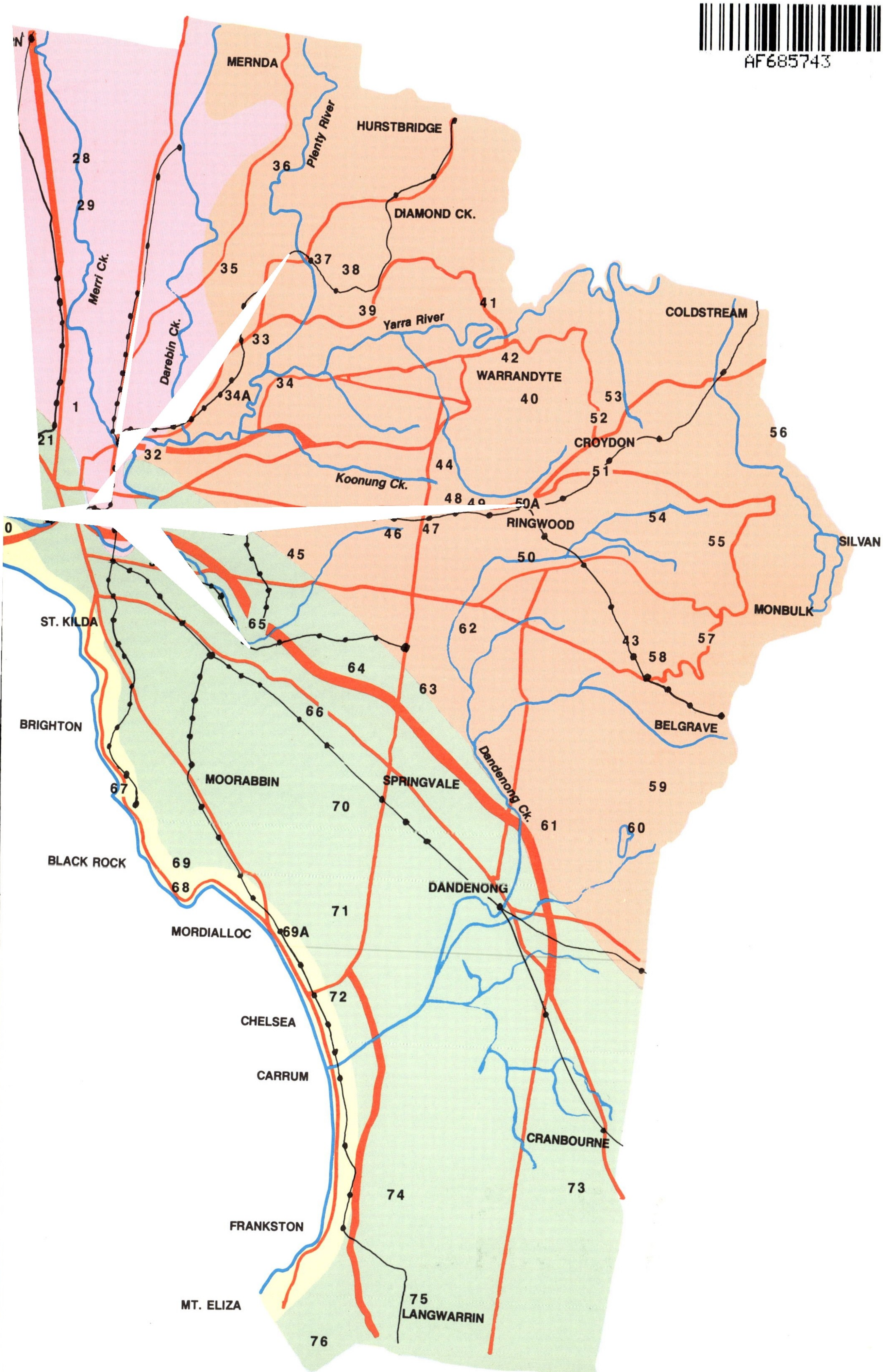
MERNDA
HURSTBRIDGE
Plenty River
DIAMOND CK.
Merri Ck.
Darebin Ck.
Yarra River
COLDSTREAM
WARRANDYTE
CROYDON
Koonung Ck.
RINGWOOD
SILVAN
MONBULK
ST. KILDA
BRIGHTON
BELGRAVE
MOORABBIN
SPRINGVALE
Dandenong Ck.
BLACK ROCK
DANDENONG
MORDIALLOC
CHELSEA
CARRUM
CRANBOURNE
FRANKSTON
MT. ELIZA
LANGWARRIN
1
21
28
29
32
33
34
34A
35
36
37
38
39
40
41
42
43
44
45
46
47
48
49
50
50A
51
52
53
54
55
56
57
58
59
60
61
62
63
64
65
66
67
68
69
69A
70
71
72
73
74
75
76

*Eucalyptus cephalocarpa*

Titles published by
S.G.A.P. Maroondah Inc.

*500 Australian Native Plants*
*Attracting Birds to Native Gardens*
*Guide to Darwinia and Homoranthus*

# Flora of Melbourne

## A Guide to the Indigenous Plants of the Greater Melbourne Area

Compiled by Society for Growing Australian Plants Maroondah, Inc.

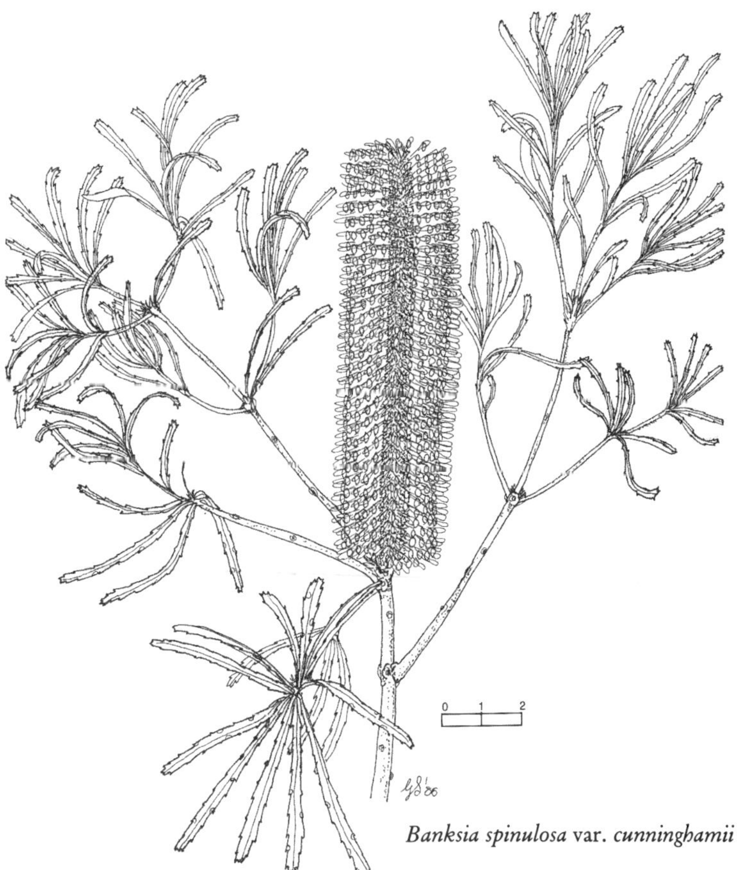

*Banksia spinulosa* var. *cunninghamii*

HYLAND HOUSE

Revised Edition published in 1993 by
Hyland House Publishing Pty Limited
(A.C.N. 005 268 208)
387-389 Clarendon Street
South Melbourne
Victoria 3205

First published 1991 by
Society for Growing Australian Plants Maroondah, Inc.
P.O. Box 33
Ringwood Victoria 3134



National Library of Australia
Cataloguing in publication data

Flora of Melbourne.

Rev. ed.
Bibliography.
Includes index.
ISBN 1 875657 07 X.

1. Botany—Victoria—Melbourne. I. Society for Growing Australian Plants. Maroondah Group.

581.99451

Editors: Marilyn Gray, John Knight *et al.*
Illustrations, George Stolfo
Cover design by Captured Concepts
Typeset by Solo Typesetting, South Australia
Printed in Australia by Brown Prior Anderson, Melbourne

Front cover: View of Melbourne from Studley Park, Kew

# *Table of Contents*

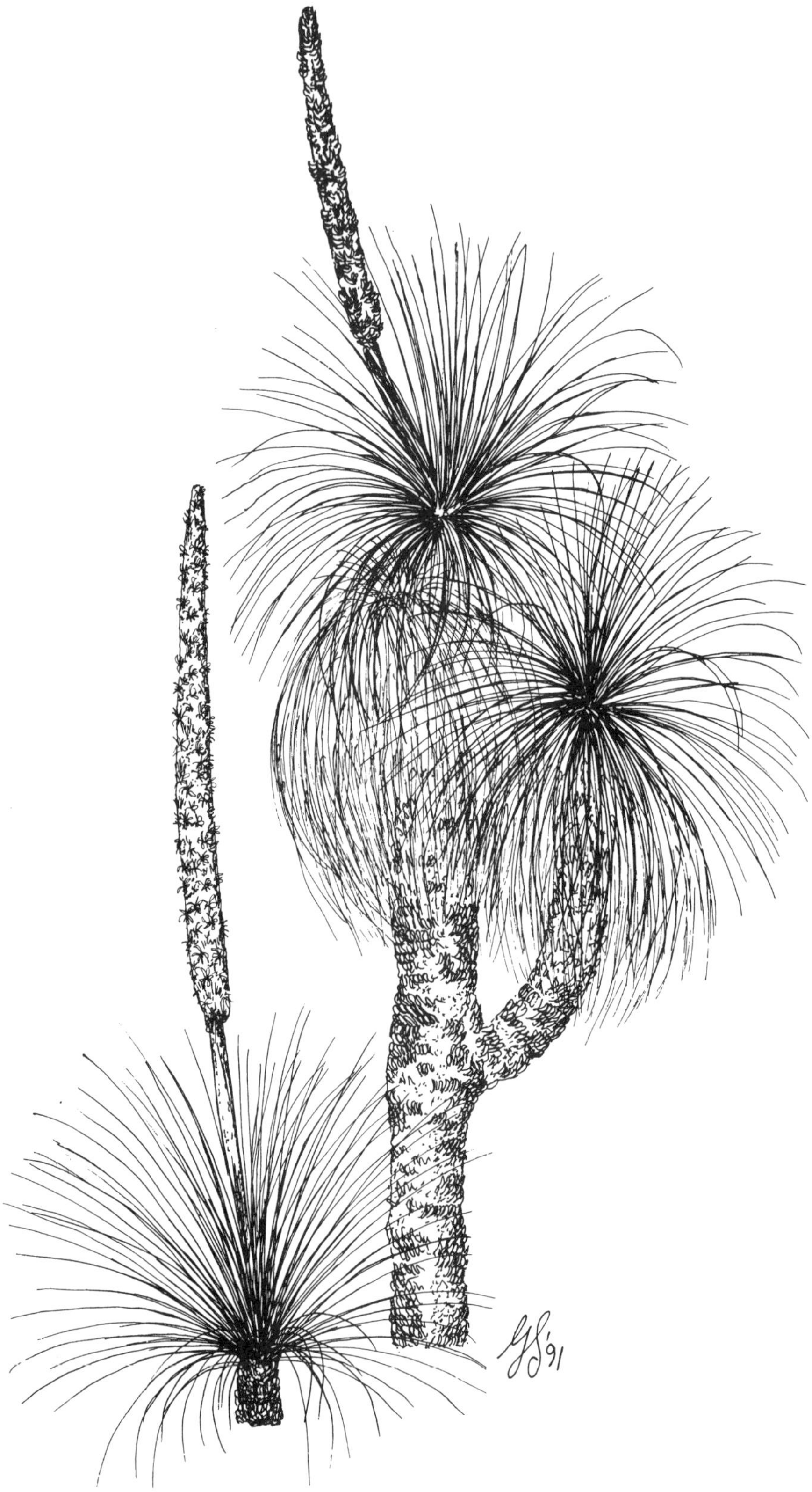

*Xanthorrhoea australis*

# *Foreword*

The *Flora of Melbourne* was more than seven years in its writing, illustration and compilation. The quality of its information, the writing of special sections, and the manner of dealing with individual species, clearly illustrate both the dedication and the expertise of the comparatively small group of people devoted to this labour of love.

*Flora of Melbourne* sets out to do a number of things: to show that we need to recognise just how fortunate we are in having such a rich flora at our back door, that we haven't been as considerate of it as we could and should have been, that the birds we all enjoy as residents or visitors in our gardens are very dependent on the bushland for food, nesting and living habitat, and that we can all do something towards reversing some of the vast damage done to this vital heritage by unchecked urban development, even within our own lifetimes.

It is not denied that we have done enormous damage to an environment that the Aborigines lived in harmony with for possibly 70,000 years or more but it is only in the last few decades that, apart from a few individuals, any number of people have made concerted efforts to bring about changes in attitudes.

We now see evidence of these changes in our national parks, our reserves, our local communities, and most importantly, in the halls of government.

But the process of awareness, training, education and implementation of conservation and revegetation policies are never ending.

Information is collected, collated and distributed through numerous publications and programs, and groups with special interest philosophies arise.

The Society for Growing Australian Plants (SGAP) is one such group. It had its beginnings in promoting the use of Australian plants in private and public gardens and over the years it has been exceedingly successful in achieving this aim. Later, broader conservation issues arose, and a different awareness evolved among certain groups of people; the ever-shrinking natural urban landscape was the focus of some of these concerns.

Many people, though, still say 'What can I do about it? I'm only one individual.' Herein lies one of the great strengths of *Flora of Melbourne*. It shows you just how involved you can become in regenerating a plot of badly treated bushland, or creating a broad habitat on your own land, including an all-important water area, and how to go about the practicalities of achieving this.

You will see how to readily identify the soil types in your area, and how best to use them for the purpose you have planned. The descriptions of nearly 1150 plants, together with over 900 line drawings, complemented by a number of colour plates will assist you in identifying any plant in your area, from the loftiest tree to the most delicate lily or ground orchid.

A section deals with the collection and storage, and the propagation of seed, and even how to recognise when it is ripe.

There is a strong and well-based philosophy expressed, that where possible . . . 'you should grow plants from seed collected locally'. While it may not always be an aim you are able to adhere to, the retention of genetic diversity is a desirable objective.

Another plant from a nursery may appear more attractive, but will it provide as adequate a habitat for the birds, animals and insects which have adapted to life in a local environment?

The commonsense, practical and easily located information in this publication will be invaluable to councils, schools, tertiary students, the horticulturalist with a special interest, or the lay person wishing to satisfy a curiosity about plants in his or her own local area, and finally to assist one's deeper involvement in the understanding of Melbourne's own floral diversity.

This is a book compiled by people who care about our native plants, who undoubtedly have extensive knowledge of and experience with them, and who wish to share this understanding and its attendant pleasures with you.

**Bill Molyneux**

Bill Molyneux is a horticultural botanist, author and nurseryman.

# *Introduction*

The publication of the *Flora of Melbourne* is the culmination of 7 years of dedication by a group committed to seeing indigenous plants appreciated as an important horticultural resource, and accepted as ecologically necessary in maintaining the survival of our local bushland environment for future generations.

It was written in response to many enquiries from a broad spectrum of the community: individuals, groups and organisations, each seeking information on local species suitable for home gardens, or the composition of the original plant communities of their area, to be used in revegetation projects. The encouragement and support of the community over this time has been the stimulus needed to keep going. Our reward will come from being involved, and encouraging others to participate, in saving what little remains of our precious remnant bushland.

Originally 'The Book', as it has become known to all involved, was planned as a horticultural guide, with details of 200 species of indigenous plants suitable for home gardens. Deciding on which species to include, and which to omit, was a much more demanding task than it might appear and was eventually abandoned in favour of a more detailed study of the known flora of Melbourne. Researching the original plant list showed just what a dearth of information existed on the plants of the Melbourne area.

Many hours of research were spent to develop a broad picture of which plants grew where. Numerous searches were made through the pressed collections at the Melbourne Herbarium. The State Library was also often visited to uncover its vast source of knowledge. Its store of older works helped us piece together an overview of Melbourne's original flora. References were made to numerous unpublished species check lists, many produced by botanists preparing management plans for specific areas. Field notes published in *Victorian Naturalist*, the journal of the Victorian Field Naturalist Club (some issues dating back over 50 years), were used to gain detailed species lists of many areas of Melbourne long since fallen victim to urban growth. Fruitful weekends were spent in the company of local experts, wandering through many of Melbourne's different habitats seeking out the many unfamiliar or little known species.

Over recent years the quest for knowledge on indigenous plants specific to local areas has exploded, as has the push for regeneration in remaining bushland areas and the revegetation of habitats in devastated areas. Work is being undertaken at a local level by concerned groups, with the publication of various booklets and pamphlets. While these works are of value to local people, they generally include only the more easily recognised plants and fail to give a picture of the overall situation in Melbourne.

With the completion of this book which contains detailed information of over 1100 species, both flowering and non-flowering, we have endeavoured to provide readers with an insight into the varied plant communities which once existed throughout the greater Melbourne area, and an idea of the original vegetation of each area. No attempt has been made to produce a genuine botanical 'flora', with its often difficult to follow identification keys. We believe that a book with simple, easily read text, and illustrations to aid identification, would far better suit the purpose of encouraging people from all walks of life to become involved in reversing the trend of the last 150 years, and hopefully rebuild the integrity and diversity of our quickly diminishing remnant bushland vegetation.

## The Benefits of Growing Local Indigenous Plants

Conservation of natural bushland habitats is a world-wide issue. The greater Melbourne area prior to European settlement contained many unique habitats. Each included a great number of different species of flora and fauna, from the insects and tiny herbs to lofty trees, which were dependent on each other for survival.

The indigenous flora of an area is a group of plants which occur naturally having evolved over thousands of years to adapt perfectly to the conditions prevailing in their environment.

Fragmenting these habitats leads to a reduction of the natural life forms found at the site.

Changes to the environment, such as clearing of land for agriculture and urban development, affect directly the viability of ecosystems. Add to this an altered fire regime, increased water runoff, erosion and nutrient build up. Each of these factors favours the establishment of weed species over the local species, ultimately leading to the destruction of the original environment.

Replanting and encouragement of regeneration of remnant bushland areas, and roadside verges, with plants indigenous to that site is the first step in arresting the destruction and brings a number of benefits.

Once established, indigenous plants, being adapted to the local climate and soil conditions, will require less maintenance and watering than exotic species. Thus the long-term cost of replanting will be less than if exotics had been used for landscape improvement.

Where possible, plants for regeneration programs should be grown from a local source to protect genetic diversity.

The use of native grasses, herbs and sedges, which produce green growth during Summer, will lessen the potential fire hazard where planted along roadsides. An understorey of introduced, weedy grasses presents a much more flammable situation as they dry out over Summer.

Using indigenous plants preserves the unique local character of an area, presenting a unified visual landscape. As the replanted areas mature, populations of native birds, and even animals, will return, occupying previously lost habitats in a relatively short time. This has been seen along the Yarra River where platypus and koalas have returned to the area.

Naturally occurring regeneration will continue to maintain the habitat, so that the bushland will ultimately become self-supporting.

This book is written for the many individuals, environmental groups and school councils who are aware of the desperate need to preserve what little remains of the natural vegetation of the greater Melbourne area. Municipal councils and government departments are charged with the responsibility of restoring the integrity of areas under their control. The *Flora of Melbourne* will assist them in making the right decisions for the long-term, despite pressure to spread the environmental improvement dollar too thinly.

It is our hope that everyone will be encouraged to be involved in reclaiming areas devastated through misuse, so that future generations may also have the opportunity to appreciate the beauty of Melbourne's unique landscapes.

If this book stimulates increased public awareness of the value of retaining and caring for our remnant indigenous plant habitats, and saves even one more bushland area from destruction, our time over the past 7 years will have been well spent.

Each of us is much richer for the experience.

**John Knight**

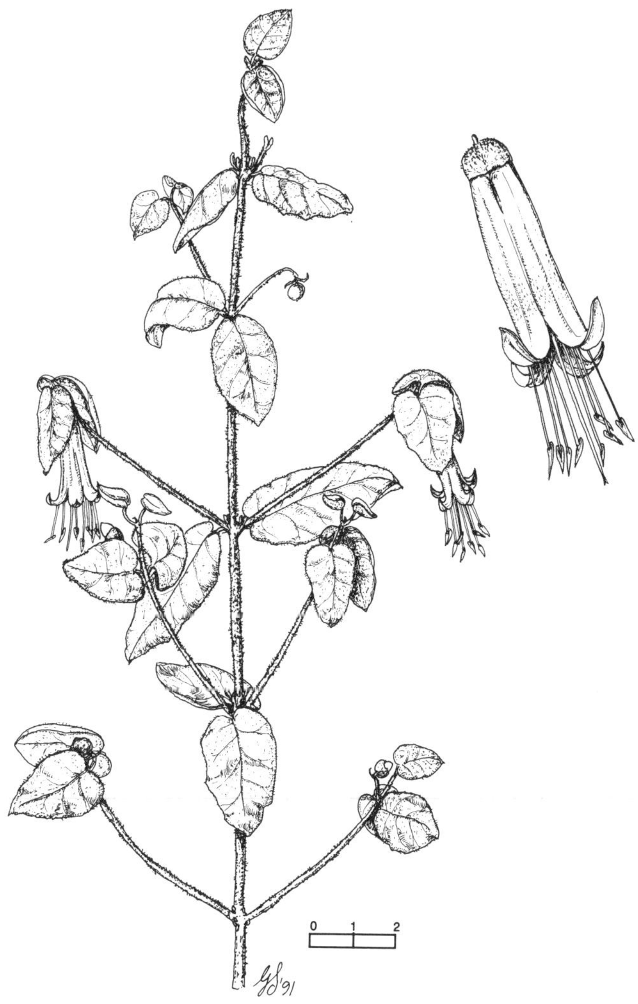


*Correa reflexa*

# Acknowledgements

We gratefully acknowledge the support and assistance of a large number of people without whom this book would never have been completed. We have met enthusiasm and encouragement from all who have been approached no matter what the query or request.

This much-needed book on Melbourne's indigenous flora was initiated by John Knight. A sub-committee was subsequently formed under Marilyn Gray's leadership, with ongoing support from John. It is the encouragement of successive Committees of SGAP Maroondah Inc. and the members plus the financial support from the annual Melbourne Wildflower Show which has made this book possible.

While we are appreciative of all who have helped along the way there are three people who deserve special recognition. We have been very fortunate to have the continued assistance and guidance of Rodger and Gwen Elliot and David Albrecht. Rodger and Gwen, as members of SGAP Maroondah Inc. and authors of numerous books on Australian plants, have shared their knowledge and experiences freely with us as we have followed the long path culminating in the publication of the *Flora of Melbourne*. Access to their botanical library has also been invaluable. David Albrecht, botanist in charge of the Identification Service at the National Herbarium of Victoria, has shared with us his amazing knowledge of the indigenous plants of Melbourne. Among other things, he has assisted with information, checked our line drawings for accuracy and identified specimens.

The immense task of editing the Flora and the articles was undertaken by David Albrecht (Dicotyledons and Conifers), Berres Colville, Tim Entwisle (Ferns), David Jones (Orchids), Graeme Lorimer (other Monocotyledons) and Gretna Weste. We appreciate the time that each put into his or her section in our pursuit of accuracy. Grant Mattingley had the unenviable task of proof reading the whole manuscript.

We are indebted to Geoff Carr, Darcy Duggan, Rodger Elliot, Vivien Freshwater, Andrew Paget, Murray Ralph, Nick Romanowski, Roslyn Savio and the Soil Conservation Authority (now part of the Dept. of Conservation & Environment) for contributing supporting articles for the project, Adrian Rigg for preparing the soil map, and to Bill Molyneux who wrote the Foreword.

The staff at the Melbourne Herbarium has kept us up to date on botanical name changes as well as allowing us access to the pressed collections for research. The Flora Branch at the Department of Conservation and Environment has allowed us the use of its yet to be published report, has verified collections and distributions from its database, and has provided us with several plant lists of remnant reserves. Both groups kindly consented to read through and comment on our draft copy.

Photographs were generously provided by Flora Anderson, Ilma Dunn, Rodger Elliott, Marion and Bill King, John Knight, Alf Salkin, Bruce Schroder and Jim Willis.

The fine-line illustrations have been painstakingly drawn by George Stolfo, with some assistance from John Armstrong who also prepared the map of the Melbourne area and the text illustrations.

The task of typing the manuscript was carried out by Marilyn Gray with assistance from Brenda Martin and David Gray.

Members of Maroondah group have assisted in many ways: providing or collecting specimens to be drawn, contributing to the early research, working bees to collate information, local knowledge of plants. We sincerely thank all who have contributed in some way and apologise for any names unintentionally omitted: Bill Aitcheson, Dave Allen, Beth Armstrong, Julie and Andrew Banks, Joan Barrett, Eva Buchanan, Trix and Lewis Chambers, Kath Deery, Millicent Denton, Tommie and Bill Ellis, Chris Fletcher, Kahn Franke, Jean Galliott, Sue Guymer, Enid and Ian Haskins, Marge and Jim Landman, Neil Marriot, Brenda Martin, Lola and Charlie Mensch, Glad Moody, Helen and Tim Morrow, Libby Powell, Arnold Romijn, Rylice Sandell, Roger and Merilyn Serong, Alice Talbot, Marie and Alf Wallbridge, Pat and Jim Watson, Gretna Weste, Mary Whitehead, Nola Wilkinson, Margaret Williams, Joe Wilson. Lorraine Marshall spent many hours researching the orchids and contacting the relevant botanists for the most up to date information. She provided us with many specimens and escorted us to sites where rare orchids could be drawn in the field.

We are grateful to the many botanists and specialists who have provided valuable information from their own research and were always willing to answer our queries: Bob Bates, Peter Branwhite, David Cameron, Roger Carolin, Geoff Carr, Margaret Corrick, Pat Coupar, Phyl Dannat, Hansjoerg Eichler, Paul Gullen, Jeff Jeanes, David Jones, Graeme Lorimer, Annette Muir, Randall Robinson, Jim Ross, Neville Scarlett, Philip Short, Thelma and George (dec.) Spice, Neville Walshe, Jim Willis, Jeff Yugovic. Beth Gott allowed us to use her unpublished database on plant species used by the local Aborigines. David Cheal checked our plant communities for accuracy.

For their assistance in taking us on field trips, providing specimens to draw, contributing information and/or plant lists from all over the Melbourne area we would like to thank: Tracey Archer, Judy Barker, Paul Barnett, Robert Bender, Robyiyn Bennison, David Boekel, Jenny Caddaye, Stewart Campbell, Jenny Casamento, Tony Chapman, Ian Clarke, Graham Clutterbuck, Scott Coutts, Jean and David Edwards, Tony Faithfull, Michael Fendley, Ferntree Gully NP rangers, Vivien Freshwater, John Galea, Laurie Gilmore, the Harpers, Robyn Heath, Don Hill, Richard Hill, Margaret Keert, Joyce Kerfoot, Chris Kenyon, Richard Leppitt, Mike Love, Lisa Milley, Adam Muyt, James O'May, George Paras, David Parkes, Loris Peggie, Murray Ralph, Anne Read, Pam Richards, Nick Romanowski, Nicky Rose, Glyn Sago, Alf Salkin, Jan Shaw, Dale Tonkinson, Eric

Ward, Neil Werrett, Mary White, Les Williams, Fiona Young, the Parks and Gardens staff of many municipal councils and shires.

Members of several groups assisted us in various ways. These include SGAP Keilor Plains, especially Carl Raynor, Ian Taylor and Barry White; SGAP Daisy Study Group; Australian Native Orchid Society (allowing our illustrator access to orchids on display at several shows) and especially Helen Richards; Bairnsdale and Latrobe Field Naturalists.

The staff of Research Publications have been extremely helpful, guiding us through the final stages of the book, offering advice and ideas and bending with us as we strove to collate the final copy ready for printing.

We gratefully acknowledge the financial support received through grants from Department of Conservation and Environment, Victoria, and Parkhills BDO, Chartered Accountants.

Finally our heartfelt thanks goes to our families who have supported us and borne the disruption of more than 7 years of research and meetings. They are Andy (dec.), Beth, David, Sue, Jill and Rene (who at last has her fridge free of specimens). Our children were all young when the project began and we thank them especially for their tolerance; Sarah, Erin (who have assisted with typing and collating) and Lauren; Darren, Keren and Liane; Michael and Naomi; Nicole and David; and Emma and Mandy (who have never known life without 'The Book').

Flora Anderson
John Armstrong
Marilyn Gray
John Knight
Roslyn Savio
Bruce Schroder
George Stolfo

Since the publication of *Flora of Melbourne* the editors have been inundated with information on the distribution of plants throughout our study area. As a result I have updated the localities listing to include as much new material as possible in the second edition. Thirty species have been added to the descriptions. Many of these have been only recently discovered within the Melbourne area.

We would be pleased to receive any additional information as it becomes available with a view to using it in future editions.

We would like to thank David Bainbridge, Bill Barker, Evan Clucas, Glen Jameson, Valentino Stajsic, Jason Stewart, Ian Taylor, Carina Watson as well as many of the previous contributors who have supplied new and updated information for this edition. The staff of the National Herbarium of Victoria have provided updated bulletins of the *Census of the Vascular Plants of Victoria* to ensure accuracy in botanical nomenclature.

The views expressed in Part 1 are not necessarily those of the Society for Growing Australian Plants Maroondah Inc.

# Photographic Acknowledgements

| | |
|---|---|
| Flora Anderson | 30 38 55 |
| Ilma Dunn | 33 45 46 |
| Rodger Elliot | 11 21 23 25 27 32 39 40 44 50 51 54 59 63 69 70 74 |
| Marilyn Gray | 1 5 6 10 12 14 16–18 20 |
| Bill King | 48 |
| Marion King | 49 58 |
| John Knight | 4 9 15 26 28 29 31 34–37 41 42 47 52 53 56 57 60–62 64 65 67 68 71–73 |
| Alf Salkin | 22 66 |
| Bruce Schroder | Front Cover 2 3 7 8 13 19 |
| Jim Willis | 24 43 |

# Part One

*Brachyscome cardiocarpa*

# *Chapter 1. Melbourne's Indigenous Plant Communities*

Roslyn Savio*

## Introduction

Plants do not grow in isolation. They grow in association with plants of the same species which interact with each other and their surrounding environment. A group of plants growing in a common environment, together with the ecological processes which determine these, is called a plant community and provides a habitat for animals, including birds, insects, spiders and reptiles. Each community can be distinguished by having more or less the same species of plants in similar proportions and the same physical vegetative structure made up of a number of intermingling layers. When these two characteristics are combined they form a description of how the plant community looks when viewed from above and in three dimensions or the side view.

A knowledge of plants in a community provides a guide as to the species of plants which may grow together naturally and the type of environment in which a particular plant can grow. The combined view is a picture of what the vegetation of Melbourne may have looked like prior to European settlement. Thus one could use this knowledge when aiming to reproduce a bird or animal habitat or simply a 'natural environment'.

It is recognised that when indigenous plants are used horticulturally or as part of a landscape design they are often not grown as part of a natural plant community. Other factors come into play such as usefulness for screening, colours of the flowers, the ability to grow quickly, shapes of the leaves, having a non-invasive root system, the ability to create shade, height and other design or horticultural requirements. It may not be possible (or even desirable) to recreate a plant community fully because of irrevocable changes to the environment that have occurred since European settlement. These changes include massive clearing, agriculture, polluted water, stormwater discharge, land fill, proximity to housing, weed invasion, the presence of exotic grasses, and also fertiliser application, incorporation of humus, Summer watering and artificial sub-surface drainage.

## 1. Influencing Factors on a Natural Community

Many factors may influence a plant community and are interdependent. Some of the most important of these are rainfall, temperature, wind, soil type and geology, topography, steepness of slopes, frequency of fire, grazing, altitude, aspect and the proximity of other plants and animals within the environment. In return the inhabitants themselves also influence these factors by altering soil, wind, sunlight and rainfall patterns, among others.

In the Melbourne area some of the major factors influencing the formation of the indigenous plant communities have been topography, geology, soil type and climate.

### 1.1 Topography

The topography of Melbourne divides into roughly four regions; the flat or undulating plains to the west and the north; the foothills of the east and north east, gradually rising to the foothills of the Dandenong Ranges and the Great Dividing Range; the coastal fringe around Port Phillip Bay, merging into the flat to undulating region to the east and south-east, and the numerous rivers and creeks flowing through the regions to create a variety of riparian and swamp environments. Topography also affects altitude, slope and aspect.

### 1.2 Soils

Soil types are the major influence on indigenous communities and follow a similar pattern to topography. The soil types are discussed in greater detail in Chapter 2, but are briefly described here as they form the basis for grouping the plants into communities.

The western plains are made up of heavy clays on younger basalt, which can extend east as far as the Plenty River and through to Richmond. In the hills of the east are the Silurian derived sedimentary clay soils of light grey loams over clay, often following rivers and creeks, with red loams on the east side of the Dandenongs. The south and south-east suburbs are a combination of deep sands free of lime around the Sandringham area, dark loams, clays and local sands. Collectively, these are known as Tertiary Sands.

### 1.3 Climate

The climate of Melbourne also varies from east to west, again affecting the formation of the plant communities. The east and northeastern hills generally have a higher rainfall than the western plains, the highest rainfall being on the Dandenong Ranges, averaging 625 mm in Spring, whereas the rainfall at Sydenham and Melton is 150 mm in Spring. The rainfall on the southern coastal fringe reflects that of the west, gradually increasing towards the east.

Temperatures are less varied with the mean maximum annual temperatures fairly uniform, the warmest mean temperatures around Melton and the coolest on Mt. Dandenong. Maritime influences are noticeable on mean minimum temperatures, the higher temperatures being around Port Phillip Bay and the near eastern and western suburbs. Cooler temperatures are found further inland, the coolest again being associated with increased altitude on Mt. Dandenong.

Wind patterns, evaporation, ground temperatures, seasonal distribution of rain are often factors influencing climate.

* Roslyn Savio is a landscape architect with working experience in the private development industry and in local and state government. Currently she has her own private practice in Croydon.

# 2. Definitions of Plant Communities

Plant community boundaries are rarely distinct lines but are more often broadly overlapping zones. Plant community descriptions relate to typical stands.

A plant community can be described in two ways. The first is a floristic description, when all the species growing at a variety of sites are identified as being inhabitants of the community. The plant community is characterised by particular species being consistently present and indicative of the local environment. It is defined by the presence of group(s) of species and the absence of other group(s). The floristic description can also include relative proportions of each species.

The second method is the structural approach, which disregards the environmental conditions and species present. This method recognises that the layers or strata within a plant community can also influence some of the environmental factors. If a slice is taken through a community, the vegetation is divided into a number of layers. The uppermost layers of vegetation form a canopy. When looked at from above, the amount of foliage cover per ground area of the uppermost layer, together with the height of the tallest layer, determines the type (and name) of the community. This is because the uppermost layer controls the solar energy used by the rest of the plant community. (Specht, R.L. 1970) This method distinguishes between the open forest or woodland with space between the treetops, and the rainforest where the canopy is closed and there is no space between the treetops.

In the Melbourne area plant communities can be divided into eighteen different structural types, listed in decreasing canopy height and density of the canopy foliage.

| | | |
|---|---|---|
| 1. Tall Open-forest | Tree height: | Greater than 30 m |
| | Foliage cover: | Density 30-70% (mid dense) |
| 2. Open-forest | Tree height: | 10-30 m |
| | Foliage cover: | Density 30-70% (mid dense) |
| 3. Low Closed-forest | Tree height: | 5-10 m |
| | Foliage cover: | Density greater than 70% (dense) |
| 4. Low Open-forest | Tree height: | 5-10 m |
| | Foliage cover: | Density 30-70% (mid dense) |
| 5. Woodland | Tree height: | 10-30 m |
| | Foliage cover: | Density 10-30% (sparse) |
| 6. Open-woodland | Tree height: | 10-30 m |
| | Foliage cover: | Density less than 10% (very sparse) |
| 7. Low woodland | Tree height: | 5-10 m |
| | Foliage cover: | Density 10-30% (sparse) |
| 8. Low-open woodland | Tree height: | 5-10 m |
| | Foliage cover: | Density less than 10% (very sparse) |
| 9. Closed-scrub | Shrub height: | 2-8 m |
| | Foliage cover: | Density 70-100% (dense) |
| 10. Open-scrub | Shrub height: | 2-8 m |
| | Foliage cover: | Density 30-70% (mid-dense) |
| 11. Closed-heath | Shrub height: | 0-2 m |
| | Foliage cover: | Density 70-100% (dense) |
| 12. Open-heath | Shrub height: | 0-2m |
| | Foliage cover: | Density 30-70% (mid-dense) |
| 13. Herbland | Herbs | |
| | Foliage cover: | Density 30-70% (mid-dense) |
| 14. Open-herbland | Herbs | |
| | Foliage cover: | Density 10-30% (sparse) |
| 15. Closed-tussock grasslands | Herbs | |
| | Foliage cover: | Density 70-100% (dense) |
| 16. Tussock grassland | Herbs | |
| | Foliage cover: | Density 30-70% (mid-dense) |
| 17. Grassland/Sedgeland | Herbs | |
| | Foliage cover: | Density 30-70% (mid-dense) |
| 18. Closed-grassland/ Sedgeland | Herbs | |
| | Foliage cover: | Density 70-100% (dense) |

# 3. Identification of Melbourne's Plant Communities

The names describing the plant communities of the Melbourne area are variable and somewhat imprecise, such as the traditional 'Dry Sclerophyll Forest'. For the sake of consistency, this book uses the names and descriptions of plant communities identified by David Cheal and others of the Department of Conservation and Environment's Flora Branch in their yet to be published report, *Vegetation Survey and Sites of Botanical Significance in the Melbourne Area*,[1] which David Cheal has generously made available, and the published *Floristic Vegetation Map of the Melbourne Area* (Cheal et al 1989). Where the boundaries of the Melbourne area of this book are not covered by this map then reference is made to another map, *Floristic Vegetation of the Upper Yarra Valley and the Dandenong Ranges*, (Parkes, D.M. 1980) and the plant communities given the names of the Flora Branch where applicable. It should be noted that the names are labels only and not a definition of the plant communities themselves.

Twenty-one communities of the Melbourne area are described in this chapter by specifying their structure, and the dominant and less frequent species for each of the different strata of the vegetation, from the upper storey or tree canopies down to the ground storey. It is important to note, however, that because the descriptions are based on site surveys of recent times, some species may be omitted because they were overlooked in the field due to seasonal differences or because they no longer exist, occur rarely or are unable to grow due to the pressure of settlement, lack of fire, grazing and weed invasion at a particular site. Many species may have also been overlooked as field surveys were not carried out at the site where they still remain. As a general rule, weed or introduced species of a particular plant community have been ignored in this description (although not in David Cheal's report), even though they may form a significant proportion of the plant community.

In order to co-ordinate the relationship between plant communities and regions of Melbourne, communities have been broadly grouped on Map 2 according to soil type and geographical area: A. Basalt Plains; B. Silurian-derived Sedimentary Hills; C. Tertiary Sands and D. Coastal areas. Within each plant community category, sub-communities have been listed according to their locality and are presented as a guide only. Readers will need to draw their own conclusion as to the particular plant community or sub-communities relevant to their chosen site. It should be remembered too that the boundaries of the areas are not distinct and that plant communities and sub-communities grade into each other.

## A. Basalt Plains

The type of soil and the topography of the western regions of Melbourne area have greatly influenced the plant communities found there. Similar soils also extend north to Mernda and east to Richmond. They are mainly fertile, heavy clay soils over younger basalts of volcanic origin which are poorly draining. Volcanic activity has also influenced the topography by forming low-lying, undulating, expansive plains with occasional rocky outcrops of earlier sedimentary or other geological origin. As mentioned previously, the rainfall is much lower than elsewhere in the Melbourne area. These are sites of rare natural grasslands, adapted to high winds and shallow soils. Trees and shrubby plants find these conditions difficult and so are not a dominant feature of the Plains Grassland and Grassy Wetlands plant communities, although they can be common on the rocky escarpments and sedimentary hills. Watercourses and drainage lines flowing through the plains have formed riparian and escarpment plant communities such as Riparian Scrub, Chenopod Rocky Open scrub and Riparian Woodland.

1 Not all communities listed are included as they are outside the boundaries of the Melbourne area of this publication, or are not listed as a community on the map by Parkes.

As the rainfall increases towards the north and east and the topography becomes more hilly, soils are of sedimentary origins or have overlying gravel deposits. Trees become more dominant and the plant communities become Red Gum Woodlands and Box Woodlands.

## 1. Plains Grasslands (Median annual rainfall 550-700 mm)

Plains Grasslands were once quite spectacular, forming waving swards of native tussock grasses interspersed with small colourful flowering herbs. The dominant grass is *Themeda triandra* but other grasses can also be present in differing numbers and species, depending on their location. Some locations where different sub-communities can be found are listed as follows:

1.1 Scattered Areas West of Darebin Creek and East of Melton
This sub-community is a Closed-tussock Grassland where *Themeda triandra* is dominant. Occasionally, other grasses such as *Danthonia caespitosa, D. setacea* and *Stipa bigeniculata* are common but not many others. *Convolvulus erubescens* and *Oxalis perennans* are characteristic herb species.

1.2 Western Area, Sunshine and Taylors Lakes (minor occurrence)
Species of this sub-community are similar to those in 1.1, but with the addition of two other tussock grasses, *Danthonia duttoniana*, instead of *D. caespitosa*, and *Chloris truncata*, but not in large numbers. The sedge, *Schoenus apogon* and rush, *Juncus bufonius* are frequent associates. Herb species include not only the characteristic *Convolvulus erubescens* but also *Acaena echinata, Asperula conferta, Eryngium ovinum, Hypericum gramineum, Solenogyne dominii* and *Plantago gaudichaudii* interspersed amongst the tussocks.

1.3 Central West—Laverton, Somerton, north of Sunbury at Jacksons Creek, formerly at Taylors Lakes
Apart from *Themeda triandra*, the tussock grasses of 1.2 are rare to absent. Instead, the associated grasses of this Closed-tussock Grassland sub-community are frequently *Stipa gibbosa* and *Danthonia caespitosa* with *Agrostis avenacea*. The herbs and rushes are similar to those of 1.2 with the addition of *Calocephalus citreus, Leptorhynchos squamatus* and *Oxalis perennans* but not *Hypericum gramineum* or *Solenogyne dominii*.

1.4 Evans Street, Sunbury
The Closed-tussock Grassland here is dminated by *Themeda triandra* and *Danthonia caespitosa*. The soils supporting this sub-community are actually derived from shale and sandstone but the community is still influenced by nearby basalt soils. Some grasses not typical of Plains Grassland are present in large numbers such as *Danthonia eriantha, Dichelachne crinita* and *Pentapogon quadrifidus* but some of the herbs of 1.3 are also typical of this community. Some additional herb species more typical of Silurian soils can be found such as *Caesia calliantha, Drosera peltata* and *Oxalis radicosa* and the small shrubs *Bossiaea prostrata, Pimelea curviflora* and *Pimelea humilis*. It is possible that this community may be transitional to Red Gum Woodland and may once have had an open tree canopy.

## 2. Grassy Wetlands (Median annual rainfall 550-700 mm)

Throughout the plains, scattered amongst the grasslands, are small depressions in the heavy fertile soils called 'gilgais', often around one hectare in size. These tend to hold water in Winter but dry out and crack deeply in Summer. Thus these depressions have supported another plant community, called Grassy Wetlands, dominated by grasses, sedges and rushes forming Grassland/Sedgelands (Herblands) or Closed-grassland/Sedgeland. Other characteristic plants include many daisy species. These communities also occur less commonly around lake margins. Significant localities are described as follows:

2.1 Scattered areas west of the Maribyrnong River.
The characteristic tussock grasses of 1.2 are present but not frequent in Grassy Wetlands. More typical are *Eleocharis acuta, Juncus flavidus* and different grasses such as *Agrostis avenacea* and *Amphibromus neesii*. Some herbs also typical of Plains Grassland grow in this sub-community such as *Acaena echinata, Asperula conferta, Eryngium ovinum* and *Convolvulus erubescens* but others are notably different, such as *Brachyscome basaltica* var. *gracilis, Calotis* spp., *Eleocharis pusilla* and *Eryngium vesiculosum*.

2.2 Western area, margins of freshwater lakes (Lake Stanley at Ardeer)
The seasonally flooded margins of freshwater lakes and swampy drainage lines also support a Grassy Wetland plant community. The structure of these communities varies from small elevated islands in areas flooded longest to cover wider areas in less flooded situations. The most typical grasses are *Agrostis avenacea* and *Amphibromus neesii* and a sedge, *Eleocharis acuta* is also common. Herbs typical of wetland areas thrive, such as *Lobelia pratioides, Lythrum hyssopifolia* and *Myriophyllum* spp.

## 3. Riparian Scrub (Median annual rainfall 500-750 mm)

Where the basalt plains are cut by watercourses and drainage lines, different plant communities have developed. The heavy loamy soils here contain more organic matter, are waterlogged for much of the Winter and are still moist during Summer.

3.1 Western area, Jacksons Ck and Werribee R.
This sub-community has a woodland structure, the tree canopy being made up of *Eucalyptus camaldulensis* over an open understorey of two wattles, *Acacia melanoxylon* and *A. mearnsii*. The dominant characteristic of this community however, is a tall dense shrub layer beside the water, made up of *Callistemon sieberi, Hymenanthera dentata* and *Leptospermum lanigerum*. Beside the water are a number of emergent aquatics and other species tolerant of prolonged waterlogging. These include *Crassula helmsii, Persicaria decipiens, Isolepis cernua, I. inundata, I. nodosa, Juncus sarophorus* and *Triglochin procera*. *Phragmites australis* forms tall reed beds amongst the aquatics. Also characteristic of this sub-community are the dense swards of *Poa labillardieri*.

3.2 Northern and Western area, Merri Ck and tributaries
The Merri Creek sub-community has a structure more of an Open-woodland but with the same tree canopy and understorey. The characteristic dense shrub layer also contains similar plant species, but with *Bursaria spinosa* instead of *Callistemon sieberi*. Similar too are the reed beds of *Phragmites australis* which also support *Calystegia sepium*. The aquatic sedges and herbs of the stream margins include large numbers of *Bolboschoenus medianus, Schoenoplectus validus, Apium prostratum, Crassula helmsii, Epilobium hirtigerum, Hydrocotyle verticillata, Lobelia alata, Lythrum hyssopifolia, Persicaria decipiens* and *Selliera radicans*.

3.3 Northern area, Jacksons Ck at Bulla, Deep Ck at Sunbury, Warringal Parklands, Heidelberg
The tree canopy of this Open-woodland is the same as 3.1 and 3.2 but the understorey has a different species of wattle—*Acacia dealbata*. The shrub layer and reed beds are similar to those of 3.1, with slightly different grasses, sedges and rushes, consisting of *Agrostis avenacea, Carex appressa, Juncus pauciflorus, Microlaena stipoides* and *Poa labillardieri*. Herb species typical of this sub-community are *Alternanthera denticulata, Rubus parvifolius* and *Dichondra repens*. Emergent aquatics beside the watercourses include *Alisma plantago-aquatica, Eleocharis* spp. and *Triglochin* spp. Dense carpets of floating plants such as *Azolla* spp. and *Spirodela oligorrhiza* may also be found in the still parts of watercourses.

### 4. Riparian Woodland (Median annual rainfall 550–600 mm)

Riparian Woodland is a very restricted plant community of the western area, with only one isolated occurrence within the study area. It occurs beside watercourses which may be dry in Summer but achieve high levels of water in Winter, and where the soils are alluvial loams of moderate depth.

4.1 Western area, occasionally along the Werribee R.
The tree canopy is formed by *Eucalyptus baueriana*, together with the occasional *E. camaldulensis* and *E. viminalis*. Beneath the canopy is a dense shrubby understorey of *Callistemon sieberi*, *Leptospermum lanigerum* and *Melaleuca lanceolata* and the taller *Acacia mearnsii*. The herbs are unusual in that they are a mixture of species from rocky sites (eg *Einadia hastata*), woodlands (*Dichondra repens*, *Rumex brownii* and *Veronica plebeia*) as well as those more typical of riparian conditions (*Poa labillardieri* and *Urtica incisa*).

### 5. Chenopod Rocky Open Scrub (Median annual rainfall 600–650 mm)

Chenopod Rocky Open Scrub is also a plant community of very restricted distribution. While it is located in the west, it is not typical of basalt-derived soils, but of Ordivician and Silurian-derived sedimentary soils on steep topographically arid slopes. The soils have a thin upper layer and a poorly permeable lower layer with a high proportion of salts. It includes the so-called 'Melton Mallee' which is just outside the study area.

5.1 Isolated Occurrences at Janefield (Plenty River) and between Sunbury and Diggers Rest (Jacksons Creek)
The tree canopy of this Open-scrub (Woodland) community is dominated by stunted *E. leucoxylon* ssp. *connata* and *E. microcarpa*, which play host to *Amyema miquelii*. The understorey is tall and shrubby and made up of *Acacia acinacea*, *A. pycnantha*, *Cassinia longifolia* and the characteristic *Rhagodia parabolica*. Characteristic sub-shrubs are *Einadia hastata*, *E. nutans* and the succulent herb *Carpobrotus modestus*. Through these shrubs scrambles *Clematis microphylla*. Grasses are not so important but typical species are *Danthonia setacea*, *Poa* spp. and *Stipa* spp.

A variation of this community occurs on dry stony western-facing slopes of deeply incised rivers and streams. *Eucalyptus leucoxylon* ssp. *connata* is the more dominant eucalypt here and is more scattered.

### 6. Box Woodland (Median annual rainfall 600–700 mm)

Box Woodland plant communities can be found on the western side of Melbourne, usually on sedimentary-derived soils but sometimes on basalt outcrops. This has a bearing on the differences between plant species found within the sub-communities. Escarpment communities are also usually classified with Box Woodlands, as a dense shrubland.

6.1 Scattered areas from Melton South to Yan Yean
This sub-community is not dominated by a eucalypt, (although there is the occasional emergent *Eucalyptus leucoxylon* ssp. *connata* or *E. melliodora*), but by *Allocasuarina verticillata*, which occurs in dense stands, either as Open-forest, Low Open-forest or Woodland. It is characterised by exposed rocky outcrops of either basalt or sedimentary origin with thin skeletal soils. *Acacia* spp., typically *A. melanoxylon*, *A. pycnantha* and *A. implexa*, form either an open understorey or link up with the canopy. There are few other shrubs except for scattered *Bursaria spinosa*, *Exocarpos cupressiformis* and *Hymenanthera dentata*, through which scrambles *Clematis microphylla*. In the past there may have been more scattered shrubs such as *Acacia acinacea*, *A. paradoxa* and *Cassinia* spp. Grasses dominate the ground layer and include a number of *Danthonia* spp. such as *D. caespitosa*, *D. racemosa* and *D. setacea*, with *Dichondra repens* and *Microlaena stipoides* in shady areas.

6.2 North Western area across to Merri Ck
A Box Woodland sub-community dominated by the tall *Eucalyptus microcarpa*, is found on gently rolling sedimentary hills, forming an Open-forest, Woodland or Low Woodland, with the occasional *E. leucoxylon* ssp. *connata*. There is generally no understorey but the field layer is dominated by the *Danthonia* spp. mentioned in 6.1, together with other grasses and tufted plants including *Elymus scabrus*, *Stipa bigeniculata*, *Themeda triandra*, *Juncus subsecundus*, *Lomandra filiformis*, *Tricoryne elatior* and *Microlaena stipoides*. Broad-leaved herbs are uncommon but may include *Oxalis radicosa* in large numbers, and *Einadia nutans*.

6.3 Gellibrand Hill
Gellibrand Hill and other low rolling hills of the central north west are made up of poorly-structured, free draining soils of Devonian granodiorite origin. The sub-community formed here is a Woodland to Open-woodland of *Eucalyptus microcarpa*. Along the drainage lines and lower slopes is *E. camaldulensis*. The shrub layer would have been open with *Acacia acinacea* and *Bursaria spinosa* plus a ground storey similar to that described in 6.2.

### 7. Red Gum Woodland (Median annual rainfall 600–650 mm)

Red Gum Woodland has also developed on a range of soil types, including those of the basalt plains, with which it shares many plant species. The community was once widespread in the north and it is likely that it intermingled with Plains Grassland communities between Craigieburn and the Plenty River, sharing many species in common. As most existing Red Gum Woodland communities are heavily disturbed, their original composition is difficult to determine. Thus some sub-communities do not reflect the 'natural environment' so much as the management history. The basalt plains sub-community is described here.

7.1 Near Northern area, Bundoora and Campbellfield
This is a Red Gum Woodland of very old *Eucalyptus camaldulensis*, forming a very open and scattered canopy. There is usually no understorey but the ground storey is very dense. The structure is of tussock grasses such as *Agrostis aemula*, *Danthonia caespitosa*, *D. laevis*, *Poa sieberiana*, *Stipa bigeniculata* and *Themeda triandra* with other herbs, some tufted, between the tussocks. These include *Acaena agnipila*, *Senecio quadridentatus* and *Tricoryne elatior* in large numbers, with lesser numbers of *Carex inversa*, *Dichelachne crinita*, *D. micrantha*, *Dichondra repens*, *Haloragis heterophylla*, *Juncus subsecundus*, *Leptorhynchos tenuifolius*, *Lomandra filiformis*, *Lythrum hyssopifolia* and *Oxalis* spp.

## B. Silurian-derived Sedimentary Hills

In the northeastern and eastern area of Melbourne, where the undulating plains give way to more hilly and mountainous topography, (often along the alignment of the Plenty River), Silurian-derived sedimentary soils are more frequently found.

These soils are generally light grey loams over clay which tend to be moist in Winter but dry out in the upper layer over Summer. Along the watercourses and drainage lines are even moister dark loams, clays and local sands. South of the Dandenong Ranges there are areas of gritty light grey loams over clay, with some red loams, while on the eastern face of the Dandenongs the soils become red loams.

As the soils change, so do the plant communities which they support. Open Woodlands on undulating hills give way to more densely forested steeper slopes. Species become much more diverse with trees and shrubs a greater feature of the structure of these communities, although grasses and herbs are still common. The main variables for the change in plant communities are soil fertility, depth of the upper layer of the soil, the amount of moisture available, rainfall and altitude.

Some plant communities are the same as for the Basalt Plains, notably Red Gum Woodland and Riparian Scrub, although the sub-communities themselves are different. Other typical plant communities of Silurian-derived soils range from Box Ironbark Woodland, the widespread Dry Sclerophyll Forest and associated Valley Sclerophyll Forest, and Swamp Scrub in the drier areas, through to Damp and Wet Sclerophyll Forests and Cool Temperate Rainforest communities of moister, higher altitudes.

## 1. Red Gum Woodland (Median annual rainfall 600–800 mm)

Bordering on the Basalt Plains and grading into sedimentary soils are large areas where Red Gum Woodland would have predominated, with a denser understorey.

1.1 Near Northeastern area, Bundoora, Yallambie, Kew, Burwood, Alamein
This sub-community is an Open-forest with dense stands of *Eucalyptus camaldulensis* and *E. melliodora*, found on low undulating hills. The understorey, if present, is made up of *Acacia implexa* and occasionally *A. mearnsii* or *A. melanoxylon*. As with Red Gum Woodland communities of the Basalt Plains, the ground layer is made up of tussock grasses, but with some changes in species. *Danthonia* spp. and *Themeda triandra* are still common, but *Poa morrisii* and *Microlaena stipoides* are also widespread. *Lomandra filiformis* is typical of the west but other species present include *Arthropodium strictum*, *Bossiaea prostrata*, *Dianella longifolia*, *Gonocarpus tetragynus*, *Opercularia ovata* and *Tricoryne elatior*.

1.2 Near Northeastern area, Lower Templestowe to Sth Morang
A sub-community of Woodland structure containing an open formation of *Eucalyptus camaldulensis* only, with an understorey of *Acacia mearnsii* and/or *A. melanoxylon*. Again there are no understorey shrubby species and grasses such as *Danthonia laevis*, *D. racemosa*, *Microlaena stipoides*, *Themeda triandra* and less commonly *Elymus scabrus* predominate the ground species with the lily *Tricoryne elatior*.

1.3 Eastern Suburbs and Bundoora
In these locations, *Eucalyptus camaldulensis* combines with *E. ovata* to form a Woodland on swamps and drainage lines. Small trees of *Acacia mearnsii* are moderately common as an understorey while *Themeda triandra* and *Danthonia geniculata* dominate the ground layer. The herbs growing between the tussocks are typical of waterlogged conditions and include *Drosera peltata* ssp. *peltata*, *Gonocarpus tetragynus*, *Luzula meridionalis*, *Schoenus apogon*, *Stylidium graminifolium* and *Wahlenbergia gracilenta*. Two orchids, typical of moist soils, are *Microtis unifolia* and *Thelymitra pauciflora*.

1.4 Near North to Northeastern area and Murrumbeena (restricted area)
This sub-community was once widespread on the low hills and plains around Melbourne, but, because of its management history of grazing, many species have disappeared. What is left is a Woodland of *Eucalyptus camaldulensis* associated with either *E. ovata* or *E. melliodora*. Again, shrubs are non-existent and the remaining ground storey is made up of *Danthonia* spp. (such as *D. laevis*, *D. setacea* and *D. racemosa*) and *Themeda triandra*, *Microlaena stipoides*, *Lythrum hyssopifolia* and *Oxalis* spp.

1.5 North Central area and Bundoora
Similar to 1.1, this Red Gum Woodland sub-community has the same structure and grows in similar soil but *Eucalyptus camaldulensis* is the only canopy tree. An additional understorey tree is *Acacia pycnantha*, while the ground storey contains the characteristic sub-shrub *Bossiaea prostrata* in addition to the grasses of 1.1. Other additional species include *Burchardia umbellata* and *Carex breviculmis*.

1.6 Near Northeastern area, Yarra River at Kew North along the Darebin Creek Valley to Bundoora
Another sub-community similar to 1.1 and 1.5 but containing *Eucalyptus ovata* as an occasional tree in association with *E. camaldulensis*. Additional tussock grasses are *Elymus scabrus*, *Stipa rudis* and *S. mollis*. Lilies are characteristic of this sub-community and *Caesia calliantha* and *Dianella revoluta* are included with those of 1.1. *Acaena agnipila* and *Convolvulus erubescens* are additional herbs.

1.7 Eastern areas (formerly), Bundoora
A sub-community of Red Gum Woodland was once widespread on drainage lines and swampy areas of the eastern suburbs, now remaining only at Bundoora, where *Eucalyptus ovata* is actually more dominant than *E. camaldulensis*, forming a Woodland structure. Except for the occasional *Acacia mearnsii* the understorey is absent, but there is a dense ground storey dominated by grasses, mainly *Themeda triandra*, but also *Danthonia geniculata*. The herbs and sedges in between the tussocks are typical of poorly drained areas and include *Drosera peltata* ssp. *peltata*, *Gonocarpus tetragynus*, *Luzula meridionalis*, *Schoenus apogon*, *Stylidium graminifolium* and the annual *Wahlenbergia gracilenta*. Two moisture-loving orchids, common in season are *Microtis unifolia* and *Thelymitra pauciflora*.

## 2. Riparian Scrub

A form of Riparian Scrub community is found on fertile heavy loams of recent alluvial soils. Now heavily weed invaded it generally forms Woodland to Open-woodland and Open-scrub with *Eucalyptus camaldulensis* as the dominant tree, where present.

2.1 Scattered areas near Streams, particularly beside the Yarra River
Beneath the tree canopy of *Eucalyptus camaldulensis* is a dense understorey of *Acacia dealbata*, *Callistemon sieberi*, *Coprosma quadrifida* and *Hymenanthera dentata*. The indigenous species of the ground storey include the rushes and sedges *Carex* spp., *Juncus* spp. and *Alternanthera denticulata*. Growing in the water is *Alisma plantago-aquatica*.

## 3. Box-Ironbark Woodland (Median annual rainfall 650–700 mm)

A plant community of very restricted distribution within the defined Melbourne region covered by this book, found only on the steep slopes of the Plenty River.

3.1 Yarrambat
The characteristic dominant tree species are *Eucalyptus leucoxylon* ssp. *connata* and *E. tricarpa*, often with *E. macrorhyncha*. These species occur elsewhere but not all together. It is an Open-forest formation with an open understorey of shrubs including *Acacia acinacea*, *A. pycnantha*, *Cassinia longifolia* and *Ozothamnus obcordatus* through which scrambles *Clematis microphylla*. Occasionally emergent from these shrubs is the taller *Exocarpos cupressiformis*. Beneath the shrubs are the tussock grasses *Danthonia geniculata*, *D. setacea*, *D. tenuior*, *Poa* spp. and *Stipa* spp. and small shrubs such as *Astroloma humifusum* and *Einadia hastata*. *Amyema miquelii* grows in the tree canopy.

## 4. Dry Sclerophyll Forest (Median annual rainfall 700–800 mm)

From Red Gum Woodland, plant communities grade into Dry Sclerophyll Forest. As the topography of Melbourne becomes more hilly towards the east, rainfall increases and the soils are of the type that are moist to near waterlogged in Winter and dry out in the upper layer in Summer. Dry Sclerophyll Forest is one of the most widespread communities of Silurian-derived sedimentary soils, containing a large number of varied sub-communities.

4.1 Eastern area from Doncaster North to Mernda

Dominating the tree canopy of this sub-community are *Eucalyptus goniocalyx*, *E. macrorhyncha* and *E. polyanthemos*, forming an Open-forest structure. *E. melliodora* is sometimes present which is indicative of deeper soils. The understorey is made up of 2 layers, with *Acacia mearnsii* and *Exocarpos cupressiformis* taller than the smaller shrubs *Acacia acinacea*, *A. genistifolia*, *Bursaria spinosa* and *Cassinia arcuata*. An even smaller shrub layer is not a common occurrence in this sub-community, but sometimes *Daviesia leptophylla* and *Dillwynia cinerascens* may be present. *Poa sieberiana* and *Themeda triandra* are widespread in the ground storey. Some herbs more characteristic of deeper soils are *Hypericum gramineum*, *Senecio quadridentatus* and *Wurmbea dioica*. Other components of the ground storey include the sedges and lilies *Arthropodium strictum*, *Burchardia umbellata*, *Dianella revoluta*, *Lepidosperma laterale* and *Lomandra filiformis*, the climber *Hardenbergia violacea* and other herbs and small shrubs such as *Astroloma humifusum*, *Drosera peltata* ssp. *peltata*, *Gonocarpus tetragynus*, *Pimelea humilis* and *Poranthera microphylla*.

4.2 East and Northeastern area, Doncaster to Yarrambat

Because of the more skeletal soils in this region, this Open-forest sub-community is more typical of Dry Sclerophyll Forest, supporting *Eucalyptus goniocalyx*, *E. macrorhyncha* and *E. polyanthemos* as the tree canopy. The understorey or shrub layer contains species of varying heights. A middle height layer of *Acacia genistifolia*, *Daviesia leptophylla* and *Dillwynia cinerascens* is interspersed with heathy species such as *Acrotriche serrulata*, *Correa reflexa*, *Hovea linearis* and *Leucopogon virgatus*. Also intermingled are taller shrubs such as *Bursaria spinosa*, *Cassinia longifolia* and *Exocarpos cupressiformis*. The ground storey is continuous beneath the shrub layer and dominated by tussock grasses. *Chionochloa pallida*, *Poa sieberiana* and *Themeda triandra* are characteristic species. Lilies and rushes similar to those in 4.1 are common, together with other herbs such as *Brunonia australis*, *Drosera peltata* and *Opercularia varia*.

4.3 Eastern area, East Burwood to Ringwood, Warrandyte and Hurstbridge

This is also an Open-forest sub-community of moister soils and grades into Damp Sclerophyll Forest. The typical trees of 4.1 are joined by *Eucalyptus obliqua* and *E. rubida*, but are not as common. The understorey is marginally different, although still in 2 layers. Beneath an open layer of taller shrubs and small trees such as *Acacia mearnsii* and *Exocarpos cupressiformis* and the occasional *Acacia melanoxylon* is an open shrub layer of mostly *Bursaria spinosa* and *Cassinia aculeata* with some *Kunzea ericoides*. The ground storey of tussock grasses is similar to that of 4.2 but also includes *Poa morrisii* and *Stipa rudis*. Some creeping grasses are also common, including *Microlaena stipoides* and *Deyeuxia quadriseta*. *Gahnia radula*, *Dianella revoluta* and *Lomandra filiformis* are found in the ground storey, together with typical moisture-loving herbs and ferns, *Acaena novae-zelandiae*, *Euchiton involucratus*, *Gonocarpus tetragynus*, *Poranthera microphylla* and *Adiantum aethiopicum* in localised areas.

4.4 Scattered Far Eastern areas, around Wattle Glen and Further Northeast

An Open-forest sub-community found on the crowns and slopes of hills is very similar to 4.2, having the same tree canopy, but with the lower storeys heathy in character. Additional species of the middle storey are *Cassinia aculeata*, *Gompholobium huegelii* and *Hardenbergia violacea* and heathy species such as *Epacris impressa* and *Pimelea linifolia*, but not *Correa reflexa*. An additional grass species to those of 4.2 is *Dichelachne micrantha*. Rushes, lilies and other herbs are also similar with the addition of *Microseris lanceolata* and the orchids *Thelymitra* spp. and *Microtis unifolia*, common in season.

4.5 Restricted areas, Mullum Creek, Donvale and the Pauline Toner Butterfly Reserve, Montmorency

On the deeper soils of the middle slopes of hills a different Open-forest sub-community is found. It contains the usual tree canopy species of 4.1, but associated eucalypt species are *Eucalyptus radiata* and *E. rubida*. Understorey shrubs are not as common as in other Dry Sclerophyll Forest sub-communities but do include species similar to those in 4.1 as well as *Correa reflexa* and *Platylobium obtusangulum*. The grasses of the ground storey consist mainly of tussocks of *Chionochloa pallida*, but also include *Danthonia* spp., *Deyeuxia quadriseta*, *Dichelachne sieberiana*, *Elymus scabrus*, *Microlaena stipoides*, *Poa sieberiana* and *Themeda triandra*.

Species uncharacteristic of Dry Sclerophyll Forests but typical of more fertile soils are *Dianella longifolia*, *Linum marginale* and *Senecio hispidulus*. *Clematis microphylla* and *Comesperma volubile* are unusually common. Heathy shrubs and other herbs include *Acrotriche serrulata*, *Astroloma humifusum*, *Burchardia umbellata*, *Dianella revoluta*, *Drosera peltata* spp. *auriculata*, *Gonocarpus tetragynus*, *Hardenbergia violacea*, *Hovea linearis*, *Hypericum gramineum*, *Lepidosperma laterale*, *Leptorhynchos tenuifolius*, *Lomandra filiformis*, *Opercularia varia*, *Senecio tenuiflorus*, *Stylidium graminifolium*, *Tricoryne elatior* and *Veronica calycina*.

At Lower Plenty near Plenty Gorge, a disturbed variation of this sub-community occurs with the same upper and lower storeys, except for *Acacia paradoxa* and *A. implexa* but missing some herb species such as *Burchardia umbellata*, *Comesperma volubile*, *Drosera peltata*, *Stylidium graminifolium* and many others.

4.6 Restricted area near Hurstbridge

This Open-forest sub-community represents a grading from Box Ironbark Woodland to Dry Sclerophyll Forest and includes *Eucalyptus tricarpa* in association with the typical trees of subcommunity 4.2. The understorey is dominated by *Acacia paradoxa*, *A. pycnantha* and *Cassinia longifolia* with the taller *Exocarpos cupressiformis* occasionally emerging. The most common grasses of the ground storey are *Chionochloa pallida* and *Poa sieberiana*, interspersed with a variety of heathy shrubs including *Acrotriche serrulata*, *Astroloma humifusum*, *Correa reflexa*, *Dillwynia phylicoides*, *Dianella revoluta*, *Gonocarpus tetragynus*, *Goodenia blackiana*, *Leucopogon virgatus* and *Stylidium graminifolium*.

## 5. Valley Sclerophyll Forest (Median annual rainfall 750–1000 mm)

Throughout the low rolling hills of the Dry Sclerophyll Forests are watercourses, gullies and drainage lines. Within these environments, often on alluvial soils, on sheltered slopes, and on slopes of north Mt. Dandenong, a related plant community known as Valley Sclerophyll Forest has developed. The species of these plant communities generally prefer moister conditions.

5.1 Far Eastern Area, from Doncaster East and Mullum Creek North to Hurstbridge

On sheltered slopes in these areas, the typical canopy is similar to those of Dry Sclerophyll Forest, forming an Open-forest of *Eucalyptus goniocalyx*, *E. obliqua*, *E. macrorhyncha* and *E. polyanthemos*. The understorey is generally two layered, with similar species but limited in diversity—a dense layer of *Acacia* spp., *Bursaria spinosa*, *Cassinia aculeata* and *C. arcuata* over a more open layer of heathy species—*Epacris impressa*, *Hardenbergia violacea*, *Hovea linearis*, *Platylobium obtusangulum* and *Pimelea humilis*. An occasional third layer is formed by the taller *Acacia melanoxylon*, *A. mearnsii* and *Exocarpos cupressiformis*. Tussock grasses are also similar and are as listed for sub-community 4.2, but *Microlaena stipoides* and the bracken *Pteridium esculentum* are included. The characteristic herb of the ground story is the common *Viola hederacea*, but the rest of the species are similar to those of many Dry Sclerophyll

Forest sub-communities: *Arthropodium strictum*, *Brunonia australis*, *Burchardia umbellata*, *Dianella revoluta*, *Lomandra filiformis*, *Euchiton involucratus*, *Gonocarpus tetragynus*, *Hypericum gramineum* and *Opercularia varia*.

5.2 Mullum Mullum Creek and Diamond Creek

The tree canopy of this Open-forest sub-community varies slightly from 5.1. *Eucalyptus goniocalyx*, *E. macrorhyncha* and *E. polyanthemos* are joined by *E. melliodora* and *E. radiata*. Again the understorey is two-layered, the tallest layer consisting of *Acacia melanoxylon* and *Exocarpos cupressiformis*. The second layer is an open combination of *Cassinia aculeata*, *C. arcuata* and *Coprosma quadrifida* mixed with smaller shrubs such as the characteristic *Spyridium parvifolium* and shrubs more typical of exposed sites: *Acrotriche serrulata*, *Correa reflexa*, *Daviesia leptophylla*, *Dillwynia cinerascens* and *Platylobium obtusangulum*. The ground storey is also similar to that of Dry Sclerophyll Forest and includes *Poa sieberiana* and *Themeda triandra* with *Dianella revoluta*, *Drosera peltata*, *D. whittakeri*, *Gonocarpus tetragynus*, *Lomandra filiformis*, *Senecio quadridentatus* and those typical of sheltered areas such as *Hydrocotyle hirta* and *Viola hederacea*. *Hardenbergia violacea*, *Comesperma volubile* and *Billardiera scandens* scramble through the shrubs.

5.3 Scattered in Eastern area from Mt. Waverley to Hurstbridge

The eucalypt species making up the tree canopy in this sub-community are the same as in 5.2 with *Eucalyptus blakelyi* also found in an area in Yarrambat. The tall understorey consists of *Exocarpos cupressiformis* with a number of wattles: *Acacia melanoxylon*, *A. mearnsii* and *A. dealbata*, and a second, dense layer of *Bursaria spinosa*, *Cassinia aculeata* and *Kunzea ericoides* with the occasional smaller shrub as for 5.2. The ground storey is slightly different, however, with *Microlaena stipoides* present amongst the *Poa* and *Danthonia* spp., *Themeda triandra*, *Arthropodium strictum* and *Lomandra filiformis*. Other moisture-loving herbs amongst the tussocks are *Acaena novae-zelandiae*, *Dichondra repens* and *Viola hederacea* with common herbs *Gonocarpus tetragynus* and *Hypericum gramineum*.

5.4 Scattered in the Northeastern area, such as Pound Bend near the Yarra River

On the small creeks and drainage lines between the low rolling hills in this area another Valley Sclerophyll Forest sub community can be found dominated by *Eucalyptus ovata*. Also present are *E. melliodora*, *E. rubida*, plus the usual *E. polyanthemos* and *E. radiata*. The understorey is a very dense layer of shrubs such as *Acacia verticillata*, *Cassinia aculeata*, *Coprosma quadrifida*, *Ozothamnus ferrugineus* and *Pomaderris aspera*, with a taller layer of *Acacia melanoxylon* and *A. mearnsii*. The only low shrub is Goodenia ovata because of the shade caused by the taller shrubs. The ground storey has a diverse range of moisture-loving plants: *Acaena novae-zelandiae*, *Adiantum aethiopicum*, *Carex appressa*, *Dichondra repens*, *Gratiola peruviana*, *Gahnia radula*, *Juncus pallidus*, *Lomandra longifolia*, *Microlaena stipoides*, *Poa labillardieri*, *P. tenera* and *Prunella vulgaris*.

5.5 Scattered in the Far Eastern area, Glen Waverley, Mt. Waverley and Nunawading

A very open sub-community for sheltered slopes with a variable combination of species: *Eucalyptus obliqua* and *E. radiata* sometimes with *E. dives*, *E. goniocalyx* and *E. ovata*. The understorey is dense, commonly containing *Acacia mearnsii* with the occasional *Exocarpos cupressiformis*, a high proportion of *Bursaria spinosa*, together with *Kunzea ericoides* and *Leptospermum continentale*. Wattles other than *Acacia paradoxa* are rare. Many grasses are present in the ground storey: *Poa morrisii*, *Stipa rudis*, *Themeda triandra*, *Deyeuxia quadriseta* and *Microlaena stipoides*. Again the herbs are typical of Dry Sclerophyll Forest: *Dianella revoluta*, *Drosera peltata* ssp. *auriculata*, *Gonocarpus tetragynus*, *Hypericum gramineum*, *Lepidosperma laterale*, *Pimelea humilis* and *Poranthera microphylla*.

5.6 Widespread from the Far East to the Far Northeastern areas, Ringwood to Hurstbridge

This is also a sheltered slope Open-Forest sub-community with a tree canopy similar to that in 5.5 but more dense. The understorey is a combination of the tall *Acacia melanoxylon* with a dense layer beneath of tall shrubs such as *A. verticillata*, *Cassinia aculeata*, *Coprosma quadrifida* and *Ozothamnus ferrugineus*. Other shrubs typical of Damp Sclerophyll Forests are also scattered throughout the understorey: *Bedfordia arborescens*, *Olearia argophylla* and *Polyscias sambucifolius*. Smaller shrubs are not a feature but there is a dense ground storey of herbs typical of sheltered areas: *Acaena novae-zelandiae*, *Adiantum aethiopicum*, *Dichondra repens*, *Euchiton involucratus*, *Lagenifera stipitata*, *Oxalis exilis* and *Viola hederacea* and the grasses *Microlaena stipoides*, *Poa labillardiera* and *P. morrisii*. Dry Sclerophyll Forest herbs are also present including *Gonocarpus tetragynus*, *Hypericum gramineum*, *Lomandra filiformis* and *Pteridium esculentum* as well as the scramblers *Billardiera scandens* and *Clematis aristata*.

5.7 Scattered in Eastern area, Glen Waverley to St. Andrews

Instead of Silurian soils, this Open-forest sub-community is found on Quaternary alluvial soils beside permanent watercourses which are moist throughout the year. Hence it has a quite different tree canopy of *Eucalyptus viminalis* with *E. melliodora* on drier soils. The understorey is dense with a tall, fairly close tree layer of *Acacia dealbata*, *A. mearnsii* and *Pomaderris aspera* over a dense shrub layer of *Acacia verticillata*, *Bursaria spinosa*, *Kunzea ericoides*, *Ozothamnus ferrugineus* and *Prostanthera lasianthos*. Smaller shrubs are resticted to *Goodenia ovata*, possibly due to the shady conditions. *Lomandra longifolia*, *Microlaena stipoides*, *Prunella vulgaris* and *Pteridium esculentum* dominate the ground storey.

5.8 Upper Western Slopes, Dandenong Ranges, Ferntree Gully

On the exposed upper western slopes of Mt Dandenong and around Ferntree Gully, an Open-forest sub-community with the dominant tree components of *Eucalyptus goniocalyx* and *E. macrorhyncha* can be found. The understorey is fairly sparse, consisting of isolated specimens of *Exocarpos cupressiformis* and *Acacia stricta*. Of the extensive heathy ground storey *Poa* spp., *Gonocarpus tetragynus*, *Lomandra filiformis*, *Pimelea humilis* and *Viola hederacea* are the most common species, but others include *Deyeuxia quadriseta*, *Chionochloa pallida*, *Glycine clandestina*, *Oxalis perennans*, *Pteridium esculentum*, *Platylobium formosum*, *Stylidium graminifolium* and *Themeda triandra*.

5.9 Northern slopes of Mt. Dandenong, Montrose, Mt. Evelyn

On the northern slopes of the Dandenongs there is an Open-forest sub-community where the dominant tree species is *Eucalyptus obliqua* together with *E. radiata*. Less common are *E. goniocalyx* and *E. macrorhyncha*. The understorey of this sub-community most commonly consists of the tall *Exocarpos cupressiformis* and *Cassinia aculeata* but also some *Acacia mucronata*, *A. stricta*, *Leptospermum continentale*, *Lomatia ilicifolia* and *Ozothamnus ferrugineus* forming an upper layer. The lower layer is heathy with similar species to those of 7.2. *Acacia myrtifolia*, *Epacris impressa*, *Goodenia lanata* and *Pultenaea gunnii* are the most common, but *Acrotriche serrulata* and *Pimelea humilis* are also included. Other climbers are *Clematis aristata*, *Comesperma volubile* and *Glycine clandestina*. The ground storey is also species-rich with the major species being *Poa ensiformis*, *Tetrarrhena juncea*, *Gonocarpus tetragynus*, *Viola hederacea* and *Pteridium esculentum*. Other herbs include *Geranium potentilloides*, *Helichrysum scorpioides*, *Hypericum gramineum* and *Oxalis perennans*, the grasses *Deyeuxia quadriseta* and *Microlaena stipoides*, the fern *Adiantum aethiopicum* and an orchid *Dipodium roseum*.

5.10 Eastern area between Dandenong and Lysterfield
Another sub-community in the eastern area not found on Silurian soils but on soils derived from Devonian volcanic rocks that are nearly waterlogged in Winter and moist at depth in Summer. This is consequently a quite different Open-forest sub-community with a low, moderately dense tree canopy of *Eucalyptus melliodora* and *E. radiata*. The moderately dense understorey is made up of *Allocasuarina littoralis*, *Acacia mearnsii* and *A. implexa*. Lower shrubs are not common but may have included *Cassinia aculeata*. The ground storey is dominated by tussock grasses: *Danthonia geniculata*, *Poa morrisii*, *Stipa pubinodis* and *Themeda triandra* and the creeping *Microlaena stipoides*. Lilies and rushes are also common amongst the grasses: *Dianella* spp., *Lepidosperma laterale*, *Lomandra filiformis* and *L. longifolia* together with the spreading, shade tolerant *Dichondra repens*, *Euchiton involucratus*, *Hypericum gramineum*, *Poranthera microphylla* and *Veronica plebeia*.

## 6. Swamp Scrub (Median annual rainfall 650–900 mm)

Swamp scrub is widespread in the eastern area where soils are poorly drained along creeks and drainage lines in lowland areas.

6.1 Eastern area
The tree canopy can be absent, but when it is present its main component is *Eucalyptus viminalis* and/or the shorter *E. ovata*. This means the structure may vary from Open-forest to Woodland to Open-woodland to Closed-scrub. The most characteristic species of this plant community is the tall dense understorey of *Melaleuca ericifolia*, and also *Acacia verticillata*, *Coprosma quadrifida*, *Ozothamnus ferrugineus* and *Leptospermum continentale*. Smaller shrubs are not a feature of the ground storey, which is composed mainly of moisture-loving herbs, grasses and sedges such as *Acaena novae-zelandiae*, *Carex appressa*, *Hydrocotyle hirta*, *Lepidosperma laterale*, *Microlaena stipoides*, *Poa tenera*, *Prunella vulgaris* and *Senecio minimus*.

## 7. Sclerophyll Woodland (Median annual rainfall approx. 900–1100 mm)

Within the study group there is one sub-community of Sclerophyll Woodland south of the Dandenong Ranges which is effectively a hilly heathland. The surface layers of the soils are shallow and nutrient deficient over a clay layer which impedes drainage and leads to waterlogging in Winter.

7.1 Low Foothills, South of Belgrave
This sub-community occurs in the low foothills south of the Dandenongs. The major tree species is *Eucalyptus cephalocarpa* in association with some *E. obliqua*. *Amyema pendulum* grows through the tree canopy. There is no tall understorey of *Acacia* spp., but there is a heathy understorey in two layers of great diversity. The dominant shrub of the upper layer is *Leptospermum continentale*, but also present are *Allocasuarina paludosa*, *Banksia marginata*, *Cassinia aculeata*, *Hakea nodosa*, *H. ulicina* and *Leptospermum myrsinoides*, with *Billardiera scandens* scrambling through them. Of the lower layer, *Platylobium obtusangulum* and *Epacris impressa* are the most common. Other species which also contribute to the heathy character of this layer are *Acrotriche prostrata*, *Acacia myrtifolia*, *Gompholobium huegelii*, *Goodenia lanata*, *Hibbertia stricta* and *Pultenaea gunnii*. Also typical of heathlands are the species found in the ground storey dominated by *Gahnia radula* and *Gonocarpus tetragynus*. Less common are *Chionochloa pallida*, *Deyeuxia quadriseta*, *Empodisma minus*, *Lepidosperma laterale*, *Lindsaea linearis*, *Poa ensiformis*, *Stipa muelleri*, *Themeda triandra* and the lily, *Burchardia umbellata*.

## 8. Damp Sclerophyll Forest (Median annual rainfall 900–1200 mm)

There are two plant sub-communities of Damp Sclerophyll Forest found on the eastern face of the Dandenong Ranges, one of which has been disturbed by frequent fire, described in 8.1.

8.1 East side of Dandenong Ranges, between Olinda and Silvan
The dominant tree of this Open-forest sub-community is *Eucalyptus obliqua* with the occasional *E. radiata*. There are no understorey trees but there are a number of shrub species. Large shrubs include *Acacia verticillata* and *Ozothamnus ferrugineus*. Smaller shrubs include large numbers of *Spyridium parvifolium* with lesser amounts of *Goodenia lanata*, *G. ovata*, *Pultenaea scabra* and *Tetratheca ciliatum*. *Tetrarrhena juncea* and *Pteridium esculentum* dominate the ground storey along with other species such as *Gonocarpus tetragynus*, *Lepidosperma elatius*, *Poa ensiformis* and *Viola hederacea*.

8.2 East side of Dandenong Ranges, West side of Silvan Reservoir
This sub-community is more diverse but has the same tree canopy as 8.1. The upper shrub layer of the understorey is composed of *Acacia mucronata*, *Cassinia aculeata*, *Leptospermum continentale* and *Ozothamnus ferrugineus*. The lower layer is composed of *Acrotriche prostrata*, *Epacris impressa*, *Goodenia lanata* and *Pultenaea gunnii* intermingled with the ground storey species, made up of rushes, ferns, sedges, grasses and herbs, which include *Gonocarpus tetragynus*, *Pteridium esculentum* and *Tetrarrhena juncea* in large numbers with fewer *Billardiera scandens*, *Deyeuxia quadriseta*, *Gahnia radula*, *Lepidosperma laterale*, *Lindsaea linearis*, *Lomandra filiformis*, *Poa ensiformis* and *Xanthorrhoea minor* ssp. *lutea*.

## 9. Wet Sclerophyll Forest

Wet Sclerophyll Forest is Tall Open-forest found on sheltered south to east facing slopes of the Dandenong Ranges, generally on deep loamy soils, and is quite different in species to those previously listed.

9.1 South and East Facing Higher Altitude Slopes, between Mt. Dandenong and Belgrave
The dominant tree is *Eucalyptus regnans*. Beneath the tree canopy is a dense understorey tree layer of *Acacia dealbata*, with *Bedfordia arborescens*, *Hedycarya angustifolia*, *Olearia argophylla* and *Pomaderris aspera*. Smaller shrubs are not common except for *Sambucus gaudichaudiana* and dense stands of *Coprosma quadrifida*, deeply shading the lower layers and making ferns the most prevalent, particularly treeferns *Dicksonia antarctica* and *Cyathea australis*, and *Polystichum proliferum*. Also present in the ground storey are *Australina pusilla* ssp. *muelleri*, *Pteridium esculentum*, the rampant grass *Tetrarrhena juncea* with *Clematis aristata* scrambling through the ferns.

9.2 South and East Facing Lower Altitude Slopes, between Mt. Dandenong and Belgrave
In this sub-community the dominant tree species is *E. cypellocarpa* instead of *E. regnans*, with some *E. obliqua*. The understorey tree layer has fewer species than in 9.1, with only *Acacia dealbata*, *A. melanoxylon*, *Ozothamnus ferrugineus* and *Pomaderris aspera*, but the dense shrub layer of *Coprosma quadrifida* is still evident, combined with some *Acacia paradoxa* and *Goodenia ovata*. Ferns are not as obvious with only a few *Cyathea australis*, *Calochlaena dubia* and *Polystichum proliferum*. *Tetrarrhena juncea* is still rampant in the ground storey, with *Viola hederacea* also a common component. Other plant species include *Clematis aristata*, *Geranium potentilloides*, *Hydrocotyle hirta*, *Lepidosperma elatius*, *Oxalis perennans*, *Poa ensiformis*, *P. tenera*, *Pteridium esculentum* and *Rubus parvifolius*.

9.3 Streams, Drainage Lines, North of Mt. Dandenong, around Silvan

This sub-community[2] is similar to 9.2 with *Eucalyptus viminalis* added to the tree canopy, *Prostanthera lasianthos* to the shrub layer, the fern *Blechnum wattsii* and the herbs *Acaena novae-zelandiae, Dianella tasmanica, Galium propinquum* and *Stellaria flaccida* added to the ground storey.

9.4 South and East Facing Slopes, between Mt. Dandenong and Belgrave

Another sub-community that is quite widespread on the sheltered slopes of the ranges contains similar species to that of 9.2, but instead of *Eucalyptus obliqua* being common, the tree canopy is formed by an association of *E. cypellocarpa* and *E. regnans*. The understorey is made up of layers, the upper layer being *Acacia dealbata*, the lower layer consisting of *Cassinia aculeata, Olearia phlogopappa* and *Polyscias sambucifolius* in fairly dense numbers. Both species of tree fern are present with *Dicksonia antarctica* the more common of the two. *Dianella tasmanica* and *Polystichum proliferum* are widespread in the ground storey, with *Poa ensiformis* and *Viola hederacea* also common. Less widespread components of the ground storey are *Acaena novae-zelandiae, Clematis aristata, Geranium potentilloides, Hydrocotyle hirta, Pteridium esculentum, Senecio linearifolius, Stellaria flaccida* and *Tetrarrhena juncea*.

## 10. Cool Temperate Rainforest

The Cool Temperate Rainforest community is located in sheltered southerly or south-easterly creek gullies on high rainfall sedimentary slopes, at mid to low altitudes.

10.1 Sheltered Gullies, Eastern Slopes of the Dandenong Ranges

Forming a Low Closed-forest, the characteristic component of this plant community is *Atherosperma moschatum* and *Acacia melanoxylon* with a few emergent Eucalyptus regnans. The fairly dense understorey is made up of tall shrubs, particularly *Olearia argophylla*, but also *Bedfordia arborescens, Coprosma quadrifida, Hedycarya angustifolia* and *Pittosporum bicolor*. Below the shrubs is a layer of *Dicksonia antarctica* and *Cyathea australis* which are host to filmy ferns and other epiphytics such as *Asplenium bulbiferum* and *Microsorum diversifolium*. The ground storey, typically low in species diversity, includes *Australina pusilla* ssp. *muelleri* and *Polystichum proliferum*.

# C. Tertiary Sands

The tertiary sands area of Melbourne contains plant communities growing on a variety of soil types derived from sandstone and wind-blown sand. These range from dark loams, clays and local sands and dark grey sands over clay, through to light grey loams over clay and deep sands free of lime. In broad geographical terms, the southeastern Tertiary sands region includes areas around Central Melbourne down through Brighton and Moorabbin to Frankston, and inland to Caulfield, Oaklcigh, Springvale and Cranbourne. In turn, these soils support a range of diverse plant communities including the very species-rich Tea-tree heath on the deep sands, Grassy Low Open Forest and Wattle tea-tree scrub on sands and sandy clays, and the ubiquitous Red Gum Woodland. Swamp scrub also occurs along drainage lines between the dunes and is referred to in detail in 6.1.

The Tea-tree heaths, or 'heathlands' were quite spectacular in Spring and once covered 'hundreds of square miles'. They used to be quite extensive around Sandringham. Swamps were common around Carrum, originally spreading to Keysborough and covering about 185 hectares.

2 On the floristic vegetation map for the Upper Yarra Valley and Dandenong Ranges, this plant community is classified as a separate community of Riparian Wet Sclerophyll Forest.

## 1. Tea-tree Heath (Median annual rainfall 700–800 mm)

Tea-tree heaths occur on nutrient-deficient deep sands on undulating dunes and are now found mainly at Langwarrin, Cranbourne, Braeside, Frankston and inland from Frankston, with odd patches at Sandringham and Springvale.

1.1 Far Southern area, Frankston, Langwarrin and Braeside

The main sub-community is an Open-woodland to Low Open-woodland where the dominant tree species is *Eucalyptus pryoriana*, forming a low open canopy over a dense shrub layer which consists mainly of *Leptospermum myrsinoides* with *Acacia oxycedrus, Banksia marginata, Leptospermum continentale* and *Ricinocarpos pinifolius* from which *Exocarpos cupressiformis* emerges, forming a very open taller understorey. Typical of heathland, smaller flowering shrubs can be found in the ground storey: *Amperea xiphoclada, Aotus ericoides, Acianthus pusillus, Bossiaea cinerea, Correa reflexa, Dillwynia glaberrima, Epacris impressa, Hibbertia prostrata, Leucopogon virgatus* and *Monotoca scoparia* to name a few. Some rushes and sedges are also typical such as *Hypolaena fastigiata, Lepidosperma concavum* and *Lomandra filiformis* as well as the grass tree *Xanthorrhoea minor* ssp. *lutea* and *Pteridium esculentum*. In the shelter provided by the taller shrubs, some herbs thrive with *Centrolepis strigosa, Dichondra repens, Drosera whittakeri, Gonocarpus tetragynus, Lagenifera stipitata, Opercularia varia, Platysace heterophylla* and *Thysanotus patersonii* being fairly common. *Trachymene anisocarpa* is commonly seen after fires or other disturbance. Many orchids may be found in season, *Pterostylis nutans, P. longifolia* being the most common.

Because the soils are generally nutrient-deficient, grasses are not characteristic of heaths, although *Microlaena stipoides* is often present in the understorey. The Dodder Laurel, *Cassytha glabella* may be found scrambling amongst the shrubs and herbs.

1.2 Far Southern area, Frankston, inland Langwarrin and Cranbourne

This sub-community is generally lacking in a tree canopy and has a structure of Low Open-woodland, Closed-heath and Open-heath. It is dominated by a dense, tall shrub layer of mainly *Leptospermum myrsinoides*, but may also include *Allocasuarina paradoxa, Banksia marginata* and *Leptospermum continentale*. Beneath this layer are the smaller heathy shrubs: *Amperea xiphoclada, Aotus ericoides, Dillwynia glaberrima, D. sericea, Epacris impressa, Hibbertia acicularis, H. prostrata, Leucopogon virgatus, Monotoca scoparia* and *Ricinocarpos pinifolius*. The ground storey is similar to that of 1.1 but is not as species-rich. It does not include grasses but may have a number of orchids.

## 2. Wattle Tea-tree Scrub (Median annual rainfall 700–800 mm)

Wattle tea-tree scrub represents a diverse array of communities some of which may be natural and others that are largely the result of disturbance since settlement of what was formerly Grassy Low Open Forest or Swamp Heaths. The 'wattle' (*Acacia longifolia*) has been introduced and is in many areas now dominating. It occurs on shallow sands and sandy clays over moderately fertile, sticky, wet, dense grey clays which are waterlogged in Winter and remain moist at a moderate depth in Summer.

2.1 Far Southern area, Langwarrin

The original community is Open-forest dominated by *Eucalyptus cephalocarpa* or, where the soils are very heavy, *E. ovata*. There is a tall understorey of *Leptospermum continentale, Melaleuca ericifolia* and *Ozothamnus ferrugineus* with *Cassytha glabella* winding through them. Typical of poorly-drained soils are the ground storey of tall sedges such as *Lepidosperma longitudinale* and *Lepyrodia muelleri*. Between the sedges are moisture-loving herbs such as *Drosera peltata* ssp. *auriculata, Gonocarpus tetragynus, Hydrocotyle hirta*,

*Isolepis marginata*, *Lagenifera gracilis*, *L. stipitata* and *Opercularia varia*. Occasionally the small herb *Xanthosia dissecta* may be found. Differing from the Tea-tree heath nearby, grasses are common and include *Microlaena stipoides* and *Poa sieberiana*.

Other likely components of this community, but now found only where the sites are often slashed, are small herbs such as *Centrolepis aristata*, *C. strigosa*, *Drosera pygmaea*, *Gonocarpus micrantha*, *Goodenia humilis*, *Hypericum gramineum* and *Schoenus apogon*.

A variation of the main community is a greater occurrence of *Lepidosperma longitudinale* and *Leptospermum continentale* and fewer herbs. Other variations have a higher proportion of *Lepidosperma longitudinale* and *Schoenus brevifolius*, *Lepyrodia muelleri* and *Gahnia radula*.

2.2 Far Southern area, Cranbourne, Langwarrin, Frankston
In between the sand dunes where there are moist, poorly-drained depressions are sub-communities of Open-forest, Low Open-forest and Closed-scrub, with a canopy, where present, of *Eucalyptus cephalocarpa* and *E. ovata*. The understorey is a combination of *Leptospermum continentale*, *Melaleuca ericifolia*, *M. squarrosa* and *Ozothamnus ferrugineus*. The ground storey is dominated by *Gahnia sieberiana* with other sedges such as *Empodisma minus*, *Lepidosperma longitudinale*, *Schoenus apogon* and *S. brevifolius*. *Gonocarpus tetragynus*, *Pteridium esculentum* and *Xanthosia dissecta* grow among them.

2.3 Far Southern area, Langwarrin
Another sub-community of Wattle tea-tree scrub, now dominated by *Acacia longifolia*, is thought to be a heathland typical of waterlogged soils which no longer exists around Melbourne. This was Closed-scrub dominated by *Allocasuarina paludosa* and *Leptospermum continentale*, and had a field layer of species such as *Drosera peltata* ssp. *auriculata*, *Epacris impressa*, *Gahnia radula*, *Gonocarpus micranthus*, *G. tetragynus*, *Patersonia occidentalis*, *Schoenus tenuissimus*, *Thelionema caespitosum* and *Xanthosia dissecta*.

### 3. Grassy Low Open Forest (Median annual rainfall 800-900 mm)

In the flat to undulating areas inland from the coast to the south east where soils are mainly sands and sandy clays, the major vegetation type is Grassy Low Open-forest. The remnants are now mainly south and east of Seaford and Frankston, but could have extended northwest of these areas, through to Braeside, Caulfield and Melbourne.

The structure of the community is Low Open-forest and Open-forest as indicated by the name.

3.1 Widespread in Far Southern area, South of Seaford
This is a very species-rich sub-community and is the most typical. The characteristic dominant tree is *Eucalyptus radiata*, sometimes with *E. ovata* and occasionally *E. cephalocarpa* and *E. pryoriana*. The tall shrub or small tree understorey includes *Acacia mearnsii*, *A. paradoxa*, *Allocasuarina littoralis*, *Cassina aculeata*, *Exocarpos cupressiformis* and *Leptospermum continentale*. The lowest storey is made up of sedges and bracken (*Gahnia radula*, *Pteridium esculentum*), small shrubs: *Acrotriche serrulata*, *Astroloma humifusum*, *Bossiaea prostrata*, *Epacris impressa*, *Hibbertia riparia*, *Leucopogon virgatus* and *Pimelea humilis*, *Xanthorrhoea minor* ssp. *lutea*, and herbs, including orchids and lilies (*Arthropodium strictum*, *Burchardia umbellata*, *Chamaescilla corymbosa*, *Dianella revoluta*, *Gonocarpus tetragynus*, *Goodenia geniculata*, *Lagenifera gracilis*, *L. stipitata*, *Opercularia varia*, *Pterostylis longifolia*, *P. nutans*, *P. pedunculata*, *Thysanotus patersonii*, *Tricoryne elatior*). Grasses such as *Microlaena stipoides*, *Poa morrisii*, *P. sieberiana* and, to a lesser extent, *Deyeuxia quadriseta* can also be dominant features. Other plants such as *Billardiera scandens*, *Comesperma volubile*, *Dichondra repens*, *Drosera peltata* ssp. *auriculata*, *D. whittakeri*, *Hypericum gramineum*, *Lepidosperma laterale*, *Lomandra filiformis*, *Schoenus apogon* and *Viola hederacea* are less common. *Amyema pendulum* is present in the tree canopy.

3.2 Far Southern area, small area South of Cranbourne
Where there are relatively heavy soils, with sand overlying heavy clay which becomes waterlogged in Winter, a slightly different sub-community of Grassy Low Open Forest exists. Here the tree canopy generally contains only one species, *Eucalyptus radiata*, and rarely, *E. cephalocarpa*. Beneath the canopy is a tall shrubby understorey of *Acacia mearnsii*, *Leptospermum continentale* and occasionally *Melaleuca ericifolia*.

There are few typical heathland species and low shrubs in the ground storey, except for *Epacris impressa*, *Pteridium esculentum*, and grass species such as *Danthonia pilosa*, *D. semiannularis*, *Deyeuxia quadriseta*, *Microlaena stipoides*, *Poa sieberiana* and *Stipa rudis*. Other herbs and rushes present amongst the grasses include *Gonocarpus tetragynus*, *Lagenifera stipitata*, *Lomandra filiformis*, *Poranthera microphylla* and *Viola hederacea*. *Arthropodium strictum* and *Tricoryne elatior* are less common. *Amyema pendulum* is occasionally present in the tree canopy.

3.3 Far Southern area, South of Frankston
A sub-community similar to 3.1 but the dominant trees are *Eucalyptus pryoriana* and *E. viminalis*, rather than *E. radiata*. *Bursaria spinosa*, *Exocarpos cupressiformis* and *Ozothamnus ferrugineus* form a dense understorey. The ground storey species are few in number but include the shade-loving plants *Adiantum aethiopicum*, *Dichondra repens*, *Microlaena stipoides* and *Poa labillardieri*, and other species, including *Acaena novae-zelandiae*, *Billardiera scandens*, *Comesperma volubile*, *Drosera peltata* ssp. *auriculata*, *Dianella revoluta*, *Gahnia radula*, *Juncus pallidus*, *Lepidosperma laterale*, *Lomandra longifolia*, *Pteridium esculentum*, *Schoenus apogon* and *Themeda triandra*.

### 4. Red Gum Woodland

In some areas of the Tertiary Sands region, where there are not sands but sedimentary clay-loams, Red Gum Woodland communities dominated. These areas are flat, broad and swampy and remain waterlogged for much of the year.

4.1 Far Southern area, Northeast of Frankston
The dominant tree of this Woodland is *Eucalyptus camaldulensis* with an understorey of *Acacia paradoxa*, *Melaleuca ericifolia* and *Ozothamnus ferrugineus*. The ground storey layer consists of the herbs *Acaena novae-zelandiae*, *Senecio glomeratus*, *S. hispidulus*, the rushes and sedges *Eleocharis acuta*, *Juncus amabilis*, *J. pallidus* and *Lomandra longifolia*, grasses such as *Poa labillardieri* and *P. sieberiana* and *Clematis microphylla*.

4.2 Southeastern area, Braeside and Noble Park
Red Gum Woodland becomes Grassy Low Open Forest in these areas, grading into other more forested communities of sedimentary-derived soils to the east. Again the dominant tree is *Eucalyptus camaldulensis* with a very open understorey of mainly *Leptospermum continentale*, and some *Kunzea ericoides*. The ground storey species are dominated by Themeda triandra with other characteristic herbs, rushes and sedges, *Carex breviculmis*, *Dichondra repens*, *Drosera peltata* ssp. *peltata*, *Hypericum gramineum*, *Hypoxis hygrometrica*, *Juncus bufonius*, *J. holoschoenus*, *J. pallidus*, *Luzula meridionalis* and *Schoenus apogon*.

## D. Coastal Areas

The coastal and near-coastal areas of Melbourne around Port Phillip Bay support a number of plant communities. East of the Maribyrnong River, the communities are Primary Dune Scrub and Coastal Banksia Woodland, while west of the Maribyrnong River the coast is host to Saltmarsh communities. The differences in soils and amount of rainfall account for the differences between east and west.

Coastal communities are amongst the most fragile, struggling for survival against continual pressure from wind, rain and tide. Destruction of even small areas of vegetation can cause rapid and severe erosion of unstable sand dunes.

### 1. **Primary Dune Scrub** (Median annual rainfall 600–800 mm)

There are two sub-communities of Primary Dune Scrub found along the coast.

1.1 Coastal Dunes Inland of Beaches, Eastern Shore of Port Phillip Bay

Because of the extreme influences of salt spray, wind exposure and early soil development from deep sands, the structure of the coastal dune plant community changes over short distances from Open-heath, Open-scrub, Open-woodland to Woodland.

On the seaward face, on the first line of defence, is a group of plants dominated by the grass, *Spinifex sericeus*, growing with *Atriplex cinerea*. As sites become more sheltered and the soils improve, shrubs are more frequent and become the second line of defence. The main shrub species is *Leptospermum laevigatum* but other species include *Acacia sophorae*, *Banksia integrifolia*, *Correa alba*, *Olearia axillaris* and *Ozothamnus turbinatus*. *Tetragonia implexicoma* scrambles through the shrubs forming dense mats. The ground storey comprises herbs such as *Actites megalocarpa*, *Apium annuum*, *Carpobrotus rossii* and *Isolepis nodosa*.

1.2 Low to High Exposed Cliffs, Eastern Shore of Port Phillip Bay

The cliffs are composed of sandy, more or less consolidated rock with the same influences as the coastal dunes, forming Closed-heath, Open-heath, Open-scrub and Woodland. Dense shrubs form the dominant vegetation, mainly *Acacia sophorae* and *Leptospermum laevigatum*. Other typical shrubs are *Leucopogon parviflorus* and *Rhagodia candolleana*, together with the distinctive *Alyxia buxifolia* and *Myoporum insulare*. Species which are also a feature in the dunes are *Tetragonia implexicoma*, *Dianella revoluta* var. *brevicaulis* and the grass *Distichlis distichophylla*.

### 2. **Saltmarsh** (Median rainfall 550–650 mm)

The Saltmarsh communities of the Melbourne area can be found on basalt-derived soils which are waterlogged for much of the year, particularly in Winter, with semi-saline to saline water. Like the Plains Grassland, the saltmarshes are treeless.

2.1 West of the Maribyrnong River

The saltmarsh community, or series of sub-communities, is quite distinctive, with a unique group of succulent or salt tolerant sedges and grasses. Its structure is of an Open-herbland, but the plant species tend to be distributed according to the depth and length of time they are inundated with saline water. The dominant species in the upper saltmarsh zones are the succulents *Sarcocornia quinqueflora* and *Suaeda australis* but the tallest plants are usually a number of salt tolerant sedge-type plants such as *Bolboschoenus caldwellii* and grasses such as *Puccinellia stricta*. *Triglochin striata* grows in depressions. Smaller growing species are *Distichlis distichophylla*, *Disphyma crassifolium* ssp. *clavellatum*, *Hemichroa pentandra* and *Samolus repens*. Closer to the beach is the shrub *Atriplex cinerea* in association with *Disphyma crassifolium* ssp. *clavellatum* and *Suaeda australis*.

The mangrove *Avicennia marina* forms an Open-scrub on intertidal mud-flats adjacent to saltmarshes, often near creek mouths. Other plant communities associated with saltmarshes are saline swamps of sedgelands composed of *Juncus krausii* and *Gahnia filum*, and on the seaward edge of the saltmarsh, a community influenced by the high tides: *Atriplex cinerea* and *Spinifex sericeus*.

### 3. **Coastal Banksia Woodland** (Median annual rainfall 800 mm)

Along the low-lying Kananook Creek is a unique community of Coastal Banksia Woodland. This is due to the environmental characteristics of this area, being flat, low, and having poorly draining, organic, slightly saline soils. Within this area are a number of sub-communities.

3.1 South of Patterson River (earlier extended north to Yarra River

The first sub-community is located on near-coastal sites south of the Patterson River where the soils are deep sand. The sites are less exposed than those of 1.1, therefore the community represents the third line of defence from salt-laden winds. This sub-community is a Woodland dominated by *Banksia integrifolia*. Beneath the tree canopy is a dense shrub layer of *Acacia sophorae*, *Leptospermum laevigatum*, *Leucopogon parviflorus*, *Myoporum insulare* and *Rhagodia candolleana*. The ground storey layer comprises rushes and herbs: *Carpobrotus rossii*, *Dichondra repens*, *Isolepis nodosa*, *Lepidosperma concavum*. Climbers and scrambling perennials through the shrubs, such as *Clematis microphylla*, *Muehlenbeckia adpressa* and *Tetragonia implexicoma*, are common.

3.2 Kananook Creek, between Patterson River and Frankston

The second sub-community is Closed-scrub and Woodland containing species typical of both Saltmarsh (*Atriplex hastata*, *Disphyma crassifolium* ssp. *clavellatum*, *Sarcocornia* spp. and *Suaeda australis*) and Swamp Scrub (*Melaleuca ericifolia* and *Phragmites australis*) because of the high water table. Species typical of free-draining coastal sites are also present: *Banksia integrifolia*, *Leptospermum laevigatum*, *Rhagodia candolleana* and *Tetragonia implexicoma*. Species more typical of this sub-community than elsewhere include *Leptinella reptans*, *Juncus krausii* and *Selliera radicans*.

## Further Reading

Australian Surveying and Land Information Group, (1990) *Vegetation Atlas of Australian Resources* Vol 6, Third Series. Department of Administrative Services, Canberra.

Cheal, D.C., et al (Oct. 1989, unpublished) *Vegetation Survey and Sites of Botanical Significance in the Melbourne Area.* Flora Branch, Department of Conservation & Environment.

Cheal, D.C., Lau, J.A., et al, (1989) *Floristic Vegetation Map of the Melbourne Area.* Department of Conservation & Environment.

Cochrane, G.R., Fuhrer, B.A., Rotherham, E.R., Willis, J.H., (rev. 1973) *Flowers and Plants of Victoria.* A.M. & A.W. Reed, Sydney.

Costermans, Leon, (rev 1991) *Native Trees and Shrubs of South-Eastern Australia.* Rigby, Adelaide.

Director of Meteorology, (Sept. 1968) *Climatic Survey Region 10—Port Phillip Victoria.* Commonwealth of Aust., Bureau of Meteorology.

Gullan, P.K., et al (July 1979) *Sites of Botanical Significance in the Upper Yarra Region.* Environmental Studies Program, Ministry of Conservation, Victoria.

McDougall, K., (1987) *Sites of Botanical Significance in the Western Region of Melbourne.* Dept. of Geography, University of Melbourne, for Dept. Conservation, Forests & Lands.

Parkes, D.M., (1980) *Floristic Vegetation of the Upper Yarra Valley and Dandenong Ranges.* Map. Upper Yarra Valley and Dandenong Ranges Authority.

Specht, R.L. (1970) 'Vegetation' in Leeper, G.W. (ed.) *The Australian Environment.* CSIRO Aust. with M.U.P.

Willis, J.H., (1965) 'Native Plants of Brighton'. *Indigenotes.* (IFFA) Vol.4 No. 8, 1991 Melb.

# *Chapter 2. Soils of Melbourne*

## Soil Conservation Authority*

In using the soil map of Melbourne, it should be remembered that builders and developers often bulldoze the topsoil, and even the subsoil, during their activities. Clean filling may be removed or brought in. Also, in Melbourne, a truly staggering amount of soil is sold to the public and spread on their gardens as: 'mountain soil' from as far away as Toolangi, 'Cranbourne loam', 'sandy loam' or just 'topsoil'. Depending on its history, the soil in a given garden may therefore vary considerably from the natural undisturbed soil profiles. The map also employs broad groupings of soils within which there is some natural variation.

The area covered by this article is larger than that of the Melbourne study area of this book. As the map also covers all sections the text has been left intact.

## 1. Red Loams

In the rolling to hilly country between Silvan and Monbulk, the upper parts of the Dandenong Ranges and the hill crests at Chirnside Park, one finds red loam soils which are deep, friable, well-structured and mildly acidic. Other features include:

- excellent drainage
- low to moderate water holding capacity
- soil digging very easy whether dry, moist or wet; non-sticky
- soil structure allows extensive root development
- shrinking and swelling is minimal
- erodibility is low
- suits wide range of plants

Example 1: MOOROOLBARK (Melway Map 37.K5)
Tertiary basalt

Soil Profile:

| | |
|---|---|
| 0–20 cm | Dark brown friable loam, grading into |
| 20–80 cm | dark reddish brown friable clay loam, over |
| 80–120 cm | reddish brown or brown strongly weathered soft basalt. |

## 2. Brown Loams over Clay

Very good soils have developed on the Older basalts of the lower Mornington Peninsula. Grey brown friable loam topsoils overlie well-structured yellow-red mottled clay subsoils. Other features include:

- good drainage
- good water holding capacity
- easily worked when moist, non-sticky, too soft when wet
- soil structure allows good root development

Example 2: BALDRY'S CROSSING (Melway Map 256.L10)
Tertiary (Older basalt)

Soil Profile:

| | |
|---|---|
| 0–30 cm | Grey brown friable loam, abruptly overlying |
| 30–300 cm | yellow-red heavy textured clay, over |
| 200+ cm | rock or decomposed basalt. |

## 3. Dark Grey Sand over Clay

In the gently undulating to flat land between Kew and Mt Waverley, and in a large triangular area bounded by Toorak, Cheltenham and Dandenong, the topsoil is a black to dark grey sand, grading into light grey sand. This topsoil overlies a brown, red and grey mottled clay. These soils are also found in the Frankston-Mornington-Balnarring-Tyabb area but the topsoils are loamier, deeper and without stones. These soils are poor in plant nutrients. Other features include:

- clay subsoil prevents too rapid drainage
- suited to a large range of plants
- topsoil easily worked whether dry, moist or wet
- non-sticky
- in winter a perched watertable may occur above the clay affecting drainage

Example 3: KEW (Melway Map 45.J7)
Tertiary sands

Soil Profile:

| | |
|---|---|
| 0–30 cm | Black to dark grey sand; grading into |
| 30–50 cm | light grey sand, with slight yellow-brown mottling and a concentration of buckshot (small iron stones) at depth; abruptly overlying |
| 50–150+ cm | brown, red and grey, mottled compact clay; strongly weathered. |

## 4. Light Grey Loams over Clay

In the north-eastern areas of Melbourne, which include: Kew to Croydon and from Bundoora to Rowville, the topsoil consists of a light grey loam with some stones or gravel. The topsoil overlies a compact yellow brown mottled clay with small and large angular stones often found at the junction. These soils are poor in plant nutrients and humus. Other features include:

- moderately well drained
- moderate to low in water holding capacity

* The Soil Conservation Authority is now part of the Deptartment of Conservation and Natural Resources.

- topsoil tends to set quite hard when dry and needs to be worked when moist as it is too soft when wet
- good for a wide range of plants
- scattered rock fragments in the subsoil can cause problems for digging

Example 4: BUNDOORA (Melway Map 10.A9)
Silurian sandstones and sandstones

Soil Profile:

| | |
|---|---|
| 0–15 cm | Light grey loam, with some stones and gravel, abruptly overlying |
| 15–60+ cm | yellow brown mottled clay. |

## 5. Gritty Light Grey Loam over Clay

Soils formed on granite rocks occur between Lysterfield and Hallam, Arthur's Seat, Mt Martha and in other areas. These soils consist of light grey gritty sandy loam, or loamy sand, over mottled yellow brown and grey clay. They are poor in nutrients and are also mildly acidic. Other features include:

- moderately well drained
- low water holding capacity in topsoil
- topsoil sets hard when dry and is moderately easy to dig when moist
- non-sticky
- waterlogging is common on lower slopes in Winter and Spring
- prone to water erosion

Example 5: MT MARTHA (Melway Map 150.G7)
Granite

Soil Profile:

| | |
|---|---|
| 0–30 cm | Light grey sandy loam; an abrupt transition to |
| 30–100 cm | yellow brown sandy clay with a blocky structure; |
| 100+ cm | decomposing granite. |

## 6. Dark Loams, Clays, Local Sands

On the flood plain and swampy areas around Melbourne the soils consist of dark loams, clays and sands. These soils can be very good for a wide range of plants if they are adequately drained (watertable kept between 1 and 2 metres in depth). They are fair to poor in plant nutrients. Other features include:

- drainage variable
- if rusty mottling along old root channels in topsoil is present it indicates very poor drainage
- mildly acidic to neutral
- sand and clay contents will vary depending upon position on the floodplain (i.e. youngest terrace or an old levee)

Example 6: BANKSIA PARK, BULLEEN (Melway Map 32.D5)
Alluvial soil on Yarra floodplain

Soil Profile:

| | |
|---|---|
| 0–20 cm | Very dark grey brown silt loam to clay loam; grading into |
| 20–60+ cm | grey brown silty clay loam to silty clay. |

## 7. Deep Sands Free of Lime

Deep grey brown sands over yellow grey, frequently with a layer of dark brown hard cemented sand ('Coffee rock'), cover the sand-dune areas from Black Rock to Brighton, and along the coast of Port Phillip Bay as far as Rye. These soils are extremely poor in nutrients and strongly to mildly acidic. Other features include:

- rapid drainage
- very low water holding capacity
- very easily worked when wet, moist or dry
- non-sticky
- prone to wind erosion
- high water tables in some areas

Example 7: FRANKSTON (Melway Map 103.K8)
Sand

Soil Profile:

| | |
|---|---|
| 0–40 cm | Light grey sand, over |
| 40–50 cm | coffee coloured sand, extending into |
| 50+ cm | yellowish brown to brown sand. |

## 8. Deep Sands with Lime

Between Cape Schanck and Sorrento, in rolling sand dune areas, the deep soils are grey to whitish sand that occasionally lie on a lime-cemented hard pan. The sandy topsoil contains a varying amount of lime and is lightly acid to alkaline, and the subsoils are alkaline. Other features include:

- excellent drainage
- very low water holding capacity
- very easily worked when wet, moist or dry
- non-sticky
- prone to wind erosion
- limited range of lime tolerant and salt wind tolerant plants

Example 8: SORRENTO (Melway Map 156.J9)
Aeolianite, calcareous and siliceous dune sand

Soil Profile:

| | |
|---|---|
| 0–90 cm | Uniform dark grey medium sand, over |
| 90–200+ cm | yellow grey uniform sand. Over sand and calcarenite. |

## 9. Heavy Clay on Younger Basalts

Over extensive areas to the north-west of Melbourne, stretching from Altona to Thomastown and from Richmond to Broadmeadows and beyond, one finds shallow dark and reddish brown heavy clays with a thin loamy topsoil. Basalt outcrops are common and basalt floaters occur extensively. Other features include:

- drainage very poor, waterlogged in wet weather
- soils very hard when dry and very sticky when wet, making them very difficult to cultivate
- slightly alkaline to slightly acidic
- clayey soil and low rainfall limits the range of plants
- clay dries very quickly in hot weather and can develop deep cracks. Large shrink-swell capacity. This can cause cracks in walls and pavements

Example 9: DAREBIN PARK, FAIRFIELD (Melway Map 31.D8)
Quaternary basalt

Soil Profile:

| | |
|---|---|
| 0–15 cm | Very dark grey brown clay, fissured and hard when dry but sticky and plastic when wet, grading into |
| 15–70+ cm | dark grey brown to brown heavy clay, coarsely angular blocky structure. |

# Notes on Soil Descriptions Used

**Acidity** Acidity is measured on the pH scale. Soils are said to be neutral if they have a pH of 7, acidic if the pH is less than 7 and alkaline if they have a pH greater than 7.

**Floaters** Rocks and stones which occur irregularly within the soil profile are referred to as floaters.

**Friable** A friable soil is one which can be easily crumbled or broken up. A friable soil allows easy cultivation.

**Hardpan** The term hardpan refers to any layer compacted or cemented within or below a soil profile.

**Humus** That part of soil organic matter which has decomposed into a dark coloured soft material without a trace of the original plant tissue.

**Mottled** Spots or blotches of different colours or shades of colour interspaced with the dominant colour.

**Organic Matter** All the constituents of soil arising from living matter: includes remains of dead plants and microfauna, both fresh and decomposed.

**Plant Nutrients** The term 'poor' used with plant nutrients means that the nutrient levels are usually enough to permit growth of most Australian plants and many exotics. Indigenous plants are happy growing without additives such as fertiliser and soil improvers.

However, to grow a range of Australian plants soil conditions may be improved by the addition of appropriate amounts of:

—**organic material** will increase the humus content of the soil. This has an excellent effect on the soil structure in loams and clays while increasing water holding capacity and nutrient storage ability of all soils

—**fertilisers** or **animal manures** will increase the nutrient level of the soil

—**lime** or **dolomite** will neutralise acidity

—**sulphur** will neutralise or acidify an alkaline soil

—**gypsum** will improve and stabilise soil structure.

**Soil Profile** A vertical section of a soil through all its layers and extending into the bedrock. A profile can be seen when the soil is exposed in a road or railway cutting.

**Structure** Structure refers to the way soil particles group together. The particles in most soils consolidate into crumbs of various sizes. Some soils, such as beach sand, have particles which do not cling together and they have no structure. In contrast, some clay soils stick together in one solid mass and also lack structure. Soils which do not readily break into crumbs are poorly structured. Well-structured soils break up easily into crumbs with definite shapes and sizes. These crumbs, being irregular in size and shape, do not completely fill the space they occupy and result in a network of open spaces and passages which are filled by air, water or plant roots. In fact, an average soil may contain forty to sixty percent of these open spaces. The structure of the soil is very important to farmers and gardeners. A well-structured soil will easily take in water through the spaces between the lumps. Excess water will drain away through the spaces and be replaced with air. Most plants need both air and water near their roots to grow well. When surface soils lose their structure they are more easily eroded.

**Texture** Texture is the feel of the moist soil when worked between the fingers and is related to the relative proportions of particle size groups present (coarse and fine sand, silt, clay), as well as being influenced by organic matter and gravel. The texture has an important bearing on how much water the soil can hold, how easy it is to dig or plough, and its fertility.

Much of our agricultural production comes from soils with friable loam surface horizons. These soils contain moderate but variable amounts of all sizes of mineral particles together with organic matter.

# *Chapter 3. Regeneration of a House Block*

Vivien Freshwater*

Why bother to retain bushland on a small block of land? Will it survive? Fortunately the move towards indigenous plants has fostered an awareness of just how important a small area of natural bush is in maintaining the local genetic diversity in plant species. Even small linear reserves are important in providing a corridor for bird and animal movement. Many species of insects have an association with a particular native plant species on which they lay their eggs, providing an immediate food source for their larvae.

Once you have acquired your bush block you should identify the indigenous and introduced plant species growing on it, as well as possible fauna habitats. The old dead tree could be an important habitat for possums, birds or bats.

Take a full year to complete your list of plant species, remembering that most orchids and lilies only appear in Spring and Summer. The identification of native grasses should be carried out in Summer, when most species are flowering. If you do not have sufficient knowledge for this task, either consult an expert, or collect plant specimens and take them to the Herbarium.

The aim of natural regeneration is to help the bush help itself. This means there should be minimal disturbance, and removal of weeds should only be carried out at the same rate as the indigenous plants are able to replace them. Experience shows the value of a long-term written plan before any revegetation work is undertaken. Natural bush regeneration takes time, so don't expect to complete the task of restoring your block in a couple of years.

It is preferable to use small hand tools when working in bushland. A kit consisting of a two-pronged daisy digger, secateurs and a pair of pliers will enable you to handle most weeds. Pliers are indispensable for removing small roots up to 10mm wide from moist soil without snapping them off. This is less destructive than digging. They are also useful when handling or moving blackberries.

Some weeds are more easily controlled than others, but if you plan your regeneration strategies to remove the introduced plants before they set seed, then you will save yourself a lot of time and effort. Probably the two introduced grasses most difficult to eliminate are Panic Veldt Grass (*Ehrharta erecta*) and Blown Grass (*Agrostis* species). The former is easily removed, but has many tiny seeds that seem to fall off the plant no matter how careful you are. The latter has a very matted root system. It is important to learn the growing and seeding habits of the weeds. With experience you learn to recognise the introduced grasses before they flower.

Each weed species has specific requirements for removal. Make a study of the various root systems, so that you will automatically know which technique and tool to use when removing them. For weeds with a more extensive root system, a garden fork is more appropriate and preferable to a mattock.

The timing for the removal of understorey trees such as Sweet Pittosporum (*Pittosporum undulatum*) can be critical. The removal of these trees will dramatically increase the light level, and in turn favour any weeds that are already established under them. It is therefore advisable to control these weeds before removing 'weedy' understorey trees. A small hot fire around the stump of *P. undulatum* will prevent regrowth and stimulate germination of soil-stored indigenous seeds.

Areas where there are few individual weeds can be restored fairly quickly. This will give you a lot of satisfaction, plus the incentive to tackle more heavily infested areas. The advantage of being 'on the spot' is that you can closely monitor the areas where you have carried out work, and constantly follow up in these areas.

The golden rule of bush regeneration is—**do not clear too large an area at one time**. Work out how much time you are prepared to spend doing follow-up and maintenance work and plan your weeding strategies accordingly. Weeds are fast colonisers of bare earth and if you overclear an area of one weed, for example Wandering Jew (*Tradescantia fluminensis*), you could find that other weed species will replace it, thus negating your work.

So start modestly. Remember that it is the quality of weed removal, rather than quantity, that will give you the best result.

You may need to reintroduce some indigenous plants to your bush block to give it the correct ecological balance. Study the surrounding bushland and make sure that you only use stock grown from local species, or better still, collect a small amount of local seed and grow your own plants. The latter at least ensures that you will not be introducing more weed species with foreign soil. Some plants such as Australian grasses can be reintroduced by direct seeding, but once again make sure that you use only local seed. It may be necessary to obtain a permit to collect seed from some areas. (See Permit Requirements in Chapter 7 Section 2).

Bush litter that lies thick on the ground forms part of the unique character of the Australian bushland but if you live in a fire-prone area it is prudent to keep the ground fuel level fairly low by removing small branches and excess bark. This needs to be carried out each Autumn and Spring, depending on the vegetation type. Some grasses such as Wire Grass (*Tetrarrhena juncea*) will need to be removed manually on a regular basis, unless you have native fauna to graze it. Other shrubs such as Hop Goodenia (*Goodenia ovata*) could be selectively thinned to encourage the germination of new

* Vivien Freshwater is a naturalist working in a voluntary capacity in bushland regeneration. She is one of the inaugural members of the Friends of Sherbrooke Forest.

plants and reduce the fire risk. Some Australian plant species such as *Senecio* spp. are early colonisers following clearing and have a 'weedy' appearance, but are in fact important food plants for the larvae of some moths. Many native sedges are food plants for butterfly larvae.

Our indigenous vegetation has a special charm and atmosphere all of its own. Its beauty is subtle, and can be fully appreciated by observing its finer details, such as grasses when in flower, or the delicate white flowers of plants such as the Bootlace Bush (*Pimelea axiflora*). It is all very precious—each small area that has survived the ravages of urbanisation.

# Further Reading

Bradley, Joan (1988) *Bringing Back the Bush: The Bradley Method of Bush Regeneration*. Lansdowne Press, Sydney.

Buchanan, Robin A, (1990) *Bush Regeneration: Recovering Australian Landscapes*. TAFE, NSW.

Friends of Sherbrooke Forest, (revised ed 1989) *Weeds of Forests, Roadsides and Gardens*. Department of Conservation & Environment, Melbourne.

Regional Pest Plant Strategy Working Group (PITE Project), (Oct. 1991) *Environmental Weeds Kit (Identification & Control)*. A Free kit C/o PO Box 104, Lilydale 3140.

*Senecio glomeratus*

# Chapter 4. *Creating or Restoring a Wetlands Area*

Nick Romanowski*

For most of the period of white settlement, wetlands throughout Australia have been drained without thought which seems ridiculous on this the driest and most drought-prone of continents. Wetlands are an important habitat for a large range of birds and other animals, and the home of many unusual waterplants quite different from the Australian plants most people are familiar with.

If you have the space for a wetland area, you will attract a surprising range of native fauna to your area, including animals which are primarily associated with water. Even a fairly small pond is worth having as it brings the distinctive feeling of water to a garden, and can add significantly to the range of Australian plants you can grow.

Many remnant bushland areas have some form of watercourse running through them. These are invariably degraded, generally choked with weeds, silted and/or eroded. Some have even been drained. Much care and sensitivity needs to be used to restore these to their original state.

This chapter is a brief guide to establishing wetlands, starting with an outline of the general requirements of waterplants. Further sections deal with the planning and preparation of new wetlands (including dealing with established weed problems) before they are filled with water, and with several potential problems.

## Conditions for waterplants

Few waterplants will do well on pure clay, although some will colonise it and gradually change conditions until they are suitable for waterplants with other needs. Planting directly with clay-colonisers and allowing a natural succession of plants is not recommended because many introduced weeds grow better under these conditions than Australian plants.

It is worth preparing soil mixes to approximate soils found in long-established wetlands, and there are additional qualities which should be incorporated for best initial growth. Wetland soils are quite different from those you are used to growing land plants in; they are usually richer, and may contain large amounts of organic matter including decayed animal materials. Under water the structure of these soils may collapse, which means they will not allow gases in and may become anaerobic (lack oxygen). As a result, plant roots may not be able to absorb nutrients from the soil. Although many water plants will absorb most of their nutrient needs from water and the thin aerobic layer on the soil's surface, most will grow faster and better if planted in a rich mix which is open and shallow enough for oxygen to circulate through it. Mixes of this type will usually become quite like soils found naturally in wetlands over a period of time, and will have given waterplants a good boost in growth by this stage.

The main component of the mix should be the topsoil of the future wetland (up to 95% by volume). To this you may need to add: organic matter (up to 10% crumbled brown coal, peat, etc.); soil-openers to improve water and gas circulation (up to 10% vermiculite, peat, or coarse sand); and fertilizer (between 1% and 2% of manures or blood-and-bone), according to the nature of the original soil. Inorganic fertilizers leach differentially, ie some chemicals will leach more quickly than others, and may cause peculiar chemical effects. A skilled operator can do the mixing with a bulldozer, and spread it as a layer 5–10 cm deep wherever you intend to plant. Sterilisation of the soil is rarely needed; if there will be a reasonable depth of water over the soil layer, most weed seeds will drown without sprouting. To further reduce weed-seed germination, and also to restrict soil or nutrients leaching into the water, spread a 2–4 cm layer of coarse sand, washed river-gravel or similar material over the soil. The pond or wetland is now ready to be filled with water, but you may want to plant into the deeper sections before much water goes in.

In small ponds, the above method can be modified for planting in ordinary pots, but lay newspaper or hessian in the bottom of the pot first to prevent leakage of soil and fertiliser. In larger ponds or dams, you should include planting-shelves or shallows. These are shelves (of any width, depth or shape) on which the layers of soil and sand are spread. To reduce loss of soil over the edge, narrower shelves should slope slightly downwards, away from deeper water. Larger areas such as planting shallows don't have to be uniformly level, but should always slope gradually between different depths. Whether you use pots or plant directly on the bottom, remember that the actual planting depth is the distance from the surface to the top of the sand layer on the soil and should be chosen to suit the particular plants you are planning to grow.

Most waterplants will grow best in full sunlight, tolerating some shade for part of the day. Those which grow fully submerged need enough light for growth. In murky water they should be planted closer to the surface. Most waterplants prefer fairly soft water which should be reasonably close to neutral. Melbourne tapwater is quite satisfactory, but should be allowed to stand whenever possible to release chlorine; this is also important for good health of any fish you may keep. Hard or saline waters reduce the range of

* Nick Romanowski is a qualified zoologist with a special interest in all aspects of freshwater biology including aquaculture. He is manager and co-owner of 'Dragonfly Aquatics', a specialist waterplant nursery.

species which can be grown. Unlike most Australian land plants you are probably familiar with, many waterplants die away for part of the cooler months. Include some evergreen ones if they are required for permanent habitat for animals.

## Planning for a wetland area

The first step in planning for a pond or wetland area is to decide how much water you can make available. (Work out the catchment of the area, and combine this with rainfall figures). If you are relying on rainfall alone to maintain the wetland, remember that a large proportion of rainfall is absorbed before any begins to run off the soil.

Australian wetlands are very varied; many dry out in warmer seasons, so a natural-looking wetland can be designed even where shortage of water is a regular Summer problem. Plants which tolerate drying out for part of the year are needed for such conditions, while others may require a minimum depth of water to grow in at all times.

The wetland does not necessarily have to be filled in just one year, but planting cannot be started until water levels are close to maximum. The larger the surface area relative to the volume of water, the greater will be the rate of evaporation; this can represent a loss of around one metre depth of water in some areas of Melbourne even in a normal Summer. If the plants you plan to grow will not tolerate large drops in water level, a reserve dam should be built at a higher point in the catchment for topping up. In the garden situation a tank may be used to collect and store water in readiness for Summer.

## Preparing the area

Whether you are building a small pond for the garden or dealing with a natural wetland which has been drained, you will effectively be damming an area to retain water. All standard damming techniques must be used including topsoil removal, tying the wall into the substrate, sheepfoot rolling (about one in three dams made with just a bulldozer will not hold water properly), and provision of an overflow capable of handling maximum expected runoff for your area. When water demands are less than 100,000 litres per day, a permanent tapped siphon is a cheap and easy option which will not weaken the wall. (If a pipe is put through the dam wall, it must have water-impervious baffles attached, and be hand compacted into place before construction of the dam wall begins.)

Once the position and water-holding height of the future wall is known, a level may be used to mark out the boundary of the wetland or pond. Alternatively, this can be marked out beforehand and the wall built to hold water to this level. Whether you are starting from scratch or reclaiming a wetland, the surface will often have been contoured to improve drainage. To correct this, future islands, peninsulas, planting shelves and shallows should be marked out at this stage. Avoid building in too many such features and follow the present slope and variation of the future wetland as much as possible, both for natural appearance and economy in earthmoving.

With the completion of contouring and rolling, topsoil can be spread where it is needed. In most cases this is the planting shelves and shallows (with organic matter and fertiliser mixed in as described), and clay surfaces above water. Don't spread topsoil in deep water areas unless you have a specific reason for putting it there. Few waterplants grow at these depths, and the deeper areas will tend to silt up anyway. Don't speed up the process unnecessarily.

The finished wetland should be a complex of shallow areas sloping gradually to meet islands and peninsulas which have been prepared for conventional planting of Australian land plants. Finally, just add water!

## Dealing with established weeds

An overgrown wetland or dam provides good habitat for some water animals such as dragonfly larvae and snails, but it is also in the last stages of turning into land, and many of the plants in it may be undesirable, introduced species. There are a number of ways to tackle cleaning out overgrown ponds or wetlands.

If you are only dealing with a small area, and are not in a hurry, it may be possible to do the work by hand and with a rake. The plants pull out of soft mud fairly easily, but you must be prepared to repeat the process regularly until new growth stops appearing. If the mud remaining is too deep it can usually be removed by a sludge pump, and is excellent used on gardens.

There are a number of types of mechanical excavator which will remove weed growth complete with mud in large-scale areas. Although this is a relatively expensive option, most of the work will be done in a single operation. Some pieces of weed will certainly remain, but these may be removed over a short period of time by hand. Alternatively, drain the entire wetland at the beginning of hot weather and allow it to dry, then remove all unwanted plants. This method is fairly quick and thorough, but also expensive, as you are effectively starting from scratch.

There are usually legal restrictions relating to the use of herbicides around and in water (for good reason—someone downstream of you will be the next person to get them!), and appropriate advice should be sought from local authorities before even considering their use. The only herbicide which is reasonably harmless to use over water is glyphosate, which is sold under a number of trade names (eg Roundup or Zero). This breaks down fairly quickly and does not appear to affect most water animals. However, many water plants are unaffected by it for at least part of their growth cycle, especially submerged ones. Other herbicides which will destroy waterplants will effectively kill most life in the water, and may still leave you with the problem of disposing of a mass of mud bound together with plant roots, if the mud is too deep for healthy underwater conditions.

## Problems

Wetlands are usually located on watercourses which may flow dramatically at some times of the year and even dry out at others. In areas permanently or frequently submerged, erosion may become a problem. It is most easily prevented by establishing a carpet of living plants. Many larger Australian plants, such as trees, will grow in areas which are only flooded occasionally. These plants are described elsewhere in this book. However there are two types of erosion which must be contained with true waterplants.

Plants which will not break up under the full force of Winter flooding must be grown in permanently moving waters. It is useful to have some idea how deep the floodwaters may run, and for how long, as not all waterplants will tolerate prolonged deep submersion. In already eroded watercourses, the same methods may be used to protect exposed soil as on land eg mesh, rock rubble, even old tyres if these can be firmly pinned down. The outside curve of bends needs particular reinforcement. Any use of rock fill must avoid causing further undercutting while allowing new plantings to take hold.

The other form of erosion occurs where water levels vary considerably over a year. Plants for these conditions must tolerate being submerged for part of the year, and dry at others, or the result is a bare band of soil which is easily eroded by heavy rains or wind. Choice of plants in this case depends on how much the water level varies, and how long plants must survive the dry conditions. For example, in areas which are only covered in water for short periods of time, land plants which will tolerate short-term flooding are the most suitable. At another extreme, when water levels

# MELBOURNE - SOIL TYPES

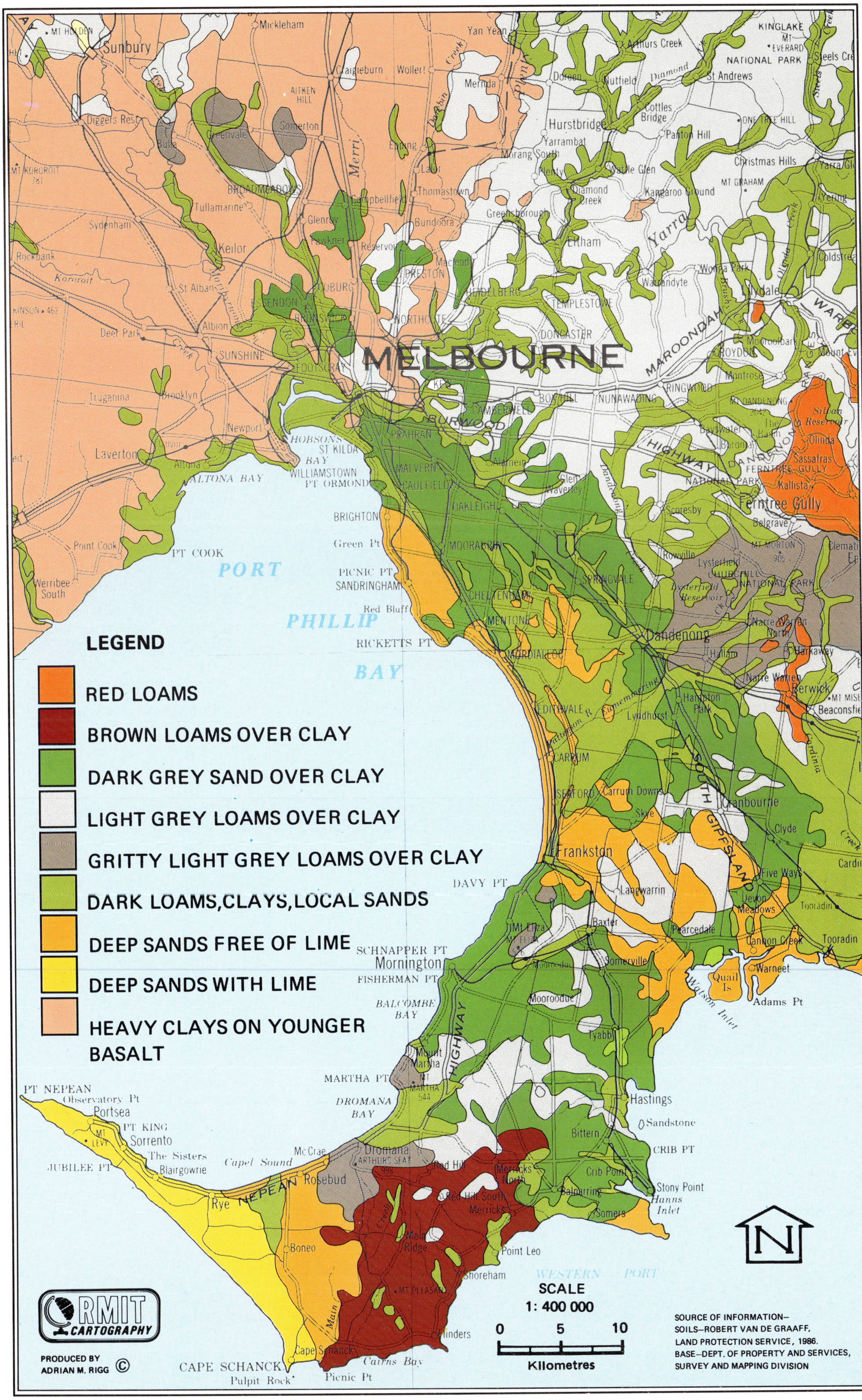

1

2

3

4

5

6

1. *Eucalyptus rubida* ssp. *rubida*
   Candlebark Gum
2. *Patersonia occidentalis*
   Long Purple-flag
3. *Podolepis jaceoides*
   Showy Podolepis
4. *Pomaderris lanigera*
   Woolly Pomaderris
5. *Stellaria flaccida*
   Forest Starwort
6. *Ajuga australis*
   Austral Bugle
7. *Stipa rudis* ssp. *rudis*
   Veined Spear-grass
8. *Pycnosorus globosus*
   Drumsticks
9. *Diuris punctata*
   Purple Diuris

7

8

9

may fall considerably over Summer, plants chosen must be able to survive dry conditions and also be able to grow in deep water.

Other problems in wetlands can be caused by too much plant growth; some plants can be very invasive, particularly *Eleocharis sphacelata*, *Phragmites australis* or either of the Australian *Typha* species. Although they make excellent and ornamental habitat for smaller water animals, waterfowl, and plants they are also fast spreading and can completely overgrow a dam or creek. The build up of silt and organic matter among their roots speeds up the ageing process of the water area, and the tall growth of *Typha* and *Phragmites* will completely screen the water from view with time. In large areas they may be kept contained by use of deep water channels, but you may need to remove seedlings once a year. In smaller ponds they are more manageable, and can be kept within bounds by planting them in a pot or other large container.

Some unwanted plants may appear around the water's edge. New plantings should be watched and seedlings removed by hand while small. Once plants are established it is usually more difficult for weeds to gain a toe-hold.

Like most Australian plants grown in their natural range, waterplants seem surprisingly trouble free. The most common problem affecting them is physical damage, and the most common culprits here are stock (particularly cattle) and yabbies. Stock of all kinds should be fenced out of wetland areas (they are a major cause of erosion and weed dispersal in any case), but yabbies require longer-term control. Some Australian fish such as cat-fish and golden perch will reduce their numbers to an acceptable level, but it may be several years before a yabby-filled dam can be planted.

One more aesthetic problem should be mentioned. Many people would like to grow native water-lilies but none of these will grow well (let alone flower) in most of Southern Australia. *Ottelia ovalifolia* with attractive, elongated floating leaves and large, showy, white flowers makes a good replacement. Other waterplants occurring naturally in the Melbourne area are described elsewhere in this book.

Water gardening adds new dimensions to an interest in Australian plants. A pond, dam or wetland draws a greater range of wildlife than land plants alone will attract, may take up surprisingly little space for the benefits it will bring, and requires as little maintenance as you would expect from any established indigenous planting.

# Further Reading

Aston, H., (1973) *Aquatic Plants of Australia*. Melbourne University Press, Melbourne. (Many of the names used have now been revised.)

Nelson, K.D. (1985) *Design and Construction of Small Earth Dams*. Inkata Press, Melbourne.

Sainty, G.R & Jacobs, S.W.L, (1981) *Waterplants of New South Wales*. Water Resources Commission, NSW.

Seagrave, C. (1988) *Aquatic Weed Control*. Fishing News, Farnham, Surrey. (An English book, but many of the common weeds, or related species, occur here also.)

*Triglochin striata*

# Chapter 5. *Planning and Design of Bushland Restoration*

Darcy Duggan*

## Introduction

An important outcome of the growing community interest in environmental issues has been an awareness of the role that native bushland and wildlife species play in defining the character of different areas around Melbourne. Little wonder then that community groups are alarmed to see the decline and loss of native vegetation as Melbourne creeps further into the surrounding countryside.

However, opportunities do exist for all residents throughout Melbourne to play a vital role in the conservation of many rare and threatened native plant and wildlife species.

This can be achieved by:

1. Encouraging the re-establishment of these species in their own backyards; and
2. Participating in the protection, management and restoration of remnant bushland.

Given the generally small and isolated nature of most urban bushland remnants, these areas will require intensive and sensitive management.

The success of any project is dependent on a commitment to funding and effort. Most important of all is the need for a well planned and structured approach, based on sound ecological guidelines and principles.

## Planning the Management of Natural Areas

The main objective of any bushland management program should be the creation of a 'manageable ecosystem', resulting in the rehabilitation of a native vegetation community and restoration of wildlife habitat.

This process of recreating the right conditions for re-growth of the desired plant community can be initiated by the removal of exotic and alien plants, and the encouragement of indigenous species to re-establish themselves.

It is vital therefore to recognise that each site will have individual and often diverse physical, chemical and biological characteristics (J. Rawlings 1990). The characteristics of a Eucalypt Woodland or Grassy Woodland community for example will differ markedly from that of a Coastal or Wet Forest community.

Understanding the history of the site and how it relates to its present condition is also critical in the planning and design of a project. There are a number of issues which must be considered such as:

- What is known about the original vegetation community structure and species diversity of a site?
- What seed or plant sources are available for revegetation works?
- What past disturbances and degree of alteration have occurred to bring about the current status of the site (eg, impact of grazing, altered fire regimes, clearing)?
- To what extent do changes in site conditions (eg filling) affect your ability to re-establish or restore a particular native vegetation community or ecosystem?

Researching historical references can often provide valuable information in reconstructing the original flora and fauna of an area, and in gaining a better understanding of the type and impact of disturbance processes. Useful sources of information include the Picture Collection (La Trobe Library), the Victorian Field Naturalists magazine, early journals including original surveyors' reports (former Dept of Lands—Victorian Archives annex—Laverton), local historical societies, and contacting older residents in an area.

Survey and study of other remnant vegetation areas nearby can be very valuable in gaining insight into the dynamics of a natural ecosystem. These sites are also extremely important for relocating species which may have been present, and in providing a source of material for re-introduction.

## Stage 1: Site Assessment

Site assessment is a critical first stage of a project, and must involve consideration of the following:

- the history of the site and how it relates to its present condition;
- overall management objectives which will influence project planning and design.

### Site Disturbance History

In determining the history of a site, it is necessary to work out the degree and type of disturbance that has occurred. Disturbance may be categorised as follows:

DEGREE—gap, patchy, marginal or all-pervasive.
FREQUENCY—one-off, periodic or continuous.
NATURAL—fire, landslips, storm damage and insect attack.
MAN-MADE—active and/or passive eg, fire, clearing, neglect.

Most bushland areas have been altered by one or more of these forms of disturbance, and will continue to fluctuate in response to environmental change. Some forms of disturbance are inevitable and part of the natural process of

* Darcy Duggan is a naturalist currently working in environmental planning. He is employed by the Yarra Bend Trust.

shaping the composition and structure of the wildlife and vegetation communities present.

Management must therefore focus on modifying the impact of disturbance initiated or magnified by human action. Major causes of site degradation include soil erosion, soil disturbance, nutrient input due to clearing and drainage discharge and rubbish dumping (hard rubbish and garden refuse).

Disturbance events causing site degradation are generally inter-related. For example, intense grazing and associated soil disturbance by rabbits will often result in soil erosion due to loss of vegetation cover. In some situations, selective grazing pressure of rabbits on vegetation may also favour the growth and spread of unpalatable weed species such as Serrated Tussock (*Nassella trichotoma*).

## Setting Management Objectives

In setting management objectives for a given site, it is essential that the area is considered and managed on a whole catchment basis wherever possible.

Account must also be taken of the irreversible changes that have occurred at each site. It may not be desirable to revegetate a swamp site which has previously been drained. Many housing and industrial estates have been established over former swamps. Another constraint is proximity to housing. Tea-tree Heaths are very flammable and revegetation would need to be modified where this would be a danger. Soil conditions were greatly altered in some coastal areas where sand dunes were bulldozed onto low-lying areas.

Current site conditions and the nature and extent of various degradation processes that have occurred will influence short term objectives. These may include the following:

**Retention** and protection of existing remnant bushland areas to prevent further degradation and loss of species;

**Restoration** of moderately degraded sites;

**Revegetation** of totally degraded or modified areas.

In defining long term objectives for a project site, consideration should be given to the following:

- full restoration to reflect the original species diversity and community structure;
- rationalisation and extension of the boundaries of the site to provide a buffer between the core bushland and urban development. This approach will enhance the overall stability of a site;
- the protection and enhancement of sites of landscape, geological, historical or archeological interest;
- planning for appropriate community use of an area consistent with preserving and maintaining its values.

## Major Factors to Consider in Site Analysis

The basic component of any site analysis is a flora and fauna survey. A good working knowledge and understanding of flora and fauna are therefore essential. It is impossible to identify problems such as weed invasion if you are unable to recognise the presence of weeds on site.

### 1. Flora survey and assessment

The initial flora survey should be carried out preferably during spring, when the majority of species including various annuals are evident. An ongoing survey throughout the year is also useful given the seasonal variations that occur within an area. This is especially the case for some sites where a large number of Winter annual and perennial species such as Angled onion (*Allium triquetrum*), Oxalis (*Oxalis pes-caprae*), Veldt grass (*Erharta longiflora*) and Quaking grass (*Briza maxima*) die back in the Summer. Additional surveys during this period allow a much greater appreciation of the extent of remnant vegetation present, without other weeds masking your view of the site.

When assessing existing vegetation, consideration must be given to the following:

- remnant indigenous species and vegetation community structure;
- native non-indigenous and ornamental/exotic species present, in particular invasive species and environmental weeds.

The relative significance of each plant species/community should be noted, for example:

- whether remnant indigenous species are of local, regional or state significance;
- historical value, landscape significance or wildlife habitat value;
- quality assessment of vegetation and extent of weed invasion. Identification of weed species present is essential in developing a management strategy for a given site (refer to Table 1 and further discussion).

### 2. Survey of existing fauna

This should include all resident and vagrant native and introduced animal and bird species. Assessment of the status or significance of each species should be made, as well as habitat needs and value of the site.

It is essential the habitat needs of wildlife are considered when developing a revegetation program. Recognition of the value of woody weeds (eg Boxthorn) in providing habitat to small bird species is important in ensuring the development of a staged program of removal of these species and replacement with suitable indigenous species. This factor is frequently overlooked in the desire to control and remove woody weeds in particular, and may result in the displacement or local extinction of remnant wildlife populations.

The presence of problem species such as rabbits must be identified and precautions taken to minimise damage. In developing a management program for these sites, priority must be given to establishing an integrated control and eradication program.

It is worth noting that the role of native fauna in ecosystem dynamics is only now beginning to be recognised. Recent observations and research suggest that the burrowing and feeding activities of small mammals such as the Bandicoot and Potoroo, may have played an important role in fire control through the reduction in ground litter along the edge of native grassland and woodland communities, in nutrient recycling and possibly in stimulating the regeneration of native herbaceous species reliant upon soil disturbance (J. Wombersley).

The extent of change in vegetation dynamics which has occurred as a result of the almost total elimination of small ground dwelling species is unknown.

In addition to a basic flora and fauna survey of a site, other factors that need to be considered during the very early phases of a project are:

**1. Soil type and structure.** May require a soil test to be carried out to determine pH, nutrient status etc. The nature of the soil influences both the dynamics and composition of a natural ecosystem and the extent to which ecological succession can occur. For example, the depth and physical, chemical and biological characteristics of topsoil will determine whether the status of a disturbed site is primary, secondary or intermediate in nature.

The biological status and 'health' of the soil is of particular importance. Only now are we beginning to realise the fundamental role that soil micro-organisms and associations between soil borne fungi and plant roots (mycorrhizas) play in nutrient cycling, water uptake and regeneration needs of native plants.

The original soils and topography may have altered resulting in the development of a completely different type of ecosystem from that previously present. This is certainly the case where these changes have been compounded by the introduction of various weed species.

**2. Landscape significance and sensitivity** of a site requiring the retention of strategic viewing areas or screening of intrusive features.

**3. Local or Regional context** of a site. Function of the site as a refuge for native plant and wildlife species, and existing or potential corridor connections with other natural areas eg road and stream reserves.

**4. Planning constraints,** ie area of land, existing underground or above ground services, ownership of land including legal status, covenants, planning controls and authorities or agencies responsible, boundaries (fencing), and planning controls affecting surrounding land use.

**5. General topography** of a site including its relationship to the surrounding landscape, levels, contours and specific features such as rocky escarpments.

**6. Existing access** into a site ie pedestrian, vehicular, parking facilities, general circulation through the area and preferred access routes which have least environmental impact.

## Survey and Mapping

To facilitate the survey and mapping of an area, it is recommended that a series of colour aerial photos (preferably 1:1000) with associated contour and grid overlay transparencies be obtained. These are particularly useful in giving an overview of the area.

Grid overlays, comprising a series of 100 m x 100 m grids or sectors, are an extremely valuable tool for preparing base maps. By accurately locating and marking the coordinates of each grid sector on the ground, it is possible to use these reference points to map in the various features for each area. By transposing this information onto transparencies, it is then possible to build up a composite picture for each sector.

This detailed approach has proved very useful in that it provides a full and comprehensive understanding of the inherent features and problems for a given site, and allows the preparation of an appropriate management program. The other major advantage is that it also ensures accurate documentation and monitoring of the site and associated management works, which can then be readily interpreted and used by others working in the field.

# Stage 2: Developing a Management Strategy and Action Plan

## Quality Assessment

The combination of each of the various features of the site will determine its overall quality. Quality assessment is therefore extremely valuable in setting both management objectives, and in developing an appropriate works program.

A useful model for assessing and mapping site criteria with regard to the quality of remnant vegetation areas, has been developed by the National Trust (NSW) Bush Regeneration program. Quality ratings range from category 1—undisturbed sites of highest quality and ecological diversity, to category 4—severe disturbance and low diversity.

A simple colour code allows ready identification of different areas.

**Red**—areas where bushland has been completely replaced by exotic plant species; or

—bushland where only overstorey species remain and regeneration is absent due to infestation of the understorey by exotics or invading native alien species.*

**Orange**—areas where bushland is severely infested by exotics and/or invading native alien species; and

—where the regeneration of the dominant species of that plant community are being significantly suppressed.

**Blue**—areas of bushland with moderate infestation of exotics and/or invading native alien species.

**Green**—areas of bushland virtually free of exotic plants where the native plant communities display the structure, species composition and diversity typical of those communities in non-urban situations.

In determining an appropriate management strategy and actions for a given area, management requirements will differ markedly. The following comments can be made with regard to the general treatment of each category:

**Red and Orange** (Category 3 & 4) sites. Totally degraded or highly modified ecosystem.

This is by far the most common situation throughout Melbourne, and is characteristic of many areas such as urban waterways which have undergone extensive filling and gross modification over the years. Associated changes in soil properties and structure as well as massive weed invasion can pose serious problems and constraints, which will substantially affect the overall approach to revegetating these areas.

An intensive weed control program is generally required prior to revegetation of these sites. Depending on the extent of disturbance and nature of the site conditions, the best approach may be by simulating ecological succession through a staged program of re-introduction of native species.

**Blue** (Category 2) sites. Moderately fragmented or degraded ecosystem.

This situation is becoming more common, and is indicative of the general decline and degradation of the environment. The increased isolation and vulnerability of flora and fauna to disturbance or local extinction is strongly influenced by factors such as reduced fitness of remnant populations due to inbreeding or inability to regenerate as a result of competition and invasion by exotic species.

Weed control is again a critical first stage in treatment of these sites to prevent further degradation and loss of species, and to encourage natural regeneration wherever possible. Supplementary planting and re-introduction of species formerly associated with the vegetation community, as well as improvement of wildlife habitat (eg hollow logs), is also valuable in re-establishing ecological diversity and stability.

**Green** (Category 1) sites. Reasonably intact or high quality ecosystem—(an increasingly rare phenomenon these days).

Vegetation is comparatively intact structurally with low to moderate levels of weed invasion and disturbance. This situation offers the greatest potential in terms of full restoration of ecological diversity and stability.

In this situation, the most appropriate strategy must be protection of the site by minimising further disturbance and degradation, and encouraging natural regeneration. This can be achieved through sensitive weed control, requiring the use of selective control techniques such as hand weeding out from the edge of intact areas (modified Bradley method), and use of spot spraying and selective herbicides.

**Table 1** Rating of Environmental Quality

| quality of understorey vegetation | | | | |
|---|---|---|---|---|
| primary physical disturbance | weed intrusion (impact of alien flora) | ecological balance | diversity of indigenous flora | rating |
| nil | nil to negligible | excellent | maximum attainable in vegetation type | 1 |
| (nil to) moderate | slight to moderate | reasonable | high to moderate | 2 |
| (nil to) severe | moderate to heavy | poor | significantly reduced | 3 |
| (nil to) severe | very heavy | drastically out of balance | low to absent | 4 |

In the case of most remnant areas where all four site conditions may be present, the overall management approach must be broad based. High priority must be given to both

* For the purposes of the above coding system, 'invading native alien plant species' is defined as any non-indigenous native plant which is capable of naturalising in a native vegetation community.

the protection of rare or significant species, and rehabilitation of high quality remnant vegetation areas, to prevent further degradation and loss of species. Equal importance should also be given to the control of specific invasive weed species, to contain their spread and prevent invasion into adjoining high quality areas.

In addition to management of high quality conservation sites, it is also good 'politics' to give preference to a few high profile areas (eg roadside reserves, picnic areas) where possible. Being seen to be doing something in the eyes of the public is critical in raising community awareness and support for your work, and increases the likelihood of gaining ongoing or additional funding.

Adoption of this approach however will ultimately be a question of balance between existing labour and resources, and other management priorities which you have identified.

# Other Planning and Design Issues

In addition to the assessment of the overall quality of a given site, there are a number of other factors which must also be considered when planning a project. These include issues such as:

## 1. Limitations relating to current information and availability of suitable revegetation materials

Lack of information regarding the original vegetation community structure and species diversity of a site, as well as a shortage of suitable seed/plant material for revegetation works, will be a major limitation to any project.

## 2. Current knowledge and level of expertise in revegetation techniques

Whilst considerable attention has been given to the development of successful strategies for the re-establishment of tree and shrub species in recent times, little attention has been given to ground flora.

There is an urgent need therefore for expanded field research and documentation in this area, particularly with regard to site and species interaction and dynamics. Wherever possible, a full complement of species must be included in any revegetation or restoration project.

In addition to encouraging natural regeneration, consideration should also be given to other techniques such as the use of tubestock, direct seeding (a potentially valuable method), transplanting material from threatened sites, and vegetative propagation, ie direct planting of cuttings and divisions of material on site.

## 3. Budget and resources

These issues are critical in influencing both the scale of the site works, and your ability to implement and maintain a project over the medium to long term. Lack of continuity and security of funding as well as over-commitment of resources is the most common cause of failure. Quality not quantity is the key to success in undertaking this work when faced with an uncertain funding future. A smaller project well done is more likely to gain support if it is seen to be successful.

# Critical Aspects of Planning and Implementation

Given the dynamic nature of revegetation and bushland restoration projects, forward planning is critical in ensuring appropriate timing and implementation of each phase of the program. Initial site preparation, for example, may need to be undertaken up to 12–18 months in advance of planting the site. This is particularly the case when dealing with heavily weed infested areas. In this situation, it will always be an advantage to expend extra time and effort in treating invasive weeds eg Fennel (*Foeniculum vulgare*) and to carry out follow up treatment to eliminate regeneration from soil-stored seed, before you attempt to revegetate an area.

Timing of post-planting maintenance is also critical to prevent re-seeding of weeds, and re-colonisation of the site. There are some situations however, such as highly exposed sites, where allowing initial recolonisation of a site by weeds can benefit young native plants, by acting as a nurse crop in providing protection from wind. In these circumstances, maintenance should be deferred until the point of flowering and seed set of weeds.

The overall importance of site preparation and follow up maintenance and their reliance on a well planned and structured approach to the project, cannot be over-emphasised.

Timing is also essential for other aspects of a project such as seed collection, where for many species, seed is only available for a few days in the year.

Finally, there is the need to maintain accurate records and to systematically document all work that is carried out. This aspect of a project is frequently ignored, resulting in much valuable information regarding the success or failure of various strategies and revegetation techniques being lost. Maintaining photographic records of a site, especially before and after photos, is extremely valuable. Photographic slides are especially useful, both as an educational resource, and to use in gaining further funding and support for projects.

# Summary

There are many clearly defined ecological principles which must be considered in developing an integrated management strategy. These include issues such as the importance of protecting existing remnant vegetation by minimising soil disturbance and damage to sites, retention and enhancement of wildlife habitat, use of local provenance genetic material in re-establishing plant species etc.

Treatment of causes of site disturbance and degradation is also critical. Weed invasion is without doubt a major environmental problem, and poses one of the greatest threats to the long term conservation of flora and fauna. This issue must therefore be given the highest priority in any project.

The extent of our ecological knowledge and understanding of native vegetation and wildlife management has expanded greatly over the last 10 years or so. We are still only scratching the surface however in many areas, and for this reason each of us has the opportunity of making an important contribution. Thorough documentation and monitoring of field research and trials undertaken, as well as exchange of information, are an essential element to the success of this process.

# Further Reading


Allen, M.F. (1988) *Reconstruction of Disturbed Arid Lands—An Ecological Approach*. Ed. E.B. Allen. American Association for the Advancement of Science (AAAS) 1988. Symposium 109.

Breckwoldt, R. (1983) *Wildlife in the Home Paddock*. Angus and Robertson, Sydney.

Buchanan, R.A. (1990) *Bush Regeneration: Recovering Australian Landscapes*. TAFE, NSW.

Carr, G.W. (1989) 'Perspectives on Environmental Weeds in Victoria'. *Weeds on Public Land—an Action Plan for Today*. Symposium presented by Weed Science Society of Victoria, Monash University, May 1988.

DePuit, E.J. & Redente, E.F., (1988) *Manipulation of Ecosystem Dynamics on Reconstructed Semi-arid Lands*. AAAS. Symposium 109.

Greening Australia Ltd, (1990) *Sowing the Seeds*. Direct Seeding and Natural Regeneration Conference Proceedings. 22–25 May, 1990.

McGee, P. (1986) *Mycorrhizal Associations of Plants in a Semi-arid Community*. Australian Journal of Botany 34, 585-593.

Rawlings, J. (1990) *Urban Bushland Management Handbook*. National Trust (NSW) unpublished Draft 3.

Stiller, D.M., Zimpfer, G.L., & Bishop, M. (1980) 'Application of Geomorphic Principles to Surface-mined Land Reclamation in Semi-arid west'. *Journal of Soil and Water Conservation*. 35: 274–277.

Wombersley, J. Director of Warrawong Wildlife Sanctuary, Adelaide Hills—South Australia. (personal communication)

# Chapter 6. Environmental Weed Invasions and their Conservation Implications

Geoff Carr*

## Introduction

The Victorian flora comprises over 3,200 indigenous species, about 30% of which occur in the Melbourne area. This is far from the only flora we have however: the exotic or introduced flora surrounds us at every turn and, more than any other factor, prejudices the survival of indigenous vegetation communities and plant species. In this chapter the size, origins, distribution and aspects of the biology and ecology of the exotic flora, as well as their significance as a threat to indigenous flora and fauna, are discussed. Particular attention will be focused on environmental weeds.

## Some Definitions

**Indigenous species** are those plants which are native to, and evolved in, a particular area or country.

**Exotic species** are those species which have been introduced from overseas. They may or may not be naturalised.

**Alien species** refer to plants which occur naturally outside the Melbourne area.

**Naturalised species** have become established outside gardens, farms or plantations and their self-maintaining populations persist without direct reliance on humans.

**Environmental weeds** are plants which aggressively invade natural bushland and displace native flora and fauna. They may be exotic, alien or ecologically 'out-of-balance' indigenous species and adversely affect survival or regeneration of indigenous species. Several categories of environmental weeds are discussed below.

**Noxious weeds** are species which affect agricultural production.

## Size of the Exotic Flora

Table 1 gives the statistics on the size of the indigenous, alien and exotic floras of Victoria. The information has been compiled from the *Census of the Vascular Plants of Victoria* (Ross 1990) and approximately 222 additional exotic species listed by others in the last 10 years. Unfortunately plant enthusiasts often neglect the weeds in favour of indigenous plants and naturalised exotic species may go unreported.

New species and new records are continually being added to the Victorian Census but the list of 'new' indigenous species is growing more slowly and should be more-or-less stable when the Australian flora is fully studied. Not so the exotic flora!

With about 1220 species (28% of the total flora), the exotic flora of Victoria is exceptionally large in proportion to the indigenous flora by both world and Australian standards.

The Melbourne region has a large indigenous flora but also an exceptionally large exotic flora. The Department of Conservation & Environment lists some 403 alien and exotic species (from quadrat data and species lists) but the real figure is much higher and few exotic species in Victoria do not occur in the Melbourne region. A tentative estimate of the number of alien and exotic species which occur in the Melbourne region is about 1000 species. This should be seen in the context of ongoing **extinction of indigenous species** and continual **addition of exotic species.**

It will be apparent to anyone who has witnessed the massive regional invasion of environmental weeds in indigenous vegetation, that these exotic species pose the greatest threat to indigenous species and the vegetation communities to which they belong.

**Table 1** Numbers of taxa in the exotic and alien vascular flora of Victoria compared with the indigenous flora. Based on Ross (1990) and 222 additional naturalised exotic taxa (Carr).

| Group | Naturalised flora | | | Indigenous flora | | |
|---|---|---|---|---|---|---|
| | Families | Genera | Species and Infraspecific taxa | Families | Genera | Species and Infraspecific taxa |
| Ferns and Fern Allies | 3 | 3 | 4 | 25 | 51 | 157 |
| Gymnosperms (Conifers) | 2 | 4 | 14 | 2 | 2 | 6 |
| Monocotyledons | 20 | 134 | 309 | 24 | 191 | 874 |
| Dicotyledons | 94 | 371 | 894 | 103 | 418 | 2107 |
| Totals | 119 | 512 | 1221 (28%)* | 154 | 662 | 3144 (72%)* |

*Percentage of total Victorian flora of approx. 4365 taxa.

* Geoff Carr, a botanical consultant, is a director of Ecological Horticulture P/L which is involved in vegetation surveys, studies for vegetation management, assessments, research and education. His special interests include research on Australian orchids and environmental weeds. He is a founder and past President of the Indigenous Flora & Fauna Association.

## Origins of the Exotic Flora

The establishment of exotic plant species began on the day that Europeans set foot in Victoria. In March 1801, Lt. Jas. Grant in the 'Lady Nelson' called at Churchill Is. in Western-port Bay and planted food plants including peaches, apples, wheat, onions and potatoes. Documentation of the exotic flora in South Australia showed that only about 20 years after founding, 101 plant species had already become naturalised. This must have been closely comparable to the development of the exotic flora of the Melbourne region—both in the methods of introduction of exotic species and the composition of the exotic floras involved.

Plants have been either deliberately or accidentally introduced, and for some species, both probably apply.

Deliberate introduction of species for ornament or utility has been more significant than accidental introductions and it is estimated that between 65 and 70% of all naturalised exotic species in Victoria have been introduced this way, mostly for ornamental horticulture. This is illustrated by Guilfoyle's *Catalogue of Plants under Cultivation in the Melbourne Botanic Gardens* (1883). Of the several thousand species listed, at least 337 (27% of the current exotic flora) have become naturalised in Victoria. Indeed many hundreds of plant species are currently available in the nursery trade and they include some of the most destructively invasive of environmental weed species. Of the 570 species of serious or potentially serious Victorian environmental weeds (listed in the recent compilation by the author and co-workers), no less than 43% are commercially available.

The horticultural industries are by no means the only source of weed species—they are routinely used for soil stabilisation and in other environmental management practices. Of the numerous plant species advocated for soil stabilisation in Victoria by government agencies, 51 are seriously weedy.

The influx of exotic plant species into Victoria and Australia goes unchecked. Agricultural scientists, for example, continue to scour the world for potentially useful species (eg for fodder or for rehabilitation of saline land) which will ideally behave as weeds. In addition, current weed species (eg *Phalaris aquatica*, Canary Grass) are being genetically 'improved' to enhance their ability to flourish in a wider range of conditions. Plant quarantine laws are outdated by contrast with animal quarantine laws.

The **geographic origins** of the exotic flora reflect regions or countries with which historically we have had greatest political, economic and cultural ties. They come from climatically similar regions, especially those with a Mediterranean climate. The main regions or countries which have provided the exotic flora are (in decreasing order of importance): Europe and the British Isles, Mediterranean Basin, South Africa, Asia, South America, North America, Western Asia, Central America and Mexico, North Africa, China and New Zealand. Naturalised species originating in Australia (including Victoria) are discussed below.

## Life Forms

A wide variety of life forms (or growth habits) is found in the exotic flora. They range from annuals (the most common life form) through perennial herbs to shrubs, vines and trees; some are purely aquatic and there is a larger range of succulents than in the indigenous flora. Life forms of the most serious environmental weed species in Victoria are given in the Appendix (Page 30).

## Conservation Implications of Environmental Weed Invasions

Invasion by environmental weeds is a world-wide phenomenon and it is without doubt the most serious conservation threat to our indigenus flora and fauna. These invasions are part of the world-wide destruction and degradation of the flora (and by implication fauna) caused by human activity. Environmental weed invasions have been documented in many regions throughout the world.

Considering the conservation and economic implications, relatively little attention has so far been given to environmental weeds within Australia. For Victoria, several generalised or anecdotal accounts have been published. More recently an overview of an environmental weed invasion in the state gave rise to the preparation of a list of some 570 species regarded as environmental weeds in Victoria.

A local weed flora has been published by the Friends of Sherbrooke Forest and several pamphlets or booklets have appeared eg. those published by the City of Frankston and the Shire of Eltham. A further publication by the Regional Pest Plant Strategy Working Group treats cultivated 'garden-escape' environmental weeds in the Dandenong Ranges. All above-mentioned publications deal with weeds contained within and/or adjoining the Melbourne region.

The most detailed and comprehensive documentation of environmental weed invasion in Victoria has come from specific studies in connection with environmental effects statements, vegetation and species management investigations, and assessments of vegetation significance. Well over 100 such studies have been carried out in the Melbourne region.

Ecological studies on specific environmental weed species have also been carried out, sometimes in the Melbourne region.

## What is the nature of the environmental problems posed by environmental weeds?

Indications are that the future of the flora and fauna of much of Victoria, and of Australia, is in jeopardy and faces annihilation at the community and species levels.

Almost all remnant bushlands are being invaded by weeds or will be invaded in the future. Without appropriate management we may expect gross changes in vegetation structure and composition. Communities will become impoverished and ultimately indigenous species will be unable to regenerate because of competition from weeds. With the destruction of the flora, the indigenous fauna, which is directly or indirectly dependent on indigenous vegetation, will also collapse. The extent and rate of this projected ecological collapse may be unprecedented in evolutionary time without appropriate management.

One of the most spectacular demonstrations of invasion, and a case which is regarded as a preview of the longer-term scenario is seen at the You Yangs, a low granite range southwest of Melbourne. Here indigenous vegetation has been substantially eliminated by invasions of *Chrysanthemoides monilifera* (Boneseed), *Myrsiphyllum asparagoides* (Bridal Creeper, Smilax Asparagus) and numerous exotic grasses, over a period of about 35 years.

The flora and fauna of the Melbourne region have already suffered severe degradation and destruction from a variety of causes related to settlement such as grazing and clearing but environmental weed invasions could now destroy most indigenous vegetation remnants. Already an alarming proportion of indigenous species in the Melbourne flora have become extinct, and many more are on the brink of extinction; the same applies regionally to vegetation communities.

## Types of Environmental Weeds

It is convenient to recognise several different types of environmental weeds according to their origins, but it is also necessary to identify another kind of invasion—that of **genes** contributed to populations of indigenous species by hybridisation with alien and exotic species. A further problem of weediness concerns naturalised indigenous species. Both are discussed below.

Three groups of environmental weeds are recognisable according to geographical origins:

- species introduced deliberately or accidentally from overseas (exotic);
- Australian plant species from outside Melbourne (alien);
- Victorian species naturalised outside their natural geographic range, including 'garden escapees' and those which have been inadvertently dispersed.

## Size of the Environmental Weed Flora

The recent list of environmental weeds in Victoria has about 570 entries. Most are species but a few hybrids and horticultural selections are included. This represents about 46% of the total exotic flora of Victoria. Almost all species on the list occur in the Melbourne region.

In any given reserve or larger area, the number of environmental weed species is invariably high, and many of these have the capacity to destroy indigenous vegetation over time. Statistics for representative areas in the greater Melbourne region are given in Table 2. The most serious environmental weed species in Victoria are listed in the Appendix.

## Australian Plants as Environmental Weeds

About 120 species of Australian plants have been recorded as naturalised in Victoria outside their natural range, as 'escapees' from cultivation. They include species from all states and are amongst the most devastatingly invasive of all weeds (see Appendix). On present trends it is clear that many additional Australian plant species will become naturalised in Victoria.

The ease with which Australian plants naturalise is demonstrated in several locations where plantations of 'native' plants have been established in sites adjoining intact or partially intact indigenous vegetation. At Anglesea where two plantations contain about 50 Australian tree and shrub species, no less than 35 of these species have become naturalised and many are exceedingly serious threats (eg. *Melaleuca armillaris*, Bracelet Honey-Myrtle, and *M. diosmifolia*, Green Honey-myrtle).

A similar response is seen at Mt. Martha Park on the Mornington Peninsula where at least 102 species of Australian plants have been cultivated in regionally significant woodland. Of these species, 38% have become naturalised and now threaten the survival of the remnant vegetation. Examples include *Hakea suaveolens* (Sweet Hakea) and *Baeckea virgata* (Tall Baeckea).

## Hybridisation between Indigenous, Alien and Exotic Species

Hybridisation between indigenous, alien and exotic species has been reported in many countries. In Victoria at least 12 such hybrids involving indigenous species have been recorded in six genera: *Acacia* (4 spp.), *Coprosma* (1), *Correa* (1), *Grevillea* (4), *Nicotiana* (1) and *Pittosporum* (1).

One very large hybrid population at West Barwon dam (Otway Ranges) involves *Acacia mucronata* x *A. longifolia* and numbers hundreds of individuals. How far or fast the genes of *A. longifolia* (naturally confined to East Gippsland and the Grampians in Victoria) will move through the extensive *A. mucronata* populations in the Otways, and the potential ecological consequences, is not known. *A. longifolia* was originally planted as garden specimens.

Of similar concern is the contamination of the gene pools of the regionally-rare *Grevillea rosmarinifolia* at Hurstbridge and Diamond Creek by alien *Grevillea* spp. Other species, involving *Grevillea baueri*, *G. juniperina* and other *G. rosmarinifolia* forms as well as hybrids, are cultivated in nearby gardens from which honey-eaters carry pollen to the local wild plants. The complex hybrid swarms resulting will cause the demise of the pure *G. rosmarinifolia* as hybrids are now

**Table 2** Statistics on the weed flora of representative biological reserves or areas in the Melbourne region.

| Reserve/area | Area (ha) | Broad vegetation communities present | Total no. of plant species recorded | Indig spp. (%) | Exotic spp. (%) | No. of environ weeds targeted for control or elimination | Refs |
|---|---|---|---|---|---|---|---|
| Gramatan Ave Heathland Sanctuary—City of Sandringham | 0.27 | Tea-tree Heath | 107 | 50 (47%) | 57 (53%) | 57 | Carr et al. 1991b |
| Hochkins Ridge Flora Reserve—Croydon Nth | 18.6 | Dry & Valley Sclerophyll Forest | 378 | 227 (60%) | 151 (40%) | 67 | Carr et al. 1991a |
| Moorooduc Quarry Reserve—Mt. Eliza | 37 | Wetland, Sclerophyll Woodland, Swamp Shrubland | 314 | 182 (58%) | 132 (42%) | 88 | Carr & McMahon 1988 |
| Plenty Gorge Metropolitan Park—Mernda to Greensborough | 1400 | Dry & Valley Sclerophyll Forests, Wetlands Grassy Woodlands Grasslands | 696 | 411 (59%) | 285 (41%) | not applic | Carr et al. 1991c |
| Puffing Billy Corridor—Shire of Sherbrooke | | Sclerophyll Forest Damp Sclerophyll Forest | 176 | 95 (54%) | 81 (46%) | 57 | McMahon et al. 1990 |
| Shire of Eltham | | Sclerophyll Forest | 696 | 348 (50%) | 348 (50%) | 80 | McMahon et al. 1989 |
| Truganina Swamp/ Cherry Lake—Altona | 350 | Saltmarsh, Grassland | 307 | 141 (46%) | 166 (54%) | 36 | Carr et al. 1989 |

more abundant and more vigorous and out-compete their wild parents.

At Mt. Dandenong, hybridisation between a rare distinctive *Grevillea alpina* population and garden-grown *G. rosmarinifolia* is also reported. A comparable outcome to the *G. rosmarinifolia* Hurstbridge hybridisation is predictable.

Hybrids of this kind will become increasingly common, as will hybrids between local and non-local provenances of the same species where the latter are cultivated near natural populations. This has been termed 'genetic pollution' and the introduction of foreign genes may have harmful ecological consequences. The most obvious potential for 'genetic pollution' is seen in ***Eucalyptus camaldulensis*** around Melbourne. Numerous wild populations of this species probably now contain 'foreign' genes because of extensive planting of Northern Victorian Red Gums, for example on the freeway around Yarra Bend Park and Latrobe University campus. Seed from these populations can no longer be used for revegetation purposes, and this is yet another convincing argument for planting **local provenances** of indigenous species wherever the latter are required.

# Ecologically 'out-of-balance' Indigenous Species

In recent decades population explosions of several Victorian indigenous species have occurred, either in vegetation communities in which the species normally occurs, or in adjoining vegetation communities. Such species may be as destructively invasive as any truly exotic environmental weed species. The most serious of these species in Victoria are *Acacia sophorae* (Coast Wattle), *Kunzea ambigua* (White Kunzea), *K. ericoides* (Burgan) and *Leptospermum laevigatum* (Coast Tea-tree).

In the Melbourne region *A. sophorae* is highly invasive in coastal and near-coastal heath, woodland and forest, as is *L. laevigatum*. *K. ericoides* has recently been reported as a serious vegetation management problem in several areas.

The reasons for the upsurge in these indigenous species are poorly understood but may involve altered fire regimes (*Leptospermum laevigatum* and *Acacia sophorae*), increased grazing pressure by kangaroos, wallabies and stock (*Kunzea ericoides*), and increased seed dispersal by vehicles and exotic animals (*A. sophorae*). A further species, *Tetrarrhena juncea* (Forest Wire-grass), is considered a problem in Sherbrooke Forest as a result of reduced grazing by wombats. It inhibits regeneration of other species and reduces availability of lyrebird feeding sites.

In future, indigenous species will increasingly require active management to prevent ecological imbalance, especially in remnant vegetation.

# Dispersal

Dispersal of weed seeds or plant parts is a key attribute determining the success of weed species. Some produce no seed, instead reproducing exclusively from bulbs, corms, fragmented stems, stolons or leaves, for example *Salix fragilis* (Crack Willow), *Oxalis pes-caprae* (Soursob) and *Vinca major* (Blue Periwinkle). The latter has dispersed along vast tracts of riverine environments in East Gippsland indicating that seed production is not a prerequisite for success.

Many different seed dispersal mechanisms are evident in the alien and exotic flora. This has important implications from a vegetation management perspective because some control may be effected to reduce dispersal of environmental weeds.

A very large proportion of species is dispersed by seed adhering to animals' fur, people's clothing, in mud on feet, and the like. This may be good reason for limiting access of people in certain areas, and is of critical importance in stock-grazing of indigenous vegetation.

Fleshy seeds and fruit (some 8% of the Victorian non-indigenous flora), as well as a range of other weeds including grasses and clovers (*Trifolium* spp.), are eaten by a large range of birds and mammals, such as gulls, silvereyes, currawongs, emus, blackbirds, starlings, cattle, horses, sheep and foxes, and even lizards, and dispersed over a wide area. Some of the most devastating weeds originate in gardens and are dispersed by birds, especially the exotic blackbird. Examples of species with fleshy fruits or seeds dispersed in this way are given in the Appendix.

There is a continual weed seed rain into indigenous vegetation from surrounding areas. In the case of wind dispersed species (notably of the family *Asteraceae*) seeds may come from a great distance.

The dumping of garden rubbish is one of the most significant of all means by which exotic plants find their way into indigenous vegetation and corms, bulbs, tubers, stem leaves and whole plants may become established in this way. Education to prevent this common practice is urgently needed.

A final but exceedingly important seed dispersal mechanism is via the motor vehicle. One Canberra scientist recovered and germinated seed of no less than 116 plant species from sludge in a car wash, including very serious environmental weeds!

The most significant management lessons from seed dispersal stories are the following:

- given the extremely effective dispersal of weeds, sometimes over long distances, there is no safe place where weeds may be tolerated as posing no threat to indigenous vegetation. When cultivation of serious environmental weeds is finally banned the ban should be universal and not restricted to certain places;
- it is pointless to control or eliminate weed populations in bushland if there is a constant rain of seed of garden-grown weed species from adjoining properties.
- restriction of access in certain areas of reserves (people, horses and/or vehicles) is often advisable to limit spread of weeds.
- stock grazing on public land and in indigenous vegetation in general is disastrous.
- drainage water from roads and tracks should be very carefully controlled because of its role in seed dispersal as well as creating physical environments conducive to weed invasions.

# The Conservation Implications

## Vegetation Management

Vegetation requires management. Gone are the days when nature 'looked after itself', when an area culd be simply fenced and left to its own devices.

The need for active vegetation management—not only that in reserves but **all** indigenous vegetation remnants—is very poorly appreciated. In the Melbourne region, as elsewhere, adequate vegetation management is a lamentably rare phenomenon, either on public or private land. If we neglect management of private land then we may consign its vegetation and fauna to the scrap heap of history. Degradation is frequently irreversible.

An array of harmful vegetation management procedures is routinely practised by public and private management agencies and individuals. Such practices include inappropriate fire regimes, vegetation clearance, mowing, altered water cycles, nutrient enrichment of soils and water, stock grazing, non-renewable resource exploitation, soil compaction and disturbance, and planting of environmental weeds.

The discipline of indigenous vegetation management is still in its infancy and little published information is available. Unfortunately there is often a lack of appreciation, even amongst practitioners, that for vegetation management 'near enough' is not good enough. A full inventory of indigenous and exotic plant species is required as well as information on the structure and composition of vegetation (ie quadrat data) and the physical environment. Management plans can be formulated on the basis of this information. Indigenous animals should not be overlooked since management for vegetation objectives may be antagonistic to faunal values.

Vegetation management plans have been prepared for many reserves (see Further Reading) but the necessary amount of management is seldom possible because of inadequate financial resources and/or available expertise. Too often the perception prevails that 'it will happen in good time'. For most regional vegetation it is now the eleventh hour; a lost opportunity or failure of timing may mean consigning vegetation and fauna to annihilation.

An excellent case in point is the phoenix-like recovery of (now) exceedingly rare heath vegetation, ostensibly eliminated by invading *Leptospermum laevigatum* (Coast Tea-tree) at Sandringham. An astonishing recovery from soil-stored seed occurred after fire destroyed the tea-tree. If this site had not been weeded in the first six months after fire, it would have been irredeemably lost. A second opportunity would **never** arise. Furthermore, ongoing maintenance of the site is mandatory.

## The Future?

The conservation scenario sketched in this chapter is disturbing but we cannot believe that it is an inevitable consequence of settlement or we will consign the results of millions of years of evolutionary history to oblivion. A great deal can be done to maintain the integrity of the flora and fauna both for its own sake (always the primary imperative) and for the benefit of future generations. To achieve this something approximating a revolution needs to occur—a major change in awareness, communication and action which needs to be Australia-wide.

The initiatives listed below are essential to tackle environmental weed invasions on the broadest front.

- overhaul of all plant importation and quarantine laws;
- horticultural, agricultural and land management practices using exotic species must be revised. It should be illegal to cultivate or trade in environmental weeds;
- the vast array of deleterious land management practices must be curtailed or revised;
- research and documentation of environmental weed invasions is required;
- education at all levels;
- action on weed monitoring and control must take place, **prevention** rather than cure being the watchword.

# Appendix: Very Seriously Invasive Environmental Weed Species in Victoria

## Life Form

A — Annual
P — Perennial
Pt — Tuberous perennial
T — Tree
Ea — Emergent aquatic herb
Pb — bulbous perennial
Rc — Root climber
V — Vine
Ls — Large shrub
Pc — Cormous perennial
S — Small to medium shrub
X — Succulent perennial herb, subshrub or shrub

## Origin

Af — Africa
C — Central
Eur — Europe
Jap — Japan
NZ — New Zealand
Tem — Temperate
Am — America
Cal — California
E/as — Eurasia
Med — Mediterranean
Qld — Queensland
Vic — Victoria
Arg — Argentina
Ch — China
Hem — Hemisphere
N — North
S — South
W — West
As — Asia
E — East
Is — Islands
NSW — New South Wales
Tas — Tasmania
WA — Western Australia

| Latin name | Common name | Family | Life form | Origin |
|---|---|---|---|---|
| *Acacia baileyana* | Cootamundra Wattle | Mimosaceae | T | NSW |
| *Acacia decurrens* | Early Black Wattle | Mimosaceae | T | NSW |
| *Acacia elata* | Cedar Wattle | Mimosaceae | T | NSW |
| *Acacia longifolia* | Sallow Wattle | Mimosaceae | Ls/T | Vic, NSW, Qld |
| *Acacia saligna* | Golden Wreath Wattle | Mimosaceae | Ls/T | WA |
| *Acacia sophorae* | Coast Wattle | Mimosaceae | Ls | Vic, NSW, Qld, Tas |
| *Acer pseudoplatanus* | Sycamore Maple | Aceraceae | T | Eur |
| *Agrostis capillaris* | Brown-top Bent | Poaceae | P | N Hem |
| *Agrostis stolonifera* | Creeping Bent | Poaceae | P | N Hem |
| *Allium triquetrum* | Angled Onion | Liliaceae | Pb | W Med |
| *Alstroemeria aurea* | Alstroemeria | Liliaceae | Pt | Chile, Arg |
| *Ammophila arenaria* | Marram Grass | Poaceae | P | Eur |
| *Anthoxanthum odoratum* | Sweet Vernal-grass | Poaceae | P | Eur |
| *Artemisia verlotiorum* | Chinese Wormwood | Asteraceae | P | Eur, As |
| *Avena fatua* | Wild Oat | Poaceae | A | Med |
| *Brassica tournefortii* | Mediterranean Turnip | Brassicaceae | A | As |
| *Briza maxima* | Large Quaking-grass | Poaceae | A | Med |
| *Bromus catharticus* | Prairie Grass | Poaceae | A | S Am |
| *Bromus diandrus* | Great Brome | Poaceae | A | Med |
| *Carpobrotus edulis* | Hottentot Fig | Aizoaceae | X | S Af |
| *Carrichtera annua* | Wards Weed | Brassicaceae | A | Med |
| *Chrysanthemoides monilifera* ssp. *m.* | Boneseed | Asteraceae | Ls | S Af |
| *Coprosma repens* | New Zealand Mirror-bush | Rubiaceae | Ls | NZ |
| *Coprosma robusta* | Karamu | Rubiaceae | Ls/T | NZ |
| *Cortaderia jubata* | Pink Pampas Grass | Poaceae | P | S Am |
| *Cortaderia selloana* | Pampas Grass | Poaceae | P | S Am |
| *Cotoneaster divaricata* | Cotoneaster | Rosaceae | Ls | As |

| Latin name | Common name | Family | Life form | Origin |
|---|---|---|---|---|
| *Cotoneaster glaucophyllus* | Cotoneaster | Rosaceae | Ls | Ch |
| *Cotoneaster pannosus* | Cotoneaster | Rosaceae | Ls | E As |
| *Crataegus monogyna* | Hawthorn | Rosaceae | Ls/T | Eur |
| *Critesion marinum* | Sea Barley-grass | Poaceae | A | Med |
| *Crocosmia* x *crocosmiiflora* | Montbretia | Iridaceae | Pc | S Af |
| *Cynara cardunculus* | Artichoke Thistle | Asteraceae | P | Med |
| *Cytisus multiflorus* | White Spanish Broom | Fabaceae | Ls | Spain, Portugal |
| *Cytisus palmensis* | Tree Lucerne | Fabaceae | Ls/T | Canary Is |
| *Cytisus scoparius* | English Broom | Fabaceae | S | Eur |
| *Delairea odorata* | Cape Ivy | Asteraceae | V | S Af |
| *Dipogon lignosus* | Dipogon | Fabaceae | V | S Af |
| *Ehrharta calycina* | Perennial Veldt Grass | Poaceae | P | S Af |
| *Ehrharta erecta* | Panic Veldt Grass | Poaceae | P | S Af |
| *Ehrharta longiflora* | Annual Veldt Grass | Poaceae | A | S Af |
| *Eragrostis curvula* | African Love-grass | Poaceae | P | S Af |
| *Erica arborea* | Tree Heath | Ericaceae | Ls | Med, SW Eur |
| *Erica baccans* | Berry-flower Heath | Ericaceae | S | S Af |
| *Erica lusitanica* | Spanish Heath | Ericaceae | S | SW Eur |
| *Foeniculum vulgare* | Fennel | Apiaceae | P | Eur, W As |
| *Fraxinus rotundifolia* ssp. *rotundifolia* | Desert Ash | Oleaceae | T | Med, SW As |
| *Freesia leichtlinii* | Freesia | Iridaceae | Pc | S Af |
| *Galium aparine* | Cleavers | Rubiaceae | A | Eur |
| *Genista linifolia* | Flax-leaf Broom | Fabaceae | S | Med |
| *Genista monspessulana* | Montpellier Broom | Fabaceae | S | Med |
| *Gladiolus undulatus* | Gladiolus | Iridaceae | Pc | S Af |
| *Hakea salicifolia* | Willow-leaf Hakea | Proteaceae | Ls | NSW, Qld |
| *Hakea suaveolens* | Sweet Hakea | Proteaceae | Ls | WA |
| *Hedera helix* | Ivy | Araliaceae | Rc | Eur |
| *Holcus lanatus* | Yorkshire Fog | Poaceae | P | Eur, As |
| *Homeria flaccida* | One-leaf Cape Tulip | Iridaceae | Pc | S Af |
| *Homeria miniata* | Two-leaf Cape Tulip | Iridaceae | Pc | S Af |
| *Hypericum androsaemum* | Tutsan | Clusiaceae | S | Eur, N Af, W As |
| *Hypericum calycinum* | Large-fl. St Johns Wort | Clusiaceae | S | SE Eur, W As |
| *Hypericum perforatum* | St Johns Wort | Clusiaceae | P | Eur, W As |
| *Ilex aquifolium* | Holly | Aquifoliaceae | T | Eur |
| *Juncus acutus* | Sharp Rush | Juncaceae | P | Med,Eur,SAm,N Am,S Af |
| *Juncus effusus* | Soft Rush | Juncaceae | P | temp N Hem |
| *Lagurus ovatus* | Hares-tail | Poaceae | A | Med |
| *Leersia oryzoides* | Cut-grass | Poaceae | Ea(P) | Eur, temp As, N Am |
| *Leptospermum laevigatum* | Coast Tea-tree | Myrtaceae | Ls/T | Vic, NSW, Tas |
| *Leycesteria formosa* | Himalayan Honeysuckle | Caprifoliaceae | Ls | Himalaya |
| *Lonicera japonica* | Japanese Honeysuckle | Caprifoliaceae | V | E As |
| *Lotus corniculatus* | Birds-foot Trefoil | Fabaceae | A | Eur |
| *Lotus uliginosus* | Greater Birds-foot Tref. | Fabaceae | P | Eur |
| *Lycium ferocissimum* | African Box-thorn | Solanaceae | Ls | S Af |
| *Marrubium vulgare* | Horehound | Lamiaceae | P | Eur |
| *Melaleuca armillaris* | Giant Honey-Myrtle | Myrtaceae | Ls/T | Vic, NSW, Qld Tas |
| *Mesembryanthemum cystallinum* | Ice Plant | Aizoaceae | A | S Af |
| *Myrsiphyllum asparagoides* | Bridal Creeper | Liliaceae | Pt | S Af |
| *Myrsiphyllum scandens* | Myrsiphyllum | Liliaceae | Pt | S Af |
| *Nassella neesiana* | Chilean Spear-grass | Poaceae | P | S Am |
| *Nassella trichotoma* | Serrated Tussock | Poaceae | P | S Am |
| *Oenothera glazioviana* | Evening Primrose | Onagraceae | P | garden origin |
| *Oxalis pes-caprae* | Soursob | Oxalidaceae | Pb | S Af |
| *Paraserianthes lophantha* | Cape Wattle | Mimosaceae | Ls/T | WA |
| *Paspalum dilatatum* | Paspalum | Poaceae | P | S Am |
| *Paspalum distichum* | Water Couch | Poaceae | Ea(P) | cosmopolitan |
| *Passiflora mollissima* | Banana Passionfruit | Passifloraceae | V | S Am |
| *Pennisetum alopecuroides* | Swamp Foxtail-grass | Poaceae | P | As, Pacific |
| *Pennisetum clandestinum* | Kikuyu | Poaceae | P | E Af |
| *Pennisetum macrourum* | African Feather-grass | Poaceae | P | S Af |
| *Phalaris aquatica* | Toowoomba Canary-grass | Poaceae | P | Med |
| *Phalaris arundinacea* | Reed Canary-grass | Poaceae | Ea(P) | N Hem, S Af |
| *Phleum pratense* | Timothy Grass | Poaceae | P | Eur |
| *Pinus pinaster* | Cluster Pine | Pinaceae | T | Med |
| *Pinus radiata* | Monterey Pine | Pinaceae | T | Cal |
| *Pittosporum undulatum* | Sweet Pittosporum | Pittosporaceae | T | E Vic, NSW, Qld |
| *Polygala myrtifolia* | Myrtle-leaf Milkwort | Polygalaceae | S | S Af |
| *Prunus cerasifera* | Cherry-plum | Rosaceae | T | E/as |
| *Prunus laurocerasus* | Cherry Laurel | Rosaceae | T | E Eur, W As |
| *Psoralea pinnata* | Blue Psoralea | Fabaceae | Ls | S Af |
| *Pyracantha angustifolia* | Orange Firethorn | Rosaceae | Ls | Ch |
| *Pyracantha crenulata* | Nepal Firethorn | Rosaceae | Ls | Himalaya |
| *Rhamnus alaternus* | Italian Buckthorn | Rhamnaceae | Ls | Eur |
| *Romulea rosea* | Common Onion-grass | Iridaceae | Pc | S Af |

| Latin name | Common name | Family | Life form | Origin |
|---|---|---|---|---|
| *Rosa rubiginosa* | Sweet Briar | Rosaceae | Ls | Eur |
| *Rubus cissburiensis* | Blackberry | Rosaceae | Ls | Eur |
| *Rubus discolor* | Blackberry | Rosaceae | Ls | Eur |
| *Rubus polyanthemus* | Blackberry | Rosaceae | Ls | Eur |
| *Rubus rosaceus* | Blackberry | Rosaceae | Ls | Eur |
| *Rubus ulmifolius* | Blackberry | Rosaceae | Ls | Eur |
| *Rubus vestitus* | Blackberry | Rosaceae | Ls | Eur |
| *Rumex sagittatus* | Climbing Dock | Polygonaceae | V | S Af |
| *Salix cinerea* | Grey Sallow | Salicaceae | Ls/T | Eur, W As, N Af |
| *Salix* x *rubens* | Crack Willow | Salicaceae | T | Eur |
| *Salpichroa origanifolia* | Pampas Lily-of-the-Valley | Solanaceae | Pt | S Am |
| *Senecio angulatus* | Climbing Groundsel | Asteraceae | V | S Af |
| *Solanum pseudocapsicum* | Madeira Winter Cherry | Solanaceae | S | S Am |
| *Sollya heterophylla* | Bluebell Creeper | Pittosporaceae | V | WA |
| *Sparaxis bulbifera* | Harlequin-flower | Iridaceae | Pc | S Af |
| *Spartina maritima* | Cord-grass | Poaceae | Ea(P) | W Eur |
| *Spartina townsendii* | Townsends Cord-grass | Poaceae | Ea(P) | Eur (hybrid) |
| *Thinopyrum junceum* | Sea Wheat-grass | Poaceae | P | W & N Eur |
| *Tradescantia albiflora* | Wandering Jew | Commelinaceae | P | S Am |
| *Trifolium arvense* | Hares-foot Clover | Fabaceae | A | Eur, W As |
| *Trifolium repens* | White Clover | Fabaceae | P | Eur, As |
| *Typha latifolia* | Great Reedmace | Typhaceae | Ea(P) | temp N Hem, N&C Af, S Am |
| *Ulex europaeus* | Gorse | Fabaceae | Ls | W Eur |
| *Vinca major* | Blue Periwinkle | Apocynaceae | P | Med |
| *Vulpia bromoides* | Squirrel-tail Fescue | Poaceae | A | Eur, W As, Af |
| *Watsonia meriana* var. *bulbillifera* | Bulbil Watsonia | Iridaceae | Pc | S Af |
| *Zantedeschia aethiopica* | White Arum Lily | Araceae | Pt | S Af |

# Further Reading


Arnold, A.H. (1987) 'Exotic Plant Invasions in Coastal Ecosystems', in proceedings *National Workshop on Coastal Management, Queenscliffe and Gosford*, Institute of Parks and Recreation, Canberra.

Audas, J.W. (1937) *The Flora of Mitcham*, Mitcham Naturalists Club, Mitcham.

Australian Institute of Agricultural Science (1976) *The Threat of Weeds to Bushland: A Victorian study*, Inkata Press, Melbourne.

Buchanan, R.A. (1989) *Bush Regeneration: Recovering Australian Landscapes*. TAFE, NSW.

Burrell, J.P. (1981) 'Invasion of coastal heaths of Victoria by *Leptospermum laevigatum* (J. Gaertn.) F. Muell.', *Australian Journal of Botany*, vol. 29.

Calder, M. (1988) 'Why bother to control alien plants in national parks and other parks?' in *Weeds on public land—an action plan for today*, (ed. R.G. Richardson), Proceedings of the Weed Science Society of Victoria and the School of Environmental Science, Monash University, Melbourne.

Carr, G.W. (1993) 'The naturalised flora of Victoria and conservation consequences of weed invasion', *Flora of Victoria* Vol 1. National Herbarium of Victoria, Melbourne.

Carr, G.W. & McMahon, A.R.G. (1988) *Vegetation and management of Quarry Reserve, Mt. Eliza, Victoria*, Report for City of Frankston, Ecological Horticulture P/L, Clifton Hill, Victoria.

Carr, G.W., McMahon, A.R.G. & Race, G.J. (1991) *The vegetation and management of Gramatan Avenue Heathland Sanctuary, City of Sandringham, Victoria*, Report for City of Sandringham, Eco. Hort. P/L, Clifton Hill, Victoria.

Carr, G.W. et al (1991) *The vegetation and management of Hochkins Ridge Flora Reserve, North Croydon, Victoria*, Report for the Hochkins Ridge Flora Reserve Committee of Management. Eco. Hort. P/L, Clifton Hill, Victoria.

Carr, G.W. et al (1989) *The vegetation, botanical significance and management of Cherry Lake, Lower Kororoit Creek and Truganina Swamp, Altona, Victoria*, Report for MMBW, Melbourne. Eco. Hort. P/L, Clifton Hill, Victoria.

Carr, G.W., Reid, J. & Albrecht, D. (1987) *The vegetation, fauna and management of Antonio Park, City of Nunawading, Victoria*, Report for the City of Nunawading, Eco. Hort. P/L, Clifton Hill, Victoria.

Carr, G.W., Todd, J.A. & Race, G.J. (1991) *Vegetation of Plenty Gorge Metropolitan Park, Greensborough—Mernda, Victoria, and its management. Part A. Significance and management issues*, Report for the MMBW, Eco. Hort. P/L, Clifton Hill, Victoria.

Carr, G.W., Yugovic, J.V. & Robinson, K. (1992) *Environmental Weeds in Victoria: an Overview*, Dept. Conservation & Environment & Eco. Hort. P/L, Melbourne.

Cheal, D.C. (1988) *Botanical Assessment of grasslands, Merri Creek-Somerton-Cooper Street*, Resource Assessment Report 88-1. Dept Conservation, Forests & Lands, Melbourne.

Cheal, D.C. et al (in prep) *Vegetation survey and sites of botanical significance in the Melbourne area*, Dept Conservation & Environment, Victoria.

City of Frankston, Shires of Hastings & Mornington (undated) *Mornington Peninsula Pest Plants*, Pamphlet, City of Frankston, Shires of Hastings & Mornington, Victoria.

Elliot, R. (1986) 'Check bushland invaders now', *Your Garden*, vol. 39(1) Melbourne.

Friends of Sherbrooke Forest and Department of Conservation, Forests & Lands (1989) *Weeds of Forests, Roadsides and Gardens*, Second edn, Department of Conservation, Forests & Lands, Victoria.

Frood, D. & Calder, M. (1987) *Nature Conservation in Victoria*, Vols 1 & 2, Victorian National Parks Association, Melbourne.

Garnett, S.T. (3 Aug 1987) 'Exotic peril to Australia's balance', *The Age*, Melbourne.

Gleadow, R.M. & Ashton, D.H. (1981) 'Invasion by *Pittosporum undulatum* of the forests of central Victoria. I. Invasion patterns and plant morphology', *Australian Journal of Botany*. vol. 29.

Gullan, P.K. (1988) 'Weeds in Victoria, where are we?', in *Weeds on public land—an action plan for today*, ed. R.G. Richardson: Proceedings of symposium presented by the Weed Science Society of Victoria and the School of Environmental Science, Monash University, Melbourne.

Holzner, W., Werger, M.G.A. & Ikasima, J. eds. (1983) *Man's impact on vegetation*, Dr W. Junk Publishers, The Hague.

Humphries, S.E., Groves, R.H. and Mitchell, D.S. (in press). *Plant invasions of Australian ecosystems: a status review and management directions*. Report to Australian National Parks and Wildlife Service. Endangered Species Program, project no 58.

Judd, T. (1990) *Invasion of native plant communities by Kunzea ambigua, Kunzea ericoides and Leptospermum laevigatum*, Ph.D. thesis, Department of Botany, University of Melbourne, Victoria.

Kloot, P.M. (1983) 'Early records of alien plants naturalised in South Australia', *Journal of the Adelaide Botanic Gardens*, vol. 6, Adelaide Botanic Gardens, S.A.

Lunt, I.D. (1988) *Derrimut Grassland Vegetation Report. Part I. The Vegetation and its Management*, Report to the Department of Conservation, Forests & Lands, Victoria.

McIntyre, S. (1990) 'Invasion of a Nation: Our Role in the Management of Exotic Plants in Australia', *Australian Biologist* Vol 3.

McMahon, A.R.G. et al (1989) *An assessment of the environmental weed problem within the Shire of Eltham and formulation of a management strategy*, Report prepared for the Shire of Eltham, Victoria.

McMahon, A.R.G. et al (1989) *An assessment of the environmental weed problem within the Shire of Eltham and formulation of a management strategy*, Report prepared for the Shire of Eltham, Victoria.

McMahon, A.R.G. et al (1989) *A review of the sites of significance in the Upper Yarra and Dandenong Ranges region.* Prepared for the Upper Yarra and Dandenong Ranges Authority. Eco. Hort. P/L, Clifton Hill, Victoria.

McMahon, A.R.G., Robinson, R.W. & Bedggood, S.E. (1990) *Preliminary assessment of the biological impacts of the proposed Puffing Billy railway extension*, Shire of Sherbrooke. Eco. Hort. P/L, Clifton Hill, Victoria.

Montrose Environmental Group (inc). *Garden Escapees: A Blot on the Montrose Landscape*, Montrose Environmental Group (inc)., Victoria.

Parsons, W.T. (1973) *Noxious weeds of Victoria*, Inkata Press, Melbourne.

Regional Pest Plants Strategy Working Group (1991) *Environmental Weed Kit*, (Identification and Control) Free kit produced by RPPSWG, Melbourne.

Robin, J.M. & Carr, G.W. (1983) 'Weed Invasions and Horticulture—an Ecological Rethink', *Australian Horticulture*, vol. 81.

Ross, J.H. (1990) *A Census of the Vascular plants of Victoria*, 3rd edition, National Herbarium of Victoria, Department of Conservation, Forests & Lands, Victoria.

Short, P.S. (1987) 'Recording and distribution of weeds in Victoria', *Victorian Naturalist*, vol. 104, Victoria.

Willis, J.H. (1965) 'Native Plants of Brighton', *Brighton Historical Society Newsletter* No. 9 (Reprinted in 'Indigenotes' Vol 4 (8) 1991). Victoria.

*Leptospermum laevigatum*

# Chapter 7. Seed Collection

## Section 1 Guidelines

Darcy Duggan

The availability of seed of local Australian species is critical for both nursery propagation and direct seeding to enable revegetation and conservation of natural areas.

Seed collecting is an interesting and rewarding way to discover and explore the local flora and fauna of an area. The techniques involved in collecting seed are easily learnt and can be carried out by almost anyone, and it provides an opportunity for an interested member of the community to play a small but vitally important part in the conservation and protection of our natural heritage.

It is essential however, that all people undertaking seed collection have a good understanding of the basic principles associated with this task, and follow a few simple guidelines.

These can be outlined as follows:

## Ethics of Seed Collecting

### 1. Respect

Most urban remnant vegetation sites are now small and isolated with often fragmented and declining populations of local native species. These sites are now extremely vulnerable to disturbances such as trampling of ground flora and excessive damage to foliage. It is absolutely fundamental therefore, that seed collectors treat all remnant natural areas with care and respect.

When working in an area:

- move carefully and watch where you walk to avoid damage to native ground flora;
- when collecting seed from shrub and overstorey species, minimise damage to vegetation by avoiding excessive pruning or breaking foliage;
- if working with a group, spread out over the site. Small groups are preferable compared with large groups of enthusiastic, well intentioned people who can inadvertently cause a lot of damage;
- carefully select sites to suit the skills level of groups. Sensitive areas should be avoided when working with inexperienced groups such as school children;
- exercise caution when disclosing the location of rare or endangered species to minimise the loss of these sites through 'poaching'. Ensure collection of seed or propagation material of these species is undertaken by experienced and trusted collectors.

### 2. Integrity

Maintaining the genetic diversity and integrity of a given species is critical to its long term protection and conservation.

The local flora of an area may comprise one or more plant communities. A plant community is made up of populations of numerous different plant species, which over a long period have evolved and adapted to suit the soils and climatic conditions of a given area. This has led to the development of local forms or varieties of particular plant species, with different genetic make-up.

The term 'provenance' is used to identify the location from where the seed or cutting material was collected. To conserve the character and genetic make-up of a local population therefore, it is necessary to collect and propagate seed from that local provenance, and make sure that propagated plants are planted back into the same area.

Seed collecting however is a subjective exercise.

There is often a tendency to either consciously or subconsciously search for superior healthy plants with a heavy seed load. This temptation must be avoided, as it is critically important to take a broad sample of the local gene pool of a species, to maintain the genetic diversity within a given population of plants. This can best be achieved by collecting seed from all individuals irrespective of seed load. It is far better to collect a small amount of seed from a large number of plants than vice versa.

In general no more than 20% of the seed present from any given plant should be collected.

There are often problems in collecting seed from rare or isolated individuals. For example the seed may have a low viability due to self pollination and/or incrossing between related individuals; and represents only a small or distorted component of the original gene pool.

### 3. Documentation

Thorough documentation is essential to ensure the integrity of seed and propagating material which has been collected.

The process of documentation is straightforward and is outlined as follows:

- locate the remnant vegetation site;
- identify species present;
- collect a plant pressing showing fruits, flowers and leaves to confirm identification. It is useful to establish a small herbarium of plants collected from a given site for research and as an educational aid.

If there is any doubt as to the identity of a plant in the field, it is essential that it be properly identified before seed is collected from it. In this way, many hours of valuable time and energy can be saved by not collecting seed from what might turn out to be either a weed or non-indigenous species.

Much of Chapter 7 is reproduced with permission from Vol 3, No. 11 (1990) of Indigenotes, the newsletter of the Indigenous Flora and Fauna Association.

## 4. Keep notes for each site

Record things such as flowering and maturation times of species present, general status of the vegetation community, environmental pressures affecting the site, eg weed invasion, rabbit grazing, disturbance etc.

These site records are valuable in developing a management strategy for the area. It is useful to record the changes in a site over a period of time, hence it is important to revisit the site regularly, but especially during Spring and Summer to monitor the rate of development and maturation of seed and the variations from year to year.

## 5. Labelling

Each collection of seeds should be immediately labelled with the name of the species, date, name of collector, location (eg Melway map reference), soil type, aspect, and any other relevant observations (eg wildlife habitat).

# Factors to Consider when Collecting Seed

There are a number of interrelated characteristics of a plant species and its seed, which influence both the techniques used to collect it, as well as the success or cost efficiency of the exercise in meeting your particular needs. For example you may require only a small amount of seed for propagation, or require bulk amounts for direct sowing and revegetation trials.

**1. The number of plants in an area** (more or less than 100 per ha). The greater the number of plants from which seed is collected, the greater the genetic diversity of the gene pool of a given collection.

**2. Height of seed pods from the ground** (more or less than 2 m). Pod or seed capsule location is an important factor when considering the ease of collection and the tools required to collect the seed e.g. hand saw, step ladder, cherry picker, long handled pruners, climbing gear.

It is often claimed that seed from the middle and upper reaches of a tree or shrub is generally of a better quality and viability than that collected lower down.

**3. The number of seed pods per plant** is both genetically and environmentally controlled. Some species (e.g. *Acacia*) exhibit distinct cycles in seed production (i.e. every 4 to 7 years a high seed yield is produced whilst in other years little or no seed matures). It is important therefore to have seed in storage to offset the lack of seed in a poor year. Environmental factors such as soil moisture and nutrient levels, temperature and rainfall will affect the health and growth of a plant and the amount of seed produced in a given year. Seed production is also clearly influenced by other factors such as strong winds which dislodge flowers and or pods, heavy frosts or attack by insect predators.

**4. Number of seeds per pod** This affects the amount of effort required to collect a required quantity of seed.

**5. Pod size** Determines the ease of collection and handling of a given species.

**6. Seed Retention** The time between flowering, seed maturation and seed shed of different species is highly variable, and may be as short as several weeks. For some plants such as *Eucalyptus*, which produce woody capsules, it may be up to one year or more after flowering before seed is mature and ready to collect (see Section 5).

A number of groups of plants retain the seed for several years within woody capsules until fire or some other environmental factor triggers seed dispersal. Genera such as *Callistemon*, *Leptospermum*, *Banksia*, *Hakea* and *Melaleuca* fall into this category. The advantage of this characteristic is that seed is generally available for collection at any time of the year.

In most other species however, seed is not retained, but dispersed fairly quickly after maturation. For most 'pea' species, e.g. *Kennedia*, *Bossiaea*, *Eutaxia*, *Pultenaea*, *Hardenbergia*, the seed is dispersed within four to five weeks after flowering, and is often forcibly ejected from the seed pods. In some species the seed may be dispersed up to 2 metres from the parent plant.

The *Acacia* group displays a fair degree of variation in the rate of seed retention which can be related to certain physical characteristics of the seed. *A. melanoxylon* and *A. implexa* both have a well developed aril (an appendage which attaches the seed to the inside of the pod), allowing seed to be retained on the plant for 4 to 6 weeks after the pods split open. *A. pycnantha* on the other hand has a greatly reduced aril resulting in seed being shed as soon as the pods split open.

Monitoring of an area from which seed is to be collected, and the timing of collection, is critical in maximising the yield of seed from many species.

**7. Distance travelled to collect seed** (travelling time more or less than 50% of time spent). This factor affects both the cost and efficiency of seed collecting, especially if collecting large quantities of seed, or a variety of different species from a site.

**8. Differential Ripening.** In some plant species such as native grasses, seed on the same plant ripens at different times. The timing of collection is less critical, as seed can be collected over a period of time.

**9. Difficulty in collecting seed.** Various native plants are 'armed' with spines or prickly foliage, making seed collection a somewhat painful and unpleasant experience. Thick protective clothing and leather gloves may therefore be required.

Some plants such as *Hakeas*, have thick woody capsules which are difficult to remove and may have to be cut off the plant.

**10. Recognising when seed is mature** and ready to collect is generally a combination of observation and common sense. The most obvious telltale sign is a change in colour and appearance of seed or seed pods from green to brown or black. Colour change is usually associated with a drying out and hardening of seed pods.

If unsure, a test sample of seed should be taken and examined. For example in *Stipa* (a native grass) if the seed is dark in colour, has a hard seed coat (requires fair pressure to split seed if squeezed between fingernails) and the seed is easily removed from the flower head, then it is mature and ready to collect.

In *Eucalyptus*, there is usually a discernible colour change in the capsules from green to red/brown. When a sample of capsules is placed in a paper bag and seed is shed within 3 to 4 days after collection (and the seed is dark in colour as opposed to pale) then it is ready to collect. You may also notice the valves in the opening of each capsule start to split open, indicating it is ready for collection (see Section 4).

# Methods of Collection

A number of different methods and a range of equipment and materials are required. The method chosen is dictated by the type of plant being collected.

Basic equipment and materials include:

- notebook and pen
- lots of large paper bags or envelopes
- plastic ground sheets
- a few plastic garbage bags
- old nylon stockings

Three broad categories of plants are annuals and herbaceous species, shrubs, and tall trees. Suggested collection methods for each category are as follows:

## 1. Annuals and Herbaceous species

This category includes native grasses, lilies and wildflower species. Collection is by hand, is intensive, time consuming and often tedious.

**Grasses.** Collect by running fingers in an upward motion along the flower head. If seed is mature, it should pull out with gentle pressure. It is important to keep awns intact as these aid germination. Place in a paper bag.

Another effective method is to grasp a handful of flower stalks and cut using secateurs or scissors. This is a particularly useful method for species such as *Poa* and *Danthonia* which have small seeds, and a distinctive tussock habit. Seed heads can then be threshed to remove seed.

***Dianella* (Flax Lilies).** Seed is contained within fleshy fruits which change colour from green to purple. Fruits drop to the ground when mature or can be gently picked off by hand. It is recommended the fruits be placed in a plastic bag as they tend to make paper bags go soggy.

***Arthropodium* (Chocolate Lilies) and Bulbine Lilies.** Seed of these species is contained in small round capsules along the stem which visibly dry out and split open. Seed disperses quickly. It is best to collect the whole stem and place it upside down in a large paper bag to allow seed to shed in the bag.

***Brunonia* (Blue Pincushions).** These have tight clustered flower heads. The seed is ready to be collected when heads can be gently picked off.

**Asteraceae** e.g. *Bracteantha*, *Helichrysum*, *Rhodanthe*, *Vittadinia*, *Senecio*: The seed of this group is small and attached to a pappus; it is generally ready to collect when the flowers start to dry and develop a fluffy appearance. Light pressure with fingers should easily dislodge seed. Remove the flower head and place in a paper bag.

It is also possible to use a small portable vacuum cleaner to collect fine herbaceous seed of this type.

**Ground cover and prostrate species**, e.g. *Eutaxia microphylla*, *Convolvulus erubescens*: collection of seed from these species can be very tedious and fiddly, hand picking individual pods. A more effective method is to gently lift the foliage and carefully slide a sheet of corrugated cardboard under the plant. Tap mature pods or seed onto the cardboard. Carefully remove the cardboard with the seed in place.

***Bossiaea prostrata*** (Creeping Bossiaea). Pods split open and forcibly eject seed. Maturation of pods therefore needs to be carefully monitored. These will first start to swell and then dry out and darken in colour. To check if they are ready, lightly squeeze pods with fingers and the pods should split open easily. Seed is light brown or tan in colour when mature. It is easier to collect the pods whole and remove the seed later.

### 2. Woody shrubs and small trees

Different methods can be employed—depending on the dispersal strategy of the species and the time lag involved. A ground sheet may be placed at the base of the plant, and seed pods can be hand picked, or seed knocked onto the ground sheet. This method can be very efficient and time saving, and is suitable for a range of species—especially *Acacia* spp., and fleshy fruited species e.g. *Hymenanthera*, *Coprosma* and *Myoporum*.

Alternatively seed may be hand picked and placed directly into a foam box or paper bag.

By far one of the most difficult groups to collect from are the 'peas' e.g. *Dillwynia*, *Pultenaea*, *Daviesia*. The seed of these species is generally ejected over a very short period (1–2 days). It is critical therefore, that these species are actively monitored to ensure correct timing of collection. Key indicators to look for are swelling of pods and colour change from green to brown.

Two methods have proven successful in collecting seed of these species:

**Selective and careful pruning of shrubs with seed pods attached**, which are then placed into large paper bags. As the material dries out, the pods open releasing seed.

This method is far more successful compared with hand stripping of pods from the plant. The attachment of pods to the branches would appear to be essential to the pods' splitting open.

When it is not possible to regularly monitor a site, branches with pods attached may be **enclosed in old nylon stockings** which are then tied firmly in place. Seed will be shed into the stockings, which must then be carefully removed from the plants.

### 3. Tall tree species

These can present problems because of the height of the trees and difficulty of access. It may be possible to collect seed from the lower branches using secateurs. For higher branches a tree lopper, commando saw, climbing gear or even a cherry picker may be required.

Wherever possible, attempt to coordinate seed collection with annual pruning work undertaken by the local council or SEC.

## Royalties

In Victoria, under the Forests Act, royalties are payable on commercial quantities of seed collected from all public land managed by the Department of Conservation and Environment. Enquiries should be made to the relevant regional office of the Department.

# *Section 2* *Permit Requirements (Vic)*

## *Department of Conservation and Natural Resources*

At present, permits under the Flora and Fauna Guarantee Act are required for collecting seed or other propagating material from protected flora on public land.

A written permit is required from the local Regional Office of the Department of Conservation and Environment, or if the collection relates to a number of regions, to the Flora Branch of the Department. The application form should be accompanied by a covering letter explaining the purpose of the seed collecting.

It is advisable to obtain verbal permission from the Department's Regional Flora and Fauna Guarantee Officer before any particular collection takes place, to regulate over-collection.

Conditions are attached to permits which are granted. These include that vegetation disturbance is minimized, that sufficient seed is left to ensure natural regeneration, and that vehicles should be confined to existing tracks.

In National and State Parks seed collection requires a research permit under the National Parks Act. Applications should be made to the National Parks and Public Land Division of the Department of Conservation and Environment phone (03) 412 4011.

On private land permission should be sought from the owner, and this is considered sufficient, except in the case of declared critical habitat (none is yet declared however).

The collection of tree ferns and grass trees requires special permission from the Department.

## Protected Flora and Communities Found within the Melbourne Region

Approximately 680 species of flora have been declared protected in Victoria under the Flora and Fauna Guarantee Act 1988. Five plant communities are also protected within Victoria. Those species and the one community found within the Melbourne region are listed below. For further information on all protected species contact the Flora & Fauna Guarantee Section of Department of Conservation & Environment, 250 Victoria Pde, East Melbourne 3002 (03) 412 4011.

### Communities

Western (Basalt) Plains Grassland Community. Every plant of that community is protected wherever it occurs within the Basalt community. The basalt community in Melbourne is shown on Map 2. It is also discussed in detail in Chapter 1.

### Species and/or Genera

*Acacia* spp. All except *A. dealbata*, *A. implexa*, *A. melanoxylon*, *A. paradoxa*.
*Agrostis adamsonii*
*Allocasuarina luehmannii*
*Boronia parviflora*
*Bracteantha* spp.
*Brunonia australis*
*Calytrix tetragona*
*Chrysocephalum* spp.
*Comesperma polygaloides*
*Correa* spp.
*Epacridaceae*—all species
*Euphrasia scabra*
*Gompholobium huegelii*
*Grevillea* spp.
*Hardenbergia violacea*
*Helichrysum* spp.
*Lepidium pseudohyssopifolium*
*L. pseudotasmanicum*
*Leucochrysum albicans*
*Orchidaceae*—all species
*Ozothamnus* spp.
*Prostanthera* spp.
*Psoralea parva*
*Psoralea tenax*
*Pteridophyta*—Clubmosses, ferns and fern allies except *Pteridium esculentum*
*Rhodanthe* spp.
*Rutidosis leptorrhynchoides*
*Senecio macrocarpus*
*Stylidium* spp.
*Thesium australe*
*Thysanotus* spp.
*Xanthorrhoea* spp.

# *Section 3* *The Separation, Cleaning and Storage of Seeds*

## Andrew N. Paget*

These notes explain first how to separate the seeds from the various types of ripe fruits that you have collected, and then how to clean the seeds. Once separated and cleaned they should be stored as described or sown immediately. The information refers mainly to the handling of seed on a large scale but is also applicable to smaller quantities.

Some of the section on separation and cleaning, and the directions for storage, have been supplied by Darcy Duggan.

## Separation

When collected, seed pods and capsules need to be dried out to fully release the seed. The method used depends on the type of seed and the ease with which it is released.

Two options are to spread pods out onto sheets on the floor, turning the pods regularly to prevent sweating and ensure even drying, or to suspend seed from the ceiling in sheets. (Duggan)

### Drying cabinet

For most fruits a drying cabinet is required to dry the fruits until seeds are released. This cabinet is best fairly well sealed and heated with a small fan heater to move the air around. The fruits can be put into brown paper bags or wooden boxes and placed on racks in this cabinet for several days until they open. Vigorous shaking when the fruits have opened will help to remove all the seed. Small quantities of seed can be separated by placing capsules into a brown paper bag and situating it in a warm position.

### Burning

For Banksias, a drying cabinet is not hot enough, and the cones must be burnt for the follicles to open and release their seeds. Either burning with a blow-torch, or on a small bonfire is effective; or you can place them in a paper bag in a warm oven.

### Bashing

Some fruits, like Blackwoods (*Acacia melanoxylon*), are reluctant to let go of their seeds. We put such fruit into a

* Andrew Paget is a botanist who has developed an interest in indigenous flora. He is a partner in 'Bushland Flora', a wholesale and retail nursery devoted to collecting and growing indigenous species.

rubbish bin and beat them with a large broom handle to separate them. It is advisable to use a dust mask when doing this to avoid the dust created.

## Cleaning

Seed cleaning is an essential process. Removal of trash is important, as this often harbours insects which may attack seed and reduce viability. For small quantities of seed you may need to use a combination of graded sieves and/or winnowing in a light breeze to remove chaff. (Duggan)

### Mush, dry and scrunch

For fleshy fruits we find that the best method to obtain clean seed is to put the fruits into a bowl and add some water. Mush all the fruits up, trying to squeeze the flesh off the seeds. Drain off as much liquid as possible and spread out the pulp onto absorbent paper. Put this into the drying cabinet or a warm oven. This is the important stage because if the pulp does not dry quickly enough or well enough it will go mouldy. Once dried the pulp can be scrunched up and all the clean seeds should separate easily. This method is suitable for *Actrotriche* spp., *Astroloma humifusum*, *Billardiera scandens*, *Coprosma* spp., *Dianella* spp., *Persoonia juniperina*, *Polyscias sambucifolius*, *Rubus parvifolius* and *Solanum* spp.

### Sieving

Sieves of various mesh sizes are essential in cleaning most seeds. In ideal situations it will be possible to select two sieves for each seed being cleaned. One will be larger than the seed, to allow the seeds to fall through but catch larger trash. The second sieve will be just smaller than the seed to catch the seeds but discard smaller trash.

It is difficult and expensive to obtain a good range of sieves. A set of 2 sieves which are useful for most seeds is available from 'Gemrocks' at Nunawading. A wide range is available through 'Greer Wire Industries', Thomastown or 'Hannafords' (manufacturers of 'Seedmaster' seed cleaning machines) of 64 East Avenue, Beverley. South Australia, 5009.

### Rubbing

Rubbing grass heads between 2 rubber mats is a method we have found useful for separating *Danthonia* seeds. These seeds will not germinate while still enclosed in the heads, but rubbing them separates them easily. Again masks are recommended to avoid inhaling the dust created.

## Storage

Low temperature and humidity conditions are important to maintain viability of seed. Satisfactory short term storage may be achieved by placing seed in paper envelopes and storing in a cool dark cupboard fumigated with moth balls (Napthalene). For longer term storage, place seed in screw top glass jars.

Clear labels with relevant data, including name of species, site and date of collection, are essential.

# *Section 4* *Recognising Ripe Indigenous Fruit and Seeds*

Andrew N. Paget

The following notes are intended as a guide to judging when fruit or seeds are ripe. They relate to the Knox area of Melbourne, but may be useful in other areas also.

## Colour changes

The fruit or seeds of most species change colour distinctly when they ripen:

*Acacia* spp.—most turn from creamy green to shiny black. Some (like *A. ulicifolia*) do not attain a high gloss finish.
*Acaena* spp.—turn from green to light grey-brown.
*Acrotriche* spp.—have a creamy bloom on the green succulent fruit.
*Alisma plantago-aquatica*—turns pale red-brown from green.
*Amperea xiphoclada*—the seeds inside the green fruit turn from cream to black and shiny.
*Amyema* spp.—turn slightly yellower when ripe, are plump, and break off easily.
*Arthropodium* spp.—seeds turn black.
*Astroloma humifusum*—fruit becomes glaucous when ripe, and seeds turn red-brown.
*Billardiera longiflora*—fruit turns purple from green, and seeds turn dark maroon.
*Billardiera scandens*—fruits turn translucent khaki, and seeds turn red-brown.
*Bulbine bulbosa*—seeds turn from creamy to grey-brown.
*Burchardia umbellata*—seeds turn from cream to red-brown.
*Bursaria spinosa*—fruits turn from green to red-brown, and the seeds also become red-brown.
*Caesia parviflora*—seeds turn black when ripe, inside green capsules.
*Cassytha glabella* and *C. pubescens*—seeds turn red-brown within green fruits.
*Cassytha melantha*—seeds turn almost black.
*Centipeda minima*—seeds turn from green to creamy yellow-green.
*Coprosma quadrifida*—fruit turns from green to red.
*Cymbonotus preissianus*—seeds turn from cream to black.
*Cynoglossum* spp.—seeds change from green to a grey-brown colour.
*Danthonia* spp.—most have seed heads which turn from green to fawn-cream when ripe.
*Dianella* spp.—fruit turns purple when ripe and seeds are shiny and black.
*Diplarrena moraea*—mature seeds are red-brown.
*Drosera* spp.—seeds are black.
*Einadia nutans*—ripe fruits are red, and the mature seeds are almost black.
*Exocarpos cupressiformis*—fruit turns red, seeds turn dark brown.
*Exocarpos strictus*—fruit turns pale silvery-pink, seeds turn dark brown.
*Gahnia sieberiana*—mature seeds turn red-orange.
*Glycine clandestina*—ripe pods turn almost black, and mature seeds inside are red-brown.
*Goodia lotifolia*—mature seeds are almost black.

*Gynatrix pulchella*—the fruits turn grey, and the ripe seeds inside are red-brown.
*Hedycarya angustifolia*—ripe fruit is orange.
*Hibbertia stricta*—ripe seeds are mottled red-brown.
*Hovea linearis*—seeds change from cream to dark grey, and the pods turn from green to grey-brown.
*Hypericum gramineum*—capsules turn from green to red-brown, and mature seeds are also red-brown.
*Indigofera australis*—seeds are red-brown.
*Lagenifera* spp.—seeds turn from green to grey-brown.
*Leptorhynchos tenuifolius*—seeds are almost black.
*Linum marginale*—seeds are shiny red-brown at maturity.
*Lomandra* spp.—creamy seeds are shiny red-brown.
*Luzula meridionalis* vars.—seeds almost black when mature.
*Pandorea pandorana*—fruit changes from green to yellowish then progressively becomes mottled with brown. At maturity the seeds are papery red-brown.
*Patersonia occidentalis*—mature seeds change from creamy to shiny red-brown.
*Pelargonium inodorum*—seeds are dark grey on maturity.
*Persoonia juniperina*—green fruit becomes yellow, and mature seeds are red-brown.
*Pimelea* spp.—most seeds are black, inside leathery fruit.
*Pittosporum bicolor*—mature fruit opens to reveal dark red sticky seeds.
*Polyscias sambucifolius*—clusters of ripe fruit are translucent grey-lilac.
*Prostanthera lasianthos*—green seeds change to grey.
*Pultenaea* spp.—green pods turn red-brown, and seeds within change from green to almost black.
*Ranunculus* spp.—green seeds turn yellowish, and fall if brushed.
*Ricinocarpos pinifolius*—green burr-like fruits turn red-brown and split open when ripe.
*Rumex brownii*—green fruit turns red-brown when ripe.
*Sambucus gaudichaudiana*—ripe fruit changes from green to a translucent creamy colour.
*Solanum aviculare*—green fruit turns dark orange-red, and seeds are red-brown.
*Solanum laciniatum*—green fruit turns yellow-orange and seeds are red-brown.
*Sphaerolobium vimineum*—green pods change through purple tinges to grey-brown.
*Stylidium graminifolium*—green capsules turn fawn-grey, and seeds change from cream to red-orange on ripening.
*Thelymitra* spp.—Most capsules turn grey-brown, and seeds from creamy to brown.
*Themeda triandra*—fruit (awn and florets) turns dark red-brown and shiny, from green-cream.
*Thysanotus* spp.—seeds turn from creamy-green to black.
*Tricoryne elatior*—seeds within green fruit turn black.
*Viminaria juncea*—pods enclosing ripe seeds turn grey-brown.
*Viola hederacea*—seeds turn dark maroon from creamy.
*Wahlenbergia* spp.—ripe seeds are red-brown, and capsules turn from green through yellow-green to grey-brown.
*Wurmbea dioica*—cream seeds mature to red-brown.
*Xanthorrhoea minor*—ripe seeds are black in colour.

## Other Changes

Other observable changes which will indicate the fruit or seeds are ripe are:

### Valves formed:

*Allocasuarina* spp., *Eucalyptus* spp., *Kunzea ericoides*, *Leptospermum* spp., and *Melaleuca* spp.

### Ease of removal:

Most fruit and seeds are easier to remove when they are ripe than when immature. This is an easy way of determining ripeness of: *Acaena* spp., *Amyema* spp., *Billardiera* spp., *Brachyscome* spp., *Carex* spp., *Centipeda minima*, *Clematis* spp., *Cynoglossum* spp., *Lagenifera* spp., *Leptorhynchos* spp., *Persoonia* spp., *Ranunculus* spp., *Triglochin* spp., and most native grasses.

### Fruit opening

This is the easiest way of telling that the seeds of the following species are ripe: *Acacia* spp., *Arthropodium* spp., *Bulbine bulbosa*, *Burchardia umbellata*, *Bursaria spinosa*, *Chamaescilla corymbosa*, *Comesperma* spp., *Cymbonotus preissianus*, *Derwentia derwentiana*, *Diplarrena moraea*, *Hypericum gramineum*, *Juncus* spp., *Kunzea ericoides*, *Linum marginale*, *Lomandra* spp., *Lomatia* spp., *Pandorea pandorana*, *Patersonia* spp., *Pittosporum* spp., *Villarsia* spp., *Wahlenbergia* spp., *Wurmbea dioica*, and *Xanthorrhoea minor*.

### Timing

It is essential to keep good records of when you collect seeds as the timing varies little from year to year. Once you have collected them one year, you can go back within a couple of weeks the following year and expect to find them ripe. To this end, Operation Revegetation has compiled a seed collecting calendar with the time of the year (in half-months) the seed of each species ripens in the Knox area.

# *Section 5* *Seed Collection Times for a Range of species in Melbourne*

## Murray Ralph*

This list has been put together from the Victorian Indigenous Nurseries Co-operative's seed collection records compiled over the last few years.

Times given are the most likely periods for seeds of a particular species to be collected. However depending on the season, local microclimate and aspect, and the area of Melbourne in which you are collecting, seed may ripen up to a few weeks earlier or later. Careful observation over time is the key to success.

The length of time for which mature seed is held on the plant varies from species to species. Some species can hold seed for up to several years, while other species release them immediately after ripening. Seed on many species ripens progressively with both mature and immature seed present on the same plant, so several visits may be required to collect quantities needed. Quantities of seed produced varies greatly from year to year with some species producing large crops of seed on a cyclical basis.

* Murray Ralph spent 12 months working with the CSIRO for the National Seed Collection Project. He became interested in the importance of propagating indigenous plants and was manager of 'Victorian Indigenous Nursery Co-operative' in East Brunswick for 4½ years. He now collects seed on contract.

Duration of time seed remains on the plant after maturity is indicated by the following:

| | | | |
|---|---|---|---|
| * | one to two days | *** | up to several months |
| ** | one to two weeks | **** | up to several years |

## Tree and Shrub Species

*Acacia* spp.—seeds of most *Acacia* spp. in the Melbourne region mature around late December to early January and generally drop soon after maturity so close attention is warranted.

| | | |
|---|---|---|
| *A. acinacea* | Late December | **** |
| *A. dealbata* | Late December | * |
| *A. genistifolia* | Late Dec–early Jan | * |
| *A. implexa* | January | * |
| *A. mearnsii* | Late December | * |
| *A. melanoxylon* | January | ** |
| *A. mucronata* | Mid Dec–Jan | * |
| *A. myrtifolia* | Mid–late Dec | * |
| *A. paradoxa* | Late Dec | * |
| *A. pycnantha* | Mid Dec–early Jan | * |
| *A. retinodes* | Late Jan–early Feb | * |
| *A. verniciflua* | Mid Dec–early Jan | * |
| *A. verticillata* | Mid December | * |

*Allocasuarina* spp.—Seed cones tend to be persistent and can be collected all year round. Late Summer to Autumn tends to be the best time as the current year's crop matures at this time.

*Allocasuarina luehmannii* seed not persistent—late Summer.

| | | |
|---|---|---|
| *Alyxia buxifolia* | Late December | ** |
| *Atriplex cinerea* | Early January | ** |
| *Banksia integrifolia* | March | *** |
| *Banksia marginata* | Late Dec–March | *** |
| *Bursaria spinosa* | March–April | ** |
| *Callistemon sieberi* | All year round | **** |
| *Callitris glaucophylla* | March | *** |
| *Cassinia* spp. | Late Jan–mid Feb | ** |
| *Coprosma quadrifida* | Late Dec–Jan | ** |
| *Daviesia* spp. | December | * |
| *Dillwynia cinerascens* | Mid Dec | * |
| *Dodonaea viscosa* ssp. *spatulata* | Mid–late Dec. | ** |

*Eucalyptus* spp.—vary throughout the year from species to species, with some having persistent seed.

| | | |
|---|---|---|
| *E. camaldulensis* | March–Nov | *** |
| *E. cephalocarpa* | April–August | *** |
| *E. cypellocarpa* | Jan–June | *** |
| *E. dives* | All year | **** |
| *E. goniocalyx* | All year | **** |
| *E. leucoxylon* ssp. *connata* | Oct–June | *** |
| *E. macrorhyncha* | All year | **** |
| *E. melliodora* | Dec–Sept | *** |
| *E. microcarpa* | Nov–August | *** |
| *E. obliqua* | All year | **** |
| *E. ovata* | All year | **** |
| *E. polyanthemos* ssp. *vestita* | Dec–June | *** |
| *E. radiata* | All year | **** |
| *E. rubida* | Feb–May | *** |
| *E. viminalis* | July–April | *** |
| *E. yarraensis* | Variable | *** |
| *Eutaxia microphylla* | Early–Mid Dec | * |
| *Goodia lotifolia* | December | * |
| *Grevillea* spp. | December | * |
| *Gynatrix pulchella* | Late December | ** |
| *Hakea* spp. | All year | **** |
| *Hymenanthera dentata* | January | ** |
| *Indigofera australis* | Late Dec–early Jan | * |
| *Kunzea ericoides* | March | *** |
| *Lavatera plebeia* | December | * |
| *Leptospermum* spp. | All year | **** |
| *Leucophyta brownii* | March | ** |
| *Melaleuca* spp. | All year | **** |
| *Olearia lirata* | December | ** |
| *Olearia ramulosa* | December | ** |
| *Ozothamnus ferrugineus* | January | ** |
| *Ozothamnus obcordatus* | Dec–Jan | ** |
| *Platylobium* spp. | December | * |
| *Pomaderris* spp. | Late Dec–early Jan | ** |
| *Prostanthera lasianthos* | January | * |
| *Pultenaea daphnoides* | December | * |
| *Rhagodia candolleana* | February | ** |
| *Rhagodia parabolica* | Mid–late Jan | ** |
| *Senna artemisioides* ssp. *filifolia* | Late Jan–early Feb | * |
| *Solanum* spp. | Jan–Feb | ** |
| *Viminaria juncea* | Jan–Feb | * |

## Tussock Habit—Native Grasses

| | | |
|---|---|---|
| *Agrostis* spp. | Late Dec–Jan | * |
| *Carex* spp. | Dec–Jan | ** |
| *Chloris truncata* | Dec–Feb | ** |
| *Danthonia* spp. | Late Nov–mid Dec | ** |
| *Dianella longifolia* | Late December | ** |
| *Dianella revoluta* | Mid–late December | ** |
| *Dianella tasmanica* | January | ** |
| *Dicanthium sericeum* | February | ** |
| *Elymus scabrus* | December | * |
| *Juncus* spp. | Dec–Jan | ** |
| *Lomandra longifolia* | Late Dec–Jan | * |
| *Microlaena stipoides* | December | * |
| *Poa* spp. | Mid–late December | ** |
| *Stipa* spp. | Late Nov–Dec | ** |
| *Themeda triandra* | Late Dec–early Jan | * |

## Groundcovers—Climbers

| | | |
|---|---|---|
| *Atriplex semibaccata* | Jan–Feb | ** |
| *Bossiaea prostrata* | Mid December | * |
| *Carpobrotus* spp. | Dec–Jan | ** |
| *Clematis aristata* | Dec–Jan | ** |
| *Clematis microphylla* | December | * |
| *Einadia nutans* | December | * |
| *Enchylaena tomentosa* | February | ** |
| *Hardenbergia violacea* | December | * |
| *Kennedia prostrata* | December | * |

## Herbs

| | | |
|---|---|---|
| *Acaena* spp. | Dec–Jan | * |
| *Brachyscome* spp. | Dec–Feb | * |
| *Bracteantha viscosa* | Mid December | * |
| *Brunonia australis* | Late Dec–mid Jan | ** |
| *Calocephalus* spp. | Feb–early March | ** |
| *Calotis* spp. | Dec–Jan | ** |
| *Chrysocephalum apiculatum* | Late Dec–early Jan | ** |
| *Chrysocephalum semipapposum* | Jan | ** |
| *Convolulus erubescens* | February | * |
| *Craspedia variabilis* | January | ** |
| *Eryngium ovinum* | February | ** |
| *Goodenia pinnatifida* | Dec–Feb | * |
| *Helichrysum rutidolepis* | Dec–Jan | ** |
| *Helichrysum scorpioides* | Dec–Jan | ** |
| *Leptorhynchos squamatus* | December | * |
| *Nicotiana* spp. | December | * |
| *Pimelea* spp. | Late Nov–Feb | * |
| *Podolepis jaceoides* | Mid Dec–mid Jan | ** |
| *Ptilotus* spp. | January | ** |
| *Pycnosorus* spp. | January | ** |
| *Rutidosis leptorrhynchoides* | Dec–Jan | * |
| *Rumex* spp. | December | * |
| *Senecio* spp. | Late Nov–Jan | * |
| *Velleia paradoxa* | Mid Dec–early Jan | * |
| *Vittadinia* spp. | Nov–early Jan | * |
| *Wahlenbergia communis* | Jan | ** |
| *Wahlenbergia stricta* | Late Dec–Jan | ** |

## Bulbs and Tuberous Species

| | | |
|---|---|---|
| *Arthropodium* spp. | Dec–Jan | ** |
| *Arthropodium minus* | Mid–Late Dec | ** |
| *Bulbine bulbosa* | Late Nov–early Dec | * |
| *Burchardia umbellata* | Dec–early Jan | * |
| *Caesia* spp. | Late Dec–early Jan | ** |
| *Stylidium graminifolium* | Mid–late Dec | * |
| *Tricoryne elatior* | Jan–Feb | * |

# Chapter 8. Propagation of Indigenous Plants of Melbourne

Rodger Elliot*

With the increasing interest in the value and importance of conserving and growing Melbourne's local plants there is also a coinciding interest in the propagation of these species.

A number of nurseries in the Melbourne region now specialise in propagating indigenous flora and they make many species available which were previously difficult to obtain. However there is a great feeling of achievement when propagating your own plants successfully. There is also the added pleasure of being able to give away any of your excess plants to friends, neighbours or local schools.

Some people may be reluctant to propagate their local plants because they believe it may be too difficult to achieve success. There is however very little difference in propagating our local plants as compared to plants from anywhere else. Generally the propagation principles are the same, with any differences in technique being only slight and of minor concern. The base line with plants, wherever they originate from, is that some are easy to propagate and others difficult and very frustrating, which hopefully adds up to interesting times if you are involved in propagation.

The information on propagation supplied here is purely an introductory exercise and there may be times when the reader will hunger for more detail. There are many excellent publications which can provide information at a greater depth. See Further Reading at the end of this chapter.

## Equipment for Propagation

Before undertaking sowing of seed, preparation of cuttings or other propagation methods, thought must be given as to the type of equipment required for success. This equipment can vary from being very simple to highly sophisticated, but most species can be propagated with the simplest of equipment, provided certain conditions are maintained, and these are discussed in more detail below.

### Containers

Any containers used must be able to hold the propagation medium and at the same time provide excellent drainage. Suitable new or used containers can be from a wide range of materials. Used containers should be thoroughly washed and sterilised prior to use. Many containers are made specifically for plant cultivation, but there are also other items that can be used such as plastic margarine or ice cream containers, if efficient drainage holes are provided.

### Propagating Structures

The next requirement is for some form of structure in which to carry out the propagation. Germination of seed can be done in the open if desired, but problems can be experienced in the form of animals, children, or heavy rain or hail and your efforts, as well as perhaps valuable seed, may be wasted.

For seedlings, one of the simplest structures is a box covered with plastic sheeting, into which seed trays or pots are placed. This can also be used for cuttings. A simple structure is also provided by polystyrene boxes which are commonly used for packaging fruit. Two wire hoops placed at each end and covered by a sheet of plastic make the unit complete.

Possibly the simplest method for propagating cuttings is to place a plastic bag or upturned bottle over a pot of cuttings. It is important however that these items, and also the fruit-box structure, should not be exposed to direct sunshine as young seedlings or cuttings may be scorched.

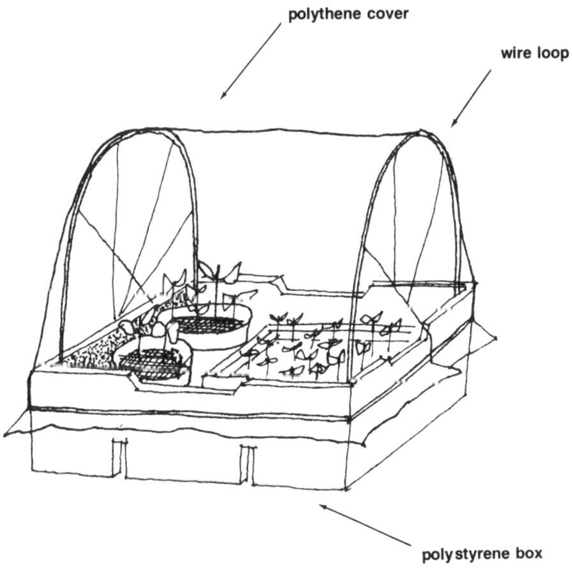

PROPAGATION BOX

A common propagating structure is the cold frame, and it is suitable for most propagation requirements. Cold frames can be easily constructed in wood or brick with a hinged upper frame covered by glass or plastic sheeting.

From cold frames the next stage is to plastic or glass houses. These should be sited in sunny areas. They will need shading during late spring to early summer. This can be done by painting with white glasshouse paint, or by covering with shadecloth. Undoubtedly these larger structures do

* Rodger Elliot is a professional horticulturist operating a wholesale nursery for Australian plants. He regularly writes articles for magazines and journals and is the author of many books on native plants and co-author of the *Encyclopaedia of Australian Plants*.

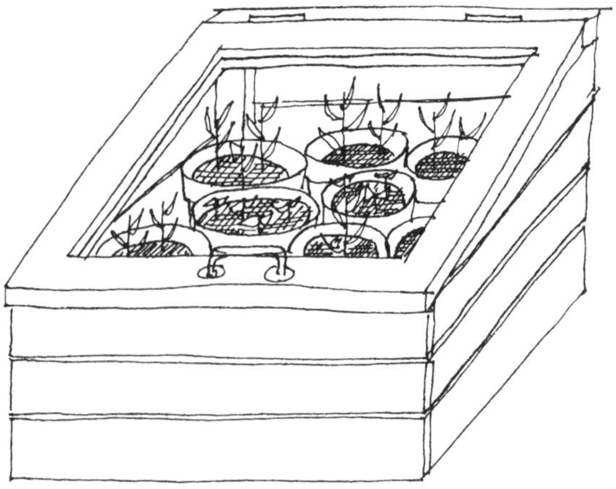

COLD FRAME

have advantages over smaller ones. You can work undercover on inclement days, and they generally will hold a greater number of seedling trays and pots, thus allowing for the production of more plants.

In propagation from cuttings often there are specific conditions required for success. Different methods and equipment can be employed to achieve those desired.

Warmth is an important aspect. If a cold frame or glasshouse is positioned well the warmth available from the sun can be all that is necessary for successful propagation. Electrically heated propagating units, as well as heating mats, are now readily available if you wish to build them into a new structure, or incorporate them into an existing one.

It is worth repeating here that in most cases you do not need sophisticated equipment to propagate plants. Plants have been propagated for thousands of years now, and in some cases equal success was achieved with early methods in comparison with today's modern technology, (eg. the propagation of grapes in the 16th century). It is therefore suggested that if you are just starting in propagation, begin with simple equipment. Then later on you can change or upgrade to different units or incorporate heating etc., if you feel the need to do so.

# Propagation Hygiene and Materials

## Hygiene

This is one of the most important aspects of propagation. It can mean the difference between success or failure. It is wise to make sure all containers have been treated in some way to eliminate all diseases or pests, especially if the containers have been used before. This can be achieved by soaking the containers in a germicide (as directed on the label), or bleach at the ratio of 1 part bleach to 19 parts water. They should then be thoroughly aired so there are no fumes left that may damage plant material. All tools, especially secateurs, should be regularly cleaned using a similar solution.

Regular treatment of propagation structures is beneficial, as there may often be a build-up of fungi and moulds. Initially they may not have much effect on cuttings and young plants, but in some cases there is a breaking point when there is an extremely quick expansion of a disease which can attack cuttings or seedlings within the structure, if not noticed and treated immediately. It is recommended that every 9–12 months structures be thoroughly washed with a germicide, then left for a day or so to enable complete airing before again filling them with propagation material.

It is also ideal that all media used for propagation, especially if it contains sand or soil, be treated to eliminate any diseases. A simple method which helps but does not entirely eliminate diseases, is to pour boiling water over the propagating mix. This has the disadvantage of making the mix very wet, and it therefore must be dried a little before use. For small quantities, moist media can be placed in a sealed plastic bag and located in a sunny site. It should be left there for 3–5 sunny days, with the bag being turned over each day.

Another method suitable for small quantities is to place the propagation mix in an oven at 95°C for 30 minutes. It is best if the mixture has been kept moist for 3 days prior to this treatment. Selected chemicals can be used but this is a somewhat more complex treatment, and the propagation mixture must be allowed to stand for some time before use, for if any trace of the chemical remains it could prove toxic to the cuttings or seedlings. Small steam sterilising units suitable for home propagators are also available.

## Propagation materials

There are many different views on which are the best materials for use in propagation. Today a wide range of materials are used with success. With propagation it is necessary to find the mixture that best suits your needs and your own situation. If you have success with a mixture there is certainly no need to change it because it differs from that used by anyone else.

Some components that are frequently used in propagation mixes are sand, peat moss, perlite, vermiculite, pinebark, sawdust and topsoil. These materials need to be mixed in different proportions, depending on the aspect of propagation for which they are to be used and the conditions under which propagation is to take place.

## Seed Raising Media

For raising seedlings there is quite a wide range of mixtures that can be used. The main pre-requisites for a sowing mixture are that it must be very well drained and at the same time it must be well aerated, i.e. it must not pack too tightly and become waterlogged. Propagation sand or coarse river sand with peat moss, pine peat and good quality topsoil can be used in varying quantities to achieve suitable germination results.

Perlite is also used and the following formula has proved successful—1 part peatmoss, 1 part perlite plus 1 part good quality sterilised loam.

## Cutting Media

For cuttings the most commonly used materials are sharp, coarse river sand and peatmoss. A widely accepted mixture is 3 parts of sand to 1 part of peatmoss. This is suitable for the majority of Australian plants. If for some reason you do not have success with this mixture, then try some variations such as a 50/50 mix. Even 100% coarse sand is worth a try, and some native plant propagators find this quite adequate. Another suitable mixture is 50% peatmoss and 50% perlite. Some growers use 1 part of peatmoss to 2 parts of perlite. Both of these ingredients are sterile, so there is no need for sterilising treatment before use.

# Propagation from Seed

For information on the collection of seed, separation, cleaning and storage, see Chapter 7 on Seed Collection.

Seed of some species can be sown immediately after it has been separated from the capsules, follicles or other fruits, eg. *Callistemon*, *Eucalyptus*, *Melaleuca* and *Stylidium*. However there are some species which need pre-sowing treatment that may help to promote seed germination. There is a range of pre-sowing treatments which can be used.

**1. Treatment of seed with hard outer coatings.**

Seeds of this type can remain viable in the wild for many years. Pre-sowing treatment is desirable to allow water to penetrate to the embryo and hasten germination. The main group of plants which respond to this treatment are the legumes, eg. *Acacia* species and all the pea plants. Some of the hard-seeded lilies also respond well. Various methods can be employed.

(a) Pour freshly boiled water onto the seed in a container and allow them to soak for 1–24 hours. Fertile seeds will swell. This is the simplest and most commonly used method. Do not boil the seeds in water as this usually damages the embryo.

freshly boiled water poured onto seeds

(b) Scarify the seeds by rubbing them between sandpaper or wet-and-dry paper.

(c) Nick the seed coat with a sharp implement (nail-clippers are ideal). This method is best used for large seeds, otherwise damage to the embryo can occur.

**2. Stratification.**
Seeds are subjected to periods of low temperatures, 0°C to -4°C for varying periods, eg. 3–10 weeks. This must be done in advance, to allow for the proposed sowing date. Some genera which are known to respond well to this treatment include *Billardiera*, *Eucalyptus* and *Wahlenbergia*.

**3. Dormancy-breaking.**
Many species will not germinate from freshly collected seed because of a dormancy period before maturity. Much better results are attained if the seed is sown 8–12 months after collection. Many of the native grasses are in this group, eg. *Themeda triandra*, Kangaroo grass. Seed needs to be stored in airtight containers at between 3°C and 15°C. It is thought that seed of *Billardiera*, *Patersonia*, some *Acacia* and *Kennedia* species also have dormancy periods.

**4. Leaching.**
Some species contain inhibitors and on occasions these can be leached from the seeds. The simplest method is to soak the seed in containers of cold water for extended periods. Another is to put the seeds in a permeable container (nylon stockings or pantyhose are useful) and place it in flowing water. Some success has been achieved by immersing such a container in a toilet cistern. *Billardiera* may respond well to this treatment.

seeds in nylon stocking

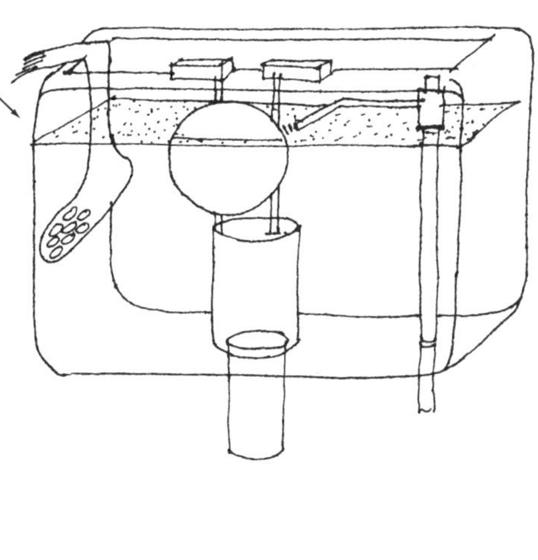

**5. Fermentation.**
Some fleshy fruits can be placed in a plastic bag and stored in a warm humid site, allowing the fruits to ferment. *Astroloma*, *Billardiera* and *Dianella* may respond favourably to this treatment.

There is still a need for much experimentation with pre-sowing treatments, and the results need to be recorded and published as they would have value for future work.

## Sowing of Seed

Containers for seed propagation should be from 4–8 cm deep. The propagation medium is filled to within about 1cm of the top. It should only be gently firmed. Compacting the medium too firmly can inhibit root growth of seedlings.

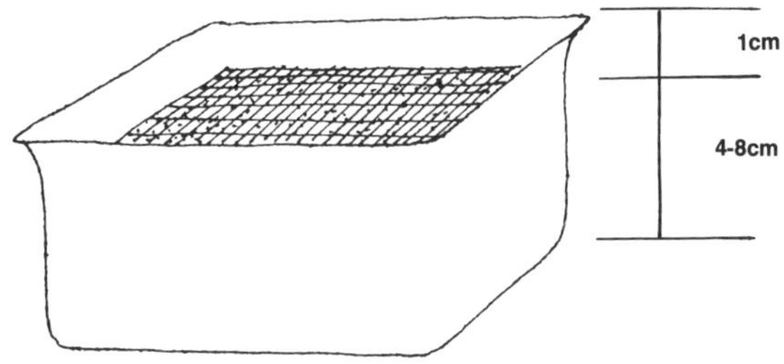


Perhaps the main rule with sowing is not to spread seed too thickly over the propagating medium. This may initially provide a lot of plants, but in most cases the end result is poor quality seedlings because of the intense competition. Many of the seedlings may die due to disease, which can spread readily among overcrowded seedlings.

Very fine seed can be difficult to spread lightly and evenly. By placing the seed and dry fine sand in a salt or pepper shaker and then dispersing the mixture over the propagation medium there is less likelihood of seedling overcrowding.

A general practice that works well is to cover the seed with propagating mixture to a depth equal to the size of the seeds. Therefore with fine seed it is only necessary for it to be barely covered. If the bog or capillary methods (see below) are used the seed does not always need to be covered.

When sowing seed, always make sure that the containers are properly labelled with the plant name and date of sowing.

The first watering must be sufficient to thoroughly soak the medium, but without dispersing the seed. A fine spray nozzle on the hose or watering can is therefore recommended. Alternatively the container should be placed in another container of water and allowed to soak. See Alternative Watering Methods, below. A fungicide can be added to the water if desired, as this can help to combat disease.

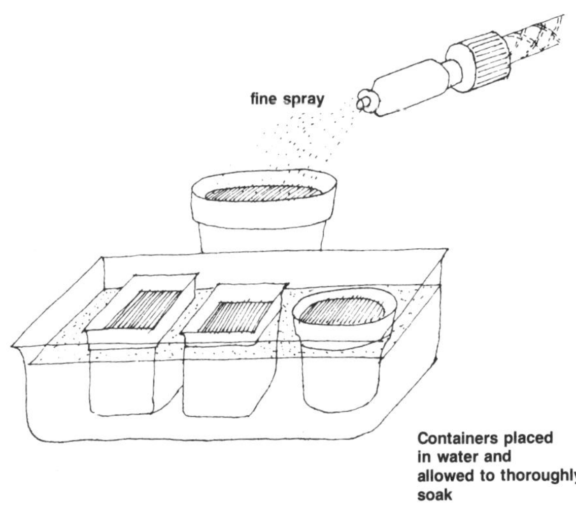


Containers placed in water and allowed to thoroughly soak

Seedling containers should be placed in a well ventilated area and generally kept moist but not wet. Germination time varies considerably from species to species. Acacias and pea plants (after treatment) are often quickest to germinate, while others such as *Epacris* spp. can take much longer. If you have any precious seeds that have not germinated, don't throw them out too soon. If the seeds don't germinate within a season put the containers aside under a tree or in a similar situation. You may be surprised to find them full of seedlings at a later date.

Some species will often not germinate until there is a continuous spell of cool weather, with plenty of rain, eg. *Billardiera* species. Seed sown in early spring may not germinate until after autumn rains.

## Alternative Watering Methods

**Bog Method.**

With many species of the Myrtle family, eg. *Callistemon*, *Kunzea*, *Leptospermum* and *Melaleuca* this method is very successful, and it is worth trying, especially with those which occur naturally in waterlogged soils or are subject to inundation for short periods. This includes aquatic species such as *Myriophyllum*.

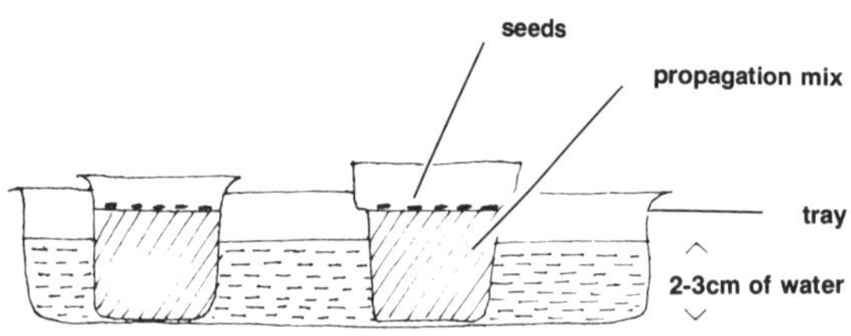


It is a simple and time-saving method. The seed should be sown in the normal manner, but it is not always necessary for it to be covered with propagating mix. The seedling container is then placed in another container that has a water level of about 2–3 cm deep. The seed is then kept continually moist by capillary action and therefore there is no need for overhead watering which can disperse fine seed and possibly damage or kill very young seedlings. It is an excellent method for keeping the seedling mix moist if you are likely to be away for a few days.

As the seed germinates and the seedlings develop they should be watched carefully. When the seedlings are about 0.5–1 cm tall the container should be removed from the water so that the seedlings do not suffer from waterlogging which can kill some of the young roots and eventually the plants.

**Capillary method.**

This is an adaptation of the Bog Method. It differs in having sand in a container with a bottle feed to keep the moisture at a constant level. The container with seeds is partially buried in the moist sand. It is ideal for people who may be away for periods of longer than 2–3 days.

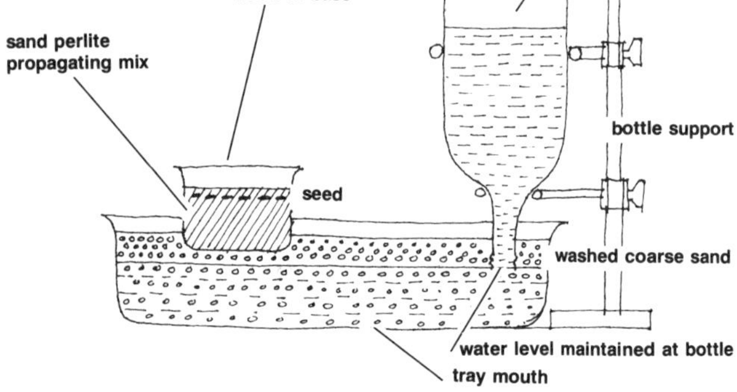


## Post-germination Treatment

As seeds germinate it is best in most cases to pot them on into individual containers, as soon as they can be handled easily. This is likely to prevent poor development of roots if seedlings are retained in smaller containers, with stronger and more healthy roots being the end product. Forestry tubes or similar sized pots, into which the seedlings can be transplanted, are ideal.

If you find the roots of the seedlings are very long, do not be afraid to prune them. Many people still believe that the roots of Australian plants should not be cut or otherwise interfered with, however experience and research have found this not to be the case when handling young seedlings. They respond well to this procedure. It prevents the roots being squeezed into a small area where they can become bent or disfigured, which can often mean that they do not develop properly and are unable when planted out to support the trunk and foliage.

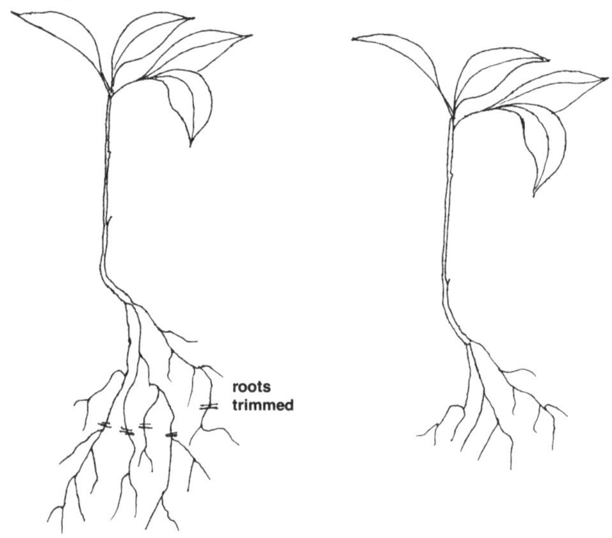


After transplanting seedlings into individual pots they should be well watered and placed in a protected shady spot for 4–7 days, to allow them to recuperate from their previous treatment. The addition of one of the commercially available plant starters and root stimulants, at the watering stage, can be beneficial.

# Propagation from Cuttings

Plants are grown from cuttings for a number of reasons including the following.

(a) There is often difficulty in obtaining seed of many species.
(b) The same forms or varieties can be perpetuated, which is not always the case in seedling grown plants.
(c) Some plants are much easier to grow from cuttings than from seed, eg. species of *Correa*, *Epacris* and *Goodenia*.

Some people are often reluctant to try growing plants from cuttings, with the idea that it is too difficult. There are many of the indigenous plants of Melbourne that are extremely easy to propagate by this method, including *Brachyscome multifida*, *Goodenia lanata*, *Mentha australis* and *Myoporum parvifolium*.

To achieve success with cuttings there are some important guidelines which should be followed, however experience is usually the best teacher, so if at first you don't succeed, have a chat with someone about what might have gone wrong, then try again.

## Collection of Cutting Material

One of the most important steps to success is to start with healthy material in good condition. The question is often asked as to when is the best time to take cuttings of Australian plants, and the answer is the nebulous 'when the material is in the optimum state for propagation'. This is usually after flowering has finished and there is new foliage growth which has begun to 'harden off' beyond the soft and floppy stage.

Cuttings can therefore be taken at varying times, depending on the plant species. Best results are usually gained between October and March, when there is warm weather and therefore more favourable propagating conditions. Artificial heating of propagation units can extend the propagation season. See below for further information on this aspect.

There are some special points worth noting while collecting material.

(1) Cutting material is best collected in the cooler times of the day, such as early morning, before the sun has been on the foliage, or in early evening, when the foliage is again cool. On collection the foliage can be cooled further by spraying with water.

(2) Cuttings are best collected from young plants, as research has shown that this type of material is usually capable of producing roots much more quickly than that from older plants.

(3) Always use disease-free material. Collect cuttings from branches that do not touch the ground, and if possible choose those that do not have black spots or markings or have not been splashed with earth and water during heavy rains. This will mean there is less likelihood of any earth-borne fungi being on the plant material.

(4) Collect material that is not too soft or too rigid, unless nothing else is available. Cuttings should have a springiness, and a useful guide is to bend the tips back until they touch the stem, then let go. If they return fairly quickly to the original position then it is likely to be good material for cuttings.

(5) At all times cutting material should be kept slightly moist, to retain it in good condition.

(6) If cuttings need to be transported for half an hour to over a period of some days, small quantities can be wrapped in newspaper and placed in an insulated cooler. Alternatively large quantities can be placed in a moistened woven polypropylene bag (often recycled fertilizers or stock food bags), which usually prevents the cuttings from sweating, as is prone to happen in clear polythene bags. Plastic garbage bags are not recommended, as they can release ethylene gas, which is detrimental to retaining cuttings in good condition.

## Preparation of Cuttings

There are a number of different methods in preparing cuttings for propagation. Stem cuttings are by far the most commonly used. Single node cuttings or single bud cuttings can also be successful but are beyond the scope of this chapter. See Further Reading for specialised publications which cover these subjects in detail.

## Preparation of Stem Cuttings

Most cuttings should be around 7–15 cm long, although there is no strict rule. Some plants, eg. *Astroloma humifusum* may not produce much material that is suitable for cuttings of this size. Therefore smaller cuttings must be used. If tip growth is too soft it will be likely to wilt and therefore should be removed by cutting or pinching just above a node. In some species the removal of the growing shoot may spoil the future shape of the plant, in which case it is best to use material with firm tip growth.

The lower cut should be just below a node, as this is the point from which roots will usually be produced. There are exceptions to most rules, and with species such as *Hardenbergia violacea* and *Pandorea pandorana* the lower cut can be done at the internode—the region of stem between the leaf nodes. Roots will develop from that cut area. It is worth experimenting with other species in the same manner.

The next step is to remove the leaves from the lower half of the cutting. In many cases they can be taken off with a quick downwards movement of the fingers, eg. *Brachyscome multifida*, *Epacris impressa* and *Myoporum parvifolium*, but once again experience will show which plants cannot be treated in this way. Some will be very prone to ripping of the outer layer of the cutting, which may cause the cutting to die. If ripping occurs, try pulling off the leaves with an upwards motion, or removal of the leaves by secateurs may be necessary.

In some plants such as those with broad-leaves, slight wounding of the base of the cutting will help to promote root growth. This can usually be done by allowing the lowest leaves to be pulled downwards, or the wound can be made with a sharp blade.

With large-leaved species it is usually desirable for the leaves to be removed with secateurs. It is the practice of some propagators to reduce the size of the remaining leaves on the cuttings. This reduces the need for the cutting to provide moisture to sustain large amounts of foliage, however in some cases cutting can also cause dieback to occur, which may spread rapidly to other parts of the cutting. **It is most important to use clean secateurs for this process, as it is in all aspects of propagation.**

If large leaves are allowed to remain on cuttings, they can prevent water from reaching the propagating medium, and extra care should be taken to prevent it from drying out.

The next step is to place the cuttings in the selected propagation medium.

The application of a hormone root stimulant can be beneficial in producing more roots of better quality, but it is not imperative that it should be used. Hormone powders or liquids for plant propagation are readily available from nurseries and garden stores.

## Management of Cuttings

From this stage there are very specific aspects which, if handled in the right manner, should produce good results. There are four major requirements for cuttings to be maintained in a living state to eventually produce roots. They are: (i) Air, (ii) Light, (iii) Water and (iv) Heat.

If these can be maintained in the right balance all the time the results will usually be excellent. There are however so many variations that can alter the environment very quickly, thus reducing the success rate for which we may hope.

**1. Air.**

A well-aerated propagation medium is important, as it allows for the provision of oxygen around the base of the cuttings, enabling callus and root production. Air movement within the propagation structure will also impede any fungal build-up on the cuttings.

**2. Light.**

Light enables the cuttings to carry out normal leaf function, eg. photosynthesis. Excessive light often leads to an increase in temperature and therefore drying out of the atmosphere and leaf surface, which is not beneficial. Lack of adequate light may lead to low temperatures and an excessive build up of moisture which can be detrimental for plant growth.

**3. Water.**

Water is of utmost importance, as there must be a humid atmosphere to reduce moisture loss from the cuttings and the propagation medium. The medium should be kept moist, but not wet, as waterlogging will reduce the access of air to the base of the cuttings. A humid atmosphere will reduce the leaf temperature, which ideally should be below that of the root area. Regular spraying of the leaf surfaces is usually beneficial in maintaining the conditions mentioned above. This action can be maintained by installing an automatic misting system, but this is not necessary for success.

**4. Temperature.**

Temperature plays an important role in the production of roots. The optimum temperature for most cuttings is in the range of 20–27°C during the daytime, and about 15°C at night. Best results are usually obtained if the top of the cutting is 5–10°C below the temperature at the base. This means that root production is not impeded by effort put into leaf-bud growth. The use of bottom heat equipment can help greatly in maintaining a suitable temperature. There should be careful control of ventilation in the propagation area, and there is usually the need for some form of shading to restrict temperatures from rising excessively during the summer months.

If you can provide suitable conditions, having regard to the above four requirements, you will be well on the way to success, especially if you have collected good quality cutting material. For further information on these subjects it is worth considering publications which are devoted entirely to propagation, see Further Reading.

# Propagation by Division

There are a number of Melbourne's indigenous plants that may be propagated by division. They are mainly monocotyledons that are clumping, tufting, have creeping rhizomes or have the capacity to form roots and self-layer at nodes on the creeping stems.

Plants in these categories include the following.

(a) Lilies and irises—such as *Dianella*, *Diplarrena*, *Libertia*, *Patersonia* and *Thelionema*.

(b) Grasses such as *Agrostis*, *Danthonia*, *Poa*, *Stipa* and *Themeda*.

(c) Rushes and sedges, such as *Carex*, *Isolepis*, *Juncus*, *Schoenus* and *Xyris*.

Although the majority of ferns are propagated from spore, a number of ferns and their allies can be grown from divisions, including some species in the genera of *Calochlaena*, *Doodia*, *Gleichenia*, *Pteris*, *Rumohra*, *Selaginella* and *Sticheris*.

There are quite a number of dicotyledons which are herbaceous perennials, and these are readily propagated by division. The following genera include species which respond well to this form of propagation: *Calotis*, *Goodenia*, *Helichrysum*, *Hydrocotyle*, *Myriophyllum*, *Pratia*, *Samolus*, *Selliera*, *Viola* and *Wahlenbergia*.

Division is a very simple method that does not need great expertise or many special tools. There are however some points which can have a significant effect on the success of plant divisions.

(1) Prevent any unnecessary damage to plant tissue.

(2) Each division must have ample healthy young roots to be able to provide food for the foliage. There should be a balance between the root system and the upper growth.

(3) Divisions must retain buds and growing points.

(4) Hygiene is very important.

As, in most cases, plants being divided have been growing in soil, either in a garden or a container, there could be diseases present that would spread readily in this propagation method. A recommended procedure is to thoroughly drench the soil area in and around the plant, 10–14 days before division, using a fungal drench and root stimulant.

The best time to divide plants is when new roots are beginning to be formed. This is usually after a growth flush. In Melbourne and its environs the period between the end of June and the middle of September has been found to be suitable. If division is done in late autumn there can be difficulty in establishing the divisions, unless they can be placed in a warm sheltered propagating structure, such as a cold frame, a glasshouse or polyhouse.

While dividing plants, always keep foliage and roots moist. Division should be carried out in a location sheltered from sun and winds which will dry the divisions.

After the plant has been removed from the ground or a container, loosen soil around the roots. If difficulty is experienced, place the plant and soil in water, agitating it as you remove the soil with the fingers. The use of water from a hose can also help to dislodge soil which is firmly attached. This also eliminates excessive damage to the root system. Some clumps can be separated with the hands, others may need to be cut with a clean sharp knife or secateurs.

For best results it is recommended that there be 3–5 growth points per division. Very small divisions initially produce more plants, but the success rate is often greatly reduced.

Any dead growth, on foliage or roots, should be removed. After balancing the amount of foliage as compared with the root system, the division is then ready for potting into containers.

# Propagation of Ferns

Ferns are usually propagated from spore, but some may be done by division, see above.

Propagation from spore is a somewhat specialised process, and is dealt with here in simple terms. If further information is desired, it can be gained from specialist fern publications and societies.

Ferns differ from the majority of other plants in producing spore rather than seeds. If propagation conditions are right these will develop on the propagation medium surface as small heart-shaped prothalli, with separate male and female organs. Sperm travels from the male to the female section to achieve fertilisation and the development of a sporeling begins.

## Collection of Spore

Spore develops on the underside of typical fern fronds or there may be special fertile fronds which are often narrower than the typical fronds.

It is best to collect mature fronds and place them in folded paper, a paper bag or envelope and store the package in a warm dry place. Never use plastic bags, as moulds can develop and ruin the spore.

The very fine, dust-like spore is shed after a few days and should be sown as soon as possible.

## Sowing of Spore

A precis of the steps for the sowing of spore.

(1) Fill the lower third of a sterile pot with coarse propagating sand.
(2) Add sowing medium, firmly packed to within about 3 cm of container rim. A mixture of thoroughly wetted peat moss and finely chopped tree fern fibre is a recommended medium for raising the spore.
(3) Pour boiling water quickly into the pot until all the medium is covered, or place the pot in a pre-heated oven 93–95°C, for 30–45 minutes.
(4) When pot and mixture have drained and cooled to be barely warm, spread the spore thinly over the surface using a spoon or a knife.
(5) Moisten the spore layer with an atomiser water spray.
(6) Cover the pot with a sheet of glass or polythene.
(7) Place the pot into another container. Add clean water to about one third of the pot's depth, (similar to the bog method of seed germination. Maintain the water level until true fronds appear.

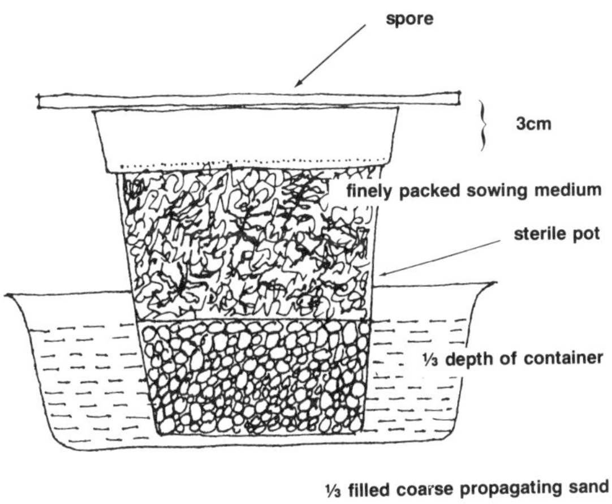


(8) Envelop the pot with a sheet of newspaper for 2–3 weeks.
(9) Leave the pot and container in a sheltered location.
(10) When sporelings develop, harden them off by removing glass or polythene.
(11) Prick out sporelings as small clusters. When further developed they can be separated into individual plants and re-potted.
(12) Ensure that young plants are given protection from sun and winds until they become established.

# Propagation of Orchids

Removal of orchids from the wild is illegal, (unless permission is granted by appropriate Government authorities), and not in the interests of conservation of our native bushland, unless some special development within an area means the plants would be destroyed anyway.

The propagation of orchids is a specialised operation. Many orchids have a symbiotic relationship with bacteria, and unless such a relationship can be maintained the orchids will die.

Orchid propagation is not covered in this chapter, but there is an excellent book on their propagation and cultivation, published by the Australasian Native Orchid Society, recommended for those with an interest in this topic. See Further Reading below. The Society also offers excellent service for members, and details of membership will be found in Appendix 3.

# Further Reading

Elliot, G. M. (1981) *Fun with Australian Plants.* Hyland House, Melbourne.

Elliot, G. M. (1985) *The Gardener's Guide to Australian Plants.* Hyland House, Melbourne.

Elliot, W.R. and Jones, D.L. (1980) *Encyclopaedia of Australian Plants Suitable for Cultivation, Vols 1–5.* Lothian, Melbourne.

Gardiner, A. (1988) *Modern Plant Propagation.* Lothian, Melbourne.

Handreck, K. & Black, N. (1984) *Growing Media for Ornamental Plants & Turf.* NSW University Press, Kensington NSW.

Hartman, H.T. & Kester, D.E. (1975) *Plant Propagation Principles and Practices.* Prentice Hall, U.S.A.

Jones, D. L. (1987) *Encyclopaedia of Ferns.* Lothian, Melbourne.

Jones, D. L. (1988) *Native Orchids of Australia.* Reed Books, Sydney.

Langkamp, P.J. (Ed.) (1987) *Germination of Australian Native Plant Seed.* Inkata Press, Melbourne.

Macdonald, B. (1986) *Practical Woody Plant Propagation for Nursery Growers.* Timber Press, Oregon U.S.A.

McMillan Browse, P.D.A. (1979) *Hardy, Woody Plants from Seed.* Grower Books, London, U.K.

Plumridge, J. (1977) *How to Propagate Plants.* Lothian, Melbourne.

Richards, H., Wootton, R. & Datodi, R. (1984) *Cultivation of Australian Native Orchids.* Australian Native Orchid Society (Vic. Group) Melbourne.

Wrigley, J. & Fagg, M. (1988) *Australian Native Plants* 3rd Ed. Collins, Sydney.

# *Part Two*

## *Plant Descriptions*

*Prostanthera lasianthos* var. *lasianthos*

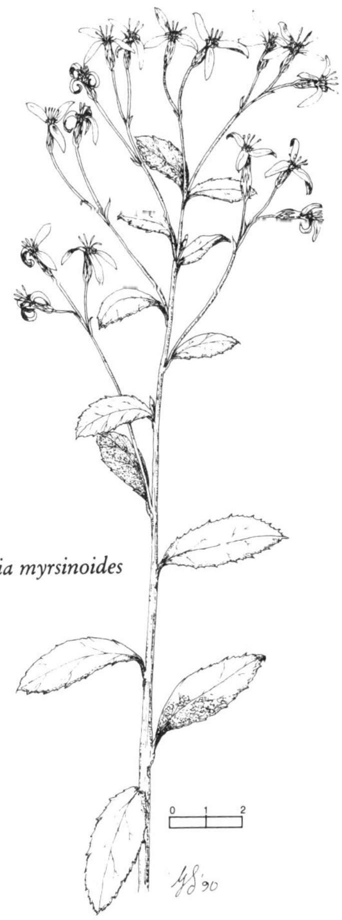


*Olearia myrsinoides*

SCALE: 0 1cm 2cm The scale representation on each plant equals 2cm.

# *Explanation of the Plant Descriptions*

Botanical Nomenclature follows '*A Census of the Vascular Plants of Victoria*', 4th Edition, J.H. Ross (January 1993).

The *Flora of Melbourne* has been listed **alphabetically** in groups according to easily recognisable flowering and non-flowering characteristics.

A total of 1152 plants indigenous to Melbourne have been described. This represents a little over ⅓ of the Victorian indigenous flora. It is estimated that about 70 species are now extinct within Melbourne, with many more known from very few plants.

No attempt has been made to describe introduced (Australian or exotic) species although botanists are still undecided about the origins of a few species included here.

**Synonyms** have been used to explain recent name changes. Botanical revision is an on-going process and many new names are found within the text. Names that have been incorrectly applied to species are also listed under the term 'synonym'.

Where several species of a genus occur in the Melbourne area, a general introduction to that genus has often been used. Features common to all the indigenous Melbourne species are discussed as well as specific generic characteristics and pertinent horticultural notes.

Botanical terms have been kept to a minimum throughout the text. An illustrated **glossary** explains all terms used.

The description of each plant refers as much as possible to local forms. Plants frequently vary in habit and foliage throughout their range. In some cases more than one form may exist within the Melbourne area. These are detailed where known.

While plants may grow in a large variety of **habitats** throughout Australia, we have chosen to list only the typical sites where one would expect to find a given species in Melbourne. These habitats are explained in Chapter 1.

The Australia-wide and world-wide **distribution** has been included to show the broader perspective of the indigenous plants of Melbourne.

The **size** usually refers to the ultimate height of the tallest part of the plant. Some plants have rosettes or tufts of leaves which are considerably shorter than the flowering stems. The foliage description will indicate the size of these plants when not in flower. Size is always given as height x width.

The size of any plant should only be used as a general guide. Cultivated plants are often larger or denser than those growing in remnant bushland. Other factors affecting ultimate size are amount of available moisture, soil type and habitat. A tree may be quite stunted on the coast where it grows on sandy, well drained soil and contends with salt-laden winds whereas the mountain form would usually be tall and straight.

Many features are common to a great many of the plants. **Foliage** may vary considerably in colour, texture and surface covering. The most common leaf colour is green and this may be assumed unless mentioned otherwise. Reference may be made to the lower surface of the leaf which is paler or brownish in appearance in some species. Leaves are also smooth and glabrous (hairless), and arranged in an alternate pattern unless otherwise stated. Leaf size is length x width.

**Flowering times** should be used as a guide only and, where possible, relate to Melbourne conditions. Flowering is often variable, and months stated indicate the period during which it is most likely to occur. Some plants also have sporadic flowering at other times of the year. Climatic conditions may influence the timing and duration of flowering. Other influencing factors are altitude and proximity to the sea. Cultivation may also affect length or time of flowering.

The **drainage requirements** of a plant refer to the amount of moisture retained in the soil over a period of time. 'Soil' is a combination of solid particles and open spaces. The solid particles, both organic and inorganic, comprise, in most soils, less than half the total volume of soil. The balance, even in poorly drained soils, is open space, known as 'pore space'. Pore space is the most important factor in determining the drainage qualities and structure of soil. It is filled with water, air and plant roots; the proportion of air:water varies with rainfall.

In well drained soils excess water does not accumulate, but moves readily through the pore spaces, allowing oxygen to become available to the plant roots quickly. Moist, well drained soils never dry out, but after wet periods excess water remains in the soil for only short periods. Moist or swampy soils are often poorly drained remaining wet for most of the year, and may retain free water for some months. Wet, boggy or waterlogged soils are highly water-retentive and remain wet all year, except that in times of drought, they may dry to varying degrees. Soils subject to periodic inundation are generally poorly drained, and free water is held above ground for varying times, particularly during Winter. However these soils may become quite dry during Summer. The section on requirements deals as much with horticultural needs as the conditions in which a plant occurs naturally. Many species will adapt to different conditions while others will succeed only in specific situations.

**Propagation** methods for each plant are mentioned in the text. Chapter 8 deals with the procedures to be used for each method.

The reader may assume that unless otherwise mentioned all plants are **frost hardy**. Plants which grow in the understorey rarely come into contact with frosts in their natural environment.

Most Australian plants benefit from pruning after flowering. This will produce more vigorous growth or a more desirable shape for a garden specimen.

The **fine-line drawings** of over 950 plants are an added identification guide. We have attempted to illustrate a species from every genus. The scale representation on each illustration is 2 cm. Live specimens have been used for most of the illustrations. Where this was not possible pressed herbarium specimens were referred to. A composite of photographs was used for a few of the rarer orchids.

The **reserves and localities** used represent the range of

habitats found within the Melbourne region, as well as examples from a broad spectrum of suburbs, enabling readers to obtain an idea of the original vegetation of their area. The key on the bookmark may be used as a quick reference, while a more detailed listing may be found in Appendix 2. All reserves and localities are shown on the Map inside the front and back covers. Suburb names have been used when locations have been obtained from herbarium specimens or personal knowledge, and do not coincide with the numbering system. At times a reserve/locality number has been used when a plant is known to come from that general area, even though it may no longer be at that specific site. Species that are known to be extinct within a particular area or reserve are marked #, eg 67#.

In order to protect species which are becoming rare or are endangered, specific locations are not given. This is the case with all orchids, where indiscriminate collection or picking has contributed to the decline in orchid populations.

We have attempted to verify the placement of species on the lists given to us. In some cases, especially with older lists, plants may have been misidentified. In other cases subsequent revision of species has made it difficult to be sure which species or subspecies was originally referred to. The authors would be pleased to receive any information regarding plants inadvertently omitted from or added to this list. Space did not permit the inclusion of every list available for the Melbourne area.

Indigenous plants should never be removed from remnant bushland sites. Seed and cuttings should only be taken with permission of the land owner and, where public land is involved, a permit may be necessary. (See Chapter 7)

**Abbreviations**

F — Full sun in an open position
P — Partial sun (full sun for only part of the day), filtered sun or dappled shade (with overhead canopy)
N — Full shade

ssp. — subspecies
var. — variety
spp. — more than one species
? — unconfirmed listing
Rges — Ranges
R — River
Pt — Point

Vic — Victoria
NSW — New South Wales
SA — South Australia
NT — Northern Territory
Qld — Queensland
Tas — Tasmania
WA — West Australia

NZ — New Zealand
PNG — Papua/New Guinea
Af — Africa
N Am — North America
S Am — South America
W Ind — West Indies
N Cal — New Caledonia
Mal — Malaysia
Poly — Polynesia
Indon — Indonesia
Eur — Europe

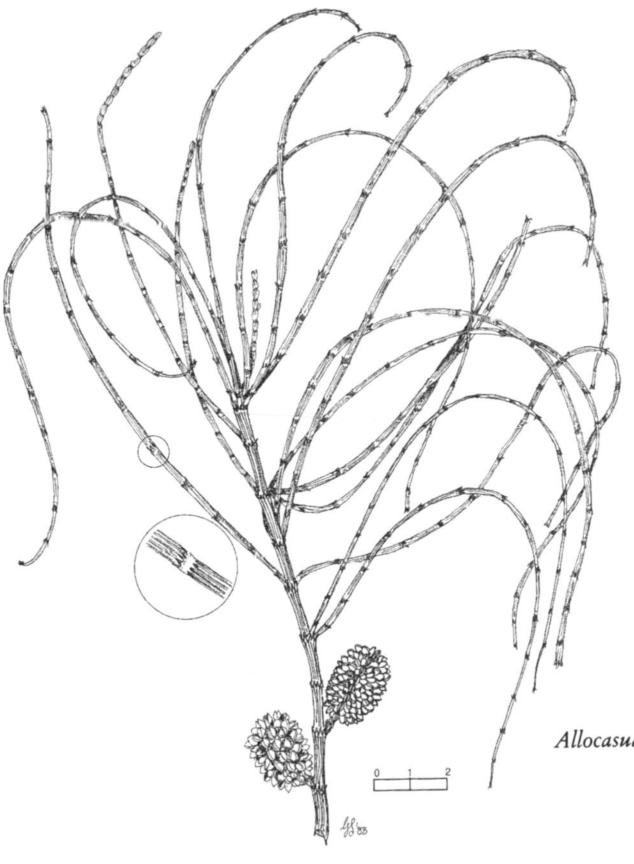

*Allocasuarina verticillata*

# Flowering Plants

# Dicotyledons

## Acacia

## Mimosaceae

The wattles are a very large genus represented by 23 species within the Melbourne area. They range in height from tall forest trees to prostrate shrubs, occurring in a wide range of habitats. Many are found as understorey shrubs in open forest. Their form and foliage are equally diverse. The true leaves of acacias are divided and fern-like and are most conspicuous when plants are seedlings. Some species retain these divided leaves while others form leaf-like structures called phyllodes. The yellow flowers are tightly clustered into globular flowerheads (or balls) or cylindrically-shaped spikes (or rods). The pods that form after fertilisation contain hard seeds which are an important food source for many birds. Pruning after flowering improves the shape of the shrubs and can extend the life of the plant. The seeds, gum, wood and bark of many species were used extensively by the local Aborigines as sources of food, fibre, medicine, implements and containers. Propagation is generally by seed, treated by immersing in boiling water, then soaking for 24 hours. (See Propagation) Some species can also be propagated by cuttings.

*Acacia acinacea*

### Acacia acinacea

Gold Dust Wattle

**Size:** 0.5–2.5 m x 2–4 m **F P**
**Habitat:** Dry sclerophyll forest, chenopod rocky open scrub, box ironbark, red gum and box woodland
**Form:** Open spreading shrub, denser in cultivation
**Foliage:** Small rounded or oblong phyllodes to 10 mm long, sometimes larger
**Flowers:** In profuse bright yellow balls along arching branches; August to November
**Requirements:** Adaptable to well drained soils
**Comments:** A good low screening plant which will often self seed in the garden. Annual pruning is beneficial.
**Localities:** Melton, 14, 16, 22, 24–26, 29, 31–39, 41, 42, 56, 58
**Distribution:** Vic, NSW, SA

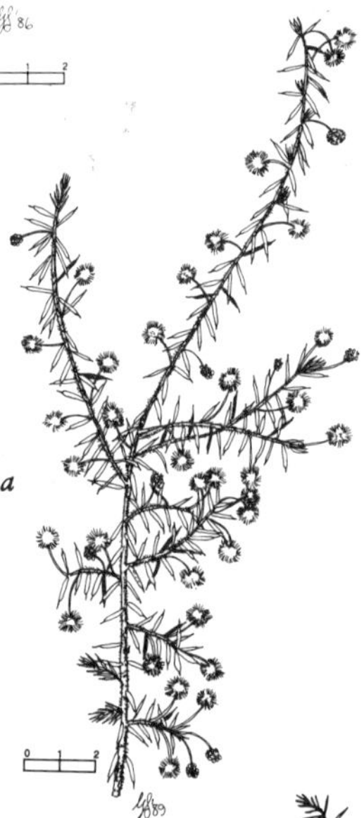

*Acacia aculeatissima*

### Acacia aculeatissima

Thin-leaf Wattle, Snake Wattle

**Size:** 0.2–0.6 m x 1–2 m **P**
**Habitat:** Dry sclerophyll forest
**Form:** Light open sprawling shrub
**Foliage:** Fine prickly phyllodes 10-15 mm long
**Flowers:** In yellow balls along wiry branches; June to December
**Requirements:** Well drained clay soils. Withstands extended periods of dryness once established.
**Comments:** Excellent for rockeries or banks. Withstands light pruning. A natural hybrid between *A. aculeatissima* and *A. verticillata* occurs in Warrandyte. It has grey phyllodes, is more upright in form than *A. aculeatissima*, and the flowerheads are intermediate between the two parents.
**Localities:** 34, 34A, 40-42, 44, 46–49, 50A, 51, 62–64
**Distribution:** Vic, NSW

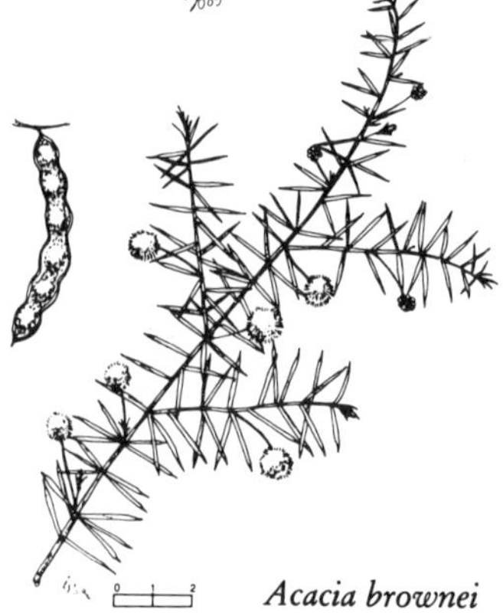

*Acacia brownei*

### Acacia brownei

Heath Wattle, Golden Prickly Wattle

**Size:** 0.5–1 m x1–2 m **P**
**Habitat:** Tea-tree heath, gravelly hillsides of dry sclerophyll forests
**Form:** Small spreading shrub
**Foliage:** Stiff narrow prickly phyllodes to 25 mm long
**Flowers:** In profuse golden yellow balls along arching stems; June to November
**Requirements:** Light soils tolerating dryness once established
**Comments:** An excellent garden plant, it differs from *A. ulicifolia* in that its phyllodes are longer and not markedly broader at base and the flowers are deeper yellow. Not common in Melbourne.
**Localities:** Warrandyte, Springvale, 50A, 51, 53, 59, 71, 74, 75
**Distribution:** Vic, Qld, NSW
**Synonym:** *A. ulicifolia var. brownei*

10

11

12

14

13

10. *Lomatia myricoides*
    River Lomatia
11. *Indigofera australis*
    Austral Indigo
12. Saltmarsh vegetation
13. *Pimelea linifolia* ssp. *linifolia*
    Slender Rice-flower
14. *Muehlenbeckia florulenta*
    Tangled Lignum
15. *Lobelia gibbosa*
    Tall Lobelia
16. *Gahnia filum*
    Chaffy Saw-sedge
17. *Hardenbergia violacea*
    Purple Coral Pea
18. *Pittosporum bicolor*
    Banyalla
19. *Swainsona lessertiifolia*
    Coast Swainson-pea
20. *Acacia verticillata*
    Prickly Moses

16

15

17

18

19

20

21

22

23

24

25

26

27

21. *Kennedia prostrata*
    Running Postman
22. *Eryngium ovinum*
    Blue Devil
23. *Dichanthium sericeum*
    Silky Blue-grass
24. *Diuris fragrantissima*
    Fragrant Doubletail
25. *Eucalytpus camaldulensis*
    River Red Gum
26. *Acacia pycnantha*
    Golden Wattle
27. *Stypandra glauca*
    Nodding Blue Lily
28. *Ottelia ovalifolia*
    Swamp Lily
29. *Leucochrysum albicans*
    Hoary Sunray

28

29

## Acacia dealbata

Silver Wattle

**Size:** 6–30 m x 5–10 m **F P**
**Habitat:** Wet, damp and valley sclerophyll forests, riparian scrub, red gum woodland, grassy low open forest
**Form:** Fast-growing open tree
**Foliage:** Bluish-green bipinnate leaves to 130 mm long
**Flowers:** Flowers in profuse lemon balls arranged in terminal racemes; July to October
**Requirements:** Easily grown, prefers deep moist soil
**Comments:** A useful tree for erosion control. An important aboriginal plant, the seeds and gum were eaten and axe handles were made from the wood. The gum was also used as an adhesive or mixed with burnt wattle bark to produce an ointment. Host to the Imperial Blue Butterfly caterpillar when less than 3 m high. Used for perfume manufacture in France where it is called Mimosa.
**Localities:** 1, 16, 19, 23–29, 32–36, 38, 39, 41, 42, 44, 46–53, 55–59, 61, 62, 64, 71, 75, 76
**Distribution:** Vic, NSW, Tas

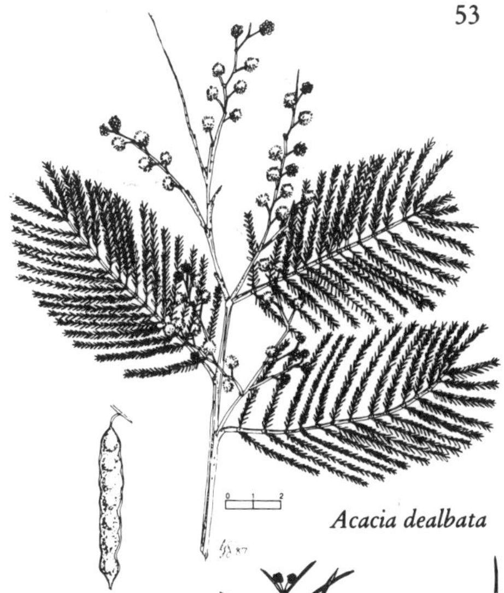

*Acacia dealbata*

## Acacia genistifolia

Spreading Wattle

**Size:** 1–3 m x 1–3 m **F P**
**Habitat:** Dry and valley sclerophyll forests
**Form:** Fast-growing, open spreading shrub
**Foliage:** Narrow prickly phyllodes to 30 mm long
**Flowers:** In perfumed lemon to cream balls. Long flowering; August to October and January to May
**Requirements:** Reliable shrub tolerating wet or dry soil
**Comments:** Because of its prickly foliage, this plant is a good refuge for small birds. Often self seeds in the garden.
**Localities:** Sth Morang, Murrumbeena[#], 33, 34A–38, 41, 42, 49, 51, 53, 58, 59, 62, 63, 67[#], 74
**Distribution:** Vic, NSW, Tas
**Synonym:** *A. diffusa*

*Acacia genistifolia*

## Acacia gunnii

Ploughshare Wattle

**Size:** 0.3–1 m x 1–2 m **F P**
**Habitat:** Exposed positions in box ironbark woodland, dry sclerophyll forest
**Form:** Prickly, low open spreading shrub
**Foliage:** Triangular phyllodes to 15 mm long, tapering to a pungent point
**Flowers:** In pale yellow balls; June to October
**Requirements:** Well drained soils tolerating dryness once established
**Comments:** A useful ground cover for erosion control. Not common in the Melbourne area.
**Propagation:** Seed, cuttings
**Localities:** Diamond Ck
**Distribution:** Vic, Qld, NSW, Tas, SA

*Acacia gunnii*

## Acacia implexa

Lightwood, Hickory Wattle

**Size:** 5–15 m x 4–7 m **F P**
**Habitat:** Common in dry or valley sclerophyll forests, red gum, box and riparian woodlands
**Form:** Fast-growing, upright small tree
**Foliage:** Sickle-shaped phyllodes to 20 cm x 20 mm
**Flowers:** In perfumed cream balls arranged in racemes; December to March
**Requirements:** Adaptable plant tolerating both wet and dry, well drained clay soil
**Comments:** An attractive, long-lived screen or shade tree, useful for erosion control. Its phyllodes are used for dyeing and the bark for tanning. The fibres produced twine.
**Propagation:** Seeds, cuttings
**Localities:** 8, 14, 16, 18, 21, 23–26, 28, 29, 31–39, 42, 44, 47–49, 50A, 51, 53, 58, 60-65, 67, 68, 69A, 71,
**Distribution:** Vic, Qld, NSW

*Acacia implexa*

## Acacia lanigera

Woolly, or Hairy Wattle

**Size:** 0.3–2 m x 1–3 m **F P**
**Habitat:** Dry sclerophyll forest, box ironbark woodland
**Form:** Small, rigid open shrub
**Foliage:** Sharp slightly curved grey-green phyllodes to 70 mm long. New growth is covered with whitish hairs.
**Flowers:** In bright yellow balls; May to October
**Requirements:** Well drained rocky and clay soils. Tolerates short periods of wetness.
**Comments:** Useful for planting under established trees and soil binding. Showy in flower.
**Propagation:** Seed, cuttings
**Localities:** Research, ?42
**Distribution:** Vic, NSW

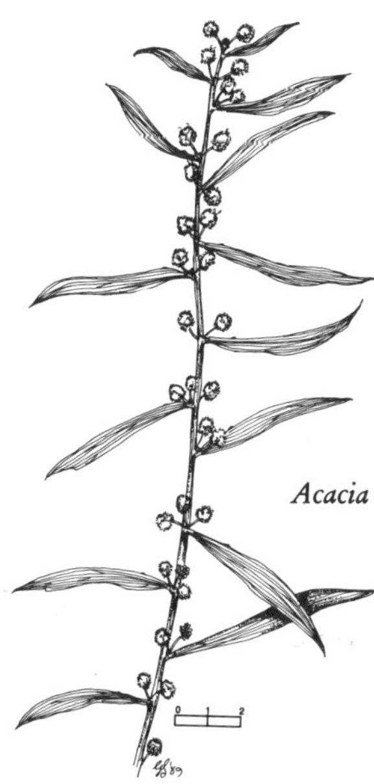

*Acacia lanigera*

SCALE: 0 1cm 2cm — The scale representation on each plant equals 2cm.

*Acacia leprosa*

## Acacia leprosa

Cinnamon Wattle

**Size:** 3–12 m x 2–6 m **P N**
**Habitat:** Damp and valley sclerophyll forests
**Form:** Fast-growing medium shrub or small tree of dense weeping habit
**Foliage:** Bright green fragrant and sticky phyllodes, variable in size from 30–150 mm long and up to 25 mm wide
**Flowers:** In abundant scented lemon to yellow balls; August to December
**Requirements:** Moist, well drained soils
**Comments:** This reliable wattle makes an excellent tall screening plant. Weeping forms are attractive planted near water. Foliage has a cinnamon fragrance on humid days or when crushed. *A. leprosa x paradoxa* occurs in dry sclerophyll forest in the Croydon/Warrandyte and Lysterfield areas.
**Localities:** 40, 42, 43, 47, 50A, 51, 53, 55–58, 60, 61
**Distribution:** Vic, NSW

*Acacia mearnsii*

## Acacia mearnsii

Black Wattle

**Size:** 8–25 m x 6–10 m **F P N**
**Habitat:** Widespread in dry and valley sclerophyll forests, grassy low open forest, red gum and riparian woodland, plains grassland, escarpment above riparian scrub
**Form:** Very fast growing open spreading tree
**Foliage:** Bipinnate dark green leaves to 20 cm long
**Flowers:** Strongly scented pale yellow balls in dense racemes; September to December
**Requirements:** Prefers well drained soils
**Comments:** This attractive tree can be short-lived and is often subject to borer attack. Its bark, seed pods, leaves and flowers can be used for dyeing. The wood was used for aboriginal weapons and bark provided twine and medicine. The gum was a source of food, drink and adhesive. Seed germinates readily after fire.
**Localities:** Tarneit, Melton, St Albans, 1, 8, 14, 16, 18, 21–23, 25, 26, 28–31, 33–51, 53, 50–76
**Distribution:** Vic, NSW, Tas, SA

*Acacia melanoxylon*

## Acacia melanoxylon

Blackwood

**Size:** 5–30 m x 4–15 m **F P**
**Habitat:** Widespread in cool temperate rainforest to dry sclerophyll forests, grassy low open forest, red gum woodland, riparian scrub, plains grassland
**Form:** Fast-growing upright tree with a heavy canopy
**Foliage:** Phyllodes are dull green and variable, to 15 cm x 10–30 mm
**Flowers:** In cream balls, arranged in short racemes; July to October
**Requirements:** Prefers deep moist soil but adaptable, tolerating dryness once established
**Comments:** A long-lived tree providing good screening and shade. It tends to sucker if its roots are damaged. Grows as a more stunted tree on the escarpments of the basalt plains. Excellent timber for furniture-making. Its fibre was used to make fishing lines by the Aborigines. Also provided weapons and medicine.
**Localities:** Sunshine, St Albans, 1, 8, 14, 16, 19, 21, 23, 25, 26, 28, 29, 32–36, 38–51, 53–69A, 71, 74–76
**Distribution:** Vic, Qld, NSW, Tas, SA

*Acacia mucronata*

## Acacia mucronata

Variable Sallow Wattle, Narrow-Leaf Wattle

**Size:** 2–6 m x 2–5 m **F P**
**Habitat:** Damp and valley sclerophyll forests
**Form:** Spreading or erect shrub of variable habit
**Foliage:** Light green linear phyllodes from 3–20 cm long
**Flowers:** In cream to yellow spikes; August to October
**Requirements:** Moist well drained soils. Drought resistant once established
**Comments:** A good screen plant but requires pruning. Suckers, especially after fires.
**Localities:** 42, 47, 49, 50A, 51, 53, 55, 56, 58, 61
**Distribution:** Vic, NSW, Tas

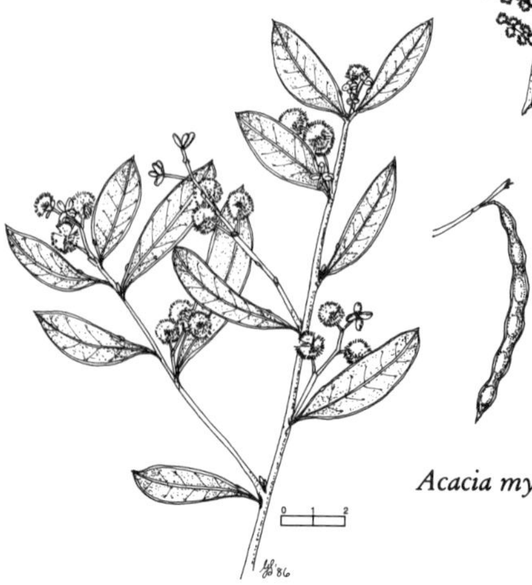

*Acacia myrtifolia*

## Acacia myrtifolia

Myrtle Wattle

**Size:** 1–3 m x 1–2 m **F P N**
**Habitat:** Sclerophyll woodland, damp and valley sclerophyll forests
**Form:** Fast-growing rounded or erect shrub
**Foliage:** Broad phyllodes often with yellow margins, to 60 mm long. New growth is bronze.
**Flowers:** Profuse, in cream to yellow balls, arranged in racemes; July to October
**Requirements:** Most soils
**Comments:** An ornamental bush, with reddish stems, it provides a low screen. A heavy prune can be beneficial. This wattle is quite variable throughout its range.
**Localities:** 34A, 35, 42, 43, 46–51, 53–56, 58–65
**Distribution:** All states

## Acacia oxycedrus

Spike Wattle

**Size:** 1–10 m x 2–5 m **F P**
**Habitat:** Tea-tree heath, dry sclerophyll forest
**Form:** Stiffly spreading shrub to small tree
**Foliage:** Rigid dark green phyllodes to 40 mm long, tapering to a sharp point
**Flowers:** In showy bright yellow spikes; June to November
**Requirements:** Adaptable, withstanding wet periods
**Comments:** A useful windbreak, its prickly foliage provides a good bird refuge or protective barrier. The local form is unlikely to grow taller than 3 m. *A. oxycedrus* x *sophorae* occurs in Oakleigh, Parkdale and Frankston.
**Propagation:** Seed, cuttings
**Localities:** 45#, 49, 50A#, 63, 67–71, 73–75
**Distribution:** Vic, NSW, SA

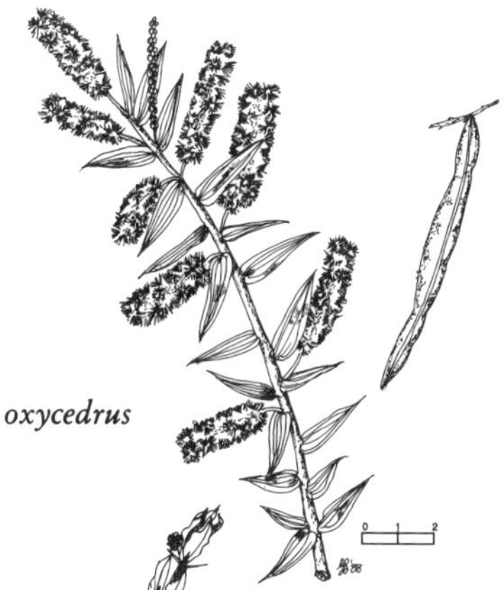

*Acacia oxycedrus*

## Acacia paradoxa

Hedge Wattle, Kangaroo Thorn

**Size:** 2–4 m x 2–5 m **F P**
**Habitat:** Widespread in dry sclerophyll forest, grassy low open forest, red gum woodland, plains grassland
**Form:** Fast-growing dense and spreading shrub covered with thorns
**Foliage:** Dark green oblong to narrow phyllodes with wavy edges, to 30 mm long
**Flowers:** Abundant in large golden yellow balls; August to November
**Requirements:** Will adapt to any situation
**Comments:** A very ornamental wattle, it has been declared a noxious weed in rural areas. Its thorny habit makes it a refuge for birds.
**Propagation:** Seed, cuttings
**Localities:** 12, 13, 16, 21, 23–26, 29, 30, 32, 33, 34A–39, 42, 44–49, 50A, 51, 53–55, 60–71, 73–76
**Distribution:** All states
**Synonym:** *A. armata*

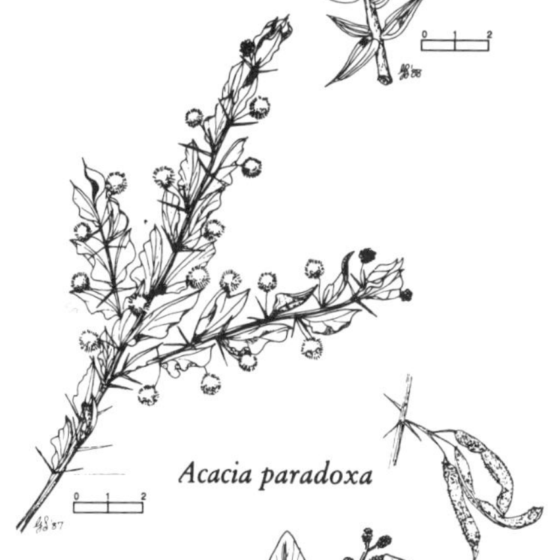

*Acacia paradoxa*

## Acacia pycnantha

Golden Wattle

**Size:** 3–10 m x 2–5 m **F P**
**Habitat:** Widespread in dry, damp and valley sclerophyll forests, plains grassland, red gum woodland
**Form:** Erect or spreading small tree with glaucous bark while young
**Foliage:** Bright green, broad curved phyllodes to 20 cm x 10–50 mm wide
**Flowers:** Profuse, in perfumed golden balls arranged in long racemes; July to October
**Requirements:** Well drained soil
**Comments:** Australia's floral emblem, this tree is fast-growing and drought resistant. Pruning whilst young encourages denser growth. It is a screening and windbreak tree and is useful for erosion control. An important aboriginal plant for food, containers, medicine and glue.
**Propagation:** Seed or cuttings
**Localities:** Montrose, 1, 4, 14, 16, 21–26, 28–30, 32–44, 47, 48, 50-53, 56, 58, 61–65, 72, 74
**Distribution:** Vic, NSW, SA

*Acacia pycnantha*

A. retinodes var. retinodes now A. provincialis

## Acacia retinodes

Wirilda, Swamp Wattle

**Size:** 4–8 m x 3–5 m **F P**
**Habitat:** Riparian scrub, coastal banksia and riparian woodlands
**Form:** Spreading tall shrub or small tree
**Foliage:** Long narrow bluish-green phyllodes to 20 cm long
**Flowers:** In lemon-yellow perfumed balls, arranged in racemes; intermittent all year, peaking November to January
**Requirements:** Adaptable to all soils and conditions including salt and moderate lime
**Comments:** This quick growing ornamental wattle can be induced to sucker by severing its roots. It can self-seed and needs to be watched near bush areas.
**Localities:** 8, 16, 18, 19, 26, 29, 33, 50A, 61, 67
**Distribution:** Vic, Tas, SA

"Swamp Form"

*Acacia retinodes*

## Acacia sophorae

Coastal Wattle

**Size:** Variable 1–8 m x 4–8 m **F**
**Habitat:** Dominant in primary dune scrub, coastal banksia woodland, invasive in grassy low open forest
**Form:** Dense mounded shrub to small tree
**Foliage:** Broad elliptical phyllodes to 15 cm long
**Flowers:** In sulphur-yellow spikes; July to October
**Requirements:** Well drained moist to dry soils
**Comments:** Taller forms make excellent screens and windbreaks. A useful soil-binding and erosion control plant, especially for front line dune stabilisation as it tolerates salt-laden winds. The green seeds were cooked in their pods and eaten by the Aborigines.
**Localities:** 32, 64, 67–69A, 71, 72, 74, 75
**Distribution:** Vic, Qld, NSW, Tas, SA

*Acacia sophorae*

SCALE: 0 1cm 2cm — The scale representation on each plant equals 2cm.

*Acacia stricta*

## Acacia stricta

Hop Wattle, Straight Wattle

**Size:** 2–5 m x 2–4 m **F P N**
**Habitat:** Valley sclerophyll forest, coastal banksia woodland, grassy low open forest, tea-tree heath
**Form:** Open upright shrub
**Foliage:** Erect narrow elliptical dull grey-green phyllodes to 15 cm long
**Flowers:** In pale yellow balls; May to October
**Requirements:** Reliable in most soils
**Comments:** A quick growing low windbreak ideal for planting under existing trees. It is drought hardy and tolerates coastal exposure.
**Localities:** 34A, 42, 44, 46, 47, 49–56, 58, 59, 61, 66, 67#, 70, 74
**Distribution:** Vic, Qld, NSW, Tas, SA

*Acacia suaveolens*

## Acacia suaveolens

Sweet Wattle

**Size:** 1–3 m x 2–5 m **F P**
**Habitat:** Tea-tree heath, wattle tea-tree scrub
**Form:** Open spreading shrub
**Foliage:** Bluish-green narrow phyllodes to 15 cm long
**Flowers:** In perfumed cream balls arranged in racemes; April to October, followed by pale bluish seed pods
**Requirements:** Well drained dry soils
**Comments:** A fast-growing wattle, it is a very ornamental low screen or windbreak, especially for coastal areas. Responds well to pruning after flowering.
**Localities:** 66–69, 74, 75
**Distribution:** Vic, Qld, NSW, Tas, SA

*Acacia ulicifolia*

## Acacia ulicifolia

Juniper Wattle

**Size:** 1–2 m x 1–2 m **P**
**Habitat:** Valley sclerophyll forest, tea-tree heath, grassy low open forest
**Form:** Open rounded shrub
**Foliage:** Very narrow dark green prickly phyllodes to 10 mm long, broader at the base
**Flowers:** Singly, in cream to pale yellow balls; March to September
**Requirements:** Prefers moist well drained soils
**Comments:** Used as a low screen it provides a prickly barrier and bird refuge.
**Propagation:** Seeds, cuttings
**Localities:** 34A, 41, 42, 50A–53, 67, 69#, 74, 75
**Distribution:** Vic, Qld, NSW, Tas

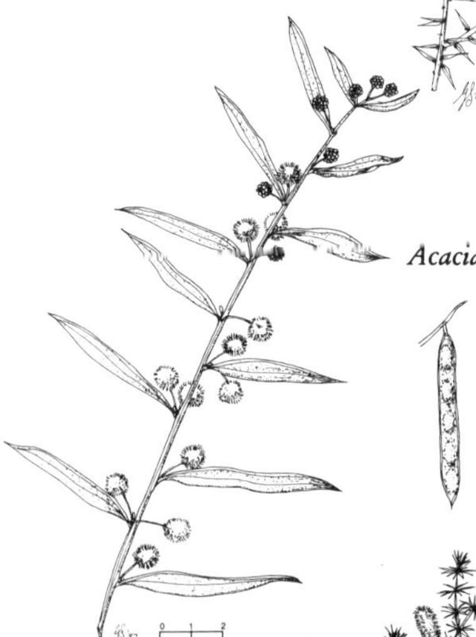

*Acacia verniciflua*

## Acacia verniciflua

Varnish Wattle

**Size:** 3–5 m x 3–5 m **F P**
**Habitat:** Damp and valley sclerophyll forests, grassy low open forest
**Form:** Slightly weeping open tall shrub or small tree
**Foliage:** Light shiny green curved phyllodes to 15 cm long
**Flowers:** Profuse, in pale to bright yellow balls; July to January
**Requirements:** Tolerates both wet and dry conditions
**Comments:** A quick growing light screening plant with sticky new growth. Seeds and pods provided food for the Aborigines.
**Localities:** Eltham, 33, 44, 47, 51, 52, 55, 59, 74, 76
**Distribution:** Vic, Qld, NSW, Tas, SA

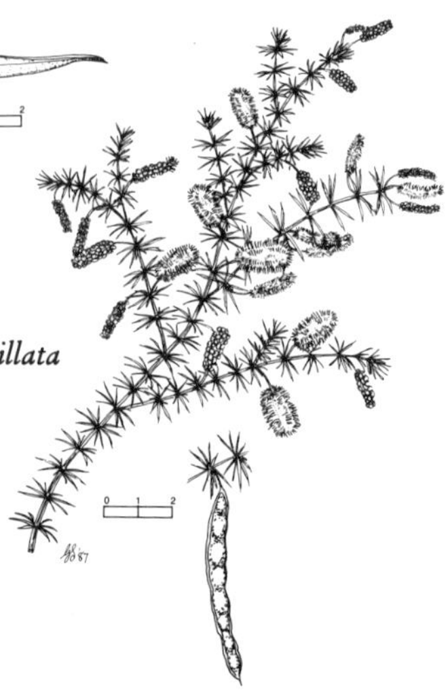

*Acacia verticillata*

## Acacia verticillata

Prickly Moses

**Size:** 2–6 m x 3–5 m **P**
**Habitat:** Widespread in wet, damp and valley sclerophyll forests, wattle tea-tree and riparian scrub
**Form:** Variable open shrub
**Foliage:** Fine prickly phyllodes to 20 mm long
**Flowers:** In light yellow spikes; June to December
**Requirements:** Tolerates most garden conditions, including alkaline soils. Withstands periods of waterlogging.
**Comments:** Pruning whilst young encourages a bushy habit. It is a useful bird refuge. Fishing line was made from the fibre.
**Localities:** 16, 26, 29, 30#, 33–36, 40–44, 46–51, 53–64, 67#, 70, 74–76
**Distribution:** Vic, NSW, Tas, SA

# Acaena — Rosaceae

*Acaena agnipila*

## Acaena agnipila — Hairy Sheep's Burr

**Size:** 20–50 cm x 30 cm — F P N
**Habitat:** Plains grassland, grassy wetland, red gum woodland, dry sclerophyll forest
**Form:** Rosette or tufted perennial herb
**Foliage:** Narrowly-obovate to oblanceolate pinnate leaves to 15 cm long, including stalk, with more than 4 pairs of leaflets. Moderately hairy upper surface, densely hairy below.
**Flowers:** Flowers are in a spike with dark purple stamens; October to November. Spines of the burr are of equal length, with slender barbs on the end.
**Requirements:** Well drained soils
**Comments:** *A. echinata* and *A. ovina* are very similar to *A. agnipila* and the 3 species may be difficult to tell apart, especially if flowerheads or burrs are not present. All 3 lack stolons and are propagated by seed.
**Localities:** St Albans, 6, 21, 25, 26, 28, 35, 36, 39, 50A, 51, 55, 56, 74, 76
**Distribution:** Vic, Qld, NSW, Tas, SA

## Acaena anserinifolia = A. novae-zelandiae

*Acaena echinata*

## Acaena echinata — Sheep's Burr

**Size:** 25–40 cm high — F P
**Habitat:** Grassy wetland, plains grassland, red gum woodland, valley sclerophyll forest
**Form:** Perennial herb with erect stems
**Foliage:** Narrowly obovate to oblanceolate, pinnate leaves 60–150 mm long, with 9–15 ovate to oblong toothed leaflets. Hairs on undersurface are confined to the major veins.
**Flowers:** Tiny flowers with dark purplish stamens form a cylindrical spike 50–150 mm long; August to November. Spines of the burr are unequal in length.
**Requirements:** Moist soil
**Localities:** Eltham, Research, 2, 4–7, 10, 12, 16–18, 20, 21, 23, 24, 26, 28, 31, 32, 36, 41, 42, 44, 59, 60, 62–65, 69A, 74
**Distribution:** All states except NT

*Acaena novae-zelandiae*

## Acaena novae-zelandiae — Bidgee Widgee

**Size:** Prostrate x 1–4 m — F P N
**Habitat:** Widespread in wet, damp and valley sclerophyll forests, swamp and riparian scrub, plains grassland, riparian and red gum woodland
**Form:** Rambling stoloniferous mat plant
**Foliage:** Oblong pinnate leaves to 40 mm. 11 oblong leaflets with toothed margins, increasing in size along the stalk to 10 mm long.
**Flowers:** Globular greenish-white flowerheads on stalks to about 20 cm long; October to January. Fruits form reddish globular heads to 25 mm covered with spines.
**Requirements:** Adaptable, tolerating wet or dry conditions
**Comments:** A useful soil-binding plant and ground cover. The burrs can be a problem on clothing. Early settlers used the leaves as a tea substitute.
**Propagation:** Seeds, cuttings, division
**Localities:** 5, 6–8, 10, 16, 21, 26–36, 38, 40–44, 46–65, 67, 69A–76
**Distribution:** Vic, Qld, NSW, Tas, SA
**Synonym:** *A. anserinifolia*

*Acaena ovina*

## Acaena ovina — Australian Sheep's Burr

**Size:** 10–50 cm high — F P
**Habitat:** Damp and dry sclerophyll forests, red gum and riparian woodland, plains grassland, grassy low open forest
**Foliage:** Tuft or rosette. Leaves to 120 mm long have 5 or more pairs of leaflets. May be glabrous or sparsely hairy above, more densely hairy below.
**Flowers:** Flowers are in a spike; September to November. Green burrs have 15–50 short spines with 3–6 much longer.
**Localities:** 4, 8, 15, 16, 26, 28, 29, 31, 32, 34A–36, 42, 44, 45, 47–49, 54, 56, 58, 63–67, 71, 74
**Distribution:** Vic, Qld, NSW, Tas, SA

SCALE: 0 1cm 2cm — The scale representation on each plant equals 2cm.

*Acrotriche prostrata*

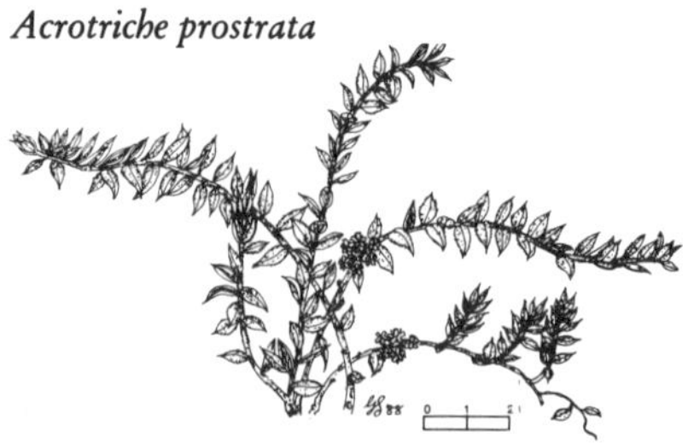

# Acrotriche — Epacridaceae

## Acrotriche prostrata

Trailing Ground-Berry

**Size:** Prostrate x 1–2 m — P N
**Habitat:** Damp and valley sclerophyll forests, sclerophyll woodland, grassy low open forest
**Form:** Open trailing plant
**Foliage:** Small dark green triangular leaves to 20 mm long
**Flowers:** Greenish translucent, tubular flowers which produce copious amounts of nectar, clustered on old wood; May to September.
**Requirements:** Moist friable soil
**Comments:** Edible fruits are globular, succulent drupes. It is ideal for growing with ferns.
**Propagation:** Cuttings
**Localities:** 55, 56, 58–62, 64, 65, 74–76
**Distribution:** Endemic to Vic

*Acrotriche serrulata*

## Acrotriche serrulata

Honey Pots

**Size:** 10-30 cm x 0.5–1 m — P
**Habitat:** Sclerophyll woodland, grassy low open forest, dry and valley sclerophyll forests
**Form:** Slow-growing, dense ground-covering shrub
**Foliage:** Hairy leaves to 10 mm long, tapering to a point
**Flowers:** Greenish translucent, tubular flowers clustered on old wood; May to October
**Requirements:** Most well drained soils tolerating dry periods
**Comments:** Flowers have a strong honey fragrance, and are rich in nectar. Used for food and drink by the Aborigines.
**Propagation:** Cuttings
**Localities:** Research, 29, 34–37, 40–44, 47–56, 58–65, 67–70, 74–76
**Distribution:** Vic, NSW, Tas, SA

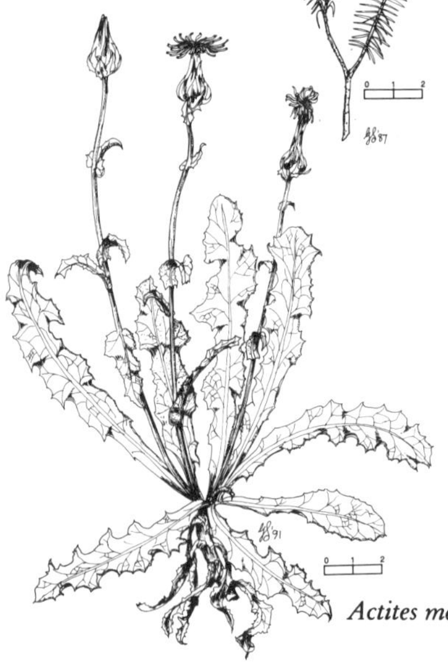

*Actites megalocarpa*

# Actites — Asteraceae

## Actites megalocarpa

Dune Thistle

**Size:** 0.1–0.6 m high — F
**Habitat:** Primary dune scrub and coastal cliffs
**Form:** Erect fleshy rhizomatous perennial herb usually forming an open clump
**Foliage:** Variable, thick glossy oblong leaves from 1.5–17 cm x 5–45 mm, stiff and stem-clasping with prickly teeth
**Flowers:** Yellow daisy flowerheads; September to June
**Requirements:** Well drained sandy soil
**Comments:** Tolerates exposure to salt spray but more stunted, reaching full height in a sheltered position. Probably extinct within the Melbourne area.
**Propagation:** Seed, division
**Localities:** Collected on Brighton Beach, 1887; 69A, 74
**Distribution:** All states

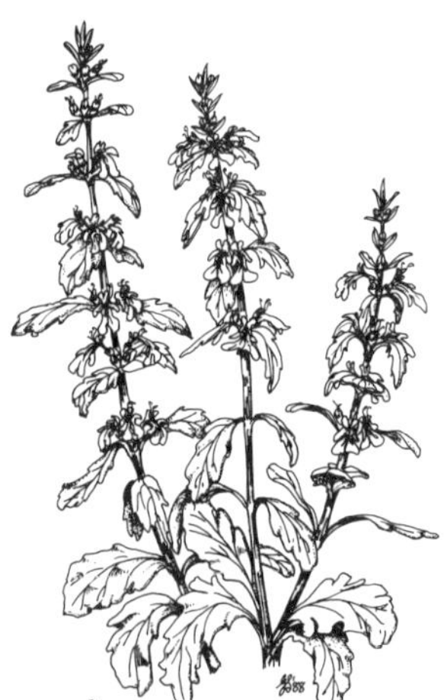

*Ajuga australis*

# Ajuga — Lamiaceae

## Ajuga australis

Austral Bugle

**Size:** 20–30 cm x 30–50 cm — F P
**Habitat:** Valley and dry sclerophyll forests, red gum woodland
**Form:** Erect or spreading perennial herb
**Foliage:** A basal rosette of velvety, grey-green toothed leaves to 120 mm long, leaves decreasing in size up the stem.
**Flowers:** Erect spikes of purple tubular flowers, sometimes blue or pink; September to February.
**Requirements:** Well drained soils, tolerates extended dry periods once established
**Comments:** A useful plant for erosion control in sandy soils. It will grow in sheltered coastal areas. Suckers vigorously in open soils. A decoction for sores was made from the leaves by the Aborigines.
**Propagation:** Seed, cuttings or division
**Localities:** 29, 34A, 35, 38, 39, 41, 42, 67#
**Distribution:** Vic, Qld, NSW, Tas, SA

# Allocasuarina — Casuarinaceae

Allocasuarinas are medium shrubs to tall trees. They are found in coastal areas with some growing inland. Most species are dioecious. Male plants have flower spikes from yellow to brown at the ends of branchlets, while female plants bear globular reddish flowers along the trunk and branches, followed by seed cones. Foliage is reduced to whorls of leaf teeth along the branchlets at the nodes. Allocasuarinas respond well to pruning if necessary. They are all ornamental and make excellent screening or windbreak plants. The timber provides excellent firewood. The local Aborigines ate the shoots and young cones and made implements and weapons from the wood. Propagation is by seed.
**Synonym:** All species in the Melbourne region were formerly placed in the genus *Casuarina*.

## Allocasuarina littoralis

Black Sheoke

**Size:** 4–8 m x 2–5 m **F P**
**Habitat:** Widespread in grassy low open forest, dry and valley sclerophyll forests, red gum woodland
**Form:** Upright small tree with fine branchlets
**Flowers:** Female flowers reddish to crimson. Male flower spikes dark brown; March to June
**Requirements:** Well drained soils
**Comments:** The bark on old trees is dark and deeply furrowed and is ideal for growing epiphytic orchids.
**Localities:** Research, 29–35, 42, 44, 47, 49, 50A, 51, 53, 54, 56, 60-67#, 69A, 71, 74–76
**Distribution:** Vic, Qld, NSW, Tas
**Synonym:** *Casuarina littoralis*

*Allocasuarina littoralis*

## Allocasuarina luehmannii

Bull Oak, Buloke

**Size:** 5–15 m x 5–10 m **F**
**Habitat:** Box woodland
**Form:** Fast-growing upright tree with fine dull green branchlets
**Flowers:** Yellow spikes to 25 mm long on male trees; October to November.
**Requirements:** Prefers heavy soil. Tolerates periods of dryness and inundation.
**Comments:** Roots may produce suckers. The cones are distinctive being squat with only 2 or 3 rows of valves. Occurring in few sites within the Melbourne area, it has become depleted in Victoria.
**Localities:** Melton, 9, 16, 26, 32–34
**Distribution:** Vic, Qld, NSW, SA
**Synonym:** *Casuarina luehmannii*

*Allocasuarina luehmannii*

## Allocasuarina paludosa

Swamp Sheoke

**Size:** 0.5–2 m x 1–2 m **F**
**Habitat:** Tea-tree heath, sclerophyll woodland, grassy low open forest
**Form:** Slow-growing open or dense shrub, branchlets grey-green, sometimes hairy in furrows
**Flowers:** Brown male flowers; March to October
**Requirements:** Moist well drained clay or sandy soils
**Comments:** Unlike most *Allocasuarina* spp., it may have both male and female flowers on the same plant. A groove down the centre of each ridge along the branchlets is a distinguishing feature.
**Localities:** 45#, 59, 63–65, 70, 71, 74–76
**Distribution:** Vic, NSW, Tas, SA
**Synonym:** *Casuarina paludosa*

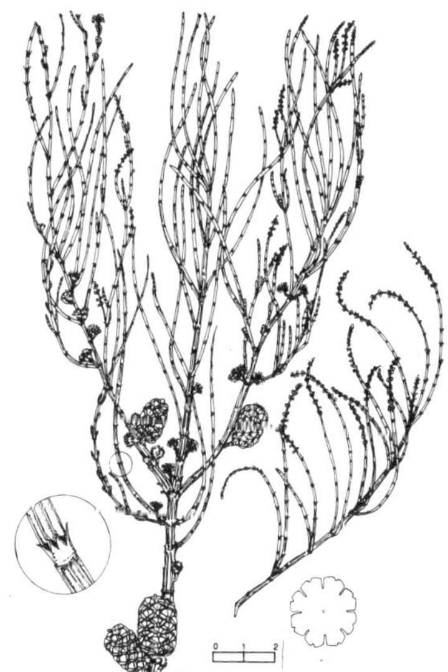

*Allocasuarina paludosa*

## Allocasuarina paradoxa

Dwarf Sheoke

**Size:** 0.5–2 m x 1–2 m **F P**
**Habitat:** Tea-tree heath
**Form:** Upright open to dense shrub
**Flowers:** Reddish-brown male flowers; September to November
**Requirements:** Well drained sandy soil
**Comments:** Commonly found with *A. paludosa* with which it may be easily confused but *A. paradoxa* prefers better drainage. It is hairless and ridges are smooth.
**Localities:** 67–70, 73–75
**Distribution:** Endemic to Vic
**Synonym:** Plants from the Melbourne region were commonly referred to as *A. pusilla*, which occurs in western Victoria.

*Allocasuarina paradoxa*

## Allocasuarina verticillata

Coast or Drooping Sheoke

**Size:** 4–11 m x 3–6 m **F**
**Habitat:** Widespread in primary dune scrub, box, red gum and box ironbark woodlands, chenopod rocky open scrub
**Form:** Small erect tree with a dense rounded crown and drooping greyish-green branchlets
**Flowers:** Yellow to brown male flowers; March to December
**Requirements:** Well drained soils
**Comments:** The golden effect provided by the dense flowers is an attractive feature of this tree. The dark furrowed bark is useful for growing epiphytic orchids.
**Localities:** Melton Sth, 14, 16, 18, 23, 25, 26, 29, 32–34A, 36, ?42, 44, 67, 69A, 71, 74, 76
**Distribution:** Vic, NSW, Tas, SA
**Synonym:** *A. stricta, Casuarina stricta*

*Illustrated p. 51*

SCALE: 0 1cm 2cm The scale representation on each plant equals 2cm.

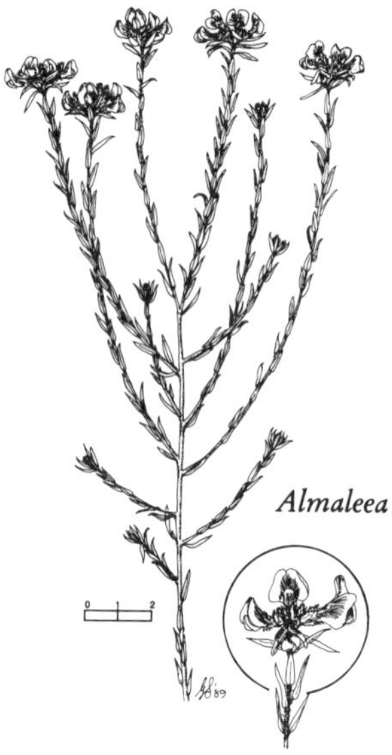

*Almaleea subumbellata*

# Almaleea — Fabaceae

## Almaleea subumbellata

Wiry Bush-pea

P

**Size:** 0.3–0.6 m high
**Habitat:** Swamp scrub
**Form:** Slender, erect, or trailing shrub
**Foliage:** Concave, lanceolate leaves to 12 mm long with minute free stipules. Leaves are usually pressed up along the stems.
**Flowers:** Terminal clusters of yellow and red flowers; October to December. Hairy oblong bracteoles.
**Requirements:** Moist to wet soils
**Comments:** An attractive low shrub with yellowish branchlets. When found in sedge swampland the plant may reach 1.5 m tall, its slender stem being supported by surrounding scrub.
**Localities:** 47, 54, 59
**Distribution:** Vic, Tas
**Synonym:** *Pultenaea subumbellata*

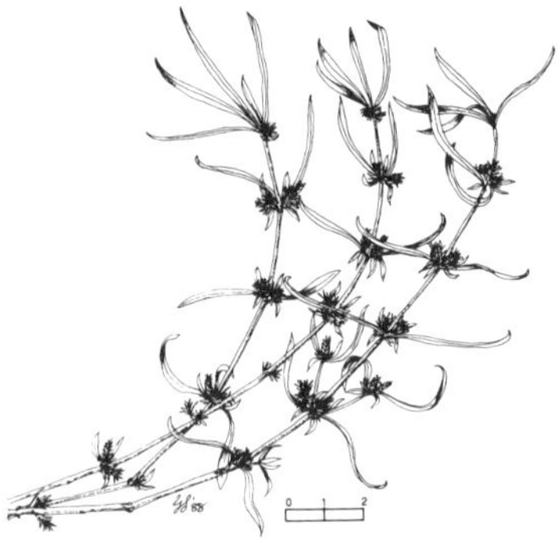

*Alternanthera denticulata*

# Alternanthera — Amaranthaceae

## Alternanthera denticulata

Lesser Joy Weed

F P

**Size:** 15–30 cm x 0.5 m
**Habitat:** Damp depressions in red gum, riparian and box woodlands, grassy wetlands, dry sclerophyll forest
**Form:** Low straggling perennial herb
**Foliage:** Narrow leaves from 20-100 mm long
**Flowers:** Short spikes of silvery-white paper flowers most of the year
**Requirements:** Moist well drained soils
**Comments:** This plant may be useful beside ponds. Prune after flowering to encourage a bushy habit. A small compact form to 15 cm wide is found on drier grasslands and box woodland of the basalt plains, rather than the usual riparian habitat. It is in the Melton area, amongst *Allocasuarina luehmannii* and has rounded leaves, white flowerheads which are twice the size of the normal form, and very short internodes.
**Propagation:** Seed, cuttings
**Localities:** Derrimut, Deer Park, Melton, 6, 8, 9, 11, 16–18, 25–30, 34A–36, 41, 42, 60, 62, 64, 67, 71, 74
**Distribution:** Vic, Qld, NSW, Tas, SA

*Alyxia buxifolia*

# Alyxia — Apocynaceae

## Alyxia buxifolia

Sea Box

F

**Size:** 1–2 m x 1–3 m
**Habitat:** Exposed coastal cliffs and primary dune scrub, grassy low open forest
**Form:** Slow-growing dense spreading shrub
**Foliage:** Leathery, dark green, glossy oval leaves to 30 mm long; paired or in 3s
**Flowers:** Sessile cluster of scented white flowers with 5 petals twisted sideways; October to February
**Requirements:** Well drained soils
**Comments:** A useful coastal plant tolerant of salt spray, its fruits are very attractive bright red drupes. It could make a decorative container plant.
**Propagation:** Seed, cuttings—slow to strike
**Localities:** Seaford, 67, 69A, 74
**Distribution:** Vic, NSW, Tas, SA, WA

*Amperea xiphoclada*

# Amperea — Euphorbiaceae

## Amperea xiphoclada

Broom Spurge

F P

**Size:** 0.3–0.8 m x 0.4–0.5 m
**Habitat:** Tea-tree heath, dry sclerophyll forest
**Form:** Wiry shrub with rigid stems arising from a woody rootstock
**Foliage:** Smooth dark green narrow leaves to 20 mm long; scale-like leaves on flowering stems
**Flowers:** Clusters of insignificant brown and cream flowers; September to February
**Requirements:** Moist well drained soils
**Comments:** Useful in landscaping for creating a natural bush appearance.
**Propagation:** Seed, cuttings
**Localities:** 49, 55, 56, 63, 67, 69–71, 73–76
**Distribution:** Vic, Qld, NSW, Tas, SA

# Amyema

**Loranthaceae**

Although parasitic, mistletoes are not a problem in forested areas because of the number of host trees available. On cleared land few trees remain. This may cause heavier infestation on each tree which may result in the tree's death. The mistletoe bird eats the berries and deposits the seeds on branches of the host trees. To propagate, the sticky seed needs to be collected and placed on the bark of a suitable host.

## Amyema linophyllum ssp. orientale

Slender-leaf Mistletoe

P

**Habitat:** Box woodland on *Allocasuarina luehmannii*
**Form:** Pendulous clump, young stems and leaves densely covered with white hairs
**Foliage:** Pairs or clusters of blue-green needle-like leaves to 120 mm long, with blunt tips
**Flowers:** 3 groups of 3 red flowers, whitish on the outside, on a common woolly stalk. The outer 2 flowers are stalked, the central flower sessile; October to February. The bud is 15–25 mm long, the globular berry is 5 mm.
**Comments:** On few sites in Melbourne, as it is dependent on *Allocasuarina luehmannii*. Also becoming rare within Victoria.
**Localities:** Melton, Werribee#, 9
**Distribution:** Vic, Qld, NSW, SA

*Amyema linophyllum* ssp. *orientale*

## Amyema miquelii

Box Mistletoe

P

**Size:** Drooping, to 3 m long
**Habitat:** Box and red gum woodlands
**Form:** Open pendulous clumps on host tree
**Foliage:** Curved, bronze-green linear leaves to 20 cm long
**Flowers:** Drooping clusters of 4–5 groups of 3 attractive, stalked orange-red flowers, each group on a common stalk. Flowering occurs sporadically throughout the year. The bud is 15–30 mm long. Fruit is a cylindrical to pear-shaped berry to 12 mm.
**Comments:** This parasitic plant may eventually kill the host branch. Its common host plant is *Eucalyptus* spp., but it is also found on *Acacia* spp.
**Localities:** Research, 25, 36, 42
**Distribution:** All states

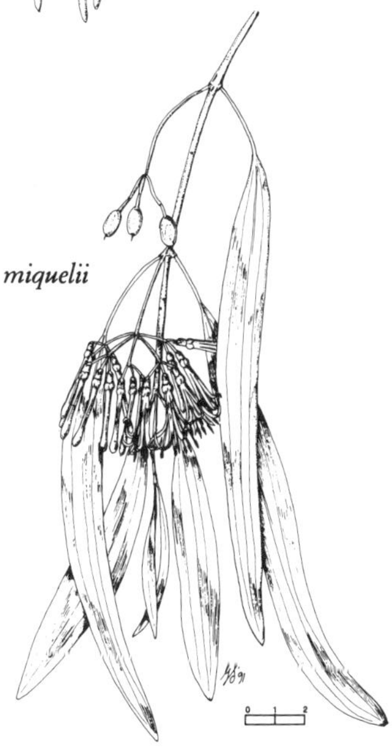

*Amyema miquelii*

## Amyema pendulum

Drooping Mistletoe

P

**Habitat:** Sclerophyll woodland, wet, damp and valley sclerophyll forest, grassy low open forest
**Form:** Very similar but larger than *A. miquelii* with lanceolate leaves to 20 cm long.
**Flowers:** The main distinguishing feature is the flower cluster which has 3–4 groups of 3 or 4 flowers. While the 2 outer flowers are stalked the central flower is sessile. The bud is 30–35 mm long and the berry elliptical to 10 mm long. Flowering mainly Summer.
**Localities:** Sth Morang, 26, 29#, 33–36, 39, 42, 44, 47–51, 53–58, 60–62, 64, 65, 67#, 71–76
**Distribution:** Vic, NSW, SA

## Amyema preissii

Wire Leaf Mistletoe

P

**Habitat:** Coastal and basalt plains
**Foliage:** Similar to *A. linophyllum* but has glabrous bright green foliage. Needle-like leaves to 100 mm long are paired, alternate or clustered.
**Flowers:** Red flowers in 2 (rarely 3) groups of 3, central flower sessile; November to April. Globular berry to 10 mm wide.
**Comments:** Host plants are *Acacia* spp., especially *A. mearnsii* and *Allocasuarina* spp.
**Localities:** Probably occurred along the Eastern coast-line of Port Phillip Bay.
**Distribution:** All mainland states *Illustrated p. 79*

## Amyema quandang var. quandang

Grey Mistletoe

P

**Habitat:** Valley sclerophyll forest
**Form:** Grey, shortly hairy plant
**Foliage:** Lanceolate to elliptical leaves to 100 mm long, often curved, in pairs or scattered
**Flowers:** Dark red flowers are held upright in 2 groups of 3; most of the year. Fruit is a pear-shaped berry to 10 mm long.
**Comments:** Always found on *Acacia* spp., especially *A. dealbata* locally.
**Localities:** Olinda Ck, Lilydale, 35, 42, 44, 61
**Distribution:** All mainland states *Illustrated p. 79*

*Amyema pendulum*

SCALE: 0 1cm 2cm The scale representation on each plant equals 2cm.

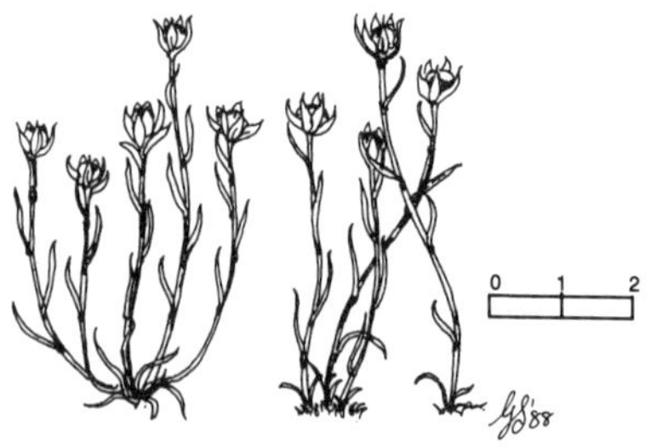

*Angianthus preissianus*

# Angianthus — Asteraceae

## Angianthus preissianus — Common Cup Flower, Salt Angianthus

F

**Size:** 10–30 cm x 15 cm
**Habitat:** Saltmarshes, primary dune scrub
**Form:** Annual matting herb
**Foliage:** Hairy, narrow silvery-grey leaves to 10 mm long
**Flowers:** In whitish subglobular terminal flowerheads; October to December
**Requirements:** Prefers moist sandy soils
**Comments:** Unknown in cultivation and locally rare, it would be worth trying.
**Propagation:** Seed
**Localities:** Williamstown, 67#
**Distribution:** Vic, SA, WA

*Aotus ericoides*

# Aotus — Fabaceae

## Aotus ericoides — Common Aotus

F P

**Size:** 0.5–1.5 m x 0.5–1.5 m
**Habitat:** Tea-tree heath
**Form:** Fast-growing bushy upright shrub
**Foliage:** Slender hairy light green leaves to 20 mm long
**Flowers:** Clusters of small yellow and red pea flowers; August to December
**Requirements:** Adaptable to well drained soils, tolerating both wet or dry periods
**Comments:** A very showy plant when in bloom, it requires pruning after flowering.
**Propagation:** Scarified seed, cuttings
**Localities:** St. Kilda#, 45#, 63, 67–69, 70, 71, 73–75
**Distribution:** All states

*Apium annuum*

# Apium — Apiaceae

## Apium annuum

F

**Size:** to 15 cm high
**Habitat:** Wet depressions in plains grassland, grassy low open forest, primary dune scrub
**Form:** Erect annual or biennial herb
**Foliage:** Broadly ovate leaves divided into 3–5 elliptic to obovate segments, 2 or 3 times longer than wide, which may be entire, lobed or further divided
**Flowers:** Sessile umbels of 4–10 tiny white flowers; September to November
**Requirements:** Moist sandy soil
**Comments:** The globular fruit have narrow ribs.
**Propagation:** Seed
**Localities:** Laverton, Altona, Seaford, 2, 3, 74
**Distribution:** Vic, SA, WA

*Apium prostratum* ssp. *prostratum* var. *prostratum*

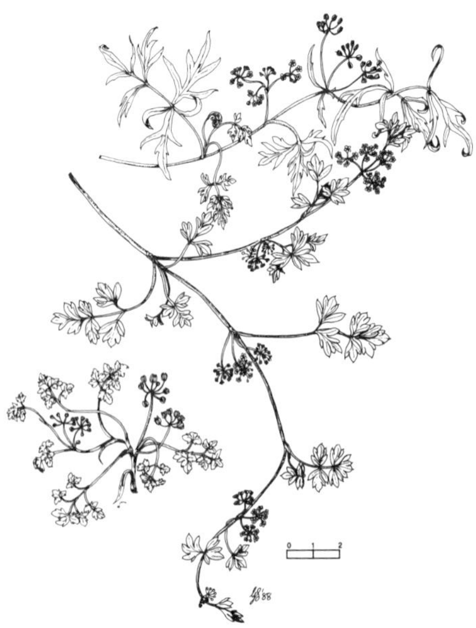

*Apium prostratum* ssp. *prostratum* var. *filiforme*

## Apium prostratum ssp. prostratum var. prostratum — Sea Celery

F

**Size:** 10–30 cm x 1–2 m
**Habitat:** Widespread in fresh or brackish swamps of coastal banksia woodland and riparian scrub
**Form:** Trailing perennial
**Foliage:** Smooth, divided or entire celery-like leaves with narrow to lanceolate segments. Leaves are 7–12 times longer than their width.
**Flowers:** Inconspicuous cream to green flowers in umbels; December to April
**Requirements:** Moist well drained sandy soils
**Comments:** Globular fruits have corky ribs. Occurs further inland than var. *filiforme*. Suitable for a rockery. It may be used as a celery substitute.
**Propagation:** Seed, division
**Localities:** Mordialloc, Carrum, 2, 16, 25, 26, 28, 35, 67, 69A, 74
**Distribution:** All states; NZ

## Apium prostratum ssp. prostratum var. filiforme

F

**Habitat:** Coastal river mouths and exposed foreshores, primary dune scrub
**Comments:** Leaves are always divided, often several times, and they are only 2-3 times longer than wide. Segments vary in shape from elliptic to cuneate. Stems are also finer than var. *prostratum*.
**Localities:** Ricketts Point
**Distribution:** All states except NT; NZ

# Asperula — Rubiaceae

## Asperula conferta — Common Woodruff

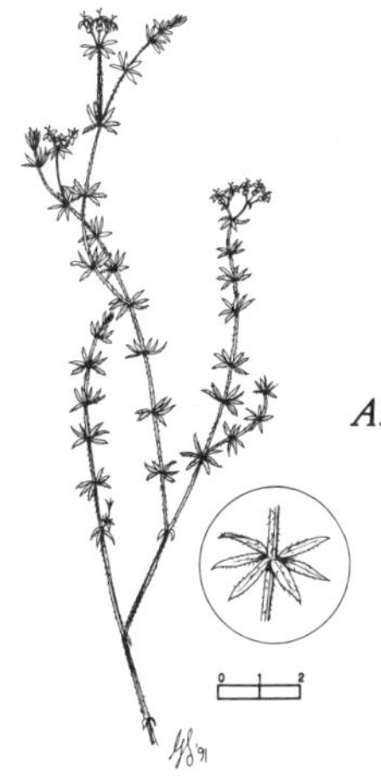

*Asperula conferta*

**Size:** to 20 cm x 0.5–1 m — P
**Habitat:** Widespread on primary dune scrub, grassy wetlands, red gum woodland, plains grassland, valley sclerophyll forest
**Form:** Low spreading perennial herb with squarish glabrous stems, becoming round with age
**Foliage:** Bright green linear pointed leaves to 10 mm long, in whorls of 6, margins and midrib on underside roughly hairy
**Flowers:** Dioecious, white terminal flowers, females in groups of 3 and shorter than males; September to December
**Requirements:** Well drained soils
**Comments:** A useful low groundcover for a moist, shady spot.
**Propagation:** Cuttings
**Localities:** Eltham, 6, 12, 13, 15–18, 25, 26, 28, 29, 31, 32, 34A, 35, 41, 42, 45, 49, 54–56, 58–67#, 74
**Distribution:** Vic, Qld, NSW, SA

## Asperula euryphylla — Broad-Leaf Woodruff

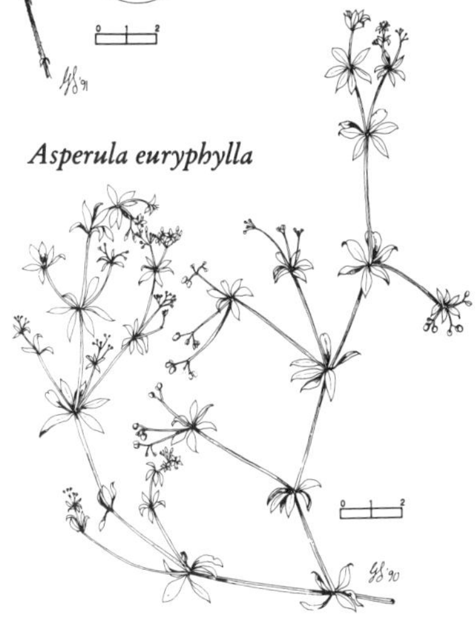

*Asperula euryphylla*

**Size:** Weak branches to 15 cm high — P
**Habitat:** Damp and wet sclerophyll forests
**Form:** Wiry perennial herb, rectangular stems covered in fine hairs, scrambling amongst other plants
**Foliage:** Whorls of 6 downy or glabrous, obovate to oblanceolate leaves to 12 mm long
**Flowers:** Dioecious, 1–3 white flowers in terminal clusters; November to December
**Requirements:** Moist well drained soils
**Localities:** 55–57, 61
**Distribution:** Vic, SA

## Asperula scoparia — Prickly Woodruff

*Asperula scoparia*

**Size:** 10 cm x 0.5–1 m — P N
**Habitat:** Damp and dry sclerophyll forests, plains grassland, grassy wetlands
**Form:** Moss-like perennial herb with rough hairy stems
**Foliage:** Whorls of 6 stiff, prickly narrow leaves
**Flowers:** Dioecious, small clusters of tiny white flowers; September to February
**Requirements:** Moist well drained soils
**Comments:** Ideal used under other plants or around rocks.
**Localities:** 2, 4–7, 10, 12, 16–18, 20, 26, 28, 34A, 36, 37, 39, ?42, 43, 49, 55, 56, 58, 71
**Distribution:** Vic, Qld, NSW, Tas, SA

# Asterolasia — Rutaceae

## Asterolasia asteriscophora — Lemon Star Bush

*Asterolasia asteriscophora*

**Size:** 1.5 m x 1 m — P
**Habitat:** Damp and valley sclerophyll forests
**Form:** Slender upright shrub
**Foliage:** Dull green soft hairy leaves to 25 mm long, locally smaller to 10 mm long
**Flowers:** Lemon-yellow star-like flowers, solitary or in clusters; September to November
**Requirements:** Moist well drained soils
**Comments:** A reliable garden plant needing annual pruning. Lasts well as a cut flower. The local form often has white flowers.
**Propagation:** Cuttings
**Localities:** Belgrave, Upper Ferntree Gully
**Distribution:** Vic, NSW

# Astroloma — Epacridaceae

## Astroloma humifusum — Cranberry Heath

*Astroloma humifusum*

**Size:** 10-50 cm x 1–1.5 m — F P
**Habitat:** Red gum woodland, dry sclerophyll forest, tea-tree heath, grassy low open forest
**Form:** Dense spreading mat-like plant
**Foliage:** Blue-green, stiff, prickly narrow leaves to 10 mm long
**Flowers:** Bright red tubular flowers along branches most of the year, especially May to September
**Requirements:** Well drained soils tolerating dry periods once established
**Comments:** The pale green globular fruit are edible. An excellent plant for rockeries, embankments, under shrubs or in hanging baskets. The flowers are generally hidden by the foliage. Responds well to pruning.
**Propagation:** Cuttings—difficult. Use very young firm growth.
**Localities:** 26, 34A–42, 44, 45#, 47–49, 50A, 51, 58, 62–69A, 74–76
**Distribution:** Vic, NSW, Tas, SA, WA

SCALE: 0 1cm 2cm — The scale representation on each plant equals 2cm.

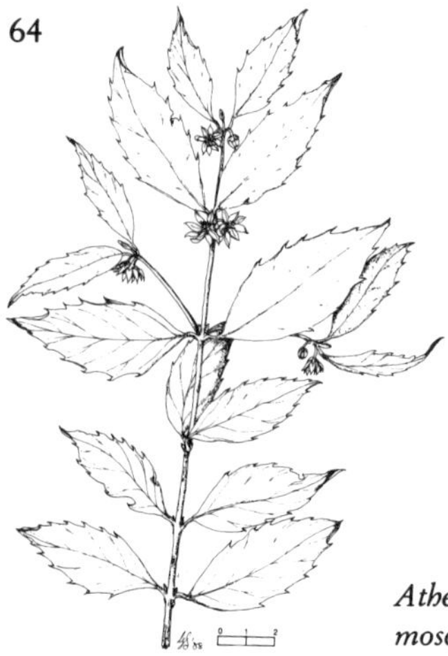

*Atherosperma moschatum*

# Atherosperma — Monimiaceae

## Atherosperma moschatum — Southern or Black Sassafras

P N

**Size:** 10–25 m x 2–5 m
**Habitat:** Cool temperate rainforest, wet sclerophyll forest
**Form:** Dense narrow conical tree
**Foliage:** Shiny green lanceolate leaves, whitish below, 40–90 mm x 15–40 mm
**Flowers:** Pendent velvety cream perfumed flowers; March to July
**Requirements:** Cool moist well drained soils
**Comments:** A slow-growing tree with nutmeg-scented bark and musky leaves.
**Propagation:** Seed, cuttings—can be very slow to strike
**Localities:** 57, 58
**Distribution:** Vic, Qld, NSW, Tas

# Atriplex — Chenopodiaceae

*Atriplex* species are commonly known as saltbushes as they can grow in saline soil. Flowers are unisexual: the males are globular forming clusters or spikes; the females form clusters in lower axils. *Atriplex* are fire-resistant, drought tolerant and excellent for soil erosion control. The seeds were ground and cooked by the Aborigines.

*Atriplex cinerea*

*Atriplex paludosa* ssp. *paludosa*

*Atriplex semibaccata*

## Atriplex billardieri — Glistening Saltbush

F

**Size:** 15–40 cm x 0.5–1 m
**Habitat:** Coastal cliffs and foreshores near high tide line, primary dune scrub
**Form:** Small succulent perennial herb
**Foliage:** Glistening, oblong yellow-green leaves to 10 mm long
**Flowers:** Small greenish flowers, male flowers in small clusters in the upper axils, females 1–2 in lower axils on the same plant
**Requirements:** Well drained sandy soil
**Comments:** An endangered species in Melbourne.
**Localities:** St. Kilda#, 67
**Distribution:** Vic, Tas
**Synonym:** *Theleophyton billardieri*

## Atriplex cinerea — Coast or Grey Saltbush

F P

**Size:** 1–2 m x 2–3 m
**Habitat:** Exposed primary dune scrub, sandy shorelines above high tide level
**Form:** Fast-growing dense spreading shrub with brittle branches
**Foliage:** Silver-grey lanceolate to oblong leaves to 80 mm long
**Flowers:** Male and female flowers usually occur on separate plants (dioecious). Male flowers are reddish-purple in dense globular clusters; female flowers are cream in axillary clusters; September to March
**Requirements:** Well drained soils
**Comments:** This saltbush has fleshy, triangular fruit. Its leaves can be eaten after cooking. A good low screen for coastal gardens.
**Propagation:** Seed, cutting
**Localities:** 2, 3, 67, 69A, 74
**Distribution:** All states

## Atriplex paludosa ssp. paludosa — Marsh Saltbush

F

**Size:** 1 m x 1–2 m
**Habitat:** Saltmarshes and tidal flats
**Form:** Open silvery-grey shrub
**Foliage:** Scaly ovate to oblong leaves to 25 mm long
**Flowers:** Dioecious, the male flowers are in slender spikes, the females in axillary clusters; most of the year
**Requirements:** Boggy saline soils
**Comments:** Useful in saline areas.
**Localities:** 2, 3
**Distribution:** Vic, NSW, Tas, SA, WA

## Atriplex semibaccata — Berry or Creeping Saltbush

F P

**Size:** 10–30 cm x 1–3 m
**Habitat:** Primary dune scrub, riparian scrub, red gum woodland, coloniser of disturbed ground
**Form:** Spreading perennial shrub
**Foliage:** Oblong toothed leaves to 30 mm long, grey-green above, mealy below
**Flowers:** Male flowers in small dense clusters in upper axils, female flowers scattered in lower axils, singly or in small clusters, on the same plant; most of the year, peaking November to February. Followed by red succulent fruit.
**Requirements:** Well drained soils. Salt tolerant
**Comments:** A useful groundcover to control soil erosion.
**Localities:** Maribyrnong, Derrimut, Melton, 2–4, 6, 16, 18, 21–29, 32, 35, 71
**Distribution:** All mainland states

### Atriplex suberecta

Sprawling or Lagoon Saltbush

**Size:** 0.3–0.6 m x 1–2 m F
**Habitat:** Saltmarshes and disturbed areas
**Form:** Sprawling annual herb with ascending stems
**Foliage:** Wedge-shaped leaves to 70 mm long are green above and mealy white below
**Flowers:** Male flowers clustered near the ends of branches, females in dense axillary clusters; all year
**Requirements:** Tolerates boggy saline soils
**Comments:** Often found alongside railway lines. May be introduced in these situations.
**Localities:** Melbourne, Footscray, Sunshine, Rockbank, Melton, 2–4, 10
**Distribution:** Vic, Qld, NSW, SA, WA
**Synonym:** Separated from *A. muelleri* which occurs in Northern Australia

*Australina pusilla* ssp. *muelleri*

## Australina — Urticaceae

### Australina pusilla ssp. muelleri

Smooth or Shade Nettle

**Size:** 20–40 cm x 0.5–1 m N
**Habitat:** Cool temperate rainforest, damp and wet sclerophyll forests
**Form:** Loosely-branched perennial herb
**Foliage:** Dark green ovate to lanceolate leaves to 50 mm long
**Flowers:** Insignificant hairy male and female flowers without petals on semi-succulent stems; November to February
**Requirements:** Moist well drained soils
**Comments:** A non-stinging herb ideal for growing with ferns.
**Propagation:** Seed, cutting, root division
**Localities:** 57, 58, 60
**Distribution:** Vic, Qld, NSW, Tas
**Synonym:** *A. muelleri* is now included in *A. pusilla*

*Avicennia marina*

## Avicennia — Verbenacaea

### Avicennia marina

White or Grey Mangrove

**Size:** 3–8 m x 2–5 m F
**Habitat:** Coastal shallows and estuaries, saltmarshes
**Form:** Slow-growing dense shrub or small tree
**Foliage:** Shiny dark green ovate-lanceolate leaves to 60 mm long
**Flowers:** Sweetly scented yellow to cream flowers; January to March
**Requirements:** Restricted to sites inundated at high tide
**Comments:** An excellent plant for stabilising mud flats, there are now only a few remnant stands left around Port Philip Bay. The Aborigines roasted and ate the fruit. The durable timber is used for marine construction.
**Propagation:** Seed—germinates on the tree before dropping. Seedlings can be transplanted but pots need flushing with brackish water.
**Localities:** Williamstown, 3#
**Distribution:** All mainland states

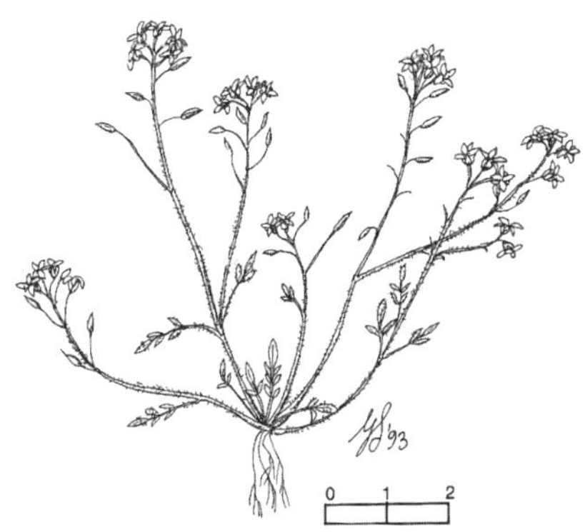

*Ballantinia antipoda*

## Ballantinia — Brassicaceae

### Ballantinia antipoda

Southern Shepherds Purse

**Size:** 2–10 cm high F P
**Habitat:** Basaltic and granitic sites in moss mats in seepage areas
**Form:** Tiny, much branched hairy annual herb
**Foliage:** Leaves pinnate, lanceolate to ovate in outline
**Flowers:** Masses of tiny white flowers with shortly-clawed petals; Spring
**Requirements:** Well drained moist soil
**Comments:** This little herb would make an interesting container plant. Extinct in Melbourne mainly due to grazing and rabbits, it is now known from 1 Victorian site.
**Propagation:** Seed
**Localities:** An early collection in Werribee
**Distribution:** Vic, Tas
**Synonym:** *Cuphonotus antipodus, Capsela antipodus, Hutchinsia australis*

SCALE: 0 1cm 2cm — The scale representation on each plant equals 2cm.

# Banksia — Proteaceae

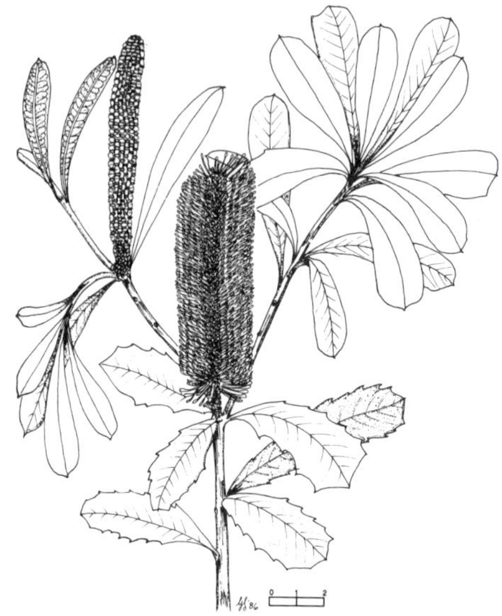

*Banksia integrifolia* var. *integrifolia*

## Banksia integrifolia var. integrifolia — Coast Banksia

F

**Size:** 10–20 m x 5–10 m
**Habitat:** Coastal banksia woodland, primary dune scrub
**Form:** Open erect or spreading tree
**Foliage:** Dark green leaves with silvery underside, linear to obovate to 15 cm long
**Flowers:** Pale yellow flowers borne on terminal spikes or brushes, 50–150 mm long. Flowering may occur at any time of the year, but mainly February to September
**Requirements:** Well drained soils responding to Summer watering
**Comments:** A sturdy ornamental windbreak with rough bark which can become gnarled and fissured with age. The seed is shed from woody cones when mature. It is useful for beach erosion control. The nectar from banksia cones was used to make drinks by the Aborigines. Paint brushes came from the stamens.
**Propagation:** Seed
**Localities:** 67, 69A, 74, 75
**Distribution:** Vic, Qld, NSW, Tas

*Banksia marginata*

## Banksia marginata — Silver Banksia

F P

**Size:** 1–10 m x 1–5 m
**Habitat:** Widespread from coast to ranges in tea-tree heath, grassy low open forest, riparian scrub, sclerophyll woodland, valley sclerophyll forest
**Form:** Variable dense to open shrub or small tree, depending on its habitat
**Foliage:** Stiff linear to obovate leaves to 100 mm long, silver below, toothed or entire, with a truncate tip.
**Flowers:** Pale to bright yellow flower spikes 40–100 mm; September to April
**Requirements:** Prefers good drainage tolerating soils which can be wet in Winter and dry in Summer
**Comments:** Bushy forms make excellent screening plants. Attractive woolly brown new growth can be encouraged by pruning. Seed is shed when mature. Cut flowers keep well. The forms presently found in the Melbourne area are unlikely to exceed 5 m. A tree form once occurred on the basalt plains in the Shire of Bulla and was referred to in early surveying reports. It now remains to the north west of the study area.
**Propagation:** Seed; specific forms must be grown from cuttings.
**Localities:** Olinda, Wonga Park, Mt. Cottrell, 1, 16, 26, 29, 43, 49, 50, 51, 54–56, 58, 59, 61–63, 67, 69–71, 73–76
**Distribution:** Vic, NSW, Tas, SA

## Banksia spinulosa var. cunninghamii — Hairpin Banksia

F P

**Size:** 2–4 m x 2–5 m
**Habitat:** Valley sclerophyll forest, sclerophyll woodland
**Form:** Dense erect or spreading shrub
**Foliage:** Narrow dark green toothed leaves to 100 mm long
**Flowers:** Golden yellow spikes to 20 cm long with black hooked styles; February to July. Seed is retained in the cone.
**Requirements:** Prefers moist well drained soils but will tolerate dryness once established
**Comments:** An excellent shrub for growing under established trees. Light pruning encourages bushiness and makes an attractive screen. This variety does not produce a lignotuber and is not appropriate for areas with alkaline soils.
**Propagation:** Seed, Cuttings
**Localities:** 47, 51, 55
**Distribution:** Vic, Qld, NSW

*Illustrated title page*

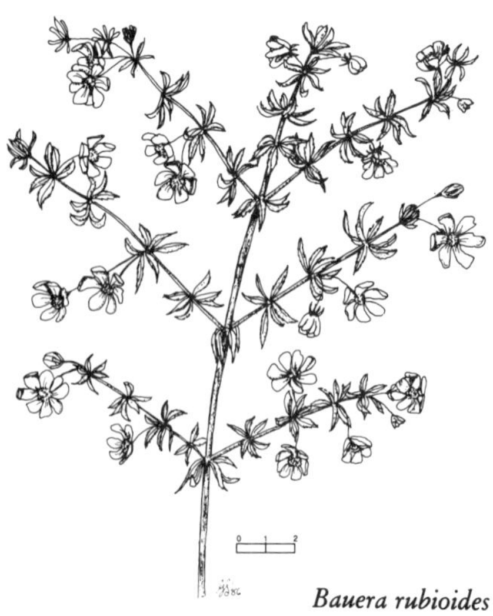

*Bauera rubioides*

# Bauera — Baueraceae

## Bauera rubioides — Wiry Bauera, River or Dog Rose

P N

**Size:** 1–2 m x 1–2 m
**Habitat:** Wet and sclerophyll forest
**Form:** Fast-growing dense scrambling shrub
**Foliage:** Pairs of leaves divided into 3 narrow sessile leaflets to 12 mm long, giving the appearance of whorls of 6 leaves
**Flowers:** Showy display of pink, sometimes white, open flowers most of the year
**Requirements:** Well drained moist soils, tolerating short periods of waterlogging
**Comments:** A hardy garden plant requiring regular light pruning.
**Propagation:** Cuttings, seeds
**Localities:** Montrose, 55–57
**Distribution:** Vic, Qld, NSW, Tas, SA

# Bedfordia — Asteraceae

## Bedfordia arborescens — Blanket Leaf

*Bedfordia arborescens*

**Size:** 3–7 m x 2–4 m — P N
**Habitat:** Cool temperate rainforest, wet and valley sclerophyll forests
**Form:** Tall shrub or small spreading open tree
**Foliage:** Soft lanceolate leaves, dark green above, woolly white below, 15–25 cm long
**Flowers:** Small yellow daisy flowerheads in loose woolly clusters; October to January
**Requirements:** Deep well drained moist soils
**Comments:** A cool root system is necessary for this undershrub. This may be attained through mulching. Its rough bark may be suitable for epiphytic orchids or ferns.
**Propagation:** Seed
**Localities:** 42, 55–57
**Distribution:** Vic, NSW

# Beyeria — Euphorbiaceae

## Beyeria viscosa — Pinkwood

*Beyeria viscosa*

**Size:** 3–6 m x 2–4 m — P
**Habitat:** Damp sclerophyll forest, along gullies
**Form:** Medium to tall sticky shrub
**Foliage:** Shiny dark green oblong leaves to 100 mm long with dense white hairs below and edges recurved
**Flowers:** Pale yellow to cream flowers; July to January. Dioecious, female flowers are solitary, male flowers in clusters of 1–3. Fruit is a rounded capsule to 8 mm long.
**Requirements:** Well drained light to medium soil
**Comments:** Typically a plant of East Gippsland, it is known in Melbourne from a single stand on Mt. Dandenong. An excellent screening plant which responds well to pruning.
**Propagation:** Seed, cuttings
**Localities:** 55
**Distribution:** Vic, NSW, Tas, SA, WA

# Billardiera — Pittosporaceae

## Billardiera longiflora — Purple Apple Berry

*Billardiera longiflora*

**Size:** Slender climber — P
**Habitat:** Cool temperate rainforest, wet and valley sclerophyll forests, grassy low open forest
**Form:** Slender quick-growing light climber
**Foliage:** Shiny narrow dark green to purplish leaves to 50 mm long
**Flowers:** Pendent tubular flowers to 30 mm, bright greenish-yellow tinged with purple; August to January
**Requirements:** Moist well drained soils
**Comments:** Grows well under established eucalypts provided it has extra summer moisture. An attractive feature is the glossy purple berries which follow the flowers. It tolerates full sun in cool areas.
**Propagation:** Seed, cuttings
**Localities:** Kalorama, ?42, 57, 74, 76
**Distribution:** Vic, NSW, Tas

## Billardiera scandens var. scandens — Common Apple Berry, Apple Dumpling

*Billardiera scandens* var. *scandens*

**Size:** Adaptable light climber — F P N
**Habitat:** Widespread in damp, dry and valley sclerophyll forests, sclerophyll woodland, grassy low open forest
**Form:** Soft climber, bushier if grown in an open aspect
**Foliage:** Dark green linear-lanceolate leaves to 50 mm long, with wavy margins. New growth is soft, light green and hairy.
**Flowers:** Greenish-yellow tubular flowers throughout the year, with the main flowering period from June to January.
**Requirements:** Well drained dry to moist soils
**Comments:** The light green berries are edible. Grows under established trees.
**Propagation:** Seed, cuttings
**Localities:** 30#, 34–38, 40–64, 67, 69–71, 74–76
**Distribution:** Vic, Qld, NSW, Tas, SA

## Billardiera scandens var. brachyantha

**Habitat:** Dry sclerophyll forest
**Comments:** A more shrubby, hairier variety with 2–3 flowers grouped together. Flower stalks and petals are shorter.
**Localities:** Eltham, 42
**Distribution:** Endemic to Vic

SCALE: 0 1cm 2cm The scale representation on each plant equals 2cm.

Boerhavia dominii

# Boerhavia

## Nyctaginaceae

### Boerhavia dominii

Tah Vine

F P

**Size:** Prostrate x 1–2 m
**Habitat:** Saltmarsh
**Form:** Trailing or climbing perennial
**Foliage:** Dark green to 25 mm long, broad basal leaves with narrower upper leaves, wavy margins
**Flowers:** Dull pink flowers, either solitary or in umbels of 2–4; August to December
**Requirements:** Dry situations
**Comments:** A useful ground cover for dry areas but could become a weed in ideal conditions. The tap root was baked and eaten by the Aborigines.
**Propagation:** Seed, cuttings
**Localities:** Altona
**Distribution:** All mainland states
**Synonym:** *B. diffusa*

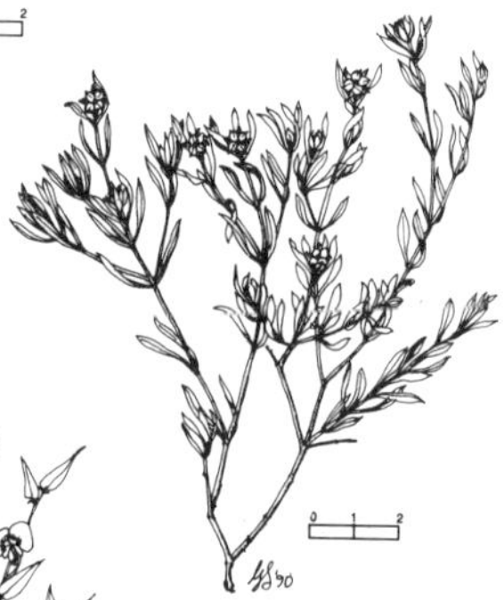

*Boronia parviflora*

*Bossiaea cinerea*

# Boronia

## Rutaceae

### Boronia parviflora

Swamp Boronia

P

**Size:** 0.3–0.5 m x 0.3–0.5 m
**Habitat:** Swamp margins of tea-tree heath
**Form:** Diffuse or trailing dwarf shrub
**Foliage:** Elliptical reddish leaves to 20 mm long
**Flowers:** White or pink flowers in clusters at the end of branchlets; August to December
**Requirements:** Moist peaty soil
**Comments:** Summer moisture is required. A light mulch will keep the roots cool.
**Propagation:** Cuttings
**Localities:** 66#, 67#, 74, 75
**Distribution:** Vic, Qld, NSW, Tas, SA

# Bossiaea

## Fabaceae

### Bossiaea cinerea

Showy Bossiaea

F P

**Size:** 1–2 m x 1–2 m
**Habitat:** Tea-tree heath, coastal banksia woodland
**Form:** Dense low, rounded or spreading shrub
**Foliage:** Alternate dull green, stalkless triangular leaves to 20 mm long, with bronze new growth
**Flowers:** Profuse yellow and red pea flowers occurring singly along stems; August to December
**Requirements:** Adapts to most well drained soils
**Comments:** An ornamental shrub which may be pruned hard. It prefers dappled shade and tolerates some coastal exposure and light frosts.
**Propagation:** Scarified seed, cuttings
**Localities:** Port Melbourne#, 42, 63, 65#–71, 73–76
**Distribution:** Vic, NSW, Tas, SA

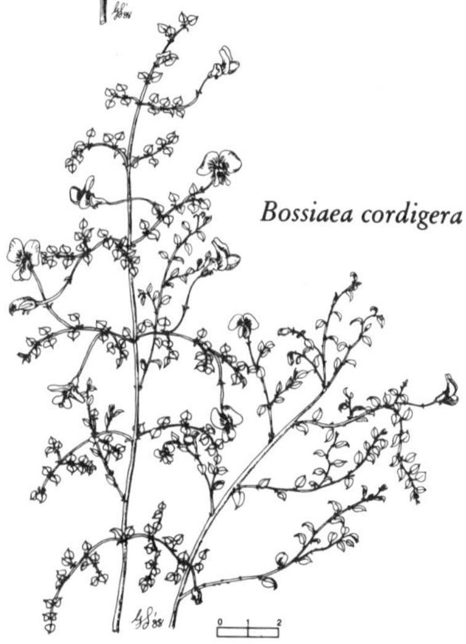

*Bossiaea cordigera*

### Bossiaea cordigera

Wiry Bossiaea

P

**Size:** Prostrate — 20 cm x 1–2 m
**Habitat:** Sclerophyll woodland
**Form:** Ornamental spreading shrub
**Foliage:** Pairs of small dark green heart-shaped leaves to 5 mm long
**Flowers:** Single yellow and red pea flowers; November to February
**Requirements:** Well drained moist soil
**Comments:** A wiry shrub which scrambles over other plants.
**Localities:** 59
**Distribution:** Vic, Tas

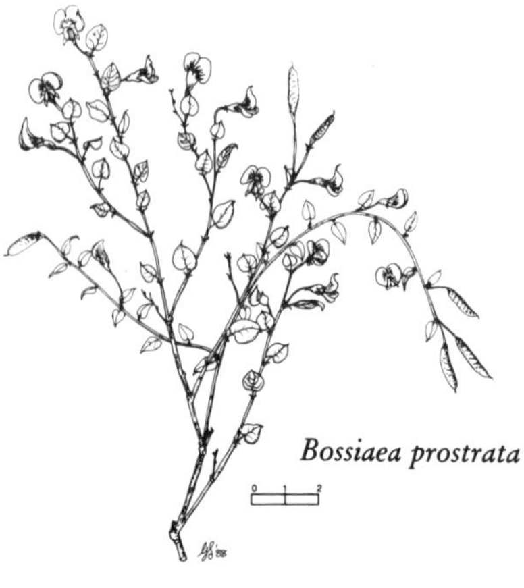

*Bossiaea prostrata*

### Bossiaea prostrata

Creeping Bossiaea

F P N

**Size:** Prostrate x 0.5–1.5 m
**Habitat:** Dry and valley sclerophyll forests, grassy low open forest, red gum and sclerophyll woodlands, plains grassland
**Form:** Spreading or matting ground cover
**Foliage:** Oval to oblong grey-green leaves to 10 mm long
**Flowers:** Yellow and brown pea flowers; October to November
**Requirements:** Easy to grow in all well drained soils
**Comments:** This attractively foliaged plant grows well under other plants. Common throughout most of Melbourne, it remains in isolated areas on the basalt plains.
**Localities:** Tottenham, 7, 12, 13, 16, 25, 26, 28–36, 38, 39, 41, 42, 44–51, 53–56, 58–67, 69–71, 74–76
**Distribution:** Vic, Qld, NSW, Tas, SA

# Brachyloma — Epacridaceae

## Brachyloma ciliatum — Fringed Brachyloma or Daphne Heath

**P**

**Size:** 0.5 m x 0.5–1 m
**Habitat:** Tea-tree heath
**Form:** Slow-growing open shrub
**Foliage:** Small dull green oval to oblong pointed leaves to 10 mm long, margins hairy
**Flowers:** Small white to cream tubular flowers, hairy inside; September to December
**Requirements:** Prefers sandy well drained soil
**Comments:** Useful as an undershrub, this plant tolerates light frosts and needs cool roots. Regenerates well after a fire. The fruit was eaten raw by the Aborigines.
**Propagation:** Cuttings
**Localities:** Baxter, 67#, 74, 75
**Distribution:** Vic, Tas, SA

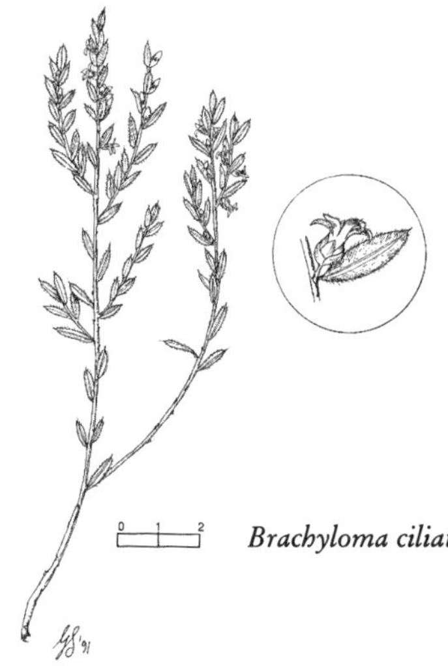

*Brachyloma ciliatum*

# Brachyscome — Asteraceae

A member of the daisy family, these plants are widespread, occurring in many habitats throughout Melbourne. Most require moist conditions or Summer watering. Seventeen species occur in the Melbourne area; several of these are already extinct while most are rare or endangered. They are perennial herbs with the typical daisy flowerhead which consists of many disc florets surrounded by petal-like ray florets. They adapt well to cultivation and are excellent for the small garden, rockeries or container growing. A heavy prune after flowering is beneficial. Propagation is by seed, cutting or division with many striking readily from cuttings.

## Brachyscome aculeata — Hill Daisy

**F P**

**Size:** 0.2–0.6 m high
**Habitat:** Riparian scrub
**Form:** Basal rosette
**Foliage:** Linear to wedge-shaped leaves to 100 mm long
**Flowers:** Branched flowering stems with numerous erect white flowerheads, 20 mm across; intermittent from late Spring to late Autumn
**Requirements:** Ample moisture
**Comments:** A very restricted range in Melbourne.
**Localities:** Lilydale#, 29
**Distribution:** Vic, Qld, NSW, Tas, SA

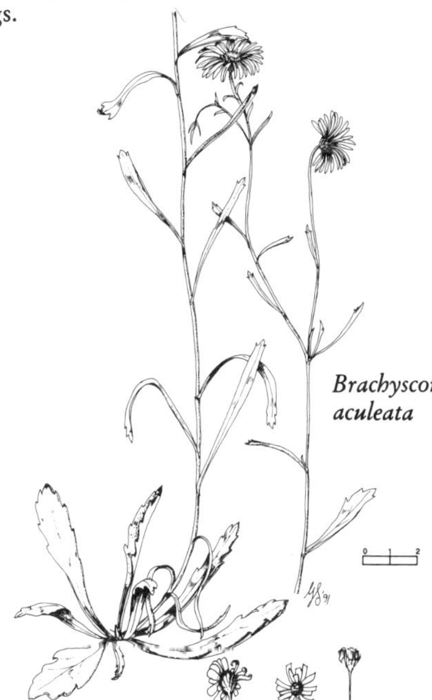

*Brachyscome aculeata*

## Brachyscome basaltica var. gracilis — Basalt Daisy

**F**

**Size:** 0.3–0.6 m high
**Habitat:** Grassy wetlands, plains grassland
**Form:** Small perennial herb
**Foliage:** Narrow-linear leaves to 90 mm long, scattered along the stems
**Flowers:** Numerous white flowerheads to 18 mm across, held erect; September to November
**Requirements:** Moist soils
**Comments:** A good bog plant from the basalt plains but now locally rare.
**Localities:** Werribee, Sydenham, Bundoora, Derrimut, 7, 11, 16, 17#, 26
**Distribution:** Vic, NSW, SA

*Brachyscome basaltica* var. *gracilis*

## Brachyscome cardiocarpa — Swamp Daisy

**F P**

**Size:** 10–30 cm x 20 cm
**Habitat:** Swamp scrub, grassy wetland, depressions in tea-tree heath
**Form:** Fast-growing tufted perennial herb
**Foliage:** Rosette of grass-like dark green or purplish leaves to 20 cm long
**Flowers:** Single white or blue flowerheads 40 mm across, on long erect stems; June to December.
**Requirements:** Moisture is essential
**Localities:** Black Rock, Scoresby, Bayswater, Kilsyth, Springvale, Montrose, 29#, 49, 58, 63, 65#, 67#, 74
**Distribution:** Vic, NSW, Tas, SA

*Illustrated p. 1*

## Brachyscome ciliaris — Variable Daisy

**F**

**Size:** 7–11 cm high
**Habitat:** Plains grassland
**Form:** Spreading annual or perennial herb
**Foliage:** Pinnate leaves to 60 mm long, leaflets narrow, covered in white woolly hairs
**Flowers:** Profuse white or blue flowerheads on wiry, down-covered stems; most of the year, peaking August to December
**Requirements:** Well drained soil
**Comments:** An excellent container plant.
**Localities:** Collected in Werribee in 1899; 74
**Distribution:** All states

*Brachyscome ciliaris*

SCALE: 0 1cm 2cm — The scale representation on each plant equals 2cm.

*Brachyscome decipiens*

*Brachyscome dentata*

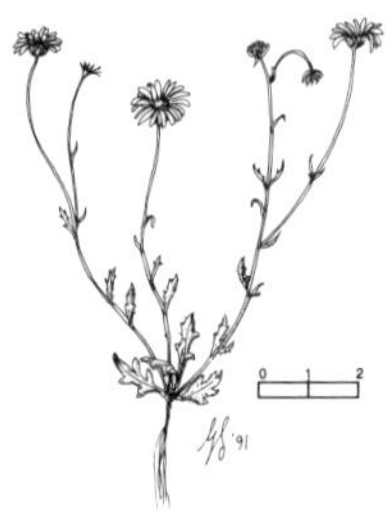

*Brachyscome exilis*

*Brachyscome diversifolia* var. *diversifolia*

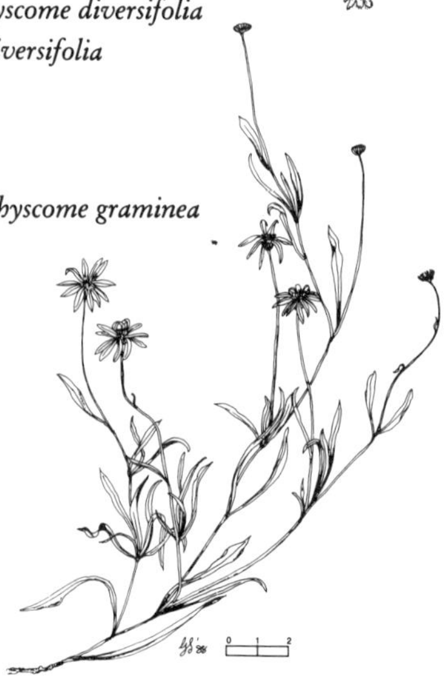

*Brachyscome graminea*

## Brachyscome decipiens

Field Daisy

**F P**

**Size:** 3–20 cm high
**Habitat:** Plains grasslands, red gum woodland, dry and valley sclerophyll forests
**Form:** Perennial herb in a basal rosette
**Foliage:** Flat tuft of broad leaves to 16 cm long,
**Flowers:** White or blue flowerheads 40 mm across on erect stalks; August to March
**Requirements:** Moist to wet soils
**Comments:** An excellent container or rockery plant. Locally rare.
**Localities:** Coldstream, Eltham–Warrandyte, 25, 26, 29#, 34A, 45, 49, 50A, 63, 64, 67#
**Distribution:** Vic, NSW, Tas, SA

## Brachyscome dentata

Golden or Lobe-seed Daisy

**F**

**Size:** 15–30 cm x 0.3–0.6 m
**Habitat:** Grassy wetlands, riparian scrub, box woodland, plains grassland
**Form:** Spreading to erect perennial herb
**Foliage:** Narrow leaves to 30 mm long
**Flowers:** White flowerheads; October to December
**Requirements:** Well drained soils and protection from slugs and snails
**Localities:** Sunbury, Tottenham, St Albans, Keilor, Williamstown, Melton Sth, Laverton Nth, 2, 4, 7, 10–12, 16, 17#, 25, 26
**Distribution:** Vic, Qld, NSW, SA
**Synonym:** *B. heterodonta*

## Brachyscome diversifolia var. diversifolia

Tall or Large-headed Daisy

**F P**

**Size:** 20–50 cm high
**Habitat:** Primary dune scrub, coastal banksia woodland, dry sclerophyll forest
**Form:** Robust tufted herb
**Foliage:** Soft, hairy, lobed bright green leaves to 100 mm long
**Flowers:** Single white flowerheads, 20–40 mm across on erect hairy stalks; intermittent, peaking from October to February.
**Requirements:** Well drained soils
**Comments:** Suitable for cut flowers. Endangered in Melbourne, it has disappeared from most of its range.
**Localities:** Warrandyte, 67#
**Distribution:** Vic, Qld, NSW, Tas, SA

## Brachyscome exilis

Slender Daisy

**F P**

**Size:** 5–15 cm high
**Habitat:** Saline areas and near brackish wetlands
**Form:** Small herb
**Foliage:** Few broad hairy leaves to 10 mm long
**Flowers:** Terminal bunches of white flowerheads; August to November
**Requirements:** Adaptable to dry or wet conditions
**Comments:** A rare Victorian plant it is now believed to be extinct within the Melbourne area.
**Localities:** Collected from the Eltham and St. Albans areas.
**Distribution:** Vic, NSW, SA, WA

## Brachyscome graminea

Grass Daisy

**F P**

**Size:** 0.3–0.7 m high
**Habitat:** Margins of brackish swamps in riparian scrub, tea-tree heath, grassy wetlands
**Form:** Sprawling perennial herb
**Foliage:** Bright green grass-like leaves to 50 mm long
**Flowers:** Numerous single-stemmed white, pink, lilac or deep blue flowerheads to 20 mm wide; September to March
**Requirements:** Moist soils
**Comments:** A good container plant but must not dry out.
**Localities:** Port Phillip Bay, 29, 30#, 34A, 67#
**Distribution:** Vic, Qld, NSW, Tas, SA

## Brachyscome leptocarpa

Downy or Small Hairy Daisy

**F**

**Size:** 2–25 cm high
**Habitat:** Plains grasslands
**Form:** Slender annual herb
**Foliage:** Hairy leaves to 30 mm long, basal leaves are narrow while stem leaves are entire or toothed
**Flowers:** Few white to mauve flowerheads; September to November
**Requirements:** Dry well drained soils
**Localities:** An early collection from Werribee
**Distribution:** Vic, NSW, SA

*Illustrated p. 79*

## Brachyscome multifida var. multifida

Cut Leaf Daisy

**F P**

**Size:** 10–40 cm x 0.2–1 m
**Habitat:** Box woodland, dry sclerophyll forest, sheltered rocky areas
**Form:** Fast growing, low spreading perennial
**Foliage:** Soft dark green pinnate leaves to 70 mm long
**Flowers:** Profuse variable lilac-blue, mauve, pink or white flowerheads, held erect. Flowers most of the year peaking in Spring and Summer.
**Requirements:** Moist clay soils, tolerating dryness once established
**Comments:** A very popular and ornamental plant. Spreads by underground suckers and roots at nodes making it a useful soil binder.
**Propagation:** Stem cutting, which strikes readily.
**Localities:** Bundoora, 36, 55, 63
**Distribution:** Vic, Qld, NSW

*Brachyscome multifida* var. *multifida*

## Brachyscome parvula var. parvula

Coast Daisy

**F**

**Size:** 10–40 cm high
**Habitat:** Marshes of primary dune scrub
**Form:** Tufted perennial herb
**Foliage:** Blunt linear to spathulate leaves to 70 mm long
**Flowers:** Single white flowerheads to 10 mm; November to January
**Requirements:** Wet soils
**Comments:** Useful as a bog plant. Locally rare.
**Localities:** Mordialloc, 67
**Distribution:** Vic, Tas, SA

*Brachyscome parvula* var. *parvula*

## Brachyscome perpusilla

Rayless Daisy

**F**

**Size:** to 10 cm high
**Habitat:** Grassy wetlands
**Form:** Tiny annual tufted herb
**Foliage:** Pinnate leaves 4–30 mm long
**Flowers:** Yellow flowerheads with minute ray florets on branched stalks; July to October
**Requirements:** Moist well drained soils
**Localities:** Werribee
**Distribution:** Vic, NSW, SA, WA

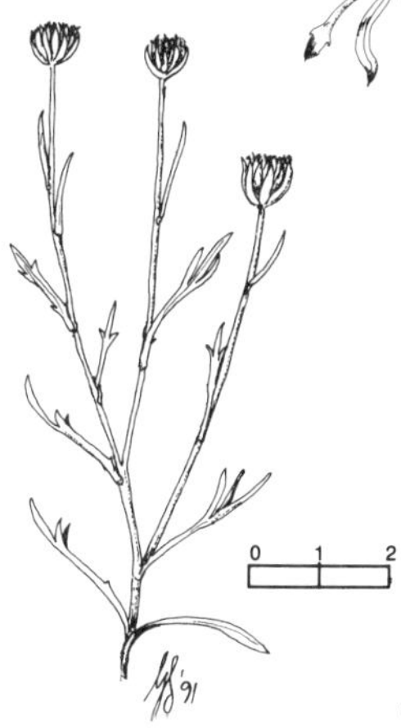

*Brachyscome perpusilla*

## Brachyscome scapigera

Tufted Daisy

**F P**

**Size:** 15–40 cm high
**Habitat:** Dry sclerophyll forest
**Form:** Densely clumping perennial herb
**Foliage:** Entire linear-lanceolate leaves to 19 cm long
**Flowers:** Few white to lilac flowerheads to 20 mm wide on unbranched stalks; September to January
**Requirements:** Moist soils, becoming dormant when dry
**Comments** Useful for rockeries and containers. May be extinct locally.
**Localities:** Collected from Research, Eltham.
**Distribution:** Vic, Qld, NSW

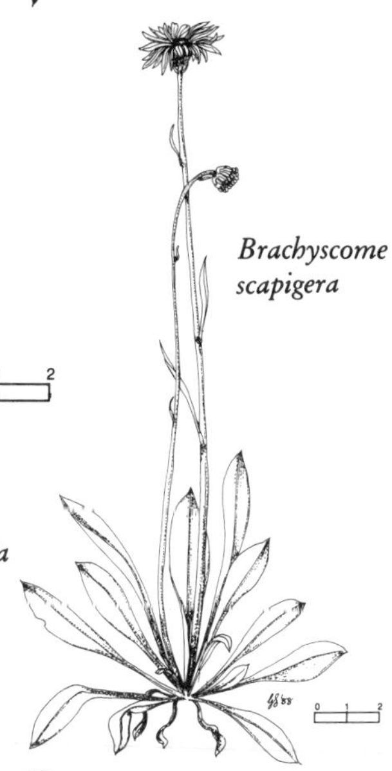

*Brachyscome scapigera*

## Brachyscome spathulata

Coarse Daisy

**F P**

**Habitat:** Riparian scrub
**Comments:** One of the *B. aculeata* group with which it is often combined, it is distinguished by the bright blue flowerheads and pointed bracts. The unbranched flowering stalks rise from a rosette of oblanceolate to broad-spathulate toothed leaves.
**Localities:** An early collection from along the Merri Creek, 1896
**Distribution:** Vic, NSW, Tas
**Synonym:** *B. scapiformis*

*Illustrated p. 79*

## Brachyscome trachycarpa

Smooth Daisy

**F**

**Size:** 10–40 cm high
**Habitat:** Sandy woodlands and rocky areas
**Form:** Small erect herb
**Foliage:** Narrow entire or lobed leaves to 35 mm long
**Flowers:** White to lilac flowerheads; most of the year
**Comments:** Appears to be locally extinct.
**Localities:** Collected at Werribee
**Distribution:** All mainland states

*Brachyscome trachycarpa*

## Brachyscome uliginosa

Small Swamp Daisy

**F**

**Size:** to 30 cm high
**Habitat:** Valley sclerophyll forest
**Form:** Creeping tufted herb
**Foliage:** Narrow pale green foliage to 60 mm long
**Flowers:** Single white or mauve flowerheads on long stems; September to January
**Requirements:** Moist soils
**Comments:** Similar to *B. scapigera*. A good container plant. Locally rare.
**Localities:** Eltham
**Distribution:** Vic, SA

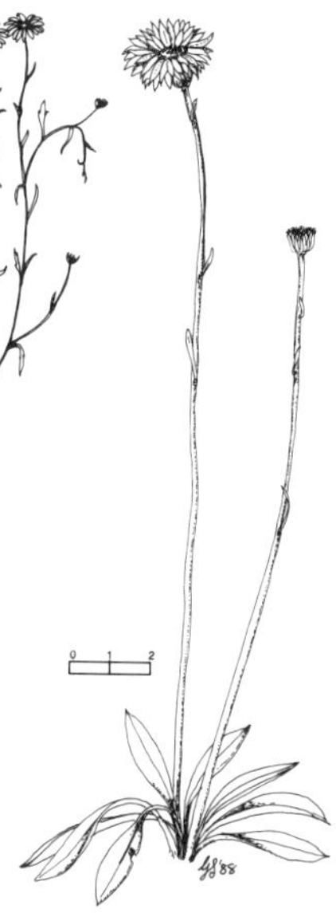

*Brachyscome uliginosa*

SCALE: 0 1cm 2cm — The scale representation on each plant equals 2cm.

# Bracteantha

## Asteraceae

*Bracteantha* are referred to as everlasting daisies. Their flowers are long-lasting and often very showy. Petal-like bracts surround many tiny flowers, giving a daisy-like appearance. Useful as cut flowers, they also dry well. *Bracteantha* are very adaptable and make excellent garden or container subjects. It is preferable to prune them back in late Winter to encourage new growth. Propagation is by seed or cuttings.
**Synonym:** Previously included in the genus *Helichrysum*

*Bracteantha bracteata*

### Bracteantha bracteata

Golden Everlasting, Straw Flower, Paper Daisy

**Size:** 0.3–1.2 m x 0.6 m — F
**Habitat:** Tea-tree heath, dry sclerophyll forest
**Form:** Fast growing erect annual or perennial herb
**Foliage:** Soft, stalkless, fresh green oblong leaves 15–120 mm long; upper leaves are narrower
**Flowers:** Flowerheads are surrounded by numerous shiny, bright yellow bracts, providing a very showy display; late Spring to Autumn
**Requirements:** Dry well drained soil
**Comments:** A spectacular daisy which is ideal for rockeries and containers. It is very variable in form and colour throughout its range. The forms generally available in nurseries are perennials from outside the Melbourne area. Hard pruning increases vigour. Locally rare.
**Localities:** Warrandyte, 30, 32, 34A, 67, 74, 75
**Distribution:** All states
**Synonym:** *Helichrysum bracteatum*

*Bracteantha viscosa*

### Bracteantha viscosa

Sticky Everlasting

**Size:** 0.5–0.8 m x 0.3–1 m — F P
**Habitat:** Dry sclerophyll forest, red gum and box woodlands, often on shallow stony soils
**Form:** Much-branched sticky biennial or perennial herb, dense or straggly
**Foliage:** Linear dark green leaves to 90 mm long, margins recurved
**Flowers:** Showy bright yellow flowerheads to 40 mm at the ends of branches; August to April
**Requirements:** Well drained soils
**Comments:** Pruning is necessary to encourage bushiness and to extend the life of this very attractive daisy. Responds to extra Summer moisture.
**Localities:** Diamond Creek, Warrandyte, Montrose, Sth of Melton, 32, 36, 67#
**Distribution:** Vic, Qld, NSW
**Synonym:** *Helichrysum viscosum*

# Brunonia

## Brunoniaceae

*Brunonia australis*

### Brunonia australis

Blue Pincushion

**Size:** 10–50 cm x 10–15 cm — F P
**Habitat:** Tea-tree heath, dry and valley sclerophyll forests, red gum woodland, plains grassland
**Form:** Perennial herb
**Foliage:** Rosette of spoon-shaped light green hairy leaves to 100 mm long
**Flowers:** Dense blue pincushion-like flowerheads on stems to 50 cm high; October to January
**Requirements:** Moist well drained soil, tolerant of extended dry periods
**Comments:** Often short-lived in cultivation, it makes an excellent container plant. Treat as an annual and collect seeds to replant. Informal drifts provide a natural look.
**Propagation:** Seed sown fresh, division—in July to August.
**Localities:** 12, 26, 29#, 34A–39, 41–44, 46–53, 55, 56, 58, 59, 62, 63, 67#, 68, 74–76
**Distribution:** All states

# Bursaria

## Pittosporaceae

*Bursaria spinosa* var. *spinosa*

### Bursaria spinosa var. spinosa

Sweet Bursaria, Blackthorn

**Size:** 2–6 m x 2–3 m — F P
**Habitat:** Widespread in grassy low open forest, riparian scrub, dry and valley sclerophyll forests, red gum and riparian woodlands
**Form:** Slender to rounded shrub or small tree with spines along the branches
**Foliage:** Shiny narrow dark green leaves, variable to 25 mm long and 10 mm wide
**Flowers:** Masses of fragrant, creamy-white bunches of flowers at the end of branches; December to March
**Requirements:** Well drained soils
**Comments:** Clusters of bronze seed capsules follow flowering on this easily grown plant. It is the food plant of the Eltham Copper Butterfly.
**Propagation:** Seed, cuttings
**Localities:** 1, 14, 16, 18, 22, 25, 26, 28, 30–56, 58–69, 74, 76
**Distribution:** All states except NT

### Bursaria spinosa var. macrophylla

**Habitat:** Plains grassland, riparian scrub
**Comments:** Leaves are longer (more than 25 mm) and wider (more than 10 mm) than *B. spinosa* var. *spinosa* and the branches are spineless.
**Localities:** Seaford, Kororoit Ck, 4, 26, 28–30, 34A, 44, 69A, 74
**Distribution:** Vic, Qld, NSW, Tas, SA

*Bursaria spinosa* var. *macrophylla*

## Calandrinia — Portulacaceae

### Calandrinia calyptrata — Pink Purslane, Small Leaf Parakeelya

**F**

**Size:** Prostrate—10 cm x 10–30 cm
**Habitat:** Primary dune scrub, coastal banksia, box and red gum woodlands, chenopod rocky open scrub
**Form:** Rosetted annual growing from Winter to Spring
**Foliage:** Succulent leaves to 35 mm long, tapered at each end, often reddish
**Flowers:** Terminal sprays of small pink flowers held high above the foliage on slender leafless stems; August to October
**Requirements:** Prefers well drained, dry soils in warm to hot areas
**Comments:** Seeds of calandrinias were ground by the Aborigines to make bread. Disappeared from much of its range.
**Propagation:** Seed, cuttings
**Localities:** Brighton 1887; 14, 16, 26, 30#, 35, 42, 67#
**Distribution:** All states except Qld

*Calandrinia calyptrata*

### Calandrinia eremaea — Small Purslane

**F**

**Comments:** Similar to *C. calyptrata*, but leaves are longer, to 40 mm, thicker, more cylindrical and stem-clasping at the base. Flowers are pink or white. May be locally extinct.
**Localities:** near Werribee R 1890
**Distribution:** Vic, NSW, SA, NT

## Callistemon — Myrtaceae

### Callistemon sieberi — River Bottlebrush

**F P**

**Size:** 3–10 m x 2–6 m
**Habitat:** Widespread in riparian scrub and woodland
**Form:** Open to dense weeping shrub
**Foliage:** Stiff narrow green leaves to 100 mm long; silky, silver new growth
**Flowers:** Scattered cream or pink bottlebrushes to 80 mm long; November to May
**Requirements:** Very adaptable, prefers very wet to moist conditions in heavy clay soil but tolerates dryness once established
**Comments:** An attractive screening shrub which benefits from pruning to encourage more prolific flowering.
**Propagation:** Seeds using bog method, cuttings
**Localities:** Keilor, 8, 14, 16, 19, 23, 26–36, 42, 47
**Distribution:** Vic, NSW, Tas, SA
**Synonym:** *C. paludosus*

*Callistemon sieberi*

## Callitriche — Callitrichaceae

### Callitriche brachycarpa — Water Starwort

**P**

**Habitat:** Moist areas of dry sclerophyll forest, riparian scrub
**Form:** Very small prostrate herb
**Foliage:** Entire oblanceolate to ovate leaves to 7 mm long
**Flowers:** Pale brown fruit with an inconspicuous wing
**Requirements:** Moist soils
**Comments:** A rare plant, with a restricted range in Melbourne. Also found north of Hurstbridge and in the Otways/Apollo Bay area.
**Localities:** Hurstbridge, 29
**Distribution:** Vic, Tas

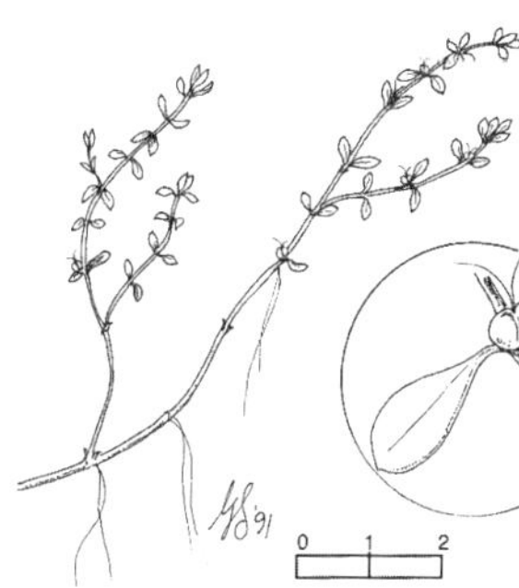

*Callitriche brachycarpa*

### Callitriche cyclocarpa — Water Starwort

**Habitat:** Floodwaters
**Form:** Tangled aquatic herb rooting from the nodes
**Foliage:** Leaves oblanceolate to 4 mm long, lower leaves occasionally more linear
**Comments:** Similar to *C. umbonata*. Distinguished by the black fruit which lacks a basal projection and is scarcely winged. An endangered plant, probably extinct in Melbourne.
**Localities:** Early collections were made in Springvale and an undefined central Melbourne site.
**Distribution:** Endemic to Vic

SCALE: 0 1cm 2cm — The scale representation on each plant equals 2cm.

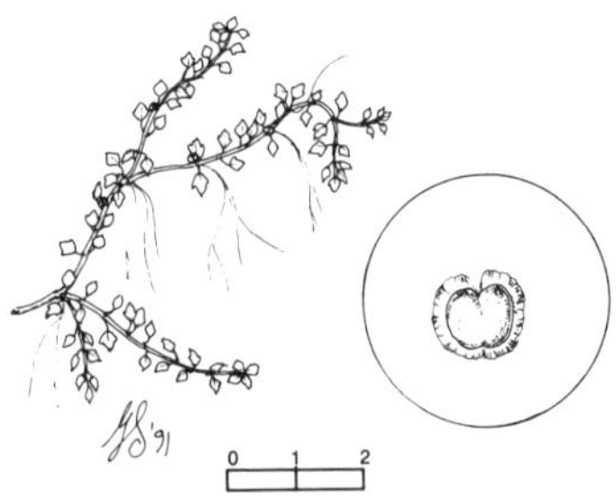

*Callitriche muelleri*

## Callitriche muelleri

Round Water Starwort

**P N**

**Size:** Prostrate x 10–20 cm
**Habitat:** Boggy sites and wet areas, also aquatic
**Form:** Creeping herb
**Foliage:** Very small dark green rhomboidal leaves, 2–7 mm long, with a single tooth on each side
**Flowers:** Insignificant, pairs of male and female flowers in each leaf axil; October to February. Tiny fruit has a conspicuous wing.
**Requirements:** Damp to wet conditions
**Comments:** It would make an attractive mat plant for bog gardens and pond margins.
**Propagation:** Division
**Localities:** Lilydale, 35, 49[#], 60
**Distribution:** Vic, Qld, NSW; NZ

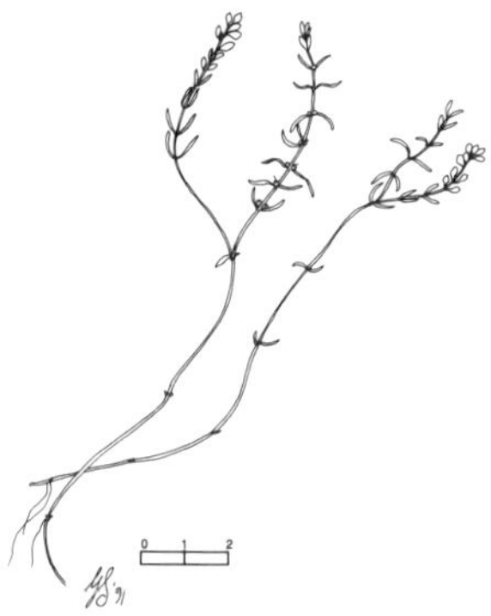

*Callitriche umbonata*

## Callitriche palustris

Water Starwort

**P**

**Habitat:** Riparian scrub, valley sclerophyll forest
**Form:** Floating or creeping stems
**Foliage:** Pairs of linear-lanceolate leaves to 3 mm long
**Comments:** Flowers have conspicuous bracts. Fruit is stalkless. Suitable for aquariums.
**Localities:** Keilor, Warrandyte North, 16, 19, 26, 32
**Distribution:** Vic, NSW; PNG, New Caledonia

## Callitriche umbonata

Water Starwort

**P**

**Habitat:** Grassy wetland
**Form:** Aquatic or terrestrial herb
**Foliage:** Lower leaves of aquatic plants linear to 10 mm, upper rosette of leaves are obovate to 8 mm long on a broad stalk; terrestrial plants lack linear leaves.
**Flowers:** Single white axillary flowers; August to December. Tiny dark brown fruit has a narrow pale brown wing projecting at the base.
**Localities:** Early collections from Darebin Creek in 1852 and unnamed inner Melbourne sites.
**Distribution:** Vic, NSW, Tas, SA

*Calocephalus citreus*

# Calocephalus — Asteraceae

## Calocephalus brownii = Leucophyta brownii

## Calocephalus citreus

Lemon Beauty Heads

**F P**

**Size:** 0.2–0.5 m x 0.3–1 m
**Habitat:** Plains grassland, red gum woodland
**Form:** Tufted perennial herb
**Foliage:** Silvery-grey pointed leaves to 100 mm long
**Flowers:** Bright yellow oblong flower heads held erect; October to February
**Requirements:** Well drained soils, tolerating dry periods once established
**Comments:** An attractive plant, suitable for mass planting in the garden or as a container plant. Flowers dry well.
**Propagation:** Cuttings, division
**Localities:** 2, 4, 5, 7, 10, 12, 13, 15–18, 20, 25, 26
**Distribution:** Vic, Qld, NSW, Tas, SA

*Calocephalus lacteus*

## Calocephalus lacteus

Milky Beauty Heads

**F P**

**Size:** 15–30 cm x 10–30 cm
**Habitat:** Plains grassland, red gum woodland, dry sclerophyll forest
**Form:** Small sprawling shrub
**Foliage:** Soft, grey, blunt leaves to 25 mm long
**Flowers:** White globular to oblong button flowerheads; September to February.
**Requirements:** Well drained soils tolerating dryness once established
**Comments:** Locally rare
**Propagation:** Seed, cuttings
**Localities:** St. Kilda[#], Sunbury, 2, 4, 5, 24–26, 28, 35, 36, 67[#]
**Distribution:** Vic, NSW, Tas, SA

# Calotis

**Asteraceae**

Low perennial shrubs with daisy flowerheads, followed by round burr-like fruiting heads. Propagate by seed or cuttings.

## Calotis anthemoides

Common Burr Daisy

**F P**

**Size:** 10–20 cm x 20–30 cm
**Habitat:** Grassy wetlands, plains grassland
**Form:** Erect glabrous perennial herb
**Foliage:** Tuft of finely divided basal leaves to 115 mm long
**Flowers:** Showy white or lilac flowerheads to 15 mm wide in Spring.
**Comments:** Locally rare plant of moist to wet areas
**Localities:** Laverton Nth, Altona, 2
**Distribution:** Vic, NSW

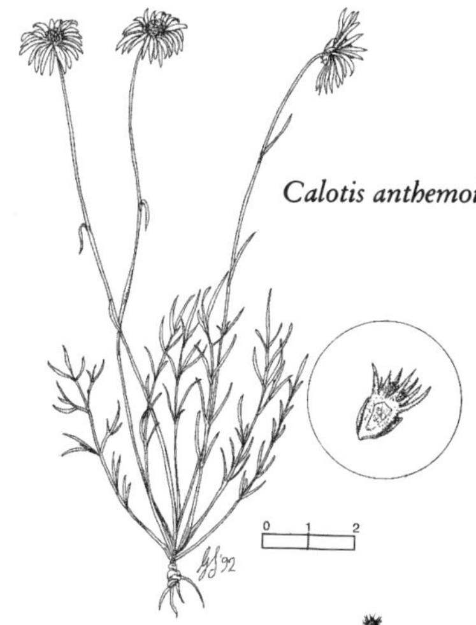

*Calotis anthemoides*

## Calotis cymbacantha

Showy Burr Daisy

**F**

**Size:** 10–30 cm high
**Habitat:** Plains grassland
**Form:** Erect, branched annual herb, covered in downy hairs
**Foliage:** Oblong to broad-linear leaves with 3–7 lobes; stalked basal leaves to 90 mm long, sessile stem leaves to 80 mm long
**Flowers:** Yellow flowerheads to 20 mm wide in loose leafy cymes; February to March, July to October
**Requirements:** Well drained soil
**Comments:** Although widespread, this plant is very rare; in Victoria it is usually found only in the drier North West region.
**Localities:** St. Albans
**Distribution:** Vic, Qld, NSW, SA, WA, NT

*Illustrated p. 79*

## Calotis lappulacea

Yellow Burr Daisy

**F P**

**Size:** 0.3–0.5 m x 0.3–1 m
**Habitat:** Box woodland and cleared land
**Foliage:** Hairy; wedge-shaped lower leaves to 60 mm long; sessile stem leaves, lobed to toothed or narrow and entire, to 25 mm long
**Flowers:** Yellow daisy flowerheads to 10 mm wide most of the year
**Requirements:** Dry, well drained soils
**Comments:** A very showy plant, ideal for rockeries. The prickly burrs limit its placement to areas away from paths.
**Localities:** Toolern Vale
**Distribution:** Vic, Qld, NSW, SA, WA

*Calotis lappulacea*

## Calotis scabiosifolia var. scabiosifolia

Rough Burr Daisy

**F P**

**Size:** 0.3–0.5 m x 0.5–1 m
**Habitat:** Plains grassland, red gum woodland
**Form:** Hairy perennial herb spreading by stolons to form tufts
**Foliage:** Basal leaves are lanceolate and toothed or lobed to 20 cm long; stem leaves are narrow, toothed or entire and sessile
**Flowers:** Showy white or lilac flowerheads to 25 mm across, singly on long stalks; August to November.
**Comments:** Suited to container growing and is drought resistant. Now locally rare.
**Localities:** Tottenham, Sydenham, Laverton Nth, 17#, 18
**Distribution:** Vic, Qld, NSW, SA

*Calotis scabiosifolia* var. *scabiosifolia*

## Calotis scabiosifolia var. integrifolia

**P**

**Comments:** Montane form. Leaves are linear and usually without teeth.
**Localities:** 55
**Distribution:** Vic, NSW

## Calotis scapigera

Tufted Burr Daisy

**F P**

**Size:** 0.1–0.5 m x 0.5–1 m
**Habitat:** Grassy wetland, swamps and depressions in red gum woodland
**Foliage:** Tufts of narrow leaves to 100 mm long
**Flowers:** White or lilac flowerheads; October to May
**Requirements:** Moist, heavy well drained soil
**Comments:** An excellent ground cover. Its suckering habit assists in controlling soil erosion.
**Propagation:** Root cuttings from runners strike readily.
**Localities:** Sydenham, 7, 12, 16–18, 26
**Distribution:** Vic, Qld, NSW, SA

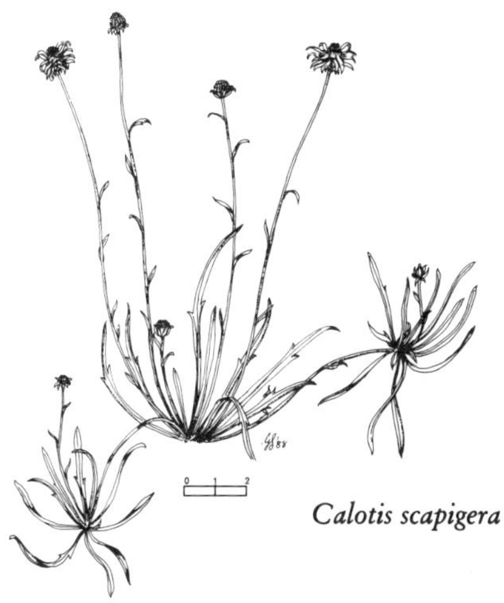

*Calotis scapigera*

SCALE: 0 1cm 2cm The scale representation on each plant equals 2cm.

*Calystegia marginata*

# Calystegia — Convolvulaceae

## Calystegia marginata — Forest Bindweed

**Size:** Twining stems to 1–2 m — **P N**
**Habitat:** Damp and valley sclerophyll forests
**Form:** Slender climber
**Foliage:** Shiny arrow-shaped leaves to 60 mm long
**Flowers:** Solitary white to pale pink funnel-shaped flowers 20–30 mm long and 20 mm across on long stalks; October to January
**Requirements:** Moist to wet areas
**Comments:** Masses of short-lived flowers are produced. Regular pruning is beneficial. Water must be available during dry periods to keep this plant growing attractively. Locally rare.
**Propagation:** Cuttings, seeds (Refer to treatment of hard coated seeds.)
**Localities:** 42, 55–58
**Distribution:** Vic, Qld, NSW

## Calystegia sepium — Large Bindweed

**Size:** Twining stems to 1–4 m long — **P**
**Habitat:** Riparian scrub, grassy wetland, valley sclerophyll forest
**Comments:** An attractive climber similar to *C. marginata* except that it has larger flowers and foliage. It may regenerate prolifically after fire but is never rampant. The roots were an important food source for the Aborigines.
**Localities:** Kororoit Ck, 4, 7, 16, 19, 26–29, 35, 42, 62, 63, 74
**Distribution:** All states except NT

*Illustrated p. 79*

*Calytrix tetragona*

# Calytrix — Myrtaceae

## Calytrix tetragona — Fringe Myrtle

**Size:** 1–2 m x 1–2 m — **F P**
**Habitat:** Riparian scrub amongst rocks
**Form:** Open erect or spreading shrub
**Foliage:** Fine, green, crowded aromatic leaves to 6 mm long
**Flowers:** Dense terminal heads of white flowers; August to November
**Requirements:** Well drained soils, tolerating extended dry periods and occasional inundation
**Comments:** Very attractive in flower. It makes an excellent container plant. Pruning promotes bushiness. An attractive feature of *Calytrix* is the deep red calyx which persists after flowering.
**Propagation:** Seed, cuttings
**Localities:** Diggers Rest#, Sunbury, 16, 26
**Distribution:** All states except NT

*Cardamine gunnii*

# Cardamine — Brassicaceae

## Cardamine dictyosperma = Rorippa dictyosperma

## Cardamine gunnii — Common Bitter Cress

**Size:** 10–30 cm high — **P**
**Habitat:** Swamps of tea-tree heath, cool temperate rainforest
**Form:** Soft, tufted perennial herb
**Foliage:** Pinnate basal leaves on long stalks, segments spathulate, sometimes only the terminal leaflet developing
**Flowers:** Tiny white flowers held above the foliage on slender stalks; Spring to early Summer. Fruit a narrow cylindrical pod to 30 mm long on elongated stalks.
**Requirements:** Moist to wet soils
**Comments:** The leaves and stems of *Cardamine* species were an important food source for the Aborigines.
**Localities:** 57, 58, 67#
**Distribution:** Vic, NSW, Tas, SA
**Synonym:** Previously included with *C. debilis*

## Cardamine laciniata = Rorippa laciniata

### Cardamine paucijuga

Bitter Cress

**Size:** to 0.4 m high — P
**Habitat:** Riparian and swamp scrub
**Form:** Erect annual herb
**Foliage:** Rosette of pinnate leaves, 1-4 pairs of linear to ovate leaf segments.
**Flowers:** Tiny white, occasionally pink, flowers in terminal clusters held above the foliage on a long flowering stem; June to November. Fruit to 30 mm long.
**Requirements:** Rich soil in moist or dry conditions
**Comments:** Similar to *C. gunnii*. Locally very rare.
**Localities:** An early collection along the Merri Creek may have been this species; 61
**Distribution:** All states except NT

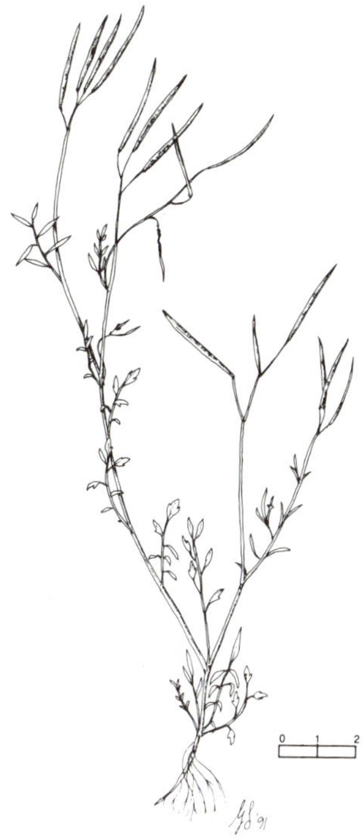

*Cardamine paucijuga*

### Cardamine tenuifolia

Slender Bitter Cress

**Size:** stems to 1 m long — P
**Habitat:** Swamp margins, plains grassland, valley sclerophyll forest
**Form:** Slender stoloniferous perennial herb with leaves scattered along the weak stems
**Foliage:** Pinnate leaves have 1-5 pairs of narrow segments
**Flowers:** Clusters of small pink or white flowers to 10 mm across; October to March. Fruit to 40 mm long.
**Requirements:** Moist to wet soils subject to inundation
**Comments:** On the basalt plains plants are protected by large boulders. Their leaves were much larger (to 70 mm long) than the finer leaves of plants east of Melbourne.
**Localities:** 18, 34A, 47
**Distribution:** All states except Qld

## Carpobrotus — Aizoaceae

### Carpobrotus modestus

Inland Noonflower or Pigface

**Size:** Prostrate x 1-3 m — F
**Habitat:** Chenopod rocky open scrub
**Form:** Spreading succulent perennial
**Foliage:** Linear 3-sided leaves to 70 mm long
**Flowers:** Profuse sessile light purple flowers to 20 mm, shading to white at the base; August to January
**Requirements:** Well drained soils
**Comments:** The flowers only open on sunny days.It is salt and drought tolerant and is useful as a soil binder. The Aborigines ate the fruit and leaves raw. The leaves were sometimes cooked.
**Propagation:** Seed, cuttings, division of layered stems
**Localities:** 1, 29#, 32, 34A-36
**Distribution:** Vic, SA, WA

### Carpobrotus rossii

Ross' Noonflower, Karkalla

**Size:** Prostrate x 2-3 m — F P
**Habitat:** Coastal banksia woodland, primary dune scrub
**Comments:** Similar to *C. modestus* but with longer leaves, to 100 mm and larger flowers to 50 mm across. The globular purplish-red fruit were eaten fresh or dried by the Aborigines. They also ate the salty leaves with meat. Grows in sandy soil.
**Localities:** Seaford, 16, 26, 67-69, 74
**Distribution:** Vic, Tas, SA, WA

*Carpobrotus rossii*

### Cassia nemophila var. nemophila = Senna artemisiodes ssp. filifolia

## Cassinia — Asteraceae

Erect shrubs found in open forest. Fast growing, they provide quick screens while other plants grow. All need pruning to maintain shape and prevent woodiness. Flowers last for a long time and can be dried. The foliage may cause skin irritations. Propagate easily from seed or cuttings.

### Cassinia aculeata

Common Cassinia, Dogwood

**Size:** 2-4 m x 1-2 m — P
**Habitat:** Widespread in wet to dry sclerophyll forests, grassy low open forest, sclerophyll woodland, from coast to foothills
**Form:** Upright open shrub
**Foliage:** Dark green linear leaves to 40 mm long
**Flowers:** Crowded, domed heads of small whitish flowers; November to March, buds often pink
**Requirements:** Heavy, moist, well drained soils
**Localities:** 29, 33-44, 46-64, 67, 68, 70-76
**Distribution:** Vic, NSW, Tas, SA

*Illustrated p. 79*

SCALE: 0 1cm 2cm The scale representation on each plant equals 2cm.

*Cassinia arcuata*

## Cassinia arcuata

Drooping Cassinia, Chinese Scrub

| | |
|---|---|
| **Size:** | 1–3 m x 1–2 m **F P** |
| **Habitat:** | Box and box ironbark woodland, dry and valley sclerophyll forests, especially on disturbed ground, grassy low open forest |
| **Form:** | Open rounded shrub |
| **Foliage:** | Drooping linear leaves to 10 mm long, with spicy aroma |
| **Flowers:** | Long, pendulous plumes of shiny pale brown flowerheads; November to February |
| **Requirements:** | Well drained soils |
| **Comments:** | This graceful plant is very easy to grow. |
| **Localities:** | Sunbury, 16, 18, 21, 25, 26, 32, 33, 35–42, 44, 46–51, 55, 56, 58, 60–64, 68, 70–72, 74–76 |
| **Distribution:** | Vic, NSW, SA, WA |

*Cassinia longifolia*

## Cassinia longifolia

Shiny Cassinia, Cauliflower Bush

| | |
|---|---|
| **Size:** | 2–4 m x 2–3 m **P N** |
| **Habitat:** | Valley and dry sclerophyll forests, grassy low open forest, box and red gum woodland |
| **Form:** | Medium to large shrub |
| **Foliage:** | Lanceolate leaves to 80 mm long. Leaves are hairy below and sticky when young. |
| **Flowers:** | Dense terminal clusters of small white flowers, to 15 cm across; November to March |
| **Requirements:** | Moist, well drained soil |
| **Comments:** | An adaptable shrub to most conditions. |
| **Localities:** | Sth Morang, Sth of Melton, 14, 16, 25, 26, 29, 32, 34–36, 38, 39, 41, 42, 44, 47, 48, 50–53, 55–60, 63, 71, 74, 76 |
| **Distribution:** | Vic, Qld, NSW, Tas |

## Cassinia trinerva

Three-Veined Cassinia

| | |
|---|---|
| **Size:** | 2–3.5 m x 1.5–3 m **P N** |
| **Habitat:** | Damp and wet sclerophyll forests |
| **Form:** | Dense rounded shrub |
| **Foliage:** | Narrow, 3-veined, dark green leaves to 80 mm long, pale underneath. |
| **Flowers:** | Broad flat clusters of whitish flowers; November to March |
| **Requirements:** | Moist, well drained heavy soil |
| **Localities:** | Wantirna, 57, 58 |
| **Distribution:** | Vic, NSW, Tas |

*Illustrated p. 79*

# Cassytha

# Lauraceae

Perennial creeping or climbing parasites, cassythas become rootless at maturity attaching to their hosts by a wedge inside attachment cups known as haustoria. Through these attachments they are able to gain some of their nutritional requirements from the host plant. Leaves are tiny and scale-like, hairy on the upper surface. Forming a tangled mass, large cassythas may kill their host. Propagation is by seed. The Aborigines ate the fruit raw.

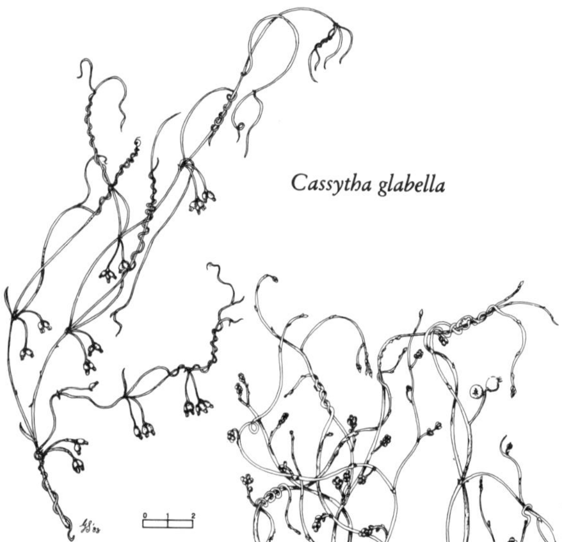

*Cassytha glabella*

## Cassytha glabella

Slender or Tangled Dodder Laurel

**F P**

| | |
|---|---|
| **Habitat:** | Tea-tree heath, wattle tea-tree scrub, damp and valley sclerophyll forests |
| **Form:** | Light parasitic creeper with fine, glabrous stems less than 1 mm thick, twining over grasses and small shrubs |
| **Flowers:** | Clusters of tiny white flowers are followed by orange or red fruit |
| **Comments:** | A most attractive species which is readily controlled by pulling the threadlike stems away from the host plant. The Aborigines infused the fruit to treat fever. |
| **Localities:** | 36, 41–43, 49, 50A, 58–61, 63, 64, 67–76 |
| **Distribution:** | All states except NT |

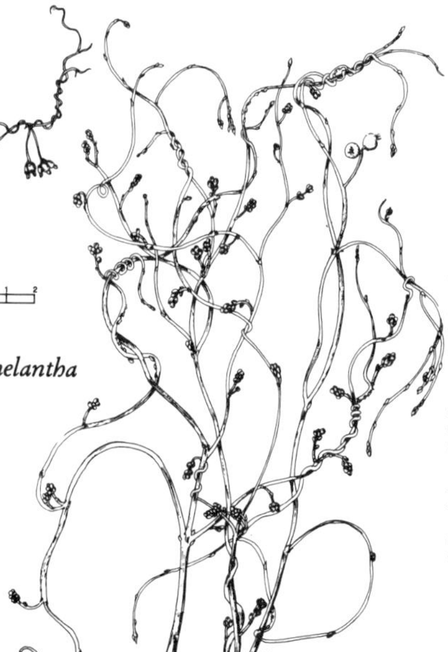

*Cassytha melantha*

## Cassytha melantha

Coarse or Large Dodder Laurel

**F P**

| | |
|---|---|
| **Habitat:** | Damp, dry and valley sclerophyll forests, coastal banksia and box woodlands |
| **Form:** | Rapidly growing parasitic creeper |
| **Flowers:** | Minute white flowers in spikes; June to October. The outside of the flowers are covered in dense black hairs. Globular green to black fruit follow the flowers. |
| **Requirements:** | Prefers light well drained soil |
| **Comments:** | Growing well above the ground, this most robust species can strangle or break its host plant, often eucalypts. |
| **Localities:** | 14, 16, 26, 27, 32, 34A–36, 40–42, 44, 47–49, 50A, 51, 54–56, 60, 62, 67, 69A, 74–76 |
| **Distribution:** | Vic, NSW, Tas, SA, WA |

*Amyema preissii*

*Amyema quandang* var. *quandang*

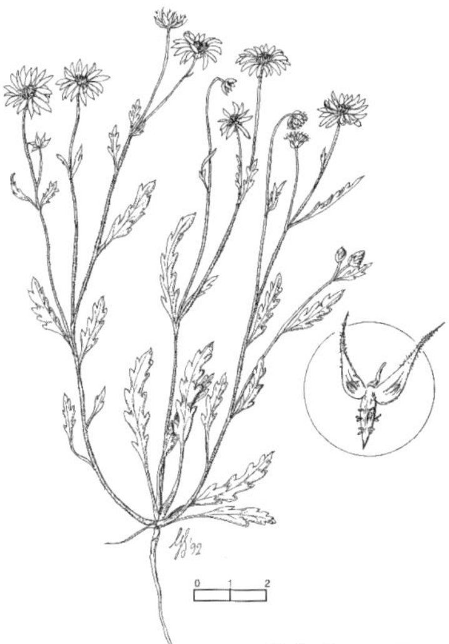

*Calotis cymbacantha*

*Brachyscome leptocarpa*

*Brachyscome spathulata*

*Cassinia aculeata*

*Cassinia trinerva*

*Calystegia sepium*

SCALE: 0 1cm 2cm The scale representation on each plant equals 2cm.

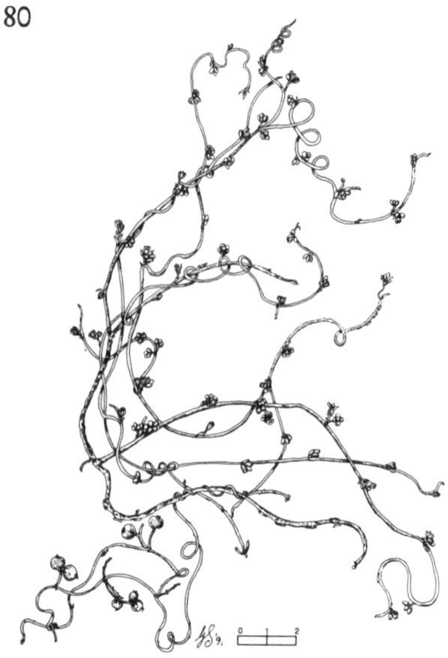

*Cassytha pubescens*

### Cassytha pubescens

Downy Dodder Laurel

**F P**

**Habitat:** Widespread in tea-tree heath, dry, damp and valley sclerophyll forests, red gum woodland
**Form:** Light, parasitic creeper, somewhat more robust than *C. glabella* with stems to about 2 mm thick. The stems, buds and fruit are usually hairy.
**Flowers:** Flowering occurs from December to April, with many tiny white flowers tightly clustered in a spike from 25–100 mm long, followed by greenish-white fruit.
**Localities:** St. Kilda#, 40–42, 44, 47, 48, 50–56, 58, 60, 62, 63, 65#–70, 73–76
**Distribution:** Vic, Qld, NSW, SA, Tas; NZ
**Synonym:** *C. phaeolasia*

### Casuarina spp. = Allocasuarina spp.

*Centella cordifolia*

## Centella — Apiaceae

### Centella cordifolia

Swamp Pennywort

**F P**

**Size:** Prostrate x 1–2 m
**Habitat:** Widespread in swamp and riparian scrub, damp, valley and dry sclerophyll forests, grassy low open forest
**Form:** Fast-growing creeping perennial herb
**Foliage:** Light green heart-or kidney-shaped leaves to 40 mm long
**Flowers:** Umbels of tiny white to pink flowers on long stalks; August to January
**Requirements:** Moist to wet soils
**Comments:** Grows well around pools and in boggy areas but may become invasive.
**Propagation:** Cuttings, division of stems
**Localities:** Montrose, 28, 29, 34A–36, 40–42, 44, 46–48, 50, 51, 53, 54, 56, 59, 60, 62–64, 70, 74–76
**Distribution:** All states except NT

*Centipeda cunninghamii*

## Centipeda — Asteraceae

### Centipeda cunninghamii

Common Sneezeweed

**F P**

**Size:** 0.1–0.5 m x 30 cm
**Habitat:** Grassy wetland, grassy low open forest, riparian scrub
**Form:** Fast-growing perennial herb
**Foliage:** Sessile coarsely-toothed oblanceolate leaves to 30 mm long, may be hairy or glabrous
**Flowers:** Sessile, globular greenish-yellow flowerheads; November to March
**Requirements:** Moist soils, tolerating inundation. Ideally suited to growing in hot inland areas, provided moisture is available.
**Comments:** Pruning is necessary to promote compact growth. The Aborigines and early settlers used this plant to treat many disorders including colds and infected eyes. Becoming rarer in Melbourne.
**Propagation:** Seed or cuttings
**Localities:** Eltham, 5, 6, 18, 30#, 34A, 35, 40, 42, 51, 60, 67#
**Distribution:** All mainland states

*Centipeda minima*

### Centipeda minima

Spreading Sneezeweed

**P**

**Size:** Prostrate—20 cm x 10–30 cm
**Habitat:** Widespread in swamp scrub, red gum woodland, valley sclerophyll forest, in flood-prone areas
**Form:** Spreading annual herb
**Foliage:** Similar to *C. cunninghamii*, but shortly stalked and more ovate, to 15 mm long, with a few teeth near the apex
**Flowers:** Flowerheads are smaller than *C. cunninghamii*
**Requirements:** Moist soils which may dry out in Summer
**Comments:** The Aborigines crushed the herb to treat colds. Becoming locally rare.
**Localities:** 12, 26, 34–36, 54, 61–63, 74, 75
**Distribution:** All states

# Ceratophyllum — Ceratophyllaceae

## Ceratophyllum demersum — Common Hornwort

**Size:** stems 0.3–0.6 m long
**Habitat:** Aquatic, submerged in 0.3–1 m water
**Form:** Rootless perennial herb
**Foliage:** Dark green dissected leaves to 40 mm long, segments linear and forked, spine-tipped teeth along margins; crowded in whorls of 7–12
**Flowers:** Inconspicuous greenish axillary flowers, male and female singly at different nodes; Summer
**Requirements:** Fresh or brackish stagnant to slow-moving water
**Comments:** Known from one area in the Melbourne area it may have once been more common. Useful in a pond providing fish habitat, but needs periodic thinning to prevent small ponds being overgrown.
**Propagation:** Stem cuttings in water
**Localities:** Werribee River
**Distribution:** Vic, Qld, NSW, SA, NT

## Chamaesyce drummondii = Euphorbia drummondii

# Chenopodium — Chenopodiaceae

Annual or perennial herbs or shrubs with small unisexual or bisexual flowers clustered in the leaf axils or in terminal panicles. Terminal clusters are often male or bisexual flowers while female flowers occur in lateral clusters. They grow in drier areas and are salt tolerant. Some are fire retardant due to the high salt content of their foliage while others provide animal fodder. The leaves and seeds of some species were eaten by the Aborigines. Propagation is by seed or cutting.

## Chenopodium desertorum ssp. microphyllum — Small-leaf Goosefoot

F

**Size:** 0.4 m x 0.8 m
**Habitat:** Chenopod rocky open scrub
**Form:** Mealy-white, much branched perennial herb
**Foliage:** Pale greenish blunt, broadly lanceolate leaves to 10 mm long
**Flowers:** Small loose clusters of sessile, mealy flowers in Winter and Spring; male and bisexual flowers terminal, female flowers lateral
**Requirements:** Moist soils
**Comments:** Locally rare.
**Localities:** Melton, Mt Cottrell, 16
**Distribution:** Vic, NSW, SA, WA
**Synonym:** *C. pseudomicrophyllum*

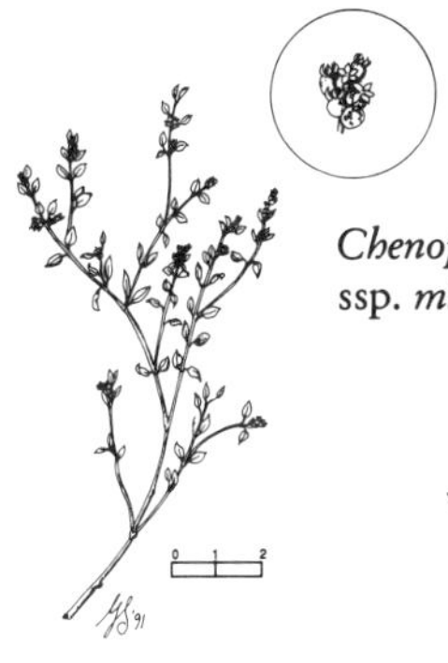

*Chenopodium desertorum* ssp. *microphyllum*

## Chenopodium glaucum — Glaucous Goosefoot

F P

**Size:** Prostrate–0.5 m x 1 m
**Habitat:** Primary dune scrub, saltmarshes, plains grassland
**Form:** Prostrate to erect branched annual herb
**Foliage:** Somewhat succulent, broadly lanceolate bright green leaves, densely mealy below, to 25 mm long
**Flowers:** Panicles of small dense clusters of mauve or green flowers in a cyme; December to June. Terminal clusters bisexual, lateral clusters female or bisexual.
**Requirements:** Moist saline soil
**Comments:** A second form is found near Rockbank in saline soils. Named *C. ambiguum* in the past, it is presently included in *C. glaucum.* It is prostrate and has small brownish leaves. There are only 2 known sites where this form still grows in Melbourne.
**Localities:** 3–5, 11, 22, 67#, 71
**Distribution:** Vic, NSW, Tas, SA, WA

*Chenopodium glaucum*

## Chenopodium pumilio — Small Crumbweed, Rough-leaved Goosefoot

F

**Size:** Prostrate–25 cm high
**Habitat:** Riparian scrub, grassy wetland
**Form:** Hairy aromatic annual herb
**Foliage:** Pale green oak-like leaves to 20 mm long
**Flowers:** Minute white flowers in compact clusters; March to July
**Requirements:** Grows in sandy soils
**Localities:** Early collections in Altona, St Kilda; Mt Cottrell, 9, 11, 16, 23, 26, 35
**Distribution:** All states

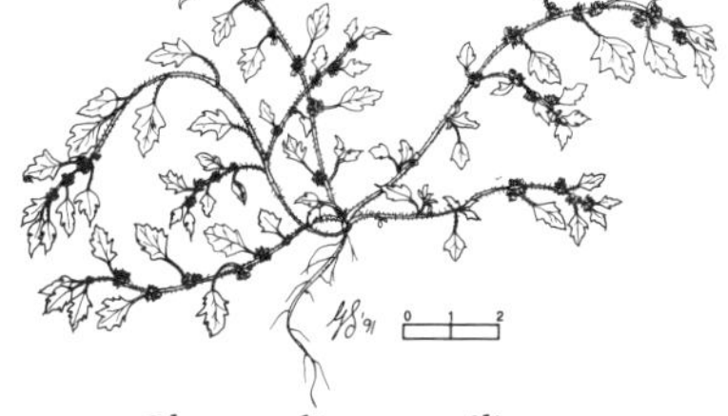

*Chenopodium pumilio*

SCALE: 0 1cm 2cm — The scale representation on each plant equals 2cm.

# Chrysocephalum

**Asteraceae**

*Chrysocephalum apiculatum*

## Chrysocephalum apiculatum

Common Everlasting, Yellow Buttons

**Size:** Prostrate-30 cm x 1–2 m **F**
**Habitat:** Widespread in plains grassland, red gum and box woodland, dry sclerophyll forest, tea-tree heath
**Form:** Variable, dense, spreading perennial herb
**Foliage:** Silver-grey obovate leaves to 60 mm long, densely hairy
**Flowers:** Terminal clusters of rounded, bright yellow flowerheads most of the year, peaking September to December
**Requirements:** Well drained soils
**Comments:** An excellent rockery plant with contrasting foliage. It should be pruned regularly to encourage new growth. The Keilor Plains form is smaller, more straggly in habit, with a finer leaf.
**Propagation:** Seed, cuttings
**Localities:** Warrandyte, 1, 4, 5, 7, 10, 12, 13, 15–18, 26, 29, 30, 35, 37, 50A, 55, 58, 67, 69, 69A, 74, 76
**Distribution:** All states
**Synonym:** *Helichrysum apiculatum*

*Chrysocephalum semipapposum*

## Chrysocephalum semipapposum

Clustered Everlasting

**Size:** 0.3–1 m x 1–3 m **F P**
**Habitat:** Box and red gum woodlands, damp to dry sclerophyll forests, plains grassland
**Form:** Variable, dense perennial herb, erect stems produced from a creeping rhizome. Individual stems live for 2 years, some forms developing secondary stems in their second year.
**Foliage:** Sticky grey-green narrow leaves 20–50 mm long
**Flowers:** Dense terminal clusters of small yellow flowerheads; October to May
**Requirements:** Variable, from moist to dry soils
**Comments:** Requires hard pruning of old growth. Three distinct forms occur within Melbourne.
**Form 1:** Large, robust grey-leafed form to 1 m high from damp sclerophyll forests and moist sites. May form very large patches to 3 m across. Stems are well spaced along the rhizome. Leaves to 50 mm long and 5 mm wide are broader at the tip. Flowers are orange yellow.
**Localities:** Baxter, 32, 34–39, 44, 49, 50A, 51, 55, 56, 58, 62, 63, 65–67
**Form 2:** Tight tussock forming clumps to 0.5 m high, from drier areas. Foliage is similar, to 30 mm long and silvery to green depending on the amount of sun received. Flowers are golden-yellow.
**Localities:** Eltham, 41, 42
**Form 3:** Drought tolerant, very tightly clumping sub-shrub to 0.4 m high. Stems are very straight in the first year, then arching with secondary stems to 100 mm high, rising erect from them. Leaves very narrow to 30 mm long, tightly revolute and falcate. Pale yellow flowers in smaller clusters on both primary and secondary stems. An excellent plant for cultivation which flowers later and responds well to Summer watering.
**Localities:** Hurstbridge, Sydenham, Melton Sth, 13–16, 25, 26, 32
**Propagation:** Seed, cuttings
**Distribution:** All states
**Synonym:** *Helichrysum semipapposum*

# Ciclospermum

**Apiaceae**

## Ciclospermum leptophyllum

Slender Celery

**Size:** 0.3–0.7 m high **P N**
**Habitat:** Riparian scrub
**Form:** Slender open annual herb
**Foliage:** Divided leaves to 40 mm, with very narrow leaf segments
**Flowers:** Umbels of tiny whitish flowers at or near the ends of branches; December to February. Fruit deeply ribbed.
**Requirements:** Moist soils
**Comments:** Can become invasive especially in disturbed areas. A cosmopolitan weed, it is considered to be introduced to Australia by some authorities.
**Propagation:** Seed
**Localities:** Early collection from Darebin Creek
**Distribution:** All mainland states; America
**Synonym:** *Apium leptophyllum*

# Clematis — Ranunculaceae

## Clematis aristata

Austral Clematis, Goat's or Old Man's Beard,
**F P N**

**Habitat:** Widespread in wet, damp and valley sclerophyll forests, grassy low open forest
**Form:** Vigorous climber
**Foliage:** Trifoliate shiny green leaves, leaflets ovate to 80 mm long. Juvenile leaves have silvery top markings, purplish below
**Flowers:** Masses of creamy white starry flowers to 60 mm across; August to March
**Requirements:** Easily grown in well drained soils
**Comments:** This very showy climber is dioecious. Attractive feathery seed heads persist on the female plant for some time. Will grow well under established eucalypts if watered during Summer.
**Propagation:** Fresh seed, stem cuttings
**Localities:** 33–34A, 40–44, 46, 47, 49, 50A, 51–53, 55–60, 62–64, 70, 74, 76
**Distribution:** Vic, Qld, NSW, Tas

*Clematis aristata*

## Clematis microphylla var. microphylla

Small-leafed Clematis
**F P**

**Habitat:** Dry sclerophyll forest, box, riparian and coastal banksia woodlands, primary dune scrub, plains grassland, grassy low open forest
**Form:** Variable medium-sized climber
**Foliage:** Oblong, dull green leaves to 30 mm long, divided 2–3 times
**Flowers:** Dioecious; masses of cream starry flowers; July to November. Feathery fruit persists on the female plant for some time.
**Requirements:** Well drained soils
**Comments:** Leaves can relieve skin irritations but may cause blisters unless used in moderation. The Aborigines cooked and kneaded the tap root to make dough.
**Propagation:** Seed, stem cuttings
**Localities:** 1, 14, 16, 18, 24–26, 28–30, 32–39, 41, 42, 44, 47, 48, 50A, 52, 53, 60–64, 67, 69A, 71, 74, 76
**Distribution:** All states except NT *Illustrated p. 106*

# Comesperma — Polygalaceae

*Comesperma* are small shrubs or twiners with attractive pea-like flowers. They prefer well drained soils with some shelter. Propagation is by seed or cuttings.

## Comesperma calymega

Blue Spike Milkwort
**F P**

**Size:** 0.3–0.5 m x 0.3–1 m
**Habitat:** Tea-tree heath
**Form:** Dwarf erect shrub
**Foliage:** Thick elliptic to lanceolate leaves to 20 mm long
**Flowers:** Dense terminal racemes of small blue flowers; October to January
**Requirements:** Well drained soil
**Localities:** Olinda, 67, 68, 71, 73–75
**Distribution:** Vic, Tas, SA, WA

*Comesperma calymega*

## Comesperma ericinum

Heath Milkwort
**P**

**Size:** 0.5–1.5 m x 1–2 m
**Habitat:** Tea-tree heath, valley sclerophyll forest
**Form:** Slender erect shrub
**Foliage:** Numerous narrow to oblong dark green leaves to 15 mm long
**Flowers:** Profuse terminal racemes of small pink flowers; October to February
**Requirements:** Adaptable, tolerating wet periods
**Comments:** This attractive shrub benefits from pruning after flowering. Suitable for growing in containers. Locally rare.
**Localities:** Montrose, 49, 50–51, 53, 54, 56, 67#
**Distribution:** Vic, Qld, NSW, Tas

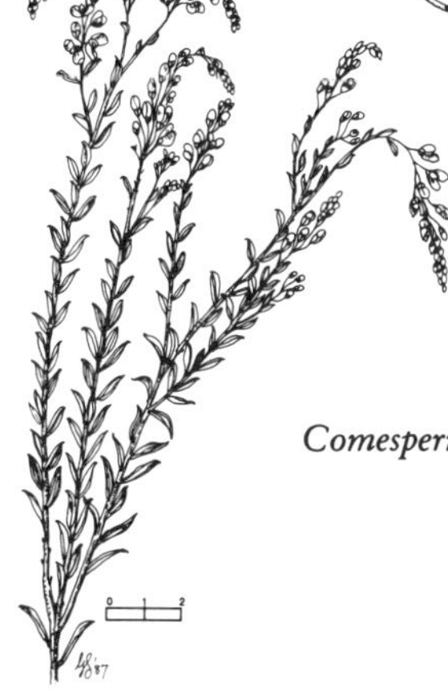

*Comesperma ericinum*

## Comesperma polygaloides

Small Milkwort
**F**

**Size:** 0.3–0.6 m x 0.3 m
**Habitat:** Plains grasslands, red gum woodland, in sandy, clay and granite areas
**Form:** Small slender shrub
**Foliage:** Narrow bluish-green leaves
**Flowers:** Profuse terminal racemes of purple or mauve flowers
**Requirements:** Well drained soils
**Comments:** Suitable for growing in containers. A vulnerable plant in Victoria, there are few sites where it remains in Melbourne. It occurred until recently at Taylors Lakes when this area was cleared for housing.
**Localities:** St Albans#, Footscray#, Port Melbourne#, Sunbury, 5, 7, 17#, 18, 26, 35
**Distribution:** Vic, SA, WA

SCALE: 0 1cm 2cm — The scale representation on each plant equals 2cm.

*Comesperma volubile*

## Comesperma volubile

Love Creeper

**P**

**Habitat:** Widespread in grassy low open forest, dry and valley sclerophyll forests, tea-tree heath
**Form:** Open slender twiner
**Foliage:** Almost leafless, with a few narrow leaves to 10 mm long
**Flowers:** Loose racemes of blue flowers; August to December
**Requirements:** Well drained soil with roots protected from drying out
**Comments:** This beautiful creeper is difficult to establish, possibly requiring a symbiotic relationship with an undetermined organism for successful growth.
**Propagation:** Seed, cuttings—difficult, needing leaves on stem
**Localities:** Research, 34, 34A, 36, 37, 41–44, 46–53, 55–64, 67, 68, 70, 71, 73–76
**Distribution:** All states except NT

# Convolvulus — Convolvulaceae

*Convolvulus erubescens*

## Convolvulus erubescens

Australian, Blushing or Pink Bindweed

**Size:** 10–30 cm x 0.5 m — **F**
**Habitat:** Tea-tree heath, red gum woodland, plains grassland, dry and damp sclerophyll forests
**Form:** Fast-growing trailing or climbing perennial
**Foliage:** Hairy grey-green leaves 15–60 mm long, with several lobes at the base, central lobe often much longer
**Flowers:** Very showy rosy pink funnel-shaped flowers, on stalks 15–35 mm long; August to February
**Requirements:** Well drained hot dry areas
**Comments:** Ideal for tubs and hanging baskets or as a ground cover. The Aborigines used a boiled extract to treat diarrhoea and stomach ache. The woody taproot was cooked in baskets, then kneaded into dough.
**Propagation:** Cuttings; woody seed coat needs to be sliced to expose seed before germination.
**Localities:** 1, 2, 4–9, 12, 13, 15–18, 20, 21, 24–26, 28–32, 34–36, 39, 42, 45#, 67#, 71, 74, 76
**Distribution:** All states

*Convolvulus remotus*

## Convolvulus remotus

**Size:** Prostrate x 0.3–0.6 m — **F**
**Habitat:** Red gum woodland, plains grassland
**Form:** Small trailing plant
**Foliage:** Entire, bright green leaves 15–70 mm long, sometimes with 2 long linear lobes at the base. Hairs are pressed against the leaf surface rather than erect as in *C. erubescens*.
**Flowers:** Pink flowers, singly or in pairs; September to December
**Comments:** Good ground cover for hot areas. Locally rare, found only on the basalt plains.
**Localities:** Bundoora, Laverton Nth, 2, 21, 28
**Distribution:** All mainland states

# Coprosma — Rubiaceae

*Coprosma hirtella*

## Coprosma hirtella

Rough Coprosma

**Size:** 1–2 m x 0.5–1.5 m — **P**
**Habitat:** Wet, damp and valley sclerophyll forests
**Form:** Slender understorey shrub
**Foliage:** Rough dull green broad elliptical leaves, pointed at ends, to 70 mm long
**Flowers:** Insignificant greenish clusters of unisexual flowers; August to October; followed by attractive reddish-orange drupes to 10 mm long
**Requirements:** Moist, well drained sheltered position
**Comments:** Easily grown. The drupes are edible but leave an unpleasant taste.
**Propagation:** Fresh seed, cuttings
**Localities:** 55–58
**Distribution:** Vic, NSW, Tas

*Coprosma quadrifida*

## Coprosma quadrifida

Prickly Currant Bush

**Size:** 2–4 m x 1–1.5 m — **P N**
**Habitat:** Cool temperate rainforest, wet and damp sclerophyll forests, sclerophyll woodland, swamp scrub
**Form:** Open, upright, spiny shrub
**Foliage:** Dull lanceolate to elliptic leaves to 15 mm long
**Flowers:** Inconspicuous single greenish flowers; September to November; followed by small edible red drupes; January to March.
**Requirements:** Moist well drained soil
**Comments:** A useful plant to grow with ferns. A form growing at the head of the Werribee River has larger leaves to 50 mm long.
**Propagation:** Seed, cuttings
**Localities:** Sth Morang, 1, 26, 29–36, 41, 42, 44, 47, 50, 51, 53, 55–59, 62–64, 70
**Distribution:** Vic, NSW, Tas

30

31

32

33

34

35

36

37

30. *Banksia marginata*
    Silver Banksia
31. *Thelionema caespitosum*
    Blue Grass-lily
32. *Acacia myrtifolia*
    Myrtle Wattle
33. *Lomandra filiformis*
    ssp. *coriacea*
34. *Wurmbea dioicia*
    Early Nancy
35. *Comesperma ericinum*
    Heath Milkwort
36. *Ricinocarpos pinifolius* with
    *Dampiera stricta*
    Wedding Bush & Blue Dampiera
37. *Calochilus robertsonii*
    Common or Purple Beard-orchid
38. *Dillwynia sericea*
    Showy Parrot-pea

38

39

40

41

42

39. *Grevillea alpina* Alpine Grevillea
40. *Acacia dealbata* Silver Wattle
41. *Olearia lirata* Snowy Daisy Bush
42. *Chiloglottis valida* Common Bird-orchid
43. *Notelaea ligustrina* Native Mock Olive
44. *Eucalyptus regnans* with *Bedfordia arborescens* Mountain Ash, Blanket Leaf
45. *Prostanthera lasianthos* Victorian Christmas Bush
46. *Lomandra filiformis* ssp. *filiformis* Wattle Mat-rush
47. *Doodia media* Common Rasp Fern
48. *Banksia spinulosa* Hairpin Banksia

43

44

45

46

47

48

# Correa — Rutaceae

Correas are widespread in varied habitats from exposed coastal to mountain areas. Nectar-producing flowers occur over a long period, attracting honeyeaters. The flowers are bell-shaped except for *C. alba* which has an open star-like flower. The leaves when crushed produce a distinctive aroma. Pruning after flowering is beneficial to maintain shape. All require a well drained position. They make attractive tub specimens. Correas are usually propagated easily from cuttings whilst seeds are very difficult to germinate.

## Correa alba — White Correa

**Size:** 0.5–2 m x 1–3 m — **F P**
**Habitat:** Primary dune scrub
**Form:** Dense spreading shrub, dwarfed by wind and salt spray
**Foliage:** Thick, round to oval grey-green leaves, pale and hairy below, to 40 mm long
**Flowers:** Waxy white starry flowers most of the year
**Requirements:** Well drained soils, tolerating moisture or extended dry periods once established
**Comments:** A useful plant for soil binding or as a low screen. The leaves were used by early settlers to make a tea substitute.
**Localities:** Seaford, 67, 74
**Distribution:** Vic, NSW, Tas, SA

*Correa alba*

## Correa glabra — Rock or Smooth Correa

**Size:** 1–3 m x 1–3 m — **F P N**
**Habitat:** Widespread in riparian scrub, on low rocky outcrops, and escarpments
**Form:** Variable dense upright to spreading shrub
**Foliage:** Dark green oblong leaves, usually with wavy edges, to 40 mm long
**Flowers:** Solitary pale green bells most of the year, peaking May to August
**Requirements:** Well drained soils
**Comments:** An attractive, easily grown shrub, ideal as a low screen. Tolerant of lime.
**Localities:** Essendon, Sunshine, Kororoit Ck, 14, 16, 26, 28, 29, 32, 33, 34A–36, 41, 42, 47, 50A, 51
**Distribution:** Vic, Qld, NSW, SA

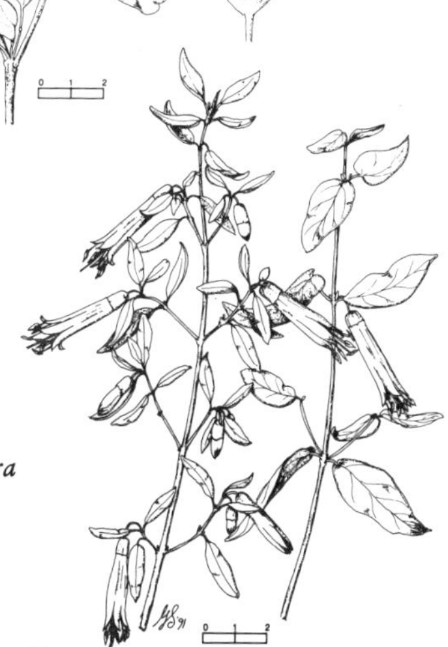

*Correa glabra*

## Correa reflexa — Native Fuchsia, Common Correa

**Size:** 0.3–2 m x 1–2 m — **F P N**
**Habitat:** Dry, damp and valley sclerophyll forests, sclerophyll woodland, tea-tree scrub, grassy low open forest
**Form:** Very variable—open upright to spreading shrub
**Foliage:** Dull green, hairy, broadly ovate leaves to 50 mm long
**Flowers:** Solitary, large, usually light green bells; March to September. A form with red flowers and yellow tips is known from Donvale, Croydon, Boronia, Dingley and Frankston.
**Requirements:** Well drained soils
**Comments:** A prostrate form has been reported from Seaford. Other Victorian forms may be prostrate or medium-sized shrubs, with flowers of cream or red with yellow-green tips. An excellent plant for dry shady positions, establishing well under existing trees. Forms collected from wetter mountain areas require some moisture.
**Localities:** St. Kilda#, Eltham, 30#, 33, 34A–37, 40–44, 46–53, 55, 56, 58, 59, 61–65#, 67–71, 73–76
**Distribution:** All states except NT — *Also illustrated p. ix*

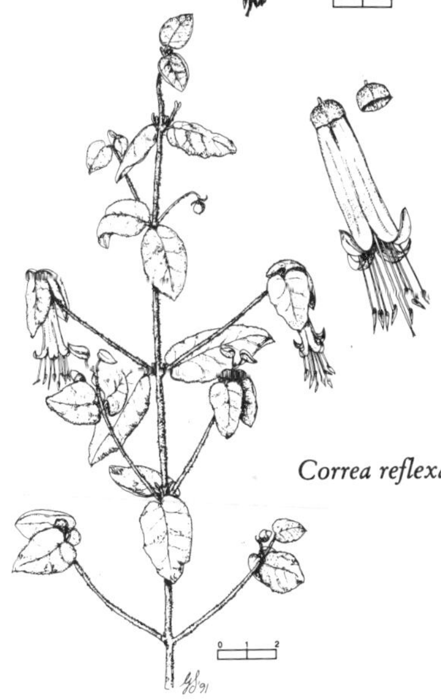

*Correa reflexa*

## Correa reflexa x glabra

**Size:** 1.5 m x 1.5 m — **F P**
**Comments:** A natural hybrid occurring in the Warrandyte/Croydon North area. Similar in appearance to *C. glabra* but with bright green wrinkly leaves, hairy below, to 25 mm long. Grows in dry, clay soil.
**Localities:** 41, 42, 51, 53

# Cotula — Asteraceae

## Cotula australis — Common Cotula

**Size:** Prostrate–20 cm x 1 m — **P N**
**Habitat:** Widespread in coastal banksia, red gum and sclerophyll woodland, riparian scrub, dry sclerophyll forest, near water
**Form:** Spreading annual or perennial herb
**Foliage:** Hairy pinnate leaves to 20 mm long
**Flowers:** Slender, long-stalked, solitary white and yellow-green flower heads; September to February
**Requirements:** Moist soils
**Comments:** A useful ground cover for difficult shady areas but may become invasive.
**Propagation:** Seed, cuttings, division
**Localities:** 16, 23, 25, 26, 29, 30, 32, 34A-36, 38, 41, 42, 44, 49, 51, 58, 63, 64, 67–71, 74
**Distribution:** All states

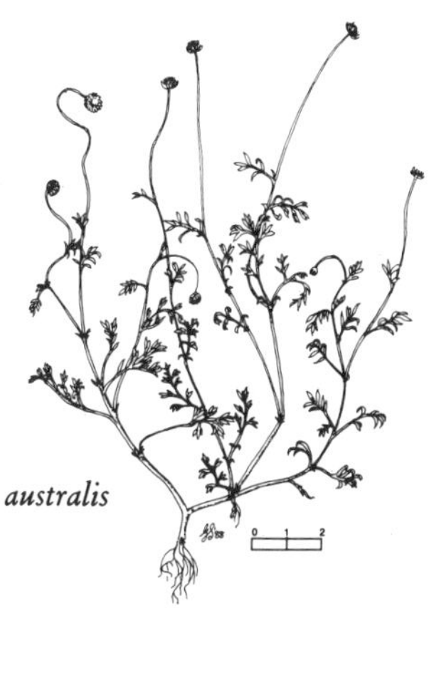

*Cotula australis*

SCALE: 0 1cm 2cm — The scale representation on each plant equals 2cm.

### Cotula reptans = Leptinella reptans

### Cotula vulgaris var. australasica

Slender Cotula
F P

**Size:** to 20 cm high
**Habitat:** Saltmarshes, moist tea-tree heath
**Form:** Slender annual herb
**Foliage:** Thread-like leaves to 40 mm long forming a tangled mass
**Flowers:** Pale yellow button flowers; August to November
**Requirements:** Moist soil, tolerating saline conditions
**Localities:** Seaholme 1942, 67#
**Distribution:** Vic, Tas, SA, WA

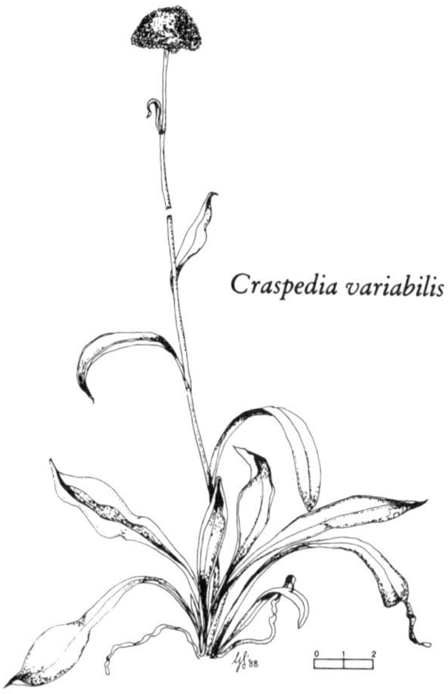

*Craspedia variabilis*

## Craspedia

Asteraceae

### Craspedia chrysantha = Pycnosorus chrysanthes

### Craspedia globosa = Pycnosorus globosus

### Craspedia variabilis

Common Billy Buttons
F P

**Size:** 30 cm x 0.5–1 m
**Habitat:** Widespread in grassy wetlands, plains grasslands, riparian scrub, tea-tree heath, grassy low open forest, dry and valley sclerophyll forest, sclerophyll woodland
**Form:** Variable tufted perennial herb
**Foliage:** Basal rosette of broad hairy leaves to 20 cm long
**Flowers:** Yellow globular heads to 20 mm held erect singly on long stalks from 15–50 cm high; September to November
**Requirements:** Moist to boggy soils
**Comments:** Spectacular in flower it may be grown in containers or rockeries or used in massed plantings. An excellent dried flower. A taller form grows in swampy areas to the north of the study area.
**Propagation:** Seed, division
**Localities:** St Albans, Sunbury, Eltham, Baxter, 2, 7, 15, 16, 26, 29#, 33–37, 42, 44, 46, 47, 49–51, 55, 58, 59, 63, 67#, 70
**Distribution:** All states
**Synonym:** *Craspedia glauca*

## Crassula

Crassulaceae

### Crassula colorata var. acuminata

Dense Crassula
F

**Size:** 1–15 cm high
**Habitat:** Plains grassland, chenopod rocky open scrub, grassy low open forests
**Form:** Erect, succulent annual herb
**Foliage:** Crowded, obovate leaves to 4 mm long, becoming red with age
**Flowers:** Dense clusters of tiny, sessile, pink 5-petalled flowers; August to October
**Requirements:** Moist soils
**Comments:** Would be attractive as a rockery plant. Distinguished from *C. sieberiana* ssp. *tetramera*, with which it often occurs, by a cluster of brown nodules near the base of each fruit. Locally rare.
**Localities:** 2, 74
**Distribution:** Vic, NSW, SA, WA; NZ
**Synonym:** *C. colorata* var. *tuberculata*

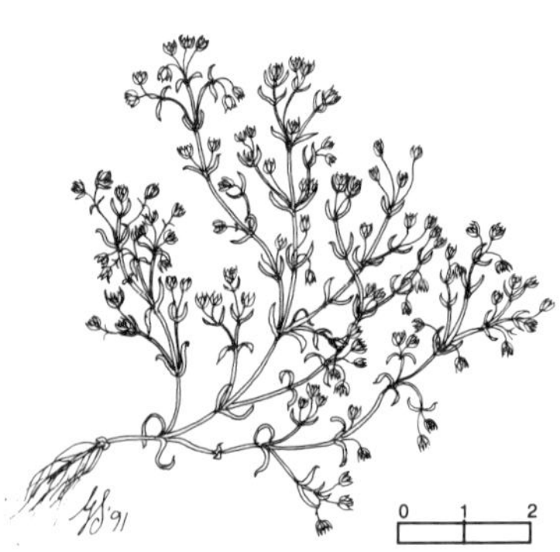

*Crassula decumbens* var. *decumbens*

### Crassula decumbens var. decumbens

Rufous or Spreading Stonecrop
P

**Size:** Prostrate–8 cm high
**Habitat:** Primary dune scrub, red gum woodland, tea-tree heath, valley sclerophyll forest
**Form:** Widespread, small reddish erect or prostrate annual herb with repeatedly branched stems
**Foliage:** Succulent, linear-lanceolate leaves to 8 mm long
**Flowers:** Stalked, tiny white or red-tinged flowers in clusters.
**Requirements:** Moist, boggy areas
**Localities:** 2, 4, 23, 25, 26, 31, 32, 35, 42, 43A, 44, 49, 51, 67#–71, 74
**Distribution:** Vic, NSW, Tas, SA, WA; S Am, S Af
**Synonym:** *C. macrantha*

## Crassula helmsii

Swamp Stonecrop

**Size:** 10-30 cm high — F
**Habitat:** Widespread in riparian scrub and woodland, valley sclerophyll forest, grassy low open forest
**Form:** Fast growing aquatic or creeping perennial herb
**Foliage:** Pairs of succulent, narrow green or reddish oblong leaves to 15 mm long
**Flowers:** Insignificant, stalked, solitary, white 4-petalled flowers along ends of branches; November to April
**Requirements:** Aquatic or bog conditions
**Comments:** A rapid grower which may become invasive. Submerged forms have longer branches that float on the surface, and broader leaves. Plants growing in moist soils form small clumps with densely clustered leaves. Makes an excellent aquarium plant.
**Propagation:** Seed, stem cuttings, division
**Localities:** Diamond Ck, 4, 8, 14, 16, 26, 28, 34A, 35, 36, 42, 44, 60, 62, 63, 71, 74-76
**Distribution:** All states except NT; NZ

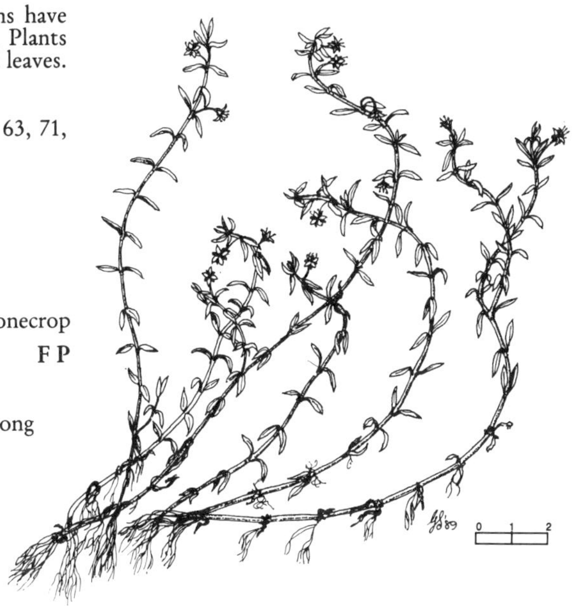

*Crassula helmsii*

## Crassula pedicellosa

Stalked Stonecrop

**Size:** to 10 cm high — F P
**Habitat:** Tea-tree heath, grassy wetlands
**Form:** Tufted erect annual herb
**Foliage:** Green to purplish-red oblanceolate to elliptic leaves to 12 mm long
**Flowers:** Tiny cream flowers in terminal clusters; August to October
**Requirements:** Moist sandy soils
**Localities:** 5, 74
**Distribution:** Vic, Tas, SA, WA

## Crassula peduncularis

Purple Stonecrop

**Size:** to 5 cm high — P N
**Habitat:** Sheltered marshy sites in box woodland and coastal banksia woodland
**Form:** Tiny upright annual herb with much-branched red stems
**Foliage:** Succulent, oval purplish leaves to 5 mm long
**Flowers:** Long-stalked, single red-tinged flowers in Summer
**Comments:** Very restricted range in moist soils
**Localities:** 24, 67#
**Distribution:** Vic, NSW, Tas, SA, WA; NZ, S Am
**Synonym:** *C. purpurata*

## Crassula sieberiana ssp. sieberiana

Austral Stonecrop

**Size:** 1-10 cm high — F P
**Habitat:** Wet crevices in red gum and box woodland, grassy wetland, plains grassland, riparian and chenopod rocky open scrub, tea-tree heath
**Form:** Succulent perennial herb forming dense colonies. Roots frequently grow from shoots. Fleshy, decumbent branches have segmented, swollen nodes, especially below the leaves, with basal internodes from 3-5 mm long.
**Foliage:** Tiny linear-lanceolate leaves to 8 mm long are green to greyish-brown
**Flowers:** Minute 4-petalled cream, pink or red cup-shaped flowers in dense, spike-like clusters; October to June
**Requirements:** Moist to wet soils
**Comments:** Grow as a rockery plant.
**Localities:** Research, Belgrave Heights, 4, 6, 12, 14, 16-18, 21, 23-26, 29, 32, 34A-36, 42, 49, 50A, 51, 55, 57-59, 67, 68, 69A, 75
**Distribution:** Vic, Qld, NSW, Tas, SA; NZ

*Crassula sieberiana* ssp. *sieberiana*

## Crassula sieberiana ssp. tetramera

**Size:** to 15 cm high — P
**Habitat:** Tea-tree heath, valley sclerophyll forest
**Form:** Erect annual with normal basal root system. Wiry-woody branches are not segmented and basal internodes are longer to 12 mm.
**Flowers:** Axillary in leaf-like bracts; August to November. The calyx is twice as long as the petals in the coastal form.
**Comments:** Often occuring with *C. colorata* var. *acuminata* which is distinguished by its 5 petals.
**Localities:** Melton, Werribee, Diamond Ck, Research, 32, 36, 69-72, 74
**Distribution:** All states

SCALE: 0 1cm 2cm — The scale representation on each plant equals 2cm.

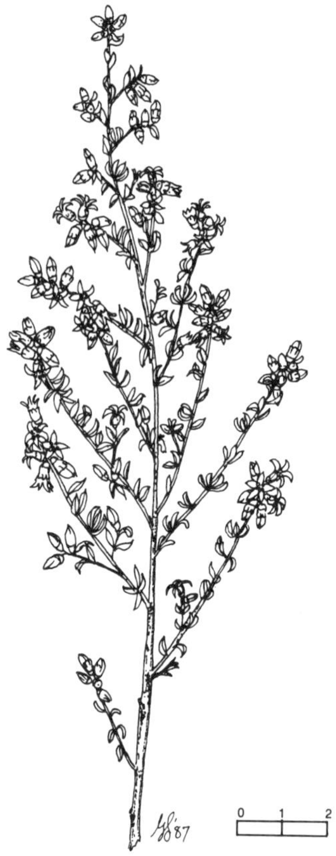

*Cryptandra amara*

# Cryptandra — Rhamnaceae

## Cryptandra amara var. amara

Bitter Cryptandra

F P

**Size:** 0.3–0.6 m x 0.5 m
**Habitat:** Box woodland
**Form:** Slow growing, wiry spreading shrub
**Foliage:** Green, narrow to obovate leaves to 6 mm long, rarely slightly hairy, on stiff, spiny branchlets
**Flowers:** Clusters of small white or reddish flowers; April to October
**Requirements:** Well drained soils, tolerating dryness once established
**Comments:** A very showy plant, especially suited to container growing or massed planting in a rockery. Requires light pruning after flowering to maintain a compact shape.
**Propagation:** Cuttings—can be slow to strike
**Localities:** Diggers Rest, Bulla
**Distribution:** Vic, Qld, NSW, Tas, SA

# Cuscuta — Cuscutaceae

## Cuscuta tasmanica

Golden Dodder

F

**Habitat:** Saltmarshes
**Form:** Leafless, rootless parasite which twines around low herbs, absorbing their nourishment through specialised attachments. Often parasitises *Wilsonia* spp. and *Chenopodium* spp. Slender stems are pale yellow.
**Flowers:** Creamy flowers are in small, dense, stalked clusters; December to May.
**Comments:** It has little ornamental value and plants may become a weed in gardens. May be extinct within Melbourne.
**Localities:** Collected in St. Kilda in 1905, Altona#; 4
**Distribution:** Vic, NSW, Tas, SA, WA

# Cymbonotus — Asteraceae

## Cymbonotus lawsonianus

Bear's Ear

F

**Size:** Prostrate x 10–30 cm
**Habitat:** Red gum woodland, plains grassland, dry or valley sclerophyll forest
**Form:** Perennial herb forming a flat rosette
**Foliage:** Ovate to lanceolate toothed leaves to 100 mm long, white and hairy underneath
**Flowers:** Single yellow daisy flowers on short woolly stalks; September to November
**Requirements:** Moist, rich well drained soils
**Comments:** Suitable for rockeries but may become a weed in cultivated or disturbed soil. Fruit are deeply curved and smooth on the convex side. Used by the early settlers in an ointment to heal wounds. Very rare.
**Propagation:** Seed, division
**Localities:** Sydenham#, St. Albans#, Greensborough, 20#, 21#, 30#, 42, 58#
**Distribution:** All states except NT

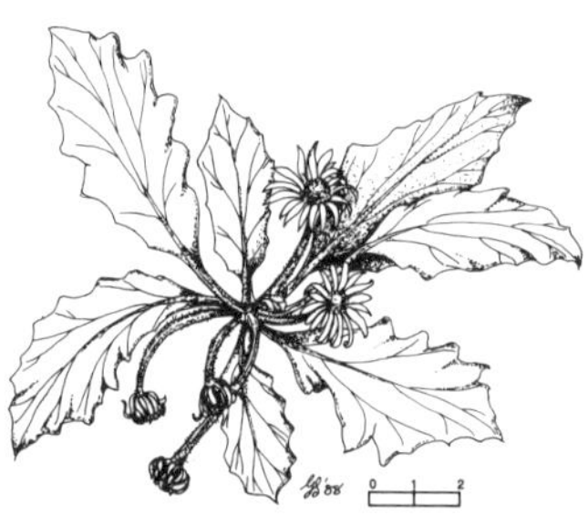

*Cymbonotus preissianus*

## Cymbonotus preissianus

Australe Bear's Ear

P N

**Habitat:** Dry and damp sclerophyll forest, red gum woodland, grassy low open forest
**Comments:** Differs from *C. lawsonianus* in preferring sheltered situations. Leaves are velvety and strap-like. Fruits are only slightly curved, and rough on the convex side.
**Localities:** 29#, 34A–39, 41, 42, 55, 58, 67#, 74, 76
**Distribution:** Vic, NSW, Tas, SA, WA

*Cynoglossum australe*

# Cynoglossum — Boraginaceae

## Cynoglossum australe

Austral Hound's Tongue

F P

**Size:** 0.3–1 m x 0.5–2 m
**Habitat:** Plains grassland, amongst rocks
**Flowers:** Similar to *C. suaveolens* but with small, slightly fragrant light blue flowers; October to February. Forked flowering stems are leafless and many-flowered.
**Requirements:** Dry, well drained soil
**Comments:** Locally endangered, it may already be extinct in the Melbourne region.
**Localities:** Early collections from Darebin Creek, Melbourne
**Distribution:** All states except Qld

## Cynoglossum latifolium

Forest Hound's Tongue

| | |
|---|---|
| Size: | 0.5–1.5 m x 0.5–2 m **P** |
| Habitat: | Damp and wet sclerophyll forests and gullies |
| Form: | Weak-stemmed prickly perennial |
| Foliage: | Oval, hairy 5-veined leaves to 80 mm long |
| Flowers: | Small bluish flowers on longer stalks (>15 mm) than *C. suaveolens* and *C. australe*, singly in axils or in few-flowered loose racemes; sporadically. Stalks do not recurve in fruit. |
| Requirements: | Moist soil in sheltered areas |
| Localities: | 56–58 |
| Distribution: | Vic, Qld, NSW, Tas |

*Cynoglossum latifolium*

## Cynoglossum suaveolens

Sweet Hound's Tongue

| | |
|---|---|
| Size: | 0.1–1 m x 0.5–1.5 m **F P** |
| Habitat: | Plains grassland, coastal banksia woodland, damp and dry sclerophyll forest |
| Form: | Erect to spreading perennial herb |
| Foliage: | Basal rosette with wavy lanceolate to oblong leaves to 120 mm long, becoming smaller and sessile on stems. Leaf surface is covered in stiff hairs. |
| Flowers: | Highly fragrant, stalked, small white flowers with yellow centres, on terminal leafy flowering stems; most of the year. Flower stalk curves downwards in fruit. |
| Requirements: | Well drained heavy soil |
| Comments: | This most desirable plant is useful for containers or on embankments where it acts as a soil binder. It responds well to light pruning. |
| Propagation: | Seed, cuttings, division |
| Localities: | Kororoit Ck, Rockbank, 18, 25, 26, 28, 30#, 35, 42, 44, 47, 49, 50A, 55, 56, 59, 67#, 69A, 74 |
| Distribution: | Vic, Qld, NSW, Tas, SA |

*Cynoglossum suaveolens*

# Dampiera — Goodeniaceae

## Dampiera stricta

Blue Dampiera

| | |
|---|---|
| Size: | 0.3–0.8 m x 0.3–2 m **F P** |
| Habitat: | Dry sclerophyll forest, tea-tree heath |
| Form: | Suckering perennial herb |
| Foliage: | Thick, elliptical leaves, sometimes hairy, to 50 mm long |
| Flowers: | Single or clustered bright blue to mauve-blue flowers with yellow centres; May to June |
| Requirements: | Moist soils with reasonable drainage |
| Comments: | Excellent garden or container plant providing a natural look. Hard pruning is beneficial. Locally rare. |
| Propagation: | Suckers, cuttings can be slow |
| Localities: | Eltham, ?42, 44 |
| Distribution: | Vic, Qld, NSW, Tas |

*Dampiera stricta*

# Daucus — Apiaceae

## Daucus glochidiatus

Austral Carrot

| | |
|---|---|
| Size: | 0.75 m high **F P** |
| Habitat: | Plains grassland, dry and valley sclerophyll forest, coastal banksia woodland |
| Form: | Slender carrot-like annual with erect stems branching at the base and covered in stiff hairs |
| Foliage: | Pinnate leaves, ovate segments have narrow lobes |
| Flowers: | Terminal umbels of tiny white or pink flowers; August to November. Fruit spiny. |
| Requirements: | Dry soils |
| Comments: | May become a weed under ideal conditions. |
| Propagation: | Seed |
| Localities: | Diamond Ck, Laverton Nth, 12, 26, 29#, 34A, 36, 38, 41, 42, 44, 53, 55, 62, 63, 67#, 75 |
| Distribution: | All states |

*Daucus glochidiatus*

SCALE: 0 1cm 2cm — The scale representation on each plant equals 2cm.

# Daviesia

## Fabaceae

These 3 species are frequently found as understorey shrubs in open forests where they will tolerate a wide range of well drained soils. Masses of pea-shaped flowers, commonly yellow and brown, make them very showy in Spring. They are followed by triangular seed pods which are a distinguishing feature. Plants will benefit from an annual pruning after flowering. Propagation is by scarified seed or cuttings.

*Daviesia latifolia*

### Daviesia latifolia

Hop Bitter-pea

**Size:** 1–3 m x 1–2 m **F P**
**Habitat:** Dry, damp and valley sclerophyll forests, grassy low open forest, disturbed areas
**Form:** Open spreading shrub
**Foliage:** Dull green ovate-elliptic leaves with wavy edges to 100 mm long and a strong network of veins
**Flowers:** Dense racemes of fragrant yellow and brown flowers; September to December
**Requirements:** Adaptable to most soils
**Comments:** Useful in massed plantings for screening or hedges. Leaves have medicinal properties and were substituted for hops.
**Localities:** Epping, Springvale#, 34A, 42–44, 47–49, 50A, 51, 54, 56, 58, 61–64, 67#, 74, 76
**Distribution:** Vic, Qld, NSW, Tas

*Daviesia leptophylla*

### Daviesia leptophylla

Narrow Leaf, or Slender Bitter-pea

**Size:** 1–2 m x 1–2 m **F P**
**Habitat:** Widespread in valley and dry sclerophyll forests, sclerophyll and coastal banksia woodland
**Form:** Open erect shrub
**Foliage:** Stiff dull green linear leaves to 100 mm long
**Flowers:** Many short racemes of yellow and red flowers; September to January
**Localities:** Hawthorn#, Hampton#, Clayton#; Epping, Research, 34, 34A, 36, 37, 41, 42, 44, 46–51, 53–56, 58, 62–64, 67#, 74, 75
**Distribution:** Vic, NSW, SA
**Synonym:** *D. virgata*

### Daviesia leptophylla x latifolia

**Comments:** A natural hybrid with characteristics mid-way between each parent. Frequently confused with *D. mimosoides* which does not occur within the Melbourne area.
**Localities:** Knox, 35, 38, 39, 42, 43, 47, 48, 50A, 74

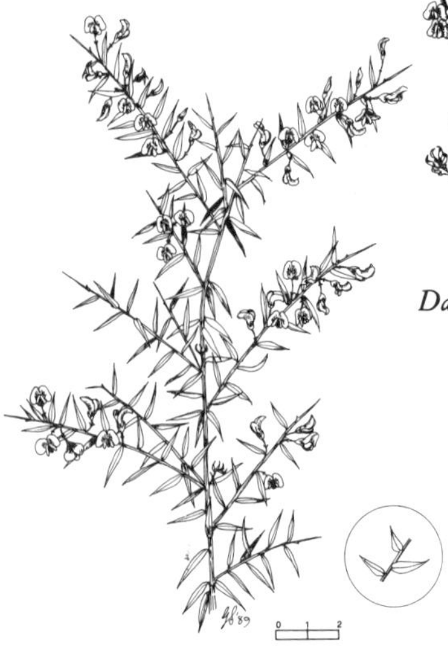

*Daviesia ulicifolia*

### Daviesia ulicifolia

Gorse Bitter-Pea

**Size:** 1.5 m x 1–2 m **P**
**Habitat:** Dry sclerophyll forest, tea-tree heath, riparian scrub
**Form:** Stiff, tangled, prickly shrub
**Foliage:** Dark green prickly, narrow-lanceolate leaves to 25 mm long, midrib prominent
**Flowers:** 1–3 flowers in leaf axils; July to November
**Localities:** 29, 49, 55, 67, 68
**Distribution:** Vic, Qld, NSW, SA, WA

# Derwentia

## Scrophulariaceae

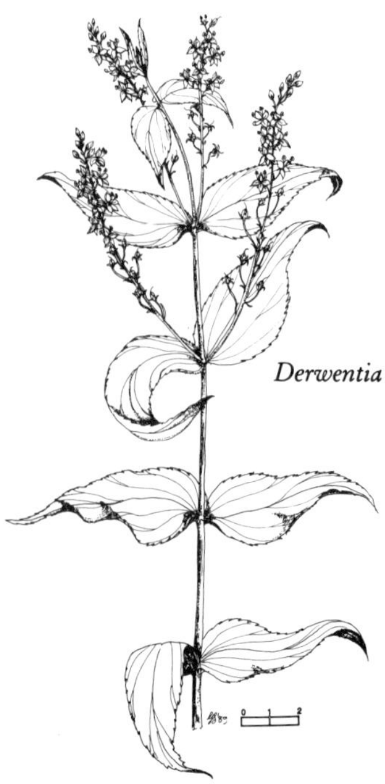

*Derwentia derwentiana*

### Derwentia derwentiana

Derwent Speedwell

**Size:** 0.6–1 m x 1 m **P N**
**Habitat:** Valley, damp and wet sclerophyll forests, riparian scrub
**Form:** Upright to straggly open perennial
**Foliage:** Dark green lanceolate leaves with toothed edges, to 100 mm long
**Flowers:** Terminal sprays of white or pale blue 4-petalled flowers; November to January
**Requirements:** Moist well drained soil
**Comments:** Fast growing perennial which may die back in Summer. Hard pruning encourages more vigorous growth.
**Propagation:** Cuttings, division
**Localities:** 29, 42, 44, 50A, 55–58, 60
**Distribution:** Vic, Qld, NSW, Tas, SA
**Synonym:** *Veronica derwentiana, Parahebe derwentiana*

# Desmodium — Fabaceae

## Desmodium gunnii — Slender Tick-trefoil

**Size:** 0.2–0.5 m x 1–2 m — P
**Habitat:** Damp and valley sclerophyll forests, grassy wetlands
**Form:** Light scrambling undershrub, occasionally rooting at the nodes
**Foliage:** Trifoliate with smooth round dark green leaflets to 25 mm long
**Flowers:** Pink pea flowers in loose racemes to 120 mm occur most of the year. Fruit is a pod to 24 mm with 3–6 tooth-like segments, covered in small bristles.
**Requirements:** Moist soils
**Comments:** An evergreen undershrub which may form large mats in moist conditions. Needs to be grown among other plants.
**Propagation:** Scarified seed, cuttings
**Localities:** 6, 36, 42, 55–58, 60
**Distribution:** Vic, Qld, NSW, Tas
**Synonym:** *D. varians* var. *gunnii*

*Desmodium varians*

## Desmodium varians — Slender Tick-trefoil

**Size:** 0.2–0.5 m x 0.2–0.5 m — F P
**Habitat:** Dry sclerophyll forest, plains grassland, crevices in escarpments
**Form:** Slender scrambling undershrub which dies back to its rootstock in Summer
**Foliage:** Yellow-green trifoliate leaves, leaflets linear with sunken veins
**Flowers:** Yellow-orange pea flowers in loose racemes; Winter to Spring
**Requirements:** Well drained soils, tolerating dry periods once established
**Comments:** Juvenile leaflets may be rounded and the plant may be easily mistaken for *D. gunnii* at this stage. When regrowth takes place a terminal flower spike develops at the end of the new shoot first before lateral shoots develop.
**Localities:** Essendon, Kororoit Ck, 17#, 34A, 41, 42, 59
**Distribution:** Vic, Qld, NSW, SA

*Desmodium gunnii*

# Dichondra — Convolvulaceae

## Dichondra repens — Kidney Weed

**Size:** Prostrate creeping herb — P N
**Habitat:** Widespread in plains grassland, red gum and coastal banksia woodland, riparian scrub and woodland, dry and valley sclerophyll forests and grassy low open forest
**Form:** Dense spreading herb rooting at nodes to form mats
**Foliage:** Hairy kidney-shaped leaves to 40 mm long
**Flowers:** Tiny greenish flowers; September to December
**Requirements:** Well drained soils
**Comments:** Alternative to grass where foot traffic is light. It grows more vigorously under cultivation.
**Propagation:** Seed, division of suckers. Seed available from most commercial outlets is not the local form but is in fact the American species *D. micrantha*.
**Localities:** 2, 4–6, 8, 9, 12, 14, 16–18, 21, 23 32, 34A–48, 50A–52, 54–71, 74–76
**Distribution:** All states

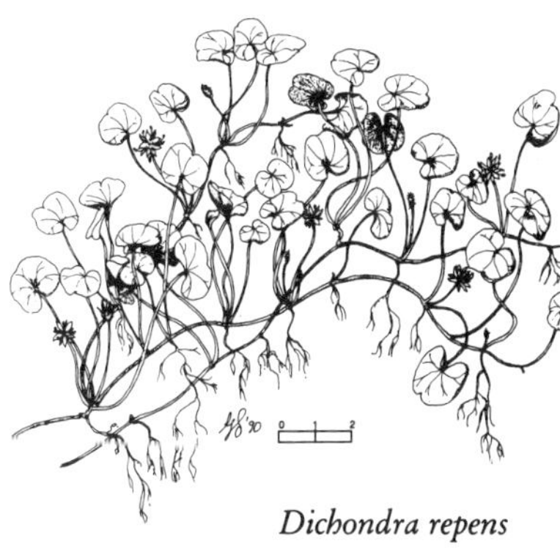

*Dichondra repens*

# Dillwynia — Fabaceae

The dillwynias are most attractive small shrubs occurring in dry sclerophyll forests or in heathlands. They are tolerant of a wide range of well drained soils. A display of showy yellow or yellow and red pea flowers occurs during Spring. They may be distinguished from the other genera with pea flowers by the groove in the upper surface of the leaf. Pruning after flowering is beneficial. Propagation is by scarified seed or cuttings.

*Dillwynia cinerascens*

## Dillwynia cinerascens — Grey Parrot Pea

**Size:** 0.6–1.5 m x 0.5–1.5 m — P
**Habitat:** Tea-tree heath, dry and valley sclerophyll forests, plains grassland
**Form:** Open, erect or spreading understorey shrub
**Foliage:** Terete greyish-green leaves to 20 mm long
**Flowers:** Terminal clusters of yellow and orange pea flowers; July to November
**Requirements:** Dry soils
**Comments:** An adaptable shrub for a shady situation.
**Localities:** Sydenham, Oaklands, Tottenham, Laverton Nth, Sunshine, 12, 15, 16, 26, 33–36, 38–44, 46–56, 58, 62–64, 67–69, 71, 73–76
**Distribution:** Vic, Tas, SA, WA

SCALE: 0 1cm 2cm — The scale representation on each plant equals 2cm.

## Dillwynia glaberrima

Heath or Smooth Parrot Pea

**F P**

**Size:** 1–2 m x 1–2 m
**Habitat:** Valley sclerophyll forest, tea-tree heath, grassy low open forest
**Form:** Wiry, open shrub
**Foliage:** Glabrous terete green leaves to 25 mm long
**Flowers:** Terminal racemes of bright yellow and red flowers; August to December.
**Comments:** Useful for planting under established trees.
**Localities:** 37, 42, 44, 47, 49, ?50A, 59, 60, 61, 67–71, 73–76
**Distribution:** Vic, Qld, NSW, Tas, SA

*Dillwynia glaberrima*

## Dillwynia hispida

Red Parrot Pea

**P**

**Size:** 0.2–0.6 m x 0.3–1 m
**Habitat:** Tea-tree heath
**Form:** Slender erect shrub
**Foliage:** Crowded terete leaves to 10 mm long with stiff spreading hairs
**Flowers:** Well displayed large red to orange flowers in terminal heads; September to November
**Requirements:** Prefers sandy, moist, well drained situations
**Comments:** Very showy plant in containers and hanging baskets but difficult to maintain in the garden.
**Localities:** The only occurrence of this plant within the Melbourne area is at Sandringham. This may be an incorrect collection.
**Distribution:** Vic, NSW, SA

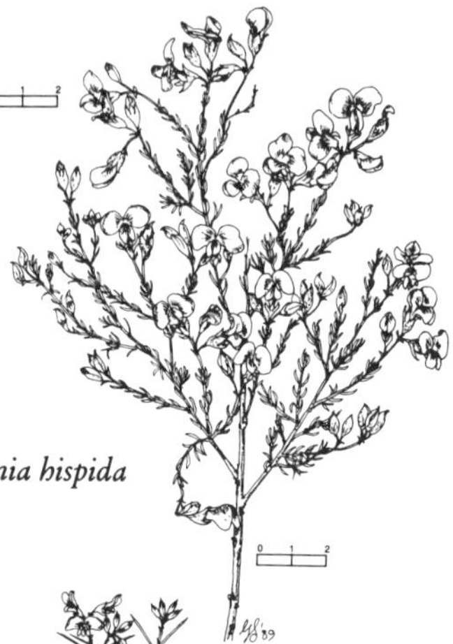

*Dillwynia hispida*

## Dillwynia juniperina

Prickly Parrot Pea or Juniper Pea Bush

**P**

**Size:** 1–1.5 m x 1–1.5 m
**Habitat:** Red gum woodland
**Form:** Prickly shrub
**Foliage:** Linear leaves to 15 mm long, providing excellent bird refuge
**Flowers:** Yellow and red flowers clustered near the ends of branches; September to November
**Requirements:** Very adaptable in well drained dry situations
**Comments:** Known from very few sites in Melbourne.
**Localities:** 29, 35
**Distribution:** Vic, Qld, NSW

*Dillwynia juniperina*

## Dillwynia phylicoides

Small-leaf Parrot Pea

**P**

**Size:** 0.5–1.5 m x 1–2 m
**Habitat:** Dry sclerophyll forest
**Form:** Spreading to erect shrub
**Foliage:** Narrow twisted leaves to 10 mm long
**Flowers:** Profuse terminal clusters of yellow and red flowers; September to December
**Requirements:** Well drained soil
**Comments:** A very hardy and ornamental species that deserves to be more widely grown. Responds very well to hard pruning.
**Localities:** Hurstbridge, 36, 41, 47#
**Distribution:** Vic, Qld, NSW
**Synonym:** *D. retorta* var. *phylicoides*

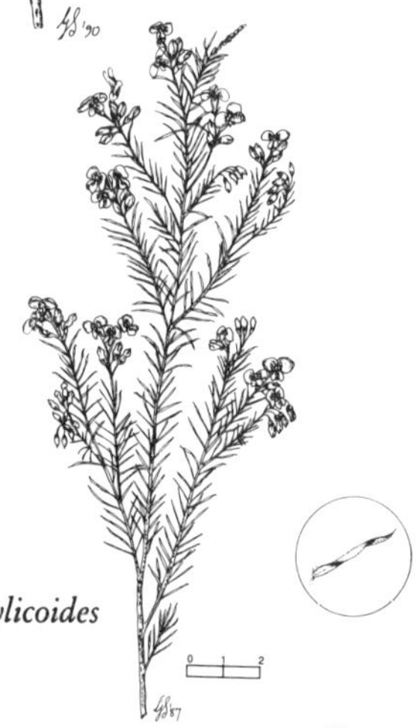

*Dillwynia phylicoides*

## Dillwynia sericea

Showy Parrot Pea

**F P**

**Size:** 0.6–1.5 m x 0.5–1.5 m
**Habitat:** Dry sclerophyll forest, sclerophyll woodland, tea-tree heath
**Form:** Small erect shrub
**Foliage:** Narrow hairy leaves to 20 mm long
**Flowers:** Cylindrical leafy spikes of yellow and red, apricot, or orange flowers; August to December
**Requirements:** Very adaptable, tolerating extended dry periods once established
**Comments:** Useful as a low screen or container plant but requires regular pruning after flowering.
**Localities:** 16, 20#, 26, 36, 42, 44, 46, 47, 49#, 50A, 67#, 71, 73–76
**Distribution:** Vic, Qld, NSW, Tas, SA

*Illustrated p. 106*

# Disphyma

**Aizoaceae**

## Disphyma crassifolium ssp. clavellatum

Rounded Moon-flower

**F**

**Size:** Prostrate x 1–2 m
**Habitat:** Coastal banksia woodland, saltmarshes
**Form:** Spreading perennial herb
**Foliage:** Succulent, cylindrical leaves to 50 mm long
**Flowers:** Daisy-like pink or magenta flowers with white centre, to 50 mm wide; October to February
**Requirements:** Very reliable in most soils. Tolerates saline conditions.
**Comments:** Useful for soil erosion on embankments. The attractive foliage varies from green through to purple and was eaten raw by the Aborigines.
**Propagation:** Stem or leaf cuttings, or seed
**Localities:** Footscray, 1–4, 32, 67, 69A, 72, 74
**Distribution:** Vic, Qld, NSW, Tas, SA
**Synonym:** *D. clavellatum*, *D. australe*

*Disphyma crassifolium* ssp. *clavellatum*

# Dodonaea — Sapindaceae

## Dodonaea viscosa ssp. spatulata — Wedge-Leaf Hop Bush

**Size:** 1-3 m x 1-3 m — F P
**Habitat:** Dry sclerophyll forest, riparian scrub, box woodland, grassy low open forest
**Form:** Open to dense spreading shrub
**Foliage:** Obovate to spathulate leaves to 75 mm long
**Flowers:** Inconspicuous unisexual flowers in clusters; August to November; followed by showy red or blackish-brown capsules with 3–4 papery wings
**Requirements:** Well drained soils
**Comments:** A very hardy shrub tolerating extended dry periods once established and moderate coastal exposure. Pruning is beneficial. Plants in the Melbourne area have been incorrectly referred to as *D. cuneata*. The typical form *D. viscosa* ssp. *viscosa* does not occur in the study area.
**Propagation:** Seed, cuttings—especially to perpetuate forms with attractively coloured capsules
**Localities:** Keilor, Greensborough, Research, 1, 9, 14, 16, 18, 22, 23, 26, 29, 31, 32, 35, 36, 74, 76
**Distribution:** All states except NT
**Synonym:** *D. viscosa* ssp. *cuneata* — *Also illustrated p. 106*

*Dodonaea viscosa* ssp. *spatulata*

# Drosera — Droseraceae

Droseras are small carnivorous perennial herbs. Known as Sundews, they trap insects on special sticky hairs on their leaves. All droseras require damp soil for at least part of the year. The main growing season is from Autumn to Spring, usually with a dormant period during Summer when plants die back to their roots. Most make excellent container plants, preferring a damp peaty soil and plenty of light. Propagation is by seed sown in Spring, or by leaf cutting.

## Drosera binata — Forked Sundew

**Size:** 0.2–0.8 m high — F P
**Habitat:** Tea-tree heath, valley sclerophyll forest
**Form:** Erect perennial herb
**Foliage:** Narrow pale green to reddish forked leaves from 5–60 cm long
**Flowers:** Flowering stem is up to 80 cm long, with many white flowers, each 25 mm across. The main flowering is from September to April.
**Requirements:** Needs continuously damp peaty soil
**Comments:** Excellent pot or basket plant. Use peat and sand, or sphagnum moss, in a shallow container.
**Propagation:** Easy from seed, leaf or stem cutting
**Localities:** Moorabbin#, 42, 67#, 74, 75
**Distribution:** All states

*Drosera binata*

## Drosera glanduligera — Scarlet Sundew

**Size:** Basal rosette to 3 cm high — F P
**Habitat:** Swamps and moist banks of red gum woodland and tea-tree heath
**Form:** Tiny rosette
**Foliage:** Elliptical leaves to 15 mm long
**Flowers:** Terminal clusters of orange flowers fading to pink; August to October
**Requirements:** Moist acidic, peaty soil
**Comments:** Usually grows in dense colonies.
**Localities:** 12, 26, 35, 45, 67#, 74, 75
**Distribution:** Vic, NSW, Tas, SA, WA

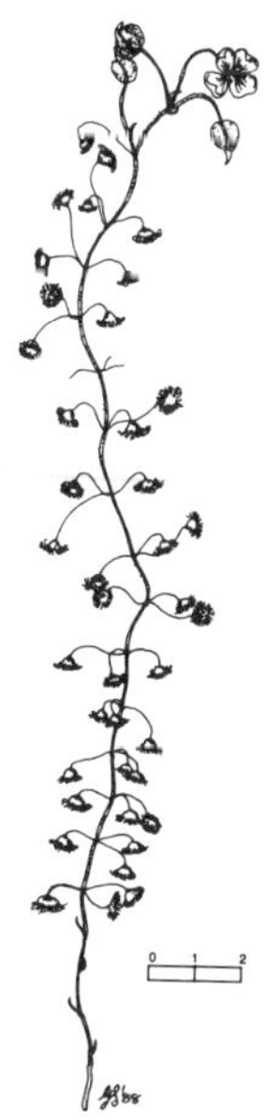

*Drosera macrantha*

## Drosera macrantha — Bridal Rainbow or Climbing Sundew

**Size:** 0.4–0.8 m high — F P
**Habitat:** Damp and valley sclerophyll forest, tea-tree heath, swamp scrub, grassy low open forest, sclerophyll woodland
**Form:** Weakly climbing perennial herb with a tuberous rootstock
**Foliage:** Rounded, cupped, yellowish-green leaves to 90 mm long
**Flowers:** Terminal clusters of white or pink sweetly perfumed flowers; June to October
**Requirements:** Sandy loam or clay soils which are moist during the growing period but dry out in Summer
**Propagation:** Seed
**Localities:** 29, 30, 32, 34A, 41, 42, 49, 53, 58, 67, 69, 70, 71, 73–76
**Distribution:** Vic, Tas, SA, WA
**Synonym:** *D. macrantha* ssp. *planchonii*, *D. planchonii*

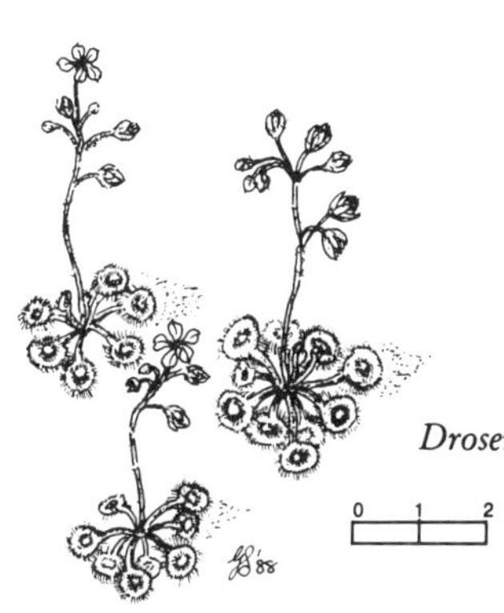

*Drosera glanduligera*

SCALE: 0 1cm 2cm — The scale representation on each plant equals 2cm.

*Drosera peltata* ssp. *auriculata*

## Drosera peltata ssp. auriculata

Tall Sundew

**Size:** 0.1–0.8 m high **F P**
**Habitat:** Widespread in wattle tea-tree scrub, tea-tree heath, grassy low open forest, dry and valley sclerophyll forest
**Form:** Upright perennial herb with a tuberous root
**Foliage:** Sparsely foliaged with few dark green spathulate leaves to 12 mm long
**Flowers:** Loose terminal cluster of white to pale pink flowers; August to December
**Requirements:** Adaptable to most soils which are damp during growth periods and dry out in Summer
**Comments:** The size of the seeds is the most reliable distinguishing feature between the 2 subspecies of *D. peltata*. They are linear, to about 1 mm long in ssp. *auriculata*.
**Propagation:** Seed
**Localities:** 29#, 32–36, 38, 41–51, 53–56, 58–64, 67–71, 73–76
**Distribution:** Vic, Qld, NSW, Tas, SA
**Synonym:** *D. auriculata*

*Drosera peltata* ssp. *peltata*

## Drosera peltata ssp. peltata

Pale Sundew

**Size:** 0.1–0.5 m high **F P**
**Habitat:** Plains grassland, red gum woodland, dry and valley sclerophyll forests, tea-tree heath
**Comments:** This subspecies differs from *D. peltata* ssp. *auriculata* in that it usually has a basal rosette of pale green leaves and usually has hairy rather than smooth buds. Seeds are ovoid and less than 0.6 mm long.
**Localities:** 7, 12, 13, 16, 25, 26, 29–31, 34A–36, 42, 44, 47, 49–52, 56, 58, 69A–71, 74–76
**Distribution:** All states except NT

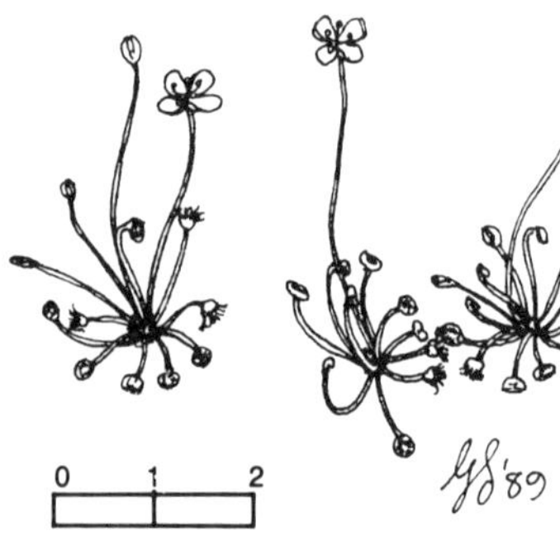

*Drosera pygmaea*

## Drosera pygmaea

Pygmy, or Tiny Sundew

**Size:** 2 cm high **F**
**Habitat:** Grassy low open forest, swamp scrub
**Form:** Tiny basal rosette
**Foliage:** Bronze round leaves to 2 mm diameter on a fine stalk to 8 mm long
**Flowers:** Single white flower 4 mm wide on an erect stalk to 20 mm; June to January
**Requirements:** Moist, sandy soil
**Propagation:** May be propagated from vegative propagules (minute bud-like, lens-shaped structures) which form in the centre of the rosette.
**Localities:** Heathmont, The Basin, Rowville, Baxter, 54, 61, 67#, 69A, 73–76
**Distribution:** All states

*Drosera spathulata*

## Drosera spatulata

Spoon-leaf or Rosy Sundew

**Size:** 5–20 cm x 2 cm **P**
**Habitat:** Tea-tree heath and swamp scrub
**Form:** Tuberous basal rosette
**Foliage:** Crowded spathulate reddish leaves to 10 mm long
**Flowers:** Small white, pink or red flowers on an erect stem to 200 mm high, throughout the year
**Comments:** An attractive container plant. Endangered locally.
**Localities:** Cheltenham#, 67#, 74, 75
**Distribution:** Vic, Qld, NSW, Tas

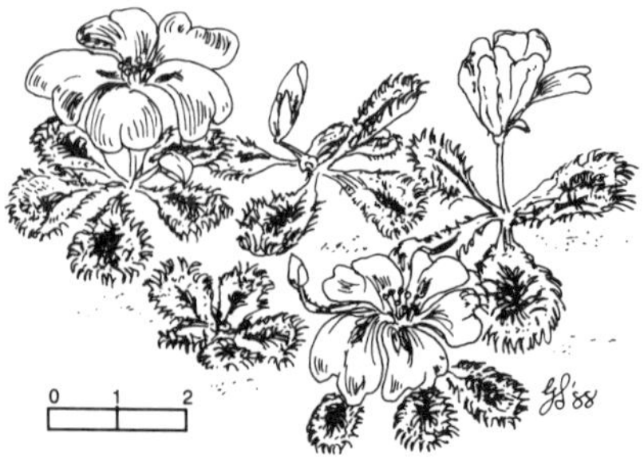

*Drosera whittakeri*

## Drosera whittakeri

Scented Sundew

**Size:** 2–4 cm x 2–4 cm **F P**
**Habitat:** Widespread in tea-tree heath, grassy low open forest, red gum woodland, plains grassland, dry and valley sclerophyll forest
**Form:** Tuberous, round flat basal rosette
**Foliage:** Green, bronze or red spathulate leaves to 25 mm long
**Flowers:** Single large, perfumed white flower to 30 mm on a thick, erect stalk to 40 mm; July to October
**Requirements:** Prefers acidic peaty soil which is moist during the growing season
**Comments:** May form large colonies in moist areas. An excellent container plant.
**Localities:** 12, 16, 25, 26, 29#, 30#, 34A–36, 38, 39, 41–44, 46–51, 53–56, 58–64, 67#, 70, 71, 73–76
**Distribution:** Vic, SA

# Eclipta **Asteraceae**

## Eclipta platyglossa

Yellow Twin-heads

**Size:** 15–25 cm high **F P**
**Habitat:** Grassy wetlands
**Form:** Hairy annual or biennial herb
**Foliage:** Pairs of lanceolate leaves to 50 mm long
**Flowers:** Tiny yellow daisy flowers with bi-lobed rays; late Spring and Summer
**Requirements:** Tolerates heavy, wet soil
**Comments:** Suitable for a bog garden. Very rare in the Melbourne area and in danger of extinction.
**Propagation:** Seed, cuttings
**Localities:** Werribee#, Derrimut on private property
**Distribution:** All states except Tas

*Einadia hastata*

# Einadia **Chenopodiaceae**

## Einadia hastata

Saloop, Berry Saltbush

**Size:** 10–20 cm x 20–50 cm **F P**
**Habitat:** Riparian and chenopod rocky open scrub, riparian, box and red gum woodland, primary dune scrub, tea-tree heath
**Form:** Many-branched rigid procumbent undershrub
**Foliage:** Green hastate leaves to 20 mm long
**Flowers:** Clusters of insignificant cream flowers followed by small succulent red berries
**Requirements:** Well drained soils, tolerating dryness
**Comments:** Regular light pruning promotes dense bushy growth. Saltbushes are useful in revegetating salt affected areas.
**Propagation:** Cuttings
**Localities:** West Melton, 9, 14, 16, 26, 29, 32, 34A–36, 67#, 69
**Distribution:** Vic, Qld, NSW
**Synonym:** *Rhagodia hastata*

*Einadia nutans* ssp. *nutans*

## Einadia nutans ssp. nutans

Nodding or Climbing Saltbush

**Size:** 30 cm x 1.2 m **F P**
**Habitat:** Dry sclerophyll forest, plains grassland, riparian and chenopod rocky open scrub, riparian, red gum and box woodland, primary dune scrub
**Form:** Sprawling or scrambling shrub
**Foliage:** Broadly lanceolate grey-green leaves to 30 mm long
**Flowers:** Insignificant clusters of greenish flowers followed by succulent red berries in Summer and Autumn
**Requirements:** Tolerates dry soils
**Comments:** A useful groundcover for dry banks and rockeries, drought resistant once established. It is also a valuable fire retardant.
**Propagation:** Cuttings
**Localities:** 2, 4, 8, 9, 12, 14–16, 18, 21–26, 28–36, 38, 39, 42, 49, 65#, 67#, 69A, 71
**Distribution:** All states except WA
**Synonym:** *Rhagodia nutans*

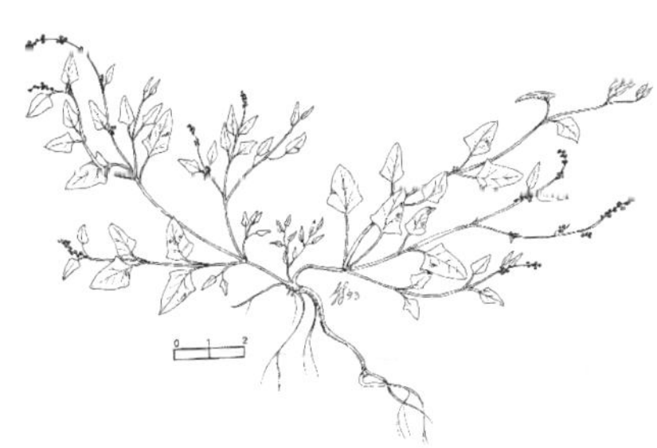

*Einadia trigonos* ssp. *trigonos*

## Einadia trigonos ssp. trigonos

Lax or Fishweed Goosefoot

**Size:** Prostrate or procumbent **P**
**Habitat:** Red gum woodland
**Form:** Low straggling herbaceous perennial
**Foliage:** Greyish-green arrow-shaped leaves to 15 mm long
**Flowers:** Insignificant flowers followed by dry black berries
**Requirements:** Moist well drained soil
**Comments:** A rare plant for Melbourne, known from few sites.
**Propagation:** Cuttings
**Localities:** 32
**Distribution:** Vic, NSW, Qld
**Synonym:** *Chenopodium trigonon*

# Elatine **Elatinaceae**

## Elatine gratioloides

Waterwort

**Size:** Prostrate x 0.3–2 m **F**
**Habitat:** Aquatic, in still water or muddy sites, of red gum and box woodland, grassy wetland and plains grassland
**Form:** Fast growing annual herb
**Foliage:** Bright green ovate to elliptic leaves 2–20 mm x 1–4 mm
**Flowers:** Solitary pinkish-green flowers in leaf axils most of the year
**Requirements:** Shallow water or boggy site subject to inundation
**Comments:** Makes a good aquarium or bog garden plant, regenerating readily in boggy conditions. Locally rare.
**Propagation:** Cuttings, seed sown in mud
**Localities:** Deer Park, Rockbank, 2, 6, 7, 17#, 18, 24, 26
**Distribution:** All states

*Elatine gratioloides*

SCALE: 0 1cm 2cm The scale representation on each plant equals 2cm.

*Enchylaena tomentosa*

# Enchylaena — Chenopodiaceae

## Enchylaena tomentosa — Barrier or Ruby Saltbush

**Size:** Prostrate–1 m x 0.5–1 m — F P
**Habitat:** Riparian and chenopod rocky open scrub, red gum and box woodland, plains grassland, primary dune scrub
**Form:** Low spreading or upright woody shrub
**Foliage:** Bluish-green to dark green succulent leaves, narrowly cylindrical to 10 mm long or linear with margins recurved, to 20 mm long
**Flowers:** Insignificant greenish flowers in Spring and early Summer, followed by succulent green berries which change to yellow or red.
**Requirements:** Very adaptable tolerating poor soils, dryness and some salinity
**Comments:** An attractive plant, useful as an undershrub in dryland plantings, and suitable for containers. Pruning is beneficial. There are two forms within the Melbourne area, an upright form with smaller and paler terete leaves and red berries favouring drier areas and a prostrate form with darker linear leaves and yellower berries, preferring alluvial soils. The Aborigines ate the sweet berries and used the leaves as a green vegetable.
**Propagation:** Cuttings, seed
**Localities:** Avondale Heights, 1, 2, 4, 12, 14–16, 21–26, 28, 29, 32, 34A, 67#, 74
**Distribution:** All mainland states

# Epacris — Epacridaceae

The 3 *Epacris* species of Melbourne commonly occur in heathlands and as understorey shrubs. They are upright plants with tubular flowers borne on the upper branches from late Autumn to Spring. All make good tub specimens, especially if massed. Often spindly, they respond to hard pruning after flowering. Root protection is necessary if grown in full sun. Propagation is by seed or cuttings which strike well provided young growth is used.

*Epacris impressa* 'Cranbourne Bells'

*Epacris gunnii*

## Epacris gunnii

**Size:** 0.5–1 m x 0.5–1 m — P
**Habitat:** Swamp scrub, valley sclerophyll forest
**Form:** Straggly, upright shrub
**Foliage:** Crowded, prickly, heart-shaped leaves to 6 mm long
**Flowers:** Very small white flowers sometimes tinged with pink; April to October
**Requirements:** Moist well drained soils which should not dry out in Summer
**Comments:** A most attractive rockery plant, also suitable for containers. It is frost and snow tolerant. Locally rare through clearing.
**Localities:** Montrose, Heathmont, Knox#
**Distribution:** Vic, Tas
**Synonym:** *E. microphylla* var. *gunnii*

## Epacris impressa — Common Heath
## Victoria's floral emblem

**Size:** 0.5–1.5 m x 0.2–0.6 m — P
**Habitat:** Widespread in tea-tree heath, wattle tea-tree scrub, low grassy open forest, sclerophyll woodland, dry, damp and valley sclerophyll forests
**Form:** Open, wiry shrub
**Foliage:** Prickly, tapering leaves to 15 mm long
**Flowers:** Masses of white, pink or red flowers; March to November
**Requirements:** Moist, well drained soil, tolerating limited dry or wet periods once established
**Comments:** An attractive rockery plant, particularly when planted in groups. Makes an excellent cut flower. A form from Cranbourne, with double white flowers, is sold as 'Cranbourne Bells'. Its only known habitat has now been destroyed.
**Localities:** St. Kilda#, Eltham, 29, 33–35, 40–44, 46–56, 58–64, 67–71, 73–76
**Distribution:** Vic, NSW, Tas, SA

*Epacris impressa*

*Epacris obtusifolia*

## Epacris obtusifolia — Blunt-leaf Heath

**Size:** 1–2 m x 0.6–1.5 m — P
**Habitat:** Moist tea-tree heath
**Form:** Narrow erect shrub
**Foliage:** Erect, blunt-tipped oblong-elliptical leaves from 5–12 mm long
**Flowers:** Masses of small white to cream perfumed flowers; June to December
**Requirements:** Moist sandy or peaty soil
**Comments:** Locally rare.
**Localities:** 66#, 67#, 73–75
**Distribution:** Vic, Qld, NSW, Tas, SA

# Epilobium — Onagraceae

## Epilobium billardierianum ssp. billardierianum

Smooth Willow Herb

*Epilobium billardierianum* ssp. *billardierianum*

**Size:** 0.1–1 m, variable depending on moisture — F P
**Habitat:** Widespread in red gum woodland, riparian and swamp scrub, valley and dry sclerophyll forest, wattle tea-tree scrub
**Form:** Erect hairy perennial herb with sprawling base and leafy stems
**Foliage:** Pairs of blunt, bluish-green ovate stem-clasping leaves to 35 mm x 6–18 mm, densely toothed with up to 40 teeth per side. Lower leaves often red.
**Flowers:** Single lilac, pink or white flowers to 15 mm long, along upper stems; December to March. Densely covered in tiny flattened hairs. Stalk lengthens to 20 mm in fruit. Capsule to 70 mm long.
**Requirements:** Will grow almost anywhere
**Comments:** The seed capsules of *Epilobium* spp. open to reveal small seeds densely covered in white hairs. May become invasive under ideal conditions.
**Propagation:** Seed
**Localities:** 2, 4–6, 11, 12, 18, 21, 26, 29#, 34A–36, 39, 41, 42, 44, 51, 52, 54, 55, 62, 63, 67, 72, 74, 75
**Distribution:** Vic, NSW, Tas, SA, WA; NZ

## Epilobium billardierianum ssp. cinereum

*Epilobium billardierianum* ssp. *cinereum*

**Size:** to 0.3 m high
**Habitat:** Damp and valley sclerophyll forests, swamp scrub
**Comments:** Leaves are narrower, to 2–7 mm, than ssp. *billardierianum*, with 0–6 coarser teeth per side. Flowers have erect as well as flattened hairs. Grows well in wet conditions.
**Localities:** 16, 53, 56, 57, 58, 60, 62, 63, 70, 74, 76
**Distribution:** All states; NZ
**Synonym:** *E. cinereum*

## Epilobium billardierianum ssp. intermedium

*Epilobium billardierianum* ssp. *intermedium*

**Comments:** Differs from the above ssp. in that lanceolate leaves have 6–20 small irregular teeth per side and margins are often wavy. Stems are often branched and have glandular and erect hairs towards the end, near the inflorescence.
**Localities:** Montrose
**Distribution:** Vic, NSW, Tas, SA, WA

## Epilobium hirtigerum

Narrow Leaf or Hoary Willow Herb

*Epilobium hirtigerum*

**Size:** 0.2–1.4 m high — F P
**Habitat:** Widespread in plains grassland, red gum and sclerophyll woodland, riparian scrub, valley sclerophyll forest, grassy low open forest
**Form:** Variable, erect perennial herb with spreading hairs
**Foliage:** Narrow, pointed sessile leaves to 60 mm long, usually alternate, with widely-spaced teeth. New growth is covered in white hairs.
**Flowers:** Single small pink or white flowers to 7 mm long; November to March. Covered in spreading hairs. Capsule to 60 mm long on stalk lengthened to 15 mm.
**Requirements:** Moist soils
**Propagation:** Seeds
**Localities:** 4, 12, 16–18, 26, 28, 34A–36, 41, 42, 53, 54, 64, 69A, 70, 74, 76
**Distribution:** All states except NT; NZ, Java, S Am

# Eremophila — Myoporaceae

## Eremophila deserti

Turkey Bush

*Eremophila deserti*

**Size:** 1–4 m x 2 m — F
**Habitat:** Box woodland
**Form:** Erect much-branched shrub
**Foliage:** Thick narrowly oblanceolate leaves to 50 mm long, with hooked tip
**Flowers:** Drooping cream bell-shaped flowers in groups of 1–4; May to October
**Requirements:** Well drained stony or sandy soils
**Comments:** The yellow drupes following flowering are eaten by emus and turkeys but are poisonous to stock. It is now very rare in the Melbourne area reproducing successfully in two sites.
**Propagation:** Cuttings
**Localities:** Spasmodic on the basalt plains, remaining in Keilor, Essendon and Melton, 16, 22, 26
**Distribution:** Vic, Qld, NSW, SA, WA
**Synonym:** *Myoporum deserti*

SCALE: 0 1cm 2cm — The scale representation on each plant equals 2cm.

*Erodium crinitum*

# Erodium — Geraniaceae

## Erodium crinitum — Blue Heron's Bill, Native Crowfoot

**Size:** 0.1–0.8 m x 0.3–1 m — F P
**Habitat:** Coastal banksia woodland, tea-tree heath, plains grassland
**Form:** Fast growing, open, spreading annual or biennial herb
**Foliage:** Deeply divided dark green leaves from 15–40 mm long
**Flowers:** Clusters of 2 to 6 blue flowers with yellow or white veins near the base; July to October
**Requirements:** Tolerates extended dry periods in sandy soil
**Comments:** It is a useful container plant, with interesting fruit. The roots are edible. Locally rare.
**Propagation:** Seed or cuttings
**Localities:** 2, 12, 16, 21, 26, 34A, 49, 67#, 71
**Distribution:** All states

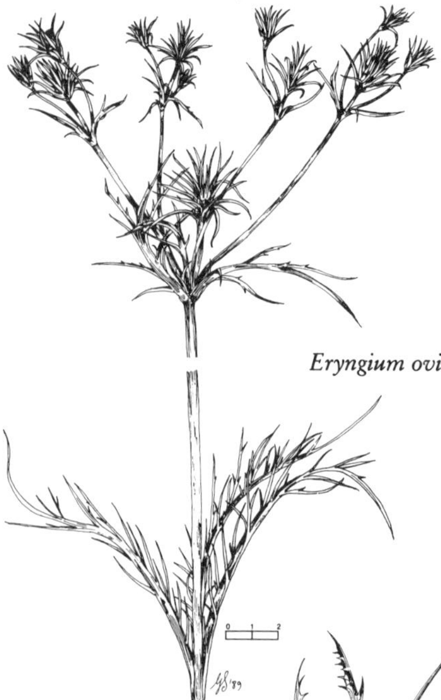

*Eryngium ovinum*

# Eryngium — Apiaceae

## Eryngium ovinum — Blue Devil, Eryngo

**Size:** 0.1–0.6 m x 0.3–0.5 m — F
**Habitat:** Plains grassland, red gum woodland
**Form:** Fast growing, stiffly erect perennial herb with ribbed stems
**Foliage:** Prickly; basal leaves finely divided from 10–25 cm long, stem leaves smaller.
**Flowers:** Globular to ovoid flowerheads to 25 mm long with rigid sharply-pointed metallic blue to purplish bracts; August to February
**Requirements:** Moist well drained soils
**Comments:** A very decorative but extremely prickly plant for a container or rockery.
**Propagation:** Seed
**Localities:** Melton, Tottenham, St Albans, Hampton#, 2, 4, 5, 7, 10, 12, 13, 15–18, 20, 24–26, 28, 29, 35, 67#
**Distribution:** All states; S Am
**Synonym:** *E. rostratum*

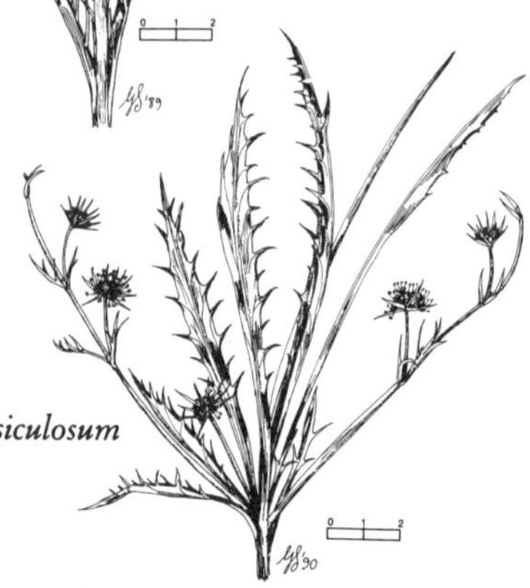

*Eryngium vesiculosum*

## Eryngium vesiculosum — Prickfoot

**Size:** 10–20 cm x 0.3–0.6 m — F P
**Habitat:** Swamp scrub, grassy wetlands, valley sclerophyll forest
**Form:** Tufted perennial herb
**Foliage:** Bright green oblanceolate leaves with sharply toothed edges, from 5–20 cm long; stem leaves shorter and lobed
**Flowers:** Small globular pale blue flower heads; October to February
**Requirements:** Moist soil, withstands wet conditions
**Comments:** It is an interesting plant for a bog garden or container, the base of which should be immersed in water.
**Propagation:** Root cuttings, seeds
**Localities:** Bundoora, Moorabbin, Carrum, 2, 4, 6, 7, 17#, 18, 28, 29, 35, 41, 42, 44, 51, 54, 67#, 71
**Distribution:** All states except NT; NZ

# Eucalyptus — Myrtaceae

Twenty-eight *Eucalyptus* species occur within the Melbourne area. They are found in most habitats covering a wide variety of soil types. All the local species have white to cream flowers. Their bark, foliage, flower size and number, and bud and capsule shape provide identifying features. The paired leaf arrangement typical of eucalypt seedlings contrasts with the alternate arrangement of adult leaves. Seedling and juvenile leaves may differ from adult leaves in a number of other features, including, size, shape, colour, and presence or absence of surface wax or hairs. Propagation is by seed collected from mature capsules.

Bark type is also a useful aid to identifying most of the indigenous eucalypts. The most common type belongs to the 'Gums'. The bark on most of the trunk and on the branches is smooth and is shed annually leaving a cream or grey trunk. Some gums retain old, darker bark around the base, sometimes extending up the tree. 'Stringybarks' are usually tall and straight and are easily identified by the thick, widely furrowed dead bark which is retained on the trunk and branches, weathering to grey or grey-brown. The long fibres can be pulled off in strands. The adult leaves of most stringybarks are asymmetrical and their intermediate leaves are noted for their larger size and variety of shapes. The outer bark of the 'Boxes' is persistent on the trunk, short-fibred, thin and flaky or scaly. The wood is close-grained providing excellent fuel. Flowers are in clusters forming a panicle. The term 'Peppermint' alludes to the strong peppermint smell of the leaves of 2 of the Melbourne species when crushed. The bark is grey and persistent, the fibres closely interlaced. There is only one 'Ironbark' in Melbourne. Its bark is hard, thick and deeply furrowed.

The Aborigines used eucalypts extensively for many products. Food was obtained from the seed, nectar and manna of many species. Many medicinal concoctions were derived from the leaves and gum. Bark provided containers, canoes and fibre for string. Many implements were made from the wood.

Although most local eucalypts may obtain their maximum size under ideal conditions it is most unlikely that these dimensions would be reached in cultivation.

## Eucalyptus albens

White Box

**Size:** 8–25 m x 10–15 m — F P
**Habitat:** Box woodland
**Form:** Medium to tall upright tree with dense spreading canopy
**Foliage:** Juvenile—glaucous ovate to rounded leaves 50–150 mm x 50–115 mm
Adult—Grey to bluish-green broadly lanceolate leaves to 15 cm x 30 mm, densely veined
**Flowers:** Terminal clusters of 3–7 sessile, long (to 18 mm), tapered glaucous buds, often ridged. Very showy creamy flowers; February to June. Glaucous barrel-shaped fruits have sunken valves.
**Requirements:** Well drained drier soils
**Comments:** An attractive fast growing tree with whitish-grey bark and hard, durable timber. Useful as a shade and windbreak tree, providing excellent honey.
**Localities:** The only occurrence within the Melbourne area is in Melton South.
**Distribution:** Vic, Qld, NSW, SA

*Eucalyptus albens*

## Eucalyptus aromaphloia = Eucalyptus ignorabilis

## Eucalyptus baueriana

Blue Box, Round-leaf Box

**Size:** 12–30 m x 8–15 m — F
**Habitat:** Riparian and box woodland
**Form:** Medium tree with dense crown
**Foliage:** Juvenile—thin, round pale to grey-green leaves 60–70 mm x 50–120 mm
Adult—thin, pale green ovate to broad-lanceolate leaves to 90 mm x 60 mm
**Flowers:** Clusters of 7 club-to diamond-shaped buds followed by profuse terminal bunches of white flowers October to January. Fruit are funnel-shaped.
**Requirements:** Well drained moist loams
**Comments:** An excellent tree for larger gardens, providing shade.
**Localities:** Mt Cottrell, 8
**Distribution:** Vic, NSW

*Eucalyptus baueriana*

## Eucalyptus baxteri

Brown Stringybark

**Size:** 3–40 m x 4–20 m — F P
**Habitat:** Damp and valley sclerophyll forests of foothills
**Form:** Variable, medium-sized tree with a rounded crown locally, smaller in exposed areas
**Foliage:** Juvenile—opposite and ovate at first, paler green and hairy below, 30–70 mm x 20–50 mm, with wavy edges
Intermediate—larger to 130 mm x 80 mm
Adult—thick, glossy green broad-lanceolate leaves, 15 cm x 30 mm
**Flowers:** Clusters of 7–15 club-shaped and warty buds on a thick stalk. Profuse white flowers December to April. Large, round fruit with the rim near the top.
**Requirements:** Well drained soils
**Comments:** A good shelter and shade tree. Timber is used for buildings, poles and fuel. The size of this tree is determined by the quality of the soil on which it is grown.
**Localities:** Montrose, 34A, 42, 43, 47, 51, 54, 55, 56, 61
**Distribution:** Vic, NSW, SA — *Illustrated p. 106*

## Eucalyptus behriana

Bull Mallee

**Comments:** After further research for the second edition it is improbable that *E. behriana* occurred within the Melbourne area. Its closest distribution is in the Bacchus Marsh/Long Forest area.

## Eucalyptus blakelyi

Blakely's Red Gum

**Size:** 15–24 m x 10–15 m — F P
**Habitat:** Box-ironbark woodland
**Form:** Medium-sized tree with a short trunk and a large dense crown
**Foliage:** Juvenile—alternate, thick, ovate to nearly round grey-green leaves to 140 mm long x 100 mm
Adult—dull green, lanceolate, falcate leaves to 16 cm x 20 mm
**Flowers:** Clusters of 5–10 yellowish-green buds, caps elongated and conical. White flowers; August to January. Small ovoid fruit with a small rim and 3–4 strongly projecting valves.
**Requirements:** Heavy, moist soils
**Comments:** The grey bark sheds in large sheets or flakes to leave a smooth surface with light patches. Found in a very localised area within Melbourne. Its durable red timber is useful for fence posts and firewood. It is also a valuable honey tree. Growing from a lignotuber it responds well to coppicing or hard pruning. Relatively slow growing, *E. blakelyi* makes a good shade and windbreak tree suited to large suburban gardens.
**Localities:** Yarrambat, Diamond Ck, 36
**Distribution:** Vic, Qld, NSW

*Eucalyptus blakelyi*

SCALE: 0 1cm 2cm — The scale representation on each plant equals 2cm.

*Eucalyptus camaldulensis*

## Eucalyptus camaldulensis

River Red Gum

**F**

**Size:** 12–50 m x 15–35 m

**Habitat:** Plains grassland, riparian scrub and woodland, red gum woodland

**Form:** Large open spreading tree

**Foliage:** Juvenile—bluish-green broad-lanceolate leaves to 26 cm x 80 mm Adult—dull narrow-lanceolate leaves to 25 cm x 20 mm. Forms from the Melbourne area generally have smaller foliage than those from inland areas.

**Flowers:** Long-stalked buds in clusters of 7–11, with caps contracting abruptly to a point. Profuse white flowers; November to March. Fruit is small with 4 strongly projecting valves.

**Requirements:** Damp alluvial soils, deep subsoils. Tolerates very dry periods and inundation once established.

**Comments:** A widespread eucalypt which has almost disappeared from the grassy woodlands of eastern Port Phillip Bay. Excellent for planting in parklands and broadacres but too large for the small suburban garden. It is very valuable as a honey tree and for its durable timber. Its bark is greyish with brown or reddish patches peeling or flaking during late Summer/Autumn. Selected forms are being planted to combat salinity in inland soils. An important species to the Aborigines, many trees are renowned as corroboree or canoe trees. Food, containers and clubs were other by-products while the gum treated burns and diarrhoea.

**Localities:** Melton, Wonga Park, Caulfield, Dandenong, Brighton, 4, 6, 8, 14, 16, 18, 19, 21, 23–36, 44, 45, 60, 61, 63, 65–69A, 71, 72, 74

**Distribution:** All mainland states

*Eucalyptus camphora*

## Eucalyptus camphora

Mountain Swamp Gum

**F P**

**Size:** 8–20 m x 5–12 m

**Habitat:** Valley sclerophyll forest

**Form:** Slow growing medium tree with short crooked trunk and rounded crown

**Foliage:** Juvenile—thick, dull to glossy green ovate to nearly round leaves to 80 mm x 40 mm, often notched at the apex; larger in intermediate stage Adult—thick, dark green elliptic to ovate leaves to 130 mm x 60 mm, densely veined.

**Flowers:** Clusters of 3–7 shiny diamond-shaped buds followed by white flowers; March to April. Funnel-shaped fruit are flat-topped with 3–4 small erectly projecting valves.

**Requirements:** Heavy wet soils, tolerating heavy frosts

**Comments:** Similar to *E. ovata* but has broader leaves and smaller fruit with exserted valves. Rough scaly bark persists at the base of the tree, peeling in long strips from most of the trunk and the branches. Useful as a shade tree in wet areas, and to control poor drainage. Locally rare.

**Localities:** Diamond Ck, Wonga Park

**Distribution:** Vic, Qld, NSW

## Eucalyptus cephalocarpa

Silver Leafed Stringbark, Mealy Stringybark

**F P**

**Size:** 8–20 m x 5–15 m

**Habitat:** Damp and valley sclerophyll forests, sclerophyll woodland, grassy low open forest, wattle tea-tree scrub

**Form:** Medium-sized straight or crooked tree with dense canopy

**Foliage:** Juvenile—sessile, round to ovate silver leaves to 110 mm x 60 mm Adult—grey-green lanceolate leaves 20 cm x 25 mm

**Flowers:** Silvery clusters of 7–11 stalkless, diamond-shaped buds. Profuse white to cream flowers March to August. Fruit is often flat topped and silvery.

**Requirements:** Most soils, withstanding periods of inundation.

**Comments:** The new foliage is an attractive feature while its juvenile foliage is used for wool dyeing and decoration. A valuable honey tree.

**Localities:** Montrose, 34, 42–51, 54, 56–64, 67#, 70, 73–76

**Distribution:** Endemic to Vic

**Synonym:** *E. cinerea* ssp. *cephalocarpa*

*Illustrated p. 106*

*Eucalyptus cypellocarpa*

## Eucalyptus cypellocarpa

Mountain Grey Gum

**F P**

**Size:** 10–65 m x 12–30 m

**Habitat:** Wet and damp sclerophyll forests of foothills

**Form:** Upright tree with dense canopy, but growth is stunted in drier soils

**Foliage:** Juvenile—broad, stalkless, 17 cm x 70 mm
Intermediate—very long to 35 cm
Adult—falcate or lanceolate to 20 cm x 25 mm

**Flowers:** Thick stalked elongated buds in clusters of 7. White flowers usually from December to July. Goblet-shaped fruit have enclosed valves.

**Requirements:** Prefers deep moist soil however it is adaptable to most conditions

**Comments:** This tree provides timber for the building industry, produces abundant honey and has potential for parkland and broadacre planting. The greyish bark sheds in strips in late Summer/Autumn to reveal yellow or whitish patches while that at the base may be rough and flaky. This species has been confused with *E. goniocalyx*.

**Localities:** Kalorama, 55, 56

**Distribution:** Vic, NSW

## Eucalyptus dives

Broad-leafed or Blue Peppermint

**Size:** 8–25 m x 6–15 m **F P**
**Habitat:** Valley sclerophyll forest, sclerophyll woodland, usually on poor soils
**Form:** Moderately dense low branching, spreading tree with large canopy
**Foliage:** Juvenile—broad, thick-textured blue to grey-green stem-clasping leaves, to 15 cm x 70 mm
Adult—aromatic, firm dull green broad-lanceolate leaves to 15 cm x 30 mm
**Flowers:** Profuse clusters of 7–15 club-shaped buds followed by white flowers September to December. Pear-shaped fruit have flat rims.
**Requirements:** Adaptable to all but wet soils, tolerating dry and stony conditions
**Comments:** Trees provide shade and windbreaks. If coppiced, new growth is useful as cut foliage.
**Localities:** 33, 44, 45, 47, 56, 60, 73
**Distribution:** Vic, NSW

*Eucalyptus dives*

## Eucalyptus globoidea

White Stringybark

**Size:** 10–30 m x 8–20 m **F P**
**Habitat:** Valley sclerophyll forest
**Form:** Upright tree with compact canopy
**Foliage:** Juvenile—ovate, hairy leaves, paler below, 30–60 mm x 20–40 mm, edges wavy
Adult—glossy and lanceolate, 60–100 mm x 25 mm, generally smaller than other stringybarks
**Flowers:** Pointed buds in groups of 7–15 on flattened stalks. White to cream flowers at various times followed by hemispherical fruit.
**Requirements:** Will adapt to a wide range of moist soils
**Comments:** A useful shade tree for parks, it has a high yield of pollen for honey production. The trunks provide straight timber for poles. Restricted within the Melbourne area.
**Localities:** Dandenong Ck valley, 47, 49–51
**Distribution:** Vic, NSW

*Eucalyptus globoidea*

## Eucalyptus goniocalyx

Long-leaved Box, Bundy

**Size:** 8–20 m x 6–15 m **F P**
**Habitat:** Dry, valley and damp sclerophyll forests
**Form:** Fast growing tree with a crooked trunk and large rounded crown
**Foliage:** Juvenile—bluish to light green and sessile, round to broad leaves to 110 mm x 100 mm
Adult—long, tapering shiny leaves to 24 cm x 30 mm
**Flowers:** Broad flat stalk has up to 7 cylindrical buds. Cap conical. White to cream flowers; March to August. Fruit cup-shaped.
**Requirements:** Tolerates dry and poor soils
**Comments:** This species is not considered a true box as its bark is thick and subfibrous and it has flowers in axillary clusters rather than in terminal panicles.
**Localities:** 33–44, 46–49, 50A-56, 58–63
**Distribution:** Vic, NSW, SA *Illustrated p. 106*

## Eucalyptus ignorabilis

Scent Bark

**Size:** 12–20 m x 7–15 m **F**
**Habitat:** Damp sclerophyll forest
**Form:** Spreading tree with open to dense canopy on open sites, taller on better soils
**Foliage:** Juvenile—grey-green, stalked, at first lanceolate, quickly becoming stalked, narrow-elliptical or broad-lanceolate, to 50 mm x 8 mm
Adult—long, narrow green to blue green pendulous leaves, 140 mm x 25 mm
**Flowers:** Egg-shaped buds in 7s. White to cream flowers from January to April are followed by small top-shaped fruit.
**Requirements:** Moist loams, tolerating wet periods but not inundation
**Comments:** A large shade tree with pinkish brown wood. The young bark is aromatic when crushed. This species is confused and currently subject to revision. The Melbourne form is now known as *E.* aff. *aromaphloia* and may become a new species.
**Localities:** Diamond Ck, Wantirna, 44, 50A, 59
**Distribution:** Vic, NSW
**Synonym:** Recently segregated from *E. aromaphloia*

*Eucalyptus ignorabilis*

SCALE: 0 1cm 2cm — The scale representation on each plant equals 2cm.

*Eucalyptus leucoxylon* ssp. *connata*

## Eucalyptus leucoxylon ssp. connata

Yellow Gum, White Ironbark, S.A. Blue Gum

**Size:** 10–20 m x 6–20 m **F P**
**Habitat:** Box ironbark, red gum and box woodlands, chenopod rocky open scrub, dry sclerophyll forest
**Form:** Upright tree with spreading crown
**Foliage:** Juvenile—sessile heart-shaped green or blue-green leaves to 90 mm x 60 mm, becoming connate (or joined at the base)
Adult—stalked olive-green to green, narrow to lanceolate leaves to 14 cm x 25 mm
**Flowers:** Conical buds in 3s on long thin stalks followed by profuse cream to white flowers; May to September. Fruit are hemispherical, or sometimes slightly subcylindrical with sunken valves.
**Requirements:** Prefers heavy soil but is very adaptable, tolerating drought once established. Grows on sandstone in the basalt plains area.
**Comments:** Bark is rough at the base, cream or light-grey above, shedding to leave yellowish patches. It is an excellent tree for larger gardens, providing shade and shelter. *E. leucoxylon* is an excellent honey-producing tree and also a valuable source of timber and fuel. The Aborigines made clubs and shields from its wood. Early settlers used oil distilled from its leaves for a variety of purposes including remedy for coughs and colds, insecticide and disinfectant. The form commonly available through the nurseries, erroneously known as *E. leucoxylon* "Rosea", is not the local Melbourne form.
**Localities:** 1, 14, 16, 26, 29, 31, 32, 34A–36, 49
**Distribution:** Endemic to Vic
**Synonym:** Segregated from *E. leucoxylon* ssp. *leucoxylon*

*Eucalyptus macrorhyncha*

## Eucalyptus macrorhyncha

Red Stringybark

**Size:** 10–35 m x 10–20 m **F P**
**Habitat:** Dry and valley sclerophyll forests
**Form:** Medium to tall upright tree with rounded canopy
**Foliage:** Juvenile—ovate, shortly stalked hairy leaves with wavy margins, 30–50 mm x 20–30 mm
Adult—glossy, lanceolate leaves 60–150 mm x 30–50 mm
**Flowers:** Diamond-shaped buds with long caps in clusters of 7–11, on a flattened stalk. Profuse white to cream flowers January to April. Globular fruit has valves projecting.
**Requirements:** Prefers well drained clay loam soil. It does not tolerate wet soils.
**Comments:** The red-brown bark is deeply fissured. Its flowers exude a sweet honey aroma, and produce excellent honey. The roots are sensitive to compaction of soil.
**Localities:** 29, 33–53, 55, 56, 58, 60, 62–66
**Distribution:** Vic, NSW, SA

## Eucalyptus melliodora

Yellow Box

**Size:** 10–30 m x 8–25 m **F P**
**Habitat:** Dry and valley sclerophyll forests, red gum and box woodlands
**Form:** Variable open to dense tree
**Foliage:** Juvenile—grey-green oval leaves to 110 mm x 50 mm
Adult—narrow leaves ranging from light green to greyish or blue, 60–140 mm x 12–25 mm
**Flowers:** Club-shaped buds in clusters of 3–7, followed by profuse, perfumed white to cream flowers; September to March. Fruit are hemispherical to sub-globular.
**Requirements:** Well drained loams and alluvial soils
**Comments:** This attractive tree is highly regarded for timber and honey. The bark is rough and brownish with yellow tinges at the base, smooth and cream on upper branches. Whilst too large for suburban gardens, it is an excellent tree for broadscape planting.
**Localities:** Mt. Cottrell, Brighton, Mooroolbark, 9, 16, 23, 25, 26, 29–32, 34–39, 41, 42, 44–49, 50A, 51, 53, 56, 58, 60–64, 67#, 69A
**Distribution:** Vic, Qld, NSW *Illustrated p. 106*

*Eucalyptus microcarpa*

## Eucalyptus microcarpa

Grey or Inland Grey Box, Narrow Leaf Box

**Size:** 10–25 m x 10–15 m **F**
**Habitat:** Box woodland, chenopod rocky open scrub
**Form:** Small to medium spreading tree, with ascending branches
**Foliage:** Juvenile—ovate to broad dull green leaves 30–120 mm x 20–50 mm
Adult—dull green narrow leaves to 15 cm x 25 mm
**Flowers:** 4–8 diamond-shaped buds with hooked ends on thin stalks. White flowers February to August. Fruit varies from cup to barrel-shaped.
**Requirements:** Heavy soils, tolerating wet winters and dry summers
**Comments:** This shade or windbreak tree is valued for its timber and honey. The scaly grey bark becomes smooth on small branches.
**Localities:** Mt Cottrell-Melton, Toolern Vale, 9, 14, 16, 24–26, 29
**Distribution:** Vic, Qld, NSW, SA

## Eucalyptus nortonii

Large-leaved Box

**Size:** 10–15 m x 5–8 m F P
**Habitat:** Red gum woodland
**Form:** Rounded, sparsely crowned small to medium tree, with a crooked trunk
**Foliage:** Juvenile—stalkless, rounded bluish-grey leaves 40–90 mm x 30–80 mm, with toothed edges
Adult—stalked, grey to light green lanceolate leaves to 25 cm x 35 mm
**Flowers:** Broad, flat, mealy stalk has up to 7 mealy cylindrical buds. White flowers in axillary clusters; August to October. Smooth or ribbed barrel-shaped fruit is mealy.
**Requirements:** Well drained soil
**Comments:** Very similar to *E. goniocalyx* but has larger, glaucous buds and fruit. An ornamental tree for home gardens and street planting. It has a disjunct distribution in Victoria with one stand occurring in Gresswell Forest.
**Localities:** 35
**Distribution:** Vic, NSW

*Eucalyptus obliqua*

## Eucalyptus obliqua

Messmate

**Size:** 4–70 m x 6–35 m F P
**Habitat:** Widespread in wet, damp, valley and dry sclerophyll forests, sclerophyll woodland, grassy low open forest
**Form:** Tall upright tree with a dense canopy
**Foliage:** Juvenile—glossy, pendent, broadly ovate leaves, 60–80 mm x 30–40 mm
Intermediate—20 cm x 14 cm
Adult—shiny oblique leaves, 90–150 mm x 30 mm
**Flowers:** Club-shaped buds without scars, in clusters of 7–15 followed by profuse white to cream flowers December to March. Wine-glass shaped fruit have sunken valves.
**Requirements:** Moist well drained soils, tolerating short dry periods
**Comments:** Important for its honey and hardwood. It is an excellent shelter and shade tree for larger areas. The bark provided string for bags and nets, and tinder for fire-making. A tall tree locally, it becomes a stunted shrub in exposed coastal areas outside the Melbourne region.
**Localities:** 34, 37, 42–44, 46–49, 50–63, 74, 76
**Distribution:** Vic, Qld, NSW, Tas, SA

*Eucalyptus ovata*

## Eucalyptus ovata

Swamp Gum

**Size:** 8–30 m x 8–20 m F P
**Habitat:** Widespread in damp and valley sclerophyll forests, riparian and red gum woodland, grassy low open forests, swamp and riparian scrub
**Form:** Fast growing upright tree with open to moderately dense canopy
**Foliage:** Juvenile—dull green elliptic leaves 30–100 mm x 30–70 mm, becoming intermediate after only a few pairs
Intermediate—larger, 110–150 mm x 50–80 mm
Adult—shiny broad-lanceolate to ovate leaves to 15 cm x 30–50 mm with wavy edges and uneven bases
Local coastal forms seem to have smaller leaves than forms further inland.
**Flowers:** Usually 7 diamond-shaped buds in a cluster followed by white to cream flowers. Flowering time can vary, usually March to June. Fruit is funnel-shaped and flat topped.
**Requirements:** Moist soils, tolerating inundation during Winter and Summer dryness
**Comments:** Rough bark persists for some way up the trunk. Long strips are shed in Summer and Autumn from the upper trunk and branches to reveal greyish new bark. It is a useful shade tree for swampy areas and creek banks. The leaves are eaten by koalas.
**Localities:** 26, 29, 32, 33, 34A, 35, 37, 38, 40–45, 47, 49–51, 53–56, 58–67, 70–76
**Distribution:** Vic, NSW, Tas, SA

## Eucalyptus ovata x camaldulensis = Eucalyptus studleyensis

**Comments:** A naturally occurring hybrid found in Studley Park/Yarra Bend and along the Yarra Valley. It has rough bark and grows to 12 m high.

*Eucalyptus pauciflora* ssp. *pauciflora*

## Eucalyptus pauciflora ssp. pauciflora

Snow Gum, White Sallee

**Size:** 8–12 m x 6–10 m F P
**Habitat:** Dry sclerophyll forest, red gum woodland
**Form:** Rounded low-branching tree locally, low stunted tree in alpine areas
**Foliage:** Juvenile—weeping blue-green ovate leaves to 18 cm x 75 mm
Adult—thick, shiny bright green leaves broadly lanceolate to 19 cm x 20–70 mm
**Flowers:** Clusters of 7–15 club-shaped buds without distinct scars, on thick stalks. White to cream flowers October to January. Fruit are cup-shaped.
**Requirements:** Well drained conditions
**Comments:** The beautiful white to cream trunk makes this tree well worth growing. It provides honey, fuel and assists with soil erosion control. Although trees may grow larger than 12 m, they are unlikely to do so under Melbourne conditions. It is tolerant of extreme exposure to cold.
**Localities:** Lilydale, Mooroolbark, 67#, 69A–71, 73–75
**Distribution:** Vic, NSW, Tas

SCALE: 0 1cm 2cm — The scale representation on each plant equals 2cm.

*Eucalyptus polyanthemos* ssp. *vestita*

## Eucalyptus polyanthemos ssp. vestita

Red Box

**F P**

**Size:** 7–25 m x 5–15 m
**Habitat:** Dry and valley sclerophyll forests
**Form:** Slow growing, small to medium tree sometimes with a crooked trunk, with compact to spreading crown
**Foliage:** Juvenile—rounded grey-green leaves, 65 mm x 30–80 mm
Adult—ovate to broad-lanceolate bluish-green leaves, 50–90 mm x 15–30 mm
**Flowers:** Terminal bluish club-shaped buds in groups of 3–7. White or cream flowers; September to January. Fruit barrel-shaped.
**Requirements:** Requires good drainage and will tolerate poor stony soils
**Comments:** An ornamental tree with attractive blue-grey foliage which is used for floral decoration. The red wood is hard and durable and makes excellent fuel. A profusely flowering tree which produces copious quantities of excellent honey.
**Localities:** Sth Morang, Research, 29, 33–36, 40–42, 44, 47–49, 50A, 51
**Distribution:** Vic, NSW

*Eucalyptus pryoriana*

## Eucalyptus pryoriana

Gippsland Manna Gum

**F**

**Size:** 8–16 m x 5–12 m
**Habitat:** Tea-tree heath, wattle tea-tree scrub
**Form:** Well-branched, spreading medium tree with a dense crown
**Foliage:** Juvenile—dark green, paler below, stalkless lanceolate leaves 80–150 mm x 25 mm
Adult—narrower dark green stalked leaves to 180 mm x 20 mm
**Flowers:** Stalkless pointed buds in 3s. White flowers March to May. Cup-shaped fruit have projecting valves.
**Requirements:** Well drained sandy soil
**Comments:** Thick fibrous bark persists on the trunk and larger branches, shedding on smaller branches. An ideal tree for larger coastal gardens, it is also a good windbreak and shelter tree providing honey and fuel. This tree is an important koala fodder tree.
**Localities:** 67–71, 74–76
**Distribution:** Endemic to Vic
**Synonym:** *E. viminalis* var. *racemosa*

## Eucalyptus radiata

Narrow-leaved Peppermint

**F P**

**Size:** 10–30 m x 6–20 m
**Habitat:** Widespread in damp and valley sclerophyll forests, sclerophyll woodland, grassy low open forests
**Form:** Attractive low branching tree with dense canopy
**Foliage:** Juvenile—lanceolate or occasionally broad-lanceolate, thin-textured and stem-clasping, becoming elongated, to 15 cm x 35 mm
Adult—aromatic, thin-textured, narrow, to 15 cm x 20 mm
**Flowers:** Numerous small club-shaped buds without distinct scars in clusters. Profuse white flowers; October to January. Small cup-shaped fruit.
**Requirements:** Well drained soils
**Comments:** This graceful upright tree provides shade and shelter. Its leaves are distilled for oil. Some forms have an attractive weeping habit.
**Localities:** 26, 33–34A, 40–51, 53–56, 58–63, 67#, 71, 73–76
**Distribution:** Vic, NSW

*Illustrated p. 106*

*Eucalyptus regnans*

## Eucalyptus regnans

Mountain Ash

**F**

**Size:** 25–95 m x 15–30 m
**Habitat:** Cool temperate rainforest, wet sclerophyll forest
**Form:** Fast growing upright tree with a small open canopy
**Foliage:** Juvenile—broad with wavy edges, 17 cm x 80 mm
Adult—lanceolate to 140 mm x 10–70 mm
**Flowers:** Club-shaped buds in paired clusters of 7–15. Profuse white to cream flowers; December to May. Fruit are shaped like a top.
**Requirements:** Deep rich loams which are moist and well drained
**Comments:** A rough finely fibrous bark is persistent to 15 m. It is smooth and greyish above with long ribbons of shedding bark often hanging from the tree. The tallest flowering plant in the world, *E. regnans* produces excellent hardwood timber and is also used in paper manufacturing. Unlike most eucalypts, individuals are killed by intense fire and regeneration is usually exclusively from seed.
**Propagation:** Seed, may germinate better if moistened and placed in the refrigerator for 3–4 weeks.
**Localities:** 55, 57
**Distribution:** Vic, Tas

## Eucalyptus rubida ssp. rubida — Candlebark Gum

**Size:** 10–25 m x 10–20 m — F P
**Habitat:** Dry and valley sclerophyll forests
**Form:** Upright tree with dense spreading to rounded crown
**Foliage:** Juvenile—sessile, silvery rounded leaves 45 mm x 30–55 mm
Adult—long narrow dark green leaves 7.5–24 cm x 10–25 mm
**Flowers:** Sessile buds with domed caps, in 3s. White flowers November to February. Domed fruit have projecting valves.
**Requirements:** Well drained drier soils
**Comments:** A decorative shade tree with a smooth white trunk which develops red or pink patches before the bark is shed in strips. It withstands very cold conditions and is a good honey producer. An important koala food tree.
**Localities:** Mooroolbark, Montrose, 33–34A, 36, 38–42, 44, 51, 53, 56, 58, 62, 63
**Distribution:** Vic, Qld, NSW, Tas, SA

*Eucalyptus rubida* ssp. *rubida*

## Eucalyptus sideroxylon ssp. tricarpa = Eucalyptus tricarpa

## Eucalyptus tricarpa — Red Ironbark, Mugga

**Size:** 10–30 m x 10–20 m — F P
**Habitat:** Box ironbark woodland, dry sclerophyll forest
**Form:** Upright to spreading tree with an open crown
**Foliage:** Juvenile—bluish, oblong leaves, 40–150 mm x 20 mm
Adult—narrow grey-green leaves to 140 mm x 18 mm
**Flowers:** Large diamond-shaped buds in 3s on long stalks. Profuse cream or pink flowers; May to December. Large barrel-shaped fruit.
**Requirements:** Well drained soils, tolerating extended dry periods
**Comments:** The rough, black bark and bluish foliage make this ornamental tree very popular. The flowers of the local form are usually cream. It is valued for its honey and durable timber. Essential oils are extracted from the foliage. The wood was used to make shields.
**Localities:** 35, 36, 42
**Distribution:** Vic, Qld, NSW
**Synonym:** *E. sideroxylon* ssp. *tricarpa*

*Eucalyptus tricarpa*

## Eucalyptus viminalis — Ribbon or Manna Gum

**Size:** 10–50 m x 8–15 m — F
**Habitat:** Widespread along watercourses of wet and valley sclerophyll forests, sclerophyll and riparian woodland, grassy low open forest
**Form:** Fast growing tall upright tree with open crown
**Foliage:** Juvenile—dark green lanceolate to broad-lanceolate leaves 35–65 mm x 10–30 mm
Adult—narrow, often sickle-shaped to 20 cm long
**Flowers:** Egg-shaped buds in clusters of 3 on a flattened stalk. White flowers January to May. Round fruit with 3 or 4 projecting valves.
**Requirements:** Very adaptable, with tallest specimens found in moist deep loam soils in mountain areas
**Comments:** Bark is shed in ribbons during Summer leaving the upper trunk and branches white. Rough bark often persists at the base. This ornamental tree is best suited to large gardens and parklands. It is used for timber, pulp, honey production and is an important koala food tree. An important source of manna to the Aborigines, it was also used to obtain medicines and make implements.
**Localities:** Malvern#, 16, 23, 25, 26, 29, 32–36, 38, 42, 44, 45#, 47–49, 50A, 53, 55–58, 60–67, 73–76
**Distribution:** Vic, Qld, NSW, Tas, SA — *Illustrated p. 106*

## Eucalyptus yarraensis — Yarra Gum

**Size:** 10–20 m x 5–10 m — F P
**Habitat:** River flats and flood plains of valley sclerophyll forest
**Form:** Small to medium tree with a short trunk and dense spreading crown
**Foliage:** Juvenile—dull green oval leaves 50–80 mm x 30–50 mm, not always opposite
Adult—elliptic or broad-lanceolate leaves 60–100 mm x 30–50 mm, with wavy edges
**Flowers:** Small diamond-shaped buds in clusters of 7. White flowers in January are followed by small hemispherical fruit.
**Requirements:** Tolerates heavy soil and limited inundation
**Comments:** Rough black fibrous bark persists on the trunk and sometimes on large branches. It makes a good shade and windbreak tree.
**Localities:** Yarra Valley between Lilydale and Healesville, Yarrambat; 47, 62, 63
**Distribution:** Endemic to Vic

*Eucalyptus yarraensis*

SCALE: 0 1cm 2cm — The scale representation on each plant equals 2cm.

*Eucalyptus baxteri*

*Clematis microphylla* var. *microphylla*

*Eucalyptus cephalocarpa*

*Eucalyptus goniocalyx*

*Dillwynia sericea*

*Eucalyptus melliodora*

*Eucalyptus radiata*

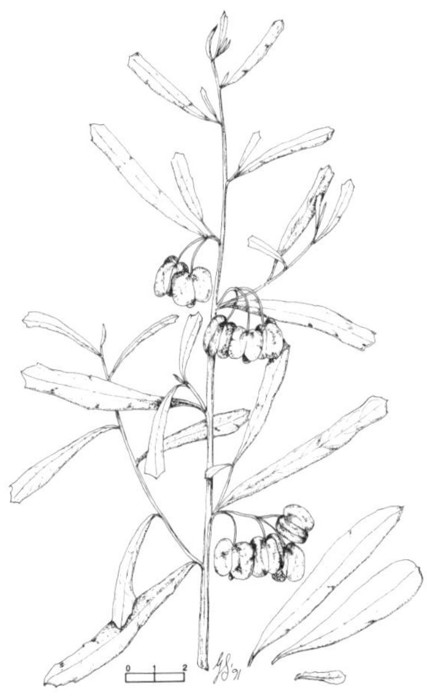

*Dodonaea viscosa* ssp. *spatulata*

*Eucalyptus viminalis*

# Euchiton **Asteraceae**

## Euchiton gymnocephalus — Creeping Cudweed

**Size:** to 30 cm high
**Habitat:** Riparian scrub, valley sclerophyll forest, wattle tea-tree scrub
**Form:** Stoloniferous perennial herb
**Comments:** Differs from *E. involucratus* in that leaves are shorter, usually less than 40 mm long, floral leaves are also shorter. Basal rosette is persistent when flowering.
**Localities:** 26, 28, 29, 35, 39, 41, 42, 47, 49, 51, 52, 56, 58, 60, 62, 63, 67, 69A, 70, 74–76
**Distribution:** ?All states
**Synonym:** *Gnaphalium gymnocephalum, G. japonicum, G. collinum.* Separated from *Euchiton involucratus.*

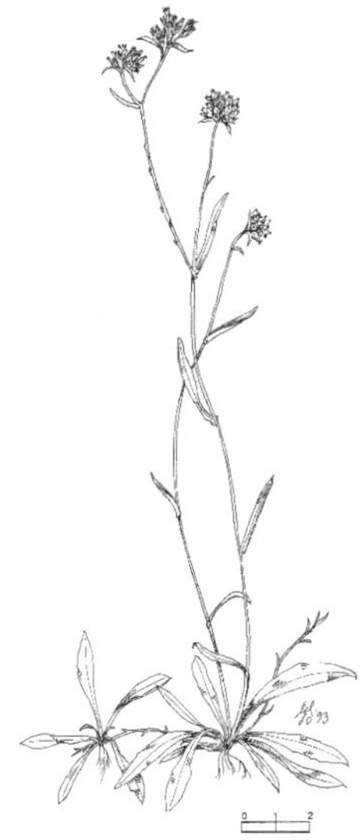

*Euchiton gymnocephalus*

## Euchiton involucratus — Common or Star Cudweed

**Size:** 10–50 cm high **F P**
**Habitat:** Widespread in dry and valley sclerophyll forests, box and coastal banksia woodland, riparian scrub
**Form:** Erect, unbranched woolly annual or perennial herb, tufted to shortly rhizomatous
**Foliage:** Leaves usually glabrous above, cottony-white underneath, to 20 cm long but usually smaller; basal rosette withering before flowering; narrowly elliptic, oblanceolate or linear stem leaves have a distinct mid-vein and may be almost stem-clasping
**Flowers:** Globular terminal clusters of brownish flowerheads to 2 mm wide, lacking ray florets, most of the year, surrounded by green leaf-like bracts much larger than the flowerheads. Flowerhead has more than one bisexual floret.
**Requirements:** Moist, well drained soils
**Propagation:** Seed, cuttings, division
**Localities:** Eltham, 6, 16–18, 28, 34A, 36, 40–42, 44, 47, 50–56, 58, 60, 62–65, 67, 69A, 74, 76
**Distribution:** All states
**Synonym:** *Gnaphalium involucratum*

## Euchiton sphaericus — Common Cudweed

**Size:** 0.3–0.6 m high
**Habitat:** Box woodland, plains grassland, valley sclerophyll forest
**Form:** Tufted annual herb, lacking a rosette
**Foliage:** Leaves similar to *E. involucratus*, to 80 mm long.
**Flowers:** Flowerheads less than 2 mm wide in dense terminal clusters to 25 mm across. Single bisexual floret per flowerhead. July to December
**Localities:** 9, 14, 25, 26, 41, 42, 56, 62
**Distribution:** All states; NZ
**Synonym:** *Gnaphalium sphaericum.* Separated from *Euchiton involucratus.*

*Euchiton sphaericus*

*Euchiton involucratus*

# Euphorbia **Euphorbiaceae**

## Euphorbia drummondii — Flat Spurge, Caustic Weed

**Size:** Prostrate x 20–30 cm **F**
**Habitat:** Box woodland, plains grassland, saltmarshes
**Form:** Annual or short-lived perennial matting plant
**Foliage:** Pairs of ovate to oblong blue-green leaves with reddish margins to 8 mm long
**Flowers:** Minute green flowers with reddish glands, singly or in small clusters; July to September, March to May
**Requirements:** Dry soils
**Comments:** Watering may encourage a longer life span. The stems contain a milky sap which was used by the Aborigines to treat snake bite, dysentery, genital diseases and wounds.
**Propagation:** Seed
**Localities:** 4, 9, 12, 14–16, 25, 26, 34A
**Distribution:** All mainland states
**Synonym:** *Chamaesyce drummondii*

*Euphorbia drummondii*

SCALE: 0 1cm 2cm — The scale representation on each plant equals 2cm.

# Euphrasia — Scrophulariaceae

*Euphrasia collina* ssp. *collina*

## Euphrasia collina ssp. collina — Purple Eyebright

**Size:** 0.2–0.8 m x 0.2–0.5 m — F P
**Habitat:** Was widespread in tea-tree heath, sclerophyll woodland
**Form:** Small, upright perennial herb
**Foliage:** Pointed, oval to oblong toothed leaves to 15 mm long x 5 mm
**Flowers:** Upper branches densely covered with white to pink and mauve hooded flowers; August to February. Flowering growth dies back each year.
**Requirements:** Well drained moist soils.
**Comments:** This very attractive small herb has disappeared from most Melbourne localities. It is semi-parasitic and has the ability to photosynthesise as well as obtain nutrients by attaching to the roots of host plants. It has proven difficult to propagate and grow. There are many subspecies and affiliated species.
**Propagation:** Seeds. Difficult
**Localities:** Dandenong, 49, 58, 59, 67, 74
**Distribution:** Vic, NSW, Tas, SA

## Euphrasia collina ssp. muelleri

**Size:** 0.4 m high
**Habitat:** Heathlands, once widespread
**Comments:** This subspecies is now very rare in Victoria. It has lilac to purple flowers; July to November.
**Localities:** Frankston[#], Mentone[#], Sandringham[#]
**Distribution:** Vic, NSW, SA

## Euphrasia collina ssp. paludosa

**Size:** 0.5 m high
**Habitat:** Swamp scrub, valley sclerophyll forest
**Comments:** Flowers are white or white with blue through to pink markings in the throat. Masses of flowers are produced from September to February.
**Localities:** Outer Eastern suburbs and Dandenong Ranges, to Dandenong
**Distribution:** Vic, Qld, NSW, SA

## Euphrasia collina ssp. trichocalycina

**Size:** 0.6 m x 0.3–0.5 m
**Habitat:** Dry sclerophyll forest
**Comments:** Prefers an open situation. Leaves narrow-ovate to oblong. Racemes of 20–50 flowers, white or white with lavendar markings; August to November. Now extinct within Melbourne and rare in Victoria.
**Localities:** Collected in Ringwood 1924
**Distribution:** Vic, SA

## Euphrasia scabra — Yellow Eyebright

**Size:** 0.1–0.5 m x 0.1–0.5 m — F P
**Habitat:** Moist grassy sites on primary dune scrub, saltmarshes
**Form:** Semi-parasitic, erect hairy annual herb
**Foliage:** Pairs of ovate to elliptic, toothed or lobed leaves to 15 mm long, with 1–5 pairs of teeth
**Flowers:** Dense terminal spikes of yellow flowers, (often mauve in alpine areas); October to April
**Requirements:** Moist well drained soils
**Comments:** Once widespread in southern Australia, it is now endangered throughout its range. It is extinct in Melbourne and survives in a few ungrazed sites in the Victorian alps. Would make an attractive container plant if problems of cultivation could be overcome.
**Propagation:** Seeds. Difficult.
**Localities:** Collected in Port Melbourne in 1850.
**Distribution:** Vic, NSW, Tas, SA, WA

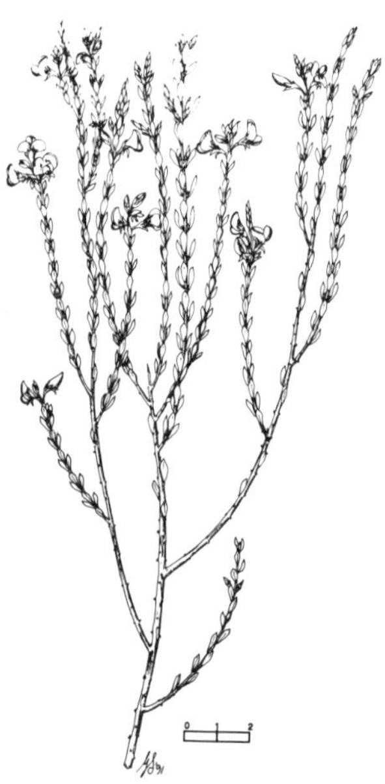

*Eutaxia diffusa*

# Eutaxia — Fabaceae

## Eutaxia diffusa

**Size:** 0.5–1.2 m x 0.8–1.2 m — F
**Habitat:** Coastal cliffs, box woodland
**Comments:** Differs from *E. microphylla* in that it is upright in form with soft branches and lanceolate leaves. Flowers are paler yellow often without any red. This species also occurs around Long Forest and Exford just outside the study area.
**Localities:** Melton area, 16, 22, 74
**Distribution:** Vic, NSW, SA
**Synonym:** *E. microphylla* var. *diffusa*

### Eutaxia microphylla var. microphylla

Small-leaved Eutaxia

**F P N**

**Size:** Prostrate or procumbent x 0.5–1.5 m
**Habitat:** Plains grassland, red gum woodland
**Form:** Variable low growing shrub
**Foliage:** Crowded, narrow, greyish-green leaves to 7 mm long. Branchlets on older growth become spiny.
**Flowers:** Profuse yellow and red pea flowers; August to October
**Requirements:** Adapts to most soils and, once established, tolerates waterlogging and extended dry periods.
**Comments:** This attractive ground cover makes an excellent container plant. It can be used on pool edges or rockeries, and may be used to control soil erosion in small areas. Responds well to pruning.
**Propagation:** Scarified seed, cuttings
**Localities:** 4, 5, 7, 12, 16, 26, 29, 35
**Distribution:** All states except NT

*Eutaxia microphylla* var. *microphylla*

## Exocarpos — Santalaceae

### Exocarpos cupressiformis

Wild Cherry, Cherry Ballart

**F P**

**Size:** 3–8 m x 3–5 m
**Habitat:** Widespread in grassy low open forest, dry, damp and valley sclerophyll forests, red gum and box woodland, tea-tree heath
**Form:** Dense, rounded shrub to small tree
**Foliage:** Leaves reduced to tiny scales on pendulous yellow-green to dark green branchlets
**Flowers:** Insignificant spikes of cream flowers are followed by globular green nuts on fleshy orange to red, edible stalks.
**Requirements:** Well drained soils, tolerating poor and dry conditions
**Comments:** Whilst a very attractive tree it needs a symbiotic relationship with another plant to grow. Plants respond well to pruning. They are brittle in the wind but coppice readily. The timber is excellent for turning and furniture. It was used by the local Yarra tribe to make spear throwers.
**Propagation:** Difficult. Minimal success with stem cuttings but root cuttings may be more successful. Suckering can be attained by damaging roots. Seed was successfully germinated after fruit was fed to hens and then sown with seeds of *Themeda triandra*. (IFFA Vol 4, No. 2 1991)
**Localities:** Melton Sth, 14, 26, 29, 30–33, 34A–69A, 71–76
**Distribution:** Vic, Qld, NSW, Tas, SA

*Exocarpos cupressiformis*

### Exocarpos strictus

Pale Fruit Ballart or Dwarf Cherry

**F P**

**Size:** 1–2.5 m x 1–3 m
**Habitat:** Damp, valley and dry sclerophyll forest, red gum woodland, grassy low open forest
**Form:** Upright to spreading broom-like shrub
**Foliage:** Light to dark green flattened branchlets, becoming pendulous. Leaves are tiny scales.
**Flowers:** Minute clusters of flowers, August to November are followed by green to purplish-black fruit on a succulent white to mauve stalk.
**Requirements:** Well drained soils
**Comments:** Forms dense thickets in its natural habitat.
**Propagation:** As *E. cupressiformis*
**Localities:** 26, 32, 33, 42, 43, 54, 55, 58, 60, 67#, 74, 76
**Distribution:** Vic, NSW, Tas, SA

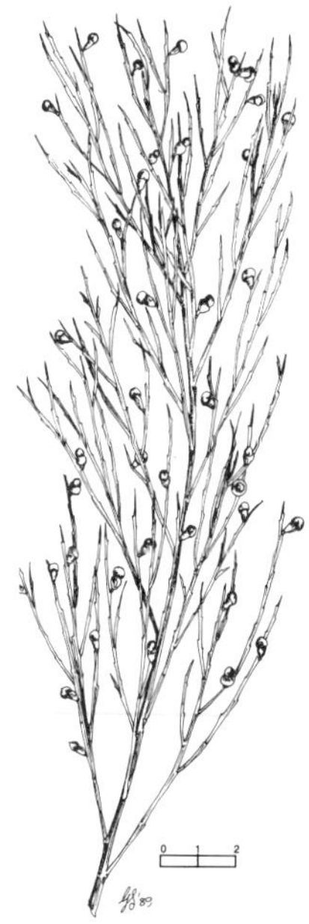

*Exocarpos stricta*

## Frankenia — Frankeniaceae

### Frankenia pauciflora var. pauciflora

Southern or Common Sea Heath

**F P**

**Size:** Prostrate-30 cm x 0.3–1 m
**Habitat:** Coastal saltmarshes, primary dune scrub
**Form:** Spreading mat-like shrub
**Foliage:** Narrow to oblong green to grey-green leaves to 7 mm long, hairy underneath
**Flowers:** Masses of small pink flowers in terminal clusters, occurring most of the year, peaking August to February
**Requirements:** Well drained soils, tolerating extended dry periods once established
**Comments:** An attractive plant for rockeries and containers, responding well to pruning to maintain denseness. It is useful for binding soil especially in saline areas. The Victorian form has been referred to as var. *gunnii*.
**Propagation:** Seed, cuttings, division of suckers
**Localities:** 2–4, 67, 74
**Distribution:** All states except Qld
**Synonym:** *F. pauciflora* var *gunnii*

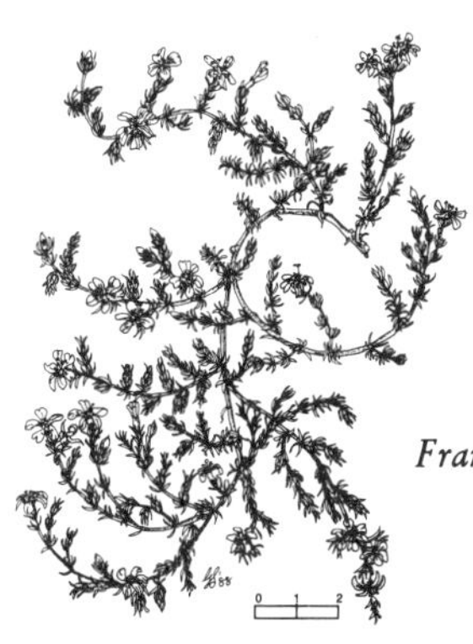

*Frankenia pauciflora* var. *pauciflora*

SCALE: 0 1cm 2cm — The scale representation on each plant equals 2cm.

# Galium

**Rubiaceae**

Spreading annual, biennial or perennial herbs found in a wide range of habitats, although generally in sheltered positions. Stems are usually quadrangular and leaves often have bristle-like hairs. Flowers are in axillary or terminal branched clusters and have a very short floral tube. Useful as ground covers in shady moist areas or for cascading over containers. Propagation is by seed, cuttings or division of creeping rhizomes. *Galium* spp. are very similar to *Asperula* spp., the latter distinguished by the dioecious nature of the plants, the longer floral tube and the fleshy fruit.

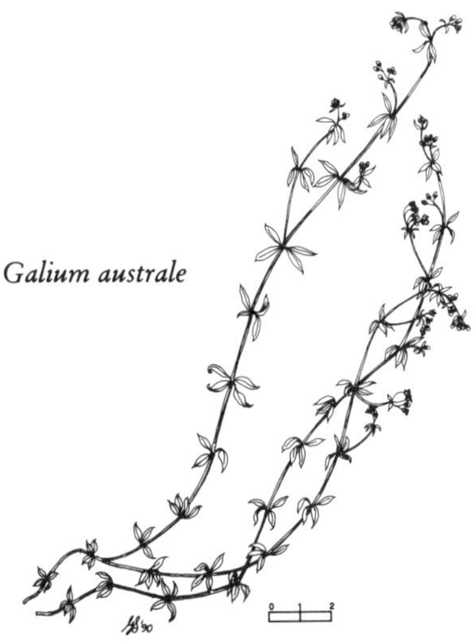

*Galium australe*

*Galium binifolium*

*Galium ciliare*

## Galium australe

Bedstraw, Tangled Bedstraw

P

**Habitat:** Grassy low open forest, riparian scrub, valley sclerophyll forest, occasionally epiphytic on tree ferns
**Form:** Often hairy perennial with tangled, weak stems
**Foliage:** Whorls of 4 dark green to brown, elliptical leaves to 12 mm long, often unequal in length
**Flowers:** Axillary clusters of 1–7 white flowers; December to February. Fruit covered in brown hooked hairs.
**Localities:** Baxter, 29#, 42, 49
**Distribution:** Vic, NSW, Tas, SA, WA

## Galium binifolium

Reflexed Bedstraw

P

**Habitat:** Damp sclerophyll forest
**Form:** Straggling annual or biennial herb
**Foliage:** Narrow elliptical leaves to 12 mm long, usually unequal in length (2 long and 2 short), especially on the lower parts of stems, often bent downwards
**Flowers:** Cream to yellowish flowers sometimes have reddish or purple tinges; September to November. 1 or 2 axillary clusters of up to 8 flowers in leaf axils.
**Localities:** 56
**Distribution:** Vic, NSW, SA

## Galium ciliare

Bedstraw

P

**Size:** 10–40 cm x 0.3–1 m
**Habitat:** Woodland, also usually on grasslands of basaltic plains
**Form:** Erect usually unbranched stems are glabrous or have spreading hairs
**Foliage:** Glossy, yellowish, ovate to round leaves to 10 mm long with recurved margins, in whorls of 4–5
**Flowers:** Yellow to cream axillary or terminal clusters on long stalks to 15 mm long; September to December
**Requirements:** Well drained soil
**Comments:** Similar to *G. propinquum* which has non-glossy leaf blades tapering to the stalk, and short or absent flower stalks.
**Localities:** Known from an early collection from the Dandenong area.
**Distribution:** Vic, Qld, NSW, Tas

## Galium compactum

Bedstraw

F

**Size:** Prostrate–30 cm x 0.2–0.6 m
**Habitat:** Dry sclerophyll forest
**Form:** Compact perennial with short hairy stems
**Foliage:** Tiny, light green pointed ovate leaves to 6 mm, in whorls of 4–5. Scattered bristles on leaves, mainly on recurved margins.
**Flowers:** 1 or 2 axillary yellowish-green flowers tinged with purple on stalks shorter than the leaves; September to December
**Requirements:** Very well drained soils, tolerating limited periods of dryness and coastal exposure
**Comments:** A useful compact plant for rockeries or a container. An endangered plant in Victoria, it is now extinct in Melbourne.
**Localities:** Research#
**Distribution:** Vic, SA

## Galium gaudichaudii

Rough Bedstraw

P

**Size:** Prostrate–30 cm x 0.5–2 m
**Habitat:** Wet sclerophyll rainforest, riparian scrub, damp and valley sclerophyll forests, grassy low open forest
**Form:** Spreading perennial herb with strong tap root
**Foliage:** Narrow leaves to 12 mm long, in whorls of 4–5, margins recurved
**Flowers:** Very small almost sessile cream flowers in terminal and axillary clusters; September to December. Buds are purplish.
**Requirements:** Moist, well drained soils
**Comments:** The most common Victorian species. It may become invasive in ideal conditions.
**Propagation:** Seed, cuttings
**Localities:** 16, 26, 29, 32, 34A, 36, 41, 42, 53, 55, 57, 58, 64, 74, 76
**Distribution:** Vic, Qld, NSW, Tas, SA

## Galium migrans

P

**Habitat:** Rock crevices in escarpment vegetation, riparian woodland and scrub, valley sclerophyll forest
**Form:** Straggling hairy perennial preferring moist situations
**Foliage:** Bright, light green elliptical to almost linear leaves to 10 mm long, upper leaves becoming narrower, in whorls of 4.
**Flowers:** Clusters of 5–15 yellowish to white flowers on long, branched thread-like stalks; September to November
**Comments:** Differs from *G. binifolium* principally in having leaves of equal length.
**Localities:** Kororoit Ck, Melton Sth, 8, 18, 28, 36, 56
**Distribution:** Vic, Qld, NSW, SA, WA

*Galium migrans*

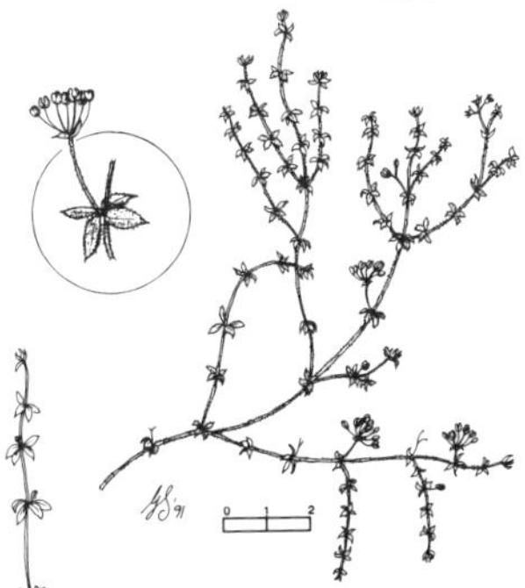

## Galium propinquum

Maori Bedstraw

P N

**Size:** Prostrate–20 cm x 0.5–1 m
**Habitat:** Coastal banksia, box and riparian woodlands, valley sclerophyll forest, grassy low open forest
**Form:** Vigorous tangled perennial herb
**Foliage:** Dull to yellowish green oval leaves to 15 mm long, in whorls of 4–5
**Flowers:** Few tiny white to cream or yellow flowers in clusters; September to December
**Requirements:** Moist well drained soils
**Comments:** Grows well in densely shaded areas, spreading by rhizomes. *G. gaudichaudii* is distinguished from *G. propinquum* by its narrower leaves and more compact habit.
**Propagation:** Seed, cuttings, division
**Localities:** 8, 14, 34A, 36, 42, 44, 58, 60, 67#, 74, 76
**Distribution:** Vic, Qld, NSW, SA; NZ

*Galium propinquum*

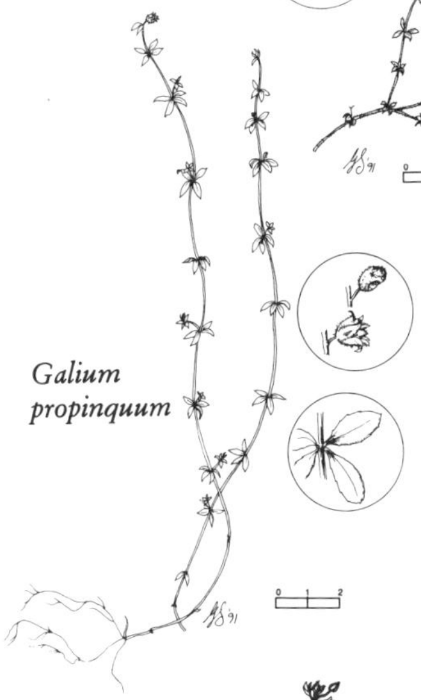

# Geranium

# Geraniaceae

## Geranium potentilloides

Crane's Bill

P N

**Size:** Prostrate x 0.5 m-1.5 m
**Habitat:** Widespread in damp and valley sclerophyll forests
**Form:** Creeping perennial herb
**Foliage:** Hairy, almost round leaves with 5–7 lobes, often purplish below. Stem leaves, 70 mm x 50 mm, on stalks to 40 mm long. The larger basal leaves are usually deciduous.
**Flowers:** Solitary open pink to almost white flowers, 10–15 mm wide; September to March. Fruit is beaked to 10 mm long.
**Requirements:** Moist shaded conditions
**Comments:** Although not ornamental it is useful for stabilising soils in shaded situations. The thick tap root of some *Geranium* species was cooked for food by the Aborigines.
**Propagation:** Seed, cuttings, division of layered stems
**Localities:** 34A, 35, 42, 44, 50A, 51, 53, 55–60, 64, 74
**Distribution:** Vic, NSW, Tas, SA; NZ, Antarctic Islands

*Geranium potentilloides*

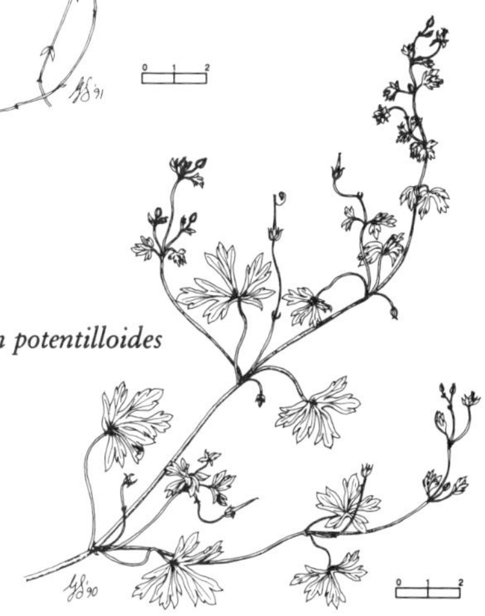

## Geranium retrorsum

Grassland Crane's Bill

P

**Size:** Prostrate–30 cm x 0.6–1 m
**Habitat:** Red gum, box and riparian woodland, plains grassland
**Form:** Creeping perennial herb
**Foliage:** Hairy, ovate to kidney-shaped stem leaves to 90 mm x 30 mm with 3–7 deeply dissected narrow lobes, held on stalks to 70 mm long. Basal leaves somewhat larger and more deeply dissected.
**Flowers:** Paired pink flowers with long hairs on sepal margins and hairs on stalks pressed backwards; July to January. Fruit beaked to 15 mm long.
**Requirements:** Most soils, preferring shade
**Comments:** Does not self layer.
**Localities:** 2, 5, 8, 10, 12, 15–18, 26, 28, 35, 44, 47, 48, 53, 60, 71, 74
**Distribution:** Vic, NSW, Tas, SA, WA; NZ

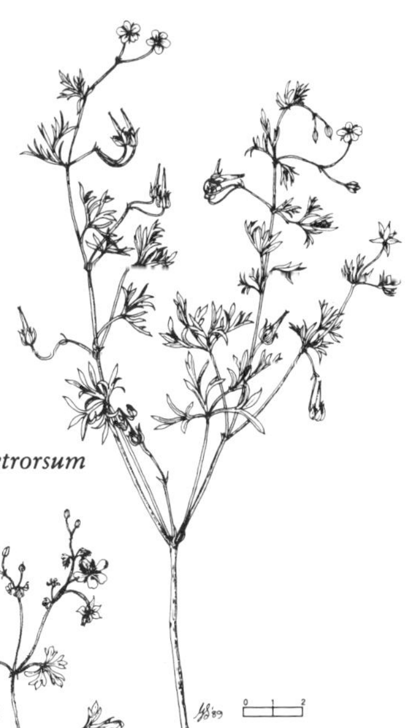

*Geranium retrorsum*

## Geranium solanderi

Austral Crane's Bill

P

**Size:** Prostrate–0.5 m x 0.6–1.5 m
**Habitat:** Widespread in red gum woodland, riparian scrub, dry and valley sclerophyll forests, grassy low open forest
**Form:** Hairy, creeping perennial herb
**Foliage:** Stem leaves 5–7 lobed, kidney-shaped to almost round to 100 mm x 40 mm, paler green below, on long hairy stalks. Basal leaves larger and more deeply dissected. Leaves age orange to red.
**Flowers:** Paired pink flowers on stalks with reflexed or spreading hairs; October to February. Fruit beaked to 12 mm long.
**Requirements:** Well drained soils but tolerates moisture
**Comments:** Similar to the other species. Cultivated plants form a dense cover in permanently damp areas. The Aborigines roasted the taproot.
**Localities:** 1, 10, 13, 18, 25, 26, 29, 31, 32, 34A–36, 39, 41, 42, 44, 46–48, 50A, 51, 53, 55–58, 60, 62, 63, 71, 74–76
**Distribution:** All states except NT; NZ

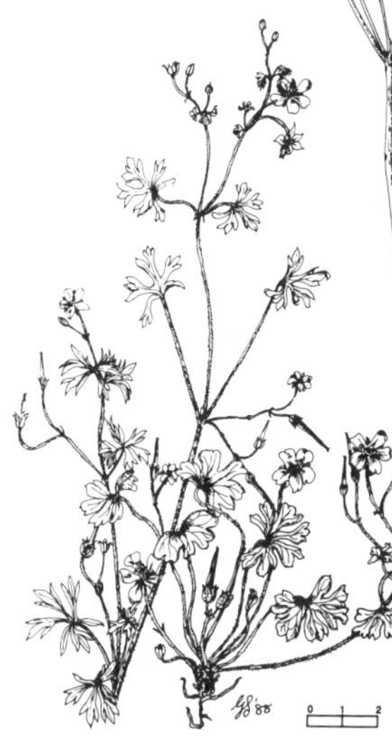

*Geranium solanderi*

SCALE: 0 1cm 2cm The scale representation on each plant equals 2cm.

*Glycine clandestina*

## Glossostigma — Scrophulariaceae

### Glossostigma elatinoides — Small Mudmat

**Size:** Prostrate x 0.3–1 m **F**
**Habitat:** In and beside freshwater swamps and shallow pools, in box woodland
**Form:** Semi-aquatic creeping perennial herb rooting at leaf nodes to form a mat
**Foliage:** Pairs of erect, stalked oblanceolate leaves to 15 mm long, along prostrate stems
**Flowers:** Tiny, single pale blue or mauve 5-petalled flowers on erect stalks shorter than the leaves; December to April
**Requirements:** Mud or shallow water to 0.1 m, on clay
**Comments:** Useful in stabilising the edges of clay-based pools where it will form green mats beneath the water. Locally rare.
**Propagation:** Fresh seed, cuttings, division
**Localities:** Melton South, 42
**Distribution:** Vic, NSW, Tas, SA; NZ

## Glycine — Fabaceae

### Glycine clandestina — Twining Glycine

**Size:** Twining 0.3–2 m tall **P N**
**Habitat:** Widespread in tea-tree heath, red gum woodland, dry and valley sclerophyll forests
**Form:** Slender open twiner
**Foliage:** Hairy, trifoliate leaves, leaflets narrow-ovate to 40 mm long, on stalks of equal length. The stalk of the terminal leaflet lacks small stipule-like structures, called stipels.
**Flowers:** Racemes of small bluish-mauve pea flowers; October to January
**Requirements:** Moist well drained soils tolerating dry periods once established
**Comments:** A dainty twiner for climbing through other plants, usually not noticed until it flowers.
**Propagation:** Scarified seed, cuttings
**Localities:** 29#, 34A–38, 41–44, 47, 49, 50A–53, 55–61, 63–67#, 71, 74, 76
**Distribution:** All states except NT

*Glycine latrobeana*

### Glycine latrobeana — Clover or Purple Glycine

**F P**
**Habitat:** Plains grassland, box woodland and dry sclerophyll forests
**Form:** A trailing rather than climbing plant with short stems
**Foliage:** Obovate leaves, 20 mm x 12 mm, broader than *G. clandestina*
**Flowers:** Racemes of 3–8 deep purple pea-like flowers; September to December
**Requirements:** Moist well drained soils
**Comments:** An endangered plant Australia-wide, it is also at risk in Melbourne.
**Localities:** Werribee#, Diggers Rest#, Melton South, Research-Eltham, 34A, 55, 58
**Distribution:** Vic, NSW, Tas, SA

### Glycine microphylla

**P N**
**Habitat:** Damp sclerophyll forest
**Form:** Stoloniferous twiner, rooting at the nodes of above ground stems
**Foliage:** Trifoliate leaves, leaflets lanceolate to 15–50 mm long, terminal leaflet often larger, on stalks of equal length. Pair of stipels on the stalk of the terminal leaflet. Leaflets of first leaves broader.
**Flowers:** Racemes of small violet flowers with a white spot; October to April
**Requirements:** Tolerates poorly drained soils in Winter
**Comments:** Similar to *G. clandestina* but smaller, the main distinguishing feature being the presence of the stipels.
**Localities:** 42, 58
**Distribution:** Vic, Qld, NSW, Tas, SA; Norfolk Is
**Synonym:** Previously included with *G. clandestina*

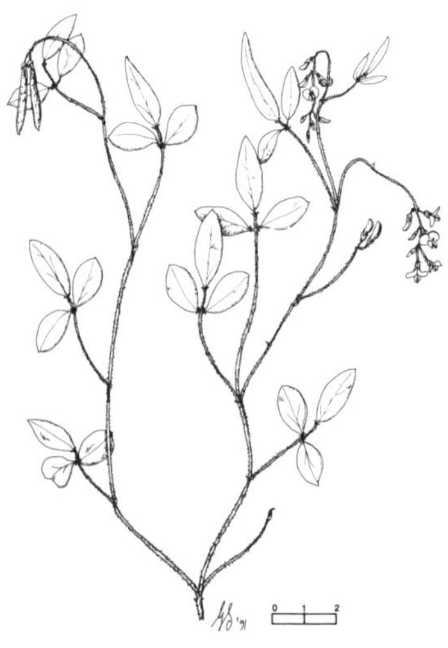

*Glycine tabacina*

### Glycine tabacina — Vanilla or Variable Glycine

**Size:** Stems 0.2–2 m long **F P**
**Habitat:** Dry sclerophyll forest, plains grassland, riparian scrub, red gum woodland
**Form:** Creeping or trailing perennial
**Foliage:** Trifoliate leaves, side leaflets elliptic to ovate to 15 mm long, terminal leaflet larger and narrower to 50 mm long. The stalk of the terminal leaflet is markedly longer than the stalks of the 2 lateral leaflets.
**Flowers:** Racemes of 4–12 blue to purple pea flowers; December to May
**Requirements:** Well drained soils, tolerating dryness once established
**Comments:** This species is not tolerant of heavy frost. The Aborigines chewed the tap root which is liquorice flavored.
**Propagation:** Seed, cuttings
**Localities:** Kororoit Ck, Melton Sth, 6, 7, 12, 16–18, 21, 23, 25, 26, 28, 29, 31#, 35, 38, 39
**Distribution:** Vic, Qld, NSW, SA, WA

# Glycyrrhiza

**Fabaceae**

## Glycyrrhiza acanthocarpa

Native or Southern Liquorice

**Size:** 0.6–1.5 m x 0.5–1 m F P
**Habitat:** Sedimentary rises
**Form:** Erect sticky shrub
**Foliage:** Aromatic pinnate leaves to 60 mm, 9–13 oblong-lanceolate leaflets to 20 mm
**Flowers:** Long axillary racemes, to 80 mm, of lilac pea flowers, along the upper part of the stalk; September to February. Fruit a rusty-brown burr-like pod.
**Requirements:** Moist soils
**Comments:** Responds well to hard pruning. Plants die back over Summer. Little known, it may be extinct within the Melbourne area.
**Propagation:** Scarified seed, cuttings
**Localities:** An early collection from Mentone. Possibly once occurred on sedimentary rises on the basalt plains.
**Distribution:** Vic, Qld, NSW, SA, WA

# Gnaphalium

**Asteraceae**

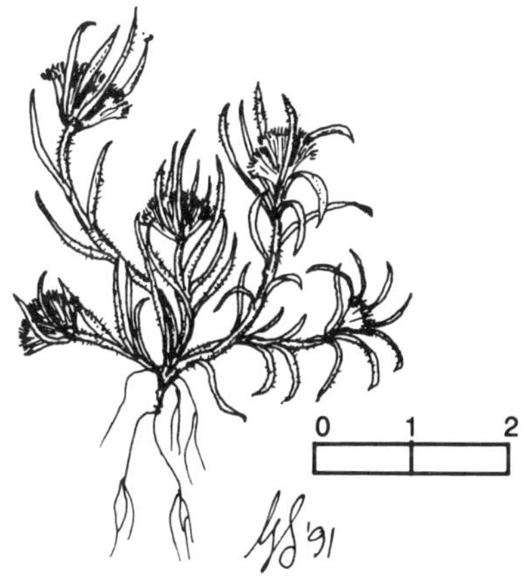

*Gnaphalium indutum*

## Gnaphalium collinum, G. gymnocephalum, G. japonicum = Euchiton gymnocephalus

## Gnaphalium indutum

Tiny Cudweed

**Size:** 10–50 mm x 150 mm F
**Habitat:** Grassy wetlands, red gum woodland, coastal banksia woodland and tea-tree heath
**Form:** Tiny wiry herb
**Foliage:** Narrow woolly grey leaves to 10 mm long
**Flowers:** Deeply clustered terminal creamy-brown flowerheads surrounded by larger bracts. Flowering occurs mainly in the warmer months.
**Requirements:** Open well drained situation
**Comments:** While rarely cultivated it has potential as a small rockery plant.
**Propagation:** Seed, cuttings
**Localities:** Seaford, 5, 7, 35, 42, 67#, 68, 70, 74
**Distribution:** All states except NT

## Gnaphalium involucratum = Euchiton involucratus

## Gnaphalium luteo-album = Pseudognaphalium luteo-album

## Gnaphalium sphaericum = Euchiton sphaericus

# Gompholobium

**Fabaceae**

*Gompholobium huegelii*

## Gompholobium huegelii

Common Wedge-pea, Karalla

**Size:** 0.3–1 m x 0.3–1 m P
**Habitat:** Tea-tree heath, sclerophyll woodland, dry sclerophyll forest
**Form:** Open spreading shrub
**Foliage:** Trifoliate leaves with narrow, pointed bluish-green leaflets to 25 mm long
**Flowers:** Large, bright yellow pea flowers; September to February. Buds and backs of petals are dark olive green.
**Requirements:** Well drained soils
**Comments:** A very attractive plant when in flower, displayed to best advantage when grouped with other plants. Tip pruning encourages bushiness. An excellent container plant.
**Propagation:** Scarified seed, cuttings
**Localities:** Hurstbridge, Warrandyte, 50–51, 53, 56, 59, 60, 67–71
**Distribution:** Vic, Qld, NSW, Tas

SCALE: 0 1cm 2cm The scale representation on each plant equals 2cm.

# Gonocarpus — Haloragaceae

*Gonocarpus elatus*

*Gonocarpus humilis*

*Gonocarpus micranthus*

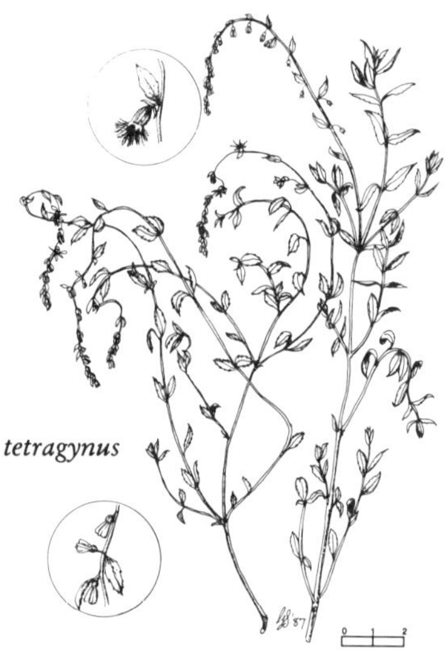

*Gonocarpus tetragynus*

## Gonocarpus elatus

Tall Gonocarpus, Raspwort

**Size:** 0.2–0.6 m x 0.3–1 m — F P
**Habitat:** Exposed rocky hillsides in box woodlands
**Form:** Dense perennial herb with hairy stems and leaves
**Foliage:** Variable green to reddish leaves, narrow to ovate, entire or toothed, 10–60 mm long
**Flowers:** Spikes of very small reddish-brown with yellow-green flowers; October to January. Small round to oval fruit after flowering.
**Requirements:** Well drained soils
**Comments:** The common name refers to the stems of raspworts, which are rough to the touch and 4-ribbed. Locally rare.
**Propagation:** Cuttings
**Localities:** Jacksons Ck Keilor, 26
**Distribution:** Vic, Qld, NSW, SA
**Synonym:** *Haloragis elata*

## Gonocarpus humilis

Raspwort

**Size:** to 0.7 m high — P
**Habitat:** Grassy low open forest, damp and valley sclerophyll forests
**Form:** Semi-prostrate to bushy shrub with minute spreading stem hairs
**Foliage:** Widely spaced decussate pairs of broad, toothed leaves to 10 mm long
**Flowers:** Terminal spikes of tiny greenish-yellow flowers; November to February
**Requirements:** Moist soils
**Comments:** Grows amongst thick undergrowth. Difficult to distinguish from *G. teucrioides* (which does not occur within the Melbourne area) but tends to have less stiffly erect stems and alternate flowers.
**Localities:** The Basin, Kalorama 41, 44, 54–56, 60, 69, 74, 75
**Distribution:** Vic, Qld, NSW, Tas, SA

## Gonocarpus micranthus

Creeping Raspwort

**Size:** Prostrate x 0.5 m — P
**Habitat:** Damp, peaty ground in wattle tea-tree scrub, tea-tree heath and swamp scrub, from coast to foothills
**Form:** Matting plant
**Foliage:** Decussate pairs of small, roundish, bright green leaves to 10 mm long
**Flowers:** Minute red flowers near ends of stems; December to February. Very small grey to reddish fruit.
**Requirements:** Boggy soils, tolerating dry periods
**Localities:** The Basin, Montrose, 47, 48, 51, 54, 56, 62–64, 67–71, 73–76
**Distribution:** Vic, Qld, NSW, Tas, SA; NZ
**Synonym:** *Haloragis micrantha*

## Gonocarpus tetragynus

Common or Poverty Raspwort

**Size:** 10–30 cm x 20–40 cm — F P
**Habitat:** Valley and dry sclerophyll forests on slopes, sclerophyll and red gum woodlands, grassy low open forest, tea-tree heaths and scrubs, and plains grassland
**Form:** Low bushy herb covered in short white appressed stem hairs
**Foliage:** Small lanceolate to elliptic serrated leaves to 12 mm long, pairs decussate
**Flowers:** Loose spikes of tiny pinkish-red flowers; December to February
**Requirements:** Moist to dry, well drained soils
**Localities:** 6, 12, 13, 16, 24–26, 28, 31, 32, 35, 36, 38–71, 73–76
**Distribution:** Vic, Qld, NSW, Tas, SA
**Synonym:** *Haloragis tetragyna*

# Goodenia — Goodeniaceae

Nine species of *Goodenia* are found in the Melbourne area. With the exception of *G. ovata*, all are perennial herbs, forming tufts or rosettes. Most grow in moist well drained habitats, tolerating periods of inundation, and are excellent used as ground covers in the garden. Easily cultivated, goodenias are mostly insignificant until in flower when they are covered with bright yellow flowers, often on long stalks, usually for lengthy periods. Propagation is usually by stem cuttings or division of stolons (where applicable) although leaf cuttings are worth trying for some species.

## Goodenia blackiana

**Size:** Prostrate–10 cm x 10–20 cm — F P
**Habitat:** Dry and valley sclerophyll forest
**Form:** Stoloniferous tufted woolly perennial herb
**Foliage:** Erect obovate to narrow-oblanceolate leaves to 60 mm long with a few scattered teeth, glabrous above, densely woolly below
**Flowers:** Single terminal yellow flowers on long erect stalks to 100 mm; September to February. Bracteoles linear to 12 mm long.
**Requirements:** Well drained soils
**Comments:** Similar to *G. geniculata* which has longer leaves with a persistent covering of stiffer hairs on both sides,. and shorter bracteoles. An excellent rockery or container plant due to its long flowering time.
**Localities:** Hurstbridge, 36, 50A
**Distribution:** Vic, SA

## Goodenia elongata

Lanky Goodenia

**Size:** 20–50 cm x 0.5–1.5 m — P
**Habitat:** Valley sclerophyll forest, swamp scrub, swampy heathland
**Form:** Creeping perennial herb
**Foliage:** Lanceolate to ovate slightly hairy leaves from 10–50 mm long, sometimes with scattered teeth. Smaller stem leaves.
**Flowers:** Large solitary yellow flowers to 25 mm on long slender stalks; October to January
**Requirements:** Damp soil, tolerating waterlogging
**Comments:** This plant may form large colonies in shaded, moist conditions. Pruning after flowering is beneficial.
**Propagation:** Cuttings—firm new growth
**Localities:** 36, 42, 47, 51, 54, 55, 67#, 74, 75
**Distribution:** Vic, NSW, Tas, SA

*Goodenia elongata*

## Goodenia geniculata

Bent Goodenia

**Size:** 5–10 cm x 10–50 cm — F P
**Habitat:** Grassy low open forest, tea-tree heath
**Form:** Perennial matting herb
**Foliage:** Rosettes of glossy oblanceolate leaves to 100 mm long, covered in stiff hairs. Young foliage is lobed.
**Flowers:** Large, solitary pale to bright yellow flowers on stalks longer than the foliage; August to February. Stalk bends after flowering.
**Requirements:** Adaptable, tolerating wet or dry periods
**Comments:** An excellent rockery or container plant due to its long flowering time. Spreads by suckering.
**Propagation:** Cuttings, division
**Localities:** Eltham, 29, 35, 36, 49, 67–71, 74–76
**Distribution:** Vic, NSW, Tas, SA

*Goodenia geniculata*

## Goodenia gracilis

Slender Goodenia

**Size:** 20–50 cm x 10–30 cm — F
**Habitat:** Grassy wetlands
**Form:** Very attractive erect perennial herb
**Foliage:** Narrow leaves from 25–150 mm long on long stalks
**Flowers:** Loose clusters of single, small yellow flowers on tall branched stalks; September to February
**Requirements:** Tolerates periodic inundation
**Comments:** Locally rare.
**Propagation:** Seed
**Localities:** Deer Park, Laverton-Werribee, 16–18, 35
**Distribution:** Vic, Qld, NSW, NT

*Goodenia gracilis*

## Goodenia heteromera

Spreading Goodenia

**Size:** Prostrate–20 cm x 0.5–2.5 m — F P
**Habitat:** Grassy wetlands
**Form:** Vigorous perennial stoloniferous herb, forming dense mats
**Foliage:** Lanceolate leaves to 80 mm long
**Flowers:** Single yellow flowers are held on long stalks which arise from a cluster of smaller stem leaves.
**Requirements:** Moist to wet heavy soil
**Comments:** Useful for soil erosion in small areas.
**Localities:** Laverton, Werribee
**Distribution:** Vic, Qld, NSW, SA

## Goodenia humilis

Swamp Goodenia

**Size:** 5–10 cm x 0.5–1.5 m — F
**Habitat:** Valley sclerophyll forest, wattle tea-tree scrub and swamp scrub
**Foliage:** Narrow lanceolate leaves to 100 mm long
**Flowers:** Profuse terminal clusters of yellow flowers; November to March
**Requirements:** Moist to wet soil
**Comments:** A very adaptable suckering, matting plant. Ideal planted around clay-based pools.
**Localities:** Montrose, 34A, 35, 50A, 54, 60–65, 67#, 69, 69A, 71, 72, 74–76
**Distribution:** Vic, NSW, Tas, SA

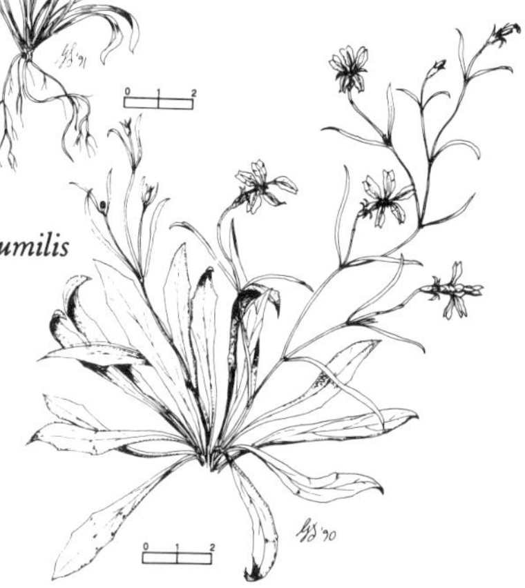

*Goodenia humilis*

## Goodenia lanata

Woolly or Trailing Goodenia

**Size:** Prostrate x 0.5–1.5 m — F P
**Habitat:** Damp and valley sclerophyll forests, sclerophyll woodland
**Form:** Hairy trailing herb, rooting at the nodes
**Foliage:** Dark green, coarsely-toothed obovate leaves to 80 mm long, paler below. Stem leaves are small and fleshy.
**Flowers:** Attractive solitary yellow flowers on long stalks; October to December.
**Requirements:** Tolerates extended dry periods once established.
**Comments:** Good in a container amongst other plants. Hard pruning will rejuvenate leggy plants.
**Localities:** 37, 41–44, 49–52, 54–56, 58, 59
**Distribution:** Vic, Tas

*Illustrated p. 172*

SCALE: 0 1cm 2cm — The scale representation on each plant equals 2cm.

*Goodenia ovata*

### Goodenia ovata — Hop Goodenia

**Size:** 1–2.5 m x 1–3 m — **F P N**
**Habitat:** Widespread from wet sclerophyll forest to grassy low open forest, box and red gum woodlands, also coastal cliffs
**Form:** Small to medium open shrub
**Foliage:** Light green ovate, toothed leaves to 100 mm long
**Flowers:** Sprays of 3–6 yellow flowers; August to February
**Requirements:** Damp soils, tolerating waterlogging
**Comments:** An understorey plant that can become straggly unless pruned. Useful as a cut flower. It is an early coloniser after fire or clearing.
**Localities:** 1, 14, 16, 22, 23, 26, 29–36, 40, 42–44, 46–51, 53–64, 67, 69A–71, 74, 76
**Distribution:** All states except WA

### Goodenia pinnatifida — Cut-leaf Goodenia, Mother Ducks, Scrambled Eggs

**Size:** 0.2–0.5 m x 0.2–0.6 m — **F**
**Habitat:** Red gum woodland, plains grassland, grassy wetlands
**Form:** Tufted annual or perennial herb
**Foliage:** Deeply and narrowly lobed oblong leaves to 80 mm long
**Flowers:** Profuse clusters of large, bright yellow flowers to 50 mm; October to November and March to May.
**Requirements:** Tolerates dryness, but Summer watering will increase flowering time
**Comments:** An excellent garden or container plant for warm areas. May completely die back during dry Summers, regenerating naturally from underground rhizomes following rain. Locally rare.
**Propagation:** Seed, cuttings, stolons
**Localities:** Tottenham, St. Albans, Nth Melbourne, Sunbury, Werribee, 2, 10, 16, 26, 28, 35, 67#
**Distribution:** Vic, Qld, NSW, SA, WA — *Illustrated p. 172*

*Goodia lotifolia*

## Goodia — Fabaceae

### Goodia lotifolia — Golden Tip, Clover Tree

**Size:** 1–5 m x 1–5 m — **F P**
**Habitat:** Damp to dry sclerophyll forests, sclerophyll woodland (on sandstone)
**Form:** Fast growing open shrub
**Foliage:** Trifoliate leaves with ovate bluish-green leaflets to 25 mm long, paler below
**Flowers:** Profuse racemes of fragrant yellow pea flowers with reddish markings; September to December
**Requirements:** Well drained soils, intolerant of alkaline soil
**Comments:** Very showy in flower. Pruning after flowering is necessary to maintain bushy growth. Heavy pruning may induce suckering.
**Propagation:** Scarified seed, stem or root cuttings
**Localities:** Sunbury, 16, 26, 42, 47, 49, 55–59
**Distribution:** All states except NT

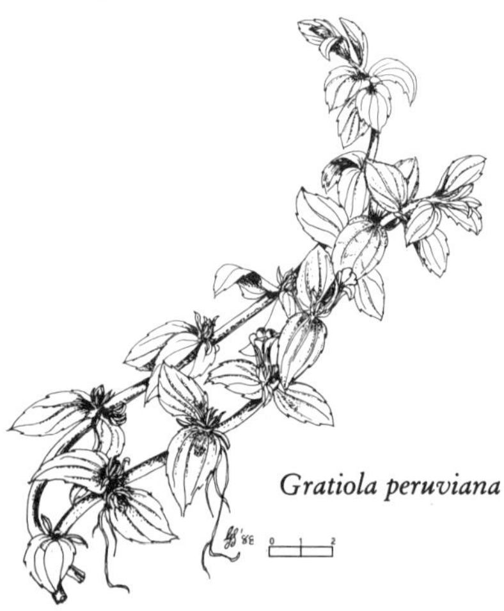

*Gratiola peruviana*

## Gratiola — Scrophulariaceae

### Gratiola peruviana — Austral Brooklime

**Size:** 30 cm x 0.5–1.5 m — **P N**
**Habitat:** Edges of swamps and creeks in riparian scrub, grassy low open forest, valley sclerophyll forests
**Form:** Sprawling or erect perennial herb
**Foliage:** Pairs of pale green, glabrous ovate stem-clasping leaves to 45 mm long x 5–27 mm with lightly-toothed margins. Sometimes leaves have sparse non-gland tipped hairs.
**Flowers:** Solitary pale pink, rarely purple or white, tubular flowers to 12 mm, either stalkless or shortly stalked, in leaf axils; October to May
**Requirements:** Easily grown in most conditions, it prefers a moist position but tolerates waterlogging and drying out once established
**Comments:** A suckering plant, it is excellent for planting beside ponds or in terrariums. The leaves have a powerful laxative effect, and were also used to treat dizziness and loss of appetite.
**Propagation:** Cuttings, division
**Localities:** 16, 26, 30, 34A, 35, 42, 44, 45, 47, 49, 56–58, 60, 67#, 71, 74, 75
**Distribution:** Vic, NSW, Tas, SA, WA; NZ, Sth Am
**Synonym:** *G. latifolia*

*Gratiola pubescens*

### Gratiola pubescens — Brooklime

**Size:** Prostrate–20 cm x 0.6–1.5 m — **P**
**Habitat:** Swamp scrub
**Flowers:** Pale pink flowers with a yellow throat, stalks to 25 mm; October to January
**Requirements:** Damp areas which dry out in Summer
**Comments:** Similar to *G. peruviana* but with densely glandular, pubescent hairs. Leaves are shorter, to 26 mm, and narrower. Locally rare due to drainage of habitat.
**Localities:** Kilsyth, Montrose, Mooroolbark, 74, 75
**Distribution:** Vic, NSW, Tas, SA, WA

49

50

51

52

53

54

55

49. *Epacris impressa* Common Heath
50. *Goodenia ovata* Hop Bush
51. *Amyema pendulum* Drooping Mistletoe
52. *Drosera peltata* ssp. *auriculata* Tall Sundew
53. *Patersonia fragilis* Short Purple-flag
54. *Bursaria spinosa* Sweet Bursaria
55. *Isopogon ceratophyllus* Horny Cone-bush
56. *Lythrum salicaria* Purple Loose-strife
57. *Lomandra longifolia* Spiny-headed Mat-rush
58. *Xanthorrhoea minor* ssp. *lutea* Small Grass Tree

56

57

58

59

60

61

62

64

59. *Themeda triandra* Kangaroo Grass
60. *Kunzea ericoides* Burgan
61. *Platylobium formosum* Handsome Flat-pea
62. *Correa reflexa* Native Fuchsia
63. *Eucalyptus polyanthemos* ssp. *vestita* Red Box
64. *Caladenia dilatata* Green Comb Spider Orchid
65. *Stackhousia monogyna* Candles
66. *Tetratheca ciliata* Pink Bells
67. *Hibbertia stricta* Erect Guinea-flower
68. *Spyridium parvifolium* Dusty Miller

63

67

65

66

68

# Grevillea — Proteaceae

## Grevillea alpina

Alpine or Mountain Grevillea, Cat's Claw

**Size:** Variable, 1–2.5 m x1–2 m — P
**Habitat:** Dry sclerophyll forest in foothills
**Form:** Dense upright shrub (Mt. Dandenong form described)
**Foliage:** Hairy oval leaves to 12 mm long, paler below
**Flowers:** Clusters of hairy, bright red and yellow flowers in terminal clusters; August to December
**Requirements:** Well drained soils
**Comments:** Flowering for a long time, this showy plant does best as an understorey shrub. Frequent pruning when young encourages bushiness in most forms. *G. alpina* is very variable over its range in habit and flower colour, with shades from white and green through yellow and gold to pinks and reds. All make attractive tub specimens. The local form makes a reliable garden plant.
**Propagation:** Cuttings
**Localities:** 55, 56
**Distribution:** Vic, NSW, Tas

*Grevillea alpina*

## Grevillea infecunda

Anglesea Grevillea

**Size:** 0.3–0.6 m x 1.5–2.5 m — F P
**Habitat:** Tea-tree heath
**Form:** Low spreading shrub, hairy young growth and branchlets
**Foliage:** Oblong, holly-like, flat to concave leaves to 60 mm long, teeth sharply pointed. Leaves paler and silky below
**Flowers:** Terminal red and green toothbrush-like flowers to 40 mm long; September to January
**Requirements:** Well drained sandy soil tolerating dryness once established
**Comments:** This unusual species never sets seed but forms colonies by suckering. Responds well to pruning and fire. Now only found in the Anglesea area it was recorded in 1950s occurring in scrub near Brighton. It is also possible that another grevillea collected near Oakleigh (1911) may have been this species.
**Propagation:** Stem and root cuttings
**Localities:** Brighton-Sandringham area#
**Distribution:** Endemic to Vic
**Synonym:** Segregated from *G. aquifolium*

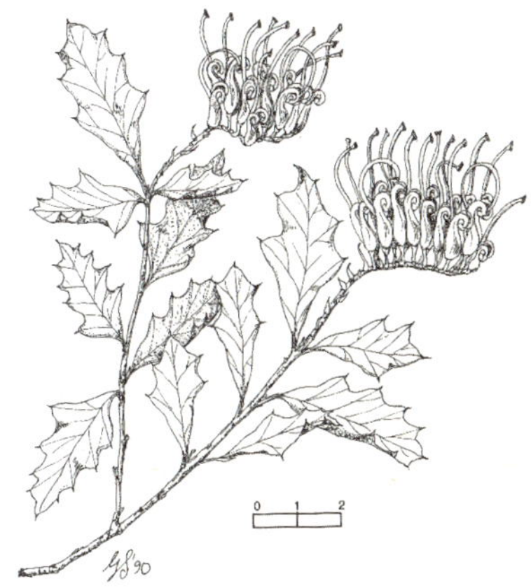

*Grevillea infecunda*

## Grevillea rosmarinifolia

Rosemary Grevillea

**Size:** Variable, 1–3 m x 2–3 m — F P
**Habitat:** Dry sclerophyll forest, plains grassland
**Form:** Dense rounded shrub
**Foliage:** Stiff, prickly needle-like or linear flat leaves to 40 mm long
**Flowers:** Racemes of red and cream, or greenish, spidery flowers drooping from the ends of branches most of the year, peaking Winter and Spring
**Requirements:** Well drained soils
**Comments:** The dense prickly foliage provides an excellent bird habitat and pruning will encourage more compact growth. Some people suffer allergic reactions to the prickly leaves. The local forms of *G. rosmarinifolia* have a very limited distribution within the Melbourne area.

The form from the basalt plains, which is sold as 'Lara Dwarf', is a small, slow-growing compact form with grey leaves and cream and pink flowers. The sole surviving plant in its natural habitat occurs just outside the Melbourne area.

The Hurstbridge/Eltham form is low growing, up to 1 m x 1 m. It has prickly bluish-green linear leaves to 15 mm long and cream flowers with pink tonings.

The Plenty Gorge form is more upright to 1.5 m x 1 m. Leaves are darker green, longer to 25 mm and broader to 3–4 mm and the flowers have stronger pink tonings.

A naturally occurring hybrid between *G. rosmarinifolia* and *G. lanigera* is found in Wattle Glen.

**Propagation:** Cuttings
**Localities:** Werribee#, Hurstbridge, Diamond Ck, Craigieburn, 26, 29#, 34, 34A, 36, 42
**Distribution:** Vic, NSW
**Synonym:** The forms to the north-east of Melbourne have previously been referred to as *G. glabella*.

*Grevillea rosmarinifolia*

SCALE: 0 1cm 2cm — The scale representation on each plant equals 2cm.

*Gynatrix pulchella*

# Gynatrix **Malvaceae**

## Gynatrix pulchella

Hemp Bush

**Size:** 2–4 m x 1.5–3 m **P**
**Habitat:** Damp and valley sclerophyll forests, riparian scrub, red gum woodland, often on rocky sites
**Form:** Open woody shrub
**Foliage:** Hairy heart-shaped leaves to 100 mm long with rounded teeth along the edges, pale green below
**Flowers:** Panicles of fragrant, small greenish-white flowers; August to October. Male and female flowers on different plants, with shorter petals on the female flowers.
**Requirements:** Well drained moist soil
**Comments:** Often profuse in flowering. Can become straggly, but regular pruning will overcome this. The Aborigines made string from the bark.
**Propagation:** Seed, cuttings
**Localities:** Diamond Ck, 23, 26–30, 32, 33, 34A–36, 41, 42, 51, 56, 62–66
**Distribution:** Vic, NSW, Tas

# Hakea **Proteaceae**

Four hakeas occur in the Melbourne region, growing in open forest and heathland. All have pungent-tipped foliage and clusters of flowers in the leaf axils. Their seeds are protected by large woody fruit which open after fire or damage. A favourite among birds, they provide food and protection from predators. Propagation is by seed.

*Hakea nodosa*

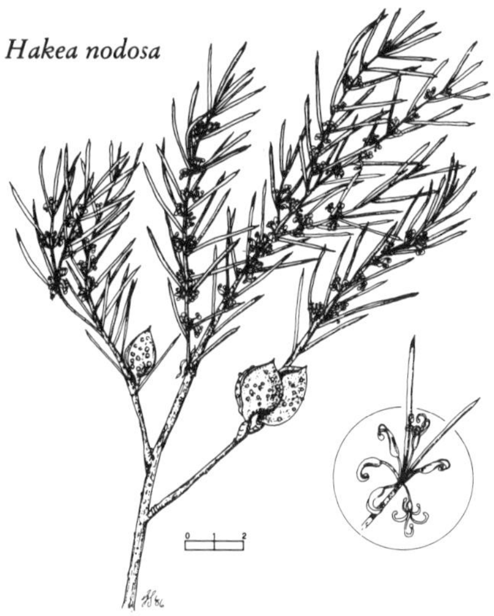

## Hakea nodosa

Yellow Hakea

**Size:** 1–3 m x 1–2 m **F P N**
**Habitat:** Swamp scrub, sclerophyll woodland, valley sclerophyll forest, from coast to foothills
**Form:** Erect, dense to open shrub
**Foliage:** Blue-green needles to 40 mm long, not as sharp as the other species
**Flowers:** Masses of small fragrant yellow flowers; April to August. Oval warty fruit to 30 mm long.
**Requirements:** Tolerates poor drainage but grows well in dry or moist conditions
**Comments:** Attractive in flower this quick growing shrub provides a good low screen. Plants may be invigorated by hard pruning.
**Localities:** 29, 47, 49–51, 54–56, 59–61, 63, 64, 67#, 70, 74, 75,
**Distribution:** Vic, Tas, SA

## Hakea ~~sericea~~ decurrens

Silky Hakea, Bushy Needlewood

**Size:** 2–5 m x 1–3 m **F P**
**Habitat:** Sclerophyll woodland, box and red gum woodland, valley sclerophyll forest
**Form:** Fast-growing erect open to dense shrub
**Foliage:** Stiff prickly needles to 60 mm long
**Flowers:** Masses of fragrant, hairy, white, or occasionally pink, flowers; May to September. Prominent rough fruit to 40 mm long have a double-pointed beak.
**Requirements:** Adaptable to any conditions
**Comments:** The prickly foliage provides a safe habitat for birds. Pruning is beneficial. Due to the indiscriminate planting of this species, and possible garden escapes, it is uncertain what the true range of this species is in Melbourne. This species is presently undergoing revision. It appears that it is different to the typical *H. sericea* which occurs in south-east NSW.
**Propagation:** Seed, cuttings
**Localities:** Montrose, 18, 29, 33–34A, 42, 47–49, 54, 56, 58, 62–64, 74, 75
**Distribution:** Vic, NSW, Tas *Illustrated p. 172*

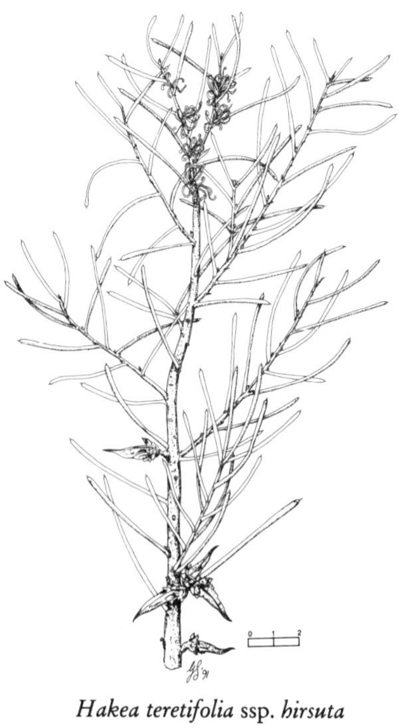

*Hakea teretifolia* ssp. *hirsuta*

## Hakea teretifolia ssp. hirsuta

Dagger Hakea

**Size:** 2–3 m x 1–3 m **F**
**Habitat:** Sclerophyll woodland
**Form:** Dense, upright to spreading shrub
**Foliage:** Stiff, sharp needle-like leaves to 50 mm long, new growth hairy and white
**Flowers:** Masses of very hairy white flowers; November to February. Fruit are narrow and dagger-shaped to 30 mm long with a sharp point.
**Requirements:** Whilst occurring naturally on moist sandy soil, it is adaptable to any well drained situation
**Comments:** Seed germinates readily after fires, forming dense stands.
**Localities:** Narre Warren East 1907; 49, 50A, 61, 74. Occurred in swamps near the Dandenong Rges in 1853.
**Distribution:** Vic, Qld, NSW, Tas

### Hakea ulicina

Furze Hakea

**Size:** 1–3 m x 1–2 m **F P**
**Habitat:** Tea-tree heath, sclerophyll woodland, dry and valley sclerophyll forest
**Form:** Rigid upright shrub
**Foliage:** Narrow, pointed leaves to 20 cm long
**Flowers:** Small clusters of white to cream flowers; July to November. Clusters of smooth short-beaked fruit to 20 mm long.
**Requirements:** Well drained dry to moist soils
**Comments:** Stands of this narrow erect hakea form an interesting landscape feature.
**Localities:** Montrose, 34, 49–51, 54, 56, 58–61, 63, 64, 67#, 70, 73–75
**Distribution:** Vic, NSW, Tas

*Hakea ulicina*

# Haloragis — Haloragaceae

### Haloragis aspera

Rough Raspwort

**Size:** 10–25 cm high **F P**
**Habitat:** Plains grassland, box woodland
**Form:** Erect perennial herb
**Foliage:** Sessile grey-green lanceolate leaves to 50 mm long, hairy and usually toothed
**Flowers:** Small clusters of red to green flowers in spikes at the end of stems; Spring to Autumn. Anthers are yellow.
**Requirements:** Moist heavy clay soils
**Comments:** Stems are rough and hairy, the hairs ending in a hook. The stems grow from underground stolons.
**Propagation:** Seed, division
**Localities:** 5, 7, 12, 17#, 18, 26, 28, 63
**Distribution:** All states

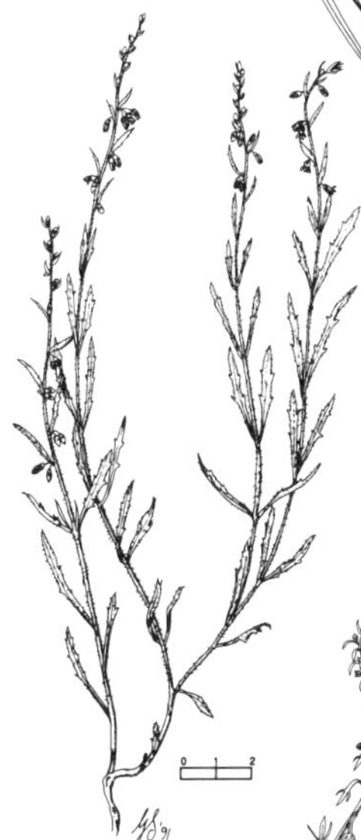

*Haloragis aspera*

### Haloragis heterophylla

Variable Raspwort

**Size:** 0.1–0.5 m high **F P**
**Habitat:** Widespread in riparian scrub and plains grassland
**Comments:** Perennial herb similar to *H. aspera* but more slender, with leaves that have fewer more deeply incised teeth. Flowers have red anthers and occur during Spring and Summer. Fruit are smaller.
**Localities:** Warrandyte, 5, 12, 18, 25, 26, 28, 35, 38, 44, 62, 71, 74
**Distribution:** Vic, Qld, NSW, Tas, SA

*Haloragis heterophylla*

# Halosarcia — Chenopodiaceae

### Halosarcia halocnemoides ssp. halocnemoides

Shrubby Samphire or Glasswort

**Size:** 5–15 cm high **F**
**Habitat:** Saltmarshes
**Form:** Erect branched sub-shrub. Stems are succulent and composed of glossy segments which are usually red or yellow.
**Foliage:** Reduced to rim on segments
**Flowers:** In short terminal spikes 5–20 mm long; November to April
**Requirements:** Withstands extreme salinity and poor drainage
**Propagation:** Seed
**Localities:** Footscray, 2–4
**Distribution:** Vic, SA, WA
**Synonym:** *Arthrocnemum halocnemoides*

*Halosarcia halocnemoides* ssp. *halocnemoides*

### Halosarcia pergranulata ssp. pergranulata

**Size:** 0.2–0.6 m high **F**
**Habitat:** Saltmarshes
**Comments:** Similar to *H. halocnemoides*, but in the Melbourne region it is a larger, erect or ascending (rarely decumbent) shrub. It has conspicuous fertile spikes 10–50 mm long. The seed is ribbed and more granular.
**Localities:** 2–4
**Distribution:** All mainland states
**Synonym:** *Arthrocnemum halocnemoides* var. *pergranulata*

*Halosarcia pergranulata* ssp. *pergranulata*

SCALE: 0 1cm 2cm — The scale representation on each plant equals 2cm.

*Hardenbergia violacea*

# Hardenbergia — Fabaceae

## Hardenbergia violacea — Purple Coral Pea, False Sarsparilla

F P

**Habitat:** Red gum woodland, dry and valley sclerophyll forests
**Form:** Fast growing dense creeper or wiry scrambler
**Foliage:** Leathery dark green ovate to lanceolate leaves to 100 mm long
**Flowers:** Sprays of mauve to purple, or sometimes pink or white, pea flowers massed along branches; July to November
**Requirements:** Well drained soils
**Comments:** A very showy scrambling climber with many forms and cultivars. The local form has dark purple flowers with a prominent green centre. It is useful for covering embankments or providing a light screen. The flowers are a source of dye. The cultivar "Happy Wanderer" is the most commonly available form of *H. violacea*. Bushy forms of *H. violacea* are also sold.
**Propagation:** Scarified seed
**Localities:** 1, 16, 21, 25, 26, 29, 30, 32–44, 46–56, 58–64, 67#, 74, 76
**Distribution:** All states

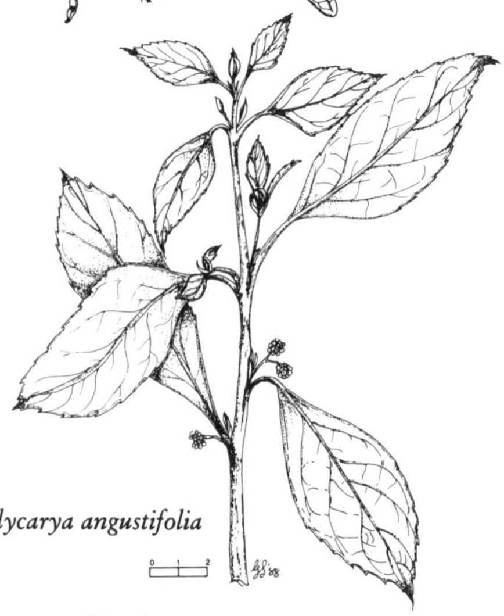

*Hedycarya angustifolia*

# Hedycarya — Monimiaceae

## Hedycarya angustifolia — Austral or Native Mulberry

**Size:** 3–7 m x 4 m — N
**Habitat:** Fern gullies of cool temperate rainforest, wet and damp sclerophyll forests
**Form:** Slender, large shrub or small tree
**Foliage:** Toothed, glossy dark green oval leaves to 120 mm long, pale green below, conspicuously veined
**Flowers:** Inconspicuous pale green male and female flowers during Spring
**Requirements:** Rich, moist well drained soil
**Comments:** The yellow fruit resembles a mulberry but is inedible. The Aborigines used the stems and wood for spear ends and firemaking.
**Propagation:** Cuttings
**Localities:** 49#, 55–58
**Distribution:** Vic, Qld, NSW

# Helichrysum — Asteraceae

Helichrysums are referred to as everlasting daisies. Their flowers are long-lasting and often very showy. They are very adaptable and brighten up a rockery or bush garden. It is preferable to prune them back in late Winter to encourage new growth. Propagation is by seed or cuttings. Due to a recent botanical revision, *Helichrysum* species have been separated into 4 genera.

*Helichrysum leucopsideum*

## Helichrysum apiculatum = Chrysocephalum apiculatum

## Helichrysum bracteatum = Bracteantha bracteata

## Helichrysum dendroideum = Ozothamnus ferrugineus

## Helichrysum leucopsideum — Satin Everlasting

**Size:** 0.1–0.5 m x 0.6 m — F P
**Habitat:** Coastal banksia and sclerophyll woodlands
**Form:** Erect hairy perennial herb
**Foliage:** Narrow leaves to 50 mm long, white and woolly below
**Flowers:** Single white paper daisy flowers with yellow centres on long stalks; October to March
**Requirements:** Adaptable but prefers well drained soil
**Localities:** Port Phillip Bay area, Belgrave, 55, 56
**Distribution:** All states except Qld

## Helichrysum obcordatum = Ozothamnus obcordatus

## Helichrysum paralium = Ozothamnus turbinatus

## Helichrysum rosmarinifolium = Ozothamnus rosmarinifolius

## Helichrysum rutidolepis

Pale Everlasting

**Size:** 0.2–0.5 m x 0.5–1.5 m **F P**
**Habitat:** Wet sites in red gum woodland, plains grassland, grassy wetland, dry sclerophyll forest
**Form:** Erect, hairy rhizomatous perennial herb with stems branching from above the base
**Foliage:** Stem-clasping and grey-green, lower leaves broadly oblanceolate, upper leaves narrower
**Flowers:** Single yellow button flowerheads 15–20 mm wide; Spring
**Requirements:** Moist well drained soil
**Comments:** An interesting rockery plant spreading by stolons. Locally rare. The species *H.* sp. aff. *rutidolepis* is now included as part of this species.
**Localities:** Diamond Ck, 2, 7, 16, 17#, 26, 61
**Distribution:** Vic, Qld, NSW, SA

*Helichrysum rutidolepis*

## Helichrysum scorpioides

Curling or Button Everlasting

**Size:** 30 cm x 20–30 cm **F P**
**Habitat:** Widespread in tea-tree heath, sclerophyll woodland, dry sclerophyll forest
**Form:** Woolly perennial herb spreading by underground stems to form a dense mat. Branched from the base.
**Foliage:** Basal clump of broad velvety leaves to 70 mm long. Stem leaves become smaller. Similar to *H. rutidolepis* but upper surface of leaves has coarse glandular hairs.
**Flowers:** Single flat yellow flowerheads 20–30 mm wide, larger than *H. rutidolepis*, surrounded by brown papery bracts, held on long stems; September to December
**Requirements:** Well drained soils
**Comments:** An attractive rockery plant which dies back after flowering. Flowers dry well.
**Localities:** 16, 26, 30, 33, 34A, 36, 41–56, 58, 59, 61–64, 67–69, 71, 74–76
**Distribution:** Vic, Qld, NSW, Tas, SA

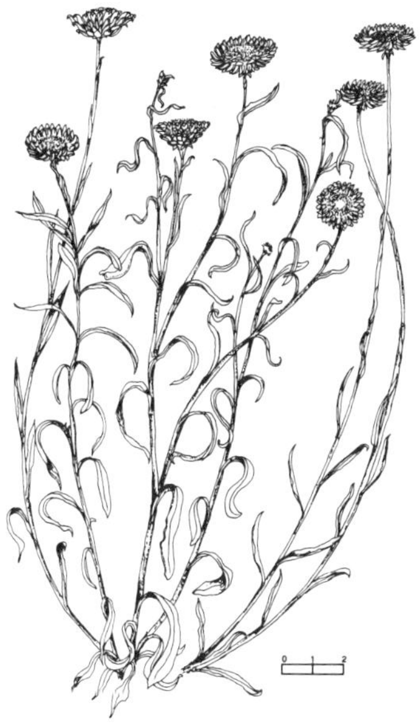

*Helichrysum scorpioides*

## Helichrysum semipapposum = Chrysocephalum semipapposum

## Helichrysum viscosum = Bracteantha viscosa

## Helipterum albicans ssp. albicans var. albicans = Leucochrysum albicans ssp. albicans var. albicans

## Helipterum anthemoides = Rhodanthe anthemoides

## Helipterum australe = Triptilodiscus pygmaeus

## Helipterum corymbiflorum = Rhodanthe corymbiflorum

# Hemichroa — Chenopodiaceae

## Hemichroa pentandra

Trailing Jointweed

**Size:** Prostrate x 30 cm **F**
**Habitat:** Saltmarshes and shorelines
**Form:** Spreading perennial herb
**Foliage:** Succulent, narrowly-ovate leaves to 12 mm long
**Flowers:** Single, tiny white flowers along the stems enclosed in papery bracts; November
**Requirements:** Wet salty soil
**Propagation:** Cuttings
**Localities:** 2–4, 67, 74
**Distribution:** Vic, NSW, Tas, SA, WA

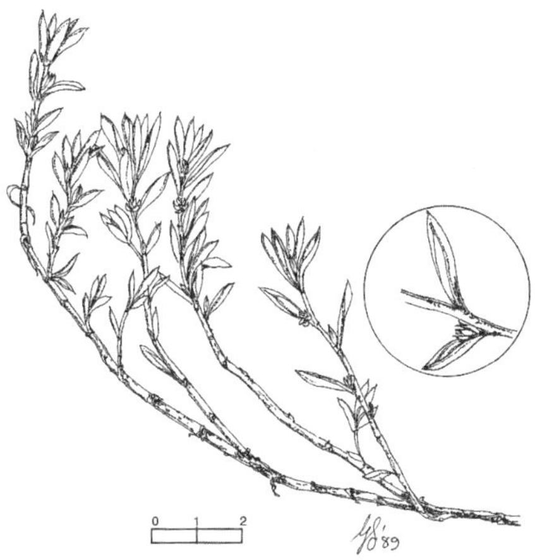

*Hemichroa pentandra*

SCALE: 0 1cm 2cm — The scale representation on each plant equals 2cm.

# Hibbertia

# Dilleniaceae

The guinea flowers are small shrubs with masses of showy yellow flowers. All make excellent garden or container plants. Pruning after flowering encourages compact growth. Propagate from cuttings.

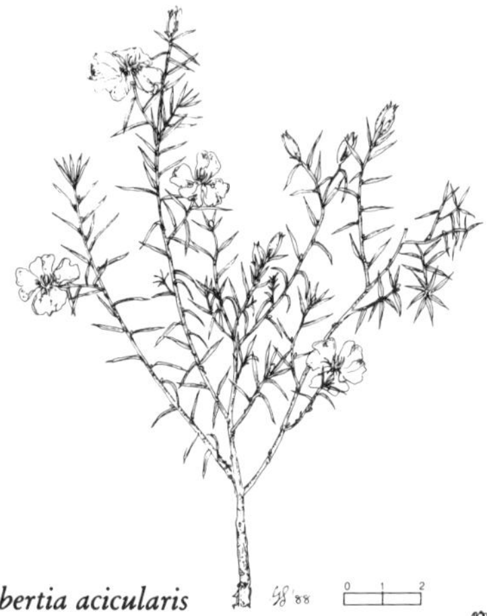

*Hibbertia acicularis*

## Hibbertia acicularis

Prickly Guinea-flower

**Size:** Prostrate-30 cm x 0.3–0.6 m **F P**
**Habitat:** Tea-tree heath, sclerophyll woodland
**Form:** Erect or sprawling small shrub
**Foliage:** Narrow pointed leaves to 12 mm, yellow below
**Flowers:** Masses of bright yellow flowers; September to February
**Requirements:** Well drained soils
**Comments:** On coastal sands it is a compact, rounded shrub, but varies according to competition from other plants. New growth is bright red and tip pruning accentuates this feature.
**Localities:** 67, 69, 70, 73–75
**Distribution:** Vic, Qld, NSW, Tas

*Hibbertia empetrifolia*

## Hibbertia empetrifolia

Trailing Guinea-flower

**Size:** 0.6–1 m x 2 m **F P**
**Habitat:** Valley sclerophyll forest
**Form:** Wiry trailing shrub covered in tiny bristly hairs
**Foliage:** Rough dark green elliptical leaves to 10 mm long, margins slightly recurved
**Flowers:** Masses of bright yellow flowers to 10 mm; August to October
**Requirements:** Well drained soils
**Comments:** Excellent scrambling shrub for embankments and under established trees and shrubs, providing a massed display. Grows more vigorously in moist soils, and may smother smaller plants. It may be encouraged to climb fences or pruned to form a low hedge.
**Localities:** 51, 53
**Distribution:** Vic, Qld, NSW, Tas
**Synonym:** *H. astrotricha, H. billardieri*

*Hibbertia obtusifolia*

## Hibbertia obtusifolia

Showy or Grey Guinea-flower

**Size:** 0.3–0.6 m x 1 m **P**
**Habitat:** Dry sclerophyll forest, box ironbark woodland
**Form:** Open rounded to spreading shrub
**Foliage:** Hairy broad to rounded grey-green leaves to 20 mm long
**Flowers:** Masses of large golden yellow flowers; August to February
**Requirements:** Well drained soil, tolerating dry shade once established
**Comments:** Very showy rockery or container plant with some flowers appearing throughout the year. The prostrate form commonly available from nurseries is not local to the Melbourne area.
**Localities:** 42, 44, 47, 49, 50A, 51,
**Distribution:** Vic, Qld, NSW, Tas

## Hibbertia procumbens

Spreading Guinea-flower

**Comments:** After further research this species was found to be incorrectly included in the first edition.

*Hibbertia prostrata*

## Hibbertia prostrata

Stalked or Bundled Guinea-flower

**Size:** 0.3–0.6 m x 0.3 m **F P**
**Habitat:** Tea-tree heaths
**Form:** Low erect sub-shrub
**Foliage:** Soft, hairy needle-like leaves to 25 mm long in dense clusters along the stems
**Flowers:** Clusters of yellow flowers for a long period, peaking September to November
**Requirements:** Moist well drained sandy soil
**Comments:** The plant is at first upright, the stems bending to the ground as they become too tall to support their own weight. Difficult in clay soils.
**Localities:** Port Melbourne#, 61, 67–71, 73–76
**Distribution:** Vic, Qld, NSW, Tas, SA
**Synonym:** *H. fasciculata*

*Hibbertia riparia*

## Hibbertia riparia

Erect Guinea-flower

**Size:** 0.3–1 m x 0.6 m **F P**
**Habitat:** Sclerophyll woodland, grassy low open forest, valley sclerophyll forest, tea-tree heath
**Form:** Open erect shrub
**Foliage:** Narrow, sometimes stiff leaves to 10 mm long, variable in hairiness
**Flowers:** Sessile yellow flowers during Spring and Summer
**Requirements:** Moist well drained soil, responding to Summer watering.
**Localities:** 47, 48, 64, 69, 70, 74, 76
**Distribution:** All states except NT
**Synonym:** Previously referred to as *H. stricta*

### Hibbertia sericea var. sericea

Silky Guinea-flower

**F P**

**Size:** 0.3–1 m x 0.6 m
**Habitat:** Tea-tree heath
**Form:** Small erect shrub covered in silky hairs
**Foliage:** Dark green oblong leaves to 25 mm long
**Flowers:** Profuse terminal clusters of yellow flowers; October to December
**Requirements:** Well drained soil
**Comments:** Cut back after flowering.
**Localities:** 33, 34A, 67–71, 73–75
**Distribution:** Vic, Qld, NSW, Tas, SA

*Hibbertia sericea* var. *sericea*

### Hibbertia stricta var. stricta

Erect Guinea-flower

**F P N**

**Size:** 0.4–1 m x 0.3–1 m
**Habitat:** Widespread in sclerophyll woodland, valley sclerophyll forest, tea-tree heath
**Form:** Showy rounded to erect shrub
**Foliage:** Narrow dull grey-green leaves to 12 mm long
**Flowers:** Light yellow distinctly stalked flowers may occur at any time, peaking August to November
**Requirements:** Well drained moist soils
**Comments:** Drought tolerant once established. Similar to *H. riparia* which has either sessile or sub-sessile flowers.
**Localities:** Newport#, 33, 34A, 35, 44, 46, 47, 49–51, 58–63, 65–68, 74, 75
**Distribution:** All states
**Synonym:** Previously referred to as *H. australis*

### Hibbertia virgata var. virgata

Twiggy Guinea-flower

**F P**

**Size:** 0.3–0.6 m x 0.6–1 m
**Habitat:** Tea-tree heath
**Form:** Open, twiggy shrub
**Foliage:** Narrow channelled leaves to 20 mm long, grouped along the stems
**Flowers:** Very showy, bright yellow flowers to 25 mm wide, produced over a long period from Winter to Spring
**Localities:** Baxter, 74
**Distribution:** Vic, Qld, NSW, Tas, SA

*Hibbertia stricta* var. *stricta*

# Hovea — Fabaceae

### Hovea linearis

Common Hovea, Blue Bonnet, Bird's Eye

**P N**

**Size:** 0.3–0.6 m x 0.3 m
**Habitat:** Dry and valley sclerophyll forest, tea-tree heath
**Form:** Open trailing sub-shrub
**Foliage:** Variable olive-green leaves, lower leaves oval, becoming narrower and longer up the stem to 65 mm long
**Flowers:** Small mauve (or rarely white) pea flowers along the stems August to October. Rounded green seed pods turn black when ripe.
**Requirements:** Dry well drained soil
**Comments:** A dainty little plant to grow amongst other plants and under eucalypts.
**Propagation:** Scarified seed, cuttings
**Localities:** 34, 34A, 36, 37, 39–44, 46–56, 58, 59, 62–64, 67#, 70
**Distribution:** Vic, Qld, NSW, Tas, SA
**Synonym:** *H. heterophylla*

*Hovea linearis*

# Hydrocotyle — Apiaceae

Hydrocotyles are either prostrate stoloniferous perennial herbs or small erect annuals. Their leaves are circular to kidney-shaped, and may be entire or lobed, usually with toothed margins. The inconspicuous flowers are grouped together in umbels. Their perfume is not always pleasant. They are useful ground covers when grown in shady, moist areas. Propagation is by seed, division of rooted nodes or cuttings.

### Hydrocotyle callicarpa

Small or Tiny Pennywort

**F P N**

**Size:** 2.5–8 cm high
**Habitat:** Valley sclerophyll forest, primary dune scrub, grassy low open forest
**Form:** Erect annual herb
**Foliage:** Kidney-shaped stalkless leaves to 10 mm across, deeply divided into 3–5 lobed segments
**Flowers:** 6–16 minute light green flowers, in umbels on long stalks; August to November
**Requirements:** Moist well drained soil
**Localities:** 35, 36, 41, 42, 44, 47, 48, 53, 55, 67#, 74–76
**Distribution:** Vic, NSW, Tas, SA, WA

SCALE: 0 1cm 2cm — The scale representation on each plant equals 2cm.

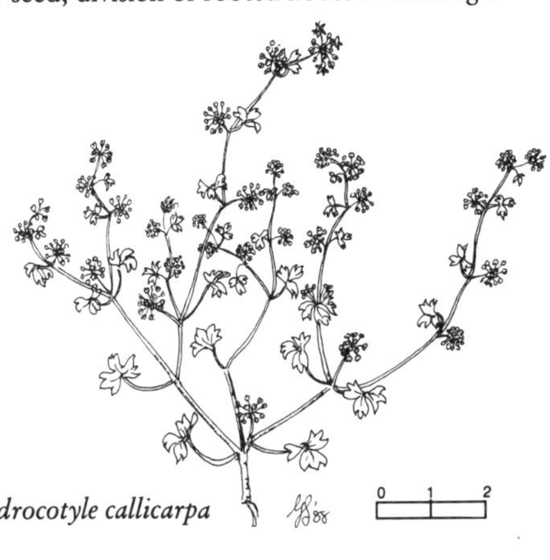

*Hydrocotyle callicarpa*

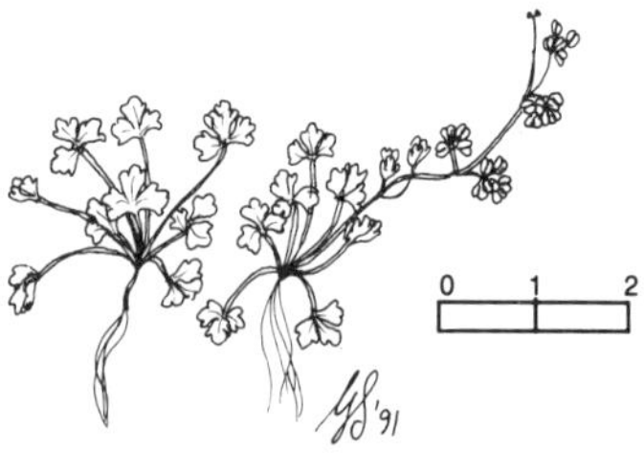


*Hydrocotyle capillaris*

## Hydrocotyle capillaris

Thread Pennywort

**F P**

**Size:** 12 mm x 25 mm
**Habitat:** Valley sclerophyll forest, grasslands, saltmarshes
**Form:** Wiry, erect annual
**Foliage:** 3-5 lobed leaves to 8 mm long
**Flowers:** Nearly stalkless umbels of tiny white flowers
**Comments:** Locally rare.
**Localities:** Werribee 1911, Research, 3, 41, 42
**Distribution:** Vic, NSW, Tas, SA, WA

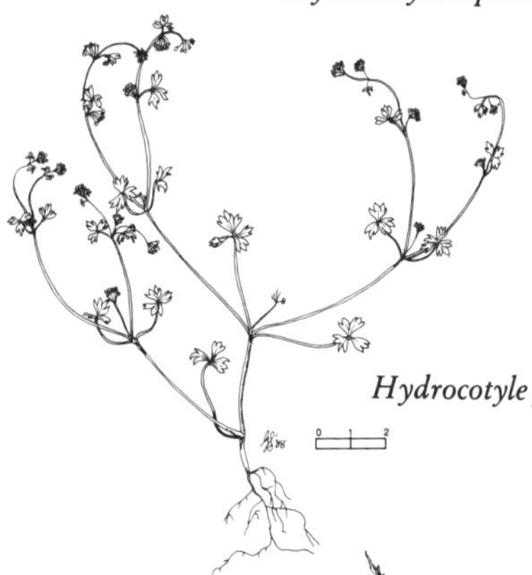

*Hydrocotyle foveolata*

## Hydrocotyle foveolata

Yellow Pennywort

**P**

**Size:** to 10 cm high
**Habitat:** Valley sclerophyll forest, tea-tree heath
**Form:** Tiny erect annual
**Foliage:** Yellowish, 3-lobed leaves with scattered long hairs, each lobe having 3 blunt teeth
**Flowers:** Stalkless clusters of 3-6 tiny flowers; September to November.
**Localities:** Warrandyte, 36, 41, 44, 50A, 51, 54, 55, 74, 75
**Distribution:** Vic, Tas, SA

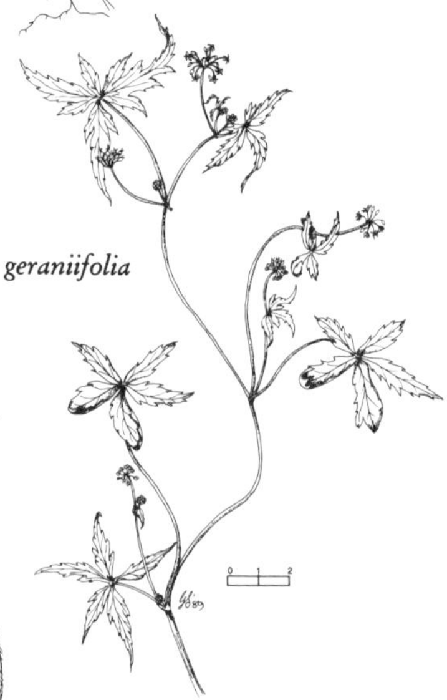

*Hydrocotyle geraniifolia*

## Hydrocotyle geraniifolia

Forest Pennywort

**P N**

**Habitat:** Fern gullies of wet and valley sclerophyll forest
**Form:** Trailing perennial herb
**Foliage:** Dark green leaves are deeply divided into 3-5 toothed, lanceolate segments
**Flowers:** Umbels of stalked small white flowers; Summer
**Requirements:** Moist soil
**Localities:** 55-58
**Distribution:** Vic, NSW

*Hydrocotyle hirta*

*Hydrocotyle hirta*

## Hydrocotyle hirta

Hairy Pennywort

**P**

**Habitat:** Damp, wet and valley sclerophyll forests, swamp and wattle tea-tree scrub
**Form:** Carpeting perennial herb
**Foliage:** Hairy stems and round or kidney-shaped leaves to 25 mm long on long stalks
**Flowers:** Tiny stalkless pale yellow flowers in a tight cluster; Summer
**Comments:** Spreads by above ground creeping stems.
**Localities:** 34A, 42, 44, 47, 54-58, 60, 71, 74-76
**Distribution:** All states

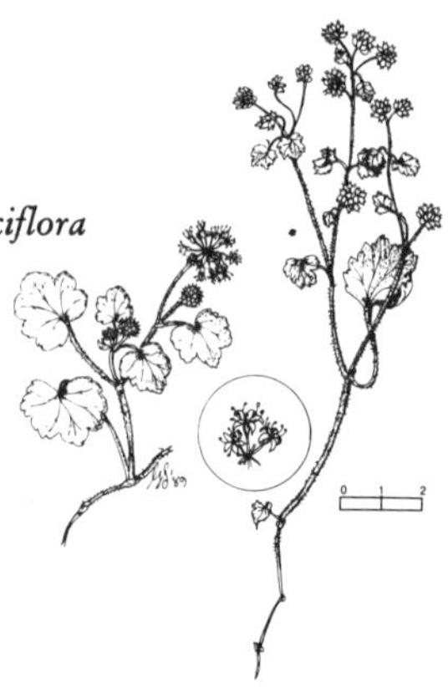

*Hydrocotyle laxiflora*

## Hydrocotyle laxiflora

Stinking Pennywort

**P N**

**Size:** 10-20 cm x 1-2 m
**Habitat:** Riparian woodland, dry and valley sclerophyll forests
**Form:** Creeping perennial herb
**Foliage:** Hairy, round or kidney-shaped leaves to 40 mm long with 5-11 scalloped lobes
**Flowers:** Globular umbels of tiny greenish-yellow flowers to 20 mm across; mainly October to December, emitting a strong offensive odour. Male flowers are on long stalks whilst female flowers occur on shorter stalks.
**Requirements:** Moist soils on poor sites
**Comments:** Spreads by underground rhizomes. A useful groundcover in shady bush areas.
**Localities:** 26, 29#, 35, 39, 41, 42, 44, 47-49, 50A, 58-60, 62-67, 69#, 71, 72, 74, 76
**Distribution:** Vic, Qld, NSW, Tas, SA

## Hydrocotyle medicaginoides

Trefoil Pennywort

**P**

**Size:** to 15 cm high
**Habitat:** Wattle tea-tree scrub
**Form:** Slender annual herb
**Foliage:** Fan-shaped leaves to 15 mm long have 3 deeply cut lobes
**Flowers:** Globular umbel of almost sessile greenish-white flowers tinged with purple; September to November
**Requirements:** Moist soil, tolerating some saltiness
**Comments:** Locally rare.
**Localities:** 74
**Distribution:** Vic, SA, WA

## Hydrocotyle muscosa

Mossy Pennywort

**Size:** to 20 cm high — P
**Habitat:** Freshwater swamps and along watercourses in grassy low open forest
**Form:** Dense matting perennial herb with thread-like stems, spreading by underground stolons
**Foliage:** Round shiny leaves to 15 mm across, deeply divided into 5 wedge-shaped leaflets, each entire or 3-toothed at the end; long hairs below.
**Flowers:** Small long-stalked umbels of 3–4 sessile deep purplish-red flowers; September to March
**Requirements:** Moist to wet soils
**Localities:** 60
**Distribution:** Vic, Tas, SA

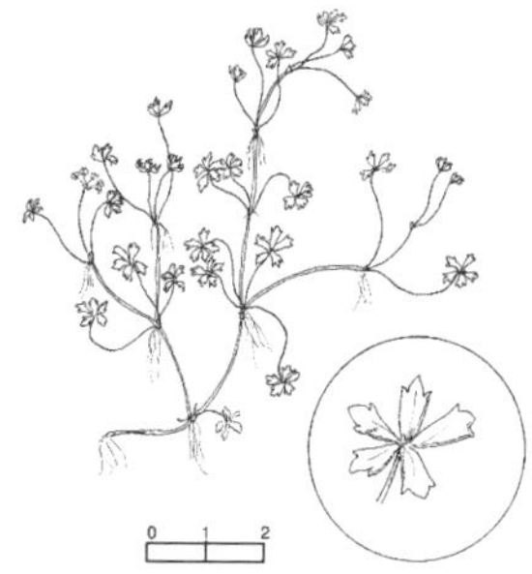

*Hydrocotyle muscosa*

## Hydocotyle sibthorpioides

Shining Pennywort

**Size:** Prostrate x 0.6–2 m — P
**Habitat:** Valley sclerophyll forest, grassy wetlands
**Form:** Trailing perennial matting herb
**Foliage:** Hairy, shiny light green circular leaves
**Flowers:** Tiny stalkless green flowers in terminal, globular umbels
**Requirements:** Moist soils, tolerating some sun
**Comments:** Often found in poorly drained lawns.
**Localities:** Laverton Nth, Rockbank, 17, 35, 41, 42, 69A
**Distribution:** Vic, NSW, Tas; PNG

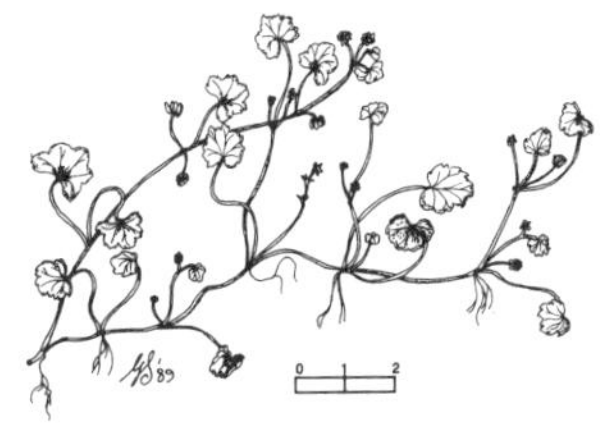

*Hydrocotyle sibthorpioides*

## Hydrocotyle tripartita

Slender Pennywort

P

**Habitat:** Grassy low open forest, box woodland
**Form:** Hairy, slender trailing perennial herb
**Foliage:** Leaves deeply divided into 3 toothed segments
**Flowers:** Minute, almost stalkless, purplish-red flowers in umbels all year, peaking November to March
**Requirements:** Moist soils
**Localities:** St. Kilda#, 17#, 18, 34A, 44, 60, 67#
**Distribution:** Vic, Qld, NSW, Tas, SA

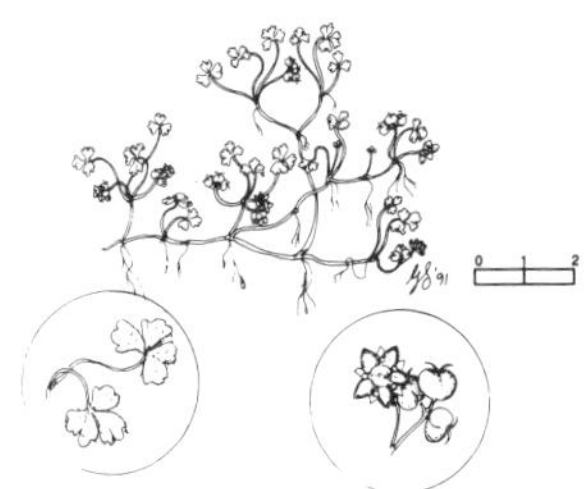

*Hydrocotyle tripartita*

## Hydrocotyle verticillata

Common or Shield Pennywort

P

**Habitat:** Freshwater swamps in riparian and wattle tea-tree scrub, valley sclerophyll forest
**Form:** Creeping perennial herb
**Foliage:** Circular light green leaves to 60 mm across, leaf attached to the stalk on its undersurface (peltate)
**Flowers:** Flower stalks rise from creeping roots, having whorls of small umbels of white flowers; November to April
**Requirements:** Wet soil
**Localities:** 28, 35, 36, 67#
**Distribution:** Vic, Qld, NSW, SA

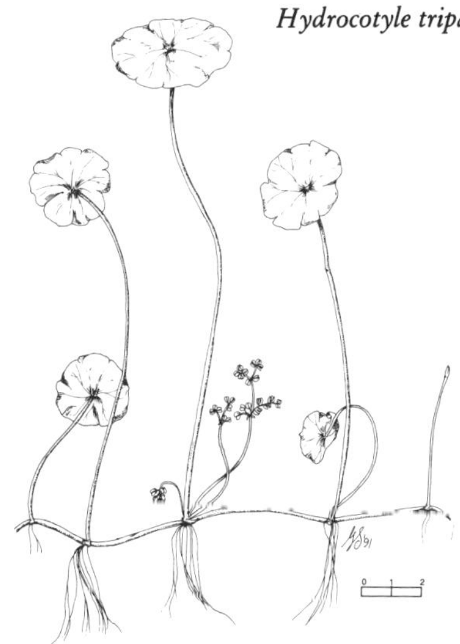

*Hydrocotyle verticillata*

Now Melicytus dentatus

# Hymenanthera — Violaceae

## Hymenanthera dentata

Tree Violet

**Size:** 2–4 m x 1–2.5 m — F P
**Habitat:** Riparian scrub, box woodland
**Form:** Rigid, often spiny, spreading shrub
**Foliage:** Dark green oblong, toothed leaves to 50 mm long, paler green below
**Flowers:** Masses of fragrant tiny cream bell flowers hanging along branches; September to November, followed by pale green to purple berries
**Requirements:** Well drained soils responding to extra watering
**Comments:** A variable shrub being lush in lightly shaded gullies but becoming stunted and spiny in exposed rocky areas.
**Propagation:** Seeds, cuttings
**Localities:** Werribee R., Seaford, 8, 14–16, 18, 23–30, 32–36, 39, 42, 44, 47, 61–64
**Distribution:** Vic, NSW, Tas, SA
**Synonym:** The variety *angustifolia* has now been combined with this species.

*Hymenanthera dentata*

## Hymenanthera sp. aff. dentata

**Size:** to 1.5 m high — F
**Habitat:** Plains grassland, escarpments
**Form:** Intricately branched shrub, branches spine-tipped
**Comments:** Differs from *H. dentata* in that leaves are smaller and dark green and bluer fruit is often spotted. Flowers slightly later in the Melbourne area. Prefers drier sites. A threatened species which occurs on private property and is therefore not protected. Confused with the weed Boxthorn by landowners.
**Localities:** Deer Park, Rockbank, St Albans, Melton, Mt Cottrell
**Synonym:** Incorrectly referred to as *H.* sp. aff. *crassifolia*.

SCALE: 0 1cm 2cm — The scale representation on each plant equals 2cm.

*Hypericum gramineum*

*Hypericum japonicum*

# Hypericum — Hypericaceae

## Hypericum gramineum — Small St. John's Wort

**Size:** to 25 cm x 5–20 cm **F P**
**Habitat:** Widespread in sclerophyll and red gum woodlands, grassy low open forest, wattle tea-tree scrub, plains grassland, dry and valley sclerophyll forests
**Form:** Erect perennial herb, stems often ridged
**Foliage:** Stem-clasping blue-green oblong leaves to 20 mm long, wavy on edges
**Flowers:** Cup-shaped orange-yellow flowers to 10 mm across in loose clusters above the leaves; October to January
**Requirements:** Most soils tolerating moisture and Summer dryness
**Propagation:** Seed
**Localities:** 5–7, 10–12, 16–18, 24, 26, 28, 32, 34A–36, 38–44, 46–68, 69A–71, 74–76
**Distribution:** All states

## Hypericum japonicum — Matted St. John's Wort

**Size:** to 10 cm x 15 cm **P**
**Habitat:** Damp sclerophyll forest, swamp and wattle tea-tree scrub
**Form:** Weak, trailing or matted herb, smaller than *H. gramineum*
**Foliage:** Oval to oblong, green leaves are not stem clasping
**Foliage:** Solitary or few yellow flowers, to 8 mm wide; Spring and Summer
**Requirements:** Moist soils, tolerating Summer dryness
**Localities:** Montrose, 29#, 40, 42, 54, 56, 57, 67, 71, 74
**Distribution:** All states except NT

*Indigofera australis*

# Indigofera — Fabaceae

## Indigofera australis — Austral Indigo

**Size:** 1–2 m x 1–2 m **F P N**
**Habitat:** Widespread in riparian scrub, valley and damp sclerophyll forests, grassy low open forest
**Form:** Open, spreading to erect shrub
**Foliage:** Pinnate blue-green leaves with elliptical leaflets 10–30 mm long
**Flowers:** Sprays of mauve, occasionally pink or white, pea flowers; September to December
**Requirements:** Adaptable to any well drained soil. Lime tolerant
**Comments:** Very attractive in flower, it is a useful shrub for understorey planting. Pruning is important to maintain bushiness and vigour. The crushed roots were used by the Aborigines to poison fish. They also obtained blue dye from the flowers.
**Propagation:** Scarified seed
**Localities:** Research, St Albans, 1, 10, 16, 26, 27, 29, 30, 32–35, 37, 41, 42, 44, 47–49, 50A, 51, 53, 55–67#, 74, 76
**Distribution:** All states

# Isoetopsis — Asteracea

## Isoetopsis graminifolia — Grass cushions

**Size:** 20–50 mm high **F**
**Habitat:** Plains grasslands, red gum woodland
**Form:** Tufted, stemless annual herb
**Foliage:** Grass-like leaves to 50 mm long
**Flowers:** In greenish-white, papery flowerheads at the base of the leaves; August to November
**Requirements:** Dry soils
**Comments:** Once more widespread in Melbourne it is now known from very few sites.
**Propagation:** Seed
**Localities:** Sydenham, Rockbank, St Albans, Sunbury, Bundoora, 26, 42, 49
**Distribution:** All states

*Isopogon ceratophyllus*

# Isopogon — Proteaceae

## Isopogon ceratophyllus — Horny Cone-bush

**Size:** 0.2–0.6 m x 0.5–1.2 m **F P**
**Habitat:** Tea-tree heath
**Form:** Stiff, rounded dwarf shrub
**Foliage:** Rigid, very prickly, finely divided light green leaves to 80 mm long, segments narrow to triangular
**Flowers:** Terminal golden-yellow flowerheads to 30 mm wide amongst the leaves; September to November. Fruiting cones to 20 mm wide.
**Requirements:** Moist well drained sandy soil, tolerating dryness once established
**Comments:** A slow growing plant, difficult to maintain for long periods unless in ideal conditions. Must be moist for extended periods. Locally rare.
**Propagation:** Seed, cuttings—using firm young growth
**Localities:** Baxter, 67–69, 73–75
**Distribution:** Vic, NSW, Tas, SA

# Isotoma — Campanulaceae

## Isotoma fluviatilis ssp. australis — Swamp Isotome

**Size:** Prostrate x 1 m — F P
**Habitat:** Plains grassland, riparian and wattle tea-tree scrub in bogs and swamps
**Form:** Dense matting plant
**Foliage:** Crowded, narrow slightly toothed leaves to 10 mm long
**Flowers:** Small blue starry flowers; October to November
**Requirements:** Moist to wet conditions
**Comments:** An excellent self-layering ground cover for damp areas and poolside plantings. Plant 30 cm apart to establish a quick cover. Restricted to very few sites.
**Propagation:** Seeds, cuttings, division
**Localities:** Knox, 17#, 26, 29#, 35, 42, 49, 67#, 75
**Distribution:** Vic, Qld, NSW, Tas, SA; NZ

*Isotoma fluviatilis* ssp. *australis*

# Ixiolaena — Asteraceae

## Ixiolaena leptolepis — Stalked Plover Daisy

**Size:** 10–40 cm x 10–40 cm — F
**Habitat:** Plains grasslands
**Form:** Upright perennial herb
**Foliage:** Rough linear to lanceolate leaves to 15 mm long, pointed at the end
**Flowers:** Long stalked, rounded flowerheads of many yellow tubular florets; August to November
**Requirements:** Well drained heavy soils, tolerating lime and short periods of wetness
**Comments:** Stems and branches are covered with loose cottony hairs. Locally rare. This species is presently undergoing revision.
**Propagation:** Seed, cuttings
**Localities:** Sydenham
**Distribution:** All mainland states

*Ixiolaena leptolepis*

## Ixiolaena sp. — Woolly Buttons

**Size:** to 0.5 m high — F
**Habitat:** Grasslands and floodplains of the basalt plains
**Form:** Perennial shrublet
**Foliage:** Soft, narrow-linear grey leaves. The leaves and branches are covered in white cottony hairs.
**Flowers:** Single yellow flowerheads are held erect on long stalks; September to January.
**Requirements:** Well drained heavy soil, tolerating dry periods but responds to Summer watering in the garden
**Comments:** An endangered plant in Melbourne. This species is undergoing review and may be found to be conspecific with *Ixiolaena leptolepis*.
**Localities:** St Albans 1938, Sydenham 1950; Laverton, Deer Park
**Distribution:** Vic, NSW, Tas
**Synonym:** *Leptorhynchos panaetioides*

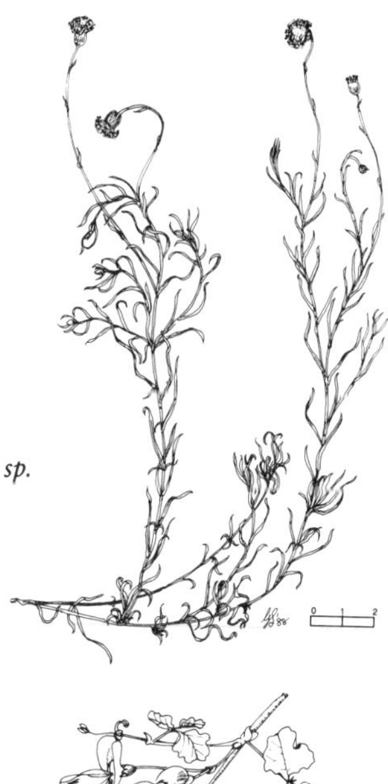

*Ixiolaena sp.*

# Kennedia — Fabaceae

## Kennedia prostrata — Running Postman, Scarlet Runner

**Size:** Prostrate x 1–2.5 m — F P
**Habitat:** Widespread in plains grassland, red gum woodland, riparian scrub, dry and valley sclerophyll forests, tea-tree heath, grassy low open forest
**Form:** Open trailing or densely matting perennial
**Foliage:** Crinkly and hairy grey-green trifoliate leaves, leaflets 6–18 mm long
**Flowers:** Single scarlet pea flowers scattered along stems; April to December
**Requirements:** Well drained soils
**Comments:** Drought tolerant once established. Recolonises in disturbed areas and after fire. Grows well in a hanging basket. The Aborigines sucked the nectar and also used the stems as twine.
**Propagation:** Scarified seed, cuttings
**Localities:** Altona, 1, 10, 12, 15, 16, 21, 26, 29, 30, 32–39, 41, 42, 44, 46–49, 50A, 51, 53, 55, 56, 58, 61–63, 65#–69A, 71, 72, 74–76
**Distribution:** All states except NT

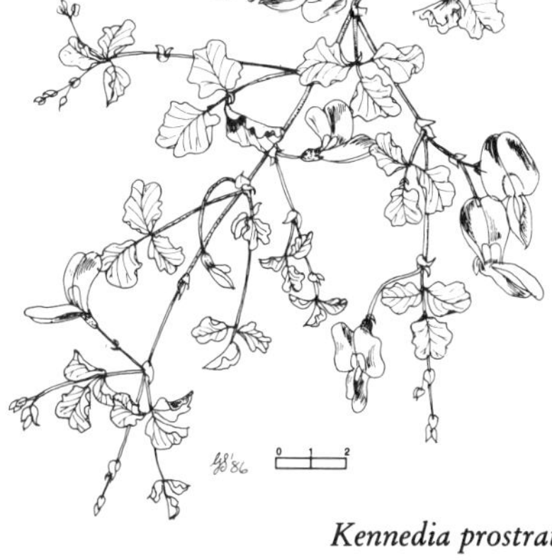

*Kennedia prostrata*

SCALE: 0 1cm 2cm — The scale representation on each plant equals 2cm.

*Kunzea ericoides*

# Kunzea — Myrtaceae

## Kunzea ericoides — Burgan

**Size:** 2–5 m x 2–4 m — F P
**Habitat:** Dry and valley sclerophyll forests
**Form:** Dense to open weeping shrub
**Foliage:** Variable, generally narrow dark green leaves from 8–25 mm long
**Flowers:** Masses of long-stalked white flowers along the stems; November to February
**Requirements:** Adaptable, tolerating wet and dry periods
**Comments:** A very attractive, fast-growing shrub, particularly the weeping form. Suitable for screening. It tolerates hard pruning. Seeds are shed annually and it may revegetate a cleared area quickly. The wood was important to the Aborigines for making fighting implements.
**Propagation:** Seed, cuttings
**Localities:** 30, 31, 33–35, 39–42, 44, 47–49, 50A–56, 59, 61–63, 67#, 71, 74
**Distribution:** Vic, Qld, NSW
**Synonym:** *Leptospermum phylicoides*

# Lagenifera — Asteraceae

Lagenifera is closely related to Brachyscome, differing in the fruit, which is bottle-shaped instead of blunt.

*Lagenifera gracilis*

## Lagenifera gracilis — Slender Bottle Daisy

**Size:** to 25 cm x 8 cm — F P
**Habitat:** Widespread in grassy low open forest, wattle tea-tree scrub, dry and valley sclerophyll forests
**Form:** Perennial herb with fleshy roots
**Foliage:** Basal rosette of dark green obovate to spathulate leaves, lighter below, to 55 mm long, with 2–4 pairs of teeth and wavy margins
**Flowers:** Small mauve or white flowerheads on smooth, often blackish, wiry stalks 8–25 cm high; October to February
**Requirements:** Damp well drained soils
**Comments:** This tiny daisy is suitable for containers and can be established under trees, provided moisture is available.
**Propagation:** Seed
**Localities:** 34A, 40–42, 44, 47, 48, 50, 51, 53–56, 59, 63, 64, 74–76
**Distribution:** All states

*Lagenifera huegelii*

## Lagenifera huegelii — Coarse Bottle Daisy

**Size:** 7–35 cm high — F P
**Habitat:** Dry sclerophyll forest, box woodland
**Form:** Small, hairy perennial herb with fleshy roots
**Foliage:** Deeply lobed oblanceolate basal leaves to 140 mm long, with 4–10 pairs of blunt teeth on wavy margins. Hairy flower stalks, 10–350 mm high, bear a few small, sessile, mostly entire, leaves, to 8 mm long.
**Flowers:** White, pink or blue flowerheads; September to November and March to June
**Requirements:** Moist well drained clay soil
**Comments:** Locally rare.
**Localities:** Eltham, Research, Sth of Melton
**Distribution:** Vic, Tas, SA, WA

*Lagenifera stipitata*

## Lagenifera stipitata — Common Lagenifera, Blue Bottle Daisy

**Size:** 5–25 cm high — F P
**Habitat:** Wet, damp and valley sclerophyll forests, grassy open forest, tea-tree heath, wattle tea-tree scrub
**Form:** Small stoloniferous perennial herb with a slender, hairy flower stalk to 25 cm high, and fibrous roots
**Foliage:** May have a basal rosette; leaves similar to *L. gracilis* but hairy
**Flowers:** Mauve flowerheads to 12 mm wide; Spring and Autumn
**Requirements:** Adaptable to moist well drained soils
**Propagation:** Seed, division
**Localities:** 34A–36, 38, 41–44, 47, 48, 50–53, 55–58, 60, 61, 63, 67#, 69A–71, 74–76
**Distribution:** All states

# Lasiopetalum — Sterculiaceae

## Lasiopetalum baueri — Slender Velvet Bush

F

**Size:** 1–1.5 m x 1 m
**Habitat:** Coastal cliffs and primary dune scrub
**Form:** Dense, erect, small shrub with a velvety covering of star-like hairs
**Foliage:** Narrow grey-green leaves, with recurved margins, to 45 mm long, hoary or rust-coloured below
**Flowers:** Small drooping clusters of hairy greyish-pink to white flowers. The petal-like organs are actually calyx lobes. October to February.
**Requirements:** Dry, well drained soil
**Comments:** An attractive plant which is now locally very rare. One plant remains in the Sandringham area.
**Propagation:** Cuttings
**Localities:** 67
**Distribution:** Vic, NSW, Tas, SA

*Lasiopetalum baueri*

# Lavatera — Malvaceae

## Lavatera plebeia var. plebeia — Australian Hollyhock

F

**Size:** 0.5–2.5 m high
**Habitat:** Coastal limestone in saltmarshes or basalt plains in riparian scrub
**Form:** Erect annual or perennial herb with downy star-shaped hairs
**Foliage:** Dark green, broadly ovate, 5–7 lobed leaves, 20–120 mm across, on long stalks
**Flowers:** Clusters of large, attractive, prominently-veined flowers, pink, white or lilac with lemon centres, petals notched; July to December
**Requirements:** Adaptable to most conditions
**Comments:** A short lived plant, it may shoot again from its woody base if pruned back hard. It is a useful background plant for a sunny rockery. The Aborigines baked the stems before chewing them. A locally rare plant of Port Phillip Bay.
**Propagation:** Seed
**Localities:** Port Melbourne 1850, Altona, 16, 18, 26, 32
**Distribution:** All states

## Lavatera plebeia var. tomentosa

F

**Habitat:** Coastal primary dune scrub
**Comments:** A stouter shrub covered in dense velvety hairs. White flowers most of the year. Remains in one location in the Melbourne area.
**Localities:** Altona
**Distribution:** Vic, Tas, SA, WA

*Lavatera plebeia* var. *plebeia*

# Lawrencia — Malvaceae

## Lawrencia spicata — Salt Lawrencia

F

**Size:** 0.3–1.8 m high
**Habitat:** Saltmarshes, peat swamps of tea-tree heath
**Form:** Erect perennial rising from a basal rosette
**Foliage:** Elliptic to narrowly ovate, toothed, long-stalked basal leaves, to 115 mm x 60 mm; stem leaves smaller, becoming sessile and entire
**Flowers:** Dense cylindrical, terminal spikes of pale yellow to white flowers; October to February.
**Requirements:** Moist salty soil
**Comments:** Flowers emit a strong odour. Locally rare and endangered within the Melbourne area.
**Propagation:** Seed
**Localities:** 2, 3, 7, 67#
**Distribution:** Vic, Tas, SA, WA

*Lawrencia spicata*

# Lepidium — Brassicaceae

## Lepidium aschersonii — Swamp or Spiny Pepper Cress

F

**Size:** 30 cm high
**Habitat:** Saltmarshes
**Form:** Branched shrubby perennial herb covered in short hairs. Branchlets end in a spine.
**Foliage:** Basal leaves bipinnate to 120 mm long; sessile upper leaves becoming smaller and lanceolate
**Flowers:** Short racemes of tiny greenish-white flowers; November to December
**Requirements:** Moist soils, salt tolerant
**Comments:** This is an endangered species Australia-wide which is now extinct in Melbourne. The leaves of *Lepidium* species were steamed by the Aborigines prior to eating.
**Propagation:** Seed
**Localities:** Was collected at Williamstown
**Distribution:** Vic, NSW, WA

SCALE: 0 1cm 2cm — The scale representation on each plant equals 2cm.

*Lepidium pseudohyssopifolium*

## Lepidium pseudohyssopifolium — Pepper Cress

**F P**

**Size:** 0.4 m high
**Habitat:** Plains grassland, riparian scrub, box woodland
**Form:** Erect, annual or perennial herb with glabrous stems
**Foliage:** Leaves are variable in size and shape from narrow and entire to 20 mm long, to broadly lanceolate and pinnate or toothed, to 90 mm long.
**Flowers:** Long spikes of tiny white flowers on glabrous stalks; June to September.
**Requirements:** Tolerates dry periods but responds well to summer watering
**Comments:** Rare throughout its range, it is endangered in Melbourne.
**Localities:** Early collections from Port Melbourne and Werribee; Sunbury, Essendon, Werribee, 16, 26, 31#, 32, 35, 36, 67#
**Distribution:** Vic, Qld, NSW, SA, WA
**Synonym:** Was previously confused with *L. hyssopifolium*

## Lepidium pseudotasmanicum — Peppercress

**P**

**Size:** 20–30 cm high
**Habitat:** Riparian scrub
**Form:** Erect, branching annual or biennial herb
**Comments:** The major feature distinguishing this species from *L. pseudohyssopifolium* is the hairy flower stalk which is shorter or equal to the length of the fruit. Also endangered in the Melbourne area.
**Localities:** Yarra R., 44, 49
**Distribution:** Vic, Qld, NSW, Tas, WA

# Leptinella — Asteraceae

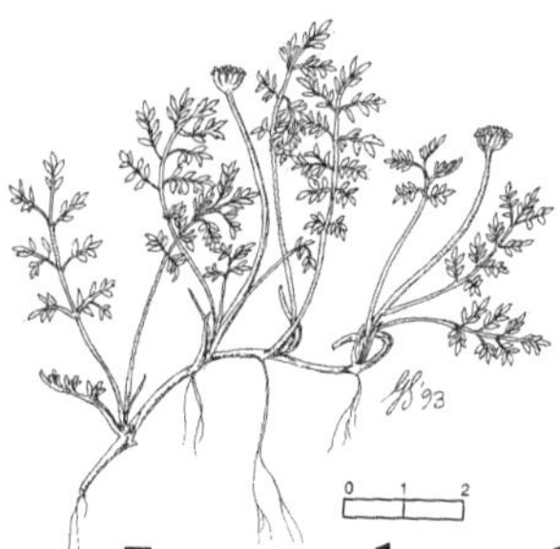

*Leptinella reptans*

## Leptinella reptans — Creeping Cotula

**F P**

**Size:** Prostrate x 1 m
**Habitat:** Bogs in riparian scrub, coastal banksia woodland
**Form:** Self-layering perennial herb
**Foliage:** Pinnate leaves to 100 mm long
**Flowers:** Yellowish button flowers throughout the year
**Requirements:** Tolerates permanently wet soil
**Propagation:** Cuttings, division
**Comments:** Previously included with *L. reptans*, *L. longipes* possibly also occurred within the Melbourne area. It may be distinguished by its almost glabrous, semisucculent leaves. Segments are larger with blunt ends.
**Localities:** 28, 30, 42, 67#, 74
**Distribution:** Vic, Qld, NSW, Tas
**Synonym:** *Cotula reptans*

# Leptorhynchos — Asteraceae

Perennial herbs or sub-shrubs. The long flowering time of *Leptorhynchos* spp. makes them useful container or rockery specimens. The yellow flowerheads are held singly above the foliage on long stems and may be used, cut or dried, for decoration. Propagation is by seed or cutting.

*Leptorhynchos elongatus*

## Leptorhynchos elongatus — Lanky Buttons

**F**

**Size:** 0.2–0.5 m x 0.2–0.6 m
**Habitat:** Dry sclerophyll forest
**Form:** Annual or perennial herb
**Foliage:** Linear-lanceolate leaves to 75 mm long
**Flowers:** White to pale yellow flowerheads; October to December.
**Requirements:** Well drained soil
**Comments:** A rare plant throughout Victoria, it is now extinct in Melbourne.
**Localities:** Early collection near Melbourne; 42#
**Distribution:** All states except WA

## Leptorhynchos panaetioides = Ixiolaena sp.

*Leptorhynchos squamatus*

## Leptorhynchos squamatus — Scaly Buttons

**F P**

**Size:** 15–30 cm x 0.4 m
**Habitat:** Plains grassland, red gum woodland, valley sclerophyll forest
**Form:** Low spreading herb
**Foliage:** Linear-lanceolate, deep green leaves to 20 mm long, woolly white below, often concentrated on one side of the stem
**Flowers:** Small yellow flowerheads held above the foliage on long scaly stalks; September to January
**Requirements:** Well drained moist soils
**Comments:** Summer watering is beneficial. The local form has flowering stems to only 15 cm high.
**Propagation:** Seed, cutting, division
**Localities:** 2, 4, 5, 7, 10, 12, 13, 16–18, 25, 26, 28, 29, 34A, 35, 38, 39, 41, 42, 45, 49, 50A, 61, 67#
**Distribution:** Vic, Qld, NSW, Tas, SA

## Leptorhynchos tenuifolius

Wiry Buttons

**Size:** 10–30 cm x 30 cm **F P**
**Habitat:** Plains grassland, red gum woodland, dry and valley sclerophyll forests, tea-tree heath, grassy low open forest
**Form:** Wiry, branched perennial
**Foliage:** Very narrow dark green leaves, woolly white below. Margins are rolled under.
**Flowers:** Single yellow flowerheads, to 10 mm wide, held erect; September to January
**Requirements:** Prefers well drained situations but accepts some moisture.
**Localities:** Kilsyth, 16, 26, 28, 35, 40–42, 44, 46–51, 54, 58–60, 62–67, 69, 69A, 71, 74–76
**Distribution:** Vic, SA

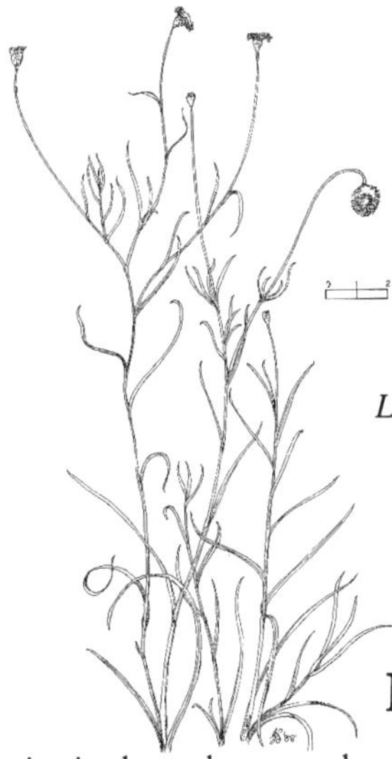

*Leptorhynchos tenuifolius*

# Leptospermum

Myrtaceae

The 6 species of *Leptospermum* local to Melbourne are shrubs or small trees. Generally growing in the understorey they are all hardy and make attractive garden specimens, most providing a good screen through their compact habit. Leaves are small and leathery. The solitary white or pink flowers cover the plant for several months. The often large petals are attached to a broad, cup-shaped floral tube. Masses of seed are stored in woody capsules. Some species drop their capsules after a year while others are only released after fire or damage to the plant. The leaves of some species were used by the early settlers as a tea substitute, hence the common name 'tea-tree'. The stems and wood were used by the Aborigines to make pegs and spears. Propagation is by seed or cuttings.

## Leptospermum continentale

Prickly Tea-tree

**Size:** 1–4 m x 1–2 m **F P**
**Habitat:** Widespread in damp and valley sclerophyll forests, sclerophyll woodland, swamp and wattle tea-tree scrub, grassy low open forest, tea-tree heath
**Form:** Rigidly upright, dense or straggling shrub or small tree
**Foliage:** Leaves rigid, prickly and concave, variable, narrow-lanceolate to lanceolate to 12 mm x 3 mm; hairy, especially on young shoots
**Flowers:** Masses of white flowers to 10 mm wide; October to March
**Requirements:** Adaptable, tolerating moisture
**Comments:** Two forms of this species exist in Melbourne. The form occurring in open forests is tall and broad with scattered leaves that are sparsely hairy below in the centre only. The heathland form is smaller, low-branching and denser with shorter and narrower leaves that are more clustered and hairier below. The heathland form also has smaller flowers and flowers earlier. The two forms co-exist in the Montrose and Langwarrin areas. *L. continentale* is often confused with *L. scoparium* which does not occur in the Melbourne area.
**Localities:** Toolern Vale, 33–35, 40, 42–51, 53–68, 69A, 71, 73–76
**Distribution:** Vic, NSW, SA
**Synonym:** A newly named species, it was known as *L. juniperinum*, a species which only occurs in NSW.

*Leptospermum continentale*

## Leptospermum grandifolium

Mountain Tea-tree

**Size:** 1.5–6 m high **F P**
**Habitat:** Damp sclerophyll forest
**Form:** Dense shrub to small rounded tree
**Foliage:** Large oblong glossy leaves to 30 mm long, paler and silky below, ending in a sharp point. New growth is silver and hairy.
**Flowers:** Large white flowers to 18 mm wide, borne at the ends of side branches; October to January. Woolly calyx tube persists as woolly base to large woody fruit
**Requirements:** Adaptable, preferring moist to wet soil
**Localities:** The Patch, Wattle Glen, 29, 47, 56
**Distribution:** Vic, NSW

*Leptospermum grandifolium*

## Leptospermum juniperinum—see L. continentale

## Leptospermum laevigatum

Coast Tea-tree

**Size:** 2–8 m x 2–4 m **F P**
**Habitat:** Primary dune scrub, tea-tree heath, coastal banksia woodland, grassy low open forest
**Form:** Tall shrub or small tree
**Foliage:** Thick, flat, dull green obovate leaves to 30 mm long
**Flowers:** Stalkless white flowers in pairs to 20 mm wide; August to October
**Requirements:** Well drained soil tolerating dryness once established
**Comments:** Attractive in flower this adaptable shrub makes a useful screen plant if pruned. Its twisted, gnarled trunk and flaking bark adds interest. It is suitable for front-line coastal planting. Great care needs to be taken when introduced outside its normal habitat as it may become an environmental weed. It has already spread further inland since European settlement. A naturally occurring hybrid with *L. myrsinoides* is found in the Mordialloc and Frankston areas.
**Localities:** 65–70, 74, 75
**Distribution:** Vic, NSW, Tas

*Also illustrated p. 33*

*Leptospermum laevigatum*

SCALE: 0 1cm 2cm The scale representation on each plant equals 2cm.

*Leptospermum lanigerum*

### Leptospermum lanigerum — Woolly Tea-tree

**Size:** 2–6 m x 1–3 m — F P
**Habitat:** Riparian scrub and woodland, valley sclerophyll forest, tea-tree heath
**Form:** Dense shrub to erect small tree
**Foliage:** Variable oblong to narrowly oblanceolate greyish leaves 5–15 mm long, ending in a point. Young growth is silver and hairy on both sides, the upper surface becoming glabrous with age.
**Flowers:** Masses of white flowers to 15 mm wide; September to January. Densely hairy floral tube and capsule.
**Requirements:** Moist soils
**Comments:** An ornamental shrub, the grey-leafed forms provide an attractive foliage contrast. Regular pruning encourages a dense shrub. The wood was an important resource for the Aborigines, providing kangaroo spears and double-barbed spears.
**Localities:** Sunshine, 1, 8, 14, 16, 26, 28–30#, 33–36, 42, 47, 49, 54–57, 59, 66#–68, 74
**Distribution:** Vic, NSW, Tas, SA

*Leptospermum myrsinoides*

### Leptospermum myrsinoides — Heath or Silky Tea-tree

**Size:** 0.5–2.5 m x 1 m — F P
**Habitat:** Sclerophyll woodland, valley sclerophyll forest, tea-tree heath
**Form:** Compact or wiry shrub
**Foliage:** Dull green oblanceolate to obovate leaves to 10 mm long, concave above
**Flowers:** White or pink flowers to 15 mm wide, singly or paired, on pinkish stems; September to November. Only the base of the calyx has silky hairs.
**Requirements:** Adaptable, tolerating poor drainage
**Comments:** An attractive shrub for understorey planting.
**Localities:** Canterbury#, 37, 49, 50A, 59, 65–71, 73–75
**Distribution:** Vic, NSW, SA

*Leptospermum obovatum*

### Leptospermum obovatum — River or Blunt-leaf Tea-tree

**Size:** 2–4 m x 1.5–2 m — P
**Habitat:** Riparian scrub
**Form:** Dense erect shrub
**Foliage:** Thick, aromatic narrow to broad leaves with blunt tip, variable from 5–20 mm long
**Flowers:** Creamy white flowers to 12 mm wide, singly or in pairs; November to January
**Requirements:** Moist soils
**Comments:** Not common in the Melbourne area.
**Localities:** Yarra R, 31, 32, 34A, 35
**Distribution:** Vic, NSW

## Leptostigma — Rubiaceae

### Leptostigma reptans — Dwarf Nertera

**Size:** Prostrate x 20 cm — P N
**Habitat:** Valley sclerophyll forest
**Form:** Hairy, small spreading herb, rooting at the nodes
**Foliage:** Pointed, ovate leaves to 10 mm long
**Flowers:** Tiny tubular cream flowers to 12 mm long covered in long hairs, between terminal pairs of leaves; November to January. Attractive orange-red drupes follow flowering.
**Requirements:** Moist to wet soil
**Comments:** Suitable for a terrarium or beside ponds. It has regional significance in Melbourne as it is now known from very few locations locally.
**Propagation:** Division
**Localities:** 44
**Distribution:** Vic, NSW
**Synonym:** *Nertera reptans*

*Leucochrysum albicans* ssp. *albicans* var. *albicans*

## Leucochrysum — Asteraceae

### Leucochrysum albicans ssp. albicans var. albicans — Hoary Sunray

**Size:** 10–30 cm x 30 cm — F P
**Habitat:** Amongst rocks in dry sclerophyll forest
**Form:** Dense clumping perennial herb
**Foliage:** Long, narrow, hairy grey leaves to 100 mm long
**Flowers:** Yellow flowerheads, surrounded by showy white or yellow papery bracts, held on stems above the leaves; November to March
**Requirements:** Well drained soils
**Comments:** A very attractive rockery or container plant, it is useful as a cut or dried flower. Prune hard after flowering to promote bushiness.
**Propagation:** Fresh seed, cuttings
**Localities:** Eltham, Diamond Ck, 34A, 42
**Distribution:** Vic, Qld, NSW, Tas, SA
**Synonym:** *Helipterum albicans* ssp. *albicans* var. *albicans*

# Leucophyta

**Asteraceae**

## Leucophyta brownii

Cushion Bush

FP

**Size:** 0.2–1 m x 0.5–2 m
**Habitat:** Coastal headlands, primary dune scrub
**Form:** Densely tangled rounded shrub
**Foliage:** Grey hairy scale-like leaves
**Flowers:** Pale yellow globular heads borne terminally; September to December
**Requirements:** Well drained dry conditions, tolerates alkaline soil
**Comments:** An attractive low shrub which withstands wind and salt spray and responds well to pruning. The grey foliage is able to reflect available light at night time making this an ideal plant for defining pathways in dimly lit areas.
**Propagation:** Seed, cuttings—without mist
**Localities:** Werribee#, St. Kilda, Williamstown, Mordialloc, Seaford, 67, 69A, 74
**Distribution:** Vic, NSW, Tas, SA, WA
**Synonym:** *Calocephalus brownii*

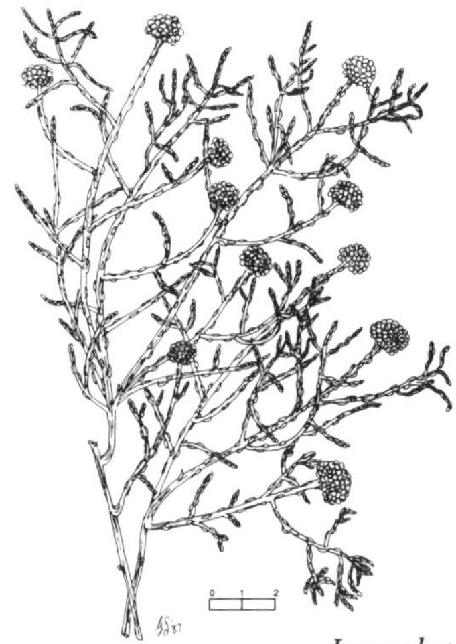

*Leucophyta brownii*

# Leucopogon

**Epacridaceae**

Understorey shrubs of heathlands, Leucopogons are called 'Beard heaths' due to the dense beard of hairs on the corolla lobes. Flowers are followed by small, succulent drupes, which were eaten by the Aborigines. They require well drained soil and a cool rootrun. Propagation is by cuttings. Care needs to be taken with their fragile roots.

## Leucopogon australis

Spike Beard Heath

P

**Size:** 1–1.5 m x 0.5–0.75 m
**Habitat:** Tea-tree heath
**Form:** Upright shrub
**Foliage:** Narrow dull green leaves to 50 mm long with parallel veins
**Flowers:** Strongly perfumed tiny white flowers, densely bearded, in rigid terminal and axillary spikes shorter than leaves; September to November. Yellow or white drupes
**Requirements:** Well drained damp sandy soil
**Comments:** The branches are pale and acutely angled in section.
**Localities:** 67#, 75
**Distribution:** Vic, Tas, SA, WA

*Leucopogon australis*

## Leucopogon ericoides

Pink Beard Heath

P

**Size:** 0.5–1.5 m x 0.7 m
**Habitat:** Tea-tree heath, dry sclerophyll forest
**Form:** Upright wiry shrub
**Foliage:** Prickly, narrow, dark green leaves to 10 mm long, margins curved under
**Flowers:** Small clusters of white to pink bearded flowers in leaf axils follow pink buds; July to November. Drupes dryish.
**Requirements:** Well drained soil, tolerating dry periods once established
**Comments:** An attractive long-flowered plant, best when intermingled with other small shrubs.
**Localities:** Eltham, 42, 44, 47, 63, 73–75
**Distribution:** Vic, Qld, NSW, Tas, SA

*Leucopogon ericoides*

## Leucopogon parviflorus

Coast Beard Heath

FP

**Size:** 1–4 m x 2–3 m
**Habitat:** Coastal banksia woodland, primary dune scrub
**Form:** Variable; much branched stunted shrub when exposed, small tree in scrub
**Foliage:** Stiff elliptic to obovate leaves, paler below, to 30 mm long; new growth bright green
**Flowers:** Masses of densely bearded white flowers tinged with maroon, in axillary spikes usually longer than the leaves; July to November. Edible round, white drupes
**Requirements:** Well drained soils
**Comments:** Bark is most attractive on old gnarled plants. It differs from *L. australis* in that leaves are broader and shorter, and the flowers lack perfume.
**Localities:** Mordialloc, 67, 74, 75
**Distribution:** All states except NT

*Leucopogon parviflorus*

SCALE: 0 1cm 2cm The scale representation on each plant equals 2cm.

*Leucopogon virgatus* var. *virgatus*

### Leucopogon virgatus var. virgatus — Common Beard Heath

**Size:** 0.3–1 m x 0.2–0.6 — **F P**
**Habitat:** Tea-tree heath, dry sclerophyll forest, grassy low open forest
**Form:** Upright wiry shrub
**Foliage:** Pointed concave, lanceolate leaves to 20 mm long, with 3 conspicuous parallel ribs on the underside. Some leaves are stem-clasping, others are spreading.
**Flowers:** Dense axillary spikes of fragrant, white bearded flowers at the ends of branches; July to December
**Requirements:** Well drained soils, tolerating some dryness once established; lime tolerant
**Comments:** An excellent garden plant, ideal for filling small gaps between other shrubs. Old plants may be rejuvenated by hard pruning.
**Localities:** Port Melbourne#, Clayton, 33, 34A, 41, 42, 44, 46–51, 53, 58, 65#, 67–70, 73, 76
**Distribution:** Vic, Qld, NSW, Tas, SA

## Levenhookia — Stylidiaceae

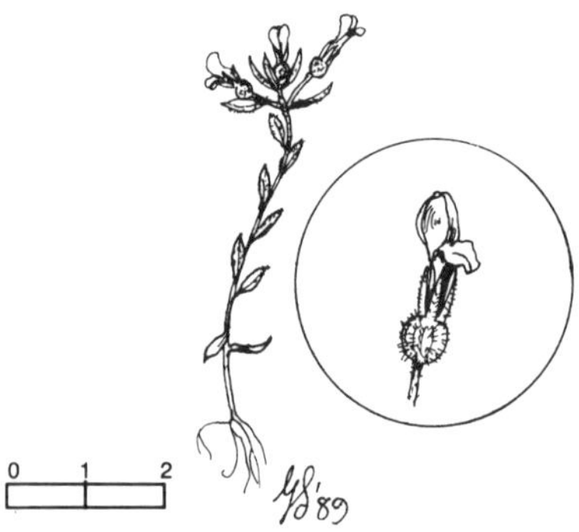

*Levenhookia dubia*

### Levenhookia dubia — Hairy Stylewort

**Size:** 15–55 mm high — **P**
**Habitat:** Grassy wetlands, valley sclerophyll forest, tea-tree heath
**Form:** Inconspicuous, erect glandular-hairy herb with reddish stems
**Foliage:** Small obovate leaves to 6 mm long, lower leaves reddish and smaller than the upper green ones
**Flowers:** Small terminal clusters of tiny whitish flowers with a pale pink lip; September to October
**Requirements:** Damp well drained soils
**Comments:** A short lived plant, which is usually abundant in times of good rainfall, but disappears during drier periods.
**Propagation:** Seed
**Localities:** Laverton Nth, Eltham, 29#, 34A, 42, 49, 58, 67#
**Distribution:** Vic, NSW, Tas, SA, WA

### Levenhookia sonderi — Slender Stylewort

**Size:** 2–6 mm high — **F P**
**Habitat:** Wet depressions in valley sclerophyll forest, grassy wetlands, tea-tree heath
**Form:** Tiny short-lived herb with green stems
**Foliage:** Round, green glandular leaves to 5 mm long
**Flowers:** Small terminal clusters of white or pinkish flowers have a purple lip
**Requirements:** Wet winter soils.
**Localities:** was in Eltham, Research; Belgrave, 7, 41, 42, 67#
**Distribution:** Endemic to Vic

## Lilaeopsis — Apiaceae

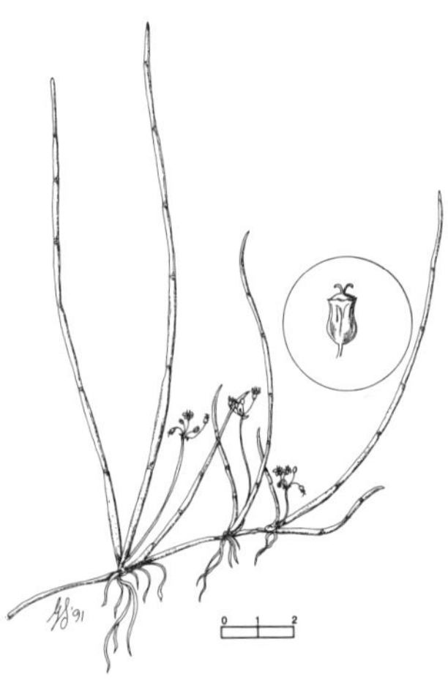

*Lilaeopsis polyantha*

### Lilaeopsis polyantha — Creeping Crantzia

**Size:** 2–25 cm high — **F P**
**Habitat:** Aquatic, in shallow water of saltmarshes, primary dune scrub
**Form:** Tangled perennial creeping herb
**Foliage:** Tufts of narrow, terete, yellow-green leaves to 25 cm long, standing erect. The leaves are hollow and partitioned.
**Flowers:** Tiny purplish flowers, 2–6 per umbel, on upright stalks to 30 mm long; October to May
**Requirements:** Usually fresh but tolerates brackish water
**Comments:** Tufts of leaves and flowers occur at nodes on the creeping rhizomes. Known at very few locations in Melbourne.
**Propagation:** Seed, division
**Localities:** Seaford, 2, 4, 74
**Distribution:** Vic, NSW, SA
**Synonym:** *L. australica*

# Limonium — Plumbaginaceae

## Limonium australe — Yellow or Native Sea Lavender

**Size:** 0.3–0.5 m x 0.2–0.5 m — F
**Habitat:** Coastal mudflats and saltmarshes
**Form:** Rosetted perennial herb
**Foliage:** Leathery, glossy green obovate-oblong leaves to 90 mm long
**Flowers:** Broad panicle of flowers on angular flowering stalks 10–40 cm high. Pale pink or white papery calyx is more conspicuous than the yellow flowers.
**Requirements:** Moist or wet soil, tolerating salt
**Comments:** An attractive plant for a rockery or container.
**Propagation:** Seed
**Localities:** Early collections around Port Phillip Bay
**Distribution:** Vic, Qld, NSW

# Limosella — Scrophulariaceae

## Limosella australis — Australian Mudwort

**Size:** Prostrate — F
**Habitat:** Swamp scrub
**Form:** Perennial, stoloniferious semi-aquatic herb forming mats
**Foliage:** Basal tufts of narrow, oblong to spathulate leaves 10–40 mm long
**Flowers:** Single, small shortly-tubular, white or pale pink 5-lobed flowers on stalks shorter than the leaves; September to November
**Requirements:** Mud or shallow water
**Comments:** A useful matting plant for bogs or ponds.
**Propagation:** Division
**Localities:** Springvale
**Distribution:** All states except NT; NZ, Af

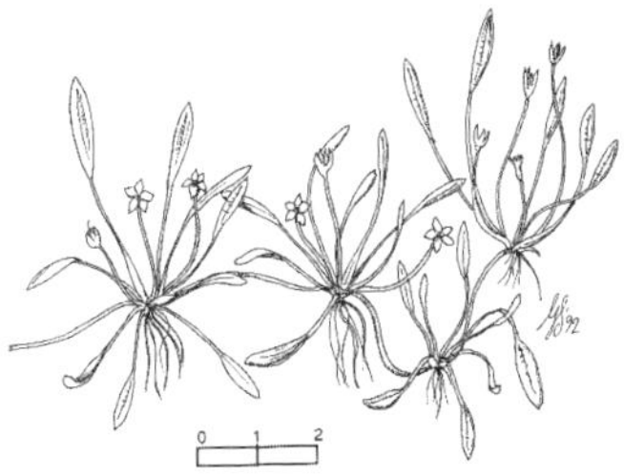

*Limosella australis*

# Linum — Linaceae

## Linum marginale — Native or Wild Flax

**Size:** 0.3–0.8 m x 0.3 m — F P
**Habitat:** Widespread in grassy wetland, red gum woodland, dry sclerophyll forest, tea-tree heath
**Form:** Slender erect perennial
**Foliage:** Ascending, narrow, sessile, bluish leaves to 20 mm long
**Flowers:** Terminal clusters of open, clear blue flowers with darker veins; most of the year, peaking September to May. The fruit is a round papery capsule.
**Requirements:** Moist well drained soils
**Comments:** The plant produces a thick tuberous rootstock, and may be cut back hard each year in late Autumn to encourage vigorous new growth. Fibre from the flax was used by Aborigines to make cord and fish nets. The seeds were used as food.
**Propagation:** Seed
**Localities:** St Albans, 2, 4, 16–18, 26, 28–30, 32, 34A, 35, 38, 40–42, 44, 47–49, 50A, 51, 55–57, 67#, 74
**Distribution:** All states except NT; NZ

*Linum marginale*

# Lissanthe — Epacridaceae

## Lissanthe strigosa — Peach Heath

**Size:** 0.2–1 m x 0.5 m — P
**Habitat:** Dry sclerophyll forest
**Form:** Rigid open shrub
**Foliage:** Stiff, narrow, green prickly leaves to 15 mm long with recurved margins. The underside is paler with conspicuous veins.
**Flowers:** Clusters of small pink or white urn-shaped flowers; August to November. Round, whitish, fleshy drupes follow flowering.
**Requirements:** Well drained soils
**Comments:** The drupes are edible.
**Propagation:** Cuttings—slow to strike
**Localities:** Research, 32, 36, 42
**Distribution:** Vic, Qld, NSW, Tas, SA

*Lissanthe strigosa*

SCALE: 0 1cm 2cm — The scale representation on each plant equals 2cm.

# Lobelia

# Campanulaceae

*Lobelia alata*

## Lobelia alata

Angled Lobelia

**Size:** Prostrate-30 cm high **F P**
**Habitat:** Damp and valley sclerophyll forests, sclerophyll woodland, riparian scrub, tea-tree heath
**Form:** Erect perennial herb spreading by layering
**Foliage:** Soft, narrow toothed leaves, the lower ones broader and longer, to 50 mm long
**Flowers:** Small pale blue fan-shaped flowers, singly in upper leaf axils; most of the year
**Requirements:** Moist soils
**Comments:** Suitable for a bog garden, this lobelia is easily recognised by its angular or narrowly winged stems.
**Propagation:** Division
**Localities:** Sunshine, Melton Sth, 14, 16, 26, 28, 35, 36, 42, 44, 46, 47, 51, 53–56, 59, 62–64, 67, 69A–71, 74–76
**Distribution:** All states except NT; NZ, S Af, S Am

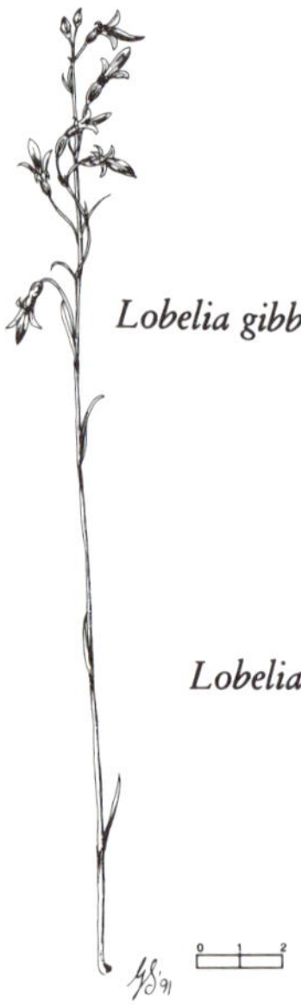

*Lobelia gibbosa*

*Lobelia simplicicaulis*

## Lobelia gibbosa

Tall Lobelia

**Size:** 10–50 cm x 10 cm **P**
**Habitat:** Tea-tree heath
**Form:** Slender erect annual herb
**Foliage:** Scattered, narrow sessile leaves to 30 mm long, usually withering during flowering
**Flowers:** One-sided spike of deep blue tubular flowers, split to the base, white markings sometimes on the lower lobes; November to March. Seeds are tiny.
**Requirements:** Moist well drained soil
**Comments:** This herb is eye-catching in a small rockery.
**Propagation:** Seed
**Localities:** 53, 55, 67[#], 74
**Distribution:** All states except NT

## Lobelia pratioides

Poison Lobelia

**Size:** Prostrate, spreading herb **P**
**Habitat:** Grassy wetlands, plains grassland, valley sclerophyll forest
**Form:** Loosely matting dioecious herb
**Foliage:** Dull green oblanceolate to cuneate leaves 10–50 mm long
**Flowers:** Single blue and white flowers are on hairy stalks longer than the leaves; November to January.
**Requirements:** Wet soils
**Comments:** An attractive plant for growing in a shaded garden, however the foliage is poisonous to stock.
**Localities:** 4, 5, 7, 17[#], 18, 24, 26, 67[#], 69A, 75
**Distribution:** Vic, NSW, Tas, SA

*Lobelia pratioides*

## Lobelia rhombifolia

Branched or Tufted Lobelia

**Size:** 10–40 cm high **P**
**Habitat:** Swampy tracts in damp sclerophyll forest
**Form:** Much branched, erect or spreading annual herb
**Foliage:** Lobed oblong leaves to 25 mm long, upper leaves smaller and narrower
**Flowers:** Single dark blue to light purple flowers with a white and yellow throat, terminal on long stalks; September to November.
**Localities:** Northern slopes of Mt Dandenong, 49, 56, 58, 59
**Distribution:** Vic, Tas, SA, WA

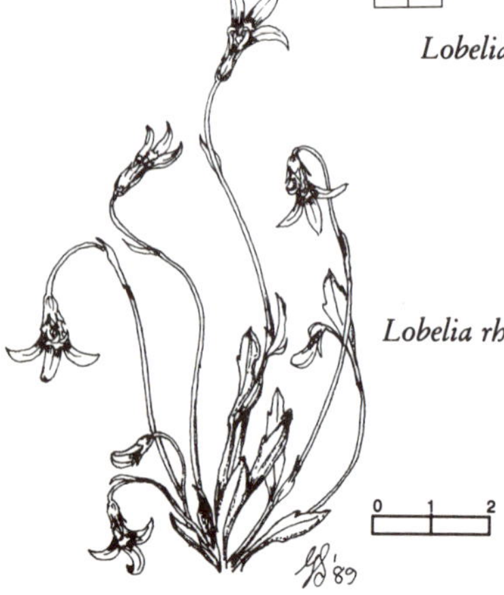

*Lobelia rhombifolia*

## Lobelia simplicicaulis

**Size:** to 0.8 m high **F P**
**Habitat:** Dry and valley sclerophyll forests. tea-tree heath
**Comments:** Similar to *L. gibbosa*. Stems are branched. Lower leaves are larger and lanceolate, to 40 mm long, and stem-clasping with toothed margins, becoming smaller and linear further up the stem. The lower leaves on coastal forms are narrow-lanceolate to 20 mm long. Seeds are much larger.
**Localities:** Cheltenham 1901; Kalorama, Olinda, 42, 49, 55–59
**Distribution:** Vic, Qld, NSW, Tas, SA
**Synonym:** *L. gibbosa* var. *simplicicaulis*

# Lomatia — Proteaceae

## Lomatia fraseri
Forest or Tree Lomatia

**Size:** 2–7 m x 1–4 m — P N
**Habitat:** Wet and damp sclerophyll forests
**Form:** Rounded shrub to small tree
**Foliage:** Stiff, leathery, narrowly elliptical leaves to 15 cm long, usually sharply and distantly toothed, silver-velvety hairs below
**Flowers:** Loose terminal or axillary racemes of creamy spider flowers, hairy on the outside; December to February
**Requirements:** Deep, moist well drained soil, tolerating some wetness
**Comments:** A very attractive small tree or shrub for a sheltered moist position. The interestingly-shaped seed follicles are also a feature. Plants are usually slow growing.
**Propagation:** Seed, cuttings
**Localities:** Mt Dandenong, 57, 58
**Distribution:** Vic, NSW

*Illustrated p. 172*

*Lomatia ilicifolia*

## Lomatia ilicifolia
Holly Leaf Lomatia

**Size:** 0.5–2 m x 1 m — F P
**Habitat:** Damp and dry sclerophyll forests, sclerophyll woodland, grassy low open forest
**Form:** Stiff erect shrub
**Foliage:** Dull green, leathery, crinkly, ovate to broad-lanceolate holly-like leaves to 15 cm long, prominent veins
**Flowers:** Racemes of glabrous cream flowers held above the foliage; December to February
**Requirements:** Well drained soil
**Comments:** A slow growing shrub regenerating vigorously from a lignotuber and flowering well after fire. It has attractive foliage and fruit. The new growth is very colourful with shades of bronze and red.
**Propagation:** Seed, cuttings
**Localities:** 42, 50A, 54–56, 58, 59
**Distribution:** Vic, Qld, NSW

## Lomatia myricoides
River or Long-leaf Lomatia

**Size:** 2–5 m x 1–3 m — P N
**Habitat:** Riparian scrub
**Form:** Open rounded shrub or small tree
**Foliage:** Dull green, narrow-lanceolate leaves to 20 cm long, veins not prominent. Margins with few distant teeth or rarely almost entire.
**Flowers:** Short axillary racemes of glabrous, perfumed, cream flowers; December to February
**Requirements:** Moist well drained soil. Tolerates short periods of wetness.
**Comments:** Attractive slow-growing small tree which flowers well in the garden with Summer watering.
**Propagation:** Seed, cuttings
**Localities:** Wonga Park, 34A, 42
**Distribution:** Vic, NSW

*Illustrated p. 172*

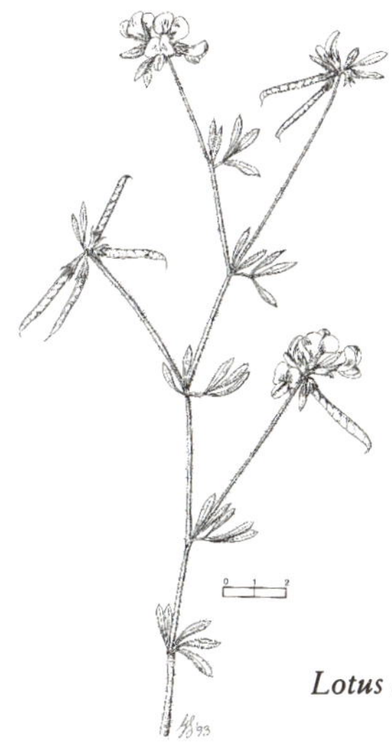

*Lotus australis*

# Lotus — Fabaceae

## Lotus australis
Austral Trefoil

**Size:** 0.3–0.6 m x 0.6 m — F
**Habitat:** Plains grassland
**Form:** Open rounded perennial herb
**Foliage:** Soft light green trifoliate leaves, leaflets narrow to 40 mm long
**Flowers:** Pale pink or white pea flowers in terminal umbels, held erect on long stalks; October to March
**Requirements:** Moist well drained soils withstanding dry periods once established
**Comments:** An attractive rockery plant resistant to salt spray.
**Propagation:** Scarified seed
**Localities:** Early collections in Braybrook 1897, Sydenham 1907; 29
**Distribution:** All states

*Lycopus australis*

# Lycopus — Lamiaceae

## Lycopus australis
Australian Gypsywort

**Size:** to 1 m high — P N
**Habitat:** Riparian scrub, valley sclerophyll forest, coastal banksia woodland
**Form:** Leafy, erect, slightly branched herb
**Foliage:** Pairs of aromatic, lanceolate leaves with jagged edges to 100 mm long
**Flowers:** Dense clusters of tiny, 4-lobed white flowers in leaf axils; December to April
**Requirements:** Moist to wet soil
**Comments:** The branches are quadrangular in cross section. An interesting foliage plant for a damp area. Not found in many locations in Melbourne.
**Propagation:** Cuttings
**Localities:** Diamond Ck, Yarra R, Seaford, 30, 35, 36, 42, 62, 63, 74
**Distribution:** Vic, Qld, NSW, Tas, SA

SCALE: 0 1cm 2cm — The scale representation on each plant equals 2cm.

*Lysiana exocarpi*

# Lysiana — Loranthaceae

## Lysiana exocarpi — Harlequin Mistletoe

**Habitat:** Box woodland
**Form:** Drooping stem parasite
**Foliage:** Narrow, curved bluish-green leaves to 150 mm long
**Flowers:** Groups of 2 or 3 curved, tubular bright red flowers to 50 mm long, greenish towards the end; January to August. Fruit is a green drupe ripening to red or purple.
**Comments:** A parasitic plant, infesting many host species in dry areas. The fruit was eaten by the Aborigines. Locally rare.
**Propagation:** Mistletoe birds eat the berries and deposit seeds on the branches of host trees.
**Localities:** Diggers Rest 1904; 9, 26
**Distribution:** Vic, NSW, SA

*Lythrum hyssopifolia*

*Lythrum salicaria*

# Lythrum — Lythraceae

## Lythrum hyssopifolia — Small or Lesser Loosestrife

**Size:** Prostrate-30 cm x 20–30 cm — **F P**
**Habitat:** Swamp and riparian scrub, grassy wetlands, red gum woodland, valley sclerophyll forest, grassy low open forest
**Form:** Trailing or erect annual herb
**Foliage:** Oblong to oblanceolate, sessile leaves to 25 mm long
**Flowers:** Single, tiny pink tubular flowers with spreading petals in leaf axils along the stem; November to April
**Requirements:** Moist soils, tolerating dry summers
**Comments:** Stems are ribbed or angular.
**Propagation:** Seed, cuttings
**Localities:** 4–7, 9, 12, 16–18, 24–26, 28, 29, 31, 34A–36, 40–42, 44, 46, 47, 49–51, 53–61, 63, 65–69A, 71, 74–76
**Distribution:** All states except NT

## Lythrum salicaria — Purple Loosestrife

**Size:** 1–2 m x 1 m — **F**
**Habitat:** Semi-aquatic in riparian scrub, valley sclerophyll forest
**Form:** Erect, hairy perennial, dying back after Summer
**Foliage:** Sessile, lanceolate leaves to 75 mm long, in pairs
**Flowers:** Clusters of tubular, purple to magenta flowers with spreading petals, crowded in terminal spikes; November to March
**Requirements:** Moist soils or shallow water
**Comments:** A very showy plant for bog gardens or beside pools, the stems should be cut back to the rootstock after flowering. Locally rare now but was collected from the banks of the Yarra River in 1856.
**Propagation:** Cuttings, division, will self-seed
**Localities:** Werribee, Warrandyte Nth, 16, 26, 30, 33, 35
**Distribution:** Vic, Qld, NSW, Tas, SA

# Maireana — Chenopodiaceae

Woody perennials or small shrubs commonly called Bluebush. All local *Maireana* spp. have hairy to quite woolly branches. Leaves are sometimes succulent. In each of these species the minute greenish flowers are bisexual and sessile. The main identifying characteristic is the shape of the fruiting body. While all have horizontal wings, the upper and lower portions are unique to each species. *Maireana* spp. are found in open flat, often saline, areas. Propagation is by seeds or cuttings, without the use of mist.

## Maireana brevifolia — Small-leaf or Short-leaf Bluebush, Cottonbush

**Size:** 1 m x 0.6–1 m — **F**
**Habitat:** Riparian scrub, box woodland
**Form:** Erect, sometimes dense shrub
**Foliage:** Succulent, obovoid glabrous leaves to 5 mm long, narrowing to a short stalk. Each leaf axil contains a tuft of wool.
**Flowers:** Solitary flowers followed by light brown fruit with 5 fan-shaped wings of equal size, inner lobes fleshy; December to July
**Requirements:** Dry well drained soils, tolerating salt
**Comments:** One small colony has recently been discovered within the Melbourne area.
**Localities:** Keilor
**Distribution:** All mainland states
**Synonym:** *Kochia brevifolia*

## Maireana decalvans
Black Cottonbush

| | |
|---|---|
| **Size:** | 0.3–0.5 m x 0.5–1 m **F P** |
| **Habitat:** | Seasonally waterlogged sites in box woodland, plains grassland |
| **Form:** | Erect or spreading bushy perennial shrub with woolly leaf axils |
| **Foliage:** | Succulent, terete, hairless blue-green leaves to 8 mm long |
| **Flowers:** | Single flowers are followed by pale green fruit. The fruit has a hemispherical base and a single wing which is flat with a radial slit; November to May |
| **Requirements:** | Prefers heavy soils, tolerating drying out in Summer. |
| **Localities:** | Laverton Nth, Deer Park, Mt Cottrell, Melton Sth, Sydenham, 16, 26 |
| **Distribution:** | Vic, Qld, NSW, SA |
| **Synonym:** | *Kochia villosa* var. *tenuifolia* |

*Maireana decalvans*

## Maireana enchylaenoides
Wingless Fissure-weed or Bluebush

| | |
|---|---|
| **Size:** | 10–20 cm x 0.3–0.6 m **F** |
| **Habitat:** | Box woodland |
| **Form:** | Slender, weak, sparsely branched perennial herb |
| **Foliage:** | Dull, narrow, oblong, hairy leaves to 10 mm long |
| **Flowers:** | Single flowers are followed by fruit with 5 small, crescent-shaped, overlapping wings which are inflated on their edges; September to March. Fruit turns black when dry. |
| **Requirements:** | Loam soil |
| **Localities:** | 9, 16, 21, 22, 26, 32 |
| **Distribution:** | Vic, Qld, NSW, SA, WA |
| **Synonym:** | *Kochia crassiloba* |

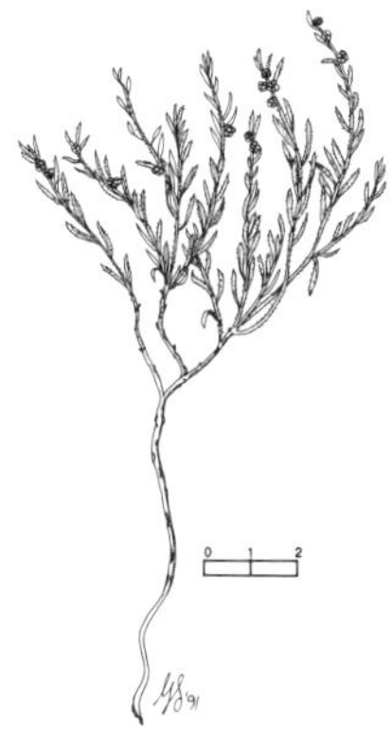

*Maireana enchylaenoides*

## Maireana pentagona
Slender Fissure-weed

| | |
|---|---|
| **Size:** | 10–30 cm x 30–50 cm **F P** |
| **Habitat:** | Plains grassland |
| **Form:** | Small woolly, prostrate to ascending perennial herb |
| **Foliage:** | Narrow leaves to 12 mm long |
| **Flowers:** | Single flowers crowded towards the end of stems. Tiny pentagonal fruit are covered in dense hairs; August to September. |
| **Requirements:** | Heavy soil |
| **Localities:** | 26 |
| **Distribution:** | Vic, NSW, SA, WA |
| **Synonym:** | *Kochia pentagona* |

# Mazus — Scrophulariaceae

## Mazus pumilio
Swamp Mazus

| | |
|---|---|
| **Size:** | 3 cm x 0.5–1 m **F P N** |
| **Habitat:** | Swamp margins of valley sclerophyll forest, tea-tree heath |
| **Form:** | Suckering mat plant |
| **Foliage:** | Glossy obovate-spathulate leaves, 8–65 mm long, with toothed, wavy margins |
| **Flowers:** | Solitary, white or mauve flowers with 2 white humps on lower lip, on stalks to 25 mm long; October to March. Several flowers on branched stalks in sheltered areas. |
| **Requirements:** | Moist to wet soil |
| **Comments:** | Suckering from stolons to form a carpet, it is useful in boggy areas and on pond edges. |
| **Propagation:** | Division |
| **Localities:** | Montrose, 34A, 51, 53, 67# |
| **Distribution:** | Vic, Qld, NSW, Tas, SA; NZ |

*Mazus pumilio*

SCALE: 0 1cm 2cm The scale representation on each plant equals 2cm.

# Melaleuca

## Myrtaceae

Fast-growing medium to large shrubs or small trees. Only two of the locally occurring species have the characteristic paper bark. The papery bark of these species was used by the Aborigines to wrap up their babies, and for blankets, bandaging and roofing. Cream to yellow bottlebrush flowers are followed by a cluster of woody seed capsules which open after fire or damage to the plant. Melaleucas may form dense copses and tolerate a wide range of conditions. Propagation is by seed, using the bog method, or cuttings.

*Melaleuca ericifolia*

### Melaleuca ericifolia

Swamp Paperbark

**Size:** 2–9 m x 3 m — F P  
**Habitat:** Red gum woodland, valley sclerophyll forest, swamp and wattle tea-tree scrub, coastal banksia woodland  
**Form:** Erect, open to bushy shrub or small tree  
**Foliage:** Narrow dark green leaves to 15 mm long, with blunt tips  
**Flowers:** Masses of terminal short cream brushes to 20 mm long, more than 10 mm wide; October to November  
**Requirements:** Moist or wet fertile soils, tolerating dry conditions once established  
**Comments:** A very adaptable plant, useful in wet garden areas, it may sucker to form a dense copse. Responds well to pruning. With age, trunks take on interesting forms. The pale papery bark is used in bark paintings. Aborigines made clubs from young stems with a root attached. The nectar produced a sweet drink and the oil from the crushed leaves treated colds.  
**Localities:** Melbourne#, Albert Park#, 1, 23, 29–35, 42, 44, 46, 47, 49–51, 54–56, 58, 60, 61–67#, 69–76  
**Distribution:** Vic, Qld, NSW, Tas

*Melaleuca lanceolata*

### Melaleuca lanceolata

Moonah

**Size:** 1–8 m x 3–6 m — F  
**Habitat:** Primary dune scrub, riparian scrub and woodland  
**Form:** Dense large shrub or small tree, stunted in exposed coastal areas  
**Foliage:** Thick, dull green, linear leaves to 12 mm long, often curved down  
**Flowers:** Cream to white terminal spikes to 50 mm long; October to March  
**Requirements:** Well drained soils, tolerating limestone and periods of inundation  
**Comments:** An adaptable species, useful as a front line coastal plant. Locally, it commonly grows in coastal areas, but may also be found scattered in riverine sites, on basalt.  
**Localities:** Werribee R, Richmond, Seaford, 69, 74  
**Distribution:** Vic, Qld, NSW, SA, WA

### Melaleuca parvistaminea

**Size:** 2–4 m x 3–5 m — F P  
**Habitat:** Riparian and swamp scrub, coloniser of disturbed areas  
**Form:** Medium shrub very similar to *M. ericifolia*  
**Comments:** Main differences include the rough bark and its non-suckering habit, although it may still form thickets. The canopy is broader and leaves smaller, to 11 mm long, with visible oil glands. The cream brushes are longer and less than 10 mm wide. Prefers infertile shallow soils in moist areas. Both species are occasionally seen growing together.  
**Localities:** Werribee R, Yarra R above Kew  
**Distribution:** Vic, NSW  
**Synonym:** Was included with *M. ericifolia*, also referred to as *M.* sp. aff. *ericifolia*.

*Melaleuca squarrosa*

### Melaleuca squarrosa

Scented Paperbark

**Size:** 2–5 m x 1–2 m — F P  
**Habitat:** Damp and valley sclerophyll forests, swamp and wattle tea-tree scrub  
**Form:** Erect, open to compact large shrub or rarely, a small tree to 10 m high  
**Foliage:** Stiff dark green ovate to triangular leaves to 18 mm long, crowded in pairs and distinctly decussate  
**Flowers:** Profuse terminal spikes of scented cream to yellow flowers; September to February  
**Requirements:** Moist to wet soils  
**Comments:** An attractive shrub responding well to pruning, it also grows well near the coast and is salt tolerant.  
**Localities:** Montrose, 33, 34A, 49, 50A, 56, 58, 59, 67, 70, 71, 73–76  
**Distribution:** Vic, NSW, Tas, SA

# Mentha
**Lamiaceae**

Small aromatic perennial herbs widespread in damp places, spreading by suckering. The mint leaves of some species have been used in drinks or as insect repellants. The Aborigines used them for medicinal purposes. All species are relatively uncommon in Melbourne. Propagation is by division or cuttings.

## Mentha australis
River Mint

**F P N**

**Size:** 0.2–0.8 m x 0.3–1 m
**Habitat:** Riparian scrub, valley sclerophyll forest
**Form:** Slender, sprawling or erect herb with square stems
**Foliage:** Pairs of lanceolate leaves to 50 mm long
**Flowers:** Clusters of small tubular white to pale mauve flowers towards the end of upright stems in whorls of 7–10 flowers; September to March. Calyx covered in grey hairs.
**Requirements:** Moist to wet soils
**Comments:** A useful plant by ponds and near pathways where the foliage, when brushed, imparts a delightful mint fragrance. The Aborigines used the leaves to flavor food and as a decoction for coughs and colds. Locally rare.
**Localities:** Yarra R., Sunbury, 16, 26, 27, 29#–31, 33, 34A–36, 42, 57, 58
**Distribution:** All states

*Mentha australis*

## Mentha diemenica var. diemenica
Slender or Wild Mint

**F P N**

**Size:** 10–30 cm x 0.5 m
**Habitat:** Grassy wetlands, red gum woodland
**Form:** Slender herb similar to *M. australis*
**Foliage:** Ovate to oblong leaves to 20 mm long, covered in downy hairs
**Flowers:** Whorls of 2–8 bright pink to violet flowers
**Requirements:** Moist soil, tolerating periodic inundation
**Localities:** Laverton Nth, Bulla, 1, 17#, 18, 35
**Distribution:** All states except NT

*Mentha diemenica* var. *diemenica*

## Mentha laxiflora
Forest Mint

**P N**

**Size:** 15–40 cm x 1 m
**Habitat:** Wet, damp and valley sclerophyll forests
**Form:** Open spreading herb with erect stems
**Foliage:** Bluntly-toothed lanceolate leaves to 50 mm
**Flowers:** Dense clusters of mauve, or rarely white, short-stalked flowers; September to March
**Localities:** 55, 57, 58
**Distribution:** Vic, NSW

## Mentha satureioides
Creeping Mint, Native Pennyroyal

**P N**

**Size:** 15 cm high
**Habitat:** Grassy wetland, red gum woodland
**Form:** Hairy stems with entire, oblong to oblong-lanceolate leaves to 30 mm long
**Flowers:** White flowers in whorls of 4–8; December to March.
**Comments:** Forms mats in sheltered areas.
**Localities:** 6, 25, 26
**Distribution:** Vic, Qld, NSW, SA, WA

*Mentha laxiflora*

# Microseris
**Asteraceae**

## Microseris lanceolata
Yam Daisy, Native Dandelion

**P**

**Size:** Prostrate-40 cm x 15–25 cm
**Habitat:** Plains grassland, red gum woodland, dry sclerophyll forest
**Form:** Tufted perennial herb, regenerating annually from a fleshy tuberous root
**Foliage:** Shiny, long narrow toothed leaves to 20 cm long
**Flowers:** Single bright yellow daisy flowerhead on a stalk 20–50 cm high; July to November. Fluffy white seed heads follow. The pappus on the narrowly-coned seed is scaly.
**Requirements:** Well drained soils
**Comments:** The flowerhead is distinguished from the dandelion as the bud droops before opening. The tuberous rootstock was a staple part of the aboriginal diet. This form was once also prevalent on the basalt plains. A second form, found in inundated areas of the basalt plains, is distinguished by the slightly waisted seed (or achene) which has a very fine, capillary-like pappus and a branched or forked taproot. Once common in moist hollows of Altona Meadows, it is still found in Laverton, although very rare.
**Propagation:** Seed
**Localities:** Eltham, 1, 7, 33–35, 37, 41, 42, 44, 45, 47–49, 50A–52, 61, 63#, 67#, 74, 76
**Distribution:** All states except NT
**Synonym:** *M. scapigera*

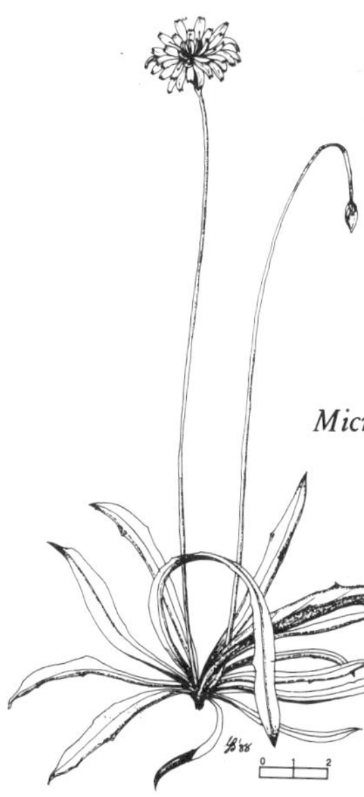

*Microseris lanceolata*

SCALE: 0 1cm 2cm The scale representation on each plant equals 2cm.

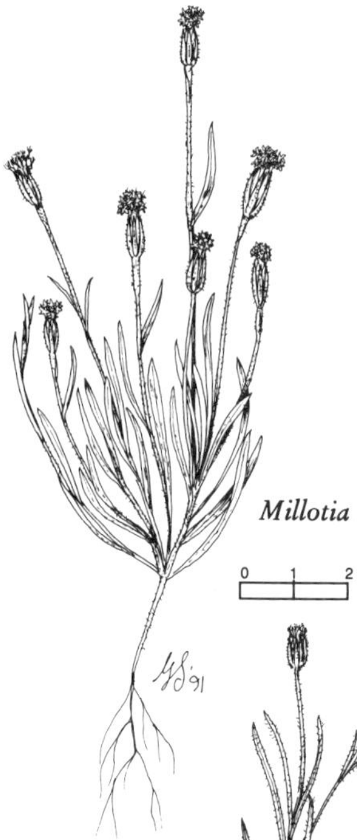

*Millotia myosotidifolia*

# Millotia **Asteraceae**

## Millotia myosotidifolia

Broad-leaf Millotia

**F P**

**Size:** 10–20 cm x 15–25 cm
**Habitat:** Red gum woodland
**Form:** Erect annual herb covered in white woolly hairs
**Foliage:** Soft, stem-clasping, narrow-linear to oblanceolate leaves to 50 mm long, at the lower end of the flower stalks
**Flowers:** Terminal creamy-white flowerheads, without ray florets, singly on long erect stalks; July to September. Bracts oblong, sparsely woolly with broad green midribs. Tiny flowers have 5 pointed petals and prominent anthers.
**Requirements:** Well drained soils
**Comments:** Probably extinct within the Melbourne area.
**Localities:** Collected in Epping 1902
**Distribution:** Vic, NSW, Tas, SA, WA

*Millotia tenuifolia*

## Millotia tenuifolia

Soft Millotia

**F**

**Size:** 25–140 mm x 10–30 cm
**Habitat:** Coastal cliff faces
**Form:** Aromatic annual herb with white-woolly and straight pale golden hairs
**Foliage:** Soft, grey, thread-like leaves to 25 mm long
**Flowers:** Single cream oblong flowerheads surrounded by narrow pointed green bracts, held above the leaves; August to November. Anthers are enclosed by 3–4 rounded petals.
**Requirements:** Well drained soil
**Propagation:** Seed
**Localities:** 67#
**Distribution:** Vic, NSW, Tas, SA, WA

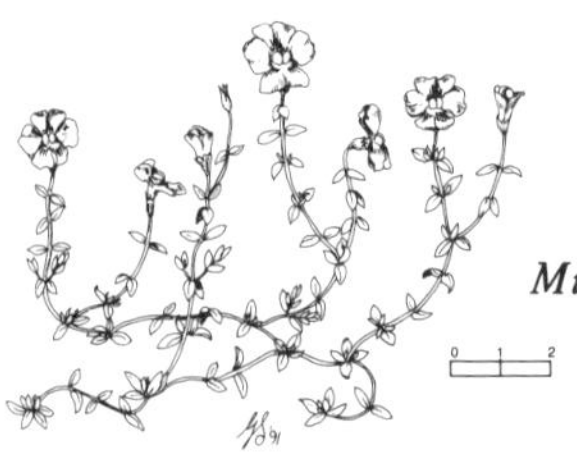

*Mimulus repens*

# Mimulus **Scrophulariaceae**

## Mimulus repens

Creeping Monkey-flower

**F P**

**Size:** Prostrate x 0.5–1 m
**Habitat:** Saltmarshes, grassy wetlands, riparian scrub
**Form:** Spreading perennial herb
**Foliage:** Pairs of succulent, stalkless, oval leaves to 6 mm long, crowded along stems
**Flowers:** Single mauve flowers with downy yellow centre on very short stalks; Spring and Autumn
**Requirements:** Moist soils tolerating salt and periods of inundation
**Comments:** Useful for planting by a water feature, it spreads by rooting at the nodes wherever it is in contact with soil.
**Propagation:** Division
**Localities:** Laverton, 4, 11, 16, 26, 67#, 72, 74
**Distribution:** All states; NZ

*Minuria leptophylla*

# Minuria **Asteraceae**

## Minuria leptophylla

Minnie Daisy

**F**

**Size:** 10–30 cm x 20–50 cm
**Habitat:** Plains grassland
**Form:** Perennial herb forming a compact rounded mound
**Foliage:** Soft narrow bright green leaves 5–25 mm long
**Flowers:** White or mauve daisy-like flowerheads with yellow centres, on short stalks above the leaves; most of the year
**Requirements:** Very well drained soil tolerating dry periods once established
**Comments:** A very showy plant, useful for rockeries or border planting in dry areas. The local form is quite small to 15 cm high with short leaves to 10 mm long. Restricted to a few sites within the Melbourne area.
**Propagation:** Seed, cuttings
**Localities:** St Albans, Sydenham, Nth Sunshine, 7, 16–18, 26
**Distribution:** All mainland states

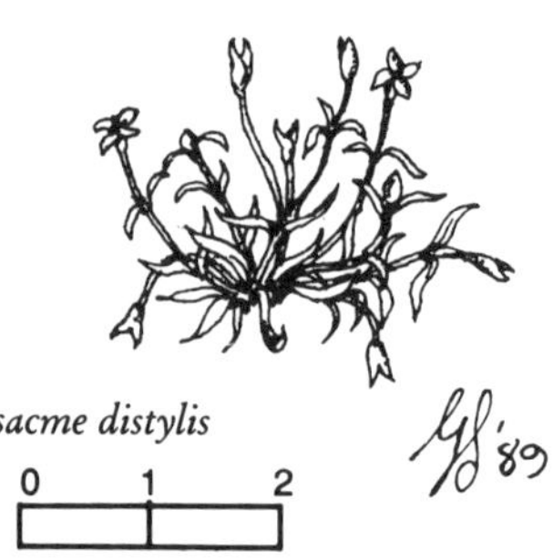

*Mitrasacme distylis*

# Mitrasacme **Loganiaceae**

## Mitrasacme distylis

Tiny Mitrewort

**Size:** to 25 mm high
**Habitat:** Coastal
**Form:** Slender erect annual herb
**Foliage:** Leaves differ from *M. paradoxa* in that they are narrow, 4 mm long, and not joined at the base
**Flowers:** Single tiny terminal flowers; October to November.
**Localities:** Collected around Port Phillip area in 1880s
**Distribution:** Vic, Tas, SA

### Mitrasacme paradoxa — Wiry Mitrewort

**Size:** 25–150 mm high — P
**Habitat:** Primary dune scrub, dry sclerophyll forest
**Form:** Inconspicuous erect annual herb with thread-like stems, and leaves mostly in a basal rosette
**Foliage:** Pairs of tiny, narrowly-oblong to ovate stem leaves (or bracts) to 8 mm long, hairy on top and joined at the base
**Flowers:** Umbels of 2–5 tiny white flowers on fine stems; September to December
**Requirements:** Well drained soils
**Comments:** An easily overlooked plant which is becoming rarer in Melbourne.
**Propagation:** Cuttings
**Localities:** Early collections—Mentone, Werribee; 41, 67#
**Distribution:** Vic, NSW, Tas, SA, WA

*Mitrasacme paradoxa*

## Monotoca — Epacridaceae

### Monotoca scoparia — Prickly Broom-heath

**Size:** 1–2.5 m x 1–2 m — F P
**Habitat:** Tea-tree heath on poor, sandy or shallow soils, dry sclerophyll forest
**Form:** Rigid, erect shrub
**Foliage:** Stiff, prickly, dark green leaves with margins curved under, whitish below, to 14 mm long
**Flowers:** Tiny cream broadly tubular flowers in small clusters; March to June; followed by small edible pale creamy-yellow drupes
**Requirements:** Moist, well drained soils
**Propagation:** Cuttings are difficult
**Localities:** 34, 51, 59, 67–71, 73–75
**Distribution:** Vic, Qld, NSW, Tas, SA

*Monotoca scoparia*

## Montia — Portulacaceae

### Montia australasica = Neopaxia australasica

### Montia fontana — Water-blinks

**Size:** to 20 cm high — P
**Habitat:** Swampy grassland
**Form:** Tiny tufted annual herb. Similar to *Neopaxia australasica* but does not spread by rhizomes.
**Foliage:** Smaller paired leaves varying from narrow to spathulate, to 20 mm long
**Flowers:** Inconspicuous pinkish flowers; September to November.
**Localities:** Croydon, 1940#
**Distribution:** Vic, NSW, SA; NZ

## Muehlenbeckia — Polygonaceae

### Muehlenbeckia adpressa — Climbing Lignum

**Size:** Climber — F
**Habitat:** Coastal banksia woodland, primary dune scrub
**Form:** Wiry trailing or climbing perennial
**Foliage:** Heart-shaped to ovate bright green leaves with wavy margins, variable 8–80 mm x 6–35 mm
**Flowers:** Axillary spikes of small greenish-yellow flowers; September to December; followed by dark reddish fruit
**Requirements:** Well drained sandy soils
**Comments:** A useful moderately vigorous climber for fences and retaining walls, tolerating salt exposure and dryness. The Aborigines ground the fruit of *Muehlenbeckia* spp. into flour.
**Propagation:** Cuttings
**Localities:** 67–69, 74
**Distribution:** Vic, NSW, Tas, SA, WA

*Muehlenbeckia adpressa*

### Muehlenbeckia florulenta — Tangled Lignum

**Size:** 1–2.5 m x 1–2 m — F
**Habitat:** Saltmarshes, grassy wetlands, riparian scrub
**Form:** Rigid, densely tangled, almost leafless, shrub
**Foliage:** Small narrow leaves found briefly on new growth. Branchlets are grey-green with parallel ridges.
**Flowers:** Spikes of small creamy-yellow flowers most of the year. Fruit is dry.
**Requirements:** Moist soil tolerating occasional inundation
**Comments:** An interesting plant which may form dense, impenetrable stands in ideal conditions. It is found on the edges of swamps on the basalt plains around Melbourne.
**Propagation:** Cuttings
**Localities:** Keilor, Laverton Nth, Sunshine, Sydenham, Melton, 4, 6, 11, 16, 26
**Distribution:** All mainland states
**Synonym:** *M. cunninghamii*

*Muehlenbeckia florulenta*

SCALE: 0 1cm 2cm — The scale representation on each plant equals 2cm.

*Muellerina eucalyptoides*

# Muellerina — Loranthaceae

## Muellerina eucalyptoides

Creeping Mistletoe

**F P**

**Habitat:** Red gum and box woodlands, valley, damp and wet sclerophyll forests, grassy low open forest
**Form:** Erect dense parasitic climber becoming pendulous under its own weight
**Foliage:** Thick, blunt pale green leaves to 75 mm long
**Flowers:** Very attractive large red and greenish flowers in loose clusters in Spring and Summer; followed by green sticky pear-shaped fruit.
**Comments:** Parasitic on a variety of hosts, but usually eucalypts, its long roots creep along the branches. Healthy host plants are usually able to cope with minor infestations. Some plants may eventually weaken to the stage where branches die and break off. Removing and burning infested branches is recommended if infestation becomes too dense. Mistletoes are the host plants of the Imperial White Butterfly.
**Propagation:** Mistletoe bird deposits seed on branches of host tree.
**Localities:** 18, 19, 21, 23, 25, 26, 28, 31, 32, 34A, 42, 44, 50, 51, 57, 58, 60, 63, 67, 69A, 71, 74–76
**Distribution:** Vic, Qld, NSW, SA

*Myoporum insulare*

# Myoporum — Myoporaceae

## Myoporum deserti = Eremophila deserti

## Myoporum insulare

Boobialla

**F**

**Size:** 1–6 m x 3 m
**Habitat:** Coastal banksia woodland, primary dune scrub
**Form:** Dense, rounded large shrub or small tree
**Foliage:** Thick, smooth lanceolate leaves, succulent beside the sea, to 100 mm long, small teeth towards the tip
**Flowers:** Clusters of fragrant white flowers with purple spots; October to November. Fruit is an edible, round purplish-green drupe.
**Requirements:** Well drained sandy soil tolerating dryness once established
**Comments:** Salt tolerant, it makes an excellent front line coastal plant. Useful for screening or a windbreak. Bark on old plants develop attractive tessellation. *Myoporum* spp. are useful fire retardant plants.
**Propagation:** Cuttings
**Localities:** Sunbury, Keilor, Seaford, 16, 26, 30#, 33, 67, 69A, 72, 74
**Distribution:** Vic, NSW, Tas, SA, WA

*Myoporum parvifolium*

## Myoporum parvifolium

Creeping Boobialla

**F P N**

**Size:** Prostrate x 2–4 m
**Habitat:** Saltmarshes, grassy wetlands
**Form:** Dense matting groundcover
**Foliage:** Thick, narrow to oblong leaves to 40 mm long, but usually smaller
**Flowers:** Masses of small, star-like white or pink flowers with purple spots; November to February. Edible green to purplish drupes
**Requirements:** Well drained soil, tolerating dryness once established
**Comments:** A very adaptable ground cover, it is excellent in rockeries. Its layering habit is useful for soil and sand binding. Other colour and leaf forms are available at nurseries. Possibly extinct within the Melbourne area.
**Propagation:** Cuttings, division at nodes
**Localities:** Early collections: Newport, Altona, Werribee, Moorabbin, 67#
**Distribution:** Vic, NSW, Tas, SA, WA

St Kilda Indig Nurses - have it - the indig form robust, dark green, 9645 2477

*Myoporum viscosum*

## Myoporum viscosum

Sticky Boobialla

**F**

**Size:** 0.5–2 m x 1.5–2 m
**Habitat:** Rocky outcrops of riparian and chenopod rocky open scrub, coastal cliffs
**Form:** Dense shrub
**Foliage:** Thin, dark green, finely toothed ovate leaves to 50 mm long, paler below. Young growth is sticky.
**Flowers:** Clusters of white flowers with purple spots and hairy centres; September to December. Edible round yellow drupes
**Requirements:** Well drained dry soils
**Comments:** A useful shrub for coastal or dry areas, tolerating exposed salty conditions and extended droughts once established.
**Propagation:** Cuttings
**Localities:** 16, 26, 29, 30, 32, 33, 34A–36, 67
**Distribution:** Vic, SA, WA

# Myosotis — Boraginaceae

## Myosotis australis — Austral Forget-me-not

**Size:** 10–30 cm high — F
**Habitat:** Dry sclerophyll forest, grassy low open forest
**Form:** Slender hairy, annual herb with 1 to several erect stems from a basal rosette
**Foliage:** Lower leaves narrowly obovate to 40 mm long, upper leaves becoming smaller and stem clasping
**Flowers:** Terminal racemes of tiny white or pale blue flowers; August to November. Calyx hairy. Bracts have upward hooked hairs.
**Requirements:** Well drained soil
**Comments:** Fast disappearing within the Melbourne area.
**Propagation:** Seed
**Localities:** 30#, 34A, 42, 63, 67#, 74, 76
**Distribution:** Vic, NSW, Tas, SA, WA; NZ, PNG

*Myosotis australis*

## Myosotis exarrhena — Sweet Forget-me-not

**Habitat:** Rocky sites
**Comments:** A perennial herb with larger flowers than *M. australis*. The fragrant white flowers have lobes as long as the tube, stamens are exserted, and calyx is glabrous. Recorded last century in the Melbourne area it may well be extinct now.
**Localities:** Frankston 1899, 29#
**Distribution:** Vic, NSW, Tas
**Synonym:** *M. suaveolens*

# Myriocephalus — Asteraceae

## Myriocephalus rhizocephalus — Woolly Heads

**Size:** 2–7 cm x 5–20 cm — P
**Habitat:** Temporarily moist depressions in red gum woodland, grassy wetlands
**Form:** Annual stemless herb forming a clump
**Foliage:** Grass-like yellowish-green leaves held vertically to 70 mm long
**Flowers:** Woolly, daisy-like flattened flowerheads, whitish bracts with yellow centres, intermixed with the leaves; September to November. Flowerheads are compound and composed of a number of partial flowerheads, each containing only one flower.
**Requirements:** Moist well drained soils
**Comments:** Short lived annual, shriveling quickly as soil dries out.
**Propagation:** Seed
**Localities:** Laverton Nth, St Albans, Werribee
**Distribution:** Vic, Qld, NSW, SA, WA

*Myriocephalus rhizocephalus*

# Myriophyllum — Haloragaceae

Annual or perennial herbs, most of the local myriophyllums, or water milfoils, are aquatic for part of the year. Others live in marshes. The plants alter their appearance as water dries up. The submerged leaves are very similar in those species with pinnate foliage. Emergent leaves are entire, toothed or lobed. Single sessile flowers occur in the upper axils of the emergent leaves, male flowers above female along the stem. Petals are absent on female flowers. The aquatic species are useful for aquariums. They may be used in and around large pools but some may become invasive. Propagation is by cuttings placed in water or a wet medium, or by division of layered stems.

## Myriophyllum amphibium — Broad Water-milfoil

P

**Habitat:** Semiaquatic in swamps and waterholes of wattle tea-tree scrub, red gum woodland and dry sclerophyll forest
**Form:** Mat-forming perennial herb with layering, prostrate stems
**Foliage:** Paired, obovate to oblanceolate leaves to 10 mm long
Submerged leaves entire or sometimes with 1 or 2 lobes
**Flowers:** Red flowers; November to February. Purplish black fruit.
**Requirements:** Moist soil
**Comments:** May be tolerant to low salinity and high levels of heavy metal pollution. Locally rare.
**Localities:** 35, 42, 75
**Distribution:** Vic, Tas, SA

*Myriophyllum amphibium*

SCALE: 0 1cm 2cm — The scale representation on each plant equals 2cm.

*Myriophyllum crispatum*

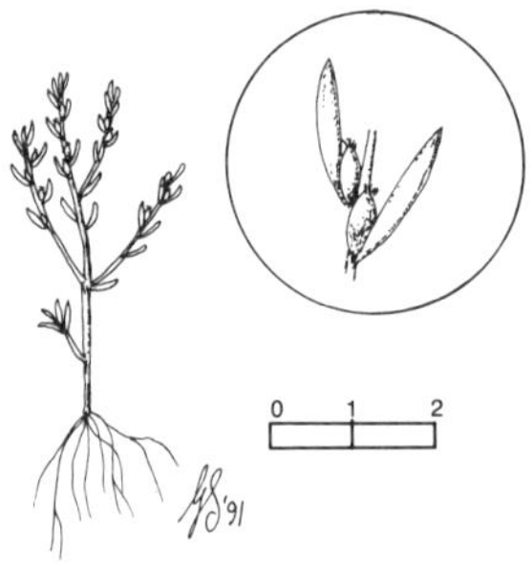

*Myriophyllum integrifolium*

*Myriophyllum muelleri*

*Myriophyllum salsugineum*

## Myriophyllum caput-medusae

Coarse Water-milfoil

**Size:** Stems 0.3–2 m long **F P**
**Habitat:** Aquatic, in riparian scrub
**Form:** Perennial herb, sparsely branched below the water line, profusely above
**Foliage:** Submerged—whorls of 3–5 pinnate leaves to 20 mm long
Emergent—glaucous, often with purplish tip, in whorls of 3–5, becoming alternate. Leaves lobed, becoming entire, ovate to oblong, to 8 mm long.
**Flowers:** Yellow to reddish male flowers above green female flowers, in simple or branched spikes arising from the upper submerged leaves. Olive-brown cylindrical fruit with dense, fine spines; November to February.
**Comments:** Plants grow in water up to 2 m deep.
**Localities:** Kew, possibly Merri Ck
**Distribution:** Vic, NSW, SA
**Synonym:** Previously included under *M. elatinoides*

## Myriophyllum crispatum

**Size:** 0.25–0.6 m high **F P**
**Habitat:** Widespread in swamps, still or slow flowing water. Will grow in water to 1 m deep.
**Form:** Erect perennial herb on short rhizomes
**Foliage:** Whorls of 5–8 leaves
Submerged—ovate in outline, pinnate to 30 mm long, segments thread-like
Intermediate—narrow to broadly lanceolate in outline, pinnate, lowest very short, upper to 17 mm long, toothed
Uppermost—terete, entire, 5–18 mm long
**Flowers:** Cream to reddish-brown flowers; October to April. Cubic fruit yellow brown to deep red.
**Requirements:** Moist to wet soil
**Comments:** This robust species can be recognised by the dense wavy hairs on the stems and on some leaves where the plant is not submerged. The emergent stem is spindle-shaped.
**Localities:** Basalt plains, 34A, 42
**Distribution:** All states except NT
**Synonym:** Previously included under *M. propinquum*

## Myriophyllum integrifolium

Tiny Water-milfoil

**Size:** 20–40 mm high **P**
**Habitat:** Valley sclerophyll forest, grassy low open forest
**Form:** Tiny, erect or trailing annual herb
**Foliage:** All leaves entire, or the lowermost rarely trifid, 1.2–3.5 mm long, upper ones shorter
**Flowers:** Red flowers are single and fruit is rounded
**Localities:** 42, 62, 66, 67, 75
**Distribution:** Vic, Tas, SA

## Myriophyllum muelleri

**Size:** to 0.6 m high **F P**
**Habitat:** Aquatic in grassy wetlands, growing in up to 60 cm of water
**Form:** Slender annual herb, rooting at the base
**Foliage:** Sessile pinnate leaves to 11 mm long, spreading at right angles from the stem. The leaves become thicker and smaller and the plant sterile if the water dries up.
**Flowers:** Single hooded red male flower at the end of flowering stem, single creamy yellow female flower in each axil immediately below. Cylindrical silvery-grey fruit.
**Localities:** Laverton, Werribee, 11
**Distribution:** Vic, Tas, SA, WA

## Myriophyllum salsugineum

**Size:** Stems 0.3–2 m long **F P**
**Habitat:** Aquatic, in slow-moving watercourses, tolerating brackish and calcareous water
**Foliage:** Similar to *M. caput-medusae* but all leaves are in whorls of 4–5.
Submerged pinnate leaves to 25 mm long
Emergent leaves—4–8 whorls, green to purplish, entire and ovate to 8 mm long, in single unbranched stem.
**Flowers:** Simple unbranched spike of green male flowers with a reddish tinge and petalless dark red and green female flowers in the axils of emergent leaves; September to February peaking in Summer. Cubic or roundish fruit is smooth.
**Comments:** Normally in water up to 2 m deep, it has been found in very deep water to 6 m.
**Localities:** Highett, Mt Eliza 1885
**Distribution:** Vic, NSW, Tas, SA, WA
**Synonym:** Previously included under *M. elatinoides*

## Myriophyllum simulans

*Myriophyllum simulans*

**Size:** to 0.4 m high — F P
**Habitat:** Aquatic, growing in water to 0.5 m deep, or in mud and seepage areas of wattle tea-tree scrub
**Form:** Perennial herb with erect, weak, often reddish stems, rooting at nodes
**Foliage:** Submerged leaves—whorls of 4–5, close together and pinnate, to 25 mm long, segments thread-like
Emergent—irregular whorls of 3–4 entire, narrow leaves to 15 mm long, fleshy and toothed
**Flowers:** Reddish male and female flowers often on different shoots; September to December. Fruit cubic, reddish purple.
**Comments:** Very variable species in both habitat and appearance, needing some drying out for good setting of fruit. Emergent leaves grow on previously submerged stems.
**Localities:** Basalt plains, 75
**Distribution:** Vic, Qld, NSW, Tas, SA
**Synonym:** Previously included under *M. propinquum*

## Myriophyllum variifolium

**Size:** stems 0.3–0.5 m long — F P
**Habitat:** Aquatic, in valley sclerophyll forest, tea-tree heath
**Form:** Aquatic perennial herb, erect in deep water to 2 m, prostrate and rooting at nodes in shallow water.
**Comments:** Very similar to *M. simulans*. All emergent leaves are whorled. Male flowers are yellow to red, female are white; fruit—cylindrical and yellow-brown.
**Localities:** Basalt plains, Montrose, 60, 61, ?67#
**Distribution:** Vic, Qld, NSW, Tas, SA
**Synonym:** Previously included under *M. propinquum*

## Myriophyllum verrucosum — Red Water-milfoil

*Myriophyllum verrucosum*

**Size:** Stems from 0.1–1.5 m — F P
**Habitat:** Aquatic, in water usually less than 2 m deep, but will grow in water up to 4 m deep. Also permanently wet or muddy swamps in grassy wetlands
**Form:** Sparsely branched perennial rooting at lower nodes, only tips emergent
**Foliage:** Whorls of 3–4 leaves
Submerged—bright to dark green pinnate leaves to 12 mm long, rounded tips
Intermediate—in 1–2 whorls
Emergent—mid green, bluish or reddish-purple leaves, dissected and lanceolate to ovate to 9 mm long
**Flowers:** Flowers may be produced throughout the year, but mostly during the warmer months. Male flowers are yellow or occasionally pink. Female flowers have deciduous petals. Cubic fruit is straw, red or grey colored.
**Comments:** The most widespread *Myriophyllum*, it can be found in deep fast flowing water as well as shallow brackish or calcareous situations. Swamp-grown plants flower and fruit prolifically as water levels begin to fall during Summer.
**Localities:** Altona, Braybrook#, Melbourne#, Caulfield#, 16, 35
**Distribution:** All mainland states

# Neopaxia — Portulacaceae

## Neopaxia australasica — White Purslane

*Neopaxia australasica*

**Size:** Prostrate creeper — F P
**Habitat:** Swamps and bogs of tea-tree heath, swamp scrub, grassy wetlands, sometimes aquatic in running water
**Form:** Spreading perennial herb, rooting at nodes
**Foliage:** Erect, narrow succulent leaves 30–100 mm x 1–4 mm, emergent in water
**Flowers:** Long-stalked, small white flowers in terminal groups of 1–4; November to March
**Requirements:** Moist to wet soils
**Comments:** A pretty little creeper for pond edges and bog gardens which may become invasive in ideal conditions. The alpine form has thicker foliage and is more floriferous.
**Propagation:** Division
**Localities:** Laverton Nth, Seaford, 33, 34A, 35, 42, 61, 67#, 69A, 71, 74, 75
**Distribution:** Vic, NSW, Tas, SA, WA; NZ
**Synonym:** *Claytonia australasica*, *Montia australasica*

## Nertera reptans = Leptostigma reptans

SCALE: 0 1cm 2cm — The scale representation on each plant equals 2cm.

*Nicotiana suaveolens*

# Nicotiana — Solanaceae

## Nicotiana maritima — Coast Tobacco

**Size:** to 0.7 m high
**Habitat:** Along watercourses in red gum woodland
**Form:** Upright annual herb with white- or grey-woolly hairs at the base of the stems
**Foliage:** Basal rosette of woolly ovate or elliptic leaves 2–22 cm long with wavy margins, stalks winged
**Flowers:** Branched flowering stem with terminal clusters of cream tubular flowers to 25 mm long most of the year, peaking in Spring. Calyx hairy.
**Requirements:** Sandy soil
**Comments:** Similar to *N. suaveolens*. Now extinct in Victoria the last collection for Melbourne was in 1883.
**Localities:** Port Phillip area, 30#
**Distribution:** Vic, SA

## Nicotiana suaveolens — Scented or Austral Tobacco

**Size:** 0.3–1.5 m **F**
**Habitat:** Rocky gullies of riparian scrub, box woodland
**Form:** Fast growing upright annual herb
**Foliage:** Basal rosette of fleshy, oblanceolate, dull green leaves 10–35 cm; few stem leaves are smaller and narrower.
**Flowers:** Terminal white to cream tubular flowers to 45 mm long in loose clusters; April to September
**Requirements:** Well drained dry sandy soil
**Comments:** Flowers are very fragrant at night. The crushed leaves were chewed as a substitute for tobacco. Locally rare.
**Propagation:** Seed, cuttings
**Localities:** Sunbury, Keilor, Essendon, Maribyrnong, Kew, 16, 26, 29
**Distribution:** Vic, Qld, NSW, SA

*Notelaea ligustrina*

# Notelaea — Oleaceae

## Notelaea ligustrina — Native or Privet Mock Olive

**Size:** 2–8 m x 3 m **P N**
**Habitat:** Cool temperate rainforest, wet sclerophyll forest
**Form:** Dense upright tall shrub or small tree
**Foliage:** Pairs of dull green, lanceolate leaves 40–90 mm long, paler below
**Flowers:** Racemes of insignificant creamy-yellow flowers; September. Oval, fleshy drupes to 10 mm long, white, pink, purple or black.
**Requirements:** Well drained, moist loamy soils
**Comments:** A useful screen plant for shady areas. Known from one location in the Melbourne area.
**Propagation:** Seed, cuttings
**Localities:** 57
**Distribution:** Vic, NSW, Tas

# Olearia — Asteraceae

Widespread through many habitats, 13 species occur in the Melbourne area. A member of the daisy family, they have flowerheads consisting of multiple tubular florets surrounded by ray florets. They are woody shrubs, varying from small shrublets to small trees, all requiring pruning to encourage bushiness. Usually flowering profusely for several months, they are an attractive addition to the garden. Propagation is by seed or cutting.

*Olearia argophylla*

## Olearia argophylla — Musk Daisy-bush

**Size:** 3–8 m x 3–5 m **P N**
**Habitat:** Moist gullies of cool temperate rainforest, wet, damp and valley sclerophyll forests
**Form:** Fast growing tall shrub or small tree
**Foliage:** Large, shiny, dark green, slightly-toothed elliptical leaves to 15 cm long, silvery below, with musky aroma
**Flowers:** Large terminal clusters of small white flowers with 3–5 ray florets; October to December
**Requirements:** Moist, rich, well drained soils
**Comments:** The largest *Olearia* sp., it is a very attractive tree for a shady area.
**Localities:** 42, 55–58, 60
**Distribution:** Vic, NSW, Tas

## Olearia asterotricha

Rough Daisy-bush

**Size:** 1–2 m x 1 m **P**
**Habitat:** Grassy low open forest, valley sclerophyll forest
**Form:** Erect shrub
**Foliage:** Variable; rough and hairy, dull green linear leaves to 30 mm long, pale below, margins rolled under and/or lobed
**Flowers:** Single blue to mauve flowerheads with 12–24 ray florets, on long stalks; November to January
**Requirements:** Moist soils
**Comments:** Branchlets and foliage are covered with star-shaped hairs. A form with larger leaves also occurs in Mallacoota. It has leaves to 60 mm long with lobed margins which are hardly or not rolled under.
**Localities:** Sth Belgrave, Dandenong, Upper Beaconsfield
**Distribution:** Vic, NSW

*Olearia asterotricha*

## Olearia axillaris

Coast Daisy-bush

**Size:** 1–2 m x 1–2 m **F**
**Habitat:** Primary dune scrub, coastal cliffs
**Form:** Erect, dense shrub
**Foliage:** Hoary narrow leaves to 25 mm long, becoming glabrous and dark green above and woolly-white below, margins rolled under
**Flowers:** Small, stalkless yellow flowerheads along upper branches, ray florets minute; February to April
**Requirements:** Well drained dry sandy soil
**Comments:** An interesting foliage plant with aromatic leaves, resistant to salt spray.
**Localities:** Seaford, 67, 74
**Distribution:** Vic, NSW, Tas, SA, WA

*Olearia axillaris*

## Olearia ciliata

Fringed Daisy-bush

**Size:** 20–40 cm x 40–60 cm **F P**
**Habitat:** Tea-tree heath
**Form:** Small dense shrub
**Foliage:** Narrow, pointed, bright green leaves with margins rolled under, to 25 mm long, crowded along stems
**Flowers:** Single bright blue to purple flowerheads to 30 mm across, with yellow centres, held above foliage on reddish stems 5–25 cm long
**Requirements:** Well drained light soils tolerating lime
**Comments:** An attractive little shrub, useful for growing amongst other plants in a rockery. An endangered Melbourne species which may already be extinct.
**Localities:** 67#
**Distribution:** Vic, NSW, Tas, SA, WA

*Olearia ciliata*

## Olearia decurrens

Clammy Daisy-bush

**Size:** 1–2 m x 1 m **F P**
**Habitat:** Box and red gum woodlands
**Form:** Erect open shrub
**Foliage:** Sticky, narrow-oblanceolate, bright green leaves 10–50 mm long, flat, usually with a few teeth or lobes
**Flowers:** Loose terminal panicles of white flowerheads with yellow centres; December to March. Ray florets 3–5.
**Requirements:** Well drained soils
**Comments:** A sticky shrub with aromatic leaves.
**Localities:** Diggers Rest 1901, Holden, Bulla
**Distribution:** Vic, NSW, SA

## Olearia erubescens

Moth Daisy-bush

**Size:** 0.3–0.5 m x 0.5 m **F P**
**Habitat:** Dry and damp sclerophyll forests
**Form:** Small spindly shrub
**Foliage:** Dark green oblong to lanceolate toothed leaves to 40 mm long, paler below with white or rusty hairs. Young growth reddish.
**Flowers:** Small loose panicles of white to pink flowerheads with yellow centres, on long stalks to 50 mm; September to November. Ray florets 4–8.
**Requirements:** Well drained soils tolerating dryness once established
**Comments:** A straggly shrub by nature, it looks good planted amongst other shrubs. This species is often confused with *O. myrsinoides* which has shorter, broader and blunter leaves and less than 4 ray florets per flowerhead.
A form resembling *O. speciosa* but believed to be a hybrid with *O. argophylla* is found in isolated patches on Mt. Dandenong and in Mt. Evelyn. It has larger, ovate, lighter green leaves and generally lacks the reddish new growth. There are also more flowers in the panicle than *O. erubescens*. Further research is required to clarify its relationship to *O. speciosa*.
**Localities:** Eltham, 49, 55, 56
**Distribution:** Vic, NSW, Tas, SA

*Olearia erubescens*

SCALE: 0 1cm 2cm The scale representation on each plant equals 2cm.

Olearia glandulosa

## Olearia glandulosa

Swamp Daisy-bush

**Size:** 1–2 m x 1 m — **F P**
**Habitat:** Swamp scrub
**Form:** Open, erect shrub
**Foliage:** Fine, sessile dark green leaves to 60 mm long, sticky to touch. The tightly rolled margins have several glandular swellings along their length.
**Flowers:** Dense terminal panicles of white or bluish flowerheads with yellow centres; November to April. Ray florets 12–15.
**Requirements:** Moist soils, tolerating some waterlogging in Winter
**Comments:** An attractive plant for a damp place in the garden. Locally rare.
**Localities:** Kilsyth
**Distribution:** Vic, NSW, Tas, SA

Olearia glutinosa

## Olearia glutinosa

Sticky Daisy-bush

**Size:** 2 m x 1.5 m — **F**
**Habitat:** Primary dune scrub
**Form:** Dense rounded shrub
**Foliage:** Sticky, narrow, greyish leaves to 30 mm long, pointed at the end
**Flowers:** Terminal clusters of blue or white flowerheads with 6–8 ray florets; December to February
**Requirements:** Well drained sandy soil
**Comments:** A useful plant for coastal gardens, resistant to salt spray.
**Localities:** Seaford, 5, 67, 74
**Distribution:** Vic, NSW, Tas

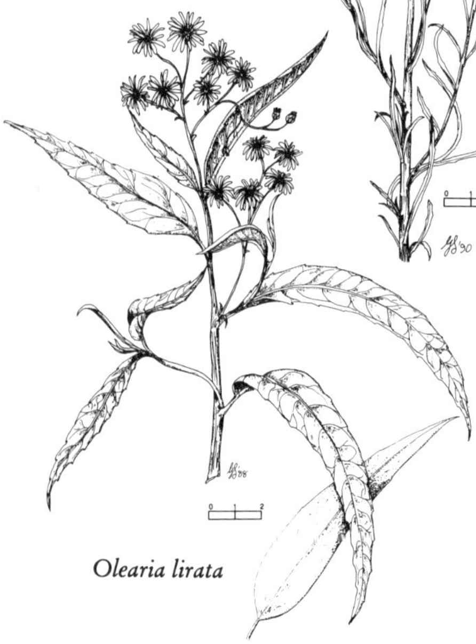

Olearia lirata

## Olearia lirata

Snowy Daisy-bush

**Size:** 2–5 m x 2–3 m — **P N**
**Habitat:** Moist gullies in damp and valley sclerophyll forests, grassy low open forest
**Form:** Soft open shrub
**Foliage:** Dark green lanceolate leaves to 16 cm long, whitish below
**Flowers:** Masses of small white flowerheads with about 15 ray florets, in large, loose panicles at or near the ends of branches; September to December.
**Requirements:** Moist well drained soil
**Comments:** An excellent shrub brightening a sheltered position in Spring.
**Localities:** 34, 34A, 42–44, 47, 49, 50A–59, 61–64, 74–76
**Distribution:** Vic, NSW, Tas

## Olearia myrsinoides

Silky Daisy-bush

**Size:** 0.3–1.5 m x 1 m — **F P**
**Habitat:** Dry and valley sclerophyll forests
**Form:** Small open spreading shrub
**Foliage:** Shiny dark green obovate leaves to 20 mm long, silvery-greyish below, with serrated margins
**Flowers:** Tight clusters of flowerheads with yellow centres, and 3 white ray florets; October to February
**Requirements:** Well drained soils
**Comments:** Similar to *O. erubescens* which has longer, narrower, more pointed leaves, reddish new growth and more than 4 ray florets.
**Localities:** Sth Morang, Park Orchards, Wantirna, Warrandyte, Montrose, 40, 43, 47, 49–53, 56, 58
**Distribution:** Vic, NSW, Tas — *Illustrated p. 49*

*Olearia phlogopappa* var. *phlogopappa*

## Olearia phlogopappa var. phlogopappa

Dusty Daisy-bush

**Size:** 1–3 m x 1–2 m — **F P**
**Habitat:** Valley sclerophyll forest
**Form:** Fast growing open to dense shrub
**Foliage:** Grey-green oblong leaves to 80 mm long, white below, with wavy, toothed margins
**Flowers:** Masses of long-stalked white, pink, blue or mauve flowerheads with yellow centres, in axillary clusters, in the upper branches; August to January. Ray florets 8–15.
**Requirements:** Well drained soils
**Comments:** One of the most attractive *Olearia* species. White is the most common colour form locally but others may be grown from selected clones.
**Localities:** 34, 47, 50, 50A, 55
**Distribution:** Vic, NSW, Tas

### Olearia ramulosa var. ramulosa

Twiggy Daisy-bush

| | |
|---|---|
| **Size:** | 0.5–2.5 m x 1 m — F P |
| **Habitat:** | Widespread in red gum woodland, damp sclerophyll forest, grassy low open forest, tea-tree heath, primary dune scrub |
| **Form:** | Open spindly, often sprawling, shrub |
| **Foliage:** | Narrow crowded leaves to 8 mm long, white below, sometimes sticky, with margins rolled under |
| **Flowers:** | Masses of small, white or mauve flowerheads along branches; September to May. Ray florets 3–15. |
| **Requirements:** | Well drained soils |
| **Comments:** | Attractive garden plant especially the mauve, suckering form. Grows well in an understorey situation. Pruning as the flowers begin to fade usually encourages a further flush of flowers. |
| **Localities:** | 30-32, 34A, 35, 42, 44, 58, 60, 66–71, 73–76 |
| **Distribution:** | All states except NT |

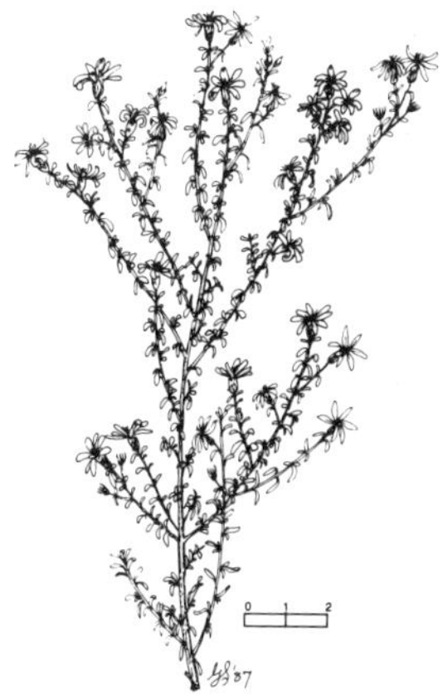

*Olearia ramulosa* var. *ramulosa*

### Olearia rugosa

Wrinkled Daisy-bush

| | |
|---|---|
| **Size:** | 1–2 m x 1 m — P |
| **Habitat:** | Damp sclerophyll forest |
| **Form:** | Upright shrub |
| **Foliage:** | Rough, deeply wrinkled, broadly-ovate leaves to 70 mm long, brown hairs below, toothed or entire on the same plant. Network of veins prominent. |
| **Flowers:** | Small panicles of white flowerheads held erect on stiff stalks; October to November |
| **Requirements:** | Moist well drained soil |
| **Comments:** | An attractive foliage plant for a sheltered position. |
| **Localities:** | Mt. Dandenong 1901, Belgrave 1932; Olinda |
| **Distribution:** | Endemic to Vic |

# Opercularia — Rubiaceae

### Opercularia ovata

Broad Stinkweed

| | |
|---|---|
| **Size:** | to 15 cm x 5–30 cm — F P |
| **Habitat:** | Red gum woodland, dry and valley sclerophyll forests, grassy low open forest |
| **Form:** | Smaller than O. *varia* with a similar densely sprawling habit. Branching occurs from several points off an underground rhizome. |
| **Foliage:** | Leaves are lighter green and ovate, on lax, somewhat angular stems. A narrow-leaf form occurs in Watsonia. |
| **Flowers:** | Spherical cluster contains 10–18 flowers. |
| **Localities:** | 34A, 35, 38, 39, 41, 42, 47–49, 50A-52, 54, 55, 59, 62–69, 74, 76 |
| **Distribution:** | Vic, Tas, SA |

*Opercularia ovata*

### Opercularia varia

Variable Stinkweed

| | |
|---|---|
| **Size:** | 5–25 cm x 5–30 cm — P N |
| **Habitat:** | Dry and valley sclerophyll forests, grassy low open forest, tea-tree heath, wattle tea-tree scrub |
| **Form:** | Sprawling or upright perennial herb forming a clump |
| **Foliage:** | Scattered pairs of sessile, narrow to lanceolate leaves to 12 mm long, with margins rolled under |
| **Flowers:** | Inconspicuous green or purple flowers in tight hemispherical clusters of 2–7; Spring. Spiky green fruit are larger than flowers. |
| **Requirements:** | Moist, well drained soils |
| **Comments:** | Branching occurs from a single short basal stem. The leaves exude an unpleasant aroma when crushed. |
| **Propagation:** | Seeds, cuttings |
| **Localities:** | Research, 34A, 36, 40–42, 44–48, 50–51, 53–56, 58–60, 62, 67#, 69A–71, 74–76 |
| **Distribution:** | Vic, Qld, NSW, Tas, SA |

*Opercularia varia*

# Oreomyrrhis — Apiaceae

### Oreomyrrhis eriopoda

Hairy Carroway

| | |
|---|---|
| **Size:** | 15–40 cm high — F P |
| **Habitat:** | Dry sclerophyll forest, box woodland |
| **Form:** | Rosetted perennial herb |
| **Foliage:** | Basal tuft of hairy, bipinnate, grey-green leaves to 20 cm long |
| **Flowers:** | Terminal umbels of 12–35 tiny white flowers with purple anthers on erect stems to 40 cm long; September to November. Large brown or purplish fruit, like carroway seeds, develop on long stalks. |
| **Requirements:** | Well drained soils |
| **Comments:** | Locally rare |
| **Propagation:** | Seed |
| **Localities:** | 26, ?42, 55 |
| **Distribution:** | Vic, NSW, Tas, SA |

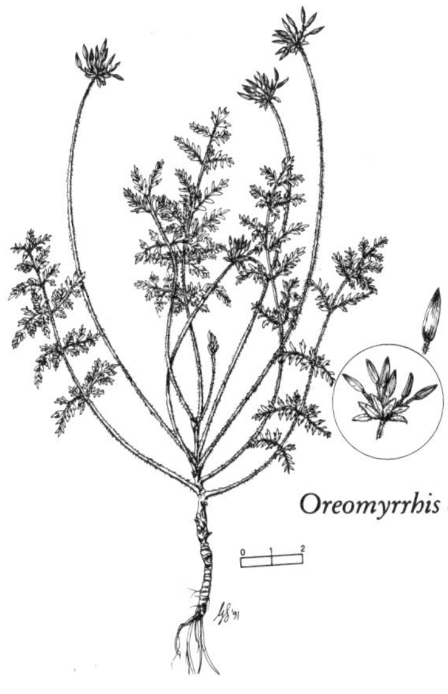

*Oreomyrrhis eriopoda*

SCALE: 0 1cm 2cm — The scale representation on each plant equals 2cm.

# Oxalis

## Oxalidaceae

*Oxalis* is a widespread genus found throughout Australia and overseas which may become invasive in the garden. The local spp. are usually hairy with trifoliate clover-like leaves and small yellow flowers which are followed by cylindrical fruit. Propagation is by seed.

*Oxalis perennans*

### Oxalis exilis

**F P**

**Habitat:** Damp and valley sclerophyll forests, box woodland, riparian scrub, grassy low open forest
**Comments:** Similar to O. *perennans* but smaller. Flowers are usually solitary, and fruit capsules shorter, to 9 mm long, and broader. Prefers rich peaty soils beside creeks. Some authors consider O. *perennans* and O. *exilis* to be conspecific.
**Localities:** Sth of Melton, 17#, 18, 21, 28, 34A, 36, 42, 44, 47, 48, 53–56, 69A, 71, 74, 76
**Distribution:** Vic, Qld, NSW, SA

### Oxalis perennans

**Size:** 30 cm high **F P**
**Habitat:** Widespread in grassy wetlands, plains grassland, red gum woodland, valley sclerophyll forest
**Form:** Erect or creeping herb
**Foliage:** Wedge-shaped leaflets; individual leaves to 15 mm long on stalks 10–40 mm long
**Flowers:** 2–3 yellow flowers on stalks as tall or taller than leaves; throughout the year. Slender hairy fruit capsules from 4–30 mm long, held erect.
**Requirements:** Heavy soil
**Comments:** Stem hairs on this species point upwards. The stout tap root is edible.
**Localities:** 2, 4, 6, 8, 9, 14–18, 21, 23, 26, 28, 31, 32, 34A–36, 42, 46–49, 50A, 54–57, 61, 65–68, 69A, 74
**Distribution:** All mainland states; NZ
**Synonym:** Has been confused in the past with the introduced species O. *corniculata* and may occur in other areas.

### Oxalis radicosa

**F P**

**Habitat:** Box woodland
**Comments:** Also similar to O. *perennans* but found on sandy soil. Dense stem hairs are widely spreading or bent backwards. Leaflets have rounded lobes. Fruit capsules are 10–20 mm long. Flowering March to November.
**Localities:** 9, 11, 12, 17#, 18, 26, 28, 71
**Distribution:** Vic, Qld, NSW, SA; Af, S Asia, Pacific

# Ozothamnus

## Asteraceae

*Ozothamnus* spp. belong to the group of daisies known as everlastings. They are all shrubby in habit. The flowerheads are in broad flat terminal clusters and are long-lasting. They are useful as fillers in dried flower arrangements. *Ozothamnus* spp. are very adaptable and make excellent garden subjects. It is preferable to prune them back in late winter to encourage bushiness. Propagation is by seed or cuttings.
**Synonym:** *Helichrysum*

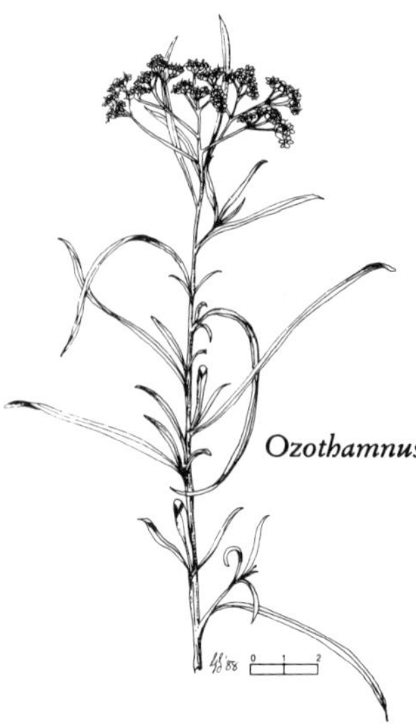

*Ozothamnus ferrugineus*

### Ozothamnus ferrugineus

Tree Everlasting

**Size:** 2–6 m x 1–3 m **F P**
**Habitat:** Wet, damp and valley sclerophyll forests, red gum woodland, swamp and wattle-tree scrub, grassy low open forest
**Form:** Open rounded shrub or small tree
**Foliage:** Narrow dark green leaves to 70 mm long, whitish below
**Flowers:** Broad clusters of white flowerheads at the ends of branches; November to February
**Requirements:** Moist well drained soil
**Comments:** The yellow-green bracts which surround the flowers before they open is a useful feature to distinguish O. *ferrugineus* from *Cassinia* spp. It is very widespread and common in moist areas. A hybrid occurs between O. *ferrugineus* and O. *rosmarinifolius* in the foothills of the Dandenongs.
**Localities:** Seaford, 29#, 30, 33–37, 41, 42, 44, 47–60, 62–67#, 69A–71, 74–76
**Distribution:** Vic, NSW, Tas, SA
**Synonym:** *Helichrysum dendroideum*

### Ozothamnus obcordatus

Grey Everlasting

F P

**Size:** 1–2 m x 1 m
**Habitat:** Dry sclerophyll forest, box and box ironbark woodland, on shallow stony soil
**Form:** Slender erect shrub
**Foliage:** Broadly elliptic, obovate or wedge-shaped, shiny dark green leaves to 15 mm long, white below
**Flowers:** Numerous dense clusters of tubular yellow flowers; October to January
**Requirements:** Well drained dryish conditions
**Comments:** A showy shrub for a difficult spot.
**Localities:** Sth of Melton, 26, 34A–36, 42, 51
**Distribution:** Vic, Qld, NSW, Tas
**Synonym:** *Helichrysum obcordatum*

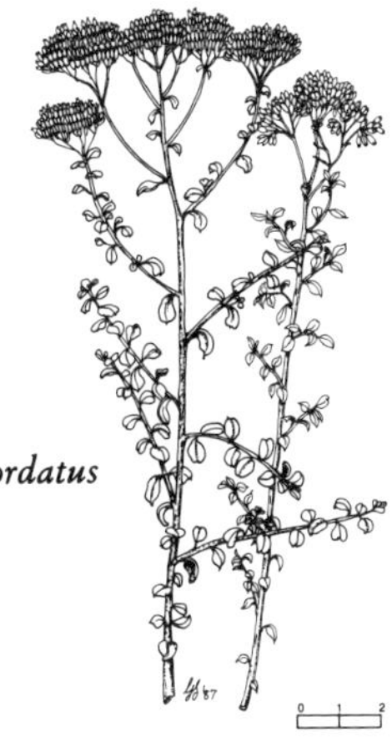

*Ozothamnus obcordatus*

### Ozothamnus rosmarinifolius

Rosemary Everlasting

F P

**Size:** 1.5–3 m x 1–2 m
**Habitat:** Swamp scrub
**Form:** Upright shrub
**Foliage:** Rough, narrow dark green leaves to 40 mm long, white below, with margins rolled under, leaves held erect
**Flowers:** Clusters of whitish flowers with yellow or crimson bracts; December to March
**Requirements:** Well drained, moist soil
**Localities:** Dingley#; Montrose, 54, 60, 67#, 74
**Distribution:** Vic, Tas
**Synonym:** *Helichrysum rosmarinifolium*

*Ozothamnus rosmarinifolius*

### Ozothamnus turbinatus

Coast Everlasting

F

**Size:** 1–3 m x 1.5 m
**Habitat:** Exposed sites on cliffs and primary dune scrub
**Form:** Dense upright to rounded hairy shrub
**Foliage:** Stiff, narrow grey-green leaves to 25 mm long, woolly white below, tinged yellow at the base, margins rolled under
**Flowers:** Dense clusters of cream tubular flowers surrounded by hairy yellow bracts; February to May
**Requirements:** Sandy, well drained soil
**Comments:** Easily grown, it benefits from hard pruning.
**Localities:** Seaford, 67#, 74
**Distribution:** Vic, Qld, NSW, Tas, SA
**Synonym:** *Helichrysum paralium*

*Ozothamnus turbinatus*

## Pandorea — Bignoniaceae

### Pandorea pandorana

Wonga Vine

F P

**Habitat:** Wet, damp, valley and dry sclerophyll forests, grassy low open forest
**Form:** Dense vigorous climber
**Foliage:** Glossy dark green pinnate leaves, 5–11 lanceolate leaflets to 50 mm long with the longest leaflet at the end
**Flowers:** Bunches of tubular creamy-white flowers with maroon markings in the throat and at the tips of the corolla lobes; September to January
**Requirements:** Moist well drained soil
**Comments:** Very showy in flower, this strong climber grows well amongst established trees or over a pergola. It prefers a shaded position. Many horticultural varieties are grown but most are not the local form.
**Propagation:** Seed, cuttings
**Localities:** 42, 43, 50A–53, 55–58, 61, 74–76
**Distribution:** Vic, Qld, NSW, Tas

*Pandorea pandorana*

### Parahebe derwentiana = Derwentia derwentiana

## Parietaria — Urticaceae

### Parietaria debilis

Shade or Forest Pellitory, Smooth Nettle

F P N

**Size:** 20–30 cm high
**Habitat:** Riparian scrub, valley sclerophyll forest, coastal banksia woodland
**Form:** Straggling annual herb
**Foliage:** Thin, flat, ovate light green leaves to 30 mm long, young growth hairy
**Flowers:** Masses of small hairy greenish flowers in clusters of 3, hanging from leaf axils; August to December
**Requirements:** Adaptable to most conditions. Tolerates limestone
**Comments:** Closely related to *Urtica incisa*, the stinging nettle, but the hairs do not sting when touched. Locally rare.
**Propagation:** Seed, cuttings
**Localities:** 28, 49, 58, 67
**Distribution:** All states

SCALE: 0 1cm 2cm — The scale representation on each plant equals 2cm.

*Parsonsia brownii*

# Parsonsia — Apocynaceae

## Parsonsia brownii

Twining Silkpod

**P N**

**Habitat:** Wet sclerophyll forest, cool temperate rainforest
**Form:** Vigorous climber
**Foliage:** Dark green; long, narrow juvenile leaves to 100 mm x 5 mm; lanceolate adult leaves yellowish below, to 120 mm x 40 mm
**Flowers:** Loose clusters of small, hairy, yellow-brown flowers; September to January. Cylindrical seed pods to 80 mm long hang vertically, producing seeds with silky hairs from August to October.
**Requirements:** Moist well drained soil
**Comments:** A good climber for a shady area but may overgrow small plants. Pruning will produce a more bushy plant.
**Propagation:** Fresh seed, cuttings
**Localities:** Rowville, 56, 57
**Distribution:** Vic, NSW, Tas

*Pelargonium australe*

# Pelargonium — Geraniaceae

## Pelargonium australe

Austral Stork's-bill

**F P**

**Size:** 0.3–0.6 m x 0.3–1 m
**Habitat:** Widespread on rocky outcrops in riparian scrub, damp and valley sclerophyll forests, coastal banksia woodland, primary dune scrub
**Form:** Soft, hairy, perennial herb forming a clump
**Foliage:** Pairs of aromatic, rounded, lobed leaves from 20–90 mm long, on stalks to 130 mm; velvety below
**Flowers:** Clusters of up to 12 pink or white 5-petalled flowers with crimson to purple veins, on long stalks; October to February. Calyx is covered in long spreading hairs.
**Requirements:** Well drained soils, tolerating dryness once established
**Comments:** An interesting rockery plant which is useful as a soil or sand binder. May die back to its rootstock during Summer, resprouting following Autumn rains.
**Propagation:** Seed, cuttings
**Localities:** Altona, St Albans, 2, 10, 16, 26, 27, 29, 30, 32, 33, 34A, 36, 42, 46, 49, 55, 58, 59, 61, 67, 72, 74–76
**Distribution:** All states except NT

*Pelargonium inodorum*

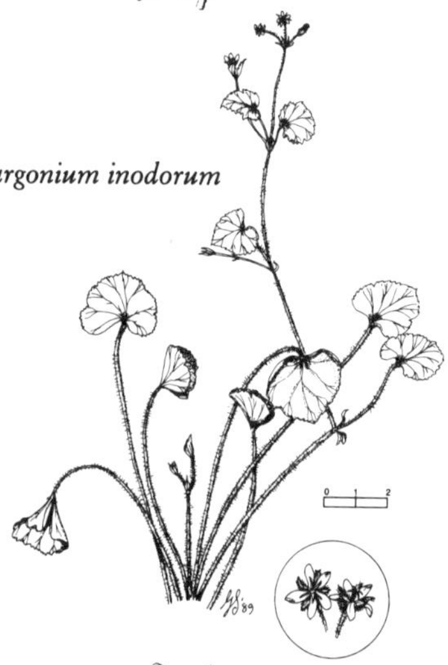

## Pelargonium inodorum

Kopata

**F P**

**Size:** 5–35 cm x 0.3–0.7 m
**Habitat:** Dry, valley and damp sclerophyll forests, grassy low open forest
**Form:** Sprawling annual or short-lived perennial herb
**Foliage:** Light green, ovate-cordate leaves to 80 mm long, with rounded lobes, on stalks to 50 mm long
**Flowers:** Clusters of tiny, deep pink flowers with darker veins; October to March
**Requirements:** Well drained soils
**Propagation:** Seed, cuttings
**Localities:** 34A, 42, 44, 54–59, 74, 75
**Distribution:** All states except NT; NZ

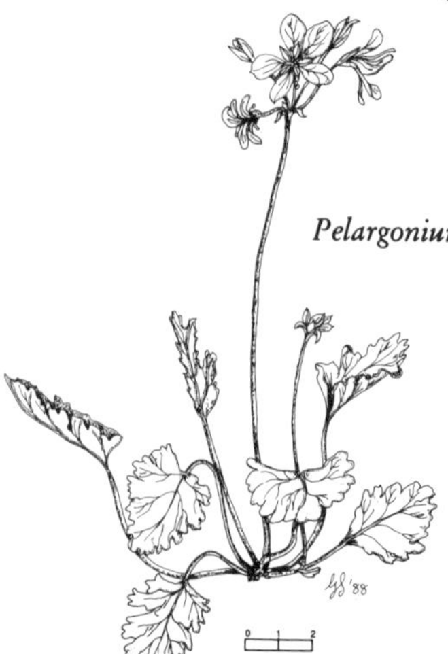

*Pelargonium rodneyanum*

## Pelargonium rodneyanum

Magenta Stork's-bill

**F P**

**Size:** 10–30 cm x 30–50 cm
**Habitat:** Rocky outcrops on plains grassland
**Form:** Perennial herb with a loose rosette of leaves growing from a tuberous root
**Foliage:** Dark green ovate leaves to 40 mm long with bluntly toothed edges, on long stalks to 100 mm long
**Flowers:** Small clusters of magenta flowers with darker veins, on long flowering stems; November to February
**Requirements:** Well drained soils
**Comments:** A very showy plant, excellent for a rockery or container, but needs root protection.
**Propagation:** Seed, cuttings
**Localities:** Melton, Rockbank, Sunbury, 16, 26, 30, 33
**Distribution:** Vic, NSW, SA, WA

# Persicaria

**Polygonaceae**

Annual or perennial herbs. *Persicaria* spp. grow in either permanently or seasonally wet sites. They are usually erect or have decumbent stems which may root at the nodes. Spikes of many small flowers occur at or near the end of branches. Propagation is by seed or division. The seeds provide an important food source for water birds.

**Synonym:** Previously included with *Polygonum*

## Persicaria decipiens

Slender Knotweed

**F P N**

**Size:** Prostrate-0.6 m x 1 m
**Habitat:** Widespread in riparian scrub and woodland, red gum woodland, valley and damp sclerophyll forests
**Form:** Aquatic to semi-aquatic; glabrous, erect to spreading annual herb
**Foliage:** Almost sessile, lanceolate leaves to 15 cm long, often with a dark blotch. Sheathing stipules are fringed with bristles.
**Flowers:** Small pink flowers in slender spikes to 50 mm long, flowering stalk sometimes branched; most of the year
**Localities:** 4, 8, 14, 21, 26–28, 30–33, 34A–36, 42, 44, 49, 53, 56–58, 60, 62, 67, 68, 74
**Distribution:** All states
**Synonym:** *Polygonum minus*, *Polygonum decipiens*

*Persicaria decipiens* (2 variants)

## Persicaria hydropiper

Waterpepper

**F P**

**Size:** 0.6–1 m high
**Habitat:** Shallow water, swamps and depressions in riparian scrub, box woodland, valley sclerophyll forest
**Form:** Upright glabrous annual herb
**Foliage:** Lanceolate leaves from 30–90 mm long, with a hot peppery taste, sheathing stipules have bristled margins
**Flowers:** Small green or white flowers scattered along pendulous spikes to 15 cm long; December to February
**Requirements:** Wet soil or shallow water at edge of ponds
**Comments:** The Aborigines roasted and peeled the stem for food and used the leaves to poison fish.
**Localities:** 14, 16, 21, 26–28, 33, 34A–36, 42, 57
**Distribution:** Vic, NSW, Qld; NZ
**Synonym:** *Polygonum hydropiper*

*Persicaria hydropiper*

## Persicaria lapathifolia

Pale Knotweed

**F P**

**Habitat:** Valley sclerophyll forest
**Form:** Branching, erect annual herb
**Foliage:** Large leaves, to 20 cm long, often have a central brown blotch and ribbed stipules without bristles
**Flowers:** Dense, thick spikes to 70 mm long, of dull pink flowers, in terminal panicles throughout the year
**Comments:** Similar to *P. hydropiper* but somewhat taller, it may form prolific stands in ideal conditions.
**Localities:** 34A–36, 42, 44, 59, 62, 63, 71
**Distribution:** All states except WA; NZ
**Synonym:** *Polygonum lapathifolium*

## Persicaria praetermissa

Spotted Knotweed

**P**

**Habitat:** Valley sclerophyll forest
**Form:** Straggling perennial herb
**Foliage:** Lanceolate leaves are lobed at the base and hairy along their midribs
**Flowers:** Pink flowers are in short spikes in leaf axils; December to April
**Comments:** Confused with *P. strigosa* which does not occur in Melbourne.
**Localities:** Knox, 30, 33, 34A–36, 42, 51
**Distribution:** Vic, Qld, NSW, Tas
**Synonym:** Segregated from *Polygonum strigosum*

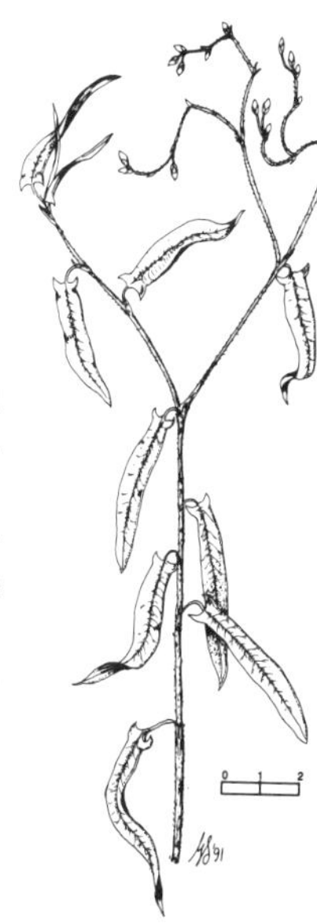

*Persicaria praetermissa*

## Persicaria prostrata

Creeping Knotweed

**F P**

**Habitat:** Grassy wetlands, plains grassland, valley sclerophyll forest
**Form:** Prostrate or decumbent perennial herb with some stems ascending, forming mats 1 m or more across
**Foliage:** Lanceolate leaves to 55 mm long are usually glabrous, but stems and leaf stalks are hairy. Hairy brownish sheathing stipule.
**Flowers:** Dense oval spikes to 20 mm long, of greenish flowers, in leaf axils; October to July
**Localities:** 9–11, 17#, 18, 26, 30, 33, 34A, 35, 42
**Distribution:** All states
**Synonym:** *Polygonum prostratum*

SCALE: 0 1cm 2cm — The scale representation on each plant equals 2cm.

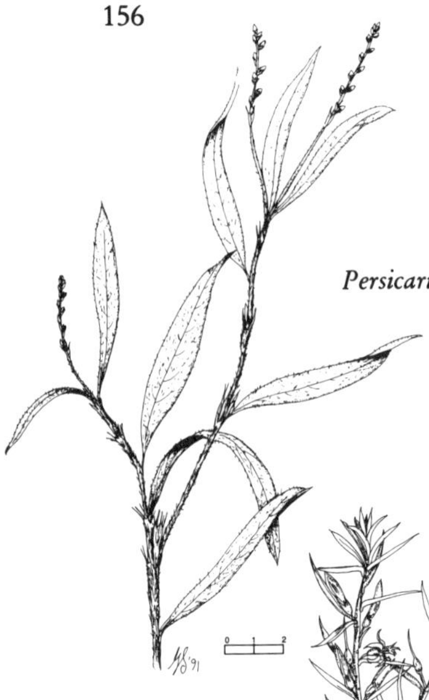

*Persicaria subsessilis*

*Persoonia juniperina*

### Persicaria subsessilis

Hairy Knotweed

**F P**

**Habitat:** Riparian scrub
**Form:** Coarsely hairy, erect, annual or perennial herb
**Foliage:** Lanceolate leaves to 100 mm long
**Flowers:** Slender spikes, to 50 mm long, of white or pink flowers, in terminal panicles
**Localities:** 19, 32, 42
**Distribution:** Vic, Qld, NSW, Tas
**Synonym:** *Polygonum subsessile*

## Persoonia — Proteaceae

### Persoonia juniperina

Prickly Geebung

**F P**

**Size:** 0.3–2 m x 0.6–0.8 m
**Habitat:** Heathy understorey of dry and valley sclerophyll forests, grassy low open forest, tea-tree heath
**Form:** Erect or rounded small shrub
**Foliage:** Fine, prickly, light green leaves to 25 mm long, crowded towards the tips of branches
**Flowers:** Single yellow tubular flowers in leaf axils; December to March
**Requirements:** Well drained soils
**Comments:** An attractive shrub which may reach 2 m high in protected coastal heaths but is usually smaller in drier areas. The fleshy green drupes that follow flowering are edible.
**Propagation:** Cuttings, which are slow to form roots
**Localities:** Clayton#, 29, 49–51, 53–56, 59, 67#, 69, 73–75
**Distribution:** Vic, NSW, Tas, SA

## Pimelea — Thymelaeaceae

Pimeleas are attractive small shrubs. Most species display striking terminal heads of flowers, often surrounded by large bracts. Leaves are almost always in pairs. Pimeleas grow in sheltered, well drained positions in the garden, some tolerating full sun. They are an important source of nectar for butterflies. Propagate from cuttings.

*Pimelea axiflora* ssp. *axiflora*

*Pimelea curviflora* var. *sericea*

### Pimelea axiflora ssp. axiflora

Bootlace Bush, Tough Rice-flower

**P N**

**Size:** 1–3 m x 1–2 m
**Habitat:** Wet, damp and valley sclerophyll forests
**Form:** Open, erect or arching shrub
**Foliage:** Narrow dark green leaves, 10–80 mm long, paler below
**Flowers:** Dioecious; sessile clusters of small creamy-white flowers in each leaf axil, surrounded by 2–4 small, brown or green papery bracts; September to November.
**Requirements:** Moist well drained soil
**Comments:** The tough bark was stripped off by the Aborigines to make nets and headbands. The early settlers used it for bootlaces.
**Localities:** Kalorama, 49, 55–58
**Distribution:** Vic, NSW, Tas

### Pimelea curviflora var. sericea

Curved Rice-flower

**F P**

**Size:** 15–30 cm x 0.2–0.6 m
**Habitat:** Widespread in plains grassland, red gum woodland, damp, valley and dry sclerophyll forests
**Form:** Erect hairy sub-shrub
**Foliage:** Dull green, elliptical leaves to 20 mm long, margins incurved, arranged alternately on stems
**Flowers:** Small terminal heads or axillary clusters of yellow-green flowers, sometimes dark red on tips, with 0–2 bracts; October to January
**Requirements:** Well drained soil
**Comments:** A variable species with many varieties, the local variety is low and spreading with upright branches.
**Localities:** Brighton#, Hawthorn#; Doncaster, Elsternwick, Montrose, 2, 7, 10, 12, 13, 15, 16, 21, 25, 26, 28, 29, 32, 34A, 35, 38–42, 44, 49, 51, 56, 58, 65–67#, 69A
**Distribution:** Vic, Qld, NSW, Tas, SA

## Pimelea flava ssp. flava

Yellow Rice-flower

**P N**

**Size:** 0.5–2 m x 0.5 m
**Habitat:** Damp and valley sclerophyll forests
**Form:** Erect, few-stemmed, slender shrub, young stems hairy
**Foliage:** Broad yellow-green leaves to 20 mm long, crowded along stems
**Flowers:** Masses of yellow flowers in terminal heads, surrounded by 2–4 ovate bracts, often larger than the leaves; August to December
**Requirements:** Moist well drained soils
**Comments:** Usually dioecious, but may be monoecious.
**Localities:** Kalorama, Montrose, 49, 50A, 51, 56, 59
**Distribution:** Vic, Tas, SA

*Pimelea flava* ssp. *flava*

## Pimelea glauca

Smooth Rice-flower

**F P**

**Size:** 0.3–0.6 m x 0.6 m
**Habitat:** Plains grassland, primary dune scrub
**Form:** Erect, many-branched glabrous shrub
**Foliage:** Narrowly-ovate, pointed, bluish-green leaves to 20 mm long
**Flowers:** Hairy; terminal, erect hemispherical heads, to 30 mm wide, of creamy-white flowers; July to February. 4 ovate bracts with pointed tips. The outer pair are glabrous but the inner pair have hairy margins.
**Requirements:** Well drained soils
**Comments:** The massed floral display makes this an excellent garden plant.
**Localities:** Laverton, Tottenham, 2, 4, 5, 10, 12, 15, 16, 18, 26, 29, 33, 34A, 41, 67#
**Distribution:** Vic, Qld, NSW, Tas, SA

*Pimelea glauca*

## Pimelea humilis

Common or Small Rice-flower

**F P**

**Size:** 10–50 cm x 0.3–1 m
**Habitat:** Widespread in red gum woodland, valley and dry sclerophyll forests, grassy low open forest, tea-tree heath
**Form:** Erect or straggling perennial sub-shrub, usually with unbranched stems and a suckering habit. Young stems are densely hairy.
**Foliage:** Elliptic leaves are green to grey-green to 15 mm long
**Flowers:** Bracts (4–6) surrounding the dense creamy-white flowerheads are larger than the leaves; September to January
**Requirements:** Moist, well drained soils
**Comments:** Generally smaller than *P. glauca* from which it is distinguished by the hairy stems and the glabrous inner floral bracts. It tolerates dryness once established but with extra Summer moisture this attractive small shrub is more vigorous and the flowers tend to last longer.
**Localities:** 12, 16, 25, 26, 28, 29, 32–36, 38–45, 47–56, 58, 59, 62–67, 69A, 71, 73–76
**Distribution:** Vic, NSW, Tas, SA

*Pimelea humilis*

*Pimelea ligustrina* ssp. *ligustrina*

## Pimelea ligustrina ssp. ligustrina

Tall Rice-flower

**P N**

**Size:** 0.5–2.5 m x 1–1.5 m
**Habitat:** Wet and damp sclerophyll forests
**Form:** Slender or bushy upright shrub
**Foliage:** Broad, thin-textured, shiny leaves, paler below, 10–75 mm long
**Flowers:** Masses of white flowers in large terminal clusters held on long, curving stalks. The yellow-tipped anthers are prominent; October to December. The 4 green bracts are broadly ovate, with pointed tips, sometimes silky hairs on the inside and margins.
**Requirements:** Moist well drained soil
**Comments:** A showy shrub for a moist, sheltered position.
**Localities:** Olinda, Mt Dandenong, 57, 58
**Distribution:** Vic, Qld, NSW, Tas, SA

## Pimelea linifolia ssp. linifolia

Slender Rice-flower

**F P**

**Size:** Prostrate-1.2 m x 1 m
**Habitat:** Riparian scrub, valley sclerophyll forest
**Form:** Variable, prostrate, erect or clump-forming, depending on habitat
**Foliage:** Narrow-elliptical dark green leaves to 30 mm long, margins curved inwards
**Flowers:** Terminal clusters of hairy white or pink flowers on long stalks. Flower-heads may be erect or nodding; July to January. The 4 broad bracts may be green or reddish.
**Requirements:** Well drained soils
**Comments:** Pruning will encourage branching. The local form is smaller, to 0.5 m high and leaves are shorter, to 15 mm long.
**Localities:** St. Albans, Hurstbridge, 16, 26, 29, 41, 42, 53, 56
**Distribution:** Vic, Qld, NSW, Tas, SA

*Pimelea linifolia* ssp. *linifolia*

SCALE: 0 1cm 2cm The scale representation on each plant equals 2cm.

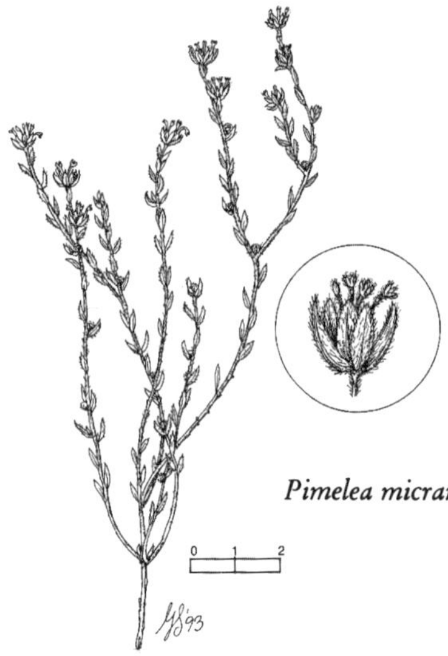

*Pimelea micrantha*

## Pimelea micrantha

Silky Rice-flower

**Size:** 10–40 cm high **F P**
**Habitat:** Riparian scrub on escarpments
**Form:** Erect sub-shrub, young stems covered in dense white hairs
**Foliage:** Hairy, pale grey-green narrowly elliptic to elliptic leaves to 10 mm long, alternate or opposite
**Flowers:** Compact terminal heads or axillary clusters of 3–8 tiny hairy flowers; bracts either absent or similar to the leaves; August to December
**Requirements:** Well drained soil
**Comments:** A species of north western Victoria which was recently discovered within the Melbourne area.
**Localities:** Taylors Lakes, Keilor Nth
**Distribution:** Vic, NSW, SA, WA
**Synonym:** *P. curviflora* var. *micrantha*

## Pimelea octophylla ssp. octophylla

Woolly Rice-flower

**Size:** 0.4–1 m x 0.5 m **F P**
**Habitat:** Tea-tree heath, sclerophyll woodland
**Form:** Slender, erect, open hairy shrub
**Foliage:** Green to dark bluish-green elliptical leaves to 20 mm long, covered in long, fine white hairs when young
**Flowers:** Masses of perfumed, woolly, creamy-yellow flowers in nodding heads; October to December. 6–12 (usually 8) densely hairy pale green bracts, no wider than the leaves.
**Requirements:** Well drained sandy soils
**Comments:** Leaves are alternate on this distinctive species.
**Localities:** Black Rock, Baxter, 59, 67, 68, 74, 76
**Distribution:** Vic, SA

*Pimelea octophylla* ssp. *octophylla*

## Pimelea pauciflora

Poison Rice-flower

**Size:** 1–3 m x 0.5–2 m **F P**
**Habitat:** Damp sclerophyll forest
**Form:** Slender, erect, many-branched shrub
**Foliage:** Dull green, narrow-elliptical leaves to 20 mm long
**Flowers:** Dioecious; few greenish-yellow flowers in clusters on short side branches, surrounded by 2–4 bracts which are longer than the flowerhead; October to December. Succulent red drupes follow flowering.
**Requirements:** Moist, well drained soils
**Comments:** A completely hairless, dioecious species which often forms dense thickets along mountain streams and moist, open areas. The tough bark was peeled and used to make nets by the Aborigines.
**Localities:** Lilydale, 50A
**Distribution:** Vic, NSW, Tas

## Pimelea phylicoides

Heath Rice-flower

**Size:** 30–50 cm x 30 cm **F**
**Habitat:** Tea-tree heath
**Form:** Erect, much branched shrub
**Foliage:** Green to blue-green elliptic leaves to 10 mm long, almost sessile, slightly toothed and hairy
**Flowers:** Small erect terminal flowerheads of hairy greenish-yellow, cream or white flowers; August to January. 3–6 hairy bracts broader than the leaves.
**Requirements:** Well drained sandy soil
**Comments:** Young stems are densely hairy.
**Localities:** Black Rock, 67[#]–69
**Distribution:** Vic, Tas, SA

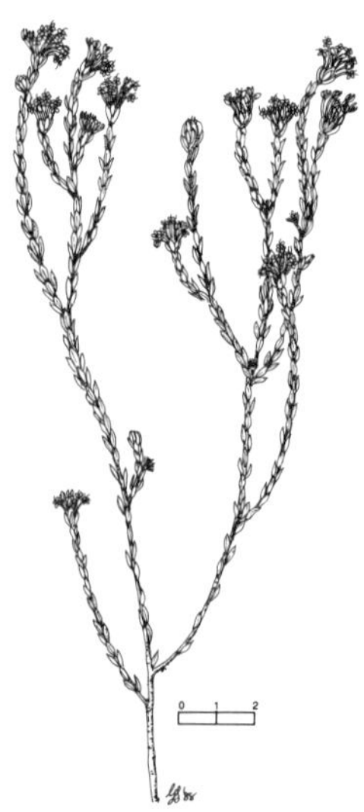

*Pimelea phylicoides*

## Pimelea spinescens ssp. spinescens

Rice-flower

**Size:** 5–30 cm high **F**
**Habitat:** Grasslands of basalt plains
**Form:** Low stunted shrub usually with leafless, spiny branchlets to 28 mm long
**Foliage:** Narrowly elliptic leaves to 10 mm long
**Flowers:** Insignificant terminal clusters of 6–12 tiny pale yellow glabrous flowers; April to October, inside 4 glabrous bracts to 7 mm long. Dioecious.
**Requirements:** Well drained soil
**Comments:** The only *Pimelea* sp. to have spines. It is smaller than *P. serpyllifolia* which grows behind sand dunes around the coast, south of the study area.
**Localities:** St Albans, Essendon, Sunshine, Deer Park, Rockbank, Altona, 4, 7, 10, 17[#], 18, 20, 26
**Distribution:** Endemic to Vic
**Synonym:** Recently separated from *P. serpyllifolia*

# Pittosporum — Pittosporaceae

## Pittosporum bicolor — Banyalla

P N

**Size:** 3–10 m x 3–4 m
**Habitat:** Sheltered gullies of wet and damp sclerophyll forests
**Form:** Upright bushy shrub or small tree
**Foliage:** Glabrous dark green lanceolate leaves to 80 mm long, paler and hairy below. New growth is much lighter than the old leaves.
**Flowers:** Single, yellow, bell-shaped flower with maroon tones on the back, in leaf axils; September to November; followed by small grey fruit and sticky red seeds.
**Requirements:** Moist, humus-rich, well drained soils
**Comments:** An interesting shrub for a shaded position although it lacks the perfume of other species. Seeds often germinate on tree fern trunks. It does not become naturalised and cause an environmental problem like *P. undulatum* from East Gippsland.
**Propagation:** Seed, cuttings
**Localities:** 55, 57, 58
**Distribution:** Vic, NSW, Tas

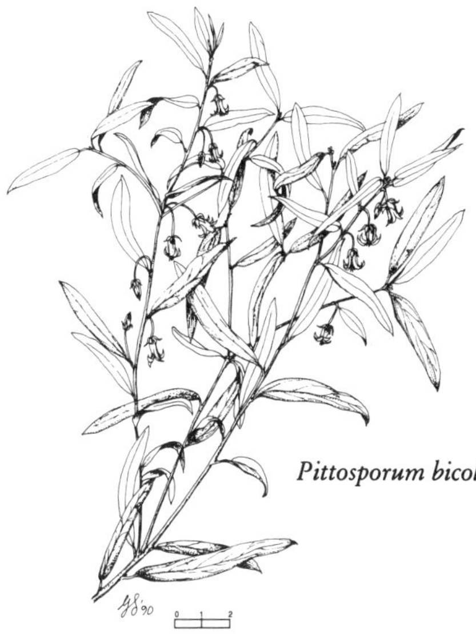

*Pittosporum bicolor*

# Plantago — Plantaginaceae

## Plantago debilis — Slender or Shade Plantain

P

**Habitat:** Damp and valley sclerophyll forests, grassy low open forest
**Form:** Rosetted perennial herb
**Foliage:** Oblanceolate leaves, to 15 cm x 30 mm are entire, broader and downy
**Flowers:** Smaller, usually glabrous, pale green flowers are widely spaced on the slender flower stalks.
**Comments:** Similar to *P. varia.* There is presently a lot of confusion within the genus *Plantago* and future revision will clarify this.
**Localities:** 34A, 36, 37, 42, ?56–58, 74
**Distribution:** All states

*Plantago debilis*

## Plantago gaudichaudii — Narrow-leaf Plantain

F

**Habitat:** Plains grassland, red gum woodland
**Foliage:** Similar to *P. varia* but has narrower, linear leaves to 30 cm x 7 mm with fine brown hairs at the base. Usually glabrous.
**Flowers:** Compact flowering spikes to 100 mm long; August to October
**Comments:** A second form from the basalt plains has erect incurved leaves to 25 cm long, with the top half distantly toothed. Central vein is dominant.
**Localities:** 2, 7, 10, 12, 13, 17#, 18, 20, 26, 58
**Distribution:** Vic, Qld, NSW, Tas, SA

*Plantago gaudichaudii*

## Plantago hispida — Plantain

F

**Habitat:** Box woodland
**Foliage:** Erect, shortly-stalked, blunt oblanceolate leaves to 145 mm long, entire or with widely-spaced lobes. Densely hairy, with a tuft of short, white or slightly tawny hairs at the base. Leaves dry to greyish-green.
**Flowers:** Compact flowering spikes to 80 mm long; August to November
**Localities:** Sunbury
**Distribution:** All states except NT

## Plantago varia — Variable Plantain

F P

**Size:** 8–30 cm high
**Habitat:** Widespread in red gum woodland, dry, damp and valley sclerophyll forests, grassy low open forest
**Form:** Rosetted perennial herb
**Foliage:** Soft, linear-elliptic to oblanceolate hairy leaves to 20 cm x 5–20 mm, with 3 veins and sometimes toothed margins, ascending or erect. Dense tuft of brown hairs at the leaf base.
**Flowers:** Dense spikes (15–140 mm long) of tiny brownish flowers with hairy bracts, on leafless, upright flower stalks; most of the year
**Requirements:** Well drained soils
**Comments:** All local species are similar. Shape of the leaves help to distinguish this species as does the darker colour of the dried leaves.
**Propagation:** Seed
**Localities:** 7, 12, 13, 16, 18, 25, 26, 28, 29, 31, 32, 34A–36, 38, 39, 41–45, 47–49, 50A–53, 55, 56, 58, 62, 63, 67#, 71, 74, 76
**Distribution:** All states; NZ

*Plantago varia*

SCALE: 0 1cm 2cm The scale representation on each plant equals 2cm.

# Platylobium **Fabaceae**

*Platylobium formosum*

## Platylobium formosum

Handsome Flat-pea

**Size:** 0.3–1.5 m x 1–1.5 m **F P**
**Habitat:** Damp and valley sclerophyll forests, tea-tree heath
**Form:** Wiry, upright or scrambling shrub
**Foliage:** Pairs of sessile, dark green, leathery, heart-shaped leaves to 60 mm long, with raised network of veins
**Flowers:** Large yellow and red pea flowers on long hairy stalks, 2–3 in leaf axils; September to December. Stalked pods to 40 mm are broad, flat and hairy with a satiny interior.
**Requirements:** Moist well drained soils
**Comments:** A useful plant for a shady area where it will scramble through other plants. The bright flowers only open on a sunny day.
**Propagation:** Scarified seed, cuttings
**Localities:** 34A, 42–45, 47, 49–51, 55, 56, 58–61, 71, 74, 76
**Distribution:** Vic, Qld, NSW, Tas

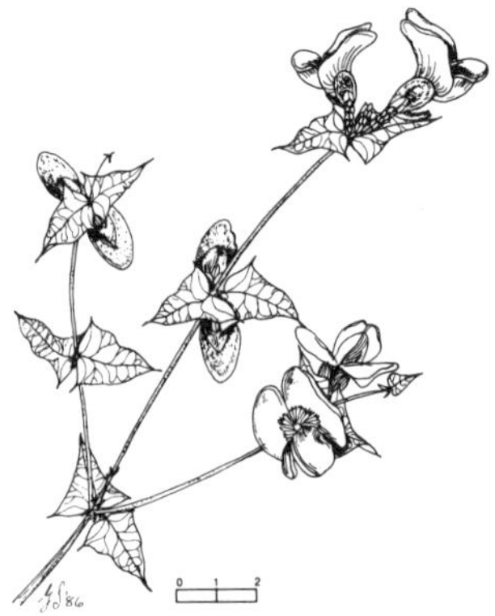

*Platylobium obtusangulum*

## Platylobium obtusangulum

Common Flat-pea

**Size:** 0.6–1 m x 1 m **F P**
**Habitat:** Dry and valley sclerophyll forests, sclerophyll and red gum woodlands, tea-tree heath, grassy low open forest
**Form:** Similar but less vigorous than *P. formosum*
**Foliage:** Leaves are triangular or arrow-shaped to 30 mm long with 1–3 sharp points
**Flowers:** Flowers (1–3 in axils) are on short stalks hidden by bracts. Pods are also shorter.
**Requirements:** Prefers drier well drained soils.
**Localities:** Highett, 32–35, 37–56, 58, 59, 62–71, 73–76
**Distribution:** Vic, Tas, SA

# Platysace **Apiaceae**

*Platysace heterophylla*

## Platysace heterophylla

Slender Platysace

**Size:** 20–35 cm x 10 cm **P**
**Habitat:** Tea-tree heath, grassy low open forest
**Form:** Trailing or tufting perennial herb
**Foliage:** Very narrow leaves to 25 mm long, sometimes dark red. Lower leaves may be divided into three.
**Flowers:** Small compound umbels of tiny, white, starry flowers on very slender stalks; October to January
**Requirements:** Well drained soils, tolerating extra moisture in Winter
**Propagation:** Seed, cuttings
**Localities:** 60, 67, 69, 70, 73–76
**Distribution:** Vic, SA

*Platysace lanceolata*

## Platysace lanceolata

Shrubby Platysace

**Size:** 0.6–1.5 m x 0.5–1 m **P**
**Habitat:** Dry sclerophyll forest
**Form:** Open to dense rounded shrub
**Foliage:** Very variable, lanceolate to nearly round, dark green leaves to 40 mm long. The local form has smaller, ovate to almost round leaves to 10 mm long.
**Flowers:** Brown or pinkish flowerbuds open to tiny white flowers in rounded compound umbels to 30 mm wide; December to February
**Requirements:** Well drained soils
**Comments:** An attractive small shrub for a lightly shaded position, tolerating periods of dryness once established. Locally rare.
**Propagation:** Seed, cuttings
**Localities:** Warrandyte Nth
**Distribution:** Vic, Qld, NSW

# Podolepis **Asteraceae**

## Podolepis jaceoides Showy Podolepis

**Size:** 0.3–0.6 m x 0.3 m **F P**
**Habitat:** Plains grassland, grassy wetland, red gum woodland, valley sclerophyll forest, tea-tree heath
**Form:** Erect tufted perennial herb
**Foliage:** Dark green lanceolate leaves to 100 mm in a basal rosette; stem leaves smaller and linear
**Flowers:** Showy, bright yellow daisy flowerheads to 50 mm wide, borne singly on long stems. Outer ligulate florets deeply 3–5-lobed; October to December
**Requirements:** Well drained soils
**Comments:** A spectacular plant in flower which grows better and flowers longer with extra Summer watering. The Aborigines baked the roots before eating. Two forms are found within the Melbourne area, both occurring together on the basalt plains. One form has few spathulate leaves and flower stalks which have 1 (or rarely) 3 flowers. The form more common in the foothills grows in clumps to 30 cm wide. It has narrow-linear leaves and branched flower stalks, each producing up to 12 flowers. The flowers are slightly smaller and more orange-yellow. Both are becoming rarer within the Melbourne area.
**Propagation:** Seed
**Localities:** Sunbury, Laverton, Werribee, Tottenham, Deer Park, Bundoora, Cheltenham#, 26, 42, 44, 47, 61, 67#
**Distribution:** Vic, Qld, NSW, Tas, SA

*Podolepis jaceoides*

# Podotheca **Asteraceae**

## Podotheca angustifolia Sticky Long-heads

**Size:** 0.2–30 cm high **F**
**Habitat:** Coastal banksia woodland
**Form:** Erect annual herb, branching from the base
**Foliage:** Sticky and hairy, narrow to oblanceolate, curved, dull green leaves 10–80 mm long
**Flowers:** Terminal, narrow, woolly yellow flowerheads, to 40 mm long, on thick stalks above the foliage. Ray florets very small; September to October
**Requirements:** Dryish, well drained soils
**Comments:** A very sticky daisy which tolerates limestone. Presumed extinct within the Melbourne area.
**Propagation:** Seed
**Localities:** Cheltenham 1909, 67# 1909
**Distribution:** Vic, NSW, Tas, SA, WA
**Synonym:** *Podosperma angustifolium*

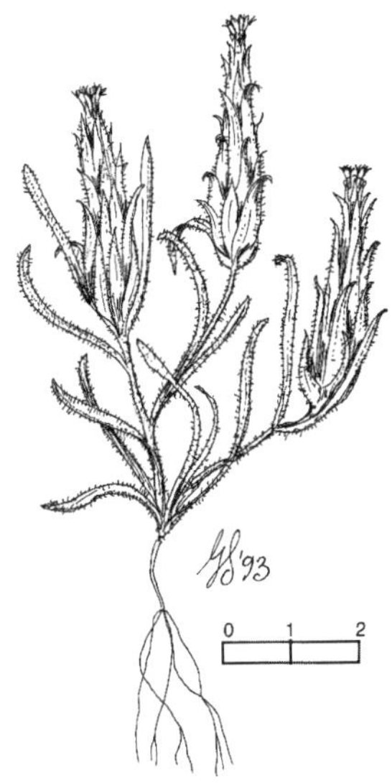

*Podotheca angustifolia*

# Polygonum **Polygonaceae**

## Polygonum hydropiper = Persicaria hydropiper

## Polygonum lapathifolium = Persicaria lapathifolia

## Polygonum minus = Persicaria decipiens

## Polygonum plebium Small Knotweed

**Size:** Prostrate x 30 cm **F P**
**Habitat:** Low-lying areas of basalt plains
**Form:** Compact, much-branched annual herb
**Foliage:** Narrow oblong leaves to 15 mm long, with a silvery sheath at the base
**Flowers:** Tight clusters of 2–5 small, white or pink flowers in leaf axils; mostly Summer
**Requirements:** Damp, muddy soils
**Propagation:** Seed
**Localities:** Probably occurred within the south west section of the study area. Known to occur at Lara.
**Distribution:** Vic, Qld, NSW, SA, NT

*Polygonum plebeium*

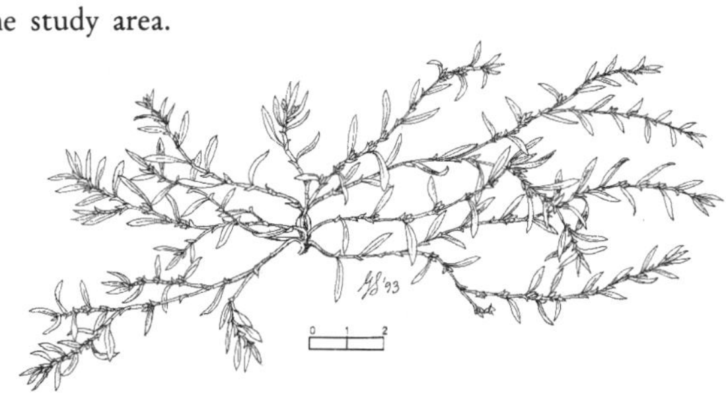

## Polygonum prostratum = Persicaria prostrata

## Polygonum subsessile = Persicaria subsessilis

## Polypompholyx tenella = Utricularia tenella

SCALE: 0 1cm 2cm The scale representation on each plant equals 2cm.

*Polyscias sambucifolius*

# Polyscias — Araliaceae

## Polyscias sambucifolius — Elderberry Panax

**Size:** 1–6 m x 1–3 m — P N
**Habitat:** Damp, wet and valley sclerophyll forests
**Form:** Very variable, slender shrub or small tree
**Foliage:** Variable pinnate leaves; local form—leaflets narrow to lanceolate, dark green, paler below, to 110 mm long
**Flowers:** Clusters of small yellow-green flowers; November to January. Succulent blue berries in Summer.
**Requirements:** Moist well drained soil
**Comments:** An interesting foliage plant suitable for indoor usage.
**Propagation:** Cuttings, root suckers
**Localities:** Kalorama, 42–44, 50A, 55–58
**Distribution:** Vic, NSW

# Pomaderris — Rhamnaceae

An attractive genus, ranging from low shrubs to small trees, often found in moist forests or near streams. Leaves are typified by the raised veins and the grey or rusty-brown star-shaped hairs on the underside. Well displayed clusters of flowers last only for a short time. Propagation is by seed or cuttings.

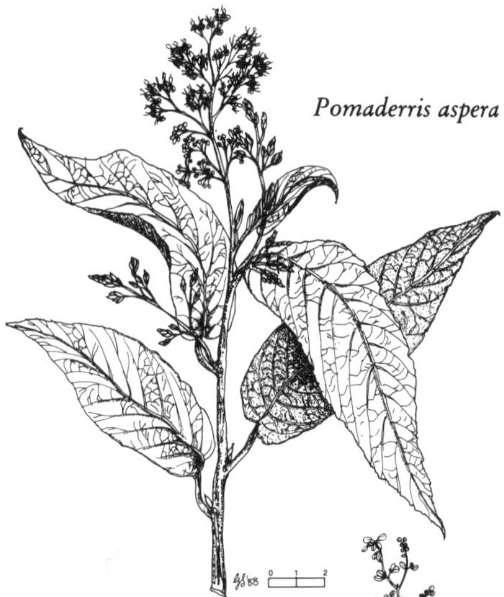

*Pomaderris aspera*

## Pomaderris aspera — Hazel Pomaderris

**Size:** 3–12 m x 2–4 m — P N
**Habitat:** Wet, damp and valley sclerophyll forests, riparian scrub
**Form:** Slender leafy shrub or small tree
**Foliage:** Large, soft lanceolate leaves to 140 mm long, with deep veins; dark green above, paler below with pinkish-brown hairs and brown veins
**Flowers:** Masses of yellow-green flowers in terminal sprays; October to December. Petals absent.
**Requirements:** Moist, well drained, humus-rich soil
**Comments:** A middle storey shrub or tree with interesting markings on the bark as the plant ages. Pegs were made from the wood by the Aborigines to stretch skins.
**Localities:** 16, 26, 29, 32–34, 36, 42, 44, 49, 50A, 51, 55–59, 62, 63
**Distribution:** Vic, NSW, Tas

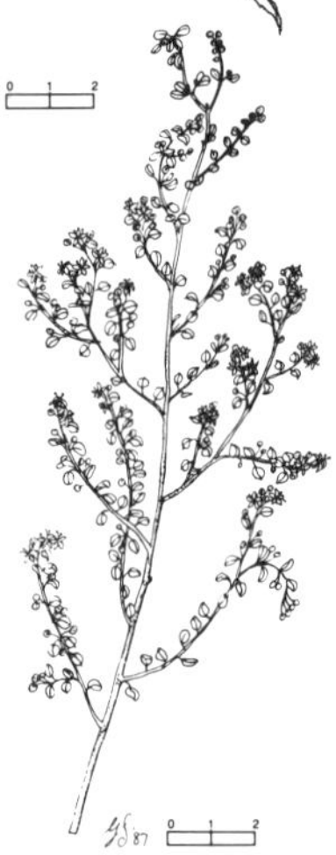

*Pomaderris elachophylla*

## Pomaderris elachophylla — Small-leaf or Lacy Pomaderris

**Size:** 1–3 m x 0.5–2 m — P
**Habitat:** Valley sclerophyll forest
**Form:** Erect wiry open shrub
**Foliage:** Tiny, elliptic dark green leaves to 6 mm long, paler below with scattered brown hairs
**Flowers:** Small axillary clusters of tiny creamy flowers; October to November; petals absent
**Requirements:** Moist well drained soils
**Comments:** Easily identified by its tiny leaves. It is rare within the Melbourne area.
**Localities:** Hurstbridge
**Distribution:** Vic, NSW, Tas

## Pomaderris elliptica — Smooth Pomaderris

**Size:** 1–4 m x 1–3 m — P
**Habitat:** Valley sclerophyll forest
**Form:** Tall erect or rounded shrub
**Foliage:** Smooth, thin, elliptic leaves to 100 mm long, pale grey below with minute hairs and raised brown veins
**Flowers:** Large rounded clusters of yellow flowers held above the foliage; September to October
**Requirements:** Moist well drained soils
**Comments:** One of the most attractive *Pomaderris* spp., easily grown in light shaded gardens. Rare within the Melbourne area.
**Localities:** Lilydale
**Distribution:** Vic, Qld, NSW, Tas
**Synonym:** *P. multiflora*

*Pomaderris ferruginea*

## Pomaderris ferruginea — Rusty Pomaderris

**Size:** 1–4 m x 1–1.5 m — F P
**Habitat:** Valley sclerophyll forest
**Form:** Open softly hairy shrub
**Foliage:** Dull green, broad-lanceolate leaves to 100 mm long, glabrous above, pale brown and softly hairy below
**Flowers:** Large terminal sprays of hairy, whitish or yellow flowers; September to October
**Requirements:** Moist, well drained soils
**Comments:** Beautiful flowering shrub which tolerates short dry periods once established. Young growth is a most attractive rusty red.
**Localities:** Along the Yarra R, Lilydale Nth, 42
**Distribution:** Vic, Qld, NSW

## Pomaderris lanigera

Woolly Pomaderris

**F P**

**Size:** 1–3 m x 1–2 m
**Habitat:** Dry and valley sclerophyll forests
**Form:** Erect hairy shrub, hairs often rusty
**Foliage:** Large, lanceolate leaves to 120 mm long, pale green below with dense brown hairs on veins; hairy on both sides of leaf
**Flowers:** Large rounded terminal clusters of yellow flowers; September to October
**Requirements:** Moist well drained soil
**Comments:** A most attractive low screening plant.
**Localities:** Mooroolbark, Wantirna, 54, 55
**Distribution:** Vic, Qld, NSW

*Pomaderris lanigera*

## Pomaderris paniculosa ssp. paralia

**F P**

**Size:** 1–1.5 m x 1–1.5 m
**Habitat:** Coastal banksia woodland
**Form:** Open rounded shrub
**Foliage:** Ovate to elliptic leaves to 50 mm long, glabrous, shiny green above, pale and densely hairy below
**Flowers:** Many small cream flowers in terminal and axillary panicles; September to October
**Requirements:** Well drained sandy soil
**Comments:** An attractive coastal shrub with reddish new growth.
**Localities:** Seaford, ?Brighton, 74
**Distribution:** Vic, Tas, SA, WA
**Synonym:** Separated from *P. oraria*

*Pomaderris paniculosa* ssp. *paralia*

## Pomaderris prunifolia

Plum-leaf Pomaderris

**F P N**

**Size:** 1–4 m x 1–4 m
**Habitat:** Valley sclerophyll forest
**Form:** Open rounded shrub
**Foliage:** Rough, dark green, elliptic to lanceolate leaves to 50 mm long, paler below
**Flowers:** Small, rounded clusters of pale yellow flowers; October to November. Petals absent.
**Requirements:** Moist well drained soils
**Comments:** More tolerant than most to drying out in Summer if it is in a shaded position.
**Localities:** 33, 34A, 36, 41, 42, 50A
**Distribution:** Vic, Qld, NSW

*Pomaderris prunifolia*

## Pomaderris racemosa

Slender or Cluster Pomaderris

**F P N**

**Size:** 2–5 m x 1–2 m
**Habitat:** Riparian scrub
**Form:** Slender shrub or, rarely, small lightly-branched tree
**Foliage:** Thin, dark green ovate leaves to 25 mm long, with margins turned down, pale green below.
**Flowers:** Small clusters of yellow flowers in leaf axils; October to November. Petals absent.
**Requirements:** Well drained soils
**Comments:** Ideal for planting in areas where space is limited, as growth is mostly erect.
**Localities:** 32, 34A, 36, 42, 50A, 62, 63
**Distribution:** Vic, Tas

*Pomaderris racemosa*

## Pomaderris vacciniifolia

Round-leaf Pomaderris

**F P**

**Size:** 2–3.5 m x 3 m
**Habitat:** Valley sclerophyll forest
**Form:** Large straggling, spreading shrub
**Foliage:** Smooth, dark green, elliptic leaves to 20 mm long, with inconspicuous veins, pale grey below
**Flowers:** Small sprays of creamy flowers in leaf axils; October to November
**Requirements:** Well drained soils
**Comments:** A rare plant with a restricted range in Melbourne.
**Localities:** 42
**Distribution:** Endemic to Vic

*Illustrated p. 172*

SCALE: 0 1cm 2cm — The scale representation on each plant equals 2cm.

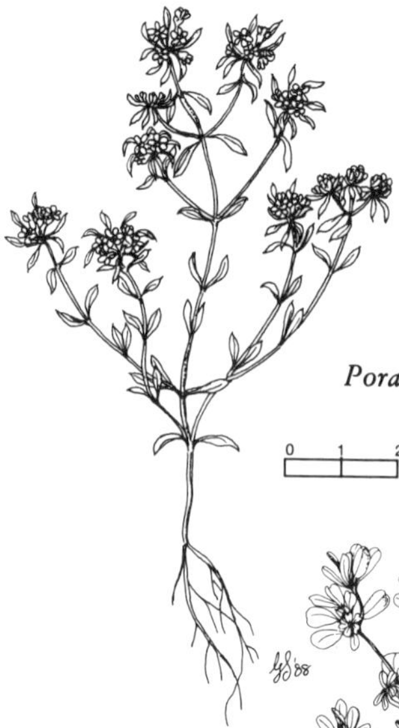

*Poranthera microphylla*

# Poranthera — Euphorbiaceae

## Poranthera microphylla — Small Poranthera

**F P**

**Size:** 3–10 cm x 10 cm
**Habitat:** Red gum and box woodland, dry and valley sclerophyll forests, grassy low open forest
**Form:** Trailing to erect annual herb
**Foliage:** Greyish, oval to spathulate leaves to 10 mm long
**Flowers:** Tiny white flowers in short, leafy heads at the end of branchlets; August to December
**Requirements:** Well drained soils
**Comments:** A relatively inconspicuous plant amongst grasses in the understorey. In moist areas, or in cultivation it becomes more vigorous and may spread 30 cm or more. It self-sows readily in the garden and can add a natural look to a low maintenance area.
**Propagation:** Seed
**Localities:** Sth of Melton, 34A–36, 38, 40–67#, 69A–71, 74–76
**Distribution:** All states; NZ

*Portulaca oleracea*

# Portulaca — Portulacaceae

## Portulaca oleracea — Common Purslane

**F**

**Size:** Prostrate x 0.4–0.7 m
**Habitat:** Wide-ranging, red gum woodland, damp sclerophyll forest, tea-tree heath, primary dune scrub, also disturbed areas
**Form:** Trailing or matting annual herb with red stems
**Foliage:** Succulent, oblong to wedge-shaped leaves to 25 mm long
**Flowers:** Sessile yellow flowers, single or clustered in leaf axils; August to March
**Requirements:** Adaptable in well drained soil
**Comments:** Salt tolerant. It may become a garden weed. Both the Aborigines and early settlers used to eat the leaves, raw or cooked, as a vegetable. The seeds were ground to make bread and cakes.
**Propagation:** Seed, cuttings
**Localities:** 25, 26, 35, 42, 58, 67, 71
**Distribution:** All states

# Pratia — Campanulaceae

Carpeting perennial herbs, pratias provide a colorful groundcover in damp to wet soils. The flowers are similar to those of *Isotoma fluviatalis* but the tubular flower is split to the base in pratias. They are useful for binding soil and can be used as a grass alternative in low traffic areas. They may become invasive under ideal conditions but are easily controlled. Propagation is by division.

## Pratia concolor — Poison Pratia

**P N**

**Size:** Prostrate–30 cm x spreading
**Habitat:** Poorly drained areas in plains grassland, grassy wetland
**Form:** Perennial herb, with trailing stems radiating from the centre
**Foliage:** Sessile, ovate-lanceolate to lanceolate grey-green leaves to 30 mm long, with finely toothed margins. Leaves are always on one plane.
**Flowers:** Small white to pale pink starry flowers; December to February; followed by succulent green fruit
**Requirements:** Moist to wet heavy soils
**Comments:** *P. concolor* has the largest leaves of the local pratias. The milky sap contains alkaloids which are suspected of stock poisoning.
**Propagation:** Cuttings, division
**Localities:** 17#, 18
**Distribution:** Vic, Qld, NSW, SA

*Pratia irrigua*

## Pratia irrigua — Fleshy or Salt Pratia

**F P**

**Size:** Prostrate or slightly ascending
**Habitat:** Saltmarshes, riparian scrub
**Form:** Weak or trailing herb
**Foliage:** Shiny, entire or inconspicuously-toothed leaves to 25 mm long, variable, often narrowly elliptic in shape. Usually (but not always) succulent.
**Flowers:** Minute white or bluish flowers on stalks shorter or longer than the leaves
**Comments:** Locally rare.
**Localities:** Rockbank, Port Melbourne, 2, 4, 7, 29, 30#, 35
**Distribution:** All states except Qld
**Synonym:** *Pratia platycalyx*

## Pratia pedunculata

Matted Pratia

**Size:** Prostrate x 2–3 m **F P**
**Habitat:** Swamp edges in plains grassland, grassy wetland, red gum woodland, valley sclerophyll forest, tea-tree heath
**Form:** Dense matting plant
**Foliage:** Tiny, toothed, oval green leaves to 8 mm long
**Flowers:** Masses of starry, blue or white flowers singly in leaf axils; November to February
**Requirements:** Moist soils, tolerating extended wetness
**Comments:** A most attractive suckering plant flowering well in full sun provided it has some protection or is kept moist.
**Localities:** 17#, 18, 35, 42, 71, 74
**Distribution:** Vic, Qld, NSW, Tas, SA

*Pratia pedunculata*

## Pratia platycalyx = P. irrigua

# Prostanthera — Lamiaceae

## Prostanthera lasianthos var. lasianthos

Victorian Christmas Bush

**Size:** 2–8 m x 2–5 m **F P N**
**Habitat:** Wet, damp and valley sclerophyll forests
**Form:** Compact tall shrub or upright small tree
**Foliage:** Pairs of dark green, toothed, lanceolate leaves to 100 mm long; minty aroma when crushed
**Flowers:** Long, loose, terminal leafless racemes of hairy, white flowers, spotted with orange and purple; November to January
**Requirements:** Moist, well drained loamy soils
**Comments:** A spectacular tree in flower, it is useful as a screen plant but needs protection from strong winds. It is very variable in form according to its origins. Mulching is beneficial and plants respond to Summer watering. A pink flowering form from Kallista is available.
**Propagation:** Fresh seed, cuttings
**Localities:** 33–34A, 42, 44, 47, 49, 50–51, 55–59, 61–63
**Distribution:** Vic, Qld, NSW, Tas

*Illustrated p. 49*

*Prostanthera melissifolia*

## Prostanthera melissifolia

Balm Mint Bush

**Size:** 1–3 m x 2–3 m **P N**
**Habitat:** Wet sclerophyll forest
**Form:** Fast-growing, dense rounded shrub
**Foliage:** Pairs of dark green, coarsely-toothed ovate leaves to 40 mm long, paler below. Strong mint aroma. Branchlets square.
**Flowers:** Loose terminal sprays of mauve to pink flowers cover the shrub; October to December
**Requirements:** Moist, well drained soils
**Comments:** An excellent shrub for a moist, sheltered position, preferably placed near a pathway as contact releases the minty fragrance. Pruning after flowering maintains bushiness and vigour.
**Propagation:** Cuttings
**Localities:** 57
**Distribution:** Vic, NSW

# Prunella — Lamiaceae

## Prunella vulgaris

Self-heal

**Size:** 7–150 mm x 10 cm **F P N**
**Habitat:** Widespread in riparian and swamp scrub, red gum woodland, valley, damp and wet sclerophyll forests, tea-tree heath
**Form:** Fast growing, spreading perennial herb
**Foliage:** Pairs of lightly hairy, dark green ovate leaves to 60 mm long; stems reddish and quadrangular
**Flowers:** Terminal heads of sessile, dark purple, rarely pink, flowers, held erect; mainly November to February
**Requirements:** Moist soils
**Comments:** A spreading groundcover for damp areas. In ideal conditions this plant may become invasive.
**Propagation:** Cuttings
**Localities:** 28, 30, 33–36, 38, 40–44, 46–51, 53–67#, 71, 74, 76
**Distribution:** All except NT; NZ

*Prunella vulgaris*

SCALE: 0 1cm 2cm — The scale representation on each plant equals 2cm.

*Pseudognaphalium luteo-album*

# Pseudognaphalium — Asteraceae

## Pseudognaphalium luteoalbum — Jersey Cudweed

**F P N**

**Size:** 20–50 cm high
**Habitat:** Widespread in saltmarshes, plains grassland, valley and damp sclerophyll forest, swamp scrub, grassy low open forest
**Form:** Slender, erect, white-woolly annual or biennial herb
**Foliage:** Densely woolly, grey leaves; lower leaves obovate to 50 mm long, stem leaves smaller, sessile and more linear
**Flowers:** Terminal clusters of creamy-yellow flowerheads, without ray florets; September to March
**Requirements:** Adaptable, tolerating dryness once established
**Comments:** An easily grown plant which is often found in disturbed soil. May become invasive. The Aborigines made a drink from the leaf to treat general sickness.
**Propagation:** Seed
**Localities:** Sth of Melton, 4, 5, 16, 29#, 34A, 35, 39, 42, 44, 48, 49, 54, 55, 57–60, 67, 68, 69A, 74, 75
**Distribution:** All states; NZ
**Synonym:** *Gnaphalium luteoalbum*

# Psoralea — Fabaceae

## Psoralea adscendens — Dusky Scurf-pea, Mountain Psoralea

**P**

**Size:** Prostrate–0.6 m x 0.5–1 m
**Habitat:** Plains grasslands
**Form:** Spreading perennial herb
**Foliage:** Dark green trifoliate leaves on long stalks to 150 mm long, leaflets lanceolate to 50 mm long
**Flowers:** Erect spike of purple, rarely pink or white, pea flowers on long stalks to 250 mm long; November to January. Calyx is covered in shaggy black hair.
**Requirements:** Moist well drained soils
**Comments:** Responds well to Summer watering. Attractive in a rockery or draping over logs. An endangered Melbourne species.
**Propagation:** Scarified seed
**Localities:** Was collected in St. Albans, Altona, Sth Morang
**Distribution:** Vic, NSW, Tas, SA

## Psoralea parva — Small Scurf-pea

**F**

**Size:** Prostrate or scrambling
**Habitat:** Plains grasslands
**Form:** Small trailing perennial herb
**Foliage:** Leaves on shorter stalks than other species with 3–5 narrow or oblanceolate leaflets to 25 mm long
**Flowers:** Dense clusters of tiny, sessile, lilac to white pea flowers; October to December. Calyx is covered in dense white hairs.
**Comments:** An endangered plant nationally, it is locally very rare.
**Localities:** Werribee, 17#
**Distribution:** Vic, NSW, SA

*Psoralea tenax*

## Psoralea tenax — Tough Scurf-pea, Emu-foot

**F P**

**Size:** 0.5 m high
**Habitat:** Plains grasslands
**Form:** Slender, trailing or erect perennial herb with angular ridges along stems
**Foliage:** Long-stalked dark green leaves have 5–7 sessile, narrow leaflets to 30 mm long, (rarely to 70 mm long) covered in small black dots
**Flowers:** Spikes, to 100 mm long, of tiny bluish pea flowers in leaf axils on long slender stalks to 20 cm long; August to March. Calyx hairs paler and shorter than *P. adscendens*.
**Requirements:** Moist soils
**Comments:** Appreciates Summer watering. Extinct throughout most of its Melbourne range and now rare on the Keilor basalt plains.
**Propagation:** Scarified seed
**Localities:** Was north east of Melbourne; St. Albans#, Sydenham, Sunshine, Werribee, 17#, 20#
**Distribution:** Vic, Qld, NSW

# Ptilotus Amaranthaceae

## Ptilotus macrocephalus

Tall Mulla Mulla, Featherheads

**Size:** 0.6–1 m x 0.6 m F
**Habitat:** Plains grassland, red gum and box woodlands
**Form:** Erect, sparsely hairy perennial herb with a long, fleshy tap root
**Foliage:** Narrow, linear to elliptic leaves 50–100 mm long, margins wavy
**Flowers:** Single, terminal, fluffy, greenish-yellow flowerheads to 120 mm x 60 mm, on stems to 0.6 m high; October to November
**Requirements:** Well drained light or heavy soils
**Comments:** A very showy plant now found in only a few isolated places locally. Useful as a cut flower.
**Propagation:** Seed
**Localities:** Werribee, Sunbury, Bundoora, Sydenham, Diggers Rest, Laverton Nth, Deer Park, Rockbank, 26
**Distribution:** All mainland states

*Ptilotus macrocephalus*

## Ptilotus spathulatus

Pussy Tails

**Size:** Prostrate x 10–40 cm F
**Habitat:** Plains grassland, primary dune scrub, box woodland
**Form:** Tufted perennial herb with a thick woody root
**Foliage:** Broad, spathulate leaves on prostrate stems, basal leaves to 60 mm long, stem leaves smaller; dark green above, paler below
**Flowers:** Single, terminal fluffy, pale green to brown erect flowerheads to 60 mm long, at the end of prostrate branches; October to November and March to June
**Requirements:** Well drained soils
**Comments:** A tiny, attractive plant restricted locally to a few remnant sites. The taproot was eaten by the Aborigines.
**Propagation:** Seed
**Localities:** Williamstown, St. Albans, Sunbury, Keilor, Bundoora, 1, 17#, 26, 29, 34A
**Distribution:** Vic, NSW, Tas, SA, WA *Illustrated p. 172*

# Pultenaea Fabaceae

A large family of shrubby understorey plants, the Pultenaeas are represented by 13 species locally. All are very showy, with bright orange-yellow and red 'egg and bacon' pea flowers during Spring. They make an attractive feature in the garden growing happily in rockeries or under trees. Planting in drifts gives the best effect. Good drainage, mulching and pruning after flowering are beneficial. *Pultenaea* spp. are distinguished from similar pea genera by the conspicuous stipules, (a pair of small leaf-like appendages) which are united behind the leaf stalk, and have rough margins. They also have a pair of bracteoles (small bracts) at the base of, or on, the flower calyx. Propagation is by scarified seed or cuttings.

*Pultenaea daphnoides*

## Pultenaea daphnoides

Large-leaf Bush-pea

**Size:** 1–3 m x 0.5–2 m F P N
**Habitat:** Dry sclerophyll forest, grassy low open forest
**Form:** Erect, branching shrub
**Foliage:** Large, flat, wedge-shaped olive green leaves to 30 mm long, paler below, with midvein produced into a slender stiff point. Stipules small.
**Flowers:** Large yellow and brown flowers in dense, terminal heads; August to November. Narrow, hairy bracteoles attached to middle of silky calyx tube.
**Requirements:** Well drained soils tolerating dryness once established, but appreciates extra moisture in full sun
**Comments:** An attractive soft shrub which, like many other species, is more stunted in coastal areas.
**Localities:** Sth Morang, 34–36, 49, 55, 56, 76
**Distribution:** Vic, NSW, Tas, SA

## Pultenaea dentata

Clustered Bush-pea

**Size:** 0.3–1 m x 0.5 m P
**Habitat:** Valley sclerophyll forest, grassy low open forest
**Form:** Hairy, wiry shrub
**Foliage:** Scattered, stiff, linear-elliptic leaves to 10 mm, margins curved inwards; stipules 1–2 mm long
**Flowers:** Compact sessile clusters of yellow and red flowers; September to December. Bracteoles 3-lobed.
**Requirements:** Moist well drained soils
**Comments:** Locally rare.
**Localities:** Clayton#, 42, 66#, 67#, 74
**Distribution:** Vic, NSW, Tas, SA

*Pultenaea dentata*

SCALE: 0 1cm 2cm The scale representation on each plant equals 2cm.

*Pultenaea gunnii*

## Pultenaea gunnii

Golden Bush-pea

**Size:** 0.5–1.5 m x 0.5 m **F P**
**Habitat:** Damp, valley and dry sclerophyll forests, sclerophyll woodland
**Form:** Wiry erect or straggling shrub
**Foliage:** Tiny ovate, dark green leaves to 6 mm long, margins recurved, concave below. Tight rosettes of leaves often at ends of branchlets. Stipules tiny and dark brown.
**Flowers:** Terminal clusters of bright yellow-orange and dark red flowers; September to October. Small lanceolate bracteoles attached in the centre of the calyx tube.
**Requirements:** Well drained soils
**Comments:** Spectacular in flower, it grows well under established trees, with other shrubs. *P. gunnii* x *P. scabra* occurs in Heathmont, Montrose and Clayton.
**Localities:** 34, 34A, 36, 41–44, 47, 49–56, 58–61, 63, 64, 67#, 73, 74
**Distribution:** Vic, Tas

*Pultenaea hispidula*

## Pultenaea hispidula

Rusty Bush-pea

**Size:** 1–1.5 m x 1 m **F P**
**Habitat:** Damp sclerophyll forest
**Form:** Hairy, much-branched shrub with arching branchlets
**Foliage:** Crowded, concave, oblong leaves to 9 mm long, hairy particularly on lower surface. Stipules reddish-brown, 2–3 mm long.
**Flowers:** Single yellow flowers in leaf axils, or terminal on short branchlets; September to December. Variable brown bracteoles are somewhat leaf-like.
**Requirements:** Well drained soil
**Comments:** A graceful small shrub with a pendulous habit. An attractive addition to the garden.
**Localities:** Kalorama, 58, 59
**Distribution:** Vic, NSW, SA

## Pultenaea humilis

Dwarf Bush-pea

**Size:** 30 cm x 1 m **P**
**Habitat:** Valley sclerophyll forest
**Form:** Hairy, erect to spreading low shrub
**Foliage:** Narrow, slightly concave, dull green leaves to 10 mm long, glabrous above and hairy below, crowded along stems. Stipules 2 mm long.
**Flowers:** Dense terminal heads of orange and yellow to red-brown flowers; October to December. Bracteoles hairy, long and narrow.
**Requirements:** Moist well drained soils. Tolerates dry shade once established but appreciates Summer watering
**Comments:** An attractive small shrub for a lightly shaded area. It will adapt to full sun if Summer watering is available. In heavy shade plants tend to grow taller and foliage is sparser. A grey-leafed form is available from nurseries.
**Localities:** 61
**Distribution:** Vic, NSW, Tas *Illustrated p. 172*

*Pultenaea juniperina* var. *mucronata*

## Pultenaea juniperina var. mucronata

Prickly Bush-pea

**Size:** 1–3 m x 1–1.5 m **F P**
**Habitat:** Valley and damp sclerophyll forests
**Form:** Erect to spreading shrub
**Foliage:** Stiff, sharply-pointed lanceolate leaves with incurved margins, to 30 mm long. Stipules erect, 2–3 mm long.
**Flowers:** Solitary fragrant yellow and red flowers in leaf axils; October to December. Short, hairy bracteoles at the base of calyx tube.
**Requirements:** Moist well drained soils tolerating some dryness once established
**Comments:** Similar in appearance to *P. muelleri*, the main difference being the axillary flowers. Seldom produces seed locally but suckers or resprouts after fire. Once common in the eastern part of the region, it is becoming locally rare.
**Localities:** Montrose, Lilydale, Kalorama, 34, 51–53, 55, 56
**Distribution:** Vic, NSW, Tas

*Pultenaea muelleri* var. *muelleri*

## Pultenaea muelleri var. muelleri

Mueller's Bush-pea

**Size:** 1–3 m x 1–2 m **P**
**Habitat:** Valley sclerophyll forest
**Form:** Tall, dense, hairy shrub
**Foliage:** Crowded, narrow lanceolate, concave leaves to 15 mm long, with pointed tip, stipules prominent
**Flowers:** Solitary yellow and red flowers at the ends of branchlets; November to January. Large oval bracteoles.
**Requirements:** Moist well drained soils
**Comments:** Similar in appearance to *P. juniperina* var. *mucronata*, but has solitary terminal flowers that are surrounded by persistent overlapping bracts, not flowers in axillary clusters. Easily grown in the garden adding colour to a shaded area. Restricted range within the Melbourne area where it is endangered and possibly extinct.
**Localities:** Dandenong Ranges#
**Distribution:** Endemic to Vic

## Pultenaea paleacea var. sericea

Chaffy Bush-pea

**Size:** 0.1–0.6 m x 0.6 m — **F P**
**Habitat:** Tea-tree heath
**Form:** Straggling to prostrate shrub
**Foliage:** Narrow grey-green leaves to 20 mm long, hairy below, with recurved margins, ending in a fine tip
**Flowers:** Dense terminal clusters of sessile gold and brown or red flowers surrounded by pale brown, papery bracts; late October to November. Brown, oblong bracteoles attached near middle of calyx tube.
**Requirements:** Moist, well drained soil
**Comments:** This variety is identified by its long, white-silky hairs on the bracts and stipules.
**Localities:** was Cheltenham, Brighton; Knox, 63, 67–69, 75
**Distribution:** Vic, Qld, NSW

*Pultenaea paleacea* var. *sericea*

## Pultenaea pedunculata

Matted Bush-pea

**Size:** Prostrate x 1–3 m — **F P**
**Habitat:** Dry sclerophyll forest, box ironbark woodland
**Form:** Layering, densely matted plant
**Foliage:** Tiny, flat, dark green elliptical leaves to 12 mm long, hairy below, margins sometimes recurved. Stipules prominent, 2–3 mm long.
**Flowers:** Masses of single yellow and red axillary flowers on long hairy stalks; October to November. Fine, long bracteoles.
**Requirements:** Well drained soils, tolerating dry periods once established
**Comments:** An excellent groundcover, useful for soil-binding, and cascading over rockeries and retaining walls. It provides an interesting base plant in a tub with other plants. The local form in Montmorency has yellow flowers and smaller leaves while a more vigorous plant occurs in the moister area of Warranwood. A pink-flowered form, while not local, is also available in nurseries.
**Localities:** 34–38, 58
**Distribution:** Vic, NSW, Tas, SA

*Pultenaea pedunculata*

## Pultenaea scabra

Rough Bush-pea

**Size:** 1–2 m x 0.5–1.5 m — **P**
**Habitat:** Damp and valley sclerophyll forests
**Form:** Hairy, erect or rounded shrub
**Foliage:** Small, rough, dark green, wedge-shaped leaves, notched at the end, to 15 mm long, paler below; stipules insignificant
**Flowers:** Masses of orange-yellow flowers in loose terminal clusters of 2–5; September to November. Lanceolate, hairy bracteoles attached at or below the centre of the calyx tube.
**Requirements:** Well drained soil which does not dry out
**Localities:** Kalorama, 49–51, 55–58, 71
**Distribution:** Vic, NSW, SA

*Pultenaea scabra*

## Pultenaea stricta

Rigid Bush-pea

**Size:** to 1 m x 0.5–1 m — **F P**
**Habitat:** Valley sclerophyll forest, tea-tree heath
**Form:** Slender upright shrub
**Foliage:** Ovate to oblong or elliptic leaves to 10 mm long, margins and short stiff point recurved, stipules insignificant (less than 1 mm long)
**Flowers:** Yellow and red flowers in loose terminal clusters; September to November. Bracteoles attached in the upper half of hairy calyx tube.
**Requirements:** Moist sandy soil
**Comments:** Now locally rare, it is still found on the Mornington Peninsula.
**Localities:** Was collected Oakleigh, Frankston 1894, Nth Croydon; 74, 75
**Distribution:** Vic, NSW, Tas, SA

*Pultenaea stricta*

## Pultenaea subumbellata = Almaleea subumbellata

## Pultenaea tenuifolia

Slender Bush-pea

**Size:** 10–50 cm x 10–30 cm — **F P**
**Habitat:** Coastal banksia woodland, primary dune scrub
**Form:** Slender, open, trailing shrub
**Foliage:** Hairy, crowded, very narrow leaves to 8 mm long, the margins so recurved that the leaves appear terete, with a prominent channelled midrib
**Flowers:** Sessile, single or paired yellow and brown flowers at the ends of branchlets, surrounded by clusters of leaves with enlarged stipules; September to October. Long hairy bracteoles attached at the base of the calyx tube.
**Requirements:** Well drained soils
**Localities:** Brighton[#], 67[#]
**Distribution:** Vic, Tas, SA, WA

*Pultenaea tenuifolia*

SCALE: 0 1cm 2cm — The scale representation on each plant equals 2cm.

*Pultenaea weindorferi*

### Pultenaea weindorferi

Swamp Bush-pea

**P N**

**Size:** 1.2 m x 0.8 m
**Habitat:** Wet depression in valley sclerophyll forest
**Form:** Glabrous upright shrub
**Foliage:** Erect, broad-linear to narrowly oblanceolate bright green leaves to 10 mm long
**Flowers:** Yellow flowers with orange-red centres in tight terminal leafy clusters; September to October
**Requirements:** Moist soil, drying out in Summer
**Comments:** A rare Victorian plant, recently discovered in a small pocket in Mt. Evelyn.
**Localities:** 56
**Distribution:** Endemic to Vic

## Pycnosorus

## Asteraceae

*Pycnosorus chrysanthes*

### Pycnosorus chrysanthes

Golden Billy Buttons

**F P**

**Size:** 30 cm x 0.3–1 m
**Habitat:** Grassy wetlands on the basalt plains
**Form:** Tufted annual or perennial herb
**Foliage:** Narrow silvery leaves to 50 mm long
**Flowers:** Solitary bright yellow globular flowerheads to 15 mm across, held above the foliage; September to December and again in March to May.
**Requirements:** Heavy moist soils
**Comments:** An excellent garden or container plant. Tolerates waterlogging for extended periods. It remains in very few sites in Melbourne.
**Propagation:** Ripe seed, cuttings, division
**Localities:** St Albans, Sunbury, Sydenham, Sunshine, 7, 16, 26
**Distribution:** Vic, Qld, NSW, SA
**Synonym:** *Craspedia chrysantha*

### Pycnosorus globosus

Drumsticks

**F**

**Size:** 10–20 cm high
**Habitat:** Moist depressions in plains grasslands
**Form:** Clumping perennial herb
**Foliage:** Erect, finely hairy, bluish-green fleshy leaves to 20 cm long
**Flowers:** Large, yellow globular flowerheads, on stems to 50 cm high; November to February
**Requirements:** Heavy moist soils
**Comments:** Found on few sites within the study area.
**Propagation:** Seed, division
**Localities:** Altona, Melton
**Distribution:** Vic, Qld, NSW, SA
**Synonym:** *Craspedia globosa*

## Ranunculus

## Ranunculaceae

Commonly known as buttercups because of their cupped, shiny yellow flowers, *Ranunculus* spp. are annual or perennial herbs of moist to wet areas. The leaves are either divided, frequently into 3 leaflets, or lobed. The fruit (achenes) are clustered together and have a short beak derived from the persistent style. They will grow in many positions in the garden if the soil is kept moist, and provide a bright splash of colour through Spring and Summer. Propagate from seeds or division.

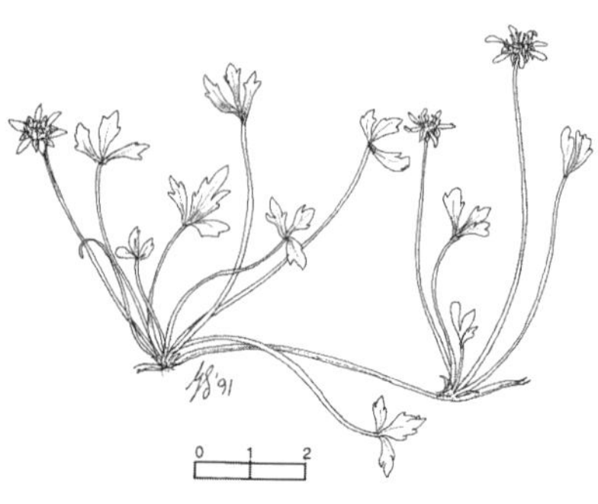

*Ranunculus amphitrichus*

### Ranunculus amphitrichus

Small River Buttercup

**F P**

**Size:** 5 cm x 20 cm
**Habitat:** Red gum woodland, valley sclerophyll forest, wattle tea-tree scrub
**Form:** Very similar in habit to *R. inundatus*. Stoloniferous, forming a mat with leaves clustered at the nodes.
**Foliage:** Leaves glabrous and shiny. Submerged plants have deeply dissected narrow-linear segments, whilst those growing in damp soil have broad, wedge-shaped, 3-lobed segments.
**Flowers:** Small pale yellow flowers to 10 mm wide with 5–9 narrow petals. Achene has slender erect beak. Flowers most of the year.
**Comments:** An attractive small plant, useful in a shallow pond, or areas with permanently wet soil. Locally rare.
**Localities:** Dandenong, 34A, 35, 49, 67#, 71, 74, 75
**Distribution:** Vic, NSW, Tas, SA; NZ
**Synonym:** *R. rivularis*

## Ranunculus glabrifolius

Shining Buttercup

**Size:** Prostrate-10 cm high **P**
**Habitat:** Swamp scrub, valley sclerophyll forest
**Form:** Slender, stoloniferous, perennial herb
**Foliage:** Leaves glossy and very variable, trifoliate with leaflets narrow and entire, or deeply dissected and toothed
**Flowers:** 5–10 narrow, yellow petals. Sepals hairy. Round, wavy and ridged achenes.
**Requirements:** Moist to wet soil
**Comments:** Grows in association with *R. amphitrichus* and *R. inundatus*. Locally rare.
**Localities:** Baxter, 41, 54, 61
**Distribution:** Vic, NSW, Tas; NZ

*Ranunculus glabrifolius*

## Ranunculus inundatus

River Buttercup

**Size:** 5–30 cm high **F P**
**Habitat:** Red gum woodland, valley sclerophyll forest, in or near fresh water
**Form:** Vigorous stoloniferous perennial often forming large mats
**Foliage:** Basal leaves, to 30 mm long, deeply divided into very narrow segments; stem leaves smaller
**Flowers:** Long, erect flowering stalks with 1–3 yellow flowers to 15 mm wide; 5–7 elliptical petals; September to January. Sepals hairless. Rounded, warty achenes have down-turned beaks.
**Localities:** Knox, 35, 42, 51, 54, 59
**Distribution:** Vic, Qld, NSW, Tas, SA

*Ranunculus inundatus*

## Ranunculus lappaceus

Common or Australian Buttercup

**Size:** 10–50 cm x 15 cm **F P**
**Habitat:** Widespread in grassy wetland, red gum woodland, wet, damp, valley and dry sclerophyll forests, wattle tea-tree scrub
**Form:** Soft, hairy, tufted perennial herb, long spreading hairs on lower stem
**Foliage:** Basal leaves divided into 3 deeply dissected or lobed leaflets to 80 mm long on stalks to 30 cm long; stem leaves fewer, narrower and less divided
**Flowers:** Single or pairs of 5-petalled golden flowers to 25 mm across, on long, erect, branched stalks; September to December. The hairy sepals are erect.
**Requirements:** Moist but not stagnant soils
**Comments:** The flattened achenes have slender coiled beaks. A bright, easy to grow herb for permanently moist areas.
**Localities:** St Albans, 12, 16, 26, 29, 34A, 35, 38, 39, 41–44, 47–49, 50A, 51, 53, 55–58, 62–64, 67#, 74, 76
**Distribution:** All states except WA; NZ

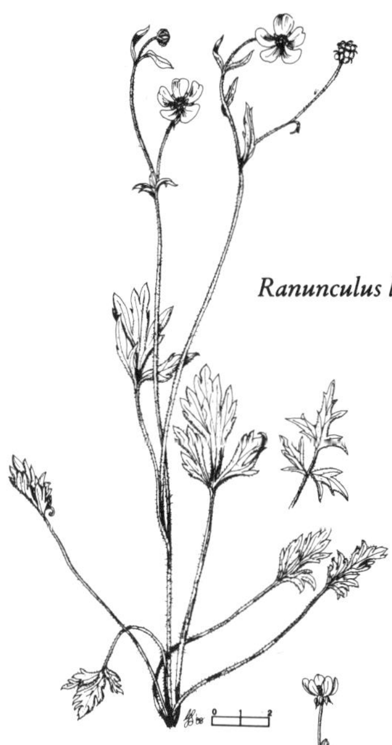

*Ranunculus lappaceus*

## Ranunculus pachycarpus

Thick-fruited Buttercup

**Size:** 6–40 cm high **F P**
**Habitat:** Valley sclerophyll forest
**Comments:** Very similar to *R. lappaceus* but the roots are tuberous rather than fibrous. The flowering stem is less branched and the achenes are plump and round with a stout beak; July to November.
**Localities:** ?42
**Distribution:** Vic, NSW, SA

## Ranunculus papulentus

Large River Buttercup

**Size:** 10–25 cm high **F P**
**Habitat:** Freshwater lagoons and along rivers in riparian scrub
**Form:** Similar to *R. inundatus*. Spreads by underground stolons.
**Foliage:** Leaves deeply dissected into several narrow segments
**Flowers:** Flowers to 20 mm wide have 9–15 petals; December to January
**Localities:** Plenty R, Yarra R
**Distribution:** Vic, Qld, NSW

*Ranunculus plebeius*

## Ranunculus plebeius

Forest or Hairy Buttercup

**Size:** 10–20 cm x 20–40 cm **P N**
**Habitat:** Riparian scrub, wet, damp and valley sclerophyll forests
**Form:** Tufted, hairy perennial herb
**Foliage:** Undivided or lobed leaves
**Flowers:** Small flowers with reflexed sepals to 15 mm wide, on long slender stalks high above the foliage; Spring to Summer. Achenes (2–3 mm long) have slender hooked beaks less than 1 mm long.
**Localities:** 29#, 42, 55–58
**Distribution:** Vic, Qld, NSW, Tas

SCALE: 0 1cm 2cm — The scale representation on each plant equals 2cm.

Goodenia pinnatifida

Goodenia lanata

Hakea sericea

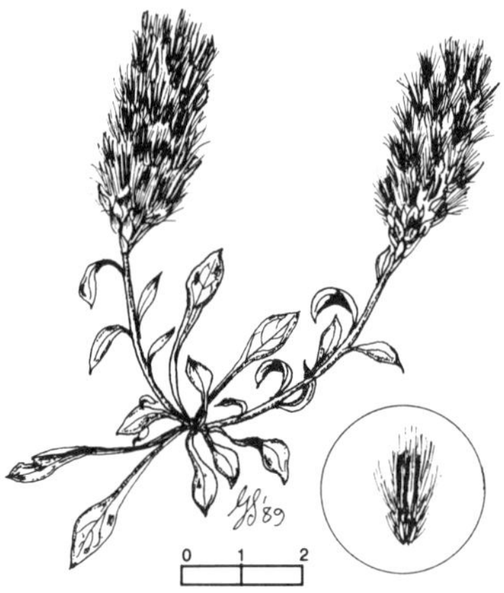

Ptilotus spathulatus

Pomaderris vacciniifolia

Lomatia myricoides

Pultenaea humilis

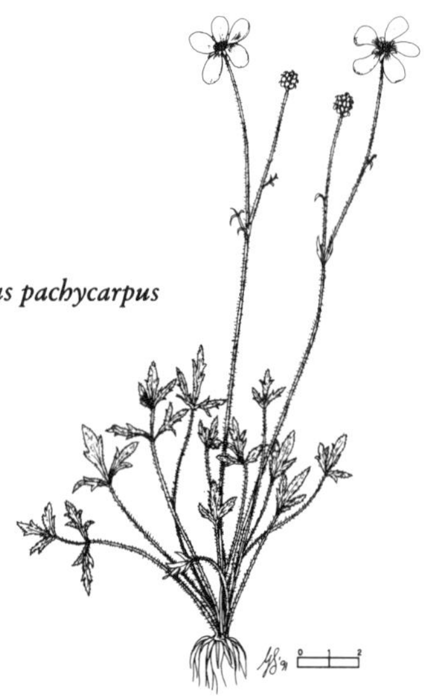

Ranunculus pachycarpus

Ranunculus papulentus

Lomatia fraseri

### Ranunculus pumilio var. pumilio
Ferny or Small-flowered Buttercup

**Size:** 1–35 cm high **F P**
**Habitat:** Floodplains and swamp margins in coastal banksia woodland, wattle tea-tree scrub and plains grassland
**Form:** Similar to *R. sessiliflorus* var. *sessiliflorus*. Densely hairy slender annual herb
**Foliage:** 3-lobed leaves up to 20 mm x 10 mm, lobes dissected on lower leaves. Longer leaf stalks to 70 mm long.
**Flowers:** Tiny stalked greenish flowers with 2–4 petals; August to November. Achenes round, short-beaked with numerous small protuberances bearing curved hairs.
**Localities:** Seaford, 5, 42, 74
**Distribution:** All states

*Ranunculus robertsonii*

### Ranunculus rivularis = R. amphitrichus

### Ranunculus robertsonii
Slender Buttercup

**Size:** 5–25 cm high **F P**
**Habitat:** Depressions in plains grassland
**Form:** Tufted, small hairy perennial herb with tuberous roots
**Foliage:** Leaves few, deeply dissected and pinnate, the 3–9 segments also deeply and narrowly lobed
**Flowers:** Bright yellow flowers with 5 petals; August to October. Recurved, strongly hooked beak on achene.
**Localities:** Melton, Sunbury
**Distribution:** Vic, SA

### Ranunculus scapiger

**Size:** to 40 cm high **P N**
**Habitat:** Dry sclerophyll forest
**Form:** Upright perennial herb
**Comments:** Differs from *R. plebeius* in that the achenes are shorter (1–2 mm), and the beaks are more than 1 mm long, straight or arching with a recurved tip.
**Localities:** Sassafras
**Distribution:** Endemic to Vic

*Ranunculus sessiliflorus var. sessiliflorus*

### Ranunculus sessiliflorus var. sessiliflorus
Australian Small-flowered Buttercup

**Size:** 2–35 cm high **P**
**Habitat:** Riparian and wattle tea-tree scrub
**Form:** Small, hairy annual herb
**Foliage:** Leaves to 20 mm wide have 3 broad, toothed lobes. Leaf stalks to 50 mm long.
**Flowers:** Tiny flowers often stalkless with 0–2 petals; July to October. Achenes ovate, long-beaked with scattered conical protuberances bearing recurved hairs.
**Localities:** St. Kilda, 34A–36, 42, 67#
**Distribution:** All states

# Rapanea — Myrsinaceae

*Rapanea howittiana*

### Rapanea howittiana
Muttonwood, Tulipwood

**Size:** 3–10 m x 2–4 m **P N**
**Habitat:** Damp sclerophyll forest, riparian scrub
**Form:** Pyramidal tall shrub or small tree
**Foliage:** Leathery, glossy dark green elliptical leaves to 90 mm long, paler green below, with wavy edges. Leaves bunched near the ends of branches. Juvenile leaves have toothed margins.
**Flowers:** Small greenish flowers clustered along older wood followed by small, round, violet to black drupes; late Winter to Summer
**Requirements:** Moist well drained soil
**Comments:** A useful screening plant for a shaded area provided it is watered during dry periods. An excellent replacement for *Pittosporum undulatum* which it superficially resembles.
**Propagation:** Fresh seed, cuttings
**Localities:** The Basin, Yarra R, 8, 16, 26, 33, 34A–36, 42, 57, 61
**Distribution:** Vic, NSW

SCALE: 0 1cm 2cm The scale representation on each plant equals 2cm.

# Rhagodia — Chenopodiaceae

*Rhagodia candolleana* ssp. *candolleana*

## Rhagodia candolleana ssp. candolleana — Seaberry Saltbush

F

**Size:** 2 m x 1–2.5 m
**Habitat:** Coastal banksia woodland, primary dune scrub, saltmarshes
**Form:** Dense sprawling or scrambling shrub
**Foliage:** Semi-succulent, dark green oblong-elliptic leaves, often broader near base, to 35 mm long, paler green and mealy below. Leaf margins often recurved.
**Flowers:** Generally dioecious; inconspicuous whitish flowers in mealy, pyramidal sprays; December to April. Small, shiny and flat dark red berries follow in Autumn.
**Requirements:** Well drained soils
**Comments:** The ribbed branchlets of this salt-tolerant shrub are mealy when young. When scrambling amongst other shrubs it may reach a height of 5 m. The Aborigines used the fruit, and possibly the leaves, for food.
**Propagation:** Cuttings
**Localities:** Altona, 4, 67–69A, 72, 74
**Distribution:** Vic, NSW, Tas, SA, WA
**Synonym:** Previously included with *R. baccata*

*Rhagodia parabolica*

## Rhagodia parabolica — Fragrant or Mealy Saltbush

F P

**Size:** 0.8–2 m x 0.5–1.5 m
**Habitat:** Sandstone cliffs of chenopod rocky open scrub, box woodlands
**Form:** Upright spindly to dense shrub
**Foliage:** Thin, broad, mealy grey-green leaves to 30 mm long, spade-like on long stalks, sometimes with broad basal lobes
**Flowers:** Dioecious; tiny, fragrant, mealy-white flowers in open sprays to 100 mm long; November to January followed by small red berries
**Requirements:** Well drained soils
**Comments:** A rare shrub of state significance in Victoria. Isolated occurrences remain in rocky gorges north-west of Melbourne.
**Propagation:** Cuttings
**Localities:** Sunbury, Melton, Werribee R, Sydenham, 26
**Distribution:** All mainland states

# Rhodanthe — Asteraceae

*Rhodanthe anthemoides*

## Rhodanthe anthemoides — Chamomile Sunray

F P

**Size:** 20–40 cm x 20–60 cm
**Habitat:** Plains grassland
**Form:** Compact perennial herb
**Foliage:** Blunt narrow bluish-green leaves to 20 mm long
**Flowers:** Terminal white papery daisies with yellow centres; October to May. Buds are deep pink to white.
**Requirements:** Moist well drained soils
**Comments:** A very attractive container or rockery plant requiring a hard prune after flowering. The variety readily available as "Paper Baby" is from NSW.
**Propagation:** Fresh seed
**Localities:** Sydenham#, 7, 16, 26
**Distribution:** Vic, Qld, NSW, Tas, SA
**Synonym:** *Helipterum anthemoides*

*Rhodanthe corymbiflora*

## Rhodanthe corymbiflora — Paper or Grey Sunray

F

**Size:** 1–30 cm x 10–20 cm
**Habitat:** Plains grassland, coastal banksia woodland
**Form:** Slender annual, usually only a single stem
**Foliage:** Soft narrow cottony grey leaves to 15 mm long
**Flowers:** Loose terminal clusters of white daisies with yellow centres; June to October.
**Requirements:** Dry well drained conditions
**Comments:** Grows well in pots. May be locally extinct.
**Localities:** Sydenham#, 67#
**Distribution:** Vic, NSW, SA
**Synonym:** *Helipterum corymbiflorum*

# Ricinocarpos — Euphorbiaceae

## Ricinocarpos pinifolius

Wedding Bush

**F P**

**Size:** 1–3 m x 1–2 m
**Habitat:** Tea-tree heath
**Form:** Dense to open rounded shrub
**Foliage:** Mostly opposite, narrow, dull green leaves to 40 mm long, margins rolled under
**Flowers:** Masses of fragrant, open, white, petalled male flowers to 25 mm wide; September to November. Clusters of male flowers surround a cone-like female flower which swells to form a round fruit.
**Requirements:** Moist, well drained sandy soil
**Comments:** Spectacular in flower but difficult to propagate.
**Propagation:** Scarified seed, cuttings
**Localities:** Port Melbourne#, St Kilda#, Caulfield, Sth Yarra, Seaford, 63, 66–71, 73–75
**Distribution:** Vic, Qld, NSW, Tas

*Ricinocarpos pinifolius*

# Rorippa — Brassicaceae

## Rorippa dictyosperma

Forest Bitter-cress

**P N**

**Size:** 0.3–0.9 m high
**Habitat:** Wet and damp sclerophyll forests
**Comments:** Differs from *R. laciniata* in that the leaves are broader and basal leaves are pinnate. The white flowers are larger with petals to 10 mm long. Each flower has 6 stamens. Seeds are pitted.
**Localities:** 32, 49, 57, 58
**Distribution:** Vic, NSW, Tas, WA
**Synonym:** *Cardamine dictyosperma*

## Rorippa laciniata

Jagged Bitter-cress, Perennial Marsh Cress

**P**

**Size:** 20–40 cm high
**Habitat:** Shaded moist slopes and river flats
**Form:** Slender, erect perennial herb
**Foliage:** Basal rosette of narrow-lanceolate, leaves to 80 mm long. They are mostly basal and may be pinnatifid, toothed or entire. Stem leaves are few, small and narrow, entire or with a few teeth or lobes.
**Flowers:** Small white flowers with 4 narrow petals to 5 mm long, clustered together at the ends of long thin stalks; most of the year. Cylindrical fruit to 40 mm long, splits to release smooth seeds.
**Requirements:** Moist soils
**Comments:** The leaves of both species were an important food source for the Aborigines. Appears to be extinct within the Melbourne area.
**Propagation:** Seed
**Localities:** Collected from Darebin Ck 1896
**Distribution:** Vic, Qld, NSW, SA
**Synonym:** *Cardamine laciniata*

*Rorippa dictyosperma*

# Rubus — Rosaceae

## Rubus parvifolius

Small-leaf Bramble, Native Raspberry

**P N**

**Size:** 0.6–1 m x 0.5–2 m
**Habitat:** Wet, damp and valley sclerophyll forests, riparian scrub and woodland
**Form:** Rambling small shrub to trailing plant covered in hooked thorns
**Foliage:** Lobed to deeply-toothed pinnate leaves, the 3–5 leaflets ovate to 60 mm long, bright green above, silvery below. The terminal leaflet is largest.
**Flowers:** Small deep pink flowers to 15 mm wide; October to December, followed by small, edible red berries (10–12 mm wide); December to April
**Requirements:** Well drained soils
**Comments:** A hardy trailer which can be trained as a climber. Similar to the introduced raspberry.
**Propagation:** Fresh seed, cuttings, root suckers
**Localities:** Sunshine, St Albans, Kalorama, 1, 8, 16, 19, 26–33, 34A–37, 41–44, 49, 50A, 55–63, 67, 69A, 71, 74
**Distribution:** Vic, Qld, NSW, Tas, SA

*Rubus parvifolius*

SCALE: 0 1cm 2cm — The scale representation on each plant equals 2cm.

# Rumex

## Polygonaceae

Widespread in Australia, these water-loving herbaceous perennials have clusters of small flowers, arranged in simple or branched spike-like or panicle-like inflorescences. Flowers age to red. Fruit are 3-sided. The valves of the fruiting bodies provide identifying features for each species. The native spp. in the Melbourne area have fruiting valves with marginal teeth. Propagation is by seed or division. Rumex reproduce freely and may become invasive in ideal conditions. Some indigenous and introduced spp. are superficially similar.

*Rumex bidens*

### Rumex bidens

Mud Dock

**Size:** 0.3–0.8 m high — **F P**
**Habitat:** Aquatic in Melbourne's rivers and creeks
**Form:** Perennial herb rooting in mud or shallow water, with hollow, swollen stems floating on or just under the water. Emergent upper stems erect.
**Foliage:** Narrow-lanceolate leaves with wavy margins to 30 cm on long stalks to 90 mm long, sheathed at base
**Flowers:** Clusters of tiny green flowers in whorls in the upper leaf axils of the upper stems, male flowers grouped above female flowers; all year, peaking Spring and Summer
**Requirements:** Shallow water or wet soil
**Comments:** The fruiting valves have 1–2 spreading basal teeth on each side. The submerged stem was an important food source for the Aborigines.
**Localities:** Maribyrnong R, 16, 26, 28, 32, 35, 42, 67#
**Distribution:** Vic, Qld, NSW, Tas, SA

### Rumex brownii

Slender or Swamp Dock

**Size:** 0.3–1 m high — **F P**
**Habitat:** Riparian scrub and woodland, box and red gum woodland, valley sclerophyll forest, wattle tea-tree scrub
**Form:** Erect perennial herb
**Foliage:** Long-stalked, oblong to lanceolate leaves to 130 mm long, mostly at the base of the stems
**Flowers:** Flowers are in widely spaced whorls in clusters of 5–16 along the leafless stems; September to December. Fruiting valves have 3–5 hooked teeth on each side and a hooked tip.
**Requirements:** Moist soils
**Comments:** Widespread low-lying areas subject to inundation. The plant may die back to a perennial rootstock during Summer. Regrowth commences the following Spring.
**Localities:** Deer Park, 8, 9, 14–18, 23–33, 34A–36, 39, 41, 42, 44, 56, 58, 60, 62, 63, 71, 74
**Distribution:** All states

*Illustrated p. 200*

*Rumex dumosus*

### Rumex dumosus

Wiry Dock

**Size:** 0.4–0.9 m x 0.4–0.9 m — **F P**
**Habitat:** Saltmarshes, grassy wetlands, plains grassland, red gum woodland
**Form:** Tangled, wiry perennial herb becoming rounded in form
**Foliage:** Lower leaves are lanceolate to 22 cm long
**Flowers:** Tiny flowers in widely spaced clusters of 1–4, each with smaller, narrow floral leaves; August to January. Fruiting valves are veined and have 2 rigid teeth on each side.
**Requirements:** Moist soils
**Comments:** When dry, the plant can break off at the base and be blown along by the wind.
**Localities:** 2, 4, 6, 7, 12, 15, 17, 18, 23, 26, 28, 34A
**Distribution:** All states

### Rumex tenax

**Size:** 10–50 cm high — **F P**
**Habitat:** Low-lying depressions in grasslands
**Form:** Erect, widely-branched perennial herb
**Foliage:** Basal and stem leaves are often linear, 90–160 mm long, with wavy margins. Leaves and stalks are of equal length.
**Flowers:** Widely spaced clusters of 12 or more small flowers in the axils of floral leaves, smaller than basal leaves; March to May and September to October. Fruiting valves are net-veined and have 2 spreading teeth on each side. They also have a callus.
**Requirements:** Moist soils
**Localities:** 7
**Distribution:** All mainland states

# Rutidosis **Asteraceae**

*Rutidosis leptorrhynchoides*

## Rutidosis leptorrhynchoides

Button Wrinklewort

**Size:** 15–30 cm x 30 cm **F P**
**Habitat:** Plains grassland, grassy wetlands of basalt plains
**Form:** Slender, glabrous perennial herb, branching from the base
**Foliage:** Sessile, linear leaves to 20 mm long, with margins curled under
**Flowers:** Yellow button flowerheads to 20 mm wide with dull green bracts, on long stalks to 20 cm; November to May; followed by fluffy seed heads
**Requirements:** Well drained heavy soils
**Comments:** An endangered species of both national and state significance, this unusual button daisy would be interesting in a container or rockery. In Melbourne, it is now restricted to only a few sites.
**Propagation:** Seed, division
**Localities:** Early collections: Footscray, Altona, 15#; St.Albans, Laverton (10 plants remain), Werribee, Deer Park, 26
**Distribution:** Vic, NSW

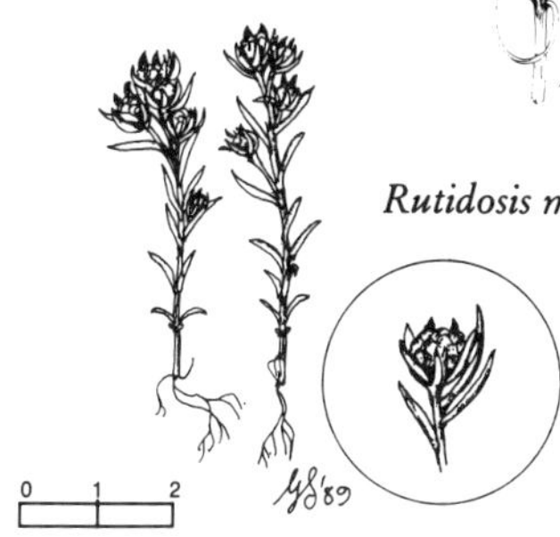

*Rutidosis multiflora*

## Rutidosis multiflora

Small Wrinklewort

**Size:** 15–50 mm high **F P**
**Habitat:** Coastal banksia woodland, primary dune scrub, dry sclerophyll forest
**Form:** Open annual herb
**Foliage:** Succulent, sessile, linear leaves to 10 mm long with long hairs near the base, lower leaves often paired
**Flowers:** Tiny, sessile, whitish-cream flowerheads grouped together in clusters among floral leaves; August to October
**Requirements:** Well drained sandy soil, tolerating dry periods
**Comments:** A rare plant in the Melbourne area resembling a tiny cauliflower, suited to containers.
**Propagation:** Seed
**Localities:** 42, 49, 67#
**Distribution:** Vic, NSW, Tas, SA, WA

# Sambucus **Caprifoliaceae**

## Sambucus gaudichaudiana

Native Elderberry

**Size:** 0.6–2 m x 0.5–1.5 m **P**
**Habitat:** Riparian scrub and woodland, wet, damp and valley sclerophyll forests
**Form:** Upright herbaceous perennial herb
**Foliage:** Pairs of light green leaves divided into 5–9 coarsely-toothed, ovate-lanceolate leaflets each 80–150 mm long; lower leaflets on larger leaves divided again
**Flowers:** Terminal clusters of small, stalkless, waxy white fragrant flowers; November to December; followed by masses of sweet, white or yellowish fleshy berries
**Requirements:** Moist well drained soil
**Comments:** A bright shrub for a shady moist area. Aborigines ate the berries.
**Propagation:** Seed, cuttings
**Localities:** Kalorama, 8, 16, 26, 29, 35, 36, 55, 57, 58
**Distribution:** Vic, Qld, NSW, Tas, SA

*Sambucus gaudichaudiana*

# Samolus **Primulaceae**

## Samolus repens

Creeping Brookweed

**Size:** 30 cm high **F P**
**Habitat:** Saltmarshes, grassy wetlands, swamps in red gum woodland, primary dune scrub
**Form:** Tufted, perennial herb, with some stems decumbent and rooting at the nodes
**Foliage:** Dull green; basal leaves thick, obovate to 40 mm long; stem leaves narrower, oblong to spathulate to 15 mm long
**Flowers:** Short terminal cluster of tiny waxy pink or white flowers on erect stems; most of the year
**Requirements:** Moist sandy soil, salt tolerant
**Comments:** A spreading herb for moist coastal areas.
**Propagation:** Cuttings, division
**Localities:** 2–4, 6, 11, 16, 25, 26, 67, 70, 74
**Distribution:** All states except NT; NZ *Also illustrated p. 343*

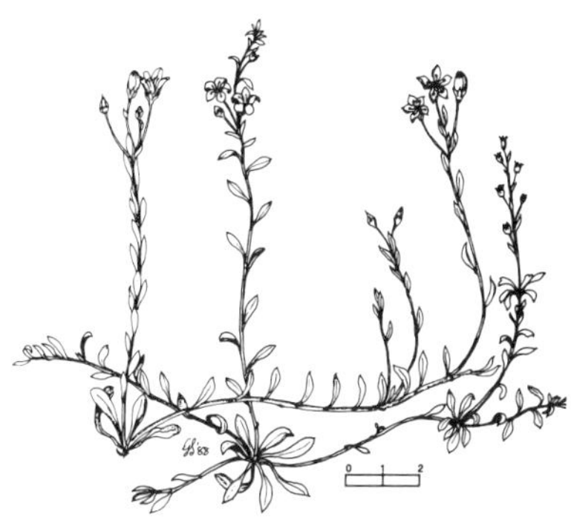

*Samolus repens*

SCALE: 0 1cm 2cm The scale representation on each plant equals 2cm.

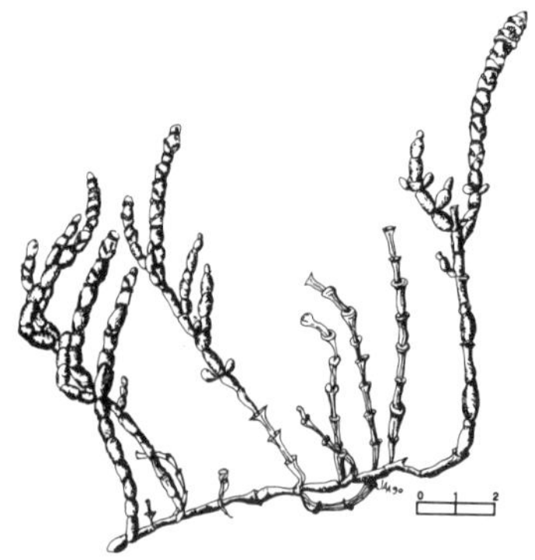

*Sarcocornia blackiana*

# Sarcocornia — Chenopodiaceae

## Sarcocornia blackiana

Thick-head Glasswort

**F**

**Size:** to 20 cm high
**Habitat:** Saltmarshes, estuaries, primary dune scrub
**Form:** Erect or spreading shrub, fleshy stems rooting at nodes
**Foliage:** Succulent, often reddish, cylindrical segments. Fruiting spikes to 50 mm long, 5–8 mm in diameter when in fruit.
**Flowers:** Rows of 5–13 green flowers embedded in the succulent axis, central flowers often in 2 rows; October to February
**Requirements:** Wet saline soil, tolerating dryness in Summer
**Comments:** Seeds are covered in small protuberances with short round hairs.
**Propagation:** Cuttings, division
**Localities:** Altona, 2, 67#, 69A, 74
**Distribution:** Vic, Tas, SA, WA
**Synonym:** *Salicornia blackiana*

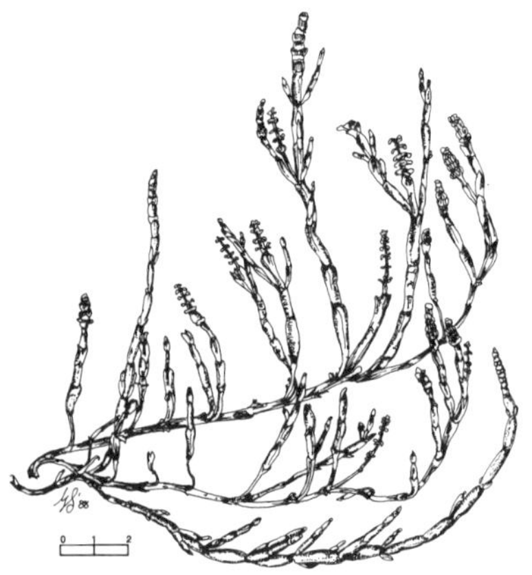

*Sarcocornia quinqueflora* ssp. *quinqueflora*

## Sarcocornia quinqueflora ssp. quinqueflora

Beaded Glasswort or Samphire

**F**

**Size:** Prostrate x 0.5 m
**Habitat:** Estuaries, saltmarshes, grassy wetlands, primary dune scrub
**Comments:** Similar to *S. blackiana* but with more slender, longer segments. Fruiting spikes are narrower, to 5 mm in diameter. 5–9 flowers are in a single row; November to March. Seeds covered with curved bristles. Prefers less salty areas. *S. quinqueflora* ssp. *tasmanica* occurs just south of the Study area, in Mt. Martha.
**Localities:** Footscray, 2–4, 7, 67, 69A, 72, 74
**Distribution:** All states except NT; NZ
**Synonym:** *Salicornia quinqueflora*

# Sclerolaena — Chenopodiaceae

## Sclerolaena diacantha

Grey Copperburr, Two-spined Saltbush

**F P**

**Size:** 10–30 cm x 20–50 cm
**Habitat:** Plains grassland
**Form:** Rounded or prostrate perennial shrub, the branches and foliage covered in densely matted hairs
**Foliage:** Fleshy, narrow grey leaves to 20 mm long
**Flowers:** Single flowers in leaf axils; most of the year. Very hairy fruiting body has 2 tiny divergent spines.
**Requirements:** Dry soils, tolerant of salt and limestone
**Comments:** A short-lived coloniser of dry areas. It seeds readily and is valuable in stabilising soils prone to wind erosion. Very restricted range in the Melbourne area, possibly occurring just outside the study area.
**Propagation:** Seed
**Localities:** Melton
**Distribution:** All states except Tas
**Synonym:** *Bassia diacantha*

*Sclerolaena muricata* var. *muricata*

## Sclerolaena muricata var. muricata

Black Roly-poly, Five-spined Saltbush

**F**

**Size:** 1 m x 1 m
**Habitat:** Saltmarshes, box woodland, riparian scrub, primary dune scrub
**Form:** Short-lived, rounded perennial shrub forming a tangled clump
**Foliage:** Dark blue-green or pale green, linear to narrowly oblong leaves to 15 mm long, hairless and slightly fleshy
**Flowers:** Single flowers in leaf axils; most of the year. Tiny, thick fruiting bodies are firmly attached to the branches. Each has 5 prominent spines, the 3 largest (to 10 mm long) being separate, the 2 smaller ones usually joined at the base.
**Requirements:** Moist soils, tolerating periodic inundation
**Comments:** A pioneer plant of disturbed or degraded areas. It has been declared a noxious weed in Victoria as the spiny fruit catch in sheep's wool making shearing and processing difficult. Dry plants break off at the base, and are blown along the ground by the wind.
**Propagation:** Seed
**Localities:** Toolern Vale, St. Albans, Sunshine, 4, 9, 12, 16, 23, 26, 67
**Distribution:** Vic, Qld, NSW, SA
**Synonym:** *Bassia quinquecuspis* var. *quinquecuspis*

### Sclerolaena muricata var. villosa

Grey Roly-poly

**F P**

**Habitat:** Riparian scrub on basalt plains
**Comments:** The branches have a covering of woolly hairs. The dark green obovate-lanceolate leaves to 10 mm long have a glabrous upper surface, sparsely downy below. Locally rare.
**Localities:** Melton, 16
**Distribution:** Vic, Qld, NSW, SA, NT
**Synonym:** *Bassia quinquecuspis* var. *villosa*

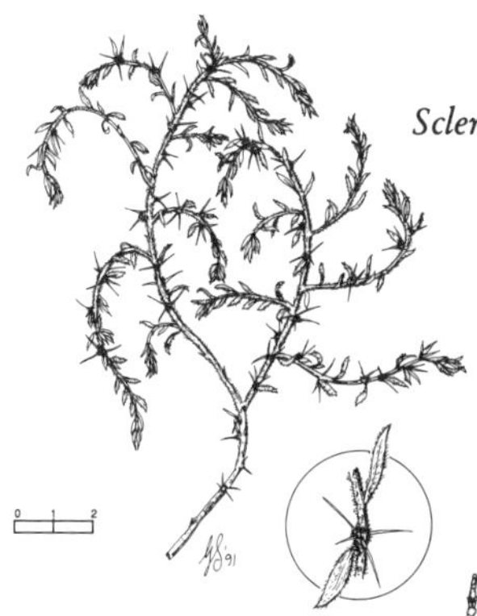

*Sclerolaena muricata* var. *villosa*

# Sclerostegia — Chenopodiaceae

### Sclerostegia arbuscula

Shrubby Glasswort

**F**

**Size:** 1–2 m x 1–2 m
**Habitat:** Saltmarshes, shoreline
**Form:** Erect, much-branched shrub
**Foliage:** Succulent stems green or reddish, with segments to 5 mm long
**Flowers:** Flowers in spikes consisting of rows of 3 flowers; all year
**Requirements:** Moist to wet saline soils
**Comments:** Restricted distribution in the Melbourne area.
**Propagation:** Seed
**Localities:** 2–4, 67
**Distribution:** Vic, NSW, Tas, SA, WA
**Synonym:** *Arthrocnemum arbuscula*

*Sclerostegia arbuscula*

# Scutellaria — Lamiaceae

### Scutellaria humilis

Dwarf Skull Cap

**F P N**

**Size:** 15 cm x 1 m
**Habitat:** Box woodland, tea-tree heath, along creeks and gullies
**Form:** Stoloniferous perennial herb with erect 4-sided stems
**Foliage:** Pairs of toothed, ovate leaves to 12 mm long
**Flowers:** Small mauve to pink flowers singly on slender stalks in leaf axils; October to February. The flowers appear to be paired as they often grow in the same direction.
**Requirements:** Moist, well drained soils, requiring Summer watering
**Comments:** Very rare small dainty plant which adds interest to a shady site. Spreads by layering.
**Propagation:** Cuttings, division
**Localities:** Mt Eliza, Sunbury#, 26, 29#, 74
**Distribution:** Vic, Qld, NSW, Tas, SA

*Scutellaria humilis*

# Sebaea — Gentianaceae

### Sebaea albidiflora

White Sebaea

**P**

**Size:** to 20 cm high
**Habitat:** Saltmarshes, valley sclerophyll forest, wattle tea-tree scrub
**Form:** Stiff, erect annual herb
**Foliage:** Pairs of erect, stem-clasping, ovate to lanceolate leaves to 10 mm long, smaller at the base and withering early
**Flowers:** Compact terminal cymes, composed of 1–5 repeatedly forked clusters with no more than 3 sessile, white to cream tubular flowers in each; September to November. Flowers divided to the base into 4 segments (4-partite). The calyx lobes are blunt, base clasped by leafy bracts.
**Requirements:** Moist to wet soil
**Comments:** Restricted distribution in Melbourne.
**Propagation:** Seed
**Localities:** 3, 4, ?42, 49, 67#
**Distribution:** Vic, Tas, SA

*Sebaea albidiflora*

### Sebaea ovata

Yellow Sebaea

**P N**

**Size:** 20 cm high
**Habitat:** Plains grassland, grassy wetlands, red gum woodland, damp and valley sclerophyll forest, tea-tree heath
**Form:** Similar to *S. albidiflora*. Stems are 4-sided with wider, ovate to almost round leaves to 12 mm long.
**Flowers:** Clusters seldom have more than 5 yellow flowers which are 5-partite and become stalked after flowering. The calyx lobes are pointed and bracts spreading. Locally rare.
**Localities:** Laverton Nth, Rockbank, 12, 17#, 18, 26, 34A, 35, 41, 42, 49, 58, 67#
**Distribution:** All states except NT; NZ

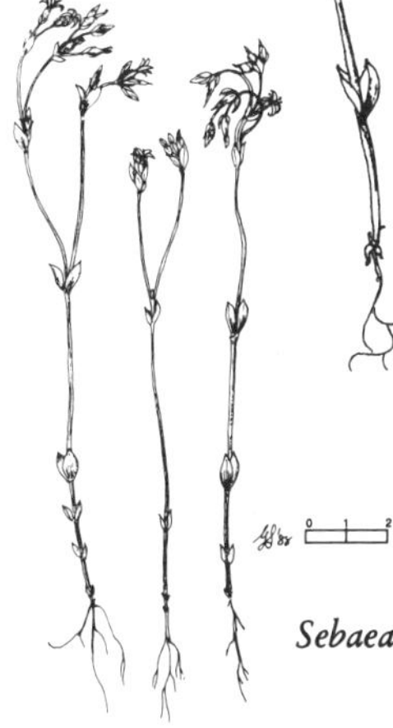

*Sebaea ovata*

SCALE: 0 1cm 2cm The scale representation on each plant equals 2cm.

*Selliera radicans*

# Selliera — Goodeniaceae

## Selliera radicans — Swampweed

**Size:** Prostrate-40 mm x 1 m — F P N
**Habitat:** Saltmarshes, riparian scrub, coastal banksia woodland, grassy wetland
**Form:** Prostrate, creeping mat plant, rooting at the nodes
**Foliage:** Succulent, bright green spathulate leaves, 10–100 mm long
**Flowers:** Small white to mauve fan-shaped flowers; October to November
**Requirements:** Moist soil, salt tolerant
**Comments:** A fast-growing creeper, useful as a soil or sand binder. Tolerates both wet and dry periods.
**Propagation:** Cuttings, division
**Localities:** Keilor, 3, 4, 16, 26, 28, 60, 67, 69A, 71, 72, 74, 76
**Distribution:** Vic, NSW, Tas, SA; NZ, S Am

# Senecio — Asteraceae

The Senecios are a very large genus of herbaceous plants represented by 16 species in the Melbourne area. Yellow flowerheads are in terminal clusters and are either radiate (having both tubular and ligulate florets) or, more frequently, discoid (having only tubular florets). The flowerhead is surrounded by a row of bracts. Fluffy seed heads follow flowering. The leaves often have ear-like growths at the base, called auricles. Senecios are pioneering plants, germinating quickly following bush-fires, and colonising disturbed areas. Generally understorey plants, some species may become invasive under ideal conditions. Propagation is by seed.

*Senecio biserratus*

## Senecio biserratus — Jagged Fireweed

**Size:** 0.6–1 m high — F P
**Habitat:** Riparian scrub, grassy low open forest, primary dune scrub
**Form:** Erect annual herb, branching below the infloresence
**Foliage:** Glabrous, ovate to broad-lanceolate toothed leaves to 140 mm long. The leaves are prominently veined, shallowly lobed and irregularly toothed; the auricles are also toothed.
**Flowers:** Dense clusters of discoid flowerheads surrounded by few, narrow green bracts; October to March
**Requirements:** Moist soils tolerant of salty winds
**Localities:** Keysborough, Chelsea, 36, 42, 71, 74, 76
**Distribution:** Vic, NSW, Tas, SA; NZ

## Senecio cunninghamii var. cunninghamii — Branching or Bushy Groundsel

**Size:** to 1 m x 1 m — F P
**Habitat:** Plains grassland
**Form:** Erect to spreading under-shrub
**Foliage:** Glabrous; sessile, usually entire, light green linear leaves 20–120 mm long, often with 2 grooves below and curled edges, auricles sometimes present. The branches are reddish.
**Flowers:** Dense flat clusters of discoid flowerheads, bracts broad and yellowish; August to April
**Requirements:** Moist soils tolerating inundation and saline conditions
**Comments:** May be useful as a coloniser of saline areas. The local form has smaller leaves to 20 mm long.
**Localities:** Sydenham, 15#, 16
**Distribution:** All states except Tas

*Senecio cunninghamii* var. *cunninghamii*

## Senecio glomeratus — Annual Fireweed

**Size:** 0.5–1.2 m high — F P
**Habitat:** Widespread in red gum woodland, wet, damp, valley and dry sclerophyll forests, grassy low open forest, tea-tree heath
**Form:** Erect, softly hairy annual herb
**Foliage:** Lanceolate-ovate leaves to 120 mm long, coarsely toothed to lobed, lobes sometimes toothed, with large, toothed auricles
**Flowers:** Densely-bunched discoid flowerheads with a few narrow, woolly bracts; November to February
**Requirements:** Moist soils
**Comments:** Similar to *S. hispidulus* var. *hispidulus* but covered in soft, cobweb-like hairs.
**Localities:** Montrose, 16, 26, 28, 34A–36, 39–42, 44, 47, 48, 50–52 54–58, 60, 63, 64, 67–72, 74–76
**Distribution:** Vic, NSW, Tas, SA, WA; NZ

*Illustrated p. 18*

## Senecio hispidulus var. hispidulus

Rough Fireweed

**Size:** 0.2–1.3 m high — F P
**Habitat:** Plains grassland, dry and valley sclerophyll forests, coastal banksia woodland, tea-tree heath
**Form:** Erect perennial, single stemmed or multi-branched from the base
**Foliage:** Sessile, lanceolate to obovate, toothed to sharply-lobed leaves to 60 mm long, erect bristly hairs below, rough on upper surface, auricles with 2 coarse teeth
**Flowers:** Branched clusters of inconspicuous spreading discoid flowerheads surrounded by green bracts; September to February
**Requirements:** Moist soils
**Comments:** Often occurs in clearings or by tracks.
**Localities:** Montrose, 28, 34A–36, 41–44, 47, 48, 50–56, 58–60, 63, 64, 67, 68, 69A–71, 74–76
**Distribution:** All states; NZ

*Senecio hispidulus* var. *hispidulus*

## Senecio hispidulus var. dissectus

Hill Fireweed

**Size:** to 0.8 m high — F P
**Habitat:** Dry sclerophyll forest, amongst rocks
**Comments:** Leaves are deeply dissected to pinnate, and margins have more rounded teeth.
**Localities:** Bulla, Diamond Ck, 32
**Distribution:** All states

*Senecio lautus*

## Senecio lautus

Variable Groundsel

**Size:** 0.1–1 m x 0.3–0.5 m — F P
**Habitat:** Riparian scrub, valley and damp sclerophyll forests, coastal banksia woodland
**Form:** Very variable. An erect or sprawling perennial herb or under-shrub.
**Foliage:** Dark green leaves, variable from narrow to pinnate, entire to toothed or lobed, 10–70 mm long, may be stem-clasping
**Flowers:** Loose clusters of radiate flowerheads with 8–14 ray florets, many disk florets; September to March. 12–20 bracts with several minute bracts at base.
**Requirements:** Well drained soils, lime tolerant
**Comments:** Some forms are particularly attractive and long flowering. The coastal form is small and spreading, with fleshy leaves to 50 mm long and well displayed flowerheads. The foothills forms range from entire, stalked and elliptic to narrow-lanceolate, toothed and stem-clasping.
**Localities:** St. Kilda#; Olinda, 14, 30#, 49, 58, 67#, 74
**Distribution:** All states; NZ

*Senecio linearifolius*

## Senecio linearifolius

Firewheel Groundsel

**Size:** 1–2 m x 1–1.5 m — P N
**Habitat:** Wet, damp and valley sclerophyll forests
**Form:** Erect perennial herb
**Foliage:** Sessile, entire or sharply-toothed, linear-lanceolate, dark green leaves to 150 mm long. The undersurface is more or less glabrous.
**Flowers:** Large flat-topped clusters of numerous, small radiate flowerheads with 5 ray florets; November to February
**Requirements:** Moist soils
**Comments:** A brilliant display in flower. Alpine forms have broader leaves than the local form. Conspicuous after fire or clearing.
**Localities:** 47, 55–58, 62, 63
**Distribution:** Vic, NSW, Tas

*Senecio macrocarpus*

## Senecio macrocarpus

**Size:** 20–40 cm high — F
**Habitat:** Plains grassland
**Form:** Straggly to erect perennial herb
**Foliage:** Downy grey narrow leaves to 60 mm long; lower leaves usually entire, stem leaves smaller, toothed, and with minute auricles
**Flowers:** Few large discoid flowerheads; September to October
**Requirements:** Well drained soils
**Comments:** Only recently rediscovered in the Melbourne area, on the basalt plains. Extinct through some of its range it is an endangered plant of both Melbourne and Victoria.
**Localities:** Laverton Nth, Sunbury, Diggers Rest, Melton
**Distribution:** Vic, Tas, SA

SCALE: 0 1cm 2cm — The scale representation on each plant equals 2cm.

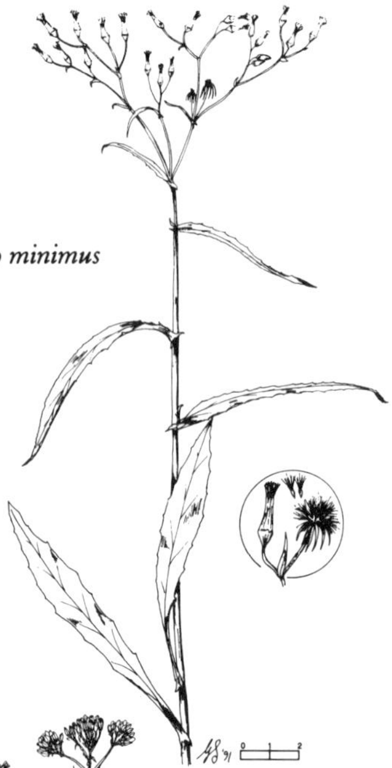

*Senecio minimus*

## Senecio minimus

Shrubby Fireweed

| | |
|---|---|
| **Size:** | 0.5–1 m high **P N** |
| **Habitat:** | Red gum and riparian woodland, swamp scrub, damp and valley sclerophyll forests, tea-tree heath |
| **Form:** | Erect annual herb often branched at the base or below flowerheads, parallel ridges along stems |
| **Foliage:** | Lanceolate, evenly-toothed leaves to 90 mm long, sometimes larger, with toothed auricles |
| **Flowers:** | Masses of discoid flowerheads, narrow green bracts; December to April |
| **Requirements:** | Moist well drained soils |
| **Localities:** | 35, 36, 42, 46, 47, 51, 54–58, 60, 71, 73–75 |
| **Distribution:** | All states except NT; NZ |

*Senecio odoratus*

## Senecio odoratus

Scented Groundsel

| | |
|---|---|
| **Size:** | 0.6–2 m x 0.6–1.5 m **F P** |
| **Habitat:** | Grassy wetlands, valley sclerophyll forest, grassy low open forest |
| **Form:** | Erect perennial undershrub |
| **Foliage:** | Coarse, sessile, usually glaucous leaves, to 120 mm long, broadly lanceolate to ovate, toothed along recurved margins, and with auricles. The veins are conspicuous. |
| **Flowers:** | Large flat-topped clusters of discoid flowerheads; October to January. 8 bracts. |
| **Requirements:** | Well drained soils, tolerating salt spray |
| **Comments:** | Coastal forms are lower and more bushy. Locally rare. |
| **Localities:** | 5, 34A, 42, 74, 76 |
| **Distribution:** | Vic, Tas, SA |

*Senecio quadridentatus*

## Senecio quadridentatus

Cotton Fireweed

| | |
|---|---|
| **Size:** | 0.4–1 m x 0.5–1 m **F** |
| **Habitat:** | Widespread in disturbed areas of plains grassland, red gum and riparian woodland, riparian scrub, dry and valley sclerophyll forests, tea-tree heath |
| **Form:** | Erect, perennial herb, much-branched from the base. Both stems and leaves are covered with dense cottony hairs, giving the plant a greyish appearance. |
| **Foliage:** | Narrow, pointed leaves to 100 mm long, becoming smaller up the stem, margins recurved. Leaves sometimes have a few teeth on the margins. |
| **Flowers:** | Small greenish-yellow discoid flowerheads in loose clusters; October to March. 11–13 bracts. |
| **Requirements:** | Very adaptable |
| **Comments:** | The most common and widely distributed *Senecio* spp. Withstands very dry conditions, but may die back to a woody rootstock until rain occurs. Major features distinguishing it from *S. tenuiflorus* are the 3 prominent veins on the floral bracts and tapering achenes (or fruit) which have an elongated beak and prominent ribs. |
| **Localities:** | 7, 8, 12, 16–18, 21, 26, 28, 31, 32, 34A–36, 39, 41–44, 46–49, 51–60, 62–68, 69A–71, 74, 76 |
| **Distribution:** | All states; NZ, Timor, N Caledonia |

## Senecio runcinifolius

Tall Groudsel

| | |
|---|---|
| **Size:** | 0.2–1 m high **F** |
| **Habitat:** | Saltmarshes |
| **Form:** | Erect annual herb |
| **Foliage:** | Lanceolate stem-clasping leaves, deeply lobed with more than 8 pairs of toothed lobes, lobes pointing towards the base of the leaf |
| **Flowers:** | Broad loosely-branched clusters of discoid flowerheads; August to November. Fruiting heads are white and very fluffy. |
| **Requirements:** | Moist soils, tolerating saline conditions |
| **Comments:** | More common after good rains, it will grow larger if it is supported by other plants. Masses of silky seeds give the plant an interesting appearance. Locally rare. |
| **Localities:** | Altona Nth |
| **Distribution:** | Vic, NSW, SA |

*Illustrated p. 200*

*Senecio spathulatus*

## Senecio spathulatus

| | |
|---|---|
| **Size:** | to 0.5 m high **F** |
| **Habitat:** | Primary dune scrub |
| **Comments:** | A weak herb often with fleshy obovate leaves to 55 mm long and few radiate flowerheads. A locally rare plant, often combined with *S. lautus.* |
| **Localities:** | Seaford, Pt. Wilson |
| **Distribution:** | Endemic to Vic |
| **Synonym:** | *S. lautus* ssp. *maritimus* |

## Senecio squarrosus

Leafy Fireweed

**Size:** 0.3–0.5 m high **F P**
**Habitat:** Grassy low open forest, plains grassland
**Form:** Erect perennial herb, branching at the flowerheads
**Foliage:** Thick linear-lanceolate leaves to 100 mm long, rough above, sparse cottony hairs below, entire or sparsely toothed or lobed; upper leaves sessile, narrower with small coarsely-toothed auricles
**Flowers:** Terminal clusters of 4–16 discoid flowerheads; December to March
**Requirements:** Well drained soils
**Comments:** A rare plant in the Melbourne area. The form found on the basalt plains may be raised to species status after further research. It is in danger of extinction due to development occurring on the only known sites.
**Localities:** Braybrook 1889, Lysterfield 1903; Werribee, Deer Park, 50A, 74
**Distribution:** Vic, Tas, SA

*Senecio squarrosus*

## Senecio tenuiflorus

**Size:** 0.3–0.6 m high **F P**
**Habitat:** Dry and valley sclerophyll forests, tea-tree heath, disturbed roadsides
**Form:** Erect annual or biennial herb branching at the base or below the flower clusters
**Foliage:** Crowded, obovate to oblanceolate lower leaves to 100 mm long, very hairy and usually purplish below, rough above. Mid-stem leaves shorter and lanceolate.
**Flowers:** Small discoid flowerheads; October to January
**Requirements:** Prefers moist soils
**Comments:** Similar to and often occurring with *S. quadridentatus* but has fewer flowerheads, one (sometimes 2) prominent veins on the bracts and a plump achene with flat ribs.
**Localities:** 34A, 36, 42, 44, 47, 48, 50–57, 70, 74
**Distribution:** All states except NT

*Senecio tenuiflorus*

## Senecio vagus

Saw Groundsel

**Size:** to 1 m high **P N**
**Habitat:** Wet sclerophyll forest
**Form:** Erect or rounded hairless perennial herb
**Foliage:** Large, dark green lanceolate leaves to 15 cm long, on long stalks. Foliage is deeply divided, with irregular lobes or teeth.
**Flowers:** Many large radiate flowerheads in large, loose clusters; sporadic, peaking September to December
**Requirements:** Moist well drained soils
**Comments:** Very restricted range in Melbourne.
**Localities:** 55, 57
**Distribution:** Vic, NSW, Tas

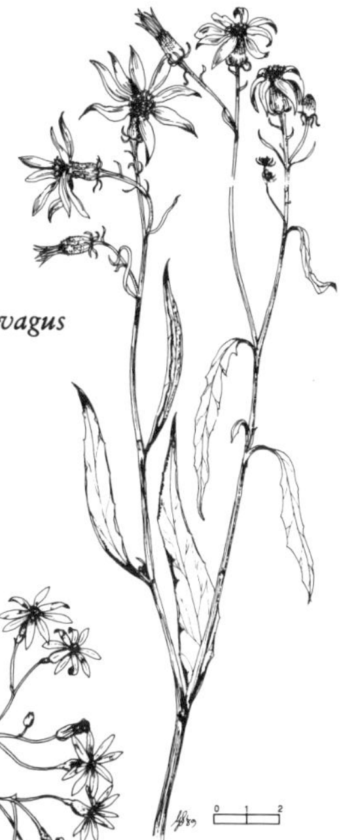

*Senecio vagus*

## Senecio velleioides

Forest Groundsel

**Size:** 0.5–1.5 m x 0.5–1 m **P N**
**Habitat:** Wet and damp sclerophyll forests
**Form:** Soft, many-branched perennial herb
**Foliage:** Stem-clasping, glaucous, heart-shaped leaves to 7–17 cm long, upper leaves almost entire while lower leaves are regularly toothed
**Flowers:** Clusters of large radiate flowerheads; November to February
**Requirements:** Moist well drained soils
**Comments:** Germinates after fire. This showy plant is becoming increasingly difficult to find locally.
**Localities:** 57, 58
**Distribution:** Vic, Qld, NSW, Tas

*Senecio velleioides*

# Senna

# Caesalpiniaceae

## Senna artemisiodes ssp. filifolia

Desert Cassia

**Size:** 1–3 m x 1 m **F P**
**Habitat:** Box woodland
**Form:** Rounded or erect shrub
**Foliage:** Pinnate leaves to 50 mm long with 1–2 pairs of terete or narrow leaflets
**Flowers:** Abundant short racemes of golden flowers; June to November
**Requirements:** Well drained soils
**Comments:** An attractive and easily grown shrub which does well under established eucalypts. It also tolerates hot dry conditions.
**Propagation:** Scarified seed, cuttings
**Localities:** Essendon, Sunbury, Tullamarine, 16, 26
**Distribution:** Vic, SA, NT
**Synonym:** *Cassia nemophila* var. *nemophila*, *C. eremophila*

*Senna artemisioides* ssp. *filifolia*

SCALE: 0 1cm 2cm — The scale representation on each plant equals 2cm

*Sigesbeckia orientalis*

## Sigesbeckia — Asteraceae

### Sigesbeckia orientalis — Indian Weed

**Size:** 0.1–0.8 m x 0.1–0.8 m — **P**
**Habitat:** Wet, damp and valley sclerophyll forests
**Form:** Stiff, erect, branched annual herb
**Foliage:** Widely-spaced pairs of soft, downy, lanceolate to arrow-shaped leaves to 120 mm long, with toothed margins
**Flowers:** Yellow daisy-like flowerheads on long stalks above the leaves; November to March. Ray florets which are tiny, number about 8, but as each is 3-lobed, it appears there are many. The 5 outer floral bracts are covered with sticky glandular hairs, somewhat like those of *Drosera* spp.
**Requirements:** Moist soils
**Comments:** Similar in appearance to *Senecio* spp., and may also be found in disturbed areas. It is not usually found in large numbers.
**Propagation:** Seed
**Localities:** Kalorama, 29, 55–58
**Distribution:** All states except Tas; NZ

## Solanum — Solanaceae

*Solanum* spp. are often colonising plants of disturbed areas and tracks. They have an important role following fires, regenerating quickly from seed or underground stems. The five fused petals are pale mauve to purplish in the Melbourne spp. The fleshy berries were eaten by the Aborigines but only when very ripe, often being placed in sand heaps to ripen. Fast growing, some species make excellent screen plants in hot, dry conditions. Propagation is by seed or cuttings.

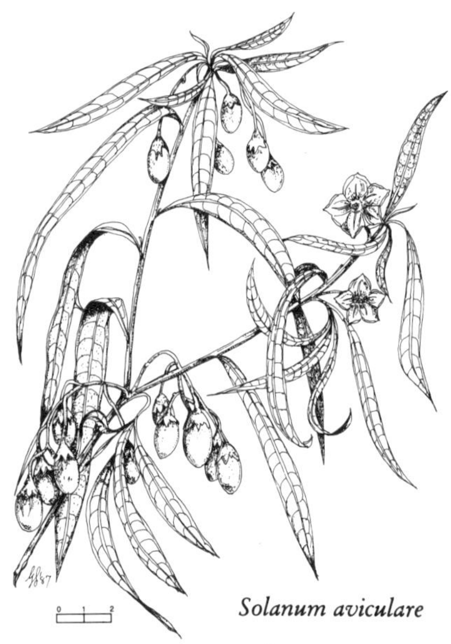

*Solanum aviculare*

### Solanum aviculare — Kangaroo Apple

**Size:** 1–3 m x 1–4 m — **F P N**
**Habitat:** Red gum woodland, valley, damp and wet sclerophyll forests, grassy low open forest
**Form:** Erect, soft woody shrub, stems angular
**Foliage:** Dark green; leaves variably lobed, broadly elliptic to obovate to 30 cm long, with 3–11 deeply cut lobes; entire lanceolate leaves 5–20 cm long often borne on the upper parts of the plant
**Flowers:** Axillary cluster of up to 10 violet flowers, darker in centre, 25–40 mm across, deeply cut lobes are sharply pointed; January. Succulent egg-shaped fruit are orange-red to scarlet when ripe.
**Requirements:** Well drained soils
**Comments:** A decorative fast growing shrub becoming straggly with age but easily rejuvenated with heavy pruning.
**Localities:** 23, 33–34A, 42, 47, 49, 55, 57, 60–63, 74, 76
**Distribution:** Vic, Qld, NSW, Tas; NZ, PNG

### Solanum laciniatum — Large Kangaroo Apple

**Size:** 1–3 m x 1–3 m — **F P N**
**Habitat:** Box and red gum woodlands, valley sclerophyll forest, riparian scrub
**Foliage:** Leaves either irregularly lobed and broadly ovate to 30 cm long with 1–7 lobes or entire and lanceolate to 100 mm long
**Flowers:** Bluish to purple flowers to 50 mm wide, September to March, have very shallow lobes. Fruit is yellow to orange-yellow when ripe.
**Comments:** Similar to *S. aviculare*. Edible when soft and orange but poisonous if eaten when green. Both species are being cultivated as a source of solasodine for making cortico-steroid drugs.
**Localities:** 1, 14, 16, 26, 29, 32, 33, 34A–36, 44, 47, 50A, 61, 64, 67–71, 74–76
**Distribution:** Vic, Tas, SA

*Solanum laciniatum*

### Solanum prinophyllum — Toothed or Forest Nightshade

**Size:** 15–30 cm x 30–50 cm — **P**
**Habitat:** Wet and damp sclerophyll forests
**Form:** Sprawling annual or short-lived prickly perennial undershrub with minute downy hairs
**Foliage:** Lustrous dark green to purplish-green lanceolate-elliptic leaves to 80 mm long, with 7–10 rounded, toothed lobes
**Flowers:** Clusters of 1–4 lilac-blue flowers to 25 mm wide, with pointed lobes; Spring to Summer. Berry green with purplish tinge when ripe.
**Requirements:** Moist soils
**Comments:** Many scattered spines to 10 mm long cover the plant. Locally rare.
**Localities:** Mt Dandenong, 57, 58
**Distribution:** Vic, Qld, NSW

*Illustrated p. 200*

### Solanum vescum

Gunyang

**Size:** 1-2 m x 1-3 m — F P
**Habitat:** Damp sclerophyll forest
**Form:** Erect or rounded shrub
**Foliage:** Variable, sessile; narrow entire leaves to 150 mm long, the leaf blades continuing down the stem as a ridge. Some leaves may have a few long lobes.
**Flowers:** Axillary cluster of pale mauve flowers to 40 mm wide, shallowly cut between broad lobes; Spring to Summer. Round, greenish-ivory berries have an unpleasant smell when ripe.
**Requirements:** Well drained soils, tolerating dryness once established
**Comments:** Restricted distribution in Melbourne.
**Localities:** 58
**Distribution:** Vic, NSW, Tas

*Solanum vescum*

## Solenogyne — Asteraceae

### Solenogyne dominii

**Size:** Prostrate-20 cm x 10 cm — F P
**Habitat:** Plains grassland, red gum woodland, dry sclerophyll forest, grassy low open forest
**Form:** Perennial herb
**Foliage:** Rosette of oblanceolate, dull green glabrous leaves to 60 mm long, paler below, 15-21 rounded marginal teeth per leaf
**Flowers:** Erect flowering stalks shorter than the leaves at flowering time, lengthening later, with a single, terminal greenish flowerhead; bracts and ray florets magenta-tinged. Most of the year.
**Requirements:** Well drained soils
**Propagation:** Seed
**Localities:** 2, 6, 12, 17#, 18, 25, 28, 34A, 35, 38, 42, 47, 48, 50A, 56, 74
**Distribution:** Vic, NSW, Tas, SA

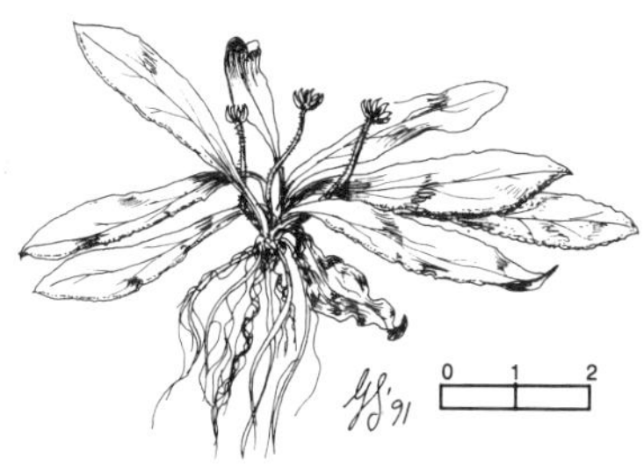

*Solenogyne dominii*

### Solenogyne gunnii

P

**Habitat:** Plains grassland, red gum woodland, dry sclerophyll forest, grassy low open forest
**Requirements:** Prefers moist, shaded positions tolerating seasonal inundation
**Comments:** Similar to *S. dominii* but leaves distinctly hairy with 9-13 marginal teeth per leaf. Bracts and ray florets rarely tinged magenta. Both species may occur together in moister areas.
**Localities:** 7, 12, 13, 20, 23, 24, 26, 31, 42, 44, 46, 47, 63, 64, 74
**Distribution:** Vic, NSW, Tas, SA
**Synonym:** *S. bellioides* var. *gunnii*

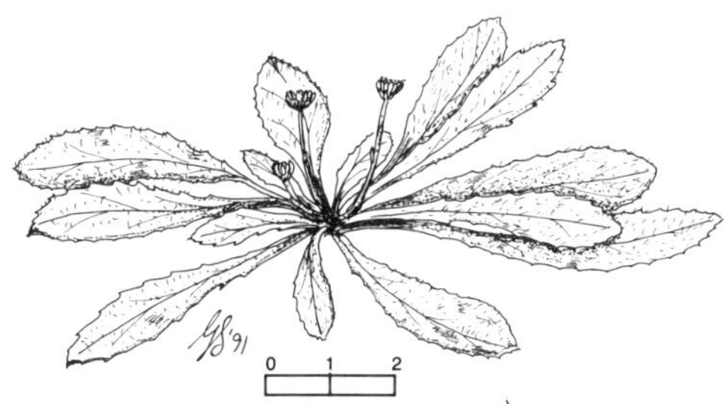

*Solenogyne gunnii*

## Sphaerolobium — Fabaceae

### Sphaerolobium vimineum

Leafless Globe-pea

**Size:** 0.3-0.6 m x 0.3-0.6 m — F P
**Habitat:** Valley sclerophyll forest, tea tree heath
**Form:** Wiry upright shrub
**Foliage:** Leafless; occasionally a few tiny scale-like leaves are found on the unbranched rounded stems
**Flowers:** Terminal spikes of small yellow pea flowers, sometimes tinged with red; September to January; seed pods small and round
**Requirements:** Moist well drained soil
**Comments:** It is attractive in a rockery or planted with other small shrubs, remaining unnoticed until it flowers. Once widespread, it is rarely seen around Melbourne now.
**Propagation:** Scarified seed, cuttings
**Localities:** Montrose, ?42, 47, 49, 50A, 54, 56, 58, 67#, 74, 75
**Distribution:** All states except NT

*Sphaerolobium vimineum*

## Sprengelia — Epacridaceae

### Sprengelia incarnata

Pink Swamp-heath

**Size:** 0.8-2 m x 0.2-0.7 m — F P
**Habitat:** Swamps in tea-tree heath, swamp scrub
**Form:** Prickly, upright shrub
**Foliage:** Stiff, concave leaves to 12 mm long with sharp points, bases sheathing the stem
**Flowers:** Dense terminal spikes of small, pink, star-like flowers; August to November
**Requirements:** Wet sandy soils, tolerating periods of waterlogging
**Comments:** A most attractive plant for a wet position but requiring special care. It performs well in the garden if given good drainage and plenty of water.
**Propagation:** Cuttings, may be difficult to strike
**Localities:** Clayton, 66#, 67#, 73, 75
**Distribution:** Vic, NSW, Tas, SA

*Sprengelia incarnata*

SCALE: 0 1cm 2cm — The scale representation on each plant equals 2cm.

*Spyridium parvifolium*

# Spyridium — Rhamnaceae

## Spyridium parvifolium

Australian Dusty Miller — **P N**

**Size:** 1–3 m x 1–2 m
**Habitat:** Damp and valley sclerophyll forests
**Form:** Dense upright shrub
**Foliage:** Heavily-veined, dark green oval leaves to 25 mm long, with notched ends, pale green and hairy below
**Flowers:** Small white flowers in flat clusters surrounded by dusty-white floral leaves; July to November
**Requirements:** Well drained soils, tolerating dryness once established
**Comments:** An interesting shrub providing a light screen in dry, shady areas, where its most attractive feature, the floral leaves, are shown to advantage.
**Propagation:** Seed, cuttings
**Localities:** 30#, 34, 34A, 41, 42, 44, 46, 47, 49–51, 55–59, 62–64
**Distribution:** Vic, NSW, Tas, SA

*Stackhousia monogyna*

*Stackhousia spathulata*

*Stackhousia viminea*

# Stackhousia — Stackhousiaceae

## Stackhousia monogyna

Candles, Creamy Stackhousiana — **P N**

**Size:** 10–30 cm x 10–30 cm
**Habitat:** Widespread in grassy wetlands, plains grassland, red gum woodland, wet, damp, valley and dry sclerophyll forests, grassy low open forest
**Form:** Erect herbaceous perennial
**Foliage:** Smooth narrow leaves to 40 mm long, often widely spaced on the stems
**Flowers:** Many tiny cream tubular flowers in a candle-like spike at the end of each stem; August to January. The perfume is especially noticeable at night.
**Requirements:** Moist well drained soil
**Comments:** Inconspicuous when not in flower, it usually grows in patches creating a massed display. An excellent rockery plant providing it is not allowed to dry out. New growth is often tinged with red.
**Propagation:** Cuttings from new leafy stems emerging from the rootstock.
**Localities:** Eltham, 4, 5, 7, 12, 16, 26, 29#, 30#, 34A–37, 41–44, 46–59, 61–64, 67#, 71, 74, 76
**Distribution:** Vic, Qld, NSW, Tas, SA

## Stackhousia spathulata

Coast Stackhousia — **F P**

**Size:** 15–30 cm x 10–30 cm
**Habitat:** Primary dune scrub
**Foliage:** Similar in form to *S. monogyna*, its leaves are thicker, broader and blunt on the ends, to 30 mm long
**Flowers:** The cream flowers are densely crowded in spikes and have a broad subtending bract; July to January
**Requirements:** Well drained sandy soil, sometimes growing on the beach
**Comments:** Now locally very rare.
**Localities:** Werribee, 67#
**Distribution:** Vic, Qld, NSW, Tas, SA

## Stackhousia viminea

Slender Stackhousia — **F P**

**Size:** 0.3–0.6 m x 0.3 m
**Habitat:** Tea-tree heath
**Form:** Similar to *S. monogyna*, spreading to form a clump
**Foliage:** Narrow, elliptic to wedge-shaped leaves to 30 mm long at the base of the stems. Leaves are sparse or non-existent on upper parts of stems.
**Flowers:** Small greenish-yellow flowers in clusters of 1–6 along the upper part of the stem, forming a dense or sparse spike. Plants with the denser spikes were known as *S. flava*.
**Localities:** Clayton, 66, 74, 75
**Distribution:** All states
**Synonym:** *S. flava*

# Stellaria — Caryophyllaceae

Stellaria is a genus of mainly herbaceous perennials, with stems which are thickened where the pairs of leaves are attached. Plants are generally very soft in their growth and sprawl along the ground unless support is available from other plants, when they may climb weakly. The white star-like flowers superficially resemble those of daisies, having 5 petals deeply lobed to the base, so that there appears to be 10. Petals are absent in some species. Each flower is 15–25 mm across, and the plants are quite showy, brightening dark shady corners.

## Stellaria flaccida
Forest Starwort

**P N**

**Size:** 10–50 cm x 0.5–1 m
**Habitat:** Wet, damp and valley sclerophyll forests
**Form:** Sprawling perennial herb
**Foliage:** Pairs of widely-spaced, bright green, ovate leaves to 25 mm long
**Flowers:** White starry flowers; September to November
**Requirements:** Moist to wet soils
**Comments:** A spreading ground cover for a poorly drained area. It may become invasive in ideal conditions.
**Propagation:** Cuttings, division
**Localities:** 55–58
**Distribution:** Vic, Qld, NSW, Tas

*Stellaria flaccida*

## Stellaria multiflora
Rayless Starwort

**F P**

**Size:** 25–75 mm high
**Habitat:** Exposed cliff tops
**Form:** Sprawling annual herb
**Foliage:** Lanceolate leaves to 5 mm long
**Flowers:** Flowers have small sepals but lack petals; September to October
**Requirements:** Grows in well drained, disturbed soils
**Comments:** Now very rare or extinct within the Melbourne area.
**Localities:** St. Kilda#, 67#
**Distribution:** Vic, NSW, Tas, SA, WA

## Stellaria palustris var. palustris
Swamp Starwort

**P**

**Size:** Sprawling to 1.2 m
**Habitat:** Grassy wetlands, tea-tree heath
**Form:** Perennial herb with weak, whitish stems
**Foliage:** Sessile, widely-spaced, narrow-lanceolate leaves to 40 mm long
**Flowers:** White axillary flowers on long stalks to 70 mm long; October to December
**Requirements:** Tolerates waterlogging. Locally rare.
**Localities:** Laverton Nth, Sunshine, 74
**Distribution:** Vic, NSW, Tas, SA

*Stellaria palustris* var. *palustris*

## Stellaria pungens
Prickly Starwort

**F P**

**Size:** 10–30 cm x 0.5 m
**Habitat:** Riparian scrub, box and red gum woodland, damp and valley sclerophyll forests, primary dune scrub
**Form:** Perennial, suckering herb with hairy, tangled stems forming a loose mat
**Foliage:** Bright green, sessile, narrow-lanceolate leaves to 15 mm long, rigid and prickly and often in clusters
**Flowers:** Single white flowers to 20 mm wide; October to December
**Requirements:** Well drained moist soil
**Comments:** May dry out and look untidy in hot weather, but is readily rejuvenated by heavy pruning and watering.
**Localities:** 14, 16, 26, 29, 32, 34A–36, 39, 41, 42, 55, 56, 58, 60, 67#, 71
**Distribution:** Vic, Qld, NSW, Tas, SA

*Stellaria pungens*

# Stenopetalum
# Brassicaceae

## Stenopetalum lineare
Narrow Thread-petal

**F**

**Size:** 0.3–0.5 m high
**Habitat:** Plains grasslands, exposed coastal cliffs
**Form:** Slender, erect, branched annual herb
**Foliage:** Basal leaves to 50 mm long with 3 or more narrow, entire or toothed lobes; stem leaves to 70 mm long, lobed, entire or thread-like
**Flowers:** Brown to yellow flowers, the petals thread-like, hanging in loose clusters along the upper stem; September to November. Oblong pods held on erect stalks.
**Requirements:** Well drained soil, tolerating dryness
**Comments:** The upper branches are covered in stellate or irregularly-branched hairs. The plants and seeds, which have a taste of mustard, were eaten by the Aborigines. Not collected for some time, it may have become extinct within the Melbourne area.
**Propagation:** Seed
**Localities:** Werribee#, 67#
**Distribution:** All states

*Stenopetalum lineare*

SCALE: 0 1cm 2cm The scale representation on each plant equals 2cm.

*Stuartina muelleri*

# Stuartina — Asteraceae

## Stuartina muelleri — Spoon Cudweed

| | |
|---|---|
| **Size:** | Prostrate-12 cm x 10–30 cm — F |
| **Habitat:** | Coastal cliffs, chenopod rocky open scrub, primary dune scrub, tea-tree heath |
| **Form:** | Erect or prostrate, spreading annual herb with woolly hairs |
| **Foliage:** | Soft, greyish-green spathulate leaves to 18 mm long, whitish below |
| **Flowers:** | Small terminal clusters of conical woolly, greenish-brown or whitish flowerheads, lacking ray florets; September to November |
| **Requirements:** | Well drained soil, responding to extra moisture |
| **Comments:** | Short-lived plants which germinate quickly following Autumn rains. Locally rare. |
| **Propagation:** | Seed, cuttings |
| **Localities:** | 34A, 67#, 74 |
| **Distribution:** | Vic, Qld, NSW, SA, WA; NZ |

# Stylidium — Stylidiaceae

Known as "trigger" plants because of their unique pollination method. The column is suddenly released when an insect lands on its base, hitting the insect and leaving pollen on its back. Flowering stems in the Melbourne spp. are held above tufts or rosettes of leaves. Each flower has 2 pairs of petals. Stylidiums make interesting container plants. Propagation is by seed, which does not set in great number.

*Stylidium beaugleholei*

## Stylidium beaugleholei — Beauglehole's Trigger-plant

| | |
|---|---|
| **Size:** | 25–100 mm high — P |
| **Habitat:** | Shallow seasonal swamps of tea-tree heath |
| **Form:** | Weak, short-lived annual with few basal leaves to 5 mm long |
| **Flowers:** | Pale pink fan-shaped flowers have a linear calyx tube and a deeper pink stripe on outside edge; October to January. |
| **Localities:** | 74 |
| **Distribution:** | Vic, SA, WA |

## Stylidium despectum — Dwarf Trigger-plant

| | |
|---|---|
| **Size:** | 15–80 mm high — F P |
| **Habitat:** | Swamp flats of riparian scrub |
| **Form:** | Short-lived erect herb |
| **Foliage:** | Tiny, narrowly-ovate leaves to 4 mm long, alternate, not in a rosette |
| **Flowers:** | Inconspicuous white or pale pink flowers on a long stalk, singly or in a raceme, petals paired vertically, calyx tube linear; October to December |
| **Requirements:** | Wet soils |
| **Comments:** | Differs from *S. inundatum* in petal orientation and lack of rosette. |
| **Localities:** | It was collected in 1896 from along the Merri Ck; 74 |
| **Distribution:** | Vic, NSW, Tas, SA, WA |

*Stylidium graminifolium* var. *graminifolium*

## Stylidium graminifolium var. graminifolium — Grass Trigger-plant

| | |
|---|---|
| **Size:** | 0.2–0.6 m x 20–30 cm — F P |
| **Habitat:** | Widespread in plains grassland, red gum woodland, dry and valley sclerophyll forests, grassy low open forest |
| **Form:** | Basally tufted perennial |
| **Foliage:** | Narrow, green to grey-green grass-like leaves 5–30 cm long, standing erect |
| **Flowers:** | Flower stem ends in a spike of many small pale to bright pink flowers; September to December |
| **Requirements:** | Moist well drained soils, tolerating both wet and dry periods once established |
| **Comments:** | A very variable plant in size, leaf and flower colour. The alpine form has deep magenta flowers. An excellent rockery plant. Plants should be used in drifts for best effect. |
| **Localities:** | 12, 26, 33–36, 38, 39, 41–44, 47–54, 59–64, 67, 69, 69A, 71, 74, 75 |
| **Distribution:** | Vic, Qld, NSW, Tas, SA |

## Stylidium graminifolium var. angustifolium

| | |
|---|---|
| **Habitat:** | Wet, damp, valley and dry sclerophyll forests |
| **Comments:** | Leaves are very narrow-linear from 15–30 cm long but no more than 2 mm wide. Flowering stem is a little shorter than above. |
| **Localities:** | 37, 55–58 |
| **Distribution:** | Endemic to Vic |

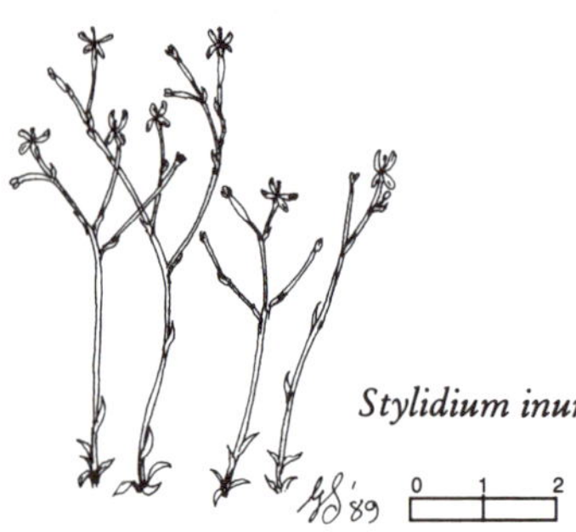

*Stylidium inundatum*

## Stylidium inundatum — Hundreds and Thousands

| | |
|---|---|
| **Size:** | 15–50 mm high — F P |
| **Habitat:** | Swamp scrub, tea-tree heath |
| **Foliage:** | Similar to *S. despectum*. When present, basal rosette of narrow leaves to 6 mm long; stem leaves ovate to 4 mm long. |
| **Flowers:** | Petals of inconspicuous flowers are paired laterally, calyx tube linear; September to November. |
| **Localities:** | 62, 70, 75 |
| **Distribution:** | Vic, Tas, SA, WA |

### Stylidium perpusillum
Tiny or Slender Trigger-plant

**Size:** 15–40 mm high **F P**  
**Habitat:** Tea-tree heath  
**Form:** Tiny slender annual herb  
**Foliage:** Few narrow leaves to 3 mm long in a basal rosette  
**Flowers:** Tiny white flowers, singly or 2–3, on a long stalk to 30 mm long, calyx tube globular; October to November.  
**Requirements:** Moist sandy soil.  
**Localities:** 67#, 74, 75  
**Distribution:** Vic, NSW, Tas, SA, WA

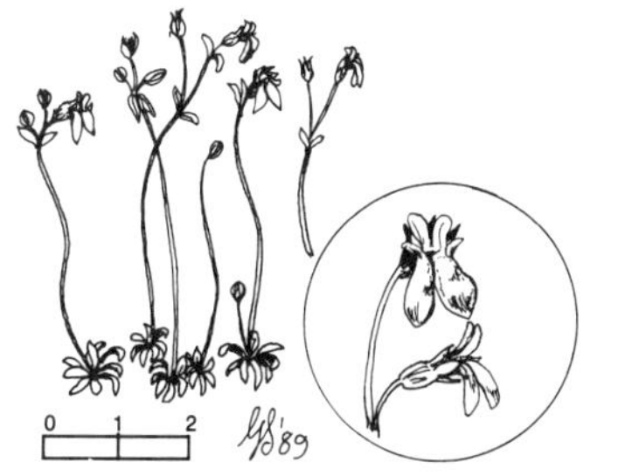

*Stylidium perpusillum*

## Suaeda
Chenopodiaceae

### Suaeda australis
Austral Seablite

**Size:** 0.15–0.8 m x 0.3–1 m **F**  
**Habitat:** Saltmarshes, coastal banksia woodland  
**Form:** Rounded perennial shrub, branching from the base  
**Foliage:** Succulent, narrow blue-green to red leaves to 30 mm long, becoming smaller up the stem  
**Flowers:** Axillary clusters of minute greenish flowers followed by succulent, round fruit. Both are found all year, peaking January to June.  
**Requirements:** Moist, saline soils  
**Propagation:** Cuttings  
**Localities:** Footscray, 2–4, 7, 67, 74  
**Distribution:** All states except NT

*Suaeda australis*

## Swainsona
Fabaceae

### Swainsona lessertiifolia
Coast or Purple Swainson-pea, Poison Pea

**Size:** 10–30 cm x 0.4–1 m **F P**  
**Habitat:** Primary dune scrub  
**Form:** Sprawling or scrambling perennial herb  
**Foliage:** Dark green pinnate leaves to 90 mm long with 13–19 oblong leaflets, softly hairy below  
**Flowers:** Spikes of bright purple pea flowers with a bright yellow mark at the base; June to October, at other times following rain  
**Requirements:** Well drained sandy soil, tolerating dryness once established  
**Comments:** Copes well with lime and salt spray but may be frost tender. It needs to be cut back each winter to encourage vigorous new growth. An endangered plant in the Melbourne area.  
**Propagation:** Scarified seed  
**Localities:** Port Phillip Bay area, Braybrook#, Seaford  
**Distribution:** Vic, Tas, SA

*Swainsona lessertiifolia*

## Templetonia
Fabaceae

### Templetonia stenophylla
Leafy Templetonia

**Size:** Prostrate-30 cm x 20–50 cm **F P**  
**Habitat:** Box woodland  
**Form:** Trailing or arching shrub with ridged to quadrangular stems  
**Foliage:** Sparse; narrow-oblong lower leaves, upper leaves linear, to 60 mm long  
**Flowers:** Showy large yellow and reddish pea flowers solitary or in pairs in leaf axils or in small terminal clusters; August to November. Stalks have 2 bracteoles near the middle. Pods flat to 20 mm long.  
**Requirements:** Well drained soils tolerating dryness once established  
**Comments:** Known to occur in the Melbourne area at two sites south of Melton, it is becoming depleted throughout Victoria.  
**Propagation:** Scarified seed  
**Localities:** Melton South, 16#  
**Distribution:** Vic, Qld, NSW, SA

## Tetragonia
Aizoaceae

### Tetragonia implexicoma
Bower Spinach

**Size:** Prostrate or straggling climber 0.3–3 m x 2 m **F P**  
**Habitat:** Coastal banksia woodland, primary dune scrub, saltmarshes  
**Form:** Succulent, trailing or scrambling shrub  
**Foliage:** Clusters of thick, bright green broadly-lanceolate to diamond-shaped leaves 20–80 mm long  
**Flowers:** Single, scented small pale-yellow flowers on long stalks; August to February. Succulent reddish berry ripens to black.  
**Requirements:** Well drained sandy soil  
**Comments:** Useful for stabilising sand dunes, it is very common along beach foreshores. Plant at 1 m intervals for dense cover. The young shoots and leaves were cooked as a vegetable by the Aborigines.  
**Propagation:** Fresh seed, cuttings  
**Localities:** Melton Sth, Werribee R, 2–5, 67, 69, 69A, 74, 76  
**Distribution:** Vic, Tas, SA, WA

*Tetragonia implexicoma*

SCALE: 0 1cm 2cm The scale representation on each plant equals 2cm.

### Tetragonia tetragonioides

New Zealand Spinach

| | |
|---|---|
| **Size:** | Prostrate x 1–2 m **F** |
| **Habitat:** | Primary dune scrub |
| **Form:** | Small spreading succulent herb similar to *T. implexicoma*. The fleshy stems become woody with age. |
| **Flowers:** | Almost sessile small green flowers, singly or in pairs in leaf axils; November to February. Longer, ribbed fruit are green and succulent but become hard and pale brown. |
| **Requirements:** | Moist, well drained soil |
| **Comments:** | An important food source for the Aborigines and early settlers. |
| **Localities:** | 67#, 69A |
| **Distribution:** | All states except NT; Pacific Ocean Countries |

*Tetratheca bauerifolia*

## Tetratheca — Tremandraceae

### Tetratheca bauerifolia

Heath Pink-bells

| | |
|---|---|
| **Size:** | 20–40 cm x 20–30 cm **P** |
| **Habitat:** | Valley sclerophyll forest |
| **Form:** | Small, erect or rounded shrub |
| **Foliage:** | Whorls of 4–6 linear to lanceolate leaves to 8 mm long, margins recurved, paler below |
| **Flowers:** | Single, rosy to dark pink bell-like flowers on long, fine red stalks in upper whorls; August to December. The stalks are glabrous and distinctly hooked above the flower. |
| **Requirements:** | Moist well drained soils |
| **Comments:** | Prefers morning sun and is attractive in flower. |
| **Propagation:** | Cuttings |
| **Localities:** | ?42 |
| **Distribution:** | Vic, NSW, SA |

*Tetratheca ciliata*

### Tetratheca ciliata

Pink Bells, Black-eyed Susan

| | |
|---|---|
| **Size:** | 0.3–0.6 m x 0.3–0.6 m **P N** |
| **Habitat:** | Damp, valley and dry sclerophyll forests, grassy low open forest |
| **Form:** | An erect or, more often, spreading clumping shrub |
| **Foliage:** | Oval leaves to 12 mm long in whorls of 3–4 distinctly fringed with hairs |
| **Flowers:** | Fragrant mauve-pink, magenta or white flowers hang profusely from short glandular hairy stalks, petals spread as they age; July to December |
| **Requirements:** | Well drained soil responding to extra moisture in Summer |
| **Comments:** | Tolerates drier and heavier soils and lime. An attractive container specimen. |
| **Localities:** | 34A, 35, 37, 38, 41–44, 47, 49–56, 58, 61, 63, 65#, 67#, 74–76 |
| **Distribution:** | Vic, NSW, Tas, SA |

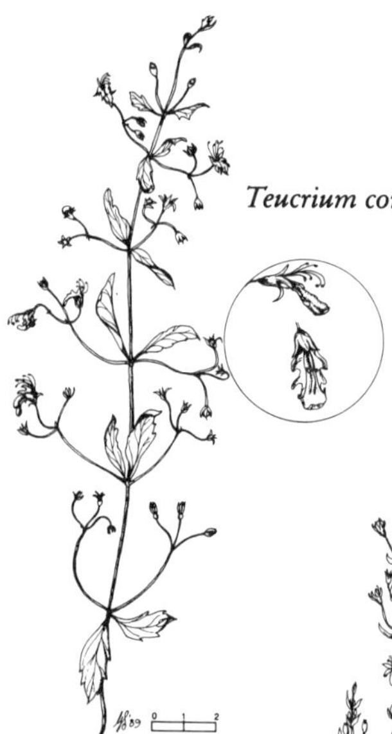

*Teucrium corymbosum*

## Teucrium — Lamiaceae

### Teucrium corymbosum

Forest Germander

| | |
|---|---|
| **Size:** | 0.3–1 m high **P N** |
| **Habitat:** | Valley sclerophyll forest |
| **Form:** | Erect perennial herb with rigid, scaly quadrangular stems |
| **Foliage:** | Pairs of ovate leaves to 50 mm long, almost glabrous above, grey hairs below, with coarsely toothed margins |
| **Flowers:** | Loose terminal spikes of white, 2-lipped flowers, the upper lip cut in two, lower lip 3-lobed, stamens are prominently exserted from the flower; December to February |
| **Requirements:** | Moist, well drained soils |
| **Comments:** | Regular pruning will maintain a tidy shape. |
| **Propagation:** | Cuttings |
| **Localities:** | Sth Morang, Bundoora, Diamond Creek, Kew, 36, 57 |
| **Distribution:** | Vic, Qld, NSW, Tas, SA |

*Teucrium racemosum*

### Teucrium racemosum

Grey Germander

| | |
|---|---|
| **Size:** | 0.4–0.6 m x 0.5–1 m **F** |
| **Habitat:** | Grassy wetlands, red gum woodland |
| **Form:** | Grey and woolly, erect perennial herb with quadrangular stems |
| **Foliage:** | Pairs of shortly-stalked, silvery, lanceolate leaves to 25 mm long, with entire, often wavy margins |
| **Flowers:** | Terminal racemes of white, 2-lipped flowers, lobed as *T. corymbosum*; November to February |
| **Requirements:** | Moist well drained soils, tolerating occasional inundation |
| **Comments:** | May sucker forming clumps. The grey foliage provides a good contrast in a damp area. |
| **Propagation:** | Cuttings, division |
| **Localities:** | Laverton, 10 |
| **Distribution:** | All mainland states |

# Thesium **Santalaceae**

## Thesium australe — Austral Toad-flax

**Size:** 30 cm high — F
**Habitat:** Plains grasslands
**Form:** Wiry, erect perennial herb
**Foliage:** Linear, yellow-green leaves to 30 mm long
**Flowers:** Minute, greenish-yellow tubular flowers singly on very short stalks in leaf axils; February to March
**Requirements:** Heavy soils
**Comments:** Extinct within the Melbourne area as a result of heavy grazing and grassland cultivation, it is known from only 3 sites in far East Gippsland. It is presumed extinct elsewhere in Australia.
**Propagation:** Seeds—difficult due to its semi-parasitic reliance on grassland herbs.
**Localities:** Once occurred on the basalt plains, Braybrook#
**Distribution:** Vic, Qld, NSW, Tas

*Thesium australe*

# Threlkeldia **Chenopodiaceae**

## Threlkeldia diffusa — Wallaby Saltbush, Coast Bonefruit

**Size:** Prostrate–30 cm x 1 m — F
**Habitat:** Saltmarshes, primary dune scrub
**Form:** Spreading, succulent perennial herb, ribbed lower branches
**Foliage:** Narrow, nearly cylindrical grey-green leaves to 15 mm long
**Flowers:** Tiny tubular flowers singly in leaf axils; May to November. Fruit are succulent and often reddish; Summer.
**Requirements:** Moist saline soils
**Comments:** A matting plant for coastal conditions.
**Propagation:** Seed
**Localities:** Seaford, 5, 74
**Distribution:** Vic, NSW, Tas, SA, WA

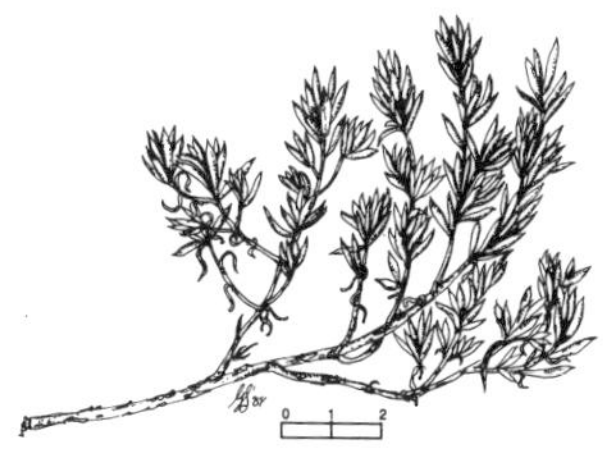

*Threlkeldia diffusa*

# Trachymene **Apiaceae**

## Trachymene anisocarpa — Wild Parsnip, Parsnip Trachymene

**Size:** stem 0.3–1.5 m high — F P
**Habitat:** Tea-tree heath
**Form:** Stoutly erect, leafy annual or biennial herb, stem finely grooved
**Foliage:** Hairy leaves 20–100 mm long, mostly in a basal rosette, deeply divided into 3 wedge-shaped, lobed segments
**Flowers:** Umbels of up to 50 lacy white flowers on long stalks to 80 mm long; September to January. Leafy flower stem held erect, up to 1.5 m high.
**Requirements:** Sandy, well drained soil
**Comments:** Fast growing, it regenerates well after fire. Most attractive in flower.
**Propagation:** Seed
**Localities:** Port Melbourne#, St. Kilda#, 67, 71, 73–75
**Distribution:** Vic, NSW, Tas, SA, WA

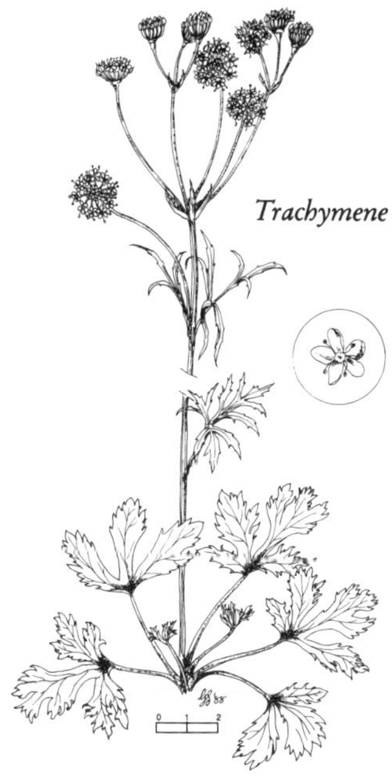

*Trachymene anisocarpa*

# Triptilodiscus **Asteraceae**

## Triptilodiscus pygmaeus — Common Sunray

**Size:** 2–12 cm x 2–12 cm — F P
**Habitat:** Red gum and box woodlands, dry sclerophyll forest
**Form:** Tiny erect annual herb
**Foliage:** Light green linear leaves to 30 mm long, covered in long scattered hairs. Leaf margins somewhat crinkled.
**Flowers:** Sessile, yellow, urn-shaped flowerheads lacking ray florets, with papery white and green bracts around the base; September to November. Flowerheads terminal or in leaf axils surrounded by leaves.
**Requirements:** Well drained soils tolerating some dryness, but responding to Summer watering, which lengthens its flowering season
**Comments:** Easily overlooked amongst herbs and grasses, it is more common in good seasons. Makes an interesting container plant.
**Propagation:** Seed, cuttings
**Localities:** Balwyn, 7, 29, 32, 35, 42, 67#
**Distribution:** Vic, Qld, NSW, SA, WA
**Synonym:** *Helipterum australe*

*Triptilodiscus pygmaeus*

SCALE: 0 1cm 2cm — The scale representation on each plant equals 2cm.

*Urtica incisa*

# Urtica — Urticaceae

## Urtica incisa

Scrub or Stinging Nettle

| | |
|---|---|
| **Size:** | 0.6–1 m high **F P N** |
| **Habitat:** | Riparian scrub and woodland, box woodland, damp sclerophyll forest |
| **Form:** | Upright, scrambling perennial herb covered with brittle stinging hairs |
| **Foliage:** | Pairs of coarsely-toothed, dark green leaves to 80 mm long, on long thin stalks |
| **Flowers:** | Loose slender spikes of small green flowers, male and female flowers in separate clusters along the spike; Winter to early Summer. Fruit is a small nut. |
| **Requirements:** | Adaptable to most sites but prefers moist, shaded conditions |
| **Comments:** | Often a coloniser of disturbed sites, particularly if plenty of moisture is available. The coarse hairs produce a very painful sting. The Aborigines used the leaves to treat rheumatism and sprains. The leaves were also cooked when food was scarce. |
| **Propagation:** | Seed |
| **Localities:** | Tullamarine, 14, 16, 26, 34A, 35, 42, 57, 58 |
| **Distribution:** | Vic, Qld, NSW, Tas, SA |

# Utricularia — Lentibulariaceae

These tiny carnivorous herbs of wet places are known as Bladderworts due to the submerged globular bladder-like trap which sucks in minute aquatic animals. The lower petal of the flower forms an enlarged lip with a pouch-like spur extending from the base of the petal downward but shorter than the lip. *Utricularia* spp. make interesting and unusual pot specimens, requiring light soil, with the pot placed in a saucer of water. Propagation is by seed.

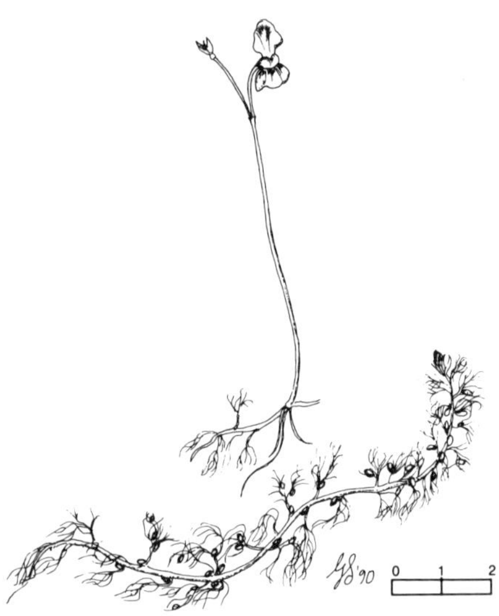

*Utricularia australis*

## Utricularia australis

Yellow Bladderwort

| | |
|---|---|
| **Size:** | Stems 0.3–0.9 m long, depending on water depth **P N** |
| **Habitat:** | Aquatic in tea-tree heath, red gum woodland |
| **Form:** | Submerged, floating rootless perennial herb with zig-zag stems |
| **Foliage:** | Pinnate leaves divided into 2 branches at the base, 10–14 alternate leaflets divided into thread-like segments; bladders to 4 mm long at the base of the leaflets |
| **Flowers:** | Spray of 2–8 yellow flowers on an erect, emergent stalk, lower lip to 18 mm wide; November |
| **Requirements:** | Fresh, still water to 0.6 m deep |
| **Comments:** | Locally rare. |
| **Localities:** | Bundoora, 74 |
| **Distribution:** | All states except NT; NZ, PNG, Af, Asia, Eur |

## Utricularia dichotoma

Fairies Aprons

| | |
|---|---|
| **Size:** | 8–30 cm high **P N** |
| **Habitat:** | Swamp margins of riparian and swamp scrub, valley sclerophyll forest, tea-tree heath |
| **Form:** | Rosette with creeping stems |
| **Foliage:** | Linear to spathulate leaves 5–25 mm long. Tiny bladders to 1.5 mm long on some modified immersed leaves. |
| **Flowers:** | Single or paired mauve to purple flowers with yellow throats on fine upright stalks; October to January. Lower lip to 20 mm wide. |
| **Requirements:** | Shallow water or mud |
| **Comments:** | Once widespread but becoming rarer in Melbourne through destruction of its habitat, it is now found in few sites. Well worth growing in a container. A yellow flowered form occurs in Frankston. |
| **Localities:** | Sunshine#, Melton#, Kilsyth, 16, 26, 49#, 61–64, 67#, 74, 75 |
| **Distribution:** | All states except NT |

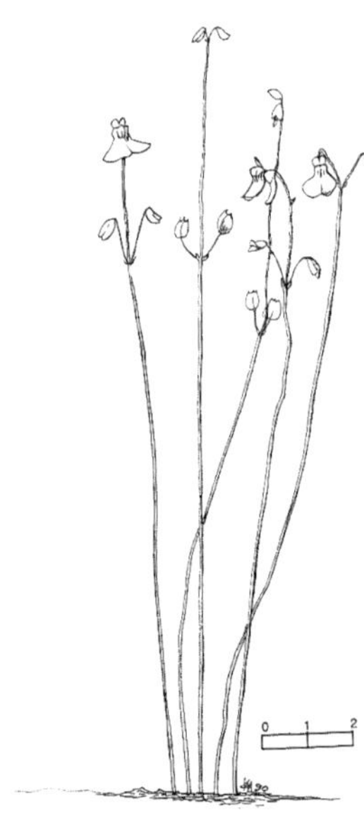

*Utricularia dichotoma*

## Utricularia sp. aff. dichotoma

| | |
|---|---|
| **Habitat:** | Grassy wetlands |
| **Requirements:** | Shallow water or mud |
| **Comments:** | An undescribed species with similarities to *U. dichotoma*. A distinguishing feature is found on the inflated section (or palate) at the base of the lower corolla lip. While this species has 4–10 slightly raised ridges on the palate, the palate of *U. dichotoma* has fewer, more strongly raised ridges. It has a disjunct distribution, mainly occurring in Western Victoria. |
| **Propagation:** | Seed |
| **Localities:** | Western suburbs |
| **Distribution:** | Vic |

## Utricularia laterifolia

Small Fairies Aprons

| | |
|---|---|
| **Size:** | to 10 cm high **P N** |
| **Habitat:** | Swamp margins of tea-tree heath |
| **Comments:** | Tiny annual herb, similar to *U. dichotoma*. The lower lip of the pale mauve flowers is 6 mm across and the flower stalks are shorter. An endangered plant in the Melbourne area. |
| **Localities:** | Moorabbin#, Clayton#, 49#, 69#, 74, 75 |
| **Distribution:** | Vic, Qld, NSW, Tas, SA; NZ |

## Utricularia tenella

Pink Bladderwort, Pink Fans

| | |
|---|---|
| **Size:** | 20–75 mm high **P** |
| **Habitat:** | Swamp margins of tea-tree heath |
| **Form:** | Tiny herb with tufted stolons |
| **Foliage:** | Narrow spathulate leaves to 6 mm long |
| **Flowers:** | 1–2 pink flowers with a yellow or white throat on a fine erect stalk, the upper lip deeply divided into tapered lobes, the lower lip to 5 mm long with 3 blunt lobes. The spur is almost as long as the lower lip. This species has 4 sepals. The other species each have 2 sepals. |
| **Localities:** | 49#, 67#, 74, 75 |
| **Distribution:** | Vic, Tas, SA, WA |
| **Synonym:** | *Polypompholyx tenella* |

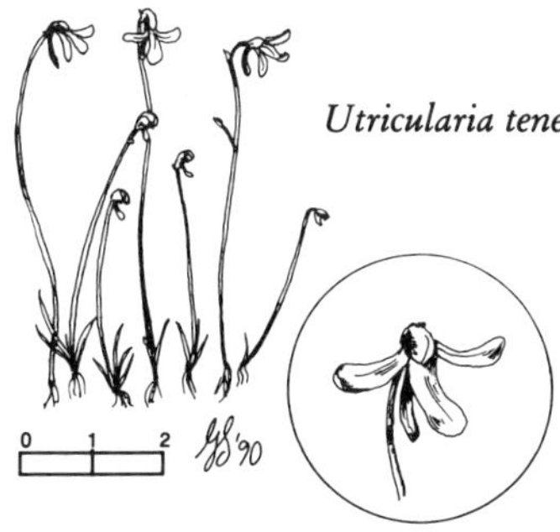

*Utricularia tenella*

# Velleia

# Goodeniaceae

## Velleia paradoxa

Spur Velleia

| | |
|---|---|
| **Size:** | Prostrate-0.6 m x 0.3 m **F P N** |
| **Habitat:** | Grassy wetlands, plains grassland, riparian scrub, red gum woodland |
| **Form:** | Erect perennial herb with forked stems |
| **Foliage:** | Basal rosette of bluntly-toothed, broadly oblanceolate light green leaves to 60 mm long; slightly hairy. Pairs of small toothed or lobed bracts at each fork. |
| **Flowers:** | Widely branched, long stalks with open, 5-petalled yellow flowers to 20 mm wide; October to November, March to May. A conspicuous spur is at the base of the flower. |
| **Requirements:** | Moist well drained soils, tolerating some wet periods |
| **Comments:** | A showy ground cover for a moist place. Will germinate readily in cultivation. It is similar to *Goodenia pinnatifida* which has smaller bracts at the forks and more deeply lobed or toothed leaves. |
| **Propagation:** | Seed |
| **Localities:** | St. Albans, Brighton#, Hampton#, 7, 10, 12, 16, 17#, 18, 25, 26, 29, 34A, 42, 67 |
| **Distribution:** | All states except NT |

*Velleia paradoxa*

# Veronica

# Scrophulariaceae

Perennial herbs of moist shady situations, *Veronica* spp. make useful groundcovers. The four blue to mauve petals join to form a tube at the base. Fruit is a flat capsule. Pruning after flowering encourages a denser habit. Propagate by division or cuttings.

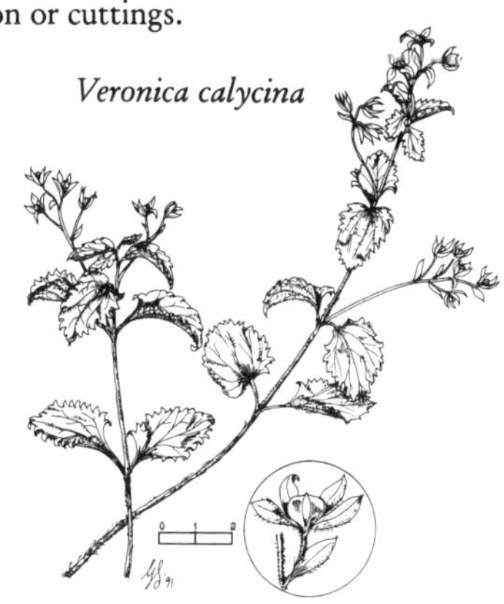

*Veronica calycina*

## Veronica calycina

Cup or Hairy Speedwell

| | |
|---|---|
| **Size:** | Prostrate–20 cm x 0.5 m **P N** |
| **Habitat:** | Widespread in wet, damp, valley and dry sclerophyll forests, riparian scrub, grassy low open forest |
| **Form:** | Erect or trailing stoloniferous herb, covered with long fine hairs |
| **Foliage:** | Pairs of dark green ovate leaves to 25 mm long with rounded teeth |
| **Flowers:** | Small, loose clusters of pale blue flowers in upper leaf axils; October to December |
| **Requirements:** | Moist soils, tolerating periods of poor drainage |
| **Comments:** | May become invasive in ideal conditions. |
| **Localities:** | 29, 34A–36, 38, 39, 41, 42, 51, 55–58, 71, 74, 76 |
| **Distribution:** | All states except NT |

## Veronica gracilis

Slender Speedwell

| | |
|---|---|
| **Size:** | 15–30 cm x 1 m **P N** |
| **Habitat:** | Widespread in grassy wetlands, riparian scrub, red gum woodland, dry, valley and damp sclerophyll forests, tea-tree heath |
| **Form:** | Slender, erect or trailing herb |
| **Foliage:** | Pairs of bright green, narrow lanceolate leaves to 20 mm long |
| **Flowers:** | Mauve to pale blue cup flowers with darker veins in small axillary clusters; September to December |
| **Requirements:** | Moist, well drained soils |
| **Comments:** | Roots freely from underground stems. |
| **Localities:** | 2, 4, 5, 16, 17#, 18, 26, 28–30, 32, 34A–36, 38, 39, 41, 42, 44, 47, 49, 51, 54, 56, 58, 59, 61–64, 67, 74, 75 |
| **Distribution:** | Vic, NSW, Tas, SA |

*Veronica gracilis*

SCALE: 0 1cm 2cm — The scale representation on each plant equals 2cm.

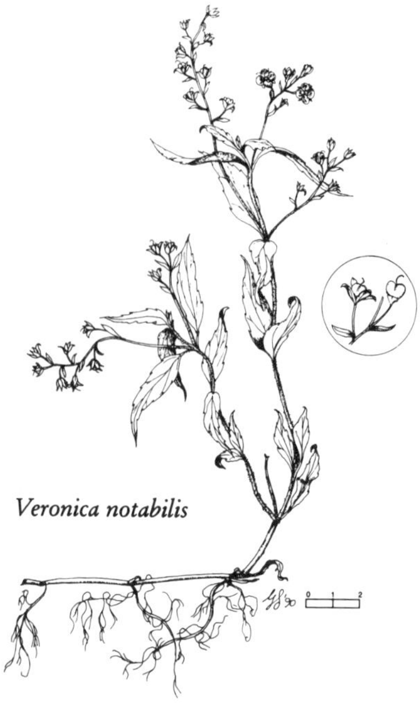

*Veronica notabilis*

## Veronica notabilis

Forest Speedwell

**Size:** Prostrate-0.8 m x 0.3-1 m **P N**
**Habitat:** Damp and wet sclerophyll forests
**Form:** Spreading perennial herb with upright stems suckering lightly from underground stems
**Foliage:** Pairs of dull green broad-lanceolate leaves to 50 mm long have impressed veins
**Flowers:** Small white to pale blue flowers with purple veins in loose terminal sprays; September to January
**Requirements:** Cool moist soils
**Comments:** The whole plant is softly hairy, especially the stems. Locally rare.
**Localities:** Kalorama, 57
**Distribution:** Vic, Qld, NSW, Tas

## Veronica plebeia

Creeping or Trailing Speedwell

**Size:** Prostrate x 2 m **F P N**
**Habitat:** Damp and valley sclerophyll forests, riparian woodland, grassy low open forest
**Form:** Creeping herb
**Foliage:** Dark green, ovate to heart-shaped coarsely-toothed leaves to 25 mm long but usually smaller
**Flowers:** Loose clusters of 2–10 violet flowers in upper leaf axils; September to January
**Requirements:** Moist well drained soils
**Comments:** Spreads readily in ideal conditions, and makes an attractive groundcover in shaded gardens.
**Localities:** Research, Kalorama, 34A, 36, 41, 42, 51, 60, 71, 74–76
**Distribution:** All states except NT

*Illustrated p. 200*

*Villarsia reniformis*

# Villarsia — Menyanthaceae

## Villarsia exaltata

Erect or Yellow Marsh Flower

**Size:** 0.3 m x 1 m, flower stalks may grow to 1.5 m **F P**
**Habitat:** Aquatic in swampy tea-tree heath
**Form:** Tufted herb, the broad basal leaves floating if plant is growing in water
**Foliage:** Shiny oval to lanceolate leaves to 80 mm long on long stalks, length determined by water depth
**Flowers:** Yellow 5-petalled flowers on long branching flowering stems held above the leaves; November to January
**Requirements:** Wet soils or water to a depth of 1 m
**Comments:** An excellent water plant for a larger pond or dam.
**Propagation:** Seed, division
**Localities:** 49#, 73–75
**Distribution:** Vic, Qld, NSW, Tas

*Illustrated p. 200*

## Villarsia reniformis

Running Marsh Flower

**Size:** 0.4 m x 0.5–1 m **F P N**
**Habitat:** Swamps or shallow water to 0.6 m deep, in swamp scrub, red gum woodland, tea-tree heath
**Form:** Tufted, or stoloniferous if growing in water
**Foliage:** Tufts of shiny dark green kidney-shaped leaves to 100 mm long, on long stalks which float if aquatic
**Flowers:** Bright yellow flowers on stems to 1 m high; September to November
**Comments:** A most attractive plant for shallow water or damp soil beside a pond.
**Localities:** Black Rock, 35, 54, 60, 66, 67#, 69–70, 75
**Distribution:** Vic, NSW, Tas, SA

*Viminaria juncea*

# Viminaria — Fabaceae

## Viminaria juncea

Native Broom, Golden Spray

**Size:** 2.5–5 m x 2 m **F P**
**Habitat:** Widespread in red gum woodland, riparian and swamp scrub, valley sclerophyll forest, tea-tree heath, grassy low open forest
**Form:** Slender erect, usually leafless shrub with pendulous branches
**Foliage:** Long flexible needle-like branchlets to 30 cm long
**Flowers:** Long drooping sprays of yellow pea flowers; October to February
**Requirements:** Adaptable to poorly drained soils, tolerating salt and lime
**Comments:** A fast-growing shrub which is most attractive in flower, providing an excellent substitute for introduced broom species. Its open weeping habit is used to best advantage beside pools, where allowed to hang over the water.
**Propagation:** Scarified seed
**Localities:** 16, 26, 28, 34A, 42, 49, 54, 60–64, 66#–68, 70, 74–76
**Distribution:** All states except NT

# Viola — Violaceae

## Viola betonicifolia

Showy or Mountain Violet

**P N**

**Size:** 15 cm x 30 cm
**Habitat:** Dry, valley and damp sclerophyll forests, grassy low open forest
**Form:** Erect tufted perennial herb
**Foliage:** Dark green triangular-ovate to lanceolate leaves to 80 mm long, on long stalks to 140 mm. Local form from the Ringwood area has much smaller, roundish to ovate leaves to 30 mm long
**Flowers:** Large purplish-blue to purple flowers to 20 mm across, borne singly on long stems to 100 mm; September to December
**Requirements:** Moist well drained soil
**Comments:** Seeds germinate freely in the garden. An attractive and adaptable plant which also thrives in protected coastal situations.
**Propagation:** Seed
**Localities:** Mordialloc 1892; 29#, 34A, 38, 42, 44, 46, 47, 49, 50A#, 58, 76
**Distribution:** Vic, Qld, NSW, Tas, SA

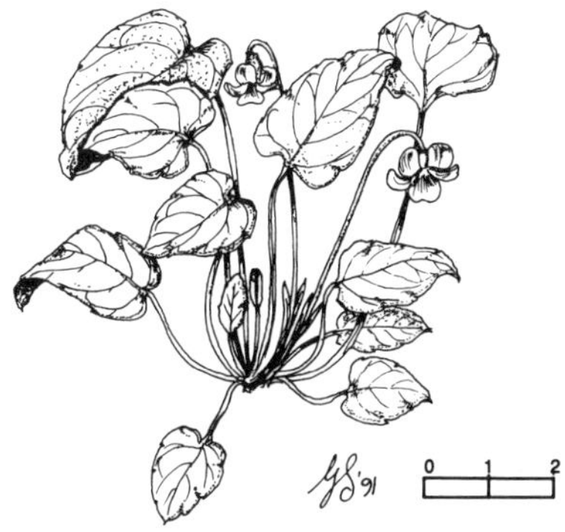

*Viola betonicifolia*

## Viola cleistogamoides

**P N**

**Habitat:** Damp tea-tree heath, swamp scrub, valley sclerophyll forest
**Form:** Tiny stoloniferous perennial herb similar to *V. sieberiana*
**Foliage:** Coarsely toothed leaves ovate to narrowly cuneate
**Flowers:** Tiny cream flowers with a purplish tinge towards the centre on shorter stalks to 25 mm long; October to January
**Requirements:** Seasonally wet, sheltered conditions.
**Localities:** Early collections from Vermont to Lilydale, 67#, 74, 76
**Distribution:** Vic, NSW, Tas, SA
**Synonym:** *Viola hederacea* ssp. *cleistogamoides*

## Viola hederacea

Native or Ivy-leaf Violet

**F P N**

**Size:** Prostrate–15 cm x 1–2 m
**Habitat:** Red gum woodland, wet, damp, valley and dry sclerophyll forests, grassy low open forest, wattle tea-tree scrub
**Form:** Fast-growing stoloniferous herb, forming a dense mat
**Foliage:** Light green kidney-shaped leaves to 30 mm long
**Flowers:** Small white flowers with purple centres held above the leaves, stalks to 100 mm long; flowers most of the year, especially June to March
**Requirements:** Moist to wet soils
**Comments:** A prolific grower once established. Larger leaves and flowers are produced when shelter and extra moisture is provided. It is an attractive hanging basket plant provided it is kept moist.
**Propagation:** Seed, division
**Localities:** 25, 26, 29#, 30, 33–36, 38, 40–45, 47–64, 67, 68, 70, 71, 73–76
**Distribution:** Vic, Qld, NSW, Tas, SA

*Viola hederacea*

## Viola sieberiana

Tiny Violet

**P N**

**Size:** 30 mm x 0.5 m
**Habitat:** Wet and damp sclerophyll forests, tea-tree heath
**Foliage:** Smaller than *V. hederacea* with denser, rhomboid to ovate, coarsely toothed leaves to 10 mm long in a tuft
**Flowers:** Tiny flowers hidden amongst the foliage on stalks to 50 mm long; October to December
**Requirements:** Moist soils
**Localities:** 58, 59, 67, 72, 74–76
**Distribution:** Vic, NSW, Tas
**Synonym:** *Viola hederacea* ssp. *sieberiana*

*Viola sieberiana*

SCALE: 0 1cm 2cm — The scale representation on each plant equals 2cm.

# Vittadinia

**Asteraceae**

Small perennial sub-shrubs found in open woodlands and grasslands, and on disturbed soil, usually on the basalt plains. All species in Melbourne are similar and have tiny mauve to blue daisy flowerheads and cuneate-spathulate to oblanceolate leaves. The main distinguishing features between the species are the fruit (or achene) characteristics. Propagation is by seed, plants often regenerating readily in the garden once established.

*Vittadinia cervicularis*

## Vittadinia cervicularis

**Size:** 15–25 cm high — F
**Habitat:** Red gum and box woodlands
**Form:** Erect perennial sub-shrub
**Foliage:** Spathulate to obovate leaves to 20 mm long, entire or with 1 pair of teeth near the middle and a fine point on the end, becoming linear. Stems and leaves are covered in bristly hairs while the floral bracts are terminated by a tuft of hairs.
**Flowers:** White or mauve flowerheads most of the year. The fruit, to 10 mm long, is flattened and covered in glandular hairs. Ribs converge into a distinct neck.
**Comments:** Only known sites in Melbourne.
**Localities:** Yarra Bend, Sth of Melton
**Distribution:** Vic, NSW, SA, WA

*Vittadinia cuneata* var. *cuneata*

## Vittadinia cuneata var. cuneata

Common or Woolly New Holland Daisy, Fuzzweed

**Size:** 10–30 cm x 30 cm — F P
**Habitat:** Red gum woodland, plains grassland, grassy wetland, tea-tree heath
**Form:** Rounded to erect perennial sub-shrub, covered in rough spreading hairs
**Foliage:** Dark green oblanceolate to cuneate leaves, 5–30 mm long, entire or with 3 lobes at the tip
**Flowers:** Narrow heads of tiny yellow flowers, the ray florets mauve to purple in 2 or more rows. The flowers are held erect in loose clusters on leafy stalks most of the year. The fruit, to 7 mm long, is wedge- to top-shaped, and sparsely to densely hairy with numerous appressed hairs on the lower part.
**Requirements:** Well drained soils
**Comments:** An attractive small plant for drier areas. The fluffy seeds are a most conspicuous feature.
**Propagation:** Seed
**Localities:** Port Melbourne, 2, 7, 12, 21, 26, 29#, 67#
**Distribution:** Vic, NSW, Tas, SA, WA

## Vittadinia gracilis

Woolly New Holland Daisy

**Size:** 10–40 cm high — F P
**Habitat:** Plains grassland, grassy wetland, red gum woodland
**Form:** Branched, erect woody perennial sub-shrub covered in fine white hairs, denser on stems
**Foliage:** Entire, linear to narrow lanceolate leaves to 40 mm long
**Flowers:** Purple flowers most of the year, bracts woolly. The fruit, to 6 mm long, has 5–7 ribs and is covered on the upper part in glandular and loose hairs.
**Localities:** Broadmeadows, Laverton, Melton, Parkville, 4, 17#, 18
**Distribution:** Vic, NSW, Tas, SA, WA

*Vittadinia muelleri*

## Vittadinia muelleri

Narrow-leaf New Holland Daisy

**Size:** 10–30 cm x 30 cm — F P
**Habitat:** Plains grassland, red gum woodland, dry sclerophyll forest
**Form:** Woody perennial subshrub, stems branching from the base
**Foliage:** Narrow to elliptical leaves to 45 mm long, entire or with a pair of pointed lobes. Minute and longer hairs are scattered along stems, but restricted to margins and lower midrib on the foliage.
**Flowers:** Dark blue flowerheads are held on stalks much longer than the foliage, most of the year. The fruit, to 5 mm long, is narrow-oblanceolate with a short tuft of appressed hairs at the base and spreading clubbed hairs on the upper part.
**Localities:** Sunbury, Eltham, Balwyn, 12, 17#, 18, 26, 38
**Distribution:** Vic, Qld, NSW, Tas

# Wahlenbergia

## Campanulaceae

Annual or perennial tufted herbs with well displayed blue or occasionally white, open, bell-shaped flowers at the ends of long slender erect stalks. The flower is 5-lobed. In most species the flowers nod in bud, becoming erect as they mature and open. Fruit is a capsule. Wahlenbergias are beautiful planted en masse throughout the garden and also make ideal container plants, flowering for a long period through Spring and Summer. Perennial species often die back in dry conditions and begin growing again during Winter. The petals are edible and are most attractive added to a salad. Propagation is by seed, which may germinate better if kept in the refrigerator at 1°C for 3 months once collected. (See stratification method.)

## Wahlenbergia communis

Tufted Bluebell

**Size:** 15–50 cm x 15 cm **F P N**
**Habitat:** Widespread on plains grassland, red gum and box woodland, riparian scrub
**Form:** Tufted or open perennial herb with sparsely branched erect flowering stems and a vigorous rhizomatous root system
**Foliage:** Small tufts or single hairy narrow leaves to 40 mm long, entire to slightly toothed
**Flowers:** Pale to bright blue flowers to 12 mm wide, singly on branching stalks to 45 cm high; November to May
**Requirements:** Moist well drained soils
**Comments:** In good seasons the foliage is denser. Often confused with *W. gracilis*, but may be easily distinguished by the flower size. In *W. gracilis*, flowers are only 6 mm wide.
**Localities:** St. Kilda, Broadmeadows, Clifton Hill, Ashburton, 2, 5, 12–14, 16–18, 20, 24–26, 28, 29, 32, 34A, 35, 63, 64, 74
**Distribution:** All mainland states

*Wahlenbergia communis*

## Wahlenbergia gracilenta

Annual Bluebell

**Size:** 10–30 cm high **F P**
**Habitat:** Plains grassland, red gum woodland, valley and damp sclerophyll forests, tea-tree heath
**Form:** Single to few-stemmed annual herb
**Foliage:** Entire, obovate hairy leaves 1–40 mm long with wavy margins, lower leaves opposite
**Flowers:** Minute pale blue flowers to about 6 mm wide on long branched stems; September to November.
**Comments:** Seeds readily in well drained soils. Withers early in Summer as the ground dries out.
**Localities:** 12, 25, 26, 35, 36, 42, 58, 67, 68, 74, 76
**Distribution:** Vic, NSW, Tas, SA, WA

*Wahlenbergia gracilenta*

## Wahlenbergia gracilis

Australian or Sprawling Bluebell

**Size:** 10–50 cm high **F P**
**Habitat:** Riparian scrub, valley, damp and wet sclerophyll forests, grassy low open forest
**Form:** Sprawling annual or perennial herb
**Foliage:** Alternate or opposite, entire narrow lanceolate leaves to 30 mm long
**Flowers:** Soft blue flowers to about 6 mm wide; September to November. Similar to *W. communis* which has larger flowers, to 12 mm wide.
**Localities:** 18, 26, 28, 29, 32, 35, 36, 41, 42, 44, 47–49, 50A, 51, 53, 55–57, 59, 74, 76
**Distribution:** Vic, Qld, NSW, Tas, SA; NG, PNZ, N Cal
**Synonym:** *W. quadrifida*

## Wahlenbergia gymnoclada

Naked Bluebell

**Size:** 10–50 cm high **F P**
**Habitat:** Valley sclerophyll and grassy low open forest
**Form:** Small erect perennial herb with spreading rhizomes
**Foliage:** Whitish-green narrow-linear, almost glabrous leaves 5–50 mm long, at the base of long unbranched flower stalks, lower leaves opposite.
**Flowers:** Single large funnel-shaped flower with long lobes, to 16 mm wide; October to February
**Requirements:** Well drained light soil
**Localities:** 31, 47, 54–56, 61, 63, 64, 74, 75
**Distribution:** Vic, Tas, SA

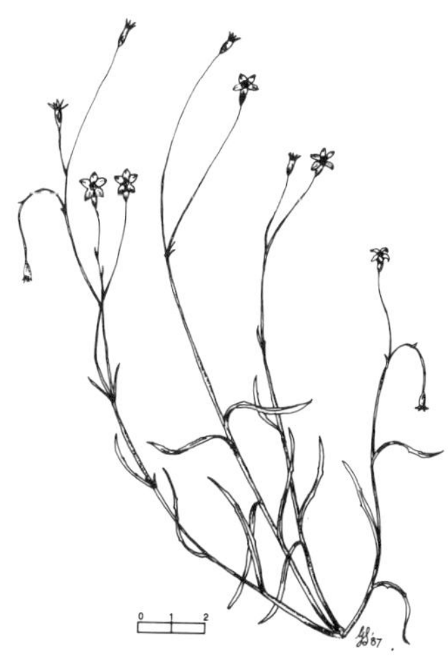

*Wahlenbergia gracilis*

## Wahlenbergia luteola

**Size:** 6–40 cm high **F**
**Habitat:** Red gum woodland, plains grassland
**Form:** Perennial tufted herb
**Foliage:** Linear leaves 4–60 mm long, lower leaves opposite
**Flowers:** Branched flowering stalk has flowers to 15 mm wide which are blue inside and yellowish-brown on the outside; most of the year.
**Localities:** Werribee 1899, 1926; Eltham, 21, 22, 25, 45
**Distribution:** Vic, NSW, SA
**Synonym:** *W. bicolor*

SCALE: 0 1cm 2cm — The scale representation on each plant equals 2cm.

*Wahlenbergia multicaulis*

### Wahlenbergia multicaulis

Tadgell's Bluebell

**F P**

**Size:** 10–50 cm high
**Habitat:** Red gum woodland, dry sclerophyll and grassy low open forest
**Form:** Tufted perennial herb
**Foliage:** Narrow strap-like, often slightly toothed alternate leaves to 60 mm long with thickened edges, lowest leaves sometimes opposite. Lower leaves and stems sometimes sparsely hairy; upper leaves smaller and linear.
**Flowers:** Flowers pale blue, with a short tube; September to May. Flower stalk may be branched.
**Requirements:** Well drained soils
**Localities:** Eltham Lower, 25, 26, 42, 44, 47, 48, 74, 76
**Distribution:** Vic, NSW, Tas, SA, WA
**Synonym:** *W. tadgellii*

### Wahlenbergia stricta

Tall Bluebell

**F P**

**Size:** 0.4–0.9 m x 0.3–0.4 m
**Habitat:** Widespread in plains grassland, red gum woodland, wet, damp, valley and dry sclerophyll forests, grassy low open forest
**Form:** Erect clumping much-branched perennial herb
**Foliage:** Hairy, ovate, opposite lower leaves 10–70 mm long with wavy edges, set at right angles to the stem. Upper leaves smaller and linear.
**Flowers:** Light blue, occasionally all white, flowers with a white throat, to 25 mm wide, tube conspicuous; August to January
**Requirements:** Well drained soils, tolerating some dryness
**Comments:** The Aborigines picked and ate the flowers.
**Localities:** 1, 12, 16, 26, 28, 34A-36, 38, 41–44, 47, 48, 50A, 51, 53–59, 61, 63, 69A, 74–76
**Distribution:** All states except NT

*Wahlenbergia stricta*

## Wilsonia — Convolvulaceae

### Wilsonia backhousei

Narrow-leaf Wilsonia

**F**

**Size:** Prostrate-20 cm x 1–1.5 m
**Habitat:** Primary dune scrub
**Form:** Perennial matting plant
**Foliage:** Succulent, glabrous dark green linear leaves to 15 mm long, sessile on woody stems
**Flowers:** Single, sessile white or cream tubular flowers to 15 mm long with greatly exserted purple-tipped stamens; October to December
**Requirements:** Moist to wet soils, tolerating salt
**Comments:** A good groundcover for coastal areas.
**Propagation:** Seed, cuttings
**Localities:** 67[#], 74
**Distribution:** Vic, NSW, Tas, SA, WA

*Wilsonia backhousei*

### Wilsonia humilis

Silky Wilsonia

**F**

**Size:** Prostrate
**Habitat:** Saltmarshes
**Form:** Very small matting plant
**Foliage:** Pairs of tiny, fleshy, silvery, ovate to oblong, concave leaves to 4 mm long, each opposite pair clasping the pair above
**Flowers:** Sessile, silky white tubular flowers longer than the leaves, to 6 mm long; September to December
**Localities:** Pt. Cook 1902, Port Melbourne[#]; 4
**Distribution:** Vic, NSW, Tas, SA, WA

### Wilsonia rotundifolia

Round-leaf Wilsonia

**F**

**Size:** Prostrate-20 cm x 0.3–0.6 m
**Habitat:** Grassy wetlands, saltmarshes, mudflats, primary dune scrub
**Form:** Small matting plant
**Foliage:** Tiny round glossy green, somewhat succulent leaves to 4 mm long on short stalks. Loose hairs cover young leaves.
**Flowers:** Tubular cream flowers with reflexed petals, very shortly stalked; September to December
**Reqirements:** Moist soils, salt tolerant
**Localities:** Laverton, 2, 4, 67[#], 74
**Distribution:** Vic, NSW, SA, WA

*Illustrated p. 200*

# Xanthosia — Apiaceae

## Xanthosia dissecta var. dissecta — Cut-leaf Xanthosia

P

**Size:** 10–30 cm x 30 cm
**Habitat:** Valley and damp sclerophyll forests, sclerophyll woodland, wattle tea-tree scrub
**Form:** Spreading perennial herb with a woody rootstock
**Foliage:** Finely divided dark green leaves to 30 mm long, on long stalks to 120 mm
**Flowers:** Axillary cluster of tiny green or reddish flowers; September to November
**Requirements:** Moist well drained soils
**Propagation:** Cuttings
**Localities:** 44, 47, 48, 50A, 51, 53–56, 58, 59, 63, 64, 66#, 67#, 70 74, 75
**Distribution:** Vic, NSW, Tas, SA

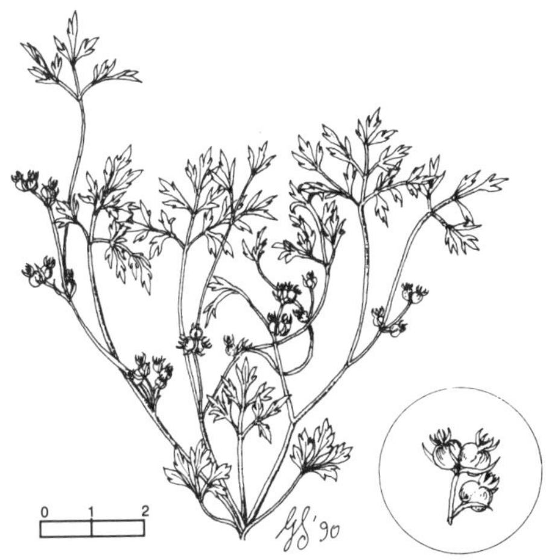

*Xanthosia dissecta*

## Xanthosia pilosa — Woolly Xanthosia

P

**Size:** 20–50 cm x 20–30 cm
**Habitat:** Grassy low open forest
**Form:** Variable erect or trailing herb
**Foliage:** Soft grey or brown hairs cover the leaves and stems. Leaves, to 30 mm long, have 3–7 broad leaflets which may be toothed or deeply lobed.
**Flowers:** Small cream flowers in long-stalked axillary umbels, often containing only 1 or 2 flowers; October to March.
**Requirements:** Moist well drained soils, preferring some protection in coastal areas
**Localities:** 74, 76
**Distribution:** Vic, Qld, NSW, Tas

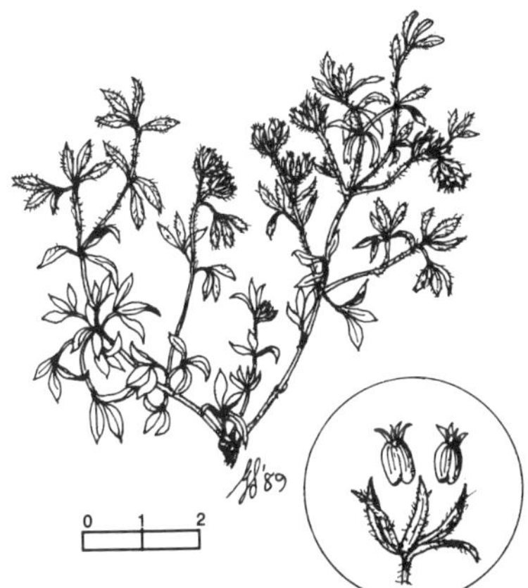

*Xanthosia pilosa*

## Xanthosia pusilla — Common or Heath Xanthosia

F P

**Size:** 5–20 cm x 20–40 cm
**Habitat:** Tea-tree heath, grassy low open forest
**Form:** Hairy, spreading or erect perennial herb
**Foliage:** Leaves divided into 3 entire, narrowly elliptic segments to 15 mm long
**Flowers:** Clusters of 1–4 sessile green flowers; October to November
**Requirements:** Well drained sandy soils
**Localities:** 67–69, 74–76
**Distribution:** Vic, Tas, SA, WA

# Ziera — Rutaceae

## Ziera arborescens — Stinkwood

P N

**Size:** 2–5 m x 3 m
**Habitat:** Wet sclerophyll forest
**Form:** Tall erect or spreading shrub or small tree
**Foliage:** Pairs of dark green glabrous leaves on long stalks divided into 3 lanceolate leaflets to 100 mm long, paler green and downy below
**Flowers:** Axillary clusters of white flowers to 10 mm wide; August to December
**Requirements:** Moist well drained soils
**Comments:** The leaves and wood have a strongly aromatic smell when touched.
**Propagation:** Cuttings
**Localities:** 57
**Distribution:** Vic, Qld, NSW, Tas

*Ziera arborescens*

# Zygophyllum — Zygophyllaceae

## Zygophyllum billardieri — Coast Twin-leaf

F

**Size:** 0.3–0.6 m x 1 m
**Habitat:** Riparian scrub, primary dune scrub
**Form:** Scrambling perennial with a single central ridge on the stem
**Foliage:** Pairs of succulent 'Y'-shaped leaves, leaflets ovate to 15 mm long
**Flowers:** Masses of bright yellow, 4-petalled cup-shaped flowers to 15 mm wide, singly in leaf axils; most of the year. Fruit a drooping, acutely 4-angled capsule to 9 mm long, triangular in outline.
**Requirements:** Sandy well drained soils, tolerating dry periods
**Comments:** Suitable for exposed coastal conditions. Salt tolerant. Appears to be resistant to fires.
**Propagation:** Cuttings
**Localities:** Port Phillip, 29#
**Distribution:** All states except NT

*Zygophyllum billardieri*

SCALE: 0 1cm 2cm — The scale representation on each plant equals 2cm.

*Zygophyllum glaucum*

## Zygophyllum glaucum

Pale Twin-leaf

**F**

**Size:** Prostrate-75 mm x 0.8–2 m
**Habitat:** Riparian scrub
**Form:** Succulent, spreading annual herb with stout ridged stems
**Foliage:** Pairs of succulent glaucous or dull green 'Y'-shaped leaves on short stalks, leaflets obovate to 30 mm long
**Flowers:** Single bright yellow flowers on short stalks in leaf axils; July to November. Drooping oblong 4-angled capsule to 20 mm long
**Requirements:** Sandy loams
**Comments:** Forming extensive mats, it dies off after flowering.
**Propagation:** Seed
**Localities:** 16, 26
**Distribution:** All mainland states

*Rumex brownii*

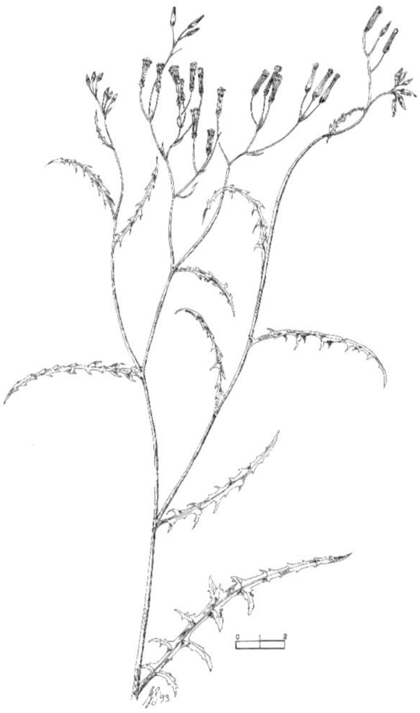

*Senecio runcinifolius*

*Solanum prinophyllum*

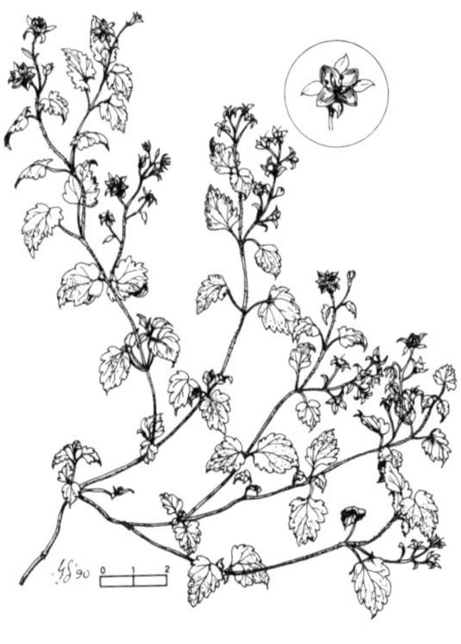

*Veronica plebeia*

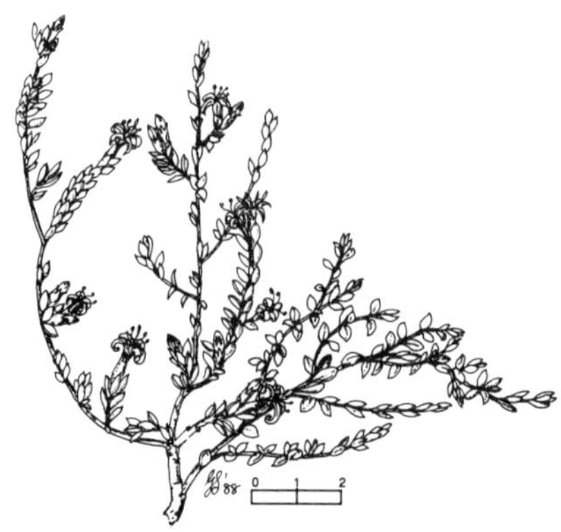

*Wilsonia rotundifolia*

*Villarsia exaltata*

# Flowering Plants — Monocotyledons

## *Lilies and Irises*

*Arthropodium fimbriatum*

## Arthropodium — Liliaceae (Anthericaceae)

### Arthropodium fimbriatum — Nodding Chocolate Lily

**F**

**Size:** 0.2–1 m x 0.2–0.8 m
**Habitat:** Box woodland, plains grassland
**Form:** Dwarf perennial herb
**Foliage:** Grass-like leaves to 40 cm long
**Flowers:** Each unbranched stem has several clusters of 1–4 blue to violet flowers held high above the foliage; September to December. Stalks nodding in fruit, and sometimes in flower.
**Requirements:** Warm, well drained soils
**Comments:** Flowers sweetly vanilla scented. Plants die back to a tuberous rootstock after flowering, and commence regrowth following Autumn rains. All species have edible tubers which were roasted by the Aborigines. Many botanists distinguish *Dichopogon* as a separate genus from *Arthropodium* because of the 2 beard-like appendages on each anther of those species.
**Propagation:** Seed, division
**Localities:** 12, 25, 26
**Distribution:** Vic, NSW, SA, WA
**Synonym:** *Dichopogon fimbriatus*

*Arthropodium milleflorum*

### Arthropodium milleflorum — Pale Vanilla Lily

**F P**

**Size:** 0.3–1 m x 30 cm
**Habitat:** Riparian scrub, damp and valley sclerophyll forests, grassy low open forest
**Form:** Grass-like tufted perennial herb
**Foliage:** Slightly fleshy dark green leaves to 30 cm x 10–30 mm
**Flowers:** Pale pink, pale mauve or white flowers in clusters of 2 or 3 on branched, leafless slender stems to 1 m tall; September to March
**Requirements:** Moist clay soils
**Comments:** Leaves die down to a dormant fleshy tuberous rootstock during the dry season. Use to good effect in mass plantings or informal drifts. Suited to container growing. Another form occurs in Melton West.
**Propagation:** Seed or division of tuberous rootstock
**Localities:** 29, 37, 42, 44, 49, 50A, 51, 55, 56, 58, 60, 74, 76
**Distribution:** Vic, NSW, Tas, SA

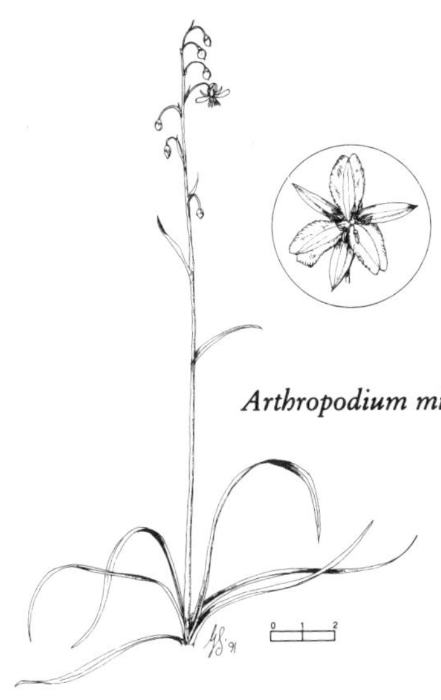

*Arthropodium minus*

### Arthropodium minus — Small Vanilla Lily

**F P**

**Size:** 20–30 cm x 10–20 cm
**Habitat:** Plains grassland, red gum woodland
**Form:** Grass-like tufted perennial herb
**Foliage:** Linear leaves to 100 mm x 5 mm
**Flowers:** Purple sweetly scented small flowers singly on slender stems to 30 cm tall; September to December
**Requirements:** Well drained soil
**Comments:** Also dormant in dry weather, this plant is well suited to an open sunny rockery situation. It tolerates dry soil conditions where overhead protection is available. Distinguish from *A. milleflorum* by the solitary flowers.
**Propagation:** Seed or division
**Localities:** Laverton Nth, Sth of Melton, 7, 12, 25, 26
**Distribution:** Vic, NSW, Tas, SA

SCALE: 0 1cm 2cm — The scale representation on each plant equals 2cm.

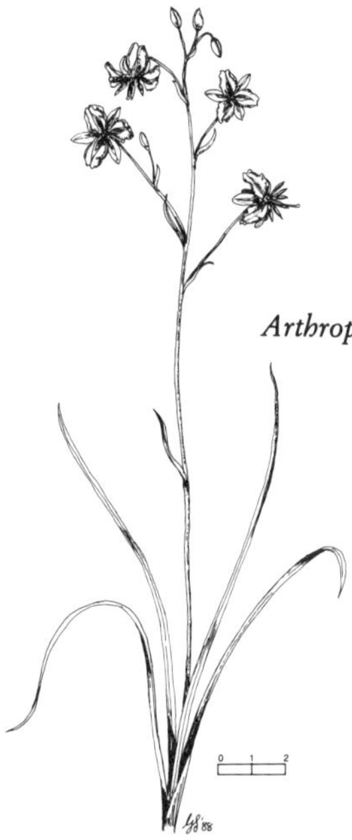

*Arthropodium strictum*

### Arthropodium strictum

Chocolate Lily

**Size:** 0.2–1 m x 0.2–0.8 m **F P**
**Habitat:** Plains grassland, red gum and sclerophyll woodland, dry, valley and damp sclerophyll forests, grassy low open forest
**Form:** Dwarf perennial herb
**Foliage:** Grass-like leaves to 40 cm long
**Flowers:** Violet, rarely white, flowers, singly on erect stalks, on branched flowering stems held high above the foliage; September to December
**Requirements:** Well drained soils
**Comments:** A most attractive and adaptable plant. Chocolate- or caramel-scented flowers brighten a rockery and add interest to natural bushland areas. Follows the same growth cycle as *A. fimbriatum* but differs in that the flowering stems are branched and flowers are solitary.
**Propagation:** Seed, division
**Localities:** St. Kilda#, 1, 2, 12, 13, 16, 21, 25, 26, 28–56, 58, 59, 61–64, 67–71, 74–76
**Distribution:** Vic, NSW, Tas, SA
**Synonym:** *Dichopogon strictus*

### Arthropodium sp. aff. strictum

**Comments:** An unnamed species similar to *A. strictum* occurring in swampy areas of plains grassland. It differs in that there are more leaves in the rosette, several flowers occur at each node and the floral bracts are larger. The flowers have darker anthers, a different fragrance and flowering time is later.
**Localities:** Sunbury

## Bulbine — Liliaceae (Asphodelaceae)

*Bulbine bulbosa*

### Bulbine bulbosa

Bulbine Lily or Wild Onion

**Size:** 0.2–0.6 m x 30 cm **F P**
**Habitat:** Widespread in plains grassland, red gum woodland, dry and valley sclerophyll forests
**Form:** Densely tufted perennial herb
**Foliage:** Erect green to grey-green succulent rush-like leaves to 40 cm long, hollow and channelled along their length
**Flowers:** Yellow star-like flowers to 20 mm wide clustered on leafless flowering stems to 60 cm high; September to January, also sporadic during the year
**Requirements:** Moist well-drained soils
**Comments:** Plants die back to a bulbous tuberous rootstock in dry weather, but may continue to produce new leaves and flower throughout the year with additional watering during Summer and Autumn. In ideal conditions they will regenerate readily, attractive informal drifts developing naturally. Differs from *B. semibarbata* in having a bulbous rootstock and all 6 staminal filaments bearded. The tubers of both species were steamed and eaten by the Aborigines.
**Propagation:** Seed or division
**Localities:** St Albans, 1, 12, 15, 16, 25, 26, 29, 33–39, 41, 42, 44, 45, 47, 49, 50A, 51, 58, 67#, 74
**Distribution:** Vic, Qld, NSW, Tas, SA

### Bulbine semibarbata

Leek Lily

**Comments:** There is an unconfirmed report that *B. semibarbata* was once found in Warrandyte. This is considered unlikely however.

## Burchardia — Liliaceae (Colchicaceae)

*Burchardia umbellata*

### Burchardia umbellata

Milkmaids

**Size:** 20–50 cm high **F P**
**Habitat:** Widespread in plains grassland, box, red gum and sclerophyll woodlands, dry and valley sclerophyll forests, grassy low open forest, tea-tree heath, coastal banksia woodland
**Form:** Small sparsely tufted perennial herb
**Foliage:** Grass-like, pale green leaves to 20 cm long and 1 or 2 narrow stem leaves
**Flowers:** Terminal umbels of 2–10 honey scented, white to light pink flowers to 30 mm wide on erect simple stems to 50 cm high; September to December. Pink triangular ovary in the centre of the flower remains after flowering as a dried capsule.
**Requirements:** Moist, well drained soils
**Comments:** This most attractive plant is best grown in informal drifts amongst other small plants, and is ideally suited to containers. Plants always die down during dry weather and regenerate from an edible, persistent fibrous rootstock.
**Propagation:** Seed
**Localities:** Somerton, 1, 12, 13, 18, 25, 26, 28–32, 34A, 35, 37–56, 58–71, 73–76
**Distribution:** All states except NT

# Caesia — Liliaceae (Anthericaceae)

*Caesia calliantha*

## Caesia calliantha — Blue Grass Lily

**Size:** 10–50 cm x 10–50 cm **F P**
**Habitat:** Plains grassland, red gum woodland, valley sclerophyll forest
**Form:** Tufted perennial herb
**Foliage:** Crowded grass-like basal leaves to 30 cm x 15 mm wide
**Flowers:** Lilac blue to deep blue star-like flowers in clusters of 2–3 along stout erect stems to 50 cm high; September to February. Petals tightly, spirally twisted after flowering.
**Requirements:** Moist soils which dry out in Summer
**Comments:** Useful rockery plant, which may grow larger and flower longer if Summer moisture is available. Best effect is gained when planted in drifts or groups. The root tubers of all species were an important part of the aboriginal diet.
**Propagation:** Seed or division
**Localities:** Doncaster, Laverton, Tottenham, Rockbank, Derrimut, DeerPark, 1, 10, 12, 15, 16, 20, 25, 26, 28, 33, 35, 39, ?42, 44, 49, 62, 63, 67#, 74, 76
**Distribution:** Vic, NSW, Tas, SA
**Synonym:** *C. vittata*

*Caesia parviflora* var. *parviflora*

## Caesia parviflora var. parviflora — Pale Grass Lily

**Size:** 10–30 cm x 10–25 cm **F P**
**Habitat:** Damp and valley sclerophyll forests, grassy low open forest, tea-tree heath
**Form:** Tufted perennial herb
**Foliage:** Grass-like basal leaves to 25 cm x 5 mm wide
**Flowers:** White to pale lavender flowers scattered singly along slender erect flowering stems to 30 cm high, branched several times; September to February.
**Requirements:** Moist to well drained soils
**Comments:** Similar to *C. calliantha* but generally smaller in all parts.
**Propagation:** Seed
**Localities:** Montrose, 42, 44, 46–51, 53, 54, 56, 58, 59, 61–64, 67–71, 74–76
**Distribution:** Vic, Qld, NSW, Tas

## Caesia parviflora var. minor

**Size:** 10–20 cm high
**Comments:** Smaller flowers and narrower leaves to 2 mm wide. Flowering stems 1-many-branched and spreading, becoming erect at the ends. Flowers are white, tinged with blue.
**Localities:** Early collections from Dandenong 1875, Oakleigh 1895
**Distribution:** Vic, Tas, SA

# Chamaescilla — Liliaceae (Anthericaceae)

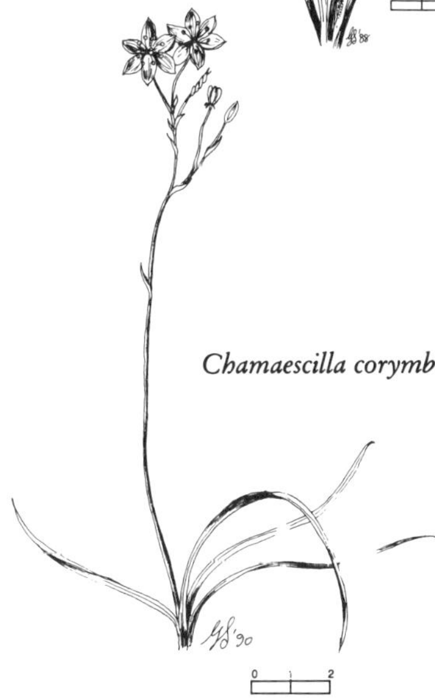

*Chamaescilla corymbosa*

## Chamaescilla corymbosa — Blue Stars

**Size:** 7–20 cm x 10–20 cm **F P**
**Habitat:** Red gum woodland, damp and valley sclerophyll forests, grassy low open forest
**Form:** Tiny perennial herb
**Foliage:** Few basal, dull green, narrow grass-like leaves to 15 cm long, often smaller, lying flat on the ground
**Flowers:** Open terminal clusters of 1–10 bright blue flowers to 15 mm wide on slender branched stems to 20 cm high; August to November. Individual flowers only last one day.
**Requirements:** Moist soils
**Comments:** Plants die back to tuberous rootstock in dry weather, and are generally short lived.
**Propagation:** Seed
**Localities:** Research, Mooroolbark, 26, 29#, 34A, 35, 42, 44, 45, 47–56, 58, 59, 63, 64, 67–69A, 71, 74–76
**Distribution:** Vic, Tas, SA, WA

# Dianella — Liliaceae (Phormiaceae)

Dianellas are most attractive garden, rockery or container plants. Their many small blue to purple flowers are in panicles held above the foliage on long flowering stems and are followed by succulent glossy blue or purple berries. The tough strap-like leaves were used by the Aborigines for basket-making and plaiting into cords. Plants are long-lived, growing in tufts or spreading colonies. Propagation is by freshly sown seed or division of clumps. Current research indicates that there may be up to 14 spp. within the Melbourne area.

SCALE: 0 1cm 2cm — The scale representation on each plant equals 2cm.

Dianella caerula

## Dianella caerula

Paroo Lily

**F P**

**Size:** 20–35 cm x 0.3–1 m
**Habitat:** Valley sclerophyll forest
**Form:** Tufted perennial clump
**Foliage:** Dark green linear leaves 15–20 cm long, on branching stems, minutely saw-toothed on edges and midrib
**Flowers:** Loose panicle of pale to dark blue flowers on erect stems to 35 cm high; August to February, followed by round blue or purple berries
**Requirements:** Well drained soils
**Comments:** An attractive long-lived accent plant. Ideal for establishing under existing trees. Other varieties of this species are taller and may develop into large colonies.
**Localities:** Belgrave, Monbulk
**Distribution:** Vic, Qld, NSW, Tas

*Dianella longifolia* var. *longifolia*

## Dianella longifolia var. longifolia

Pale or Smooth Flax-lily

**F P**

**Size:** 0.3–0.8 m x 0.5 m
**Habitat:** Plains grassland, box woodland, valley and dry sclerophyll forests, grassy low open forest
**Form:** Tufted perennial clump with thick short underground rhizome and tuberous roots
**Foliage:** Soft, linear glaucous, or sometimes light green, leaves to 80 cm x 12 mm wide, edges and midrib smooth
**Flowers:** Loose to dense panicles of pale blue flowers in groups of 2–12, on robust stems to 1 m high; August to January, followed by dark blue or purple berries
**Requirements:** Moist well drained soils
**Comments:** An attractive, easily maintained clumping plant, ideal for growing under trees. The leaves were favoured by the Aborigines for basket making. At least 3 varieties occur in the Melbourne area. The typical green form of the foothills from Hurstbridge to the Dandenong Ranges forms an evergreen tussock. The form from the basalt plains is blue with short very wide leaves that die off in Summer. It is rhizomatous and has only 2–3 'fans' if it clumps. Another form, found on the silurian sandstone around Greenvale and eastward, is also bluish, rhizomatous and summer-deciduous, but the foliage is very thin and has short tooth-like projections along the margins.
**Localities:** 2, 4, 12, 13, 16, 23, 25, 26, 28, 29, 32, 33–36, 39, 40–42, 44, 46–48, 50–56, 58, 60–64, 67, 69, 69A, 74, 76
**Distribution:** All states
**Synonym:** *D. laevis*

*Dianella revoluta* var. *revoluta*

## Dianella revoluta var. revoluta

Black Anther, or Spreading Flax-lily

**F P**

**Size:** 0.3–1 m x 0.5–2.5 m
**Habitat:** Widespread in plains grassland, box and red gum woodland, dry, valley and damp sclerophyll forests, tea-tree heath
**Form:** Robust open tufted perennial, spreading by vigorous, branched rhizomes
**Foliage:** Dark green linear leaves to 70 cm long with recurved leaf margins which are finely serrated
**Flowers:** Loose to dense panicles of blue to whitish flowers on branched stems to 1 m high; August to May, followed by small, shiny dark blue berries
**Requirements:** Well drained soils
**Comments:** A very tolerant plant once established. Ideal for growing close to trees. More vigorous if moisture is available. The Aborigines ate the berries and also used them to obtain blue dye.
**Localities:** 1, 2, 5, 7, 10, 12, 13, 15–18, 24–26, 28, 29, 31–39, 41, 42, 44–56, 58–71, 74–76
**Distribution:** All states except NT

*Dianella revoluta* var. *brevicaulis*

## Dianella revoluta var. brevicaulis

**F P**

**Size:** 30–50 cm x 30–50 cm
**Habitat:** Primary dune scrub and cliffs
**Comments:** Smaller tufts with slightly narrower yellow-green leaves. Flowering stems are amongst the foliage rather than above it, flowers on very short stalks. Prefers sandy soil.
**Localities:** 4, 69, 69A, 74, 76
**Distribution:** Vic, Tas, SA, WA

### Dianella tasmanica — Tasman Flax-lily

**Size:** 0.6–1.5 m x 0.5–2 m — **P N**
**Habitat:** Wet, damp and valley sclerophyll forests
**Form:** Very robust tufted perennial, usually a clump but may spread vigorously by strong rhizomes
**Foliage:** Broad linear leaves up to 1 m x 40 mm, distinctly serrated along the margins and underside of midrib
**Flowers:** Panicles of blue flowers with yellow anthers on strong many-branched stems to 1.5 m high; August to February, followed by large violet to blue berries
**Requirements:** Moist soils, preferably in a cool position
**Comments:** Plants are very tolerant once established and will adapt to most situations including snow cover. A most attractive addition to a fern garden. Seedlings often appear in gardens.
**Localities:** Floodplains west of Dandenong Rges, Kalorama, 42–44, 47, 49, 50A, 51, 54–60, 71, 74, 76
**Distribution:** Vic, NSW, Tas

*Dianella tasmanica*

### Dichopogon fimbriatus = Arthropodium fimbriatum

### Dichopogon strictus = Arthropodium strictum

## Diplarrena — Iridaceae

### Diplarrena moraea — Butterfly Flag

**Size:** 0.3–0.6 x 0.5–1 m — **F P**
**Habitat:** Damp sclerophyll forest, sclerophyll woodland
**Form:** Tussock-forming perennial herb with rhizomatous rootstock
**Foliage:** Strap-like, bright green leaves to 60 cm x 1 cm wide, with 20–30 parallel veins
**Flowers:** Large flowers with 3 white outer petals enclosing 3 smaller white and yellow, and sometimes purple, petals; October to December. Flowers on robust stems as long or longer than the leaves.
**Requirements:** Well drained soils
**Comments:** Very attractive accent plant suitable for container cultivation. Each honey-scented flower lasts one day, but they are produced in profusion over a long period. Plants respond favourably to extra moisture in Summer. Restricted range in Melbourne.
**Propagation:** Seed, division
**Localities:** Belgrave, 67#
**Distribution:** Vic, NSW, Tas

*Diplarrena moraea*

## Hypoxis — Liliaceae (Hypoxidaceae)

### Hypoxis glabella var. glabella — Tiny Star

**Size:** 5–20 cm high — **P**
**Habitat:** Plains grasslands, red gum woodland, dry and valley sclerophyll forests, grassy low open forest
**Form:** Sparse, glabrous perennial tufting herb
**Foliage:** Few narrow dark green grass-like leaves to 20 cm long, usually less, and 1 mm wide
**Flowers:** Single, or pairs of bright yellow star-like flowers to 20 mm wide, on stems as long as the leaves; July to March. Flowering stems have a pair of fine bracteoles 2–18 mm long near the middle.
**Requirements:** Moist, well drained soil
**Comments:** An attractive long flowering little herb which should be planted en masse for best effect. Plants regenerate from an edible bulbous tuber as soon as regular Autumn rains set in, then they flower quickly. The flowers close up in dull weather. Snails and slugs cause serious damage.
**Propagation:** Seed
**Localities:** 7, 12, 25, 26, 29, 34A, 35, 37, 38, 41, 42, 45–47, 49–51, 58, 59, 62, 63, 75
**Distribution:** Vic, NSW, Tas, SA, WA
**Synonym:** *H. hookeri*, *H. pusilla*

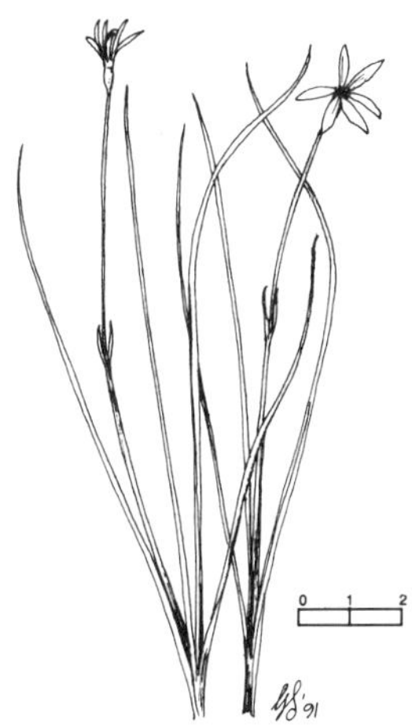
*Hypoxis glabella* var. *glabella*

### Hypoxis hygrometrica var. hygrometrica — Golden Weather-glass

**Size:** 5–20 cm high — **F P**
**Habitat:** Swamp scrub
**Flowers:** Usually a single yellow star-like flower. Differs from *H. hygrometrica* var. *villosisepala* in that the ovary and sepals are glabrous. The single bracteole is less than 4 mm long.
**Comments:** Rare within Melbourne, occurring in low-lying areas at the base of the foothills. More common just outside the study area.
**Localities:** 54
**Distribution:** Vic, NSW, Tas, SA

SCALE: 0 1cm 2cm — The scale representation on each plant equals 2cm.

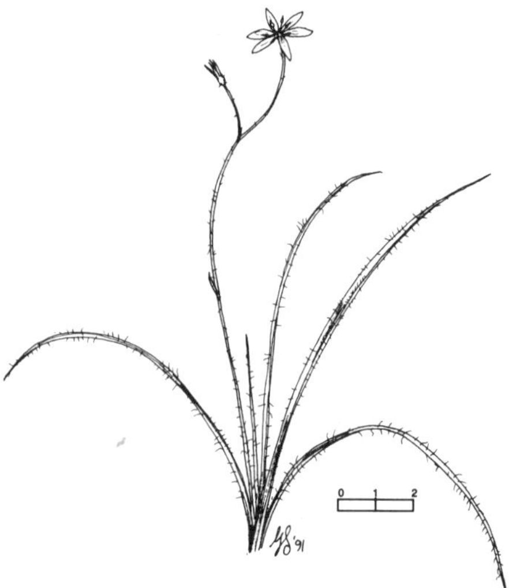

*Hypoxis hygrometrica* var. *villosisepala*

### Hypoxis hygrometrica var. villosisepala

F P

**Size:** 5–30 cm high
**Habitat:** Plains grassland, valley sclerophyll forest
**Form:** Sparsely hairy perennial tuft
**Foliage:** Few, narrow, slightly fleshy grass-like leaves to 20 cm long
**Flowers:** Pairs of yellow star-like flowers to 30 mm wide with a few hairs on the ovary and ends of the sepals; August to December. The flowering stem has a single bracteole to 10 mm long below the flowers.
**Requirements:** Moist well drained soils
**Comments:** May be distinguished from *H. glabella* var. *glabella* by the presence of tiny hairs.
**Propagation:** Seed
**Localities:** Yarra, Montrose, 25, 26, 28, 42, 50A, 54, 55, 62, 63, 67, 68, 70, 74
**Distribution:** Vic, NSW, Tas

### Hypoxis vaginata var. vaginata

Yellow Star

F P

**Size:** 5–35 cm high
**Habitat:** Valley sclerophyll forest, grassy low open forest
**Comments:** A variable species similar to and often confused with *H. glabella*. Leaves are longer, to 35 cm, and wider, to 3 mm, and flowers are larger, to 40 mm wide. It has a single sheathing bracteole to 70 mm long on the flowering stem.
**Localities:** Eltham, Doncaster, Boronia, 7, 26, 34A, 44, 55, 64, 66, 69, 74–76
**Distribution:** Vic, NSW, Tas, SA, WA
**Synonym:** Previously referred to as *H. glabella*

*Hypoxis vaginata* var. *vaginata*

### Hypoxis vaginata var. brevistigmata

**Habitat:** Tea-tree heath
**Comments:** Smaller herb to 10 cm high. Flowers occur singly on the flowering stem and are smaller in all aspects. The main distinguishing feature is the tiny size of the stigmatic lobes. Grows in damp situations.
**Localities:** An early collection in Melbourne; 69
**Distribution:** Vic, NSW, Tas

## Laxmannia

## Liliaceae (Anthericaceae)

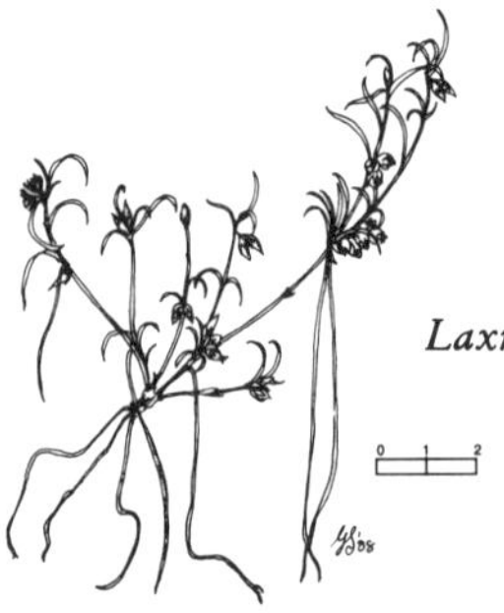

*Laxmannia orientalis*

### Laxmannia orientalis

Dwarf Wire Lily

F P

**Size:** 2–5 cm x 5–15 cm
**Habitat:** Tea-tree heath
**Form:** Tiny wiry tufting plant, often with aerial roots descending from nodes on stems
**Foliage:** Narrow, stem-sheathing, needle-like leaves 5–25 mm long, with white, woolly hairs along sheath margins, borne at regular intervals
**Flowers:** Nodding clusters of 3–7 tiny stalkless white flowers with red-lined sepals in each tuft of leaves; September to November.
**Requirements:** Well drained soils in an open sunny position
**Comments:** Particularly suited to container cultivation. This slow growing plant is more of interest to collectors than regenerators. Locally rare.
**Propagation:** Division
**Localities:** Baxter, 67–69, 74
**Distribution:** Vic, Tas, SA
**Synonym:** Previously referred to as *L. sessiliflora*, which is now regarded as endemic to WA.

## Libertia

## Iridaceae

*Libertia pulchella*

### Libertia pulchella

Pretty Grass-flag

P N

**Size:** 7–30 cm x 30 cm
**Habitat:** Damp sclerophyll forest
**Form:** Sparsely tufting perennial herb
**Foliage:** Narrow, flat, overlapping leaves, 10–20 cm long, often in fan-shaped tufts
**Flowers:** Loose clusters of small white fragrant flowers on stems to 30 cm high; October to January, followed by interesting seed heads.
**Requirements:** Moist well drained soil, rich in humus, tolerating boggy periods
**Comments:** A dainty, long-lived plant but difficult to flower unless growing conditions suit. Ideal container plant, or used as a border in fern gardens. Requires Summer watering. Restricted range in Melbourne.
**Propagation:** Seed, division
**Localities:** Ferntree Gully
**Distribution:** Vic, NSW, Tas

# Patersonia — Iridaceae

## Patersonia fragilis — Short Purple-flag

**Size:** 10–20 cm x 40 cm **F P**
**Habitat:** Tea-tree heath, swamp scrub
**Form:** Tufting rhizomatous perennial herb
**Foliage:** Thick and fleshy, stiff, narrow dull green to blue-green leaves to 20 cm long
**Flowers:** Bright, deep purple iris-like flowers on stems shorter than the leaves; September to January. Flowers are in terminal clusters, enclosed by 2 large bracts, the floral tube extending beyond the bracts. They have 6 petals: 3 large, rounded and spreading outer petals and 3 tiny, erect inner ones.
**Requirements:** Moist open position tolerating inundation for short periods
**Comments:** An attractive plant even when not in flower especially planted in a rockery. Flowers are short lived, but are produced in great numbers on sunny days.
**Propagation:** Seed, division
**Localities:** Montrose, 54, 67, 69, 74, 75
**Distribution:** Vic, Qld, NSW, Tas, SA
**Synonym:** *P. glauca*

*Patersonia fragilis*

## Patersonia occidentalis — Long Purple-flag

**Size:** 20–40 cm x 0.3–0.6 m **F P**
**Habitat:** Valley sclerophyll forest, swamp and wattle tea-tree scrub
**Form:** Compact clumping perennial herb
**Foliage:** Long, flat leaves to 60 cm x 5 mm, convex on one surface
**Flowers:** The tube of the bluish-purple, or occasionally white, flowers is almost hidden by the bracts. Flowering stems are as long as, or longer than, the foliage, to 85 cm long; September to January.
**Requirements:** Tolerates inundation during Winter and drying out in Summer
**Comments:** Suitable for bog gardens or pond edges but also tolerant of much drier positions. Distinguished from *P. fragilis* by the wider leaves, longer flowering stems and hidden floral tube.
**Propagation:** Seed or division
**Localities:** Montmorency, Montrose, 49, 51, 54, 62, 63, 67–71, 73–76
**Distribution:** Vic, Tas, SA, WA
**Synonym:** *P. longiscapa*

*Patersonia occidentalis*

# Stypandra — Liliaceae (Phormiaceae)

## Stypandra glauca — Nodding Blue Lily

**Size:** 0.5–1.5 m x 1.5 m **F P**
**Habitat:** Riparian and swamp scrub
**Form:** Soft, sprawling clump. Stems often twisted or bent, open-branched.
**Foliage:** Glaucous, lanceolate, sessile leaves to 20 cm long, but usually less, arranged alternately along the stems
**Flowers:** Bright blue nodding star-shaped flowers with bright yellow anthers, in open branching terminal clusters; July to November, also sporadic throughout the year
**Requirements:** Well drained, moist to dry soil
**Comments:** An attractive plant, especially in flower. Plants are at first herbaceous but become woody with age. Pruning encourages new growth from the centre of the clump and increases flower production. Regenerates vigorously after fire.
**Propagation:** Seed, division
**Localities:** 29, ?42, 61
**Distribution:** Vic, Qld, NSW, SA, WA

*Stypandra glauca*

# Thelionema — Liliaceae (Phormiaceae)

## Thelionema caespitosum — Blue Grass-lily

**Size:** 0.2–0.6 m x 0.3–0.6 m **F P**
**Habitat:** Valley sclerophyll forest, tea-tree heath, generally on damp peaty soil, swamp and wattle tea-tree scrub
**Form:** Tufted herbaceous perennial
**Foliage:** Blue-green to pale green linear leaves to 30 cm long, "V"-shaped in cross-section
**Flowers:** Blue, or sometimes white, star-shaped flowers on leafless wiry branched stems to 60 cm high; September to January
**Requirements:** Moist soils
**Comments:** An attractive, easily grown plant for a moist rockery. A white-flowered form occurs in the Frankston area.
**Propagation:** Division
**Localities:** Montrose, ?42, 44, 49, 54, 63, 64, 67, 70, 73–76
**Distribution:** Vic, Qld, NSW, Tas, SA
**Synonym:** *Stypandra caespitosa*

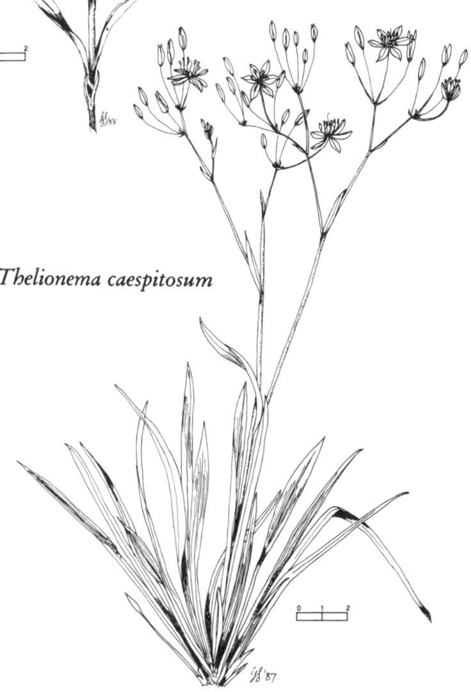

*Thelionema caespitosum*

SCALE: 0 1cm 2cm — The scale representation on each plant equals 2cm.

*Thysanotus patersonii*

# Thysanotus — Liliaceae (Anthericaceae)

## Thysanotus patersonii — Twining Fringe-lily

**F P**

**Size:** Stems 0.1–1 m long
**Habitat:** Red gum woodland, plains grassland, riparian scrub, valley and dry sclerophyll forests, grassy low open forest, tea-tree heath
**Form:** Light perennial creeper or climber, twining weakly on other plants; usually a single stem, occasionally 2, terete and hairy at the base, glabrous and quadrangular towards the tips
**Foliage:** 1–2 annual short fine leaves, withering when the leafless flowering stem develops
**Flowers:** Numerous mauve to violet flowers 15 mm wide, consisting of three distinctly fringed petals, borne singly at the end of short branches on the flowering stem; August to November
**Requirements:** Well drained soils
**Comments:** A delightful little climber, not noticeable until in flower. Stem and leaves annual, emerging from the dormant tuber in Winter.
**Propagation:** Seed
**Localities:** 12, 16, 26, 29#, 34A–37, 39, 41–44, 47–51, 53–56, 58, 59, 61, 63, 64, 67–71, 74–76
**Distribution:** Vic, NSW, Tas, SA, WA

*Thysanotus tuberosus* ssp. *tuberosus*

## Thysanotus tuberosus ssp. tuberosus — Common Fringe-lily

**F P**

**Size:** 15–30 cm x 15–20 cm
**Habitat:** Riparian scrub, red gum woodland, valley and dry sclerophyll forests
**Form:** Slender, erect herbaceous perennial tuft
**Foliage:** Few narrow greyish leaves to 20 cm long, channelled for most of length
**Flowers:** Umbels of 1–8 fringed bright mauve flowers to 35 mm wide; November to January. Flowers held above the foliage on leafless many-branched wiry stems.
**Requirements:** Open well drained soils, tolerating dryness once established
**Comments:** An attractive herb which is usually not noticed unless it is flowering. The flowers only last 1 day, generally opening early in the morning and closing by midday in hot weather, but plants continue to produce flowers over some months. The leaves shoot annually from the tuberous rootstock and may persist until after flowering or wither as flowering commences. The Aborigines ate the tubers and cooked and ground the leaves and flowers. While widespread it is becoming less common locally.
**Propagation:** Seed
**Localities:** Cheltenham#, 29, 35, 42, 47–51, 53–56, 58, 59, 61, 67–69, 71, 74–76
**Distribution:** Vic, Qld, NSW, SA

*Tricoryne elatior*

# Tricoryne — Liliaceae (Anthericaceae)

## Tricoryne elatior — Yellow Rush-lily

**F P**

**Size:** 30–50 cm x 30–50 cm
**Habitat:** Widespread in plains grassland, red gum and box woodland, dry and valley sclerophyll forests, grassy low open forest, tea-tree heath
**Form:** Wiry, many-branched, grass-like rhizomatous perennial herb
**Foliage:** Few, widely-spaced grass-like linear grey-green leaves to 90 mm long, often withering at flowering, reduced and scale-like on upper branches
**Flowers:** Many bright yellow star-shaped flowers in open clusters, held above the foliage; September to February
**Requirements:** Moist or dry soils
**Comments:** An attractive, long flowering little plant for the garden, which is not noticed until flowering commences. Used to good effect in mass plantings or informal drifts. Petals and sepals twist spirally after flowering.
**Propagation:** Seed, division
**Localities:** Laverton Nth, 5, 6, 12, 15–18, 21, 24–26, 28, 31, 32, 34A–42, 44, 45, 47–55, 58, 59, 61–68, 69A, 71, 74–76
**Distribution:** All states

# Wurmbea Liliaceae (Colchicaceae)

## Wurmbea dioica Early Nancy

**Size:** 5–30 cm x 10 cm **F P**
**Habitat:** Red gum and sclerophyll woodland, dry, valley and damp sclerophyll forests, grassy low open forest, tea-tree heath
**Form:** Small glabrous perennial herb
**Foliage:** 3 fleshy, narrow stem-sheathing leaves to 100 mm long
**Flowers:** 1–11 white starry 6-petalled flowers, usually with a purplish band (or nectary) toward the base of each sepal and petal. Most plants are either male or female. Males have red or purple anthers. Flowers July to October. Flowers spaced one above the other along the single zig-zagging flowering stem.
**Requirements:** Moist well-drained soils
**Comments:** The honey-scented flowers may be monoecious or dioecious. Dormant over Summer, it shoots again from its bulbous rootstock as soon as Autumn rains fall. It takes its common name from being the first lily to flower each year. Tubers were eaten by the Aborigines.
**Propagation:** Seed
**Localities:** 16, 25, 26, 29, 30, 32, 34A, 35, 37, 38, 41, 42, 44–51, 53–56, 58, 59, 61–64, 67, 68, 69A, 70, 74–76
**Distribution:** All states except NT
**Synonym:** *Anguillaria dioica*

*Wurmbea dioica*

SCALE: 0 1cm 2cm The scale representation on each plant equals 2cm.

# *Orchids* *Orchidaceae*

Orchids occur in a number of diverse habitats, ranging from the coast to the mountains and may be found in heaths, scrub, open forests and woodlands. All orchids in Melbourne are terrestrial with the exception of *Sarcochilus australis*, the most southerly occurring epiphytic orchid in Australia.

The orchids which occur locally are perennial herbs and may be distinguished from other monocotyledons by their floral structure. All have 6 perianth segments (3 petals and 3 sepals). The labellum is the modified lowest petal and is usually different to the two side petals. The male and female parts are united in a structure known as the column. The labellum is modified in some genera to sexually attract the males of specific insects which then pollinate the flower. The fruit is a capsule which, in most species, dehisces (splits open at maturity) releasing a large number of minute seeds.

Most Australian terrestrial species grow from a tuberoid which is replaced annually. Some species also produce daughter tuberoids and may form extensive colonies. Other species including the saprophytic, *Dipodium roseum*, grow from rhizomes. *Cryptostylis* spp. and *Spiranthes sinensis* also grow by this means but, unlike other terrestrial orchids, these species do not undergo a Summer dormancy period.

The majority of Australian orchids are difficult to propagate as they rely on a symbiotic association (or mycorrhiza) between a soil-borne fungus and their seeds for at least some stages of their development. The fungus penetrates the seed and provides the necessary nutrients for the developing orchid.

Fire plays an important role in the regeneration of orchids and some species such as *Burnettia cuneata* may not be seen for many years, only reappearing after a major fire. Others will flower more prolifically, gradually having less blooms with each subsequent year after the bushfire.

The tuberoids of many indigenous orchid spp. provided a staple diet for the local aboriginal tribes. However continued development and human pressures in the Melbourne region have led to many species becoming rare and locally endangered. Some species are currently verging on extinction while others have already disappeared. Specific localities have not been given for the orchids as a means of protecting the remaining local populations.

Orchids must never be transplanted from the bush as, without specialist knowledge, they generally do not survive in cultivation. At the same time removal will lead to their ultimate extinction in their own habitat. Orchids may be obtained from time to time from specialist Australian plant nurseries or from the Australian Native Orchid Society during their sales and exhibitions. The more reliable species, plus advice, are available where orchids are offered for sale.

Currently the Australian Orchidaceae is undergoing revision. This has already led to many new species being described and many more changes are likely to occur in the future as the taxonomists continue their work.

## ORCHIDS

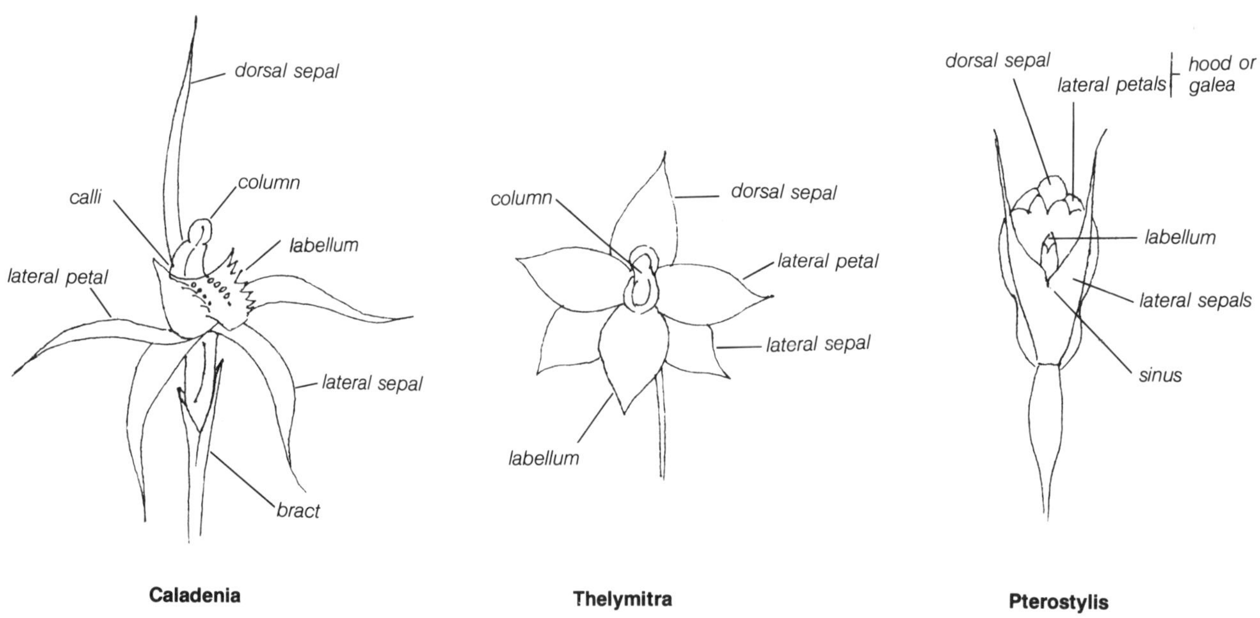

# Acianthus

## Acianthus caudatus

Mayfly Orchid

**Size:** Stem 5–25 cm high **P**
**Habitat:** Dry and valley sclerophyll forests, tea-tree heath, grassy low open forest
**Form:** Slender erect tuberous terrestrial herb
**Foliage:** Single heart-shaped dark green basal leaf to 30 mm long, purple below; held on stem, above ground level
**Flowers:** 1–9 crimson to dark purple-red flowers clustered at the end of the stem. Sepals long and thread-like, dorsal to 30 mm long, laterals to 15 mm long; petals to 5 mm long curve downwards.
Labellum is smooth, ovate and recurved, with 2 prominent glands at the base. May to October
**Requirements:** Well drained clay loam or sandy soils, remaining moist during the growing season
**Comments:** On warm days the flowers exude an unpleasant odour. While suited to a container this colony-producing species is difficult to keep growing. *Acianthus* spp. are pollinated by small flies.
**Propagation:** Seeds, increase in tuberoids
**Localities:** Eltham, Heathmont-Bayswater, Dandenong Rges, Belgrave, Langwarrin, Cranbourne, 34A, 42, 50A, 54, 74
**Distribution:** Vic, NSW, Tas, SA

*Acianthus caudatus*

## Acianthus pusillus

Gnat or Mosquito Orchid

**Size:** stem 5–20 cm high **P**
**Habitat:** Widespread in dry, damp and valley sclerophyll forests, tea-tree heath
**Form:** Slender erect tuberous herb
**Foliage:** Single green heart-shaped leaf to 40 mm long, purplish below, margins entire, held up to 30 mm above the ground
**Flowers:** Terminal raceme of 3–20 greenish to pinkish flowers. Sepals small to 8 mm long, tapering and gland-tipped; petals smaller and bent backwards.
Labellum is narrow with 2 glands at the base. March to August
**Requirements:** Moist soils, drying out over Summer
**Comments:** Easily grown in a pot where it will flower well and form colonies if watered from late February. Mix must be well drained. Wild populations may form large colonies in ideal conditions. Distinguished from *A. exsertus* by its paler flower colour and smaller labellum.
**Propagation:** Seed, increase in tuberoids
**Localities:** Eltham, Heathmont-Boronia, Montrose, Hampton, Cheltenham, Seaford, Keysborough, Cranbourne, 34A, 42, 50A, 51, 56, 58, 63, 65#–67, 71, 74
**Distribution:** Vic, Qld, NSW, Tas, SA
**Synonym:** Previously included with *A. exsertus.*

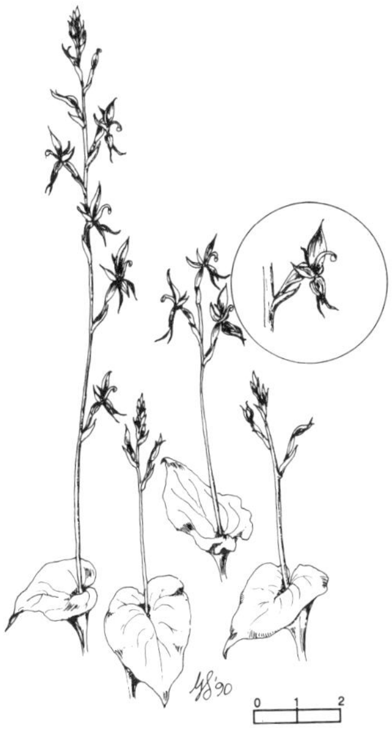


*Acianthus pusillus*

## Acianthus reniformis = Cyrtostylis reniformis

# Arthrochilus

## Arthrochilus huntianus

Elbow Orchid

**Size:** Stem to 15 cm high **P**
**Habitat:** Scattered in dry sclerophyll forest, often on ridges
**Form:** Leafless, erect tuberous herb with a reddish, wiry stem
**Flowers:** Terminal spike of 2–10 tiny, reddish-green insect-like flowers. Slender, reflexed sepals and petals to 5 mm long.
Labellum, to 6 mm long, is hinged at the base to the column. Its mid-section is fringed with long purplish-red hairs with 2 rough, hairy tails from the lower end. November to March
**Requirements:** Well drained soil
**Comments:** A saprophytic orchid feeding off dead and decaying organic matter. Locally rare and difficult to see but in good seasons it is more abundant.
**Propagation:** Impossible to cultivate
**Localities:** Dandenong Rges, Langwarrin, 42#
**Distribution:** Vic, NSW, Tas

*Arthrochilus huntianus*

SCALE: 0 1cm 2cm — The scale representation on each plant equals 2cm.

*Burnettia cuneata*

## Burnettia

### Burnettia cuneata

Lizard Orchid

**Size:** Stem 8–15 cm high **P**
**Habitat:** Wet peaty heathland, tea tree swamp
**Form:** Terrestrial saprophytic herb with purplish-brown fleshy but brittle stem
**Foliage:** Reduced to brownish fleshy sheathing scales, sometimes with a well developed basal bract to 20 mm long
**Flowers:** Clusters of 1–7 flowers to 25 mm across, white inside, greenish-brown to purplish-brown with darker veins on the outside. Fleshy, widely-spreading sepals and petals to 13 mm long; dorsal sepal erect broad and deeply hooded. September to December
**Requirements:** Wet peaty soils
**Comments:** Rarely seen, remaining dormant for years and flowering for one season after a bushfire. Rare in Victoria, it may be extinct within the Melbourne area.
**Propagation:** Impossible to cultivate
**Localities:** Cheltenham#, 50A#, 67#, 74#
**Distribution:** Vic, NSW, Tas

# Caladenia

Caladenias are dainty terrestrial orchids growing from a round tuberoid and are common in the grassy understorey of heathlands and open forests. Most species in Melbourne are now rare and in danger of extinction due to loss of habitat and weed invasion. They have a single basal leaf which is usually hairy. The hinged labellum is 3-lobed, usually fringed or toothed and bears rows of sessile or stalked calli. The column is erect, incurved and usually winged in the upper part.

*Caladenia* spp. may be divided into 2 groups. The first group has relatively small flowers and 4 of the perianth segments projecting forward like fingers. The Fairy or Finger Orchids usually occur in shades of pink, white or blue. The second group has larger flowers which are often suffused with red, and thread-like perianth segments. This group is commonly referred to as 'Spider Orchids'.

Species of *Caladenia* are difficult to grow and maintain in cultivation with most only reproducing by seed. Some will grow and reproduce through the production of daughter tuberoids, but these may be shy to flower. Bees pollinate the colourful species while the maroon and green flowered species sexually attract male thynnid wasps.

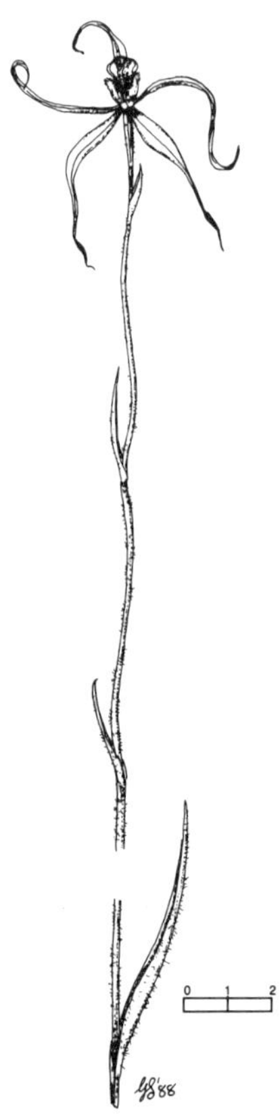

*Caladenia australis*

### Caladenia aurantiaca

Orange-tip Caladenia

**Size:** stem to 15 cm high **P**
**Habitat:** Beside swamps in tea-tree heaths
**Form:** Erect herb with thin wiry stem
**Foliage:** Narrow linear sparsely hairy leaf to 80 mm long
**Flowers:** 1–2 white flowers to 20 mm across. Perianth segments to 10 mm long. Labellum—3-lobed, lateral lobes erect, mid-lobe of labellum is orange with toothed margins. Stalked calli are clubbed and orange. September to November
**Requirements:** Moist peaty sands
**Comments:** Related to *C. carnea*, with which it is often confused. However, *C. aurantiaca* may be readily distinguished by the orange-tipped labellum. Impossible to cultivate. Known from one small colony in the Melbourne area.
**Localities:** Cranbourne
**Distribution:** Vic, NSW, Tas

### Caladenia australis

**Size:** stem to 30 cm high **P**
**Habitat:** Dry sclerophyll forest
**Form:** Erect herb with hairy, wiry stem
**Foliage:** Leaf hairy, to 120 mm long
**Flowers:** 1–2 creamy-yellow or reddish flowers to 60 mm across. Perianth segments to 50 mm long, with long dark clubbed tips.
Labellum is red, lateral lobes erect, mid-lobe recurved, margins shortly toothed. September to November
**Requirements:** Well drained soil
**Comments:** Differs from *C. reticulata* in that the labellum is red rather than cream with red stripes and margins.
**Localities:** Hurstbridge, Templestowe, Dandenong Rges, Frankston-Cranbourne, 42
**Distribution:** Vic, Tas, SA
**Synonym:** Previously confused with *C. reticulata* which does not occur further east than Stawell.

## Caladenia caerula
Blue Fingers

**Size:** stem to 15 cm high — P
**Habitat:** Dry sclerophyll forest, dry stony ground in box ironbark woodland, tea-tree heath
**Form:** Erect herb with hairy, wiry stems
**Foliage:** Oblong to linear-lanceolate, sparsely hairy leaf to 70 mm long
**Flowers:** Single blue flower to 25 mm across, paler on the outside. Perianth segments, to 20 mm long, have scattered tiny dark blue glands.
Labellum—3-lobed with dark blue transverse bands, lateral lobes are entire and erect. July to September
**Requirements:** Well drained soil
**Comments:** Difficult to maintain in cultivation.
**Localities:** Sth Morang, Diamond Ck, Eltham, Warrandyte, Heathmont, Belgrave, Black Rock#, 50A, 74
**Distribution:** Vic, Qld, NSW, Tas

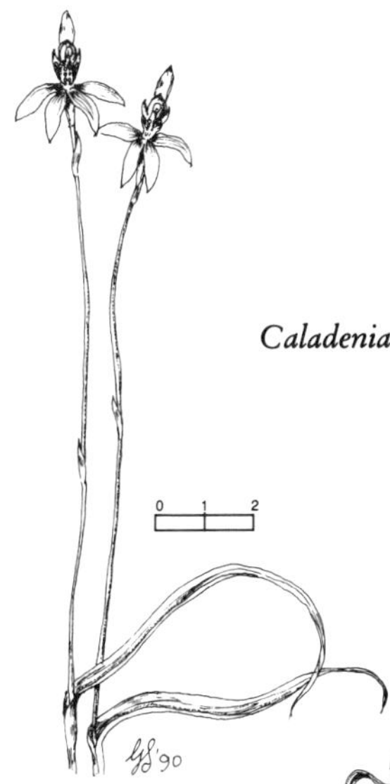

*Caladenia caerula*

## Caladenia cardiochila
Fleshy-lipped Caladenia

**Size:** stem to 30 cm high — P
**Habitat:** Coastal scrub and heathlands
**Form:** Erect herb
**Foliage:** Thick, hairy, linear-lanceolate leaf to 110 mm long
**Flowers:** Single yellowish flower to 30 mm across, each segment with a brown-red central stripe. Perianth segments spreading, to 20 mm long ending in glandular tips which are sometimes club-shaped; dorsal sepal hooded; lateral sepals wider than petals.
Labellum cordate to cordate-ovate, the tip thickened and slightly crenulate, very dark red stalked calli densely crowded in a narrow band from the base to about ½ its length. September to October
**Requirements:** Well drained sandy soils
**Localities:** Collected in Sandringham 1939; 76
**Distribution:** Vic, SA
**Synonym:** Previously included with *C. tessellata* from NSW, which also has isolated occurrences in East Gippsland.

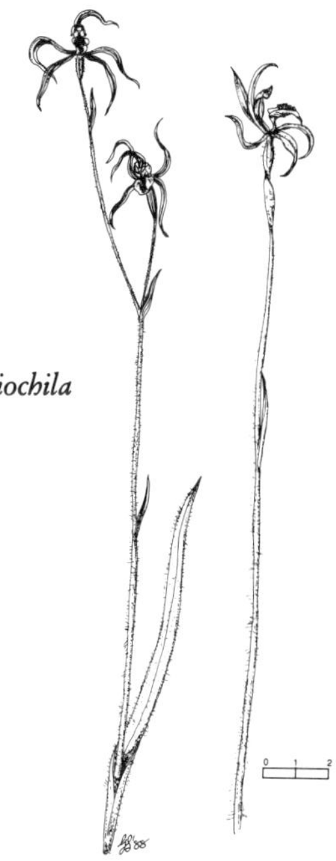

*Caladenia cardiochila*

## Caladenia carnea var. carnea
Pink Fingers

**Size:** stem 8–25 cm high — F P
**Habitat:** Widespread in dry sclerophyll forest, sclerophyll and red gum woodland, tea-tree heath
**Form:** Erect hairy herb with a wiry stem
**Foliage:** Narrow-linear leaf 8–20 cm long
**Flowers:** 1–3 pink flowers to 30 mm across, less commonly white to greenish-white. Labellum and column have dark red bars. Perianth segments to 15 mm long.
Labellum to 8 mm long, 3-lobed, lateral lobes erect, yellowish mid-lobe recurved, triangular, rarely with short marginal teeth. 2 rows of yellow calli. August to December
**Requirements:** Well drained soil
**Comments:** One of the easiest Caladenias to grow. White forms predominate in the Dandenongs.
**Localities:** Somerton, Bundoora, Hurstbridge-Eltham, Templestowe, Park Orchards, Box Hill, Heathmont, Dandenong Rges, Noble Park, Cranbourne, 33, 34A, 42, 44, 47, 49, 50A, 56, 58, 61, 65#, 67#, 74
**Distribution:** Vic, Qld, NSW, Tas, SA

*Caladenia carnea* var. *carnea*

## Caladenia catenata
White Fingers, White Caladenia

**Size:** stem to 30 cm high — P
**Habitat:** Dry and valley sclerophyll forests, red gum woodland, grassy low open forest
**Form:** Erect herb with hairy wiry stem
**Foliage:** Sparsely hairy linear leaf to 120 mm long
**Flowers:** 1–2 white flowers to 30 mm across. Perianth segments to 15 mm long, lateral sepals deflexed
Labellum—3-lobed, lateral lobes entire and erect, the longer mid-lobe recurved with marginal teeth, tip orange-yellow and calli club-like. April to October
**Requirements:** Well drained soils
**Comments:** Confused with *C. carnea* var. *carnea* but readily distinguished by the longer, toothed orange mid-lobe and lack of red bars on the labellum. The flowers are also more open. Seldom seen in Melbourne now, except after fire. Readily cultivated.
**Localities:** Bundoora, Eltham, Dandenong Rges, Belgrave, 42, 50A, 55, 63, 71, 74
**Distribution:** Vic, Qld, NSW
**Synonym:** *C. alba*

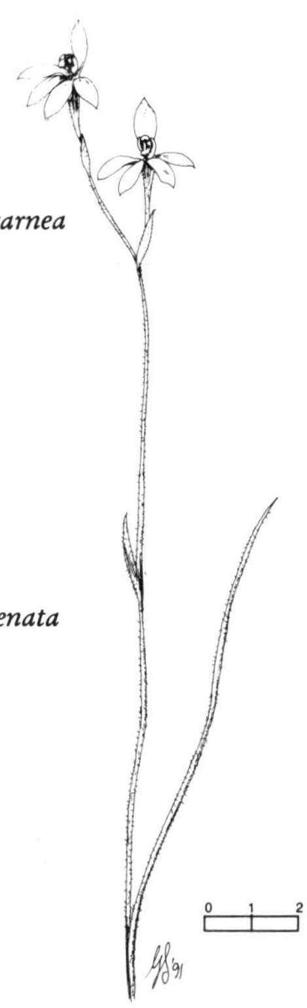

*Caladenia catenata*

SCALE: 0 1cm 2cm — The scale representation on each plant equals 2cm.

*Caladenia clavigera*

## Caladenia clavigera

Clubbed or Plain-lipped Spider Orchid

**Size:** stem 10–40 cm high **F P**
**Habitat:** Widespread but scattered occurrences in dry and valley sclerophyll forests
**Form:** Erect herb with hairy, wiry reddish stem
**Foliage:** Hairy lanceolate leaf to 15 cm long, red-spotted at the base
**Flowers:** 1–2 greenish-yellow flowers to 60 mm across. Perianth segments have a prominent red central stripe, sepals, to 50 mm long, taper to clubbed tips; petals are shorter and not clubbed.
Labellum to 12 mm long, cordate, greenish with dark red recurved mid-lobe and 4–6 rows of clubbed calli. September to December
**Requirements:** Moist well drained soils
**Comments:** Locally rare.
**Localities:** Heathmont-Boronia, Belgrave, 42, 44, 50A
**Distribution:** Vic, NSW, Tas, SA

## Caladenia concinna

**Size:** stem to 20 cm high **P**
**Habitat:** Sparse scrub in dry sclerophyll forest
**Form:** Erect herb with a wiry, hairy stem
**Foliage:** Hairy oblong to elliptic leaf to 120 mm long, red-spotted at the base
**Flowers:** 1–2 pale yellowish-green flowers with red markings, to 50 mm across. Perianth segments bent backwards, with faint red stripe, to 35 mm long; dorsal sepal erect, sepals ending in a poorly developed, yellowish club.
Labellum—broadly flared, to 18 mm across, green with maroon in the upper half. Lateral lobes have few broad teeth, mid-lobe tip is recurved, dark red calli are club-like in 2 rows. September to October
**Requirements:** Well drained soils
**Comments:** Locally rare. Has been confused with *C. toxochila* in the past.
**Localities:** Wattle Glen, Greensborough, Belgrave
**Distribution:** Vic, NSW
**Synonym:** *C. dilatata* var. *concinna*

*Caladenia deformis*

## Caladenia congesta

Black-tongue Caladenia

**Size:** stem to 0.6 m high **P**
**Habitat:** Valley sclerophyll forest, swamp margins of wattle tea-tree scrub
**Form:** Erect slightly hairy herb with slender stem
**Foliage:** Linear leaf to 18 cm long
**Flowers:** 1–3 bright pink or reddish flowers to 30 mm across, with a conspicuous black labellum. Perianth segments spreading and flat to 20 mm long; dorsal sepal erect, shorter and hooded.
Labellum—3-lobed, lateral lobes erect, mid-lobe longer and very narrow, completely covered with 2 rows of flat, sessile blackish calli. October to December
**Requirements:** Intolerant of high rainfall and high humidity
**Comments:** Very difficult to cultivate, requiring very well drained mixture.
**Localities:** Ringwood-Bayswater, Dandenong Rges, Belgrave, Lysterfield, Moorabbin#, 33, 34A, 67#
**Distribution:** Vic, NSW, Tas, SA

## Caladenia deformis

Blue Fairies, Bluebeard Caladenia

**Size:** stem 10–20 cm high **F P**
**Habitat:** Widespread in dry sclerophyll forest, grassy low open forest, tea-tree heath
**Form:** Erect herb with wiry stem, growing in tufts
**Foliage:** Bright green linear leaf to 100 mm long
**Flowers:** Single dark blue faintly perfumed flower to 40 mm across, with scattered tiny purple glands. Perianth segments lanceolate to 25 mm long, spreading; dorsal sepal erect.
Labellum—3-lobed, lateral lobes erect, entire, mid-lobe triangular, recurved, deep purple and fringed. 4–8 dense rows of calli. June to October
**Requirements:** Well drained soils
**Comments:** Now locally rare. More commonly found in shrubby coastal heaths.
**Localities:** Campbellfield, Bayswater#, Belgrave, Cranbourne, 41, 42, 47, 49, 50A, 58, 67#, 74
**Distribution:** Vic, NSW, Tas, SA, WA

## Caladenia dilatata
Green Comb Spider Orchid

**Size:** stem to 45 cm high **F P**
**Habitat:** Widespread in dry, valley and damp sclerophyll forests, red gum woodland, tea-tree heath
**Form:** Tall hairy herb with wiry stems
**Foliage:** Hairy oblong to elliptical leaf to 120 mm long
**Flowers:** 1–2 large green or yellowish-green flowers to 100 mm across. Perianth segments to 70 mm long, each having 3 red stripes; sepals tipped with yellowish hairy clubs, petals thread-like.
Labellum—green, 3-lobed, maroon towards the tip, lateral lobes have linear teeth on margins, mid-lobe is recurved with short, blunt teeth. Clubbed, reddish-purple linear calli crowded in 4 rows. Column has 2 yellow basal glands. August to January
**Requirements:** Well drained soils
**Comments:** One of the more common *Caladenia* species found in Melbourne. Variable in form, the species is currently undergoing revision.
**Localities:** Bundoora, Wattle Glen, Greensborough, Eltham, Belgrave, 34A, 42, 47, 49, 50A, 51, 55, 56, 58, 67#, 74, 76
**Distribution:** Vic, Qld, NSW, Tas, SA

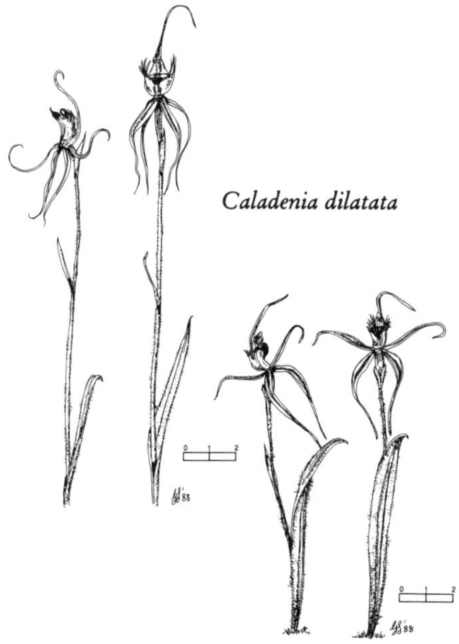

*Caladenia dilatata*

## Caladenia filamentosa var. filamentosa
Daddy Long Legs

**Size:** stem 10–40 cm high **F P**
**Habitat:** Valley and damp sclerophyll forests
**Form:** Erect herb covered in soft long hairs with a thin wiry stem
**Foliage:** Narrow-linear leaf to 18 cm long
**Flowers:** 1–4 deep crimson, reddish, or occasionally greenish-white flowers, to 80 mm across. Perianth segments to 60 mm long, flat, becoming thread-like, droopy and entangled.
Labellum—narrow with darker veins, curved, with calli on margins, tip entire, 2 rows of calli on back half of tongue. September to October
**Requirements:** Moist, well drained soils
**Comments:** Usually grows in small colonies. Locally rare.
**Localities:** Yarrambat, Boronia, The Basin, 58
**Distribution:** Vic, NSW, Tas, SA

*Caladenia filamentosa* var. *filamentosa*

## Caladenia flavovirens
Summer Spider Orchid

**Size:** stems to 40 cm high **P**
**Habitat:** Dry and valley sclerophyll forests with grassy understorey
**Form:** Robust, erect herb
**Foliage:** Coarsely-haired lanceolate leaf to 18 cm long
**Flowers:** 1–2 bright yellow-green flowers to 70 mm across. Perianth segments more than 50 mm long and drooping, sepals longer ending in club-like tips, dorsal sepal erect.
Labellum broadly ovate, undivided, sometimes with dark red tip. 4–6 rows of calli. November to January
**Requirements:** Well drained soil
**Comments:** Regenerates well after fire. Its habitat locally is being threatened by weed invasion and encroaching housing development.
**Localities:** Montrose, Boronia, The Basin, 42, 50A, 55, 58
**Distribution:** Vic, NSW, possibly Tas
**Synonym:** Formerly referred to as *C. pallida* which is endemic to Tasmania.

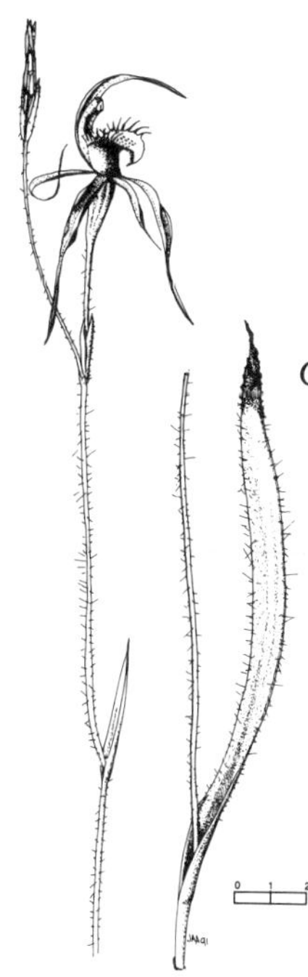

*Caladenia flavovirens*

## Caladenia gracilis
Musky Caladenia

**Size:** stem 10–40 cm high **P**
**Habitat:** Dry and valley sclerophyll forests, grassy low open forest
**Form:** Sparsely hairy erect herb with thin, wiry stem
**Foliage:** Thick linear leaf to 30 cm long
**Flowers:** Raceme of 1–6 white musk-scented flowers to 30 mm across, outer surfaces purple or greenish-brown. Spreading perianth segments to 15 mm long; dorsal sepal erect, hooded.
Labellum fringed along the end with purple, red or yellow calli. 4 rows of stalked yellow or purple calli also along tongue. September to November
**Requirements:** Well drained soil
**Comments:** The hooded dorsal sepal may cause confusion with *C. cucullata* which grows north of Melbourne. *C. gracilis* may be distinguished by the strong musky scent exuded on warm days. Locally rare.
**Localities:** Park Orchards, Doncaster, Heathmont-Boronia, Belgrave, Montrose, 42, 51, 55, 56, 58, 74
**Distribution:** Vic, Qld, NSW, Tas, SA
**Synonym:** *C. angustata*

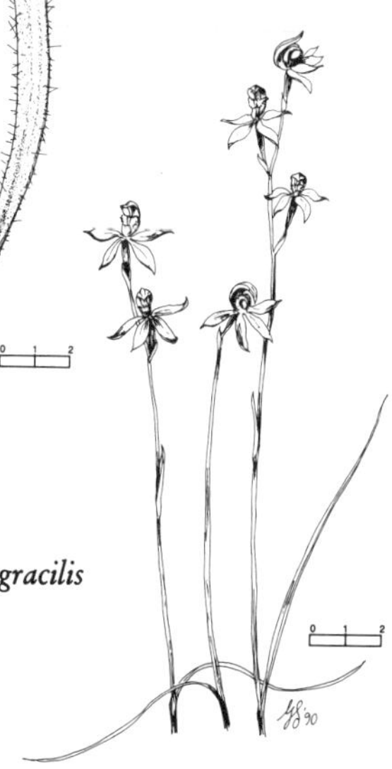

*Caladenia gracilis*

SCALE: 0 1cm 2cm — The scale representation on each plant equals 2cm.

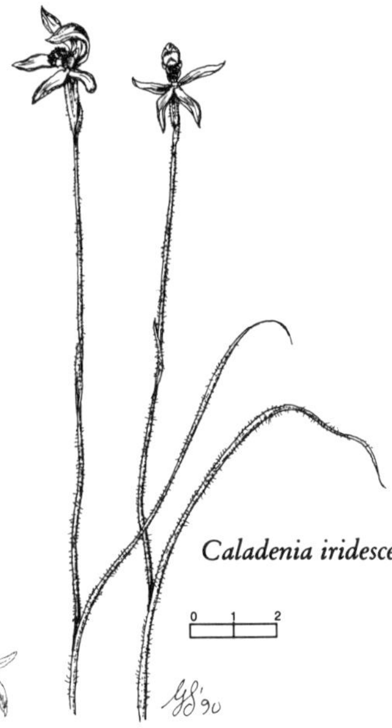

*Caladenia iridescens*

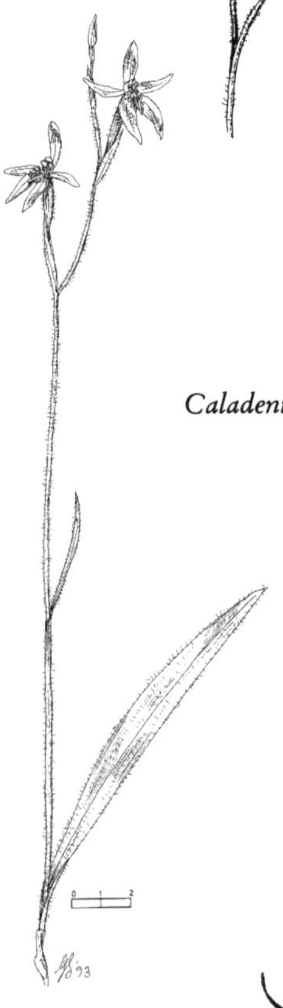

*Caladenia latifolia*

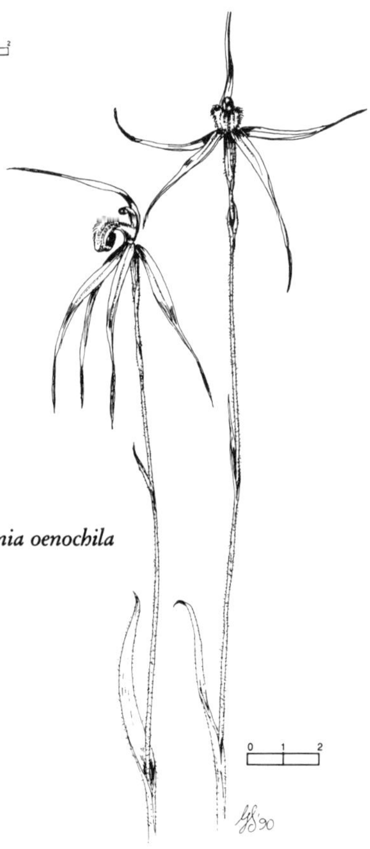

*Caladenia oenochila*

## Caladenia iridescens

Bronze Caladenia

**Size:** stem 10–25 cm high **P**
**Habitat:** Valley sclerophyll forest, grassy low open forest
**Form:** Erect herb with wiry purplish stems
**Foliage:** Narrow-linear leaf to 100 mm long
**Flowers:** 1–4 dusky red, purplish-crimson to greenish flowers to 20 mm across. Narrow, widely spaced perianth segments to 10 mm long.
Labellum 3-lobed, white with red or purple transverse bars, lateral lobes erect, purplish mid-lobe recurved with glandular-headed calli. September to November
**Requirements:** Well drained soils
**Comments:** Locally rare.
**Localities:** Heathmont-Boronia, Lysterfield, Montrose, Kilsyth, 50A, 51, ?55, 74
**Distribution:** Vic, NSW, Tas

## Caladenia latifolia

Pink Fairies

**Size:** stem to 40 cm high **F P**
**Habitat:** Tea-tree heath, coastal banksia woodland
**Form:** Hairy erect herb with wiry stems
**Foliage:** Bright green, very hairy oblong-lanceolate leaf to 20 cm long
**Flowers:** 1–4 bright pink, rarely white, flowers to 35 mm across. Perianth segments spreading; dorsal sepal erect, sepals to 18 mm long, petals shorter.
Labellum 3-lobed, narrow and pink-striped, lateral lobes erect, mid-lobe recurved with 3–4 pairs of long teeth on margin, calli in 2 converging rows. August to December
**Requirements:** Well drained sandy soil
**Comments:** Easily grown but rarely flowers in cultivation. Water regularly during growing season. Forms clumps which need to be separated to maintain vigour.
**Localities:** Hampton, Cheltenham, Beaumaris, Seaford, 67#, 74
**Distribution:** Vic, NSW, Tas, SA, WA

## Caladenia menziesii

Hare or Rabbit Orchid

**Size:** stem 5–25 cm high **P**
**Habitat:** Damp and valley sclerophyll forests, tea-tree heath, grassy low open forest
**Form:** Slender erect herb with glabrous, wiry stem
**Foliage:** Bright green glabrous ovate to ovate-lanceolate leaf to 90 mm long
**Flowers:** 1–3 pink and white flowers to 15 mm across. White lateral sepals spreading, to 10 mm long; dorsal sepal reddish, hooded; dark red petals held erect like ears, to 20 mm long.
Labellum white with transverse pink bars, sides erect, tip white and recurved, 2–4 rows of yellow calli. August to November
**Requirements:** Moist well drained soils
**Comments:** Flowers profusely after fires forming dense colonies. Easy to grow but shy to flower under cultivation, it reproduces by tuberoids, forming a clump. Banana skins placed beside the plants produce ethylene gas which may induce them to flower.
**Localities:** Belgrave, Cheltenham, Beaumaris, 34A, 42, 47, 49, 55–58, 67#, 74, 76
**Distribution:** Vic, NSW, Tas, SA, WA
**Synonym:** *Leptoceras menziesii*

## Caladenia oenochila

**Size:** stem to 40 cm high **P N**
**Habitat:** Damp and valley sclerophyll forests
**Form:** Hairy erect herb with wiry stem
**Foliage:** Linear-lanceolate leaf to 15 cm long
**Flowers:** 1–3 sweetly perfumed pale yellow-green flowers to 80 mm across, with very dark red hairs on the outer surface.
Labellum is undivided and usually deep red, paler towards the base and ovate-cordate. Marginal teeth and calli are deep red. August to November
**Requirements:** Moist well drained soil
**Comments:** Locally very rare. Will sometimes grow and flower in a pot for a few years but will not increase in number. Similar to *C. patersonii* from which it differs mostly in flower colour.
**Localities:** Nunawading-Boronia, The Basin, Lysterfield, 33, 34A, 42, 55, 56, 58, 63
**Distribution:** Endemic to Vic
**Synonym:** Previously known as *C. patersonii* which is endemic to Tas, and *C.* sp. aff. *patersonii* (foothills)

## Caladenia praecox

Early Caladenia

| | |
|---|---|
| **Size:** | stem to 15 cm high **P** |
| **Habitat:** | Dry and valley sclerophyll forests |
| **Form:** | Erect herb with thin wiry stem |
| **Foliage:** | Sparsely hairy very narrow leaf to 15 cm long |
| **Flowers:** | 1–4 white, or rarely pink, flowers to 30 mm across, with pink or green markings. Perianth segments, to 15 mm long, are dark red on the outside.<br>Labellum—white spotted with red, lateral lobes erect, mid-lobe crimson or purple, fringed white, yellow or purple. Stalked, clubbed calli white or yellow. August to October |
| **Requirements:** | Well drained soil |
| **Comments:** | Occurs in open forests where it is protected by grassy tussocks. |
| **Localities:** | Diamond Ck, Eltham, Heathmont-Boronia, Mooroolbark, Montrose, The Basin, Belgrave, Lysterfield, 34A, 41, 49, 51, 56, 58 |
| **Distribution:** | Vic, NSW, Tas |

## Caladenia pusilla

Tiny Caladenia

| | |
|---|---|
| **Size:** | stem to 9 cm high **P** |
| **Habitat:** | Valley sclerophyll forest, grassy low open forest |
| **Form:** | Tiny erect herb |
| **Foliage:** | Linear, sparsely hairy leaf to 60 mm long |
| **Flowers:** | Tiny single cream to deep pink flower to 15 mm across, with a central stripe, reddish or dark brown on outside. Perianth segments to 7 mm long; lateral sepals differ from *C. carnea* var. *carnea* in that they are relatively broad and are joined together in the basal half.<br>Labellum—3-lobed, white, transversely barred with red; lateral lobes oval, erect, mid-lobe narrow-triangular, fringed and yellow tipped. 2 rows of yellow calli. September to October |
| **Requirements:** | Well drained soils |
| **Localities:** | Croydon, Belgrave, 42#, 67#, 74, 76 |
| **Distribution:** | Vic, NSW, Tas, SA |
| **Synonym:** | *C. carnea* var. *pygmaea* |

## Caladenia robinsonii

| | |
|---|---|
| **Size:** | stem to 35 cm high **P** |
| **Habitat:** | Eucalyptus viminalis woodland with grassy understorey |
| **Form:** | Small erect hairy herb |
| **Foliage:** | Narrow-lanceolate leaf to 110 mm long |
| **Flowers:** | 1–2 cream or greenish flowers to 60 mm across. Perianth segments to 35 mm long, becoming thread-like.<br>Labellum—small with short lateral teeth. September to November |
| **Requirements:** | Well drained sandy soil |
| **Comments:** | This species is possibly extinct as the only known recent occurrence has been destroyed for housing. |
| **Localities:** | Frankston |
| **Distribution:** | Endemic to Vic |
| **Synonym:** | Previously known as C. sp. aff. *reticulata* |

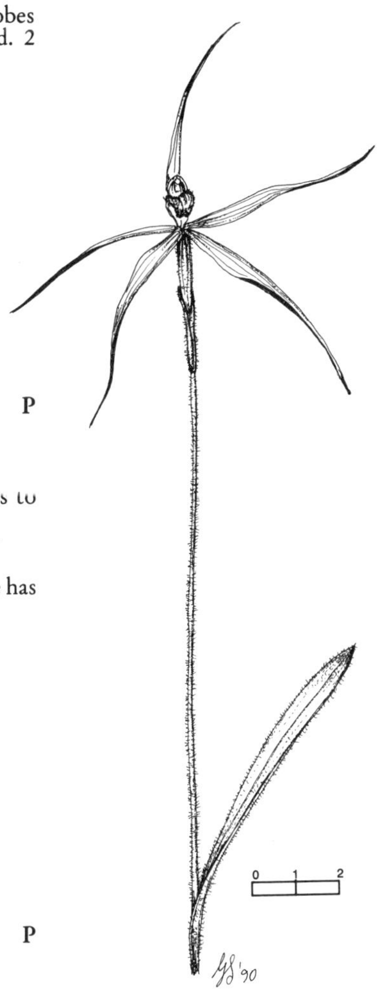


*Caladenia rosella*

## Caladenia rosella

| | |
|---|---|
| **Size:** | stem 10–17 cm high **P** |
| **Habitat:** | Dry sclerophyll forest |
| **Form:** | Small erect hairy herb with a greenish to reddish-purple stem |
| **Foliage:** | Hairy, linear-lanceolate leaf to 100 mm long. |
| **Flowers:** | Single brilliant pink flower to 60 mm across, with musky fragrance.<br>Labellum has 6 widely-spaced rows of finger-like light rose pink or yellowy-pink calli. July to October |
| **Requirements:** | Well drained soil |
| **Comments:** | Similar to *C. patersonii* but has pink flowers and is smaller in all its parts. An endangered species with a very restricted occurrence. |
| **Localities:** | Hurstbridge, Eltham |
| **Distribution:** | Endemic to Vic |

SCALE: 0 1cm 2cm — The scale representation on each plant equals 2cm.

*Caladenia tentaculata*

## Caladenia tentaculata

Spider Orchid

**Size:** stem to 40 cm high — P
**Habitat:** Valley sclerophyll forest, grassy low open forest
**Form:** Erect hairy herb with a thin wiry stem
**Foliage:** Lanceolate leaf to 15 cm long, often red-spotted near the base
**Flowers:** 1–2 yellowish-green flowers to 100 mm across, with crimson stripes and markings. Perianth segments to 70 mm long, becoming thread-like, with 3 reddish stripes; sepals tipped with yellowish hairy clubs, lateral sepals projecting forward, dorsal sepal erect; petals thread-like, bent backwards.
Labellum—3-lobed, as broad as it is long, to 20 mm across, green with prominent white patch and maroon tip; lateral lobes have blunt marginal teeth to 10 mm long. Crowded reddish calli in 4 rows. August to October
**Requirements:** Moist, well drained soils
**Comments:** Similar to *C. dilatata*, but with a broader labellum.
**Localities:** 42, 74, 76
**Distribution:** Vic, SA

## Caladenia thysanochila

**Size:** stem to 25 cm high — F P
**Habitat:** Eucalyptus viminalis woodland
**Form:** Hairy erect herb
**Foliage:** Linear-lanceolate stem-clasping leaf to 20 cm long
**Flowers:** Small white flowers with reddish-purple tinges. Light purplish-brown club-like tips on petals and sepals.
Labellum densely fringed with reddish-purple teeth. Up to 4 rows reddish-purple calli.
**Requirements:** Well drained sandy soil
**Comments:** Known from 2 plants, it is on the verge of extinction.
**Localities:** Mt. Eliza
**Distribution:** Endemic to Vic

## Caladenia venusta

**Size:** stem to 45 cm high — P
**Habitat:** Tea-tree heath
**Comments:** Similar to *C. patersonii* but distinguished by its larger flowers which are white except for the marginal teeth and calli. Teeth of the labellum are longer and its fragrance resembles mouldy oranges. A rare plant with one small colony within the Melbourne area. A new orchid found in Kilsyth in 1992 closely resembles this species.
**Localities:** Dingley#; Frankston
**Distribution:** Vic, Tas
**Synonym:** Previously known as *C.* sp. aff. *patersonii* (large white)

*Caleana major*

# Caleana

## Caleana major

Large Duck Orchid, Flying Ducks

**Size:** stem 15–40 cm high — F P
**Habitat:** Dry sclerophyll forest, grassy low open forest
**Form:** Glabrous erect herb with thin, wiry, green or reddish-brown stem
**Foliage:** Single, narrow-lanceolate, reddish basal leaf to 120 mm long. Leaf often withers by flowering time.
**Flowers:** 1–5 shiny reddish-brown flowers, displayed upside down to resemble a duck flying. Perianth segments to 15 mm long, narrow to linear with pointed tips.
Labellum ovoid, to 8 mm long, hollow underneath. The labellum resembles a duck's head and is attached to the base of the winged column by a broad sensitive strap. September to February
**Requirements:** Well drained sandy soils
**Comments:** Often grows in association with *Eucalyptus macrorhyncha*. It is difficult to keep growing in cultivation. Pollinated by sexually attracted male sawflies.
**Propagation:** Seed
**Localities:** Moorabbin#, 42#, 50A, 63, 67#, 74
**Distribution:** Vic, Qld, NSW, Tas, SA

# Calochilus

## Calochilus campestris

Copper Beard-orchid

**Size:** stem 0.3–0.6 m high **P**
**Habitat:** Damp and valley sclerophyll forests, grassy low open forest
**Form:** Glabrous, erect herb with slender to stout, often greyish, stem
**Foliage:** Single rigid dark green linear-lanceolate leaf to 30 cm long, channelled with fleshy triangular cross-section
**Flowers:** 5–15 pale yellow-green flowers to 20 mm across, with reddish-brown markings and coppery beard. Erect dorsal sepal ovate and concave to 9 mm long, lateral sepals narrower and longer spreading below the labellum; petals erect, shorter and narrower than the dorsal sepal.
Labellum to 15 mm long, ovate, curved, basal part glabrous, with 2 metallic blue plates, central part covered with purple bristly hair-like calli resembling a beard. Labellum ends in a short glabrous tail. September to January
**Comments:** All *Calochilus* species are difficult in cultivation, dying out after a few years. They require moist, well drained soils. Most species can self-pollinate if not insect pollinated. This species remains in few Melbourne sites.
**Localities:** Dandenong Rges, Boronia, Belgrave, Lysterfield, Caulfield#, Cheltenham#, 57, 61, 67#, 74
**Distribution:** Vic, Qld, NSW, Tas, SA; NZ

*Calochilus campestris*

## Calochilus paludosus

Red Beard-orchid

**Size:** stem to 35 cm high **P**
**Habitat:** Dry and valley sclerophyll forests
**Form:** Slender herb with wiry stem
**Foliage:** Thin, ribbed linear-lanceolate leaf to 18 mm long, triangular in cross-section, often reddish at base
**Flowers:** 1–9 reddish-bronze flowers to 25 mm across, with a coppery beard. Broadly elliptic hooded dorsal sepal to 15 mm long, lateral sepals smaller, narrow-lanceolate; petals to 8 mm long, triangular with pointed tip.
Labellum—to 30 mm long, basal part covered in calli, central section has a beard of coarse reddish hairs. Labellum ends in a glabrous strap-like tail to 15 mm long. September to January
**Localities:** Hurstbridge, Boronia, Bayswater, Belgrave, Kilsyth, Lysterfield, 41, 42, 50A, 51, 74
**Distribution:** Vic, Qld, NSW, Tas, SA; NZ

## Calochilus robertsonii

Common or Purplish Beard-orchid

**Size:** stem to 0.45 m high **P**
**Habitat:** Widespread in sclerophyll woodland, dry, valley and damp sclerophyll forests, grassy low open forest
**Form:** Glabrous erect herb with a slender stem
**Foliage:** Single fleshy, ribbed, dark green linear-lanceolate leaf to 40 cm long, triangular in cross-section and reddish at the base
**Flowers:** 1–9 green flowers to 30 mm across, with reddish or purple stripes and a bronze-purple beard. Erect dorsal sepal, ovate to 25 mm long, lateral sepals as long but narrower, petals to 10 mm long
Labellum—ovate to 35 mm long, basal part covered in purple callus-like glands, central part has a beard of purple to bronze-brown hair-like calli. Ends in a short glabrous tail, longer than *C. campestris*.
The base of each column wing bears a bead-like gland, the wings connected by a prominent purplish ridge. August to October
**Comments:** The most widespread and common beard orchid in Melbourne. Plants flower freely and are most prominent on warm sunny days when the flowers open widely to fully display their beards.
**Localities:** Hurstbridge, Bayswater, Dandenong Rges, 42, 44, 47, 49–51, 54–56, 58, 66#, 67#, 74, 76
**Distribution:** All states except NT; NZ

*Calochilus robertsonii*

SCALE: 0 1cm 2cm The scale representation on each plant equals 2cm.

# Chiloglottis

Small terrestrial orchids with a single terminal flower and a pair of basal leaves. A floral bract surrounds the flower stalk. The dorsal sepal is erect and curved inwards. *Chiloglottis* spp. prefer sheltered areas and spread by increasing their tuberoid numbers vegetatively to form colonies. They are commonly called Bird Orchids, alluding to the resemblance of the flowers in some species to a baby bird with its beak open, waiting to be fed.

Some species may be grown, and flower well, in pots provided the mixture is well drained and includes well rotted organic material. They need watering from Autumn to Spring while in active growth and should dry out over Summer when they are dormant. Protection is needed from slugs and snails. Pollinated by sexually attracted male thynnid wasps. Propagate by division of tuberoids.

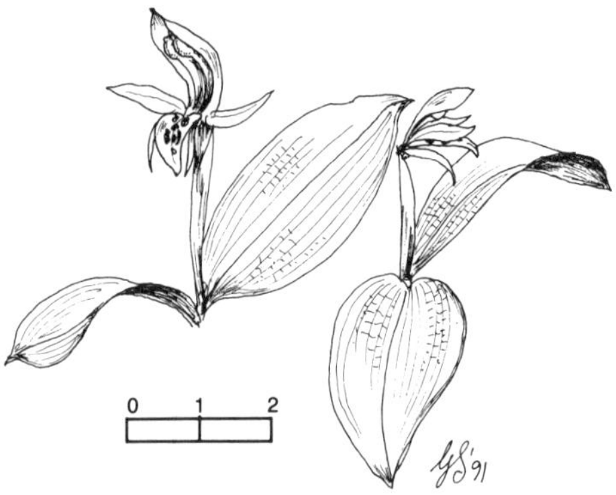


*Chiloglottis cornuta*

## Chiloglottis cornuta

Green Bird-orchid

**Size:** stem to 4 cm high — P N
**Habitat:** Wet sclerophyll forest, along stream banks
**Foliage:** Bright green broadly-ovate leaves 30–100 mm long
**Flowers:** Pale green flower to 30 mm across, with reddish markings. Perianth segments to 20 mm long, lateral sepals curved downwards, petals erect. Labellum ovate, green to reddish with a few dark green to reddish sessile calli in the centre. October to January
**Requirements:** Moist soils in high rainfall areas
**Localities:** 57
**Distribution:** Vic, NSW, Tas, SA; NZ

## Chiloglottis reflexa

Autumn Bird-orchid

**Size:** stem 8–18 cm high — P
**Habitat:** Dry sclerophyll forest, tea-tree heath, grassy low open forest
**Foliage:** Dark green elliptical stem-clasping leaves to 60 mm long with wavy margins
**Flowers:** Narrow greenish-bronze or purplish flower to 15 mm across, on a pinkish stem. Perianth segments to 20 mm long, petals bent backwards. Labellum brownish, obovate, to 12 mm long with reddish-black calli in the centre, ending in a prominent short point. March to August
**Requirements:** Well drained sandy soils
**Comments:** Flowering is sporadic, poor in some seasons, prolific in others, and may be stimulated by Summer fires. Grows well in a pot but is prone to rotting, so the mix must be perfectly drained.
**Localities:** Heathmont, Lysterfield, Cranbourne, 42, 67#, 74
**Distribution:** Vic, NSW, Tas

*Chiloglottis trapeziformis*

## Chiloglottis trapeziformis

Broad Lip or Dainty Bird-orchid

**Size:** stem to 8 cm high — P
**Habitat:** Dry sclerophyll forest, grassy low open forest
**Foliage:** Dark green ovate-lanceolate, shortly stalked leaves to 80 mm long, entire or with wavy margins
**Flower:** Brownish-green to reddish flower to 20 mm across. Perianth segments to 15 mm long.
Erect labellum broadly rhomboid (or trapeziform) to 12 mm long with a compact group of sessile black calli towards the base. September to November
**Requirements:** Well drained soils
**Comments:** In cultivation, the hardiest of the local *Chiloglottis* species. Closely related to the earlier-flowering *C. reflexa*, but is readily distinguished by the stalked leaves and more compact calli. Locally threatened and already extinct in most Melbourne sites.
**Localities:** Mt. Eliza, 34A, 51#, 74#, 76
**Distribution:** Vic, Qld, NSW, Tas

*Chiloglottis valida*

## Chiloglottis valida

Common Bird-orchid

**Size:** stem to 40 mm high — P
**Habitat:** Widespread in damp and valley sclerophyll forests, grassy low open forest
**Foliage:** Dark green ovate leaves to 100 mm long
**Flowers:** Dark reddish-brown to purplish-green flower to 40 mm across. Perianth segments to 20 mm long.
Labellum purplish, broadly ovate to 16 mm long with dark red to black calli. September to January
**Requirements:** Moist well drained humus-rich soils
**Comments:** Widely distributed in cooler forests often forming very large colonies. Readily distinguished from *C. cornuta* by the calli, some of which are on short stalks.
**Localities:** Heathmont-Boronia, Kalorama, Belgrave, 34A, 42, 44, 47, 49, 50A, 51, 55–58, 63, 74, 76
**Distribution:** Vic, NSW, Tas; NZ
**Synonym:** Previously included with *C. gunnii*

# Corybas

Dwarf terrestrial orchids of cool, sheltered areas reproducing by tuberoids to form colonies which are often hidden beneath shrubs and grassy tussocks. They have a single, flat, ground-hugging leaf and a single sessile or shortly stalked flower conspicuously hooded by the large erect dorsal sepal. Lateral sepals and petals are reduced and narrow, the large tubular labellum is erect at the base, expanding then curving downwards. Pollination is by small fungus gnats. Propagate by division.

## Corybas aconitiflorus

Spurred Helmet-orchid

**P N**

**Habitat:** Damp sclerophyll forest
**Foliage:** Dark green cordate leaf to 35 mm long, purplish below
**Flowers:** Erect purplish flower to 30 mm long on a short slender stalk. Hooded dorsal sepal encloses most of the labellum; lateral sepals and petals are thread-like. Lower half of labellum is tubular, upper part widely flared. March to July
**Requirements:** Moist well drained soils
**Comments:** Forms large colonies under ideal conditions. Successful under cultivation.
**Localities:** Belgrave
**Distribution:** Vic, Qld, NSW, Tas

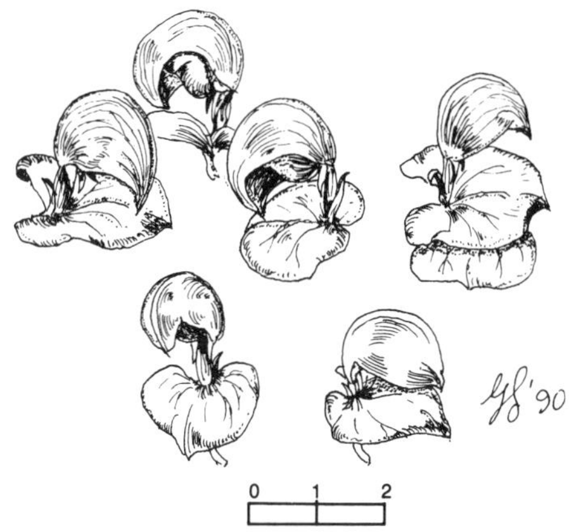

*Corybas aconitiflorus*

## Corybas diemenicus

Veined Helmet-orchid

**N**

**Habitat:** Fern gullies in wet and valley sclerophyll forest, grassy low open forest
**Foliage:** Circular to cordate leaf to 25 mm diameter with a circular marginal vein
**Flowers:** Large dark reddish-purple flower to 25 mm long, erect or recurved on a short slender stalk. Lower part of labellum is erect, upper part is sharply recurved and widely flared, with prominent veins, a central white patch and short fringed marginal teeth pointed forward. June to September
**Requirements:** Moist soil with plenty of humus
**Comments:** Seldom flowers in cultivation. In fern gullies, often colonises treefern trunks.
**Localities:** Montrose, Dandenong Rges, Belgrave, 42, 57, 74
**Distribution:** Vic, NSW, Tas, SA, WA
**Synonym:** *C. dilatatus* has been included with *C. diemenicus*. The species previously known as *C. diemenicus* has now become *C. incurvus*.

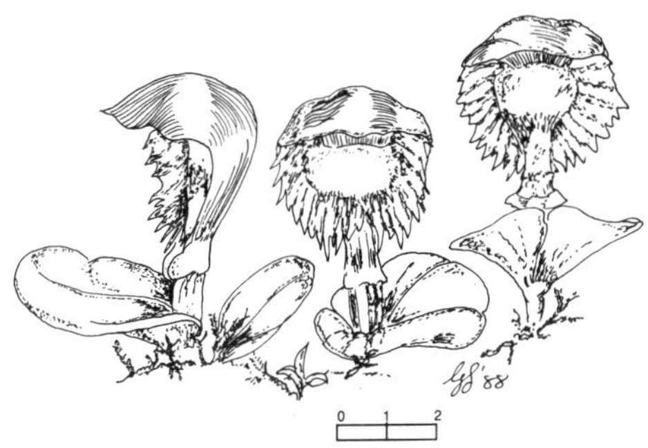

*Corybas diemenicus*

## Corybas incurvus

Slaty or Toothed Helmet-orchid

**P**

**Habitat:** Moist sheltered slopes and gullies in valley sclerophyll forest, tea-tree heath, grassy low open forest
**Foliage:** Cordate to ovate leaf to 30 mm long
**Flowers:** Erect, sessile dark purple flower to 15 mm long. Crimson-veined labellum differs from *C. diemenicus* in that the white central spot is channelled and the fringed margins of the upper part are curved inwards. June to September
**Requirements:** Moist soils
**Localities:** Bundoora, Wattle Glen, Eltham, Bayswater, Heathmont, Montrose, Springvale, Cranbourne, Cheltenham, Hampton, Beaumaris, 42, 65–67, 74–76
**Distribution:** Vic, NSW, Tas, SA
**Synonym:** *C. diemenicus*

## Corybas unguiculatus

Small Helmet-orchid, Pelicans

**P N**

**Habitat:** Protected sites in moist tea-tree heath, grassy low open forest
**Foliage:** Cordate grey-green, often 3-lobed, leaf to 30 mm long, reddish below
**Flowers:** Reddish-purple to purple-black flower to 15 mm long, held erect on a short stalk. Dorsal sepal narrow at base, expanded and hooded in upper half. Labellum is tubular with margins incurved and the tip flared. May to November
**Requirements:** Sandy or peaty soils, moist to wet in Winter
**Comments:** A solitary species which is difficult to grow. The flower profile resembles a duck's head.
**Localities:** Belgrave, Cranbourne, Beaumaris, Cheltenham, 74
**Distribution:** Vic, NSW, Tas, SA

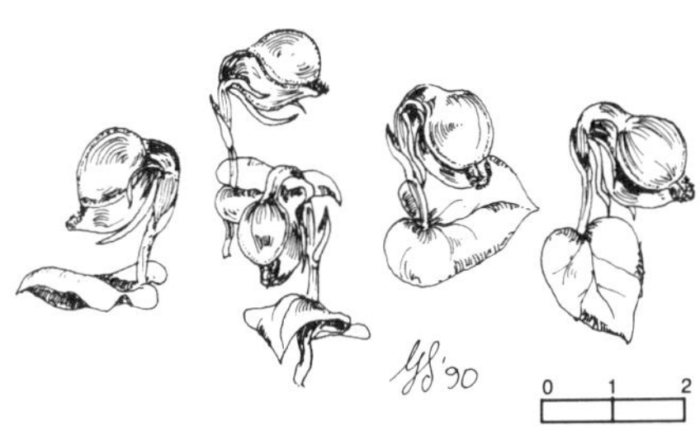

*Corybas unguiculatus*

SCALE: 0 1cm 2cm The scale representation on each plant equals 2cm.

*Cryptostylis subulata*

# Cryptostylis

## Cryptostylis leptochila

Small Tongue-orchid

**Size:** stem 20–40 cm high **P**
**Habitat:** Damp and valley sclerophyll forests, grassy low open forest
**Form:** Stiff, erect glabrous herb with tuberous rhizomes
**Foliage:** Stalked, leathery dark green ovate leaves to 120 mm long, held stiffly erect. Back of leaf brownish-purple.
**Flowers:** Terminal raceme of 9–12 brownish-purple flowers, each to 20 mm long. Thread-like green perianth segments to 20 mm long, stiffly spreading with margins rolled in.
Softly hairy labellum to 20 mm long with rounded, shiny black calli, projects forward before sharply curving up, margins rolled inwards. December to April
**Requirements:** Well drained soils
**Comments:** Forms small colonies and is slow to re-establish once disturbed. Pollinated by sexually attracted male ichneumonid wasps. Both *Cryptostylis* spp. grow actively over Spring and Summer, retaining leaves all year unless very dry.
**Propagation:** Division of rhizomes
**Localities:** Heathmont-Boronia, Dandenong Rges, Lysterfield, 42, 47, 49, 51, 56, 74
**Distribution:** Vic, Qld, NSW, Tas

## Cryptostylis subulata

Large Tongue-orchid

**Size:** stem 0.3–0.8 m high **P**
**Habitat:** Moist depressions in valley and damp sclerophyll forests, wattle tea-tree scrub, grassy low open forest
**Form:** Same as *C. leptochila*
**Foliage:** 2 or 3 stalked leathery dark to yellow green lanceolate leaves to 15 cm long, held stiffly erect
**Flowers:** Terminal raceme of 3–20 yellow and reddish-brown flowers to 25 mm long. Green to yellow-green thread-like perianth segments to 30 mm long, margins inrolled.
Labellum oblong to 30 mm long, held rigidly forward, yellowish towards the base, margins rolled up to show the reddish underside; 2 dark purple ridges run along the labellum ending in a 2-lobed callus. August to April
**Requirements:** Moist soil
**Comments:** Easy to grow, it may be possible to establish in a bog garden or terrarium.
**Propagation:** Division of rhizomes
**Localities:** Heathmont-Boronia, Belgrave, Lysterfield, Noble Park, Springvale, Montrose, 42#, 47, 49, 50A, 51, 54, 56, 63–65#, 67#, 73, 74
**Distribution:** Vic, Qld, NSW, Tas, SA; NZ

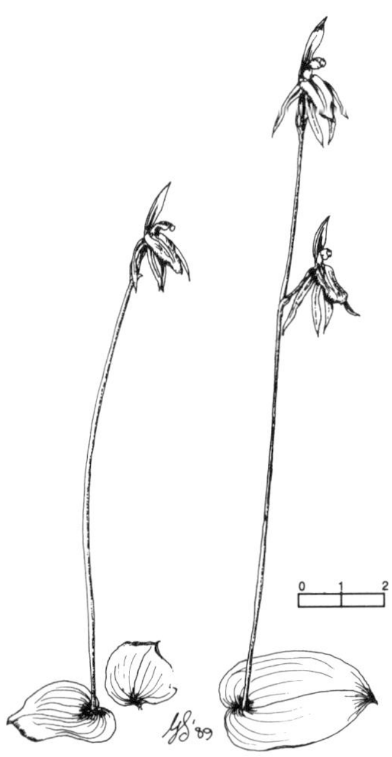

*Cyrtostylis reniformis*

# Cyrtostylis

## Cyrtostylis reniformis

Gnat Orchid

**Size:** stem to 10 cm high **P**
**Habitat:** Dry sclerophyll forest, tea-tree heath, coastal woodland
**Form:** Slender erect herb on wiry stem
**Foliage:** Single roundish ground-hugging grey-green leaf to 40 mm long with whitish veins, crystalline below
**Flowers:** 1–6 reddish-brown flowers to 10 mm long. Slender glabrous perianth segments to 8 mm long, dorsal sepal erect.
Labellum oblong with a conspicuous longitudinal vein along each edge. May to October
**Requirements:** Well drained soil
**Comments:** May form large dense colonies in moist shaded sites but is less prolific on drier forest slopes. Grows well in a pot and is often available through specialist nurseries. Dormant over Summer.
**Propagation:** Division of tuberoids
**Localities:** Dandenong Rges, Belgrave, Black Rock, Cheltenham, 32#, 33, 34A, 35, 42#, 49, 51, 67, 74, 76
**Distribution:** All states except NT; NZ
**Synonym:** *Acianthus reniformis*

# Dipodium

## Dipodium roseum

Hyacinth Orchid

**Size:** stem 0.3–0.9 m high **P**
**Habitat:** Widespread in dry and valley sclerophyll forests, coastal woodland, grassy low open forest
**Form:** Erect, leafless saprophytic herb with thick, fleshy purplish-red stem
**Flowers:** Long terminal raceme of 6–50 pale pink flowers to 40 mm across, with small pink spots. Reflexed oblong-lanceolate perianth segments to 17 mm long.
Labellum erect to 13 mm long, 3-lobed, lateral lobes much shorter, tip of mid-lobe hairy. December to April
**Requirements:** Well drained soils, growing in close association with ***Eucalyptus*** spp.
**Comments:** Being saprophytic, it is impossible to cultivate. The fleshy roots were eaten by the Aborigines. Distinguished from *D. punctatum*, which does not occur in Melbourne, by the paler pink flowers, finer spots, and reflexed perianth segments.
**Localities:** Montrose-Kilsyth, Belgrave, Heathmont-Boronia, Springvale, 35–37#, 42–44, 47–51, 55–58, 63, 67#, 71, 74, 76
**Distribution:** Vic, NSW, Tas, SA
**Synonym:** Previously included with *D. punctatum*

*Dipodium roseum*

# Diuris

Glabrous terrestrial orchids growing in colonies. Diuris are readily recognised by their erect, ear-like petals, composed of a blade and a claw, giving them the common name of 'Donkey Orchid'. The lateral sepals are narrow-linear and pendent while the dorsal sepal is erect. The labellum is entire and 3-lobed, the mid-lobe being contracted at the base. The yellow or purple flowers are held in a terminal raceme of 2 or more and are often blotched or spotted with brown. The 2 or more grass-like leaves are in a basal clump. *Diuris* spp. hybridise readily. They increase in number by producing daughter tuberoids as well as by seeds and propagation can take advantage of both methods. Seeds should be sprinkled around the base of the parent plant. Pollination is by native bees. They grow well in pots using a well drained medium, in bush houses or cool glasshouses. Plants are dormant over Summer.

## Diuris brevissima

Short-tailed Leopard Orchid

**Size:** stem to 35 cm high **F P**
**Habitat:** Dry sclerophyll forest, usually in hilly country
**Foliage:** 2 narrow channelled leaves to 20 cm long
**Flowers:** 6–9 yellow flowers to 30 mm across, with distinct red-brown blotches. Dorsal sepal ovate; lateral sepals parallel, to 10 mm; petals to 12 mm long, with a short dark brown claw.
Labellum bent downwards, lateral lobes spreading. October to November
**Requirements:** Well drained soil
**Comments:** Similar to *D. pardina* with which it is often confused. However *D. brevissima* is later flowering, has broader lateral lobes on the labellum and broader lateral sepals. The claws of the petals tend to be shorter.
**Localities:** 42
**Distribution:** Vic, NSW, SA

## Diuris corymbosa

Wallflower or Donkey Orchid

**Size:** stem 10–50 cm high **P**
**Habitat:** Dry and valley sclerophyll forests, coastal woodland, grassy low open forest
**Form:** Slender erect herb
**Foliage:** 2–3 deeply channelled linear leaves to 25 cm long
**Flowers:** 1–8 yellow, brownish-red and purple flowers to 40 mm across, the colours suffusing together. Dorsal sepal very broad and rounded to 10 mm long; narrow, green lateral sepals to 20 mm long, hang parallel; stalked petals to 20 mm long, blade oval.
Labellum has broad, spreading or recurved lateral lobes as long as or longer than the mid-lobe, to 12 mm long; mid-lobe notched on the end with a single raised ridge down the centre. July to November
**Requirements:** Well drained soils
**Comments:** A widespread and variable species in Melbourne, with increased flowering after fire. Some forms have much larger flowers, to 70 mm across. Pure yellow forms are also found.
**Localities:** Greensborough, Eltham, Doncaster, Blackburn-Boronia, The Basin, Belgrave, Baxter, 34A, 41, 42, 44, 51, 55, 56, 58, 63, 67, 71, 74–76
**Distribution:** Vic, NSW, Tas, SA, WA
**Synonym:** Previously confused with *D. longifolia*

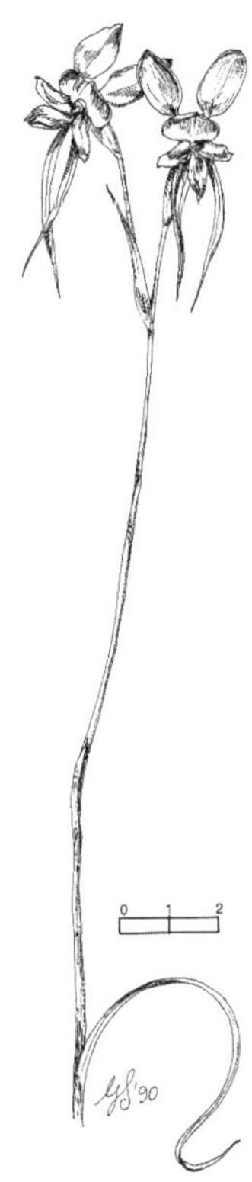

*Diuris corymbosa*

SCALE: 0 1cm 2cm The scale representation on each plant equals 2cm.

*Diuris fragrantissima*

## Diuris fragrantissima

Fragrant Doubletail

**Size:** stem to 0.2 m high **F**
**Habitat:** Plains grasslands
**Comments:** Similar to *D. punctata* but shorter and stouter. Leaves paired to 18 cm long. 2–7 highly perfumed white flowers to 50 mm across, with purple suffusions and streaks; October and November.
This orchid was once quite widespread on the Keilor Plains but is now endangered and has not been seen recently at some known sites. It has however been successfully propagated and has been reintroduced into sites where it would have previously occurred.
**Localities:** Tottenham#, St. Albans, Sydenham#, Williamstown#, Sunshine, Laverton, Keilor#
**Distribution:** Endemic to Vic
**Synonym:** *D. punctata* var. *albo-violacea*

## Diuris lanceolata

Golden Moths, Snake-orchid

**Size:** stem 10–30 cm high **F P**
**Habitat:** Moist depressions in plains grasslands, red gum and sclerophyll woodland, dry and valley sclerophyll forests, tea-tree heath
**Foliage:** 3-6 linear leaves to 15 cm long
**Flowers:** 1-5 nodding, bright yellow flowers to 25 mm across. Broadly ovate dorsal sepal to 15 mm long; narrow channelled lateral sepals to 25 mm long, spreading below the labellum; petals shorter, spreading to the side. Mid-lobe of labellum broadly ovate to 20 mm long with 2 fleshy ridges, lateral lobes to 5 mm long, spreading, ends toothed. September to October
**Requirements:** Moist soils
**Comments:** Plants of dwarf populations on the plains west of Melbourne have deep orange labellums.
**Localities:** Sunbury, Essendon, Tottenham, St. Albans, Laverton, Camberwell-Mitcham, Scoresby, Belgrave-Rowville, Lyndhurst, Black Rock, 29, 33, 34A, 35, 38, 41, 42, 58, 63, 67, 74
**Distribution:** Vic, NSW, Tas, SA
**Synonym:** Was previously known as *D. pedunculata*, a species which only occurs in NSW.

## Diuris x fastidiosa

Proud Doubletail

**Size:** stem to 20 cm high **F P**
**Habitat:** Moist grasslands
**Foliage:** Loose tussock of 7–8 grass-like, slightly twisted leaves to 100 mm long
**Flowers:** Similar to *D. lanceolata* but paler, smaller and not nodding, with darker striations on the outer edges and brown stripes at the base of the dorsal sepal. Dorsal sepals and petals are nerved; lateral sepals obliquely erect. Labellum has brown markings on outside. August to September
**Comments:** A naturally occurring hybrid between *D. palustris* and *D. lanceolata* with intermediate characteristics. The only known site in Melbourne has been destroyed. It also occurs in Stawell.
**Localities:** Tottenham#
**Distribution:** Endemic to Vic

*Diuris lanceolata*

## Diuris lanceolata x pardina

**Comments:** A naturally occurring hybrid found in the dry sclerophyll forests of north east Melbourne. Intermediate between and found with both parents.
**Localities:** Heidelberg, Eltham, Templestowe, Warrandyte
**Distribution:** Endemic to Vic
**Synonym:** Previously confused with *D.* x *palachila* which does not occur in Melbourne.

## Diuris palustris

Swamp diuris

**Size:** stem 10–25 cm high **F P**
**Habitat:** Wet depressions in plains grassland, tea-tree heath
**Foliage:** Tuft of 8–10 channelled, narrow spirally-twisted leaves to 80 mm long
**Flowers:** 1–4 yellow flowers to 20 mm across, blotched with brown, especially on the outside. Dorsal sepal ovate to 8 mm long, recurved; lateral sepals green, parallel, to 16 mm long; petals to 10 mm long have a dark brown vertical stripe.
Labellum lobes similar in length with 2 thick, fleshy raised lines from the base to the middle of the mid-lobe, merging into one towards the tip. July to December
**Requirements:** Moist to wet soils
**Comments:** The flowers release a spicy perfume on warm days. This species is slow to establish in cultivation.
**Localities:** Laverton#, Tottenham, Keilor Plains, St. Albans, 67#
**Distribution:** Vic, Tas, SA

## Diuris pardina

Leopard Orchid

**Size:** stem 20–30 cm high **P**
**Habitat:** Dry and valley sclerophyll forests, tea-tree heath, grassy low open forest
**Foliage:** 2–3 narrow channelled leaves to 20 cm long
**Flowers:** 2–6 yellow flowers to 30 mm across, distinctly and heavily blotched with brown. Dorsal sepal ovate; green lateral sepals to 12 mm long, recurved and crossing each other; petals to 15 mm long, covered in dark brown blotches.
Lateral lobes of labellum as long as the mid-lobe. August to November
**Requirements:** Well drained soils
**Comments:** Confused with *D. maculata* which is not as heavily blotched and flowers earlier than *D. pardina*.
**Localities:** Greensborough-Eltham, Park Orchards, Blackburn-Boronia, Belgrave, Springvale, 34A, 35, 37, 41, 42, 44, 51, 58, 67#, 71, 74
**Distribution:** Vic, Tas, SA
**Synonym:** Previously confused with *D. maculata*, an orchid endemic to NSW.

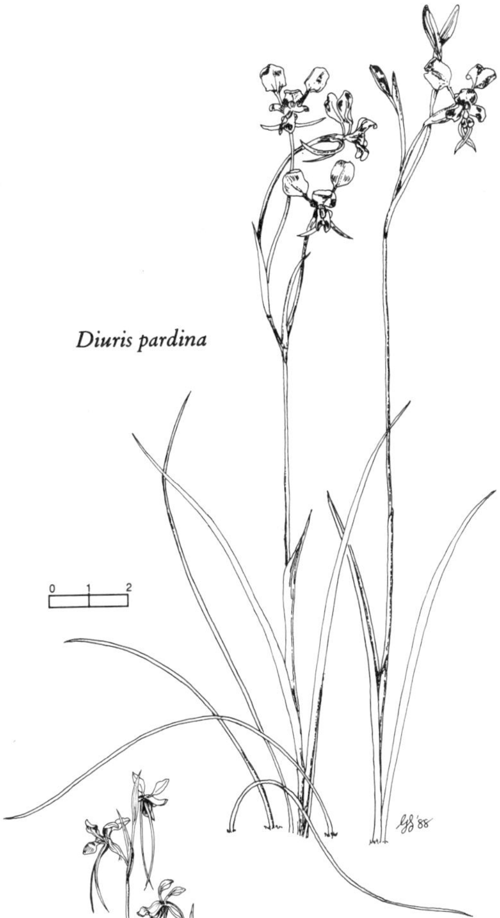

*Diuris pardina*

## Diuris punctata

Purple Diuris

**Size:** stem 20–60 cm high **F P**
**Habitat:** Moist areas in box, red gum and sclerophyll woodlands, grassy low open forest
**Foliage:** Pair of grass-like channelled leaves to 15 cm long
**Flowers:** 2–10 large purple flowers to 50 mm across. Broadly ovate dorsal sepal to 15 mm long; lateral sepals brownish or green, thread-like 40–80 mm long; petals elliptic to 30 mm long.
Mid-lobe of labellum fan-shaped to 20 mm long, with prominent yellow central blotch and 2 raised yellow ridges, lateral lobes much smaller and lobed to the base. October to November
**Requirements:** Moist soils
**Comments:** A very showy orchid which is now restricted to a few plants in the Melbourne area.
**Localities:** Werribee, Bundoora, Mentone, Hampton#, Mt. Eliza, Cranbourne, 26, 49, 50A#, 74
**Distribution:** Vic, Qld, NSW, SA

*Diuris punctata*

## Diuris sulphurea

Tiger Orchid

**Size:** stem to 40 cm high **P**
**Habitat:** Red gum, box and coastal woodland, around swamp margins in valley and dry sclerophyll forests, tea-tree heath
**Foliage:** 1–3 channelled grass-like leaves to 35 cm long
**Flowers:** 2–7 yellow flowers to 30 mm across, with brown markings. Ovate dorsal sepal to 20 mm long has 2 dark brown to blackish spots near the base; lateral sepals to 22 mm long; petals, to 30 mm long, ovate on short stalks.
Lateral lobes of labellum short, to 5 mm long, wide and blunt, mid-lobe spade-shaped with a single raised ridge extending to the middle. October to December
**Requirements:** Moist well drained soils
**Comments:** Often growing in association with *Eucalyptus baxteri*. Rare hybrids with *D. corymbosa* have been found in the Dandenong Rges.
**Localities:** Tullamarine, Bundoora, Nunawading-Boronia, Lysterfield, Springvale, 26, 42, 44, 51, 55, 56, 58, 67, 74
**Distribution:** Vic, Qld, NSW, Tas, SA

*Diuris sulphurea*

# Eriochilus

## Eriochilus cucullatus

Parson's Bands

**Size:** stem 8–25 cm high **F P**
**Habitat:** Widespread in red gum and sclerophyll woodlands, dry and valley sclerophyll forests, grassy low open forest
**Form:** Glabrous erect herb with thin wiry stem
**Foliage:** Shiny dark green, ovate basal leaf to 35 mm long, becoming fully developed after flowering finishes
**Flowers:** 1–5 white, or sometimes pink, flowers to 20 mm across. Perianth segments to 15 mm long; greenish dorsal sepal erect to 8 mm long, hooded; lateral sepals spreading, elliptical-lanceolate, petals erect, narrower and shorter.
Labellum cream or yellowish with dark spots, upper half recurved and covered in stiff hairs. December to May
**Requirements:** Dry to moist well drained soils
**Comments:** Dormant over Summer. May be successful in a container provided it is left undisturbed.
**Propagation:** Seed, scattered around established plants.
**Localities:** Basalt plains, Bundoora, Heathmont-Boronia, Kalorama, Belgrave, Baxter, 42, 45#, 50A, 51, 56, 61, 67#, 71, 74–76
**Distribution:** Vic, Qld, NSW, Tas, SA

*Eriochilus cucullatus*

SCALE: 0 1cm 2cm The scale representation on each plant equals 2cm.

# Gastrodia

## Gastrodia procera

**Size:** stem 0.7 m to over 1 m high P N
**Habitat:** Damp sclerophyll forest
**Form:** Tall leafless saprophytic herb with erect fleshy brown stem
**Flowers:** Terminal raceme of up to 75 tubular flowers to more than 20 mm long, cinnamon brown on outside, creamy inside.
Callus plate on upper part of labellum quite elevated, dividing into 2 widely spaced ridges from the middle. Fruits are larger and paler brown than *G. sesamoides*. November to January
**Requirements:** Moist soils with high organic content and high rainfall
**Comments:** Similar to *G. sesamoides*. Both species may occur together usually growing close by Eucalypt species. *G. procera* has a strong, sweet perfume, is more floriferous, with larger, darker flowers, and prefers moister habitats. Both are pollinated by native bees. Aborigines roasted and ate the tubers.
**Propagation:** Impossible
**Localities:** Dandenong Ranges, Olinda
**Distribution:** Vic, Qld, NSW, Tas
**Synonym:** Previously included with *G. sesamoides*

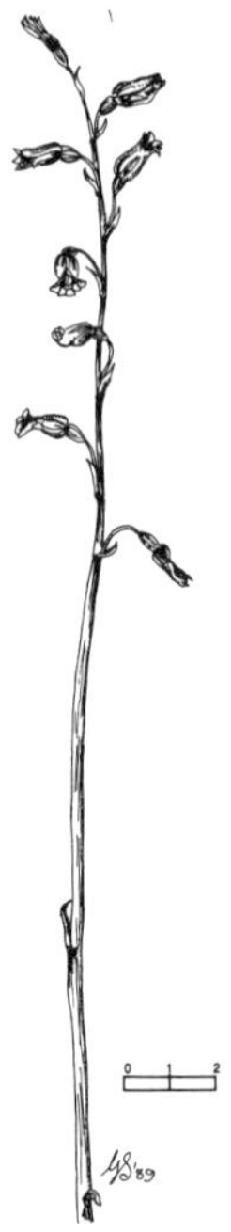

*Gastrodia sesamoides*

## Gastrodia sesamoides

Potato Orchid, Cinnamon Bells

**Size:** stem 0.3–0.8 m high P
**Habitat:** Widespread in damp and valley sclerophyll forests, grassy low open forests
**Form:** Leafless erect saprophytic herb with a fleshy brown stem, nodding before flowering
**Flowers:** Terminal raceme of up to 25 drooping, tubular flowers to 20 mm long, rough and light grey-brown on outside, creamy inside.
Labellum yellowish-orange, calli plate less elevated than *G. procera* with dividing ridges closer together. October to December
**Requirements:** Moist humus-rich soils, in high rainfall areas
**Comments:** Flowers emit a sweet perfume on warm days.
**Localities:** Eltham, Heathmont-Kalorama, Belgrave, 42, 50A, 51, 55–58, 61, 63, 74
**Distribution:** Vic, Qld, NSW, Tas, SA; NZ

# Genoplesium

*Genoplesiums* are terrestrial orchids with few to many very small dull-coloured flowers in short dense terminal spikes. Each flower is reversed with the labellum uppermost and hinged on a small claw. Flowering occurs during Summer and Autumn. Each orchid has a single, erect terete leaf which clasps the stem for most of its length. Pollination is by tiny flies. While propagation is by seed they are very difficult to grow in cultivation. *Genoplesium* spp. have been separated from *Prasophyllum* spp., which differ in details of the labellum and column.

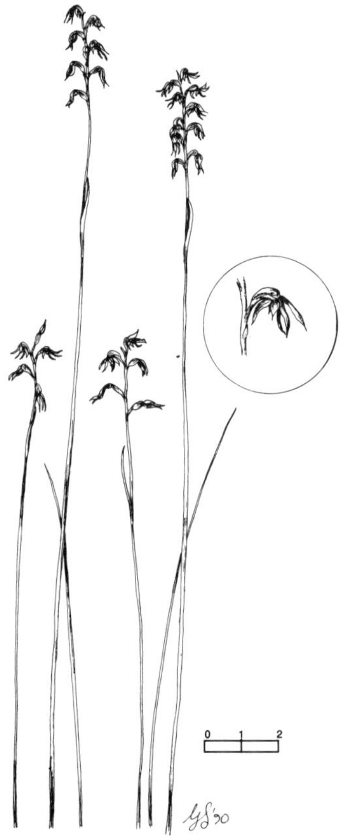

*Genoplesium despectans*

## Genoplesium archeri

Variable Midge-orchid

**Size:** stem 10–20 cm high P
**Habitat:** Valley sclerophyll forest, primary dune scrub, grassy low open forest
**Foliage:** Leaf 15 cm long, the top 25 mm free of the stem
**Flowers:** Squat spike of 2–15 pale yellowish-green or purplish flowers to 10 mm across, with reddish stripes and markings. Dorsal sepal hooded, ovate; lateral sepals and petals spreading, to 7 mm long.
Tiny dark purple labellum has hairy margins, 2 parallel ridges and upturned tip. December to June
**Requirements:** Well drained soils
**Comments:** Uncommon, it remains in few sites in Melbourne.
**Localities:** Bayswater, Dandenong Ranges, Belgrave, Lysterfield, 67#, 74
**Distribution:** Vic, Qld, NSW, Tas
**Synonym:** *Prasophyllum archeri*

## Genoplesium despectans

Sharp Midge-orchid

**Size:** stem to 40 cm high P
**Habitat:** Dry and valley sclerophyll forests, grassy low open forest
**Foliage:** Leaf to 25 cm long, free for 15 mm
**Flowers:** Dense pyramidal spike of 1–45 tiny, glabrous, nodding, dark purplish-brown or green flowers with purple markings, to 3 mm across. Dorsal sepal is hooded; lateral sepals and petals very pointed.
Labellum also sharply pointed; January to June
**Requirements:** Well drained clay loam soils
**Comments:** Often found with *Eucalyptus baxteri*.
**Localities:** Eltham, Heathmont-Boronia, Belgrave, Lysterfield, 34A, 41, 42, 44, 51, 71, 74
**Distribution:** Vic, NSW, Tas, SA
**Synonym:** *Prasophyllum despectans*

## Genoplesium morrisii
Bearded Midge-orchid

**P**

**Size:** stem to 30 cm high
**Habitat:** Valley sclerophyll forest, grassy low open forest, on heavier soil
**Foliage:** Leaf to 30 cm long, free for 20 mm
**Flowers:** Dense spike of 3–25 dark purple or greenish flowers to 8 mm across with purple and reddish markings. Hooded dorsal sepal, petals and labellum fringed with long dark purple hairs. Lateral sepals glabrous, spreading, to 6 mm long. December to May
**Requirements:** Damp clay soils
**Comments:** Readily recognised by the fringed labellum.
**Localities:** Montrose, Belgrave, Eastern Bayside Suburbs#, 56, 74
**Distribution:** Vic, Tas
**Synonym:** *Prasophyllum morrisii*

*Genoplesium morrisii*

## Genoplesium rufum
Red Midge-orchid

**F P**

**Size:** stem to 15 cm high
**Habitat:** Tea-tree heath, red gum woodland
**Foliage:** Leaf to 15 cm long, mostly sheathing
**Flowers:** Open spike of 3–20 tiny black to dark greenish flowers. Deeply hooded dorsal sepal; lateral sepals spreading, broad-lanceolate and often gland-tipped, petals toothed.
Labellum margins toothed, callus plate raised, very dark. Column is winged and lightly coloured. January to May
**Requirements:** Moist, well drained sandy soil
**Localities:** Noble Park, Sth Oakleigh, 67#
**Distribution:** Vic, Qld, NSW, Tas, SA
**Synonym:** *Prasophyllum rufum*

# Glossodia

## Glossodia major
Wax-lip Orchid

**F P**

**Size:** stem to 30 cm high
**Habitat:** Widespread in box ironbark woodland, dry and valley sclerophyll forests, coastal woodland, tea-tree heath
**Form:** Erect, slender hairy herb
**Foliage:** Single, hairy, dark green oblong to oblanceolate leaf to 100 mm long
**Flowers:** 1–2 large purple to mauve flowers to 45 mm across, with a prominent white base on labellum. Perianth segments glabrous, broad, spreading, to 25 mm long.
Labellum cordate, hairy at the base. Notched, erect yellow basal calli. September to November
**Requirements:** Well drained soils
**Comments:** Plants usually occur singly or in scattered groups, but are much more abundant after fire. On warm days flowers exude a sweet perfume. They are difficult to cultivate, usually persisting only for a short while. All-white variants are found occasionally. Pollinated by native bees.
**Localities:** Eltham, Belgrave-Rowville, 34A, 36, 41, 42, 44, 47–51, 55, 56, 58, 61, 64, 67, 74, 76
**Distribution:** Vic, Qld, NSW, Tas, SA

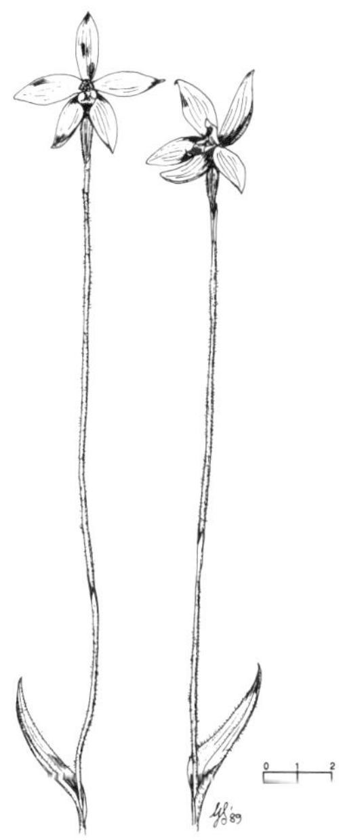

*Glossodia major*

# Leporella

## Leporella fimbriata
Fringed Hare-orchid

**F P**

**Size:** stem to 25 cm high
**Habitat:** Tea-tree heath, grassy low open forest
**Form:** Slender erect glabrous herb
**Foliage:** 1, sometimes 2, bluish-green to yellowish-green, sessile, basal, broadly-lanceolate leaves to 30 mm long, developing 3–5 prominent parallel red veins after flowering.
**Flowers:** 1–3 reddish and yellow-brown flowers to 15 mm across. Perianth segments spreading; dorsal sepal wide, erect, to 10 mm long; lateral sepals very narrow, petals longer, erect, clubbed tips covered in black glandular hairs.
Labellum fan-shaped, greenish with hairy purple spots, recurved with deeply and coarsely fringed margins. March to August
**Requirements:** Well drained sandy soil
**Comments:** Once formed extensive colonies, especially prominent after fire. Pollinated by sexually attracted winged male ants. Difficult to maintain in cultivation. As it has not been collected recently it is presumed extinct within the Melbourne area.
**Localities:** Cheltenham#, Black Rock#, Beaumaris#, 67#, 74
**Distribution:** Vic, Tas, SA, WA
**Synonym:** *Leptoceras fimbriatum*

*Leporella fimbriata*

## Leptoceras menziesii = Caladenia menziesii

SCALE: 0 1cm 2cm — The scale representation on each plant equals 2cm.

*Lyperanthus nigricans*

*Lyperanthus suaveolens*

# Lyperanthus

## Lyperanthus nigricans

Red Beaks

**Size:** stem to 25 cm high **P**
**Habitat:** Coastal banksia woodland, tea-tree heath, grassy low open forest
**Form:** Erect fleshy herb with leaf-like stem bracts to 40 mm long
**Foliage:** Bright green cordate basal leaf, variable to 90 mm long, often with red or black markings, ground-hugging
**Flowers:** 2–8 large lightly perfumed flowers to 40 mm across, white with red stripes, each with a large floral bract. Perianth segments curved forward, to 35 mm long, spreading, red-tipped; dorsal sepal hooded, curved over column.
Labellum white with crimson lines, erect, deeply fringed. August to November
**Requirements:** Well drained soils
**Comments:** Flowers freely after fires but difficult to maintain in cultivation.
**Propagation:** Division of tuberoids
**Localities:** Springvale, Cheltenham#, Sandringham, Black Rock, Beaumaris, 74
**Distribution:** Vic, NSW, Tas, SA, WA

## Lyperanthus suaveolens

Brown Beaks

**Size:** stem to 45 cm high **P**
**Habitat:** Dry and valley sclerophyll forests, tea-tree heath
**Foliage:** Stiff leathery dark green linear-lanceolate leaf to 20 cm long
**Flowers:** 2–8 yellowish-brown flowers to 30 mm across. Perianth segments narrow, stiff and leathery, to 30 mm long, spreading with margins curved downwards; petals erect.
Bright yellow labellum is curved and 3-lobed. August to November
**Requirements:** Well drained soils
**Comments:** Flowers are fragrant in warm weather. Plants display considerable variety in flower colour, from greenish, yellow-brown to dark brown or black. Grows and flowers well in a container. Locally very rare.
**Propagation:** Division of tuberoids
**Localities:** Nunawading-Bayswater, Kilsyth, Belgrave, 34A, 42, 44, 51, 58, 67#, 74
**Distributin:** Vic, Qld, NSW, Tas

# Microtis

Slender glabrous terrestrial orchids which die back to their tuberoids over Summer and Autumn. A single, erect, terete, hollow leaf sheathes the stem. Tiny green flowers, in dense or open spikes, are on the end of larger ovaries. Each has a broad, erect, hooded dorsal sepal. Lateral sepals and petals are much narrower. Flowers are self-pollinated if not visited by insects. Propagation is by division of tuberoids. Easily overlooked in the bush due to their colour and size.

*Microtis atrata*

*Microtis parviflora*

## Microtis atrata

Yellow or Tiny Onion-orchid

**Size:** stem 5–15 cm high **P N**
**Habitat:** Low-lying depressions and swamps of valley sclerophyll forest, tea-tree heath, grassy low open forest
**Foliage:** Leaf to 90 mm long, often yellow-green
**Flowers:** Dense spike of 2–40 minute semi-erect flowers less than 2 mm wide. Lateral sepals spreading, oblong and blunt-tipped, and not hidden by labellum.
Labellum—ovate and entire, bent sharply downwards, with no calli. September to December
**Requirements:** Boggy soils
**Comments:** The whole plant is yellowish-green drying to black with age. It forms dense colonies and is easy to cultivate tolerating inundation for some time. Locally rare.
**Localities:** Belgrave, 42, 47, 49, 50A, 67#, 74
**Distribution:** Vic, Tas, SA, WA

## Microtis parviflora

Slender Onion-orchid

**Size:** stem to 0.5 m high **F P**
**Habitat:** Widespread in moist areas of dry and valley sclerophyll forests, grassy low open forest, tea-tree heath
**Foliage:** Leaf to 40 cm long
**Flowers:** Dense spike of 10–80 tiny erect flowers to 3 mm wide. Flowers similar to *M. rara*. Lateral sepals not as tightly curved back.
Labellum shorter, cordate, not pressed against the ovary, margins entire or wavy with a small recurved tip. 2 calli at the base. October to February
**Requirements:** Moist soils in Winter
**Comments:** A common orchid colonising lawns and disturbed areas, forming large colonies.
**Localities:** Bundoora-Warrandyte, Nunawading-Dandenong Rges, Belgrave, 63, 67, 74–76
**Distribution:** All states except NT; NZ, N Cal

## Microtis rara

Scented Onion-orchid

**Size:** stem to 0.6 m high **F P N**
**Habitat:** Beside permanent swamps in areas of high rainfall and valley sclerophyll forest
**Foliage:** Leaf to 60 cm long
**Flowers:** Spike of 5–50 well-spaced fragrant flowers to 4 mm wide. Dorsal sepal only narrowly hooded; lateral sepals linear-oblong, spreading, curved backwards; petals erect.
Labellum oblong, pressed closely to and about the same length as the ovary, margin wavy and tip notched; 2 large dark green calli at the base and 1 at the tip. October to January
**Requirements:** Moist soils
**Comments:** Forms small clumps, flowering more freely after fires which may produce robust plants over 1 m high. Plants prefer sunny sites protected by grasses. A rare plant, first collected in Ringwood in 1920.
**Localities:** Mitcham-The Basin, Belgrave, 58, 74
**Distribution:** Vic, NSW, Tas, SA
**Synonym:** *M. oblonga*

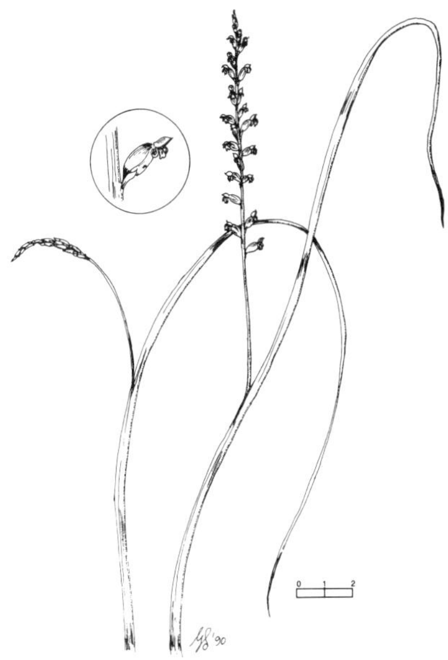

*Microtis unifolia*

## Microtis unifolia

Common Onion-orchid

**Size:** stem 15–80 cm high **F P**
**Habitat:** Widespread, favouring swamps in heaths, dry and valley sclerophyll forests from coast to foothills, but also colonising drier soils throughout its range; grassy wetland, red gum woodland
**Foliage:** Leaf to 70 cm long
**Flowers:** Dense spike of 6–100 pale to golden-green erect flowers to 4 mm wide. Dorsal sepal is deeply hooded, partly hiding petals.
Labellum less than ½ as long as ovary with wrinkled margins, 2-lobed; 2 basal calli and a narrow central callus band. October to January
**Requirements:** Well drained to moist situations
**Comments:** May form extensive colonies.
**Localities:** Throughout the Greater Melbourne area
**Distribution:** All states except NT; countries of western Pacific

# Orthoceras

## Orthoceras strictum

Horned or Crow Orchid

**Size:** stem 25–60 cm high **F P**
**Habitat:** Swamp edges in valley sclerophyll forest, wattle tea-tree scrub, grassy low open forest
**Form:** Glabrous herb with a rigid stem
**Foliage:** Erect clump of 2–5 channelled, linear leaves to 30 cm long
**Flowers:** 2–9 green to dark brown flowers to 10 mm across. Dorsal sepal to 10 mm long, hooded and pointed at tip, hiding small petals; lateral sepals almost terete, erect and spreading on each side of hood, to 25 mm long.
Labellum 3-lobed, lateral lobes erect, mid-lobe elliptical with central yellow spot, curved downwards, large yellow callus at base. November to January
**Requirements:** Moist to wet soils
**Comments:** Plants are easily overlooked in grassy habitats, but are locally common.
**Propagation:** Not cultivated
**Localities:** Dandenong Rges, Belgrave, Lysterfield, Port Phillip, Cheltenham#, 47, 49, 67#, 74
**Distribution:** Vic, Qld, NSW, Tas, SA; NZ, N Cal

*Orthoceras strictum*

# Prasophyllum

Terrestrial orchids with a single, erect, terete basal leaf which is stem-clasping only up to two thirds of its length. Prasophyllums have a spike of many small to medium, dull-coloured, reversed flowers in which the labellum is uppermost and usually a different colour to the perianth segments. The labellum is immovable and has a central callus which can occupy most of the surface. Margins are often crisped or wavy.

*Prasophyllum* spp. flower in Spring and Summer and are often strongly perfumed. They are pollinated by nectar-seeking wasps. Propagation is by seed, but all are difficult to grow. Autumn flowering species with movable labella and almost fully stem-clasping leaves are now called *Genoplesium* spp.

## Prasophyllum affine = P. fuscum

## Prasophyllum archeri = Genoplesium archeri

SCALE: 0 1cm 2cm The scale representation on each plant equals 2cm.

*Prasophyllum colemanae*

## Prasophyllum australe

Austral Leek-orchid

**Size:** stem to 0.75 m high **P**
**Habitat:** Damp and valley sclerophyll forests
**Foliage:** Slender to stout, erect leaf to 35 cm long, free for 30 cm of its length, reddish at base
**Flowers:** Dense spike of 15–60 sweetly perfumed flowers to 15 mm across, white with reddish-brown and greenish stripes on long slender ovaries, appressed against the stem. Perianth segments lanceolate, sharply pointed, to 8 mm long; lateral sepals usually united almost to the tip.
Labellum is white with curly margins and a swollen base, sharply recurved from the middle with a callus ridge ending at the bend with 2 raised lumps. September to January
**Requirements:** Moist to wet soils
**Comments:** Grows in colonies amongst grasses and sedges.
**Localities:** Bayswater, Scoresby, Lysterfield, 42, 58
**Distribution:** Vic, Qld, NSW, Tas, SA

## Prasophyllum brevilabre

Short-lip Leek-orchid

**Size:** stem to 50 cm high **P**
**Habitat:** Damp and valley sclerophyll forests
**Foliage:** Slender, fleshy leaf to 35 cm long, often reddish at base
**Flowers:** Loose spike of 8–30 green flowers to 10 mm across with dark red to dark purplish-green markings. Perianth segments to 8 mm long, lateral sepals usually joined throughout.
White labellum curves sharply back on itself from the middle with green callus extending just beyond the bend, margins curly. August to January
**Requirements:** Moist soil
**Comments:** The ovaries in this and most other species stand out from the stem and are swollen, unlike *P. australe* and *P. elatum* which are pressed against the stem.
**Localities:** Nth Warrandyte, Bayswater, 55, 58
**Distribution:** Vic, Qld, NSW, Tas

## Prasophyllum colemanae

**Size:** stem 0.25–0.9 m high **P**
**Habitat:** Valley sclerophyll forest
**Comments:** Very similar to *P. odoratum* but more robust with leaf often longer than the flowering stem. The lavender flowers are smaller, to 10 mm across, with up to 20 per spike. The dorsal sepal is greenish and petals have a narrow green central stripe. The labellum is flat and only slightly flexed. Last collected in 1922 in Bayswater.
**Localities:** Heathmont-Bayswater
**Distribution:** Endemic to Vic
**Synonym:** Previously included with *P. odoratum*

## Prasophyllum despectans = Genoplesium despectans

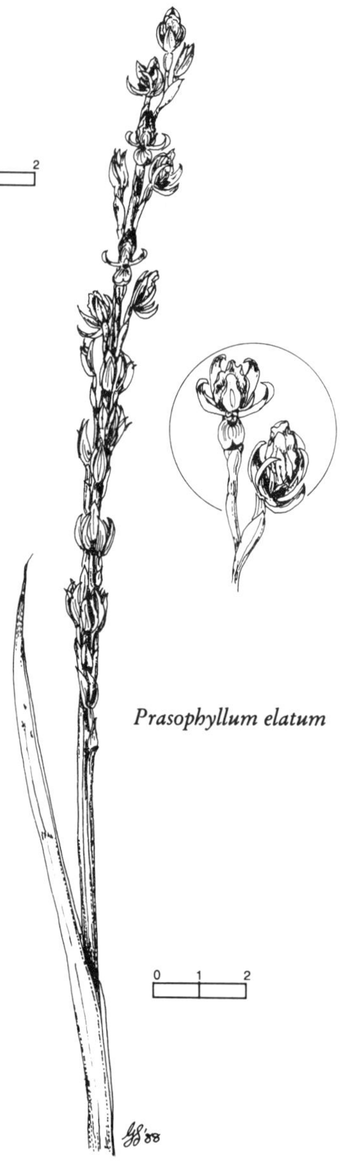

*Prasophyllum elatum*

## Prasophyllum elatum

Tall Leek-orchid

**Size:** stem 0.6–1.2 m high **F P**
**Habitat:** Damp sclerophyll forest, wattle tea-tree scrub, moist tea-tree heath
**Form:** Robust herb varying from pale green to dark purple
**Foliage:** Thick fleshy leaf to 1 m long
**Flowers:** Dense spike of 15–60 relatively large, lightly perfumed yellow-green to brown or purple flowers to 15 mm across. Flowers and ovaries similar to *P. australe*. Perianth segments longer, to 10 mm long, and often curved, tips moderately pointed.
White labellum has a green callus ridge almost to the tip. It is slightly recurved and base is not swollen, margins are wavy. August to December
**Requirements:** Moist, well drained soils
**Comments:** A variable species which flowers well after fire. One of the tallest orchids in Melbourne.
**Localities:** Belgrave, Lysterfield, Beaumaris[#], Cheltenham[#], 67[#], 74
**Distribution:** All states except NT

## Prasophyllum flavum

Yellow Leek-orchid

**Size:** stem to 0.9 m high **P N**
**Habitat:** Wet sclerophyll forest
**Foliage:** Closely sheathing purplish leaf to 60 cm long, sometimes yellow at the base
**Flowers:** Crowded spike of 6–50 yellow to yellowish-green flowers to 10 mm across, sometimes with reddish-brown markings. Perianth segments to 10 mm long, lateral sepals usually fully united.
Labellum green with white or cream wavy margins, gently recurved; callus extending halfway. October to February
**Requirements:** Moist humus-rich soils
**Comments:** Saprophytic herb growing in dense undergrowth.
**Localities:** Sherbrooke, Olinda
**Distribution:** Vic, Qld, NSW, Tas

## Prasophyllum frenchii

Maroon or Slaty Leek-orchid

**Size:** stem 30–60 cm high **F P**
**Habitat:** Valley sclerophyll forest, tea-tree heaths, wattle tea-tree scrub
**Foliage:** Slender to stout leaf to 50 cm long, free part withering at flowering time
**Flowers:** Dense spike of 30–60 flowers to 8 mm across, grey-brown or green with red markings. Perianth segments to 10 mm long, tips bent backwards; lateral sepals free.
Labellum broad and fleshy in basal section, tip erect and pointed, margins whitish and slightly wavy. Low broad callus with thickened margins, extends to bend. October to December
**Requirements:** Moist, well drained soils
**Comments:** Known from few Melbourne sites. Also rare throughout Victoria.
**Localities:** Bayswater#, Dandenong Rges, Mt. Eliza, 35, 51, 67#
**Distribution:** Vic, SA
**Synonym:** *P. hartii*

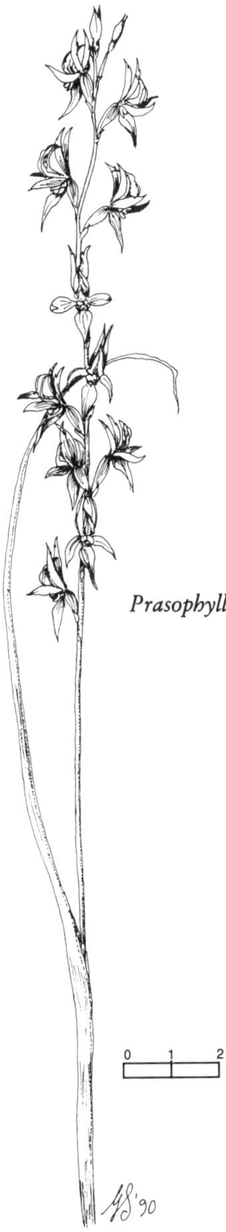

*Prasophyllum fuscum*

## Prasophyllum fuscum

Tailed or Tawny Leek-orchid

**Size:** stem 25–45 cm high **F P**
**Habitat:** Moist grasslands of basalt plains, damp sclerophyll forest, wattle tea-tree scrub
**Foliage:** Slender to stout leaf as long as the flowering spike, free part often withering at flowering time
**Flowers:** Loose to dense spike of 10–30 perfumed green, greenish-brown or purplish flowers, to 7 mm across. Perianth segments to 8 mm long, lateral sepals joined in the basal half, becoming free with maturity, margins curved inwards.
Labellum erect from above the middle or near the tip, margins flat or wrinkled; callus narrow, raised, extending almost to the tip. September to November
**Requirements:** Moist soils
**Comments:** A widespread orchid, becoming increasingly restricted throughout Victoria. Flowers well after fire.
**Localities:** Keilor Plains, St. Albans, Belgrave, 67#
**Distribution:** Vic, NSW, Tas, SA
**Synonym:** Confused with *P. affine*

## Prasophyllum lindleyanum

Green Leek-orchid

**Size:** stem to 35 cm high **P**
**Habitat:** Damp and valley sclerophyll forests, grassy low open forest
**Foliage:** Slender yellowish-green leaf to 35 cm long, free part usually withering at flowering time
**Flowers:** Loose slender spike of 12–50 yellowish-green flowers, to 7 mm across. Perianth segments to 8 mm long, curved back; lateral sepals spreading, free with margins inrolled.
White or pink s-shaped labellum with curly, hairy margins; narrow green callus ridge extending towards the tip. September to January
**Requirements:** Well drained soils
**Comments:** Vulnerable in Victoria. Flowers have a strong, sweet fragrance.
**Localities:** Mitcham-Bayswater, Dandenong Ranges, Belgrave, Black Rock, Cheltenham
**Distribution:** Vic, Tas
**Synonym:** *P. brainei*

*Prasophyllum lindleyanum*

## Prasophyllum morrisii = Genoplesium morrisii

## Prasophyllum odoratum

Scented or Sweet Leek-orchid

**Size:** stem to 0.9 m high **P**
**Habitat:** Damp and valley sclerophyll forests
**Foliage:** Slender or stout leaf to 70 cm long
**Flowers:** Spike of 10–40 perfumed green and white flowers to 15 mm across, sometimes with reddish stripes. Perianth segments to 12 mm long, pointed; lateral sepals free, widely spreading.
White labellum sharply curved back below the middle, tip passing through the lateral sepals, margins wavy; green callus extends just beyond the bend. August to February
**Requirements:** Moist, well drained soils
**Comments:** One of the most widespread *Prasophyllum* spp., prominent after fires.
**Localities:** Knox, Ringwood-Belgrave, Ferny Ck, 55, 56, 58
**Distribution:** Vic, Qld, NSW, Tas

*Illustrated overleaf*

SCALE: 0 1cm 2cm — The scale representation on each plant equals 2cm.

*Prasophyllum odoratum*

### Prasophyllum patens

Broad-lip Leek-orchid

**Size:** stem 25–45 cm high **F P**
**Habitat:** Primary dune scrub, tea-tree heath
**Foliage:** Slender leaf to 30 cm long, free part often withering at flowering time
**Flowers:** Loose or dense spike of 15–50 pinkish-green flowers, to 12 mm across. Perianth segments spreading, to 8 mm long, dorsal sepal ovate-lanceolate; lateral sepals lanceolate, free; petals narrower.
White or pink labellum ovate-lanceolate with wavy margins, curved backwards; green callus ridge extends just beyond the bend, edges not thickened. September to December
**Requirements:** Well drained sandy soil
**Comments:** Rare in Victoria, it may already be extinct in Melbourne.
**Localities:** Sandringham#
**Distribution:** Vic, Qld, NSW, Tas, SA; NZ
**Synonym:** Was confused with *P. truncatum* which is endemic to Tas.

### Prasophyllum pyriforme

Graceful Leek-orchid

**Size:** stem to 0.5 m high **P**
**Habitat:** Valley sclerophyll forest
**Foliage:** Slender leaf to 30 cm long, free part often wilting at flowering time
**Flowers:** Dense spike of 15–40 fragrant yellowish-green, brownish or purplish flowers, to 12 mm across. Perianth segments to 15 mm long, lateral sepals free or joined in the basal third.
White, pink or purplish labellum with wavy margins, upper half erect, fleshy green callus raised and extending almost to the tip. September to January
**Requirements:** Well drained soil
**Comments:** The original specimen was collected in Wonga Park. Plants favour lightly timbered sites, growing among grass tussocks.
**Localities:** Doncaster, Wonga Park, Belgrave, 51, 58
**Distribution:** Endemic to Vic
**Synonym:** Previously known as *P.* sp. aff. *rostratum*. Wrongly named *P. gracile*.

### Prasophyllum rufum = Genoplesium rufum

### Prasophyllum spicatum

**Size:** stem to 25 cm high **P**
**Habitat:** Tea-tree heath
**Comments:** Similar to *P. odoratum*. A very robust orchid with many green and white flowers in a dense spike to 15 cm long.
**Localities:** Cranbourne, 74
**Distribution:** Endemic to Vic
**Synonym:** *P. odoratum* var. *album*

## Pterostylis

A well represented genus of terrestrial orchids in Melbourne, common in cool moist forests, with a few species occurring on the basalt plains. Most *Pterostylis* spp. are glabrous and have a basal rosette of leaves which may either encircle the flower stem or appear to one side. The rosette may only be present in non-flowering plants. In those species where the rosette is absent from flowering plants, the leaves are scattered up the stem.

*Pterostylis* spp. are commonly known as Greenhoods because flowers are generally green, sometimes with brown or reddish tinges, and the dorsal sepal and petals are united into a hooded structure called the galea, which encloses the winged column. The galea is usually erect for about ¾ of its length before curving forward. The labellum is variable in shape and is attached by a movable claw to the foot of the column. The lateral sepals are united for part or most of their length to form a lower lip which may be erect, embracing the galea, or, in many species, bent downwards. The sinus, or recess, between the lateral sepals differs in shape and size and acts as a further distinguishing feature.

Many of the colony-forming species are amongst the easiest of the terrestrial orchids to grow. They are often available from specialist nurseries or at Native Orchid displays and shows and should not be removed from the bush. The main requirement is a well drained mix and the observance of a Summer dormancy period, when the pots are permitted to dry out. Solitary species can generally be grown but are sensitive to overwatering. Propagation is by division of tuberoids. Small flies or fungus gnats are the pollinators.

*Pterostylis alata*

### Pterostylis alata

Striped Greenhood

**Size:** stem 8–25 cm high **P**
**Habitat:** Widespread in dry and valley sclerophyll forests, tea-tree heath
**Form:** Erect slender herb on thin stem; non-flowering plants form leafy rosette
**Foliage:** Rosette of 3–11 long-stalked, ovate dark green leaves with wavy margins, to 20 mm long; flowering plants have 3–4 sheathing sessile, lanceolate stem leaves to 40 mm
**Flowers:** Single, occasinally 2, pale green or translucent white flowers, with dark green stripes, to 25 mm long. Galea ends in a fine point. Labellum green to brown, erect, narrowed in upper half.
Lateral sepals erect, sinus broad and notched, free points to 25 mm long and thread-like. April to August
**Requirements:** Moist, well drained soils
**Comments:** Forms colonies but difficult to establish in cultivation. Similar to *P. robusta*, but taller and more slender with smaller leaves and flowers. Grows in higher rainfall areas. Becoming locally rare.
**Localities:** Greensborough, Hurstbridge-Warrandyte, Eastern heathlands of Port Phillip Bay, 65#–67
**Distribution:** Vic, NSW, Tas, SA

## Pterostylis alpina

Mountain Greenhood

**Size:** stem to 30 cm high **P**

**Habitat:** Damp and valley sclerophyll forests

**Form:** Stout herb with a rough fleshy stem, non-flowering plants form extended leafy rosette

**Foliage:** 3–5 sessile, dark green, fleshy, ovate to elliptic leaves to 60 mm long scattered up the stem, margins entire or wavy

**Flowers:** Single translucent white flower to 30 mm long with dark green lines and tinges. Dark green labellum curved near end to protrude through the sinus.
Lateral sepals erect, sinus broad with notch, free points to 15 mm long swept back behind galea. August to October

**Requirements:** Moist, well drained soil

**Comments:** An easily grown colonising *Pterostylis* which usually occurs on southern slopes growing amongst grassy tussocks. Often confused with *P. scabrida* which is endemic to Tasmania.

**Localities:** Montrose, Belgrave, 34A#, 41, 42, 47, 49, 51, 55–58

**Distribution:** Vic, NSW Tas

*Pterostylis alpina*

## Pterostylis concinna

Trim Greenhood

**Size:** stem 8–30 cm high **F P**

**Habitat:** Widespread in dry and valley sclerophyll forests, tea-tree heath, grassy low open forest, coastal woodland

**Form:** Wiry erect herb

**Foliage:** Rosette of 4–6 dark green ovate to oblong stalked leaves to 30 mm long, margins entire or wavy

**Flowers:** Single, sometimes 2, translucent pale green flowers to 15 mm long with dark green stripes and brown tinges near sepal points and sinus. Galea ends in a fine point. Brown labellum slightly curved, ending in 2 teeth.
Lateral sepals erect, with a very wide sinus and fine erect points to 20 mm long held above the galea. Tips of the points appear clubbed. May to October

**Requirements:** Moist, well drained soils

**Comments:** May form dense colonies, usually in shaded areas, but will tolerate full sun in moist protected positions.

**Localities:** Eltham, Belgrave, Dingley, Cheltenham, Hampton, Beaumaris, 33, 34A, 42, 47, 49, 58, 65#–67, 69A, 74–76

**Distribution:** Vic, NSW, Tas, SA

*Pterostylis concinna*

## Pterostylis x toveyana

Mentone Greenhood

**Size:** stem to 20 cm high **P**

**Habitat:** Dry sclerophyll forest, tea-tree heath, grassy low open forest

**Foliage:** Non-flowering plants have a rosette of 2–4 ovate leaves to 20 mm long similar to both parents. Flowering plants also have 2–3 lanceolate sheathing stem leaves to 30 mm long.

**Flowers:** The single flower, to 15 mm long, is translucent white with broad green stripes and brown tinges on the tips.
Lateral sepals tightly embrace the galea. Free points erect to 20 mm long, tips expanded and clubbed. May to October

**Requirements:** Moist, well drained soil

**Comments:** A naturally occurring hybrid between *P. concinna* and *P. alata*. First collected in Mentone.

**Localities:** Greensborough, Mentone, Aspendale, Frankston

**Distribution:** Vic, SA, Tas

*Pterostylis x toveyana*

## Pterostylis cucullata

Leafy Greenhood

**Size:** stem 8–15 cm high **P**

**Habitat:** Tea-tree heath

**Foliage:** Rosette of 5–7 sessile, oblong to elliptical dark green leaves to 100 mm long, with 2 or 3 leaf-like bracts scattered up the stem, the large, uppermost bract sheathing the base of the flower.

**Flowers:** Single, large softly hairy green flower to 30 mm long, petals and lower lip reddish-brown. Tip of galea curved down. Dark brown labellum, curved, blunt end not protruding.
Lateral sepals erect with a very narrow sinus, points to 10 mm long, tapering, just exceeding the galea. August to October

**Requirements:** Sandy soil

**Comments:** Very restricted in Melbourne, it is also vulnerable throughout its range in Australia. Forms small colonies.

**Localities:** Hampton 1948; 29#, 67#, 74#

**Distribution:** Vic, NSW, Tas, SA

*Pterostylis cucullata*

SCALE: 0 1cm 2cm — The scale representation on each plant equals 2cm.

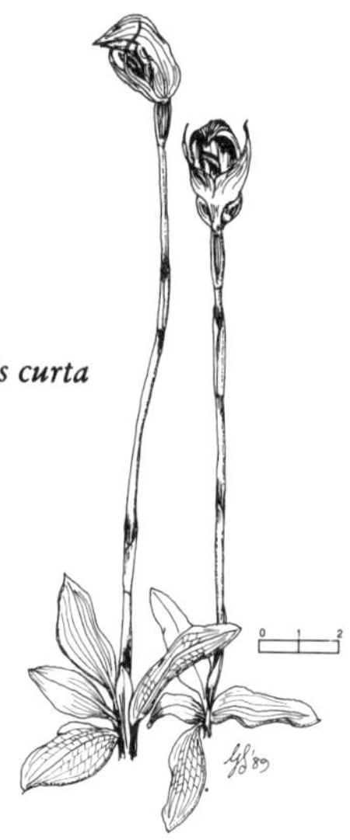

*Pterostylis curta*

## Pterostylis curta

Blunt Greenhood

**Size:** stem 10–30 cm high **P N**

**Habitat:** Widespread in valley and damp sclerophyll forests, riparian scrub, tea-tree heath

**Foliage:** Rosette of 2–6 dark green ovate or oblong leaves to 100 mm long, on long stalks, margins wavy or flat

**Flowers:** Single green flower with darker stripes and brown tinges, to 35 mm long. Brown labellum has a raised central ridge and a distinctly twisted tip protruding through the sinus.
Lateral sepals erect, loosely embracing the galea, sinus very broad with free points to 12 mm long, not above the galea. July to October

**Requirements:** Moist, well drained soils

**Comments:** An easily grown free flowering plant.

**Localities:** Widespread from Heidelberg to Belgrave, 29#, 67, 69A, 74, 76

**Distribution:** Vic, Qld, NSW, Tas, SA

*Pterostylis cycnocephala*

## Pterostylis cycnocephala

Swan Greenhood

**Size:** stem to 20 cm high **F P**

**Habitat:** Box woodland, dry sclerophyll forest

**Foliage:** Rosette of 6–12 sessile, ovate leaves to 25 mm long, with a pointed tip, crowded and overlapping. 2 or more closely sheathing stem bracts

**Flowers:** Spike of 2–24 tiny bright green and white flowers to 9 mm long. Galea curved throughout with a short, blunt, downward pointing tip. The labellum has a very dark shallowly notched tip, and the dark green basal appendage is curved towards the tip.
Lateral sepals, to 6 mm long, are bent downwards, concave, with free or united tips. August to January

**Requirements:** Well drained soils

**Comments:** Similar to the closely related *P. mutica*, but may be distinguished by the dark labellum tip, and the basal appendage curving towards the tip. Solitary plants often grow amongst rocks, sometimes occurring with *Allocasuarina* species. A drought tolerant plant.

**Localities:** Greensborough#, Keilor Plains

**Distribution:** Vic, Qld, NSW, Tas, SA

*Pterostylis decurva*

## Pterostylis decurva

Summer Greenhood

**Size:** stem to 30 cm high **P N**

**Habitat:** Valley and damp sclerophyll forests

**Foliage:** Rosette of 2–5 long-stalked, dark green oblong leaves to 30 mm long, with a blunt or rounded tip; 4–5 sheathing stem leaves to 40 mm long on flowering plants

**Flowers:** Single yellowish-green and white flower to 25 mm long, with brown tinges especially on the petals. Galea bent abruptly downwards at the tip, ending in a thread-like point to 20 mm long. Blunt reddish-brown labellum is sharply curved down in the upper third, protruding through the sinus.
Lateral sepals tightly embrace the galea with wide sinus slightly protruding, free points to 40 mm long, extending well above the galea. October to March

**Requirements:** Moist soils

**Comments:** Plants are usually solitary, but may form small colonies amongst grassy tussocks. Easily grown in a container.

**Localities:** Knox, Boronia, Belgrave, 57, 58

**Distribution:** Vic, NSW, Tas

## Pterostylis foliata

Slender Greenhood

**Size:** stem to 30 cm high **P N**

**Habitat:** Damp and valley sclerophyll forests

**Foliage:** 3–6 sessile, elliptical dark green leaves to 80 mm long, scattered along the stem, margins wavy; sometimes a few basal leaves form a loose rosette

**Flowers:** Single translucent white flower to 20 mm long with broad green stripes and brown tinges towards the tip. Galea at first erect, then curved forward. Brown labellum is curved in the upper third, protruding through the sinus.
Lateral sepals erect with a narrow sinus and free points to 20 mm long, exceeding the galea. August to January

**Requirements:** Moist soil

**Comments:** Locally rare in Melbourne.

**Localities:** Wattle Glen, Greensborough, Dandenong Rges

**Distribution:** Vic, NSW, Tas; NZ

## *Pterostylis furcata* — Sickle Greenhood

**P N**

**Size:** stem to 30 cm high
**Habitat:** Swamps and moist depressions in damp and valley sclerophyll forests
**Foliage:** Rosette of 4–5 bright green lanceolate leaves to 80 mm long
**Flowers:** Single large green and white flower to 60 mm long. Galea prominently curved and sickle-shaped. Curved, narrow dark green to brown labellum protruding through the sinus.
Lateral sepals erect, widely separated from the galea with very narrow free points to 30 mm long, curving forward above the galea. September to January
**Requirements:** Moist to wet soils
**Comments:** Needs to remain very wet during growth period.
**Localities:** Bayswater, Mooroolbark, Belgrave, The Basin, Wantirna South, Rowville, 33, 34A, 51
**Distribution:** Vic, Qld, NSW, Tas, SA
**Synonym:** *P. falcata*

*Pterostylis furcata*

## *Pterostylis grandiflora* — Cobra or Superb Greenhood

**P**

**Size:** stem 10–40 cm high
**Habitat:** Valley sclerophyll forest, grassy low open forest, tea-tree heath
**Foliage:** Flowering plants have 6–9 dark green, lanceolate sheathing stem leaves to 40 mm long. After flowering, plants develop a rosette of 3–5 shortly stalked, ovate leaves to 20 mm long, with wavy margins. A similar rosette is borne on non-flowering plants.
**Flowers:** Single translucent white flower to 35 mm long with green stripes and red-brown tinges. Galea is at first erect, then curves forward, then down with a fine tip to 10 mm long. Bronze-brown petals flared at right angles on the curved sepal. Erect red-brown labellum tapers sharply in the upper half and protrudes through the sinus.
Lateral sepals erect, sinus wide and protruding, free points very long, to 50 mm, extending high above the galea. May to August
**Requirements:** Moist soils
**Comments:** Very restricted now in Melbourne and rare throughout Victoria.
**Localities:** Sth Belgrave, Lysterfield, Eastern Port Phillip Bay, 67#
**Distribution:** Vic, Qld, NSW, Tas

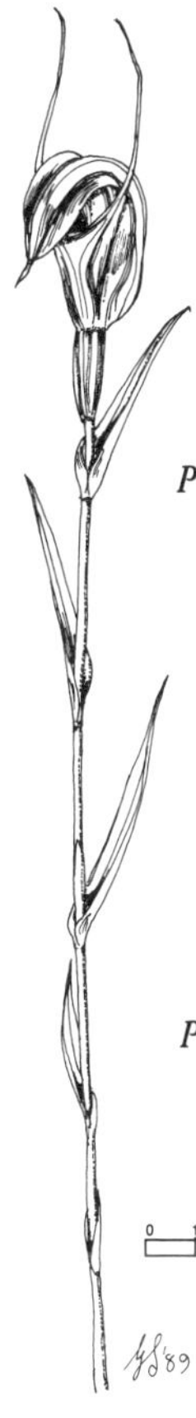

*Pterostylis grandiflora*

## *Pterostylis longifolia* — Tall Greenhood

**P**

**Size:** Variable, stem 0.15–0.9 m high
**Habitat:** Widespread in dry and valley sclerophyll forests, tea-tree heath, grassy low open forest
**Foliage:** Basal rosette of 3–6 long-stalked, linear dark green leaves to 35 mm long on non-flowering plants only; 5–8 stem-clasping, dark green lanceolate stem leaves to 100 mm long, held erect
**Flowers:** Raceme of 1–15 green and white flowers with brown tips, to 15 mm long, widely spaced along the stem. Galea curved from the base, ending in a short point. Brown to dark green hairy labellum is narrowly 3-lobed with a notched tip turned upwards.
Lateral sepals bent downwards, joined for ½ their length then tapering to a short point. April to November
**Requirements:** Moist, well drained soils
**Comments:** Plants usually grow as solitary specimens, often close to Eucalypt trunks or through grass tussocks. Because of their height they are easily recognised. Although the most common species in Melbourne it can be difficult to maintain in cultivation.
**Localities:** Heidelberg to Dandenong Rges, 61, 63, 67#, 74–76
**Distribution:** Vic, Qld, NSW, Tas, SA

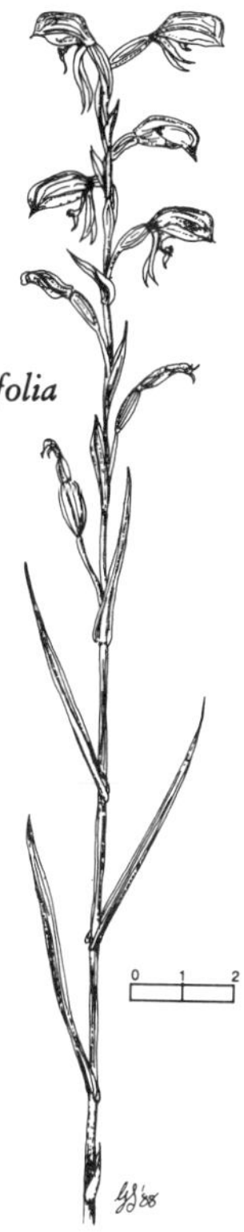

*Pterostylis longifolia*

## *Pterostylis* sp. aff. *longifolia*

**Size:** stem 0.4 m high
**Habitat:** Box ironbark woodland, dry sclerophyll forest
**Comments:** Similar to *P. longifolia* but more robust with larger flowers which are crowded towards the top of the stem rather than spaced along it. Labellum differs in that it is larger, with a slightly broader mid-lobe. Erect lateral lobes are tinged brown. Leaves are longer and curved rather than erect. An orchid known from 3 sites one of which has since disappeared for housing development. The other 2 sites are currently at great risk.
**Localities:** Eltham#, Greensborough, Diamond Creek
**Distribution:** Endemic to Vic

*Pterostylis* sp. aff. *longifolia*

SCALE: 0 1cm 2cm — The scale representation on each plant equals 2cm.

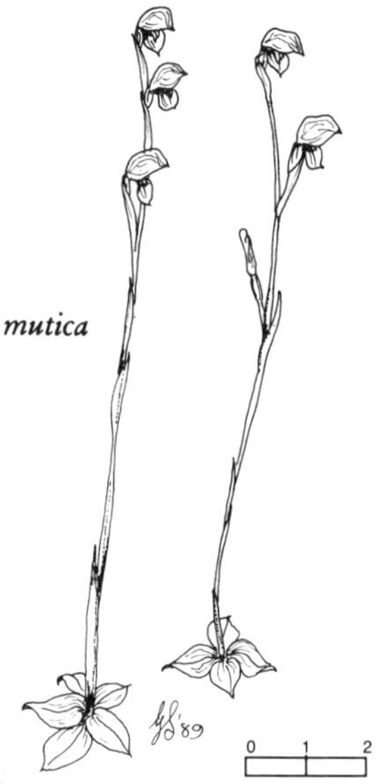

*Pterostylis mutica*

## Pterostylis mutica

Midget Greenhood

**Size:** stem 8–25 cm high **F P**
**Habitat:** Woodlands of basalt plains, dry sclerophyll forest, tea-tree heath
**Foliage:** Rosette of 6–12 sessile, ovate dark green leaves to 25 mm long, with flat margins; 1–9 pointed sheathing stem bracts
**Flowers:** Dense spike of 2–10 tiny waxy green flowers to 8 mm long. Galea curved throughout, blunt-tipped. Nearly round pale green labellum has a notched tip and a fleshy appendage at the base curved like an 'S' away from the tip.
Lateral sepals bent downwards, concave and united almost to the tip. July to January
**Requirements:** Well drained soils
**Comments:** A solitary species propagated by seed. It usually grows in drier situations, often protected only by rocks or small grasses. When found on the basalt plains it commonly occurs with *Callitris glaucophylla*. May be extinct within the Melbourne area.
**Localities:** Melton#, Tottenham#, Werribee#, Greensborough#, 29#, 34A, 67#
**Distribution:** All states except NT; NZ

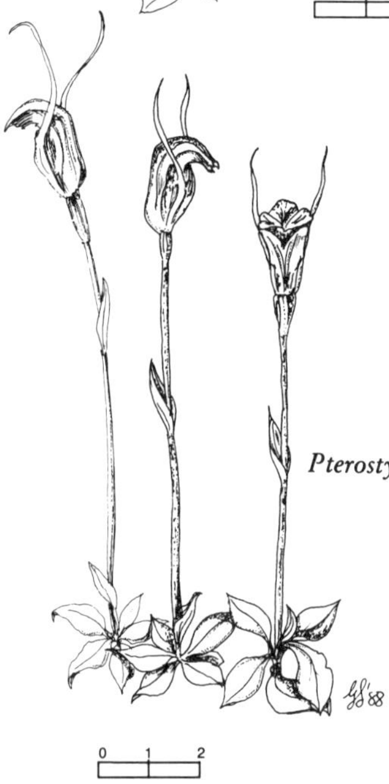

*Pterostylis nana*

## Pterostylis nana

Dwarf or Snail Greenhood

**Size:** stem 5–15 cm high **F P N**
**Habitat:** Dry sclerophyll forest, grassy low open forest, primary dune scrub
**Foliage:** Rosette of 4–8 dull green, shortly-stalked ovate leaves to 30 mm long, margins wavy; flowering plants also have 2–3 sheathing stem bracts
**Flowers:** Single bright green flower to 15 mm long with dark green stripes. Erect, thick greenish-brown labellum is hidden.
Lateral sepals erect with a central incurved tooth in the large sinus, free points to 25 mm long above the galea, tips often clubbed. July to October
**Requirements:** Well drained soils
**Comments:** A very adaptable orchid tolerating a wide range of soils and habitats. Plants exhibit a remarkable degree of variation, and may be solitary or grow in small colonies. In dry open situations the stem may be as small as 2 cm in length, but in moist protected sites, slender vigorous plants to 30 cm high have been observed. A white form is known from 1 site in Eltham.
**Localities:** Bundoora, Wattle Glen, Eltham, Belgrave, 34A, 42, 51, 58, 67#, 74–76
**Distribution:** Vic, NSW, Tas, SA, WA; NZ

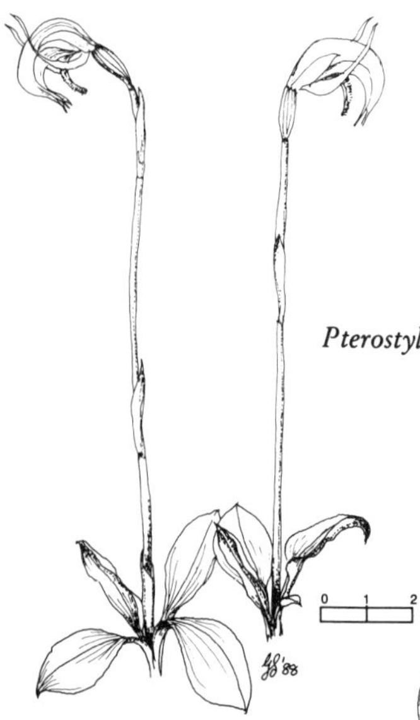

*Pterostylis nutans*

## Pterostylis nutans

Nodding Greenhood or Parrot's Beak Orchid

**Size:** stem 10–30 cm high **P N**
**Habitat:** Widespread in dry and valley sclerophyll forests, red gum woodland, tea-tree heath, grassy low open forest
**Foliage:** Rosette of 3–6 dark green shortly-stalked ovate to oblong leaves 25–80 mm long with wavy margins; 3 stem leaves on flowering plants
**Flowers:** Single nodding translucent flower to 25 mm long with green stripes, sometimes reddish towards the tip. Galea curved forward ending in a sharp point. Very hairy green labellum with reddish-brown central ridge, curved to protrude through the sinus.
Lateral sepals with narrow sinus, loosely embracing the galea, curved downwards at the base then curving forwards, free points extending just beyond the galea. March to October
**Requirements:** Moist soils
**Comments:** A locally common plant forming large colonies.
**Localities:** Bundoora, Greensborough-Dandenong Ranges, Hampton-Frankston, Cranbourne, 36, 63, 65#, 71, 74–76
**Distribution:** Vic, Qld, NSW, Tas, SA; NZ

*Pterostylis x ingens*

## Pterostylis x ingens

Large Pointed Greenhood

**Size:** stem to 30 cm high **P**
**Habitat:** Valley sclerophyll forest
**Foliage:** Rosette of 3–7 ovate leaves to 35 mm long, with wavy margins
**Flowers:** Displays similarities to both parents. Flower large to 35 mm long. Galea is humped near the base before bending forwards and then downwards from halfway. The greenish-brown labellum is hairy, with a raised central ridge and is prominent as in both parents.
Lateral sepals loosely embrace the galea with free points to 25 mm long, just beyond it. The sinus is narrow and does not protrude.
**Comments:** A naturally occurring hybrid between *P. nutans* and *P. furcata* with intermediate features. Hybrid populations are likely where the parents occur in close proximity in moister habitats of their range. Easily grown in cultivation, producing large colonies and flowering freely.
**Localities:** Bayswater, Wantirna South, Mooroolbark, Wonga Park, Blackburn, 61
**Distribution:** Vic, NSW, Tas

## Pterostylis obtusa

Blunt-tongue Greenhood

**Size:** stem 10–25 cm high **P**
**Habitat:** Valley and damp sclerophyll forests, often growing with ferns or amongst grasses, tea-tree heath
**Foliage:** Rosette of 3–6 ovate to oblong, shortly-stalked leaves to 25 mm on non-flowering plants; flowering plants have 3–5 lanceolate stem leaves to 40 mm long
**Flowers:** Single translucent white flower to 30 mm long with bright green stripes and brown tinges. Galea is erect for most of its length, then curves forward, ending in a short fine tip. Dark brown oblong labellum slightly protruding through sinus.
Lateral sepals erect with very broad bulging sinus, free points to 25 mm long held high above the galea. February to June
**Requirements:** Moist soils
**Comments:** Plants may form extensive colonies when protected by dense understorey. Colonies are generally sparse in more open situations. Locally endangered.
**Localities:** Montrose, The Basin, 58, 67#, 74
**Distribution:** Vic, Qld, NSW, Tas, SA

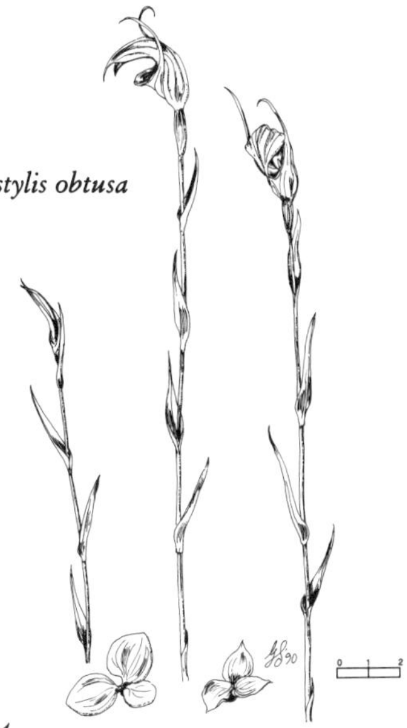

*Pterostylis obtusa*

## Pterostylis parviflora

Tiny Greenhood

**Size:** stem 15–30 cm high **P**
**Habitat:** Widespread in dry sclerophyll forest, tea-tree heath, grassy low open forest
**Foliage:** Rosette of 3–10 long-stalked, bright green to bluish-green ovate, pointed leaves to 15 mm long on non-flowering plants only; several small, closely sheathing stem bracts on flowering plants, with rosettes on side shoots off the flowering stem
**Flowers:** Spike of 1–13 green and maroon flowers to 10 mm long. Galea blunt ended. Blunt tip of labellum just visible.
Lateral sepals erect with broad sinus and very short free points. February to August
**Requirements:** Moist soils
**Comments:** Flowers are unusual in that they face towards the stem. A solitary species which is difficult to maintain in cultivation.
**Localities:** Eltham, Montrose, Nunawading-Bayswater, Belgrave, Springvale, Cheltenham, 42, 51, 56, 58, 67, 73, 74
**Distribution:** Vic, Qld, NSW, Tas, SA

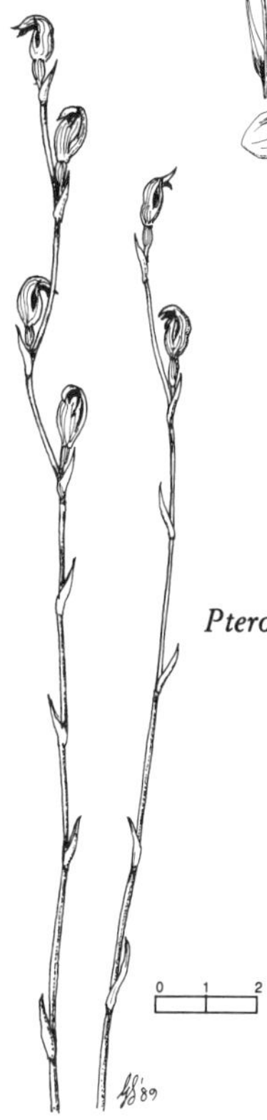

*Pterostylis parviflora*

## Pterostylis pedoglossa

Prawn Greenhood

**Size:** stem 10–15 cm high **P N**
**Habitat:** Peaty tea-tree heath, grassy low open forest
**Foliage:** Rosette of 3–6 dull greyish-green, ovate leaves to 25 mm long. The very thin flowering stem bears 2–4 sheathing bracts.
**Flowers:** Single translucent white flower to 15 mm long with green stripes at the base, brownish at the top. Galea ends in a fine point to 30 mm long. Labellum is tiny and hidden.
Lateral sepals erect, sinus narrow with free points to 30 mm long, held high above the galea. March to July
**Requirements:** Moist sandy soil
**Comments:** Plants are difficult to find, usually being well hidden in thick undergrowth or sedges and grasses, even when in large colonies. Readily identified by the long thread-like tips on all sepals. Difficult to cultivate. Vulnerable in Victoria and restricted to few sites in Melbourne.
**Localities:** Clayton#, Springvale#, Cheltenham#, 67#, 74
**Distribution:** Vic, NSW, Tas

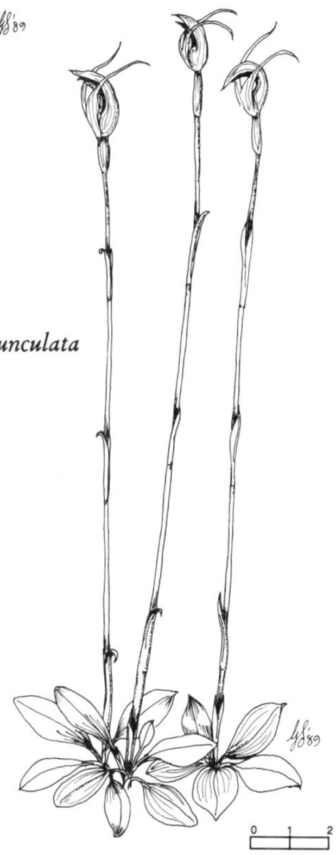

*Pterostylis pedunculata*

## Pterostylis pedunculata

Maroon Hood

**Size:** stem to 25 cm high **P N**
**Habitat:** Widespread in valley and dry sclerophyll forests, red gum woodland, tea-tree heath, grassy low open forest
**Foliage:** Rosette of 4–6 long-stalked, prominently veined, ovate to oblong dark green leaves to 40 mm long, margins sometimes wavy; 2–3 small pointed bracts closely sheathing the stem
**Flowers:** Single flower to 20 mm long, green and white at the base, the top tinged dark reddish-maroon. Tip of maroon labellum just visible.
Lateral sepals loosely erect, sinus narrow, free points to 30 mm long, held high above the galea. July to October
**Requirements:** Moist soil amongst leaf litter
**Comments:** Forms small dense colonies beneath low shrubs and grasses, or protected by tree trunks. In gullies, plants may grow on logs or fern trunks. Easily grown and flowers reliably.
**Localities:** Bundoora, Wattle Glen, Eltham, Park Orchards, Belgrave, Dingley, 34A, 36, 41, 42, 47, 51, 54, 57, 58, 61, 67, 69A, 74–76
**Distribution:** Vic, Qld, NSW, Tas, SA

SCALE: 0 1cm 2cm The scale representation on each plant equals 2cm.

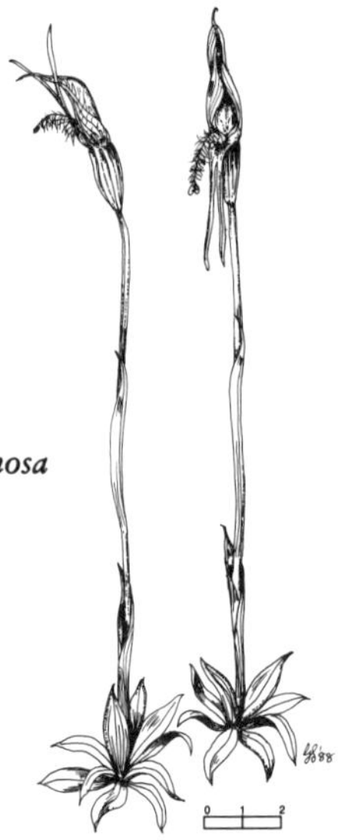

*Pterostylis plumosa*

## Pterostylis plumosa

Bearded Greenhood

**Size:** stem 20–30 cm high **P**
**Habitat:** Dry and valley sclerophyll forests, grassy low open forest
**Foliage:** Clump of 10–20 crowded, pale green or yellowish lanceolate leaves to 35 mm long, prominently veined, often extending for some distance up the stem
**Flowers:** Single, translucent, shiny green flower to 35 mm long, with dark veined markings. Galea is erect, ending in a short fine tip. Prominent thread-like labellum to 25 mm long, covered in long yellow hairs, ending in a dark knob.
Narrow lateral sepals bent downwards with inrolled margins, joined in lower third, free points nearly parallel. August to November
**Requirements:** Moist, well drained soils
**Comments:** Easily distinguished by its narrow, long yellow-haired tongue with an apical knob, and see-through galea. It usually occurs as solitary specimens but is sometimes found in small colonies. Difficult to cultivate.
**Localities:** Greensborough and Hurstbridge-Warrandyte, Heathmont, Bayswater, Belgrave, Moorabbin, 58, 67#, 74
**Distribution:** Vic, NSW, Tas, SA, WA; NZ

*Pterostylis revoluta*

## Pterostylis revoluta

Autumn Greenhood

**Size:** stem to 25 cm high **P**
**Habitat:** Dry sclerophyll forest, box ironbark woodland
**Foliage:** Rosette of 5–10 long-stalked, bluish-green ovate leaves to 20 mm long, only on non-flowering plants; 3–5 small pointed sheathing stem bracts on flowering plants
**Flowers:** Single translucent white flower with dark green stripes to 45 mm long, often with reddish-brown tinges. Galea leaning forward in a gentle curve, ending in a fine tip to 20 mm long. Curved pointed labellum protrudes through the sinus.
Lateral sepals loosely erect with narrow slightly protruding sinus, thread-like points to 35 mm long held high above galea. February to June
**Requirements:** Well drained soils
**Comments:** Plants generally occur in drier regions in the shelter of grass tussocks and small shrubs, and rely on good Summer/Autumn rains for flowering. Large free-flowering colonies are found in wetter years. Easy to cultivate. Locally rare.
**Localities:** Greensborough, Hurstbridge, Wonga Park, Park Orchards, 42, 49#, 50A#, 65#, 66
**Distribution:** Vic, Qld, NSW

*Pterostylis robusta*

## Pterostylis robusta

Large Striped Greenhood

**Size:** stem 5–15 cm high **P**
**Habitat:** Dry and damp sclerophyll forests
**Foliage:** 6–7 ovate to elliptic basal leaves only present on non-flowering plants; flowering plants bear 4–5 broad lanceolate stem leaves to 50 mm long
**Flowers:** Large single flower to 40 mm long, bright green and white with deep green stripes. Labellum dark green.
Lateral sepals are held erect or curved, free points to 30 mm long, sinus deeply notched. June to September
**Requirements:** Well drained soil
**Comments:** Similar to *P. alata* but shorter on a more robust stem, sinus more deeply notched. Easy to cultivate but shy to flower.
**Localities:** Greensborough, Dandenong Rges, 42
**Distribution:** Vic, NSW, SA, WA

*Pterostylis rufa*

## Pterostylis rufa

Rusty Hood

**Size:** stem to 30 cm high **P**
**Habitat:** Dry sclerophyll forest
**Foliage:** Rosette of 5–12 thick, dull green oblong to lanceolate leaves to 30 mm long, often withering by flowering time; 2–6 closely sheathing stem bracts
**Flowers:** Spike of 3–15 reddish flowers to 10 mm long. Galea leaning forward with a narrow opening and short tip, petal edges end before the base. Many tiny white hairs on margins of fleshy brown labellum.
Lateral sepals bent downwards, free points curved forwards from a narrow sinus. September to December
**Requirements:** Well drained gravelly soils
**Comments:** Widespread but not common, usually growing as solitary plants. They are not easy to cultivate as the tuberoids tend to rot.
**Localities:** Eltham, Park Orchards, 33#, 34A, 42, 50A, 51
**Distribution:** All states except NT

### Pterostylis sanguinea

Banded Greenhood

**Size:** stem 10–40 cm high F P
**Habitat:** Coastal banksia woodland, grassy low open forest
**Foliage:** Rosette of 3–6 ovate grey-green leaves to 30 mm long on non-flowering plants; 5–8 sessile, sheathing lanceolate leaves to 80 mm long, becoming gradually smaller down the flowering stem
**Flowers:** Compact raceme of 1–10 nodding reddish-brown flowers with greenish-white stripes, to 15 mm long. Galea and lateral sepals very wide. Fleshy brown to purple labellum has hairy margins and ends in upturned notched tip.
Lateral sepals bent downwards, tapering, with a narrow sinus, ending in short pointed tips. April to September
**Requirements:** Well drained sandy soil
**Comments:** Solitary species which is never common. Grows in protection of shrubby plants, often in peaty soils, and is difficult to locate. Unreliable in cultivation.
**Localities:** Cheltenham#, Baxter, 67#, 74, 76
**Distribution:** Vic, Tas, SA, WA
**Synonym:** Previously known as *P. vittata* which is endemic to WA.

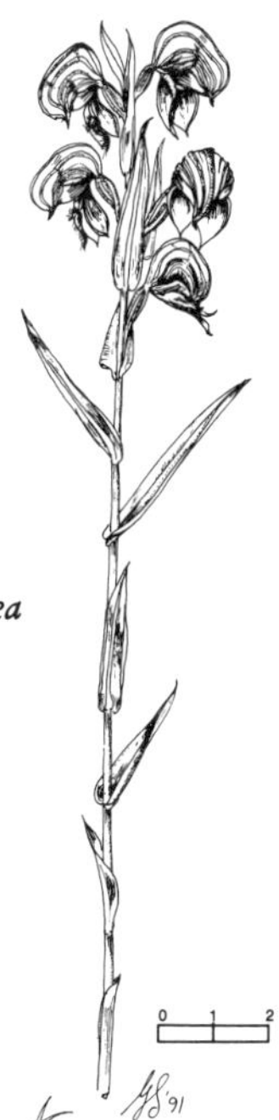

*Pterostylis sanguinea*

### Pterostylis truncata

Little Dumpies, Brittle Greenhood

**Size:** stem to 15 cm high P
**Habitat:** Rocky or grassy sites on plains grassland, valley sclerophyll forest, sclerophyll woodland
**Foliage:** Rosette of 2–6 bright green ovate leaves to 30 mm long on non-flowering plants; 2–5 closely sheathing tapering stem leaves to 30 mm long, on flowering plants
**Flowers:** Single large greenish-white flower to 50 mm long with dark green or reddish stripes. Galea curved forward ending in a point, petals broadly dilated. Labellum tapered to a blunt point, protruding slightly through the sinus.
Lateral sepals erect, sinus narrow, free points to 30 mm long, held erect or recurved above the galea. February to July
**Requirements:** Moist, well drained soil
**Comments:** Grows in dense colonies. Easy to cultivate but rarely flowers well. Most of its habitat has been destroyed by housing development leaving it endangered throughout Victoria.
**Localities:** Tottenham#, Sunshine#, Sth Belgrave, Beaumaris#, 5
**Distribution:** Vic, NSW

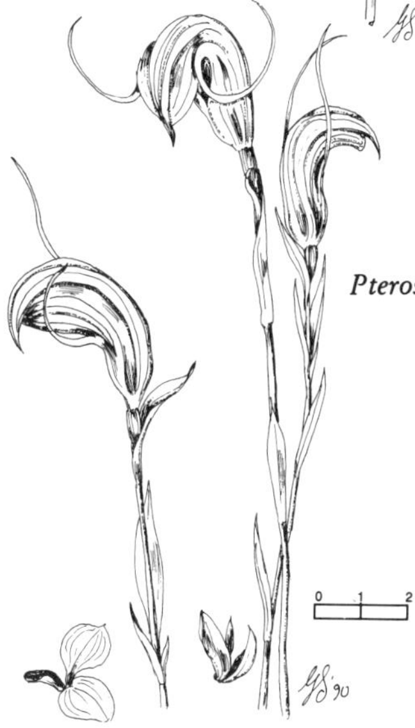

*Pterostylis truncata*

# Sarcochilus

### Sarcochilus australis

Butterfly Orchid, Gunn's Orchid

**Size:** 5–15 cm across, stems to 5 cm long P N
**Habitat:** Fern gullies in wet sclerophyll forest, cool temperate rainforest
**Form:** Semi-pendulous epiphytic orchid
**Foliage:** Thin, leathery, curved dark green linear-lanceolate leaves to 80 mm long
**Flowers:** Pendulous racemes of 2–14 fragrant green flowers to 15 mm across. Spreading perianth segments are narrow and fleshy, dorsal sepal incurved. Labellum is white with purple markings. October to December
**Requirements:** Moist humid conditions
**Comments:** The only epiphytic orchid in the Melbourne area, it lives on mossy branches and trunks of trees and shrubs such as *Coprosma quadrifida*, *Hedycarya* and *Bedfordia*, often occurring close to the ground. It is very restricted in its range in Melbourne.
**Propagation:** Impossible to cultivate
**Localities:** Dandenong Ranges
**Distribution:** Vic, Qld, NSW, Tas

*Sarchochilus australis*

# Spiranthes

### Spiranthes sinensis

Austral Ladies Tresses

**Size:** stem 25–50 cm high F P
**Habitat:** Swamps and bogs of wattle tea-tree scrub, valley sclerophyll forest, growing amongst rushes and sedges
**Form:** Erect tufted evergreen herb with a thin wiry stem
**Foliage:** Tuft of 3–5 erect, dark green narrow-lanceolate leaves to 15 cm long with 3–5 stem bracts
**Flowers:** Dense spirally-arranged spike of tiny pink and frosty white flowers. Perianth segments to 5 mm long, petals and dorsal sepal erect, lateral sepals spreading. Labellum is white with fringed margins. October to March
**Requirements:** Moist to wet soil
**Comments:** A very distinctive plant which is easily grown in a container. The flowers produce a sweet honey perfume, especially in the evenings. Very restricted in the Melbourne area and possibly already extinct.
**Propagation:** Seed, division
**Localities:** Bayswater#, Sandringham#, 74
**Distribution:** Vic, Qld, NSW, Tas, SA; NZ, Asia

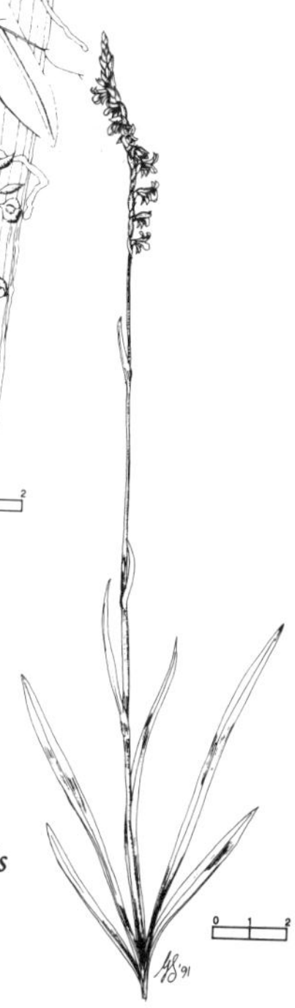

*Spiranthes sinensis*

SCALE: 0 1cm 2cm The scale representation on each plant equals 2cm.

# Thelymitra

A colourful, glabrous group of terrestrial orchids. Flowers open wide on warm sunny days, giving them the common name of Sun Orchids. They have a solitary, erect, channelled leaf sheathing at the base of the stem and 1–2 leafy stem bracts. The flowers are simple, all perianth segments including the labellum similar in appearance. Identification is made through the column which varies considerably. Important features include the 2 column arms (or lateral lobes) found on either side of the anther, which have different forms, and may be covered in many hairy tufts; the post-anther (or mid) lobe—an erect projection between the column arms and behind the anther which sometimes forms a fleshy hood overhanging the anther; and the side lobule—a fleshy projection behind the column arm but distinct from the mid lobe, with jagged margins. Propagation is by seed. They are difficult to cultivate.

*Thelymitra antennifera*

## Thelymitra antennifera

Rabbit's Ears, Vanilla Orchid

**Size:** stem 10–20 cm high **F P**
**Habitat:** Sclerophyll woodland, damp and valley sclerophyll forest, tea-tree heath, grassy low open forest
**Foliage:** Long grass-like leaf to 120 mm long, often reddish at the base. The pinkish flexuose flower stem bears 2 leaf-like bracts
**Flowers:** Zig-zagging terminal raceme of 1–3 large fragrant yellow flowers to 40 mm across; September to November
Yellow column is erect with broad wings, not hoded; dark brown column arms are 2-lobed and ear-like, held high above the column.
**Requirements:** Moist, well drained soils
**Comments:** Widespread, fairly common orchid, growing in colonies and appearing in numbers after fire.
**Localities:** Bundoora, Montrose, Belgrave, Cheltenham, 42, 47, 49, 58, 67#, 74
**Distribution:** Vic, NSW, Tas, SA, WA

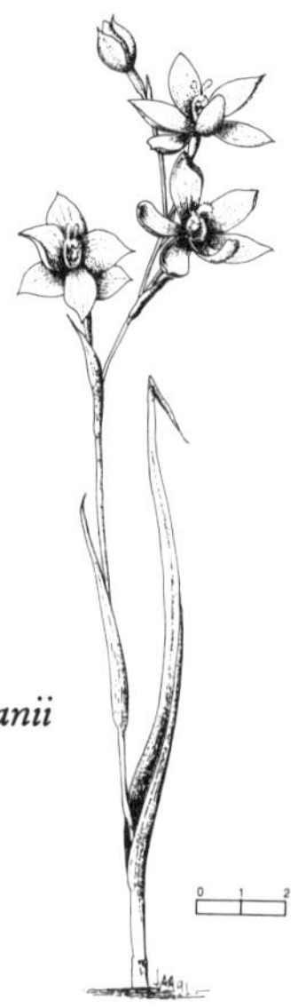

*Thelymitra x macmillanii*

## Thelymitra x macmillanii

Crimson Sun-orchid

**Size:** stem 15–20 cm high **F P**
**Habitat:** Grassy low open forest
**Foliage:** Thick, fleshy linear leaf, channelled inside.
**Flowers:** Slender wavy flower stem has 1–6 pink to crimson flowers to 40 mm across; September to October.
Column pink to crimson; post-anther lobe short, narrow and toothed; yellow to reddish column arms held high above anther, densely covered in small rounded teeth. Yellow anther is broad, thick, and hairy, curved horizontally forward.
**Comments:** A natural hybrid between *T. antennifera* and *T. nuda* occurring wherever the parents grow in close proximity. Because the parent plants are variable, the form of this hybrid also varies. In Melbourne it is only recorded for the eastern coastal fringe, but is more widespread in other areas. Flowers open freely on hot days, but in cooler weather the buds close and remain inflated as do those of *T. antennifera*.
**Localities:** Belgrave, 67#, Mt. Eliza
**Distribution:** Vic, Tas, SA, WA

*Thelymitra aristata*

## Thelymitra aristata

Scented Sun-orchid

**Size:** Robust stem to 0.8 m high **F P**
**Habitat:** Plains grassland, red gum woodland, dry sclerophyll forest
**Foliage:** Thick broad-lanceolate leaf to 35 cm long with 3–4 large leafy stem bracts
**Flowers:** Long terminal raceme of up to 30 blue to mauve flowers to 40 mm across; October to November.
White or bluish column erect, wings inflated; hood curved, 2-lobed and toothed; column arms erect, ending in a tuft of white hairs.
**Requirements:** Moist soils
**Comments:** An impressive orchid, very showy in flower. Plants appear in large numbers in moist to wet heathlands. They are generally more scattered in forested areas. Locally rare, flowering well after fire.
**Localities:** Werribee#, St. Albans, Sydenham, Eltham, 25, 33
**Distribution:** Vic, NSW, Tas, SA
**Synonym:** *T. grandiflora*

## Thelymitra carnea

Pink Sun-orchid

**Size:** stem 20–35 cm high **F P**
**Habitat:** Valley sclerophyll forest, grassy low open forest
**Foliage:** Narrow, terete or channelled leaf to 15 cm long, 2–3 stem bracts
**Flowers:** Zig-zagging raceme of 1–4 pinkish, rarely cream flowers, to 15 mm across; September to November.
Column cream to reddish, post anther lobe yellow, short and narrow, not hooded, with notched end; yellow column arms erect, margins covered in blunt teeth.
**Requirements:** Soils wet in Winter, drying out in Summer
**Comments:** Flowers only open during hot humid weather and are self-pollinated. Distinguished from *T. rubra* by the smaller flowers, shorter column, narrower column arms and post anther lobe. Fairly common, growing in moist or wet soils among sedges and low shrubs of heathlands, less conspicuous in forests.
**Localities:** Blackburn, Bayswater, Belgrave, 34A, 47, 49, 67, 74
**Distribution:** All states except Qld; NZ

*Illustrated p. 241*

## Thelymitra epipactoides

Metallic Sun-orchid

**Size:** stem 20–50 cm high **F P**
**Habitat:** Marshes in tea-tree heath
**Foliage:** Thick fleshy lanceolate leaf to 25 cm long. Stout flower stem bears 1 or 2 large sheathing leaf-like bracts.
**Flowers:** Raceme of 6–20 large blue-green or bronze flowers to 35 mm across, with a metallic shine; September to November.
Column widely winged, post-anther lobe incurved, end fringed with teeth; erect column arms end in dense white hair tufts.
**Requirements:** Moist to wet soils
**Comments:** Once widespread in Melbourne, mainly on near-coastal tracts, it is now presumed extinct due to destruction of its habitat. An endangered species throughout its Australian range.
**Localities:** Black Rock#, 67#
**Distribution:** Vic, SA

*Thelymitra carnea*

## Thelymitra flexuosa

Twisted Sun-orchid

**Size:** stem 10–25 cm high **F P**
**Habitat:** Moist depressions and swamp margins in valley sclerophyll forest, wattle tea-tree scrub, grassy low open forest
**Foliage:** Linear terete leaf to 100 mm long. The reddish flowering stem is flexuose, bearing 2 stem bracts.
**Flowers:** Zig-zagging raceme of 1–3 pale yellow or cream flowers to 14 mm across; September to November.
Erect, white to cream widely winged column with 4 rows of short hairs down the back. Column arms reduced to small lumps, post-anther lobe not hooded, slightly notched and shorter than anther. Anther large, swollen and protruding, densely hairy.
**Requirements:** Winter wet soils
**Comments:** Flowers only open on hot days and are self-pollinated. Preferring wet sites, plants sometimes occur in shallow water. Similar to *T. antennifera*, the flowers are much smaller and lack the 'rabbit's ear' appendages or the column arms.
**Localities:** Blackburn, Belgrave, 33, 34A, 42, 67#, 74
**Distribution:** Vic, Tas, SA, WA

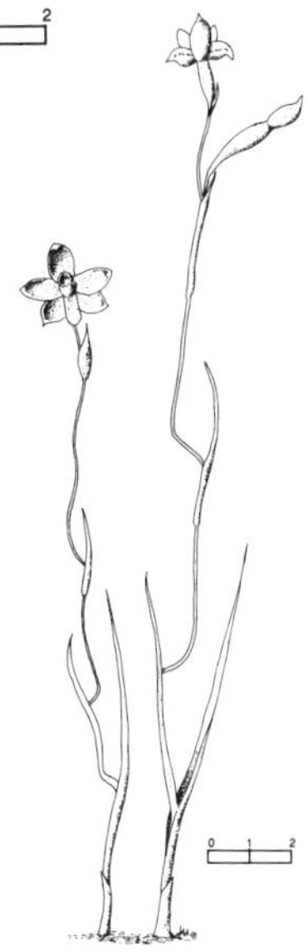

*Thelymitra flexuosa*

## Thelymitra ixioides var. ixioides

Dotted Sun-orchid

**Size:** stem 20–60 cm high **F P**
**Habitat:** Wet grassy flats of valley and dry sclerophyll forests, grassy low open forest
**Foliage:** Dark green linear to lanceolate leaf to 20 cm long, often reddish at base
**Flowers:** Raceme of 3–9 mauve to blue flowers to 20 mm across, with darker spots on dorsal sepal and petals, but occasionally absent; October to November.
Column erect, white; erect column arms projected forward, ending in tufts of white, pink or mauve hairs; short post-anther lobe not hooded, end covered in rows of finger-like orange calli; side lobules fringed, often orange.
**Requirements:** Wet soils
**Comments:** Beautiful orchid which is quite variable. It may be robust and many-flowered, or slender with few flowers. Widespread throughout Melbourne favouring moist, open sites.
**Localities:** Blackburn-Boronia, Scoresby, Belgrave, Springvale, Moorabbin, 34A, 41, 42, 44, 51, 55, 56, 58, 67#, 74, 76
**Distribution:** Vic,Qld, NSW, Tas, SA; NZ, N Cal

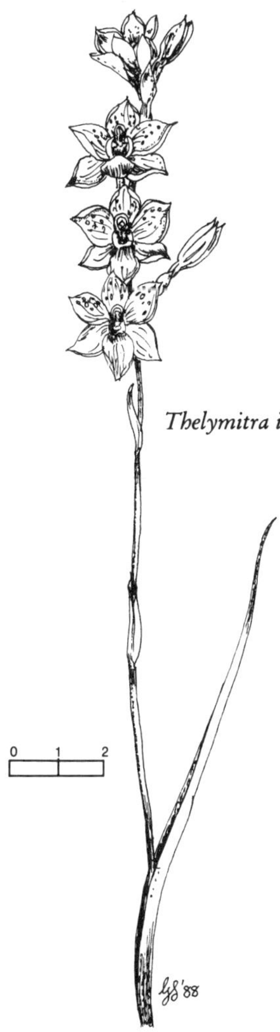

*Thelymitra ixioides* var. *ixioides*

## Thelymitra ixioides var. subdifformis

**Comments:** Differs from *T. ixioides* var. *ixioides* in flower colour. Sepals are green, petals are pale blue to lavender.
**Localities:** Blackburn
**Distribution:** Endemic to Vic

## Thelymitra luteocilium

Fringed or Pink Sun-orchid

**Size:** stem to 35 cm high **F P**
**Habitat:** Moist depressions in valley sclerophyll forest, wattle tea-tree scrub
**Foliage:** Fleshy, shallowly channelled dark green lanceolate leaf to 20 cm long. Slender, reddish-brown flower stem with 2–3 stem bracts.
**Flowers:** Raceme of 2–5 pale pink flowers to 20 mm across; September to December.
Column erect with narrow wings; post-anther lobe short, yellow, slightly hooded and densely fringed, behind the dense terminal tuft of yellow hairs of the short column arms.
**Requirements:** Moist, well drained soils
**Comments:** An inconspicuous orchid growing in the shelter of trees and shrubs, opening only on very hot humid days. Self-pollinating.
**Localities:** Ringwood, Bayswater, Langwarrin#, 34A
**Distribution:** Vic, NSW, Tas, SA

SCALE: 0 1cm 2cm The scale representation on each plant equals 2cm.

*Thelymitra nuda*

*Thelymitra pauciflora*

## Thelymitra media var. media

Tall Sun-orchid

**Size:** stem to 0.9 m high **F P**
**Habitat:** Valley, damp and wet sclerophyll forests
**Foliage:** Thick lanceolate leaf to 30 cm long. Robust or slender stem, with bracts similar to *T. aristata*.
**Flowers:** Dense spike of 5–25 blue flowers to 30 mm across; October to January. Cream or bluish column with wide wings; post-anther lobe is not hooded and has a prominent band, fringed with short erect lemon calli; column arms extend forward ending in a tuft of white hairs.
**Requirements:** Moist soils
**Comments:** A beautiful orchid, of high rainfall areas. Flowers open freely on hot days, and are generally long-lasting. Similar to *T. aristata*, distinguished by the horizontal hair tufts on the column and a distinctive collar-like band of colour below the column lobes.
**Localities:** Boronia, Montrose, The Basin, Belgrave, Lysterfield, 55, 57, 58
**Distribution:** Vic, Qld, NSW, Tas *Illustrated p. 243*

## Thelymitra megcalyptra

Scented Sun-orchid

**Size:** stem 20–40 cm high **F P**
**Habitat:** Plains grassland, coastal banksia woodland
**Foliage:** Very thick dark green lanceolate leaf 10–30 cm long, reddish at the base
**Flowers:** Raceme of 2–25 large fragrant mauve to deep blue flowers to 45 mm across; Octber to November.
White or bluish column; dark brown to blackish post-anther lobe with greatly expanded hood, deeply cleft; column arms projected forward ending in long dense white hairs.
**Requirements:** Well drained soil
**Comments:** Grows in large colonies. Flowers open readily in the sun producing an impressive display. Similar to *T. nuda* but more robust with larger flowers and a greatly swollen post-anther lobe.
**Localities:** Keilor, Sydenham, Toolern Vale, Diggers Rest, 67
**Distribution:** Vic, NSW, SA
**Synonym:** *T. aristata* var. *megcalyptra*

## Thelymitra nuda

Scented or Plain Sun-orchid

**Size:** stem to 0.6 m high **F P**
**Habitat:** Dry sclerophyll forest, tea-tree heath
**Foliage:** Variable, thick, linear to lanceolate, dark green with a red base, or bluish-green, to 30 cm long. Flowering stem stout.
**Flowers:** Raceme of 2–20 large fragrant blue flowers to 40 mm across, mauve on outside, occasionally white, pink or mauve inside; August to November. Column erect, white or bluish; dark brown post-anther lobe with large, rounded hood, often yellow margins; column arms bent inward, with short dense white, rarely pink or lilac hairs in terminal tufts, held horizontally.
**Requirements:** Well drained soils
**Comments:** A beautiful orchid with flowers opening readily. Closely related to *T. pauciflora* which has smaller flowers that seldom open, and *T. aristata* which has a yellow, toothed post-anther lobe.
**Localities:** Montrose, 34A, 41, 74
**Distribution:** All states except NT

## Thelymitra pauciflora

Slender Sun-orchid

**Size:** stem 20–50 cm high **F P**
**Habitat:** Plains grassland, dry and valley sclerophyll forests, red gum woodland, tea-tree heath, grassy low open forest
**Flowers:** Loose raceme of 3–15 blue flowers 10–20 mm across; August to January. Hood is smaller and narrower.
Column arms are erect with less dense white tufts of hair.
**Requirements:** Moist, well drained soils
**Comments:** Similar to *T. nuda* but smaller in all aspects. Flowers only open in very hot weather. Self-pollinating, it often matures without opening at all. This is the most common and widespread of Australia's *Thelymitra* species and is found in most temperate habitats. Populations often occur in large colonies, and show marked variations in flower size and colour. The plant is easily raised from seed and is reliable in cultivation.
**Localities:** Sunbury, Bundoora, Kew, Eltham, Montrose, Scoresby, Belgrave, Springvale, 25, 26, 33, 34A, 36, 42, 44, 47, 50A, 51, 54–56, 58, 63, 67, 69A, 74–76
**Distribution:** All states; NZ

## Thelymitra rubra

Salmon or Pink Sun-orchid

| | |
|---|---|
| **Size:** | stem 15–40 cm high **P** |
| **Habitat:** | Dry and valley sclerophyll forests, wet tea-tree heath, grassy low open forest |
| **Foliage:** | Deeply channelled linear leaf to 20 cm long, reddish at base. 2–3 closely-sheathing bracts on the reddish, slightly flexuose stem |
| **Flowers:** | Raceme of 1–4 pink flowers to 20 mm across; September to December. Column cream to pinkish; post-anther lobe above anther, rounded but not hooded, with a densely toothed yellow end; broad erect column arms yellow, covered in wrinkled, distinctive teeth. |
| **Requirements:** | Moist soils |
| **Comments:** | Opens readily in warm sun but is usually self-pollinated. Similar to *T. carnea* but flowers are larger and column arms are broader and warty. |
| **Localities:** | Eltham, Nunawading-Bayswater, Montrose, Belgrave, Cranbourne, Cheltenham, 35, 42, 44, 51, 55, 56, 58, 67, 74–76 |
| **Distribution:** | All states |
| **Synonym:** | *T. carnea* var. *rubra* |

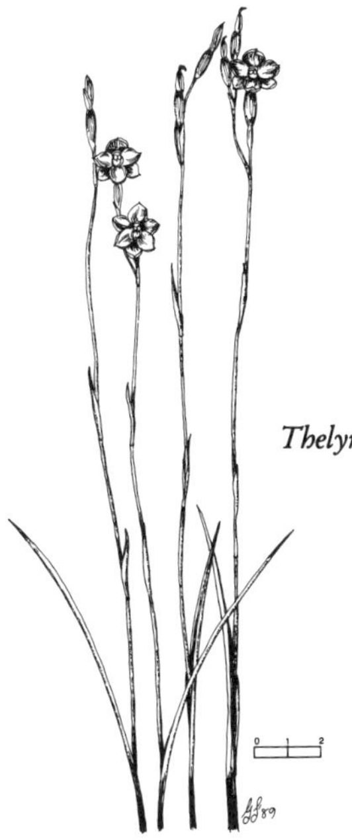

*Thelymitra rubra*

*Thelymitra media* var. *media*

SCALE: 0 1cm 2cm The scale representation on each plant equals 2cm.

# *Grasses, Rushes and Sedges, Aquatic and Semi-aquatic Herbs*

## Agropyron scabrum = Elymus scabrus

## Agrostis

**Poaceae**

Annual or perennial grasses, generally tufted and with rhizomes, sometimes forming low spreading patches. A world-wide genus with 5 species indigenous to the Melbourne area, the majority of which are annuals. The inflorescence is an open or contracted, usually much branched, panicle of small single-flowered spikelets. The lemma (or lower bract) is toothed along the top and has an awn arising dorsally from near the middle or base. Propagation is generally by seed but division of the rhizome is also successful, however few of the Australian species have horticultural merit.

*Agrostis aemula* var. *aemula*

*Agrostis avenacea*

*Agrostis billardieri* var. *billardieri*

### Agrostis adamsonii

Adamsons Bent

**Comments:** Similar to *A. billardieri* but spikelets are smaller, to 3.5 mm long, awns are straight and only shortly exserted. Previously known only from one collection, it has recently been rediscovered at another Victorian site. Extinct in the Melbourne area.
**Localities:** Studley Park#
**Distribution:** Endemic to Vic

### Agrostis aemula var. aemula

Blown Grass

**F P**

**Size:** to 10 cm high; stems 0.25–0.6 m high
**Habitat:** Plains grassland, grassy wetland, valley sclerophyll forest
**Form:** Compact annual tuft
**Foliage:** Rough, linear, flat leaves to 7 mm wide
**Flowers:** Panicle is open and broadly spreading. Spikelets to 5 mm long, purplish at maturity. Lemma, to 3 mm long, is hairy on the back. September to January
**Requirements:** Heavy clay and basalt soils
**Comments:** A widespread grass.
**Localities:** Campbellfield, St. Albans, 4, 12, 26, 28, 34A, 41, 42, 44, 47, 48, 50A, 54, 55, 65# 70, 74
**Distribution:** All states except NT

### Agrostis aemula var. setifolia

Blown Grass

**Habitat:** Grassy wetland
**Requirements:** Seasonally wet soils
**Comments:** Differs from *A. aemula* var. *aemula* in that the lemma is slightly longer, to 3.5 mm, and leaves are narrower, to 1 mm wide and inrolled.
**Localities:** Craigieburn
**Distribution:** Vic, Qld, NSW, Tas

### Agrostis avenacea

Common Blown Grass

**F P**

**Size:** stems to 0.6 m high
**Habitat:** Plains grassland, grassy wetlands, box woodland, valley sclerophyll forest, grassy low open forest, primary dune scrub
**Form:** Annual, biennial or perennial tuft
**Foliage:** Linear leaves to 20 cm long, tapering to a fine point
**Flowers:** Grey-green to pale green spikelets to 4 mm long in spreading panicles to over 30 cm long, branches often drooping when young. Lemma hairy on back. October to February
**Requirements:** Moist clays or clay loams
**Comments:** Very similar to *A. aemula* and widespread throughout the region. Readily naturalises in areas of light disturbance. A useful volunteer in rough lawns.
**Localities:** Footscray, Research, 2, 4–9, 11, 12, 15–18, 24–29, 32, 34A–36, 42, 44, 47, 48, 50A, 51, 54–56, 58, 60, 61, 64, 67#, 69A–72, 74–76
**Distribution:** All states; NZ

### Agrostis billardieri var. billardieri

Coast Blown Grass

**F P**

**Size:** to 15 cm high; stems 0.15–0.6 m high
**Habitat:** Primary dune scrub, riparian scrub
**Form:** Annual grassy tuft
**Foliage:** Flat linear leaves to 20 cm x 3–7 mm
**Flowers:** Very loose, spreading, straw-coloured panicles to 30 cm long, spikelets to 7 mm long. Lemma glabrous, 4-toothed, outer pair longer. October to February
**Requirements:** Moist soils
**Comments:** Usually confined to coastal tracts.
**Localities:** 5, 67#, 74
**Distribution:** Vic, NSW, Tas, SA; NZ

### Agrostis billardieri var. robusta

**Size:** to 0.6 m high
**Comments:** Differs from *A. billardieri* var. *billardieri* in that the leaves are narrower and inrolled and the plant is usually taller. It does not occur in coastal areas.
**Localities:** Collected in Warrandyte, 1853
**Distribution:** Vic, Tas, SA

### Agrostis aff. hiemalis

Winter Bent

**Size:** to 30 cm high — P N
**Habitat:** Damp sclerophyll forest
**Requirements:** Moist, well drained soils
**Comments:** Tufted perennial grass, sometimes with a short rhizome. The main distinguishing features are the small spikelets to 2 mm long and the awnless lemma.
**Localities:** 57
**Distribution:** Vic, Qld, NSW, Tas; Am, Asia

*Agrostis* aff. *hiemalis*

# Alisma — Alismataceae

### Alisma plantago-aquatica

Water Plantain

**Size:** 0.5–1 m x 0.5 m — F P
**Habitat:** Aquatic in riparian scrub, grassy wetlands
**Form:** Erect perennial semi-aquatic herb
**Foliage:** Erect long-stalked ovate to ovate-lanceolate leaves, blade to 30 cm long, with prominent network of veins
**Flowers:** Much-branched, whorled panicle, of many small, long-stalked, pale pink 3-petalled flowers, each lasting a single day, borne on rigid leafless stems to 1 m high or more; Summer
**Requirements:** Moist to wet soils, tolerating poorly drained sites
**Comments:** A most attractive, easily grown plant for bog gardens and shallow ponds. The closely related *A. lanceolatum*, from Europe and Asia, can often be distinguished by its lanceolate leaves. *Damasonium minus* is similar but may be distinguished by the lesser number of carpels (up to 10).
**Propagation:** Seed or division of bulbous rootstock
**Localities:** Laverton Nth, 4, 14, 16, 19, 26, 27, 32, 34A, 36, 42, 44, 51, 53, 54, 56, 62, 63, 65, 74, 76
**Distribution:** Vic, NSW; NZ

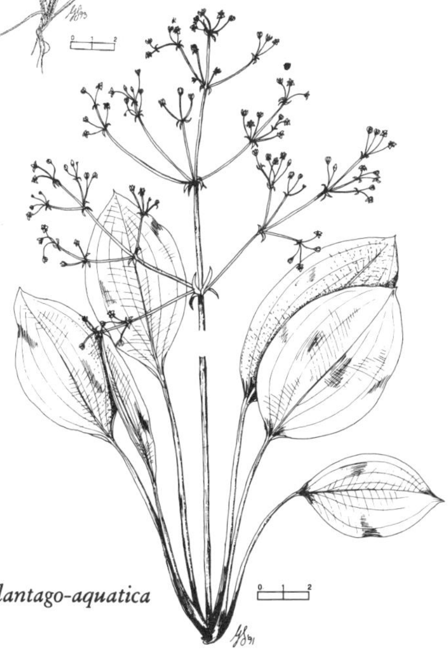

*Alisma plantago-aquatica*

# Amphibromus — Poaceae

Slender tufted perennial grasses growing up to 1.5 m high, usually found growing in wet places beside pools and streams. Leaf blades are generally linear, either flat or inrolled. The inflorescence is a narrow, loose, elongated panicle with slender, erect or flexuose branches. Spikelets have 3 or more florets. The lemma of each floret ends in bristle-like teeth and has a dorsal awn arising from above the middle. A prominent hairy tuft surrounds the base of the lemma. Propagation is by seed or division.

### Amphibromus archeri

Pointed Swamp Wallaby Grass

**Size:** to 30 cm high; stems to 1.2 m high — F P
**Habitat:** Riparian and swamp scrub
**Form:** Tufted perennial clump
**Foliage:** Greenish deeply ribbed leaves 30 cm x 2.5–4 mm
**Flowers:** Loose erect panicles to 35 cm long, spikelets to 18 mm long contain 2–4 florets. Glumes are very unequal. Lemma to 10 mm long, is 4-toothed (2 of which are much longer than the others), each ending in a bristle; awn to 15 mm long, is dark, bent, and arises well above the middle of the lemma, hair tuft large. October to January.
**Requirements:** Most soils with ample moisture
**Comments:** Useful grass for the bog garden but it is not tolerant of drying out. Very rare in Melbourne.
**Localities:** Oakleigh#, Kilsyth, 28, 69A, 71
**Distribution:** Vic, Tas, SA

*Amphibromus archeri*

### Amphibromus fluitans

Graceful Swamp Wallaby Grass

**Size:** 0.15–0.7 m high — P
**Habitat:** Beside swamps in grassy low open forest, riparian scrub
**Foliage:** Leaf sheaths are rough
**Flowers:** Spikelets 7–10-flowered, glumes more or less equal, lemma 2-toothed, awn pale, straight, rising from the middle of the lemma; hair tuft tiny or absent.
**Requirements:** Moist soils, tolerating inundation
**Comments:** It was first collected along the Yarra R. flats. Probably extinct in the Melbourne area.
**Localities:** Camberwell#; Kew#, Balwyn#, 33#, 60#
**Distribution:** Vic, NSW; NZ
**Synonym:** *A. gracilis*

*Illustrated p. 255*

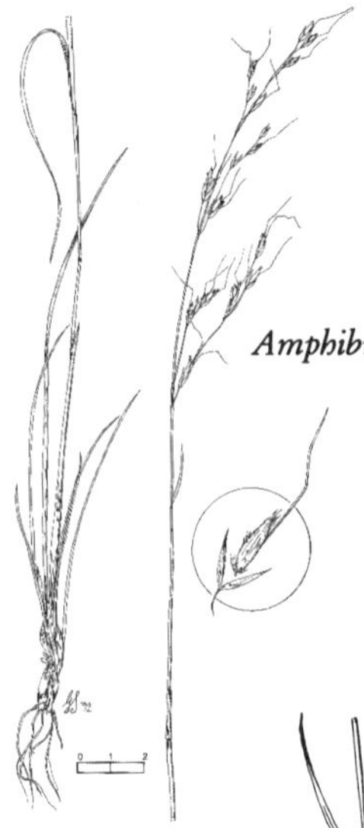

*Amphibromus macrorhinus*

## Amphibromus macrorhinus
Long-nosed Swamp Wallaby Grass

**Size:** to 25 cm high; stems to 1 m high **F**
**Habitat:** Grassy wetlands
**Foliage:** Flat or inrolled leaves to 25 cm long
**Flowers:** Green or purplish 4–5-flowered spikelets in erect, often spreading panicle to 40 cm long. Lemma has a drawn out, papery apex, very rough on lower back with awn rising from or just below middle. October to November
**Requirements:** Moist to wet soils. Plants do not tolerate drying out.
**Comments:** A rare plant in the Melbourne area.
**Localities:** Laverton
**Distribution:** Vic, NSW, Tas, SW, WA
**Synonym:** Previously included with *A. neesii*

*Amphibromus neesii*

## Amphibromus neesii
Swamp Wallaby Grass

**Size:** to 25 cm high; stems 0.4–1 m high **F P**
**Habitat:** Grassy wetlands, swamps of plains grasslands, wattle tea-tree scrub
**Form:** Densely tufted perennial grass
**Foliage:** Flat narrow leaves to 25 cm long, with rough, glabrous surface
**Flowers:** Very loose panicle to 40 cm long on slender erect stems to 1 m high, with reddish, 4–7-flowered spikelets. Glumes (or bracts) at base of spikelet narrow. Lemma irregularly toothed, pale brownish awn slightly bent, rising just above the middle, small hair tuft. Spring to Autumn
**Requirements:** Moist to wet soils
**Comments:** A common grass of low-land wet areas with an attractive weeping habit. It is readily grown but does not tolerate drying out. Until recently this species has often been confused with *A. nervosus*.
**Localities:** 2, 7, 12, 18, 26, 34A, 35, 71, 75
**Distribution:** Vic, NSW, Tas, SA, WA

## Amphibromus nervosus

**Size:** to 25 cm high; stems to 1.5 m high **F P**
**Habitat:** Grassy wetland
**Foliage:** Pale green linear leaves 5–25 cm x 1.5–6 mm, flat or loosely folded
**Comments:** Differs from *A. neesii* in that glumes are broad and lemma tapers away evenly.
**Localities:** Greenvale, Laverton, 4, 28, 34A, 47, 70
**Distribution:** Vic, NSW, SA, WA *Illustrated p. 255*

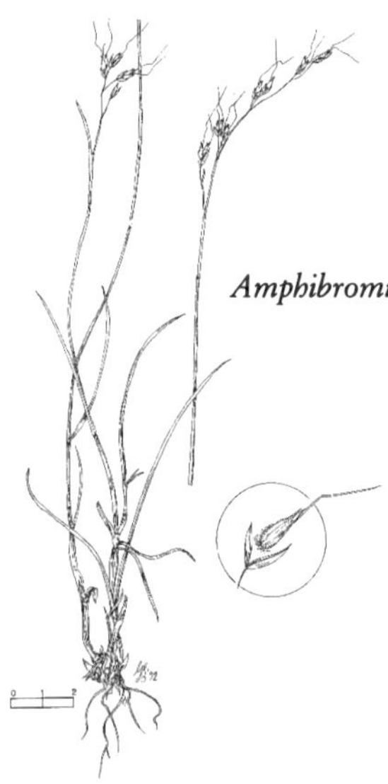

*Amphibromus pithogastrus*

## Amphibromus pithogastrus
Swollen or Plump Swamp Wallaby Grass

**Size:** to 20 cm high; stems to 1 m high **F P**
**Habitat:** Grassy wetlands
**Form:** Tufted perennial with swellings on lower internodes of stems
**Foliage:** Narrow, flat to inrolled leaves to 20 cm long
**Flowers:** Green 2–6-flowered spikelets in narrow, erect panicle to 25 cm long. Lemma is smooth or minutely warty and swollen towards the base with an awn rising shortly below the apex.
**Requirements:** Moist to wet soils
**Comments:** An endangered plant in Victoria, it has recently been found on a few scattered sites west and north of Melbourne.
**Localities:** Basalt plains
**Distribution:** Vic, NSW

# Aphelia — Centrolepidaceae

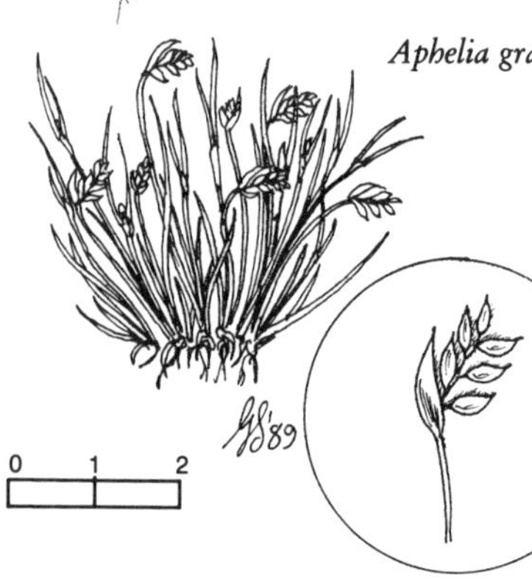

*Aphelia gracilis*

## Aphelia gracilis
Slender Aphelia

**Size:** 20–40 mm high **F P**
**Habitat:** Moist tea-tree heath, swamp scrub
**Form:** Tiny annual tuft
**Foliage:** Filiform leaves to 25 mm long
**Flowers:** Spike to 4 mm long, turned to one side, with 6–8 blunt bracts, the lowest one longest; September to November
**Requirements:** Moist open soil on temporarily wet ground
**Comments:** The flowerhead of Aphelia spp. is a spike with 6–16 paired bracts and unisexual flowers while the closely related Centrolepis spp. have flowerheads of bisexual flowers enclosed by a pair of bracts. Locally very rare.
**Propagation:** Seed
**Localities:** Dandenong, St. Kilda#, 49#, 61, 63, 67#, 74, 75
**Distribution:** Vic, NSW, Tas, SA, WA

### Aphelia pumilio

Dwarf Aphelia

**Size:** 10–30 mm high **F P**
**Habitat:** Moist tea-tree heath
**Form:** Tiny annual
**Foliage:** Linear leaves to 20 mm long
**Flowers:** Spike erect, to 6 mm long; 8–16 tapering bracts, the lowest 2 longer and leaf-like; October to November
**Requirements:** Seasonally wet ground
**Propagation:** Seed
**Localities:** Heathmont#, St. Kilda#, 67#, 74, 75
**Distribution:** Vic, NSW, Tas, SA

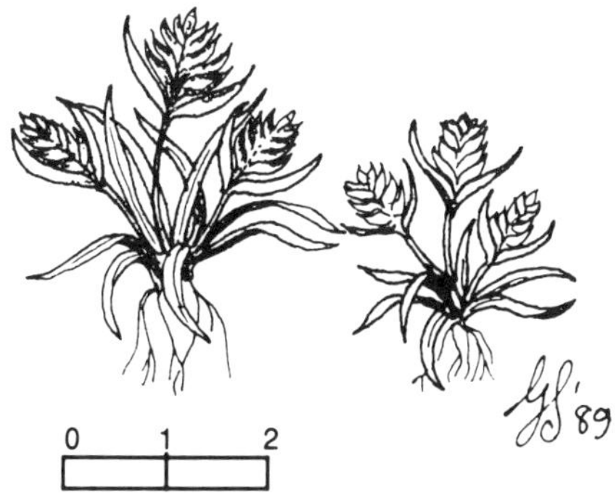


*Aphelia pumilio*

# Aristida

## Poaceae

### Aristida ramosa

Purple or Cane Wiregrass

**Size:** 0.3–1 m high **F**
**Habitat:** Plains grassland, box woodland, often in rocky areas
**Form:** Slender perennial grass tussock with stiff, wiry stems, branching at the upper nodes
**Foliage:** Leaves tightly inrolled
**Flowers:** Very narrow, loose panicle to 15 cm long with purplish 1-flowered spikelets. Distinguishing feature is the 3-branched awn on the lemma. August to October
**Requirements:** Well drained soils
**Comments:** An attractive summer-growing grass for a sunny position, tolerating poor soils. For best results and longer flowering, Summer moisture is desirable.
**Propagation:** Seed
**Localities:** 16, 26
**Distribution:** Vic, Qld, NSW, WA

*Aristida ramosa*

# Austrofestuca

## Poaceae

### Austrofestuca hookeriana

Hooker's Fescue

**Size:** 10–30 cm high; stems to 1.8 m high **F P**
**Habitat:** Swamp and wattle tea-tree scrub, valley sclerophyll forest
**Form:** Coarse, dense tussock-forming perennial grass
**Foliage:** Rough, bright green, flat linear leaves to 15 cm long, finely ridged, tips shaped like the bow of a boat
**Flowers:** Greenish 4–5-flowered spikelets to 20 mm long, ageing to straw coloured, in open conical panicles. Tiny awn to 3 mm long, lemma shortly hairy at base. November to January. Culms much taller than the leaves.
**Requirements:** Moist soils, tolerating Winter inundation and drying out in Summer
**Comments:** Being tolerant of heavy shade, this grass is ideal for use as a foliage contrast in fern gardens. Often confused with a serious introduced weed, *Festuca arundinacea*, which is distinguished by the panicles which tend to fall to one side, the glabrous lemmas and the different leaf tips.
**Propagation:** Seed, division
**Localities:** Bayswater, Montrose, 53, 54
**Distribution:** Vic, NSW, Tas
**Synonym:** *Festuca hookeriana* *Illustrated p. 255*

### Austrofestuca littoralis

Coastal Fescue

**Size:** 0.3–0.8 m high **F**
**Habitat:** Primary dune scrub
**Form:** Tussock-forming perennial grass
**Foliage:** Rigid, sharp, narrow straw-coloured leaves to 45 cm long, often longer than the culms
**Flowers:** Dense yellowish spike-like panicle held rigidly erect, with overlapping 4–6-flowered spikelets. Lemma awnless and glabrous at base.
**Requirements:** Well drained sandy soils
**Comments:** Important plant for beach erosion control.
**Propagation:** Seed or division
**Localities:** Seaford, 2, 67#, 74
**Distribution:** Vic, NSW, Tas, SA, WA
**Synonym:** *Festuca littoralis*

*Austrofestuca littoralis*

# Baumea

## Cyperaceae

Baumeas are slender, rush-like, rhizomatous perennial sedges. The flowerhead is a terminal panicle. Leaves and flowering stems (or culms) are similar in shape and usually in size. They are ideal as accent plants in a bog garden, occurring naturally along stream banks and in low-lying swampy areas. Propagation is by division of the rhizome or seed.
**Synonym:** Previously included with *Cladium*

SCALE: 0 1cm 2cm The scale representation on each plant equals 2cm.

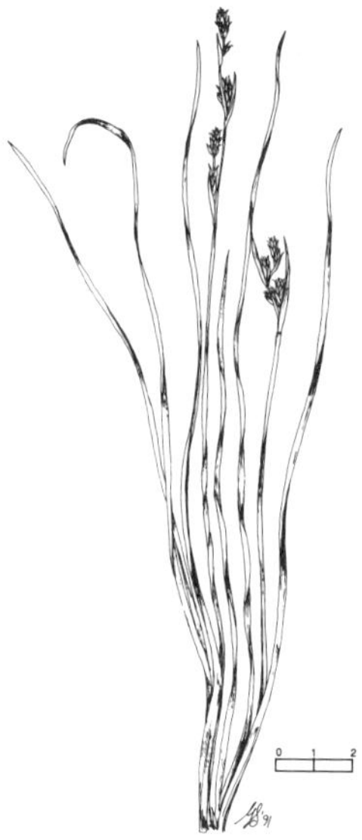
Baumea acuta

## Baumea acuta

Pale Twig-rush **F**

**Size:** 15–30 cm high
**Habitat:** Tea-tree heath
**Foliage:** Flattened leaves and flowering stems to 2 mm wide with blunt margins
**Flowers:** Reddish-brown 1-flowered spikelets to 5 mm long, densely congested in short panicles to 50 mm long; October to April
**Requirements:** Moist sandy soils
**Localities:** 50, 67#, 70, 71, 74, 75
**Distribution:** Vic, NSW, Tas, SA, WA

Baumea articulata

## Baumea articulata

Jointed Twig-rush **F P**

**Size:** 1–2 m high
**Habitat:** Riparian scrub
**Form:** Tall erect tussock
**Foliage:** Stout hollow cylindrical stems and leaves with pungent tips and prominent transverse partitions
**Flowers:** Numerous brown 3–5-flowered spikelets in large drooping panicles 20–40 cm x 3–6 cm; December to April. Nut light brown to whitish.
**Requirements:** Moist soils, tolerating inundation
**Comments:** Useful as a shelter plant for waterbirds. Very effective beside ponds.
**Localities:** 17#, 18, 35, 72, 75
**Distribution:** All states except NT; PNG, NZ, N Cal, Poly

Baumea gunnii

## Baumea gunnii

Slender Twig-rush **F P**

**Size:** 0.3–1 m high
**Habitat:** Swamp and wattle tea-tree scrub
**Foliage:** Leaves reduced to basal sheaths, sometimes with 1 or 2, very short and pointed blades; stems slender with longitudinal furrow and several fine striations
**Flowers:** Loose elongated panicles to 20 cm long, branches erect, lower branches distant. Few brownish or greyish 1-flowered spikelets, distantly scattered. Flowers most of the year.
**Requirements:** Adaptable to any soils providing ample moisture is available.
**Comments:** Interesting bog garden plant.
**Localities:** 61, 74
**Distribution:** All states except NT; PNG, NZ

## Baumea juncea

Bare Twig-rush **F P**

**Size:** 0.3–1 m high
**Habitat:** Coastal banksia woodland, tea-tree heath
**Form:** Rush-like clump with creeping rhizomes
**Foliage:** Terete greyish stems with 2 or 3 nodes, leaves reduced to sharply pointed sheaths
**Comments:** Differs from *B. gunnii* in that the spike-like panicle is smaller, to 50 mm long, and the few reddish-brown spikelets are crowded together. September to April.
**Localities:** 28, 67, 68, 69A–71, 74
**Distribution:** All states except NT; NZ

*Illustrated p. 255*

Baumea rubiginosa

## Baumea rubiginosa

Soft Twig-rush **F**

**Size:** 0.3–1 m high
**Habitat:** Swamp scrub
**Form:** Rhizomatous clump with round or slightly angled to flattened stems, usually with one node
**Foliage:** Basal leaves wider than stem; stem leaves with long sheaths and short channelled blades, both sharply pointed
**Flowers:** Narrow dense panicle to 12 cm long with reddish or brown, 3–5-flowered spikelets in erect clusters. Nut reddish-brown. Flowers most of the year.
**Requirements:** Ample moisture
**Comments:** Locally common along watercourses.
**Localities:** 51, 54, 63, 64, 69A, 70, 72, 74
**Distribution:** All states; NZ, Asia
**Synonym:** *B. glomerata*

Baumea tetragona

## Baumea tetragona

Square Twig-rush **F**

**Size:** 0.3–1 m high
**Habitat:** Swamp and wattle tea-tree scrub, tea-tree heath
**Form:** Erect rush-like clump, stems and leaves strongly 4-angled
**Foliage:** Leaves basal with pointed tips, wider than stems
**Flowers:** Dense, erect reddish-brown panicles to 100 mm long. Spikelets 1-flowered, close together. Most of the year
**Requirements:** Ample moisture, but tolerates drying out in Summer
**Comments:** Attractive accent plant for bog gardens.
**Localities:** Knox, Clayton, 54, 67#, 69, 71, 73, 74, 75
**Distribution:** Vic, Qld, NSW, Tas, SA

# Bolboschoenus — Cyperaceae

## Bolboschoenus caldwellii — Sea Club-rush

**Size:** 0.3–0.9 m high **F P**
**Habitat:** Saltmarshes, grassy wetland, wet depressions in red gum woodland
**Form:** Erect rhizomatous semi-aquatic perennial herb with leafy triangular stems
**Foliage:** Flat grass-like leaves of varying length, to 4 mm wide
**Flowers:** Flowerheads a terminal compound cluster of 1–3 many-flowered, golden or red-brown spikelets with 1–3 floral bracts. Glumes downy, keel ending in a point. Nut straw-coloured to golden-brown. October to April
**Requirements:** Fresh or brackish water on heavy clay or sandy soils
**Comments:** *Bolboschoenus* species are distinguished by the leaflike bracts which are much longer than and surround the flowerheads. The tubers of both species were roasted by the Aborigines, then pounded to make thin cakes.
**Propagation:** Seed, division
**Localities:** Rockbank, Sunshine, 2, 4, 25, 26, 72, 74
**Distribution:** Vic, NSW, Tas, SA, WA; NZ
**Synonym:** *Scirpus caldwellii, S. maritimus*

*Bolboschoenus caldwellii*

## Bolboschoenus medianus — Marsh Club-rush

**Size:** 0.7–2 m high **F P**
**Habitat:** Swamps in riparian scrub
**Comments:** Similar to but usually taller than *B. caldwellii.* The foliage is darker green, broader and more prominent to 50 cm x 6–11 mm. Clusters of 4–6 dull red-brown spikelets, surrounded by 2–6 floral bracts. Glume ends in a rough, recurved awn. Nut is red-brown to black. Also occurs in shallow brackish water.
**Localities:** 4, 28, 29, 34A, 35, ?42, 74, 75
**Distribution:** Vic, NSW, Tas, SA; NZ
**Synonym:** *Scirpus medianus, S. fluviatilis* *Illustrated p. 255*

# Bothriochloa — Poaceae

## Bothriochloa macra — Redleg Grass

**Size:** 0.3–0.6 m high **F P**
**Habitat:** Box and red gum woodland, plains grassland
**Form:** Tufted perennial grass with tough reddish-purple flowering stems
**Foliage:** Flat, linear, pointed leaves to 20 cm long
**Flowers:** Erect unbranched panicle of 2–5 racemes, to 60 mm long. Spikelets in very unequal pairs. Long white silky hairs on stalks of spikelets; whitish glume is 2-keeled, fertile lemmas are dimpled and have a brown awn to 20 mm long; anthers purple. December to April, August to September
**Requirements:** Well drained clay loam soils
**Comments:** An attractively flowered summer-growing grass, tolerant of extended dry conditions but responding well to extra moisture.
**Propagation:** Seed, division
**Localities:** Sunbury, Keilor, Laverton Nth, 2, 9, 16–18, 21, 25, 26, 32
**Distribution:** Qld, NSW, Tas, SA

*Bothriochloa macra*

# Carex — Cyperaceae

*Carex* spp. are common in moist to wet sites, usually occurring along watercourses and in swamps. They are perennial, grass-like sedges forming tufts to 1 m high (often smaller), often with rhizomes. Flowers are unisexual, in plume-like terminal spikes with glumes spirally overlapping in several rows. Bracts at the base of the spike are either large and leaf-like, or small, inconspicuous and glume-like. The nut is enclosed in a loose covering. Some species are suited to bog gardens or moist sites in a garden as accent plants. Propagated by division or seed, the more attractive *Carex* spp. are becoming readily available through the specialist nurseries.

## Carex appressa — Tall Sedge

**Size:** 0.5–1.2 m x 0.5–1 m **F**
**Habitat:** Widespread in riparian scrub, dry and valley sclerophyll forests, red gum woodland, grassy low open forest, tea-tree heath
**Form:** Robust dense tuft, stems rough, sharply triangular and solid, the same length as leaves
**Foliage:** Slender, bright green, arching leaves to 6 mm wide, with rough edges; bracts inconspicuous
**Flowers:** Long narrow dense to loose panicle 5–25 cm long, with 20 or more spikes of brownish or dull yellow spikelets. Spikes bisexual (male florets above female florets), the few female florets inconspicuous. August to January
**Requirements:** Ample moisture, tolerating periods of inundation
**Comments:** Although slow growing, this tough sedge is suitable for stabilising eroded stream banks. It may be used as either an aquatic or bog garden plant. The Aborigines used the tough leaves in basket-making.
**Localities:** 7, 26, 27, 29, 31, 34A–36, 39, 42, 44, 46, 47, 49, 50A, 51, 54–57, 60, 62–66, 70, 71, 74–76
**Distribution:** All states except NT; NZ *Illustrated p. 255*

SCALE: 0 1cm 2cm The scale representation on each plant equals 2cm.

*Carex breviculmis*

## Carex bichenoviana

Sedge

**F P**

**Size:** stems 25–50 cm high
**Habitat:** Plains grassland, red gum woodland
**Form:** Tufted grass-like sedge, with long-creeping rhizome
**Foliage:** Narrow grass-like leaves and bracts shorter than the triangular stems
**Flowers:** Heads of 9–23 dense spikes with purple-brown to blackish glumes; terminal cluster of 6–20 sessile male spikes, few female spikes lower, sometimes stalked. October to February.
**Requirements:** Moist depressions on heavy clay
**Comments:** May form dense carpets in shady situations. Very rare in the Melbourne area.
**Localities:** St Albans, Deer Park
**Distribution:** Vic, Tas, SA

## Carex breviculmis

Short-stem Sedge

**F P**

**Size:** to 15 cm high
**Habitat:** Widespread in saltmarshes, grassy wetland, plains grassland, red gum woodland, damp, valley and dry sclerophyll forests, grassy low open forest
**Form:** Small, densely tufted sedge with short woody rhizomes; stems triangular with rough margins, base covered with brown fibrous remains of old leaf sheaths
**Foliage:** Flat leaves, longer than flowering stems
**Flowers:** 2–5 cylindrical spikes, terminal spike male 4–25 mm long, others female to 20 mm long. Covering of nut often downy. Bracts long. August to December
**Requirements:** Very adaptable, from exposed slopes to moist depressions
**Comments:** Flower spikes hidden inside leaves. This plant is often overlooked because of its grass-like habit and stature.
**Localities:** 2, 4, 6, 8, 15, 17#, 18, 26, 28, 34A–36, 38, 39, 41, 42, 44, 45, 47, 48, 50–52, 54–56, 63, 64, 69A, 74–76
**Distribution:** Vic, Qld, NSW, Tas, SA; NZ

*Carex fascicularis*

## Carex fascicularis

Tassel Sedge

**P**

**Size:** 0.5–1 m high
**Habitat:** Valley sclerophyll forest
**Form:** Coarse tufted plant; triangular stems with rough margins
**Foliage:** Bright green partitioned leaves almost as long as stems, to 8 mm broad; leafy bracts much longer than the flowering spikes
**Flowers:** 3–7 long-stalked, cylindrical, drooping spikes 30–60 cm long, borne near the end of the stem. Terminal spike male. Glumes roughly awned. October to April.
**Requirements:** Moist soil, tolerating inundation
**Comments:** A common sedge found along watercourses and near swamps. This graceful tussock looks most attractive beside a pool. Avoid planting on heavy basalt soils.
**Localities:** Montrose, 35, 36, 56, 62, 63
**Distribution:** All states; NZ

*Carex gaudichaudiana*

## Carex gaudichaudiana

Tufted Sedge

**F P**

**Size:** stems 0.1–0.6 m high
**Habitat:** Valley sclerophyll forest, swamp scrub
**Form:** Coarsely tufted plant sometimes covering large areas, with woody ascending rhizomes and triangular stems with rough margins
**Foliage:** Flat, erect, dark bluish-green leaves to 4 mm wide, often longer than the stems, lower bracts long
**Flowers:** 3–8 stalked, cylindrical spikes 1.5–6 cm long, at the top of stem, terminal spike male. Glumes almost black. Nut covering is distinctly 4–6-veined. September to February
**Requirements:** Grows in gravel or mud at the water's edge
**Comments:** The bluish foliage of this attractive low-growing sedge provides relief from the typical greens of other bog plants. It is easily grown and controlled. Although tolerant of drying out once established, it looks best in permanently moist soils.
**Localities:** 34A, 35, 42, 54, 56, 62, 63
**Distribution:** Vic, Qld, Nsw, Tas, SA

*Carex incomitata*

## Carex incomitata

**F**

**Size:** stems 0.5–0.7 m high
**Habitat:** Plains grassland
**Form:** Differs from *C. appressa* in that it is a diffuse rather than dense tuft with a stout, short-creeping rhizome
**Foliage:** Leaves are paler green and wider, to 8 mm.
**Flowers:** Shorter spike-like panicle is always dense and narrow-oblong, to 80 mm long and mostly unbranched. The many female florets are widely reflexed, hiding the male section.
**Comments:** Unlike most other *Carex* spp. it prefers well drained soils and tolerates saline conditions.
**Localities:** Altona, Deer Park, 26, 36
**Distribution:** Vic, NSW
**Synonym:** Recently segregated from *C. appressa* as forma *minor*

## Carex inversa — Knob Sedge

**Size:** stems 10–30 cm high
**Habitat:** Widespread in plains grassland, grassy wetlands, box and red gum woodland, valley sclerophyll forest, grassy low open forest, tea-tree heath
**Form:** Small tufted or spreading clump with shortly creeping rhizome, stems slender, flat and smooth, usually shorter than the leaves
**Foliage:** Bright green, grass-like leaves to 1–3 mm wide; 2 leaf-like bracts to 15 cm long
**Flowers:** Terminal cluster of 2–4 sessile, bisexual, ovoid spikes to 10 mm long, (sometimes one a little lower down). Florets mostly female, few male florets at the base. Glumes pale or greenish, midrib extending into an awn. September to April.
**Requirements:** Moist well drained soils
**Comments:** A variable species which sometimes becomes a weed in gardens.
**Localities:** Montrose, 6, 7, 17#, 18, 23–26, 29, 31, 32, 34A–36, 39, 41, 42, 44, 47, 49, 50A, 54–56, 62, 63, 69A–71, 74, 76
**Distribution:** All states except NT; NZ

*Carex inversa*

## Carex iynx

**Size:** to 40 cm high — **P**
**Habitat:** Valley sclerophyll forest
**Form:** Coarse tufted herb with short, ascending, woody rhizomes, forming large clumps; stems at base covered with fibrous remains of old leaf sheaths
**Foliage:** Bracts short, leaves similar in length to stems
**Flowers:** Densely-flowered spikes to 8 mm thick, in tufts of 1–5 at nodes, upper 1–4 spikes mostly male and shortly-stalked. Lowermost spikes drooping on long slender stalks. Yellow-chestnut glumes of female florets awned.
**Requirements:** Moist soils
**Localities:** Eltham, 34A, 41, 42
**Distribution:** Vic, NSW, Tas, SA

*Illustrated p. 255*

*Carex pumila*

## Carex pumila — Strand Sedge

**Size:** stems 10–30 cm high — **F**
**Habitat:** Coastal grassy wetlands, primary dune scrub
**Form:** Tufted herb with a creeping rootstock
**Foliage:** Flat leaves, keeled in the lower part, coiled at the tip; leaf bracts long, sometimes hiding flowerheads which are held on triangular stems shorter than the leaves
**Flowers:** 4–7 spikes, 1–4 terminal male spikes, close together, female spikes distant, reddish-brown, the lowest shortly-stalked. August to January. Nut covering is plump and yellowish.
**Requirements:** Moist sandy soil
**Comments:** Once locally frequent but remaining in few sites.
**Localities:** Port Melbourne#, Brighton#; Maribyrnong R, 67#, 69A
**Distribution:** All states except NT; NZ, S Am, Asia

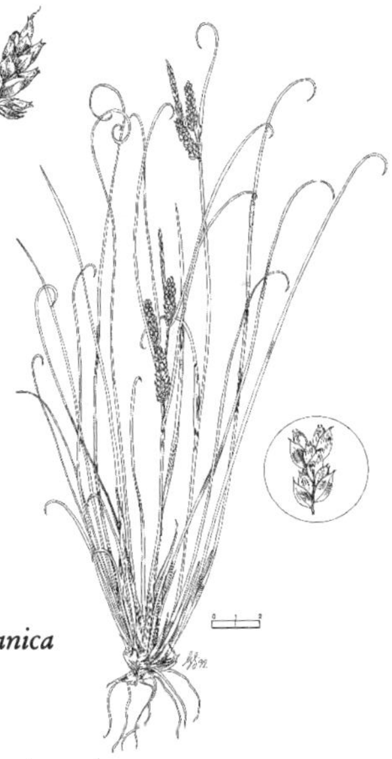

*Carex tasmanica*

## Carex tasmanica

**Size:** stems to 0.4 m high — **F P**
**Habitat:** Grassy wetlands
**Form:** Tufted perennial sedge
**Foliage:** Leaves and stem bracts strongly veined with coiled tips, basal sheaths with papery margins
**Flowers:** Raceme, to 90 mm long of 2–4 short dense spikes, each to 25 mm long. Terminal spike of male flowers is narrower than spikes of female flowers.
**Requirements:** Heavy wet soils
**Comments:** The Melbourne occurrence is a major disjunction from its normal distribution in Southwest Victoria.
**Propagation:** Division, seed
**Localities:** Craigieburn
**Distribution:** Vic, Tas

*Carex tereticaulis*

## Carex tereticaulis — Common Sedge

**Size:** 1 m x 1 m — **F P**
**Habitat:** Plains grassland, riparian scrub, valley sclerophyll forest
**Form:** Densely tufted clumps with terete, hollow, smooth stems which are finely striated
**Foliage:** Leaves rudimentary, to 4 mm wide, or reduced to sheaths
**Flowers:** Narrow, dense, spike-like panicle usually less than 10 cm long, with more than 20 spikes; August to April.
**Requirements:** Moist soils, tolerating occasional inundation
**Comments:** Similar in habit and appearance to *C. appressa*, but readily distinguished by the terete stems. The stems of *C. appressa* are distinctly triangular. The Aborigines split the stems while still green and used the fibre for string and baskets.
**Localities:** Rockbank, Altona, 4, 6, 16–18, 24–26, 33, 34A–36, 74
**Distribution:** Vic, NSW, Tas, SA, WA

SCALE: 0 1cm 2cm — The scale representation on each plant equals 2cm.

*Caustis restiacea*

# Caustis — Cyperaceae

## Caustis restiacea — Slender Twist-rush

**Size:** 0.3–0.6 m x 10–20 cm — F P
**Habitat:** Swamp scrub
**Form:** Erect perennial herb with spreading rhizomes, leaves reduced to sheathing bracts, stems slender, terete and slightly twisted
**Flowers:** Flowers borne in unisexual spikelets to 4 mm long; August to January
**Requirements:** Moist open soil
**Comments:** Known from one site within the Melbourne area, also occurring around Gembrook. Distribution is highly restricted within Victoria.
**Propagation:** Seed, division
**Localities:** Kilsyth
**Distribution:** Vic, NSW

# Centrolepis — Centrolepidaceae

Small or minute annual or perennial sedge-like or moss-like herbs with linear leaves forming a basal tuft. Flowers are sessile in a terminal head enclosed by 2 erect sheathing bracts which may continue upwards, resembling leaves. The flowering stem is erect and leafless. Found in a variety of habitats but usually in swampy ground or in moist depressions, and often quite common. Propagate by seed.

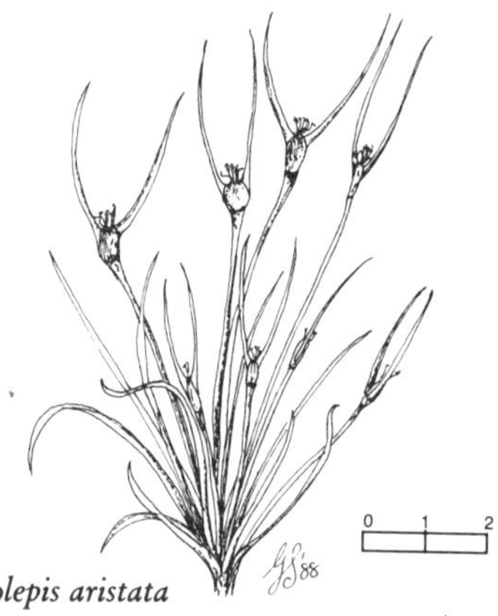

*Centrolepis aristata*

## Centrolepis aristata — Pointed Centrolepis

**Size:** 2–10 cm high — P
**Habitat:** Red gum woodland, swamp scrub, valley sclerophyll forest, wattle tea-tree scrub
**Form:** Rigidly erect annual herb, bright green or becoming reddish after flowering, stems flattened
**Foliage:** Shiny leaves paired and folded with pointed tip, 10–60 mm long
**Flowers:** Flowering head flattened, to 3 mm wide with 10–40 flowers per head; September to December. Bracts with leaf-like awns, the lower awn to 40 mm long, upper one shorter.
**Requirements:** Moist infertile soils
**Localities:** 12, 26, 28, 35, 42, 44, 53–55, 60, 67#, 69A–71, 73–75
**Distribution:** Vic, NSW, Tas, SA, WA

## Centrolepis fascicularis — Tufted Centrolepis

**Size:** 2.5–6 cm x 3–12 cm — P
**Habitat:** Swamp and wattle tea-tree scrub
**Form:** Bright green perennial herb forming dense cushions; stems terete and thread-like
**Foliage:** Many fine soft leaves to 45 mm long, lower half with sparse soft hairs
**Flowers:** Ovoid flowerhead to 3 mm long; November to February. Bracts hairy, just longer than flowerhead.
**Requirements:** Moist open soils
**Localities:** Dandenong Rges, 61, 74
**Distribution:** All states except NT; PNG

*Illustrated p. 255*

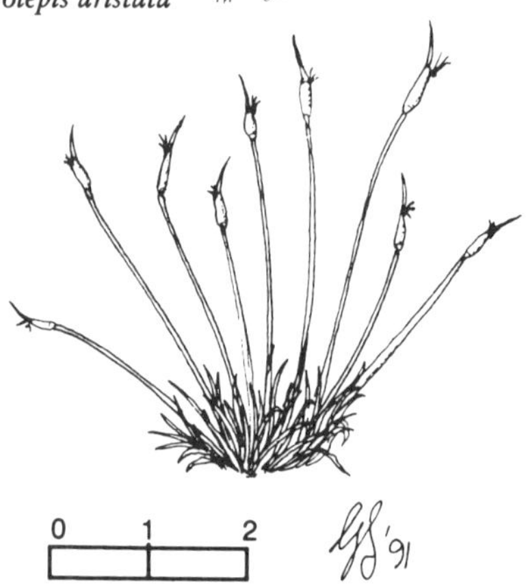

*Centrolepis polygyna*

## Centrolepis polygyna — Wiry Centrolepis

**Size:** 1.5–6 cm high — F P
**Habitat:** Saltmarshes, tea-tree heath
**Form:** Dull green annual tufted herb, becoming red-brown after flowering, stems terete
**Foliage:** Narrow crowded glabrous leaves 4–12 mm long, rigidly recurved
**Flowers:** Cylindrical flowerhead 3–5 mm long, with a single flower; July to November. Lower bract has a leaf-like curved awn to 10 mm long, upper bract awnless.
**Requirements:** Moist sandy soil, tolerating saline conditions
**Localities:** 67#
**Distribution:** Vic, NSW, Tas, SA, WA

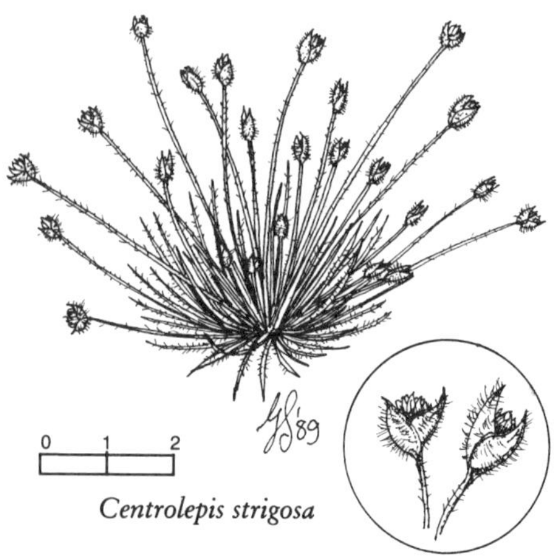

*Centrolepis strigosa*

## Centrolepis strigosa — Hairy Centrolepis

**Size:** 2–7 cm high — F P
**Habitat:** Red gum woodland, valley sclerophyll forest, tea-tree heath, wattle tea-tree scrub
**Form:** Small hairy bright green annual herb forming a hemispherical tuft
**Comments:** Similar to *C. fascicularis*. Soft spreading leaves to 25 mm long are softly hairy all over. The flowerheads, to 4 mm long, are sheathed in hairy bracts which end in a short point no longer than the heads. September to November.
**Propagation:** Seed
**Localities:** St. Kilda, 35, 42, 44, 60, 63, 67#, 69–72, 74, 75
**Distribution:** All states except NT; NZ

# Chionochloa — Poaceae

## Chionochloa pallida — Silver Top Wallaby Grass, Red Anther Wallaby Grass

**Size:** to 0.3 m x 1 m; stems to 1.5 m high — F P
**Habitat:** Sclerophyll woodland, dry and valley sclerophyll forests
**Form:** Spreading tufted perennial grass forming a tussock, or, when very large, a hummock with a sparse centre
**Foliage:** Fine green or bluish leaves to 70 cm long, with margins inrolled
**Flowers:** Spikelets ageing to straw-coloured, with prominent red anthers. Panicles to 35 cm long, held high above the leaves; October to January (mostly after fire)
**Requirements:** Well drained soils, tolerating periods of wetness
**Comments:** A most attractive grass, especially in flower, forming large clumps. Tolerates *Phytophthora cinnamomi* fungus, and poorer soils low in nutrient.
**Propagation:** Seed, division
**Localities:** Research, 25, 26, 33–36, 40–42, 44, 47, 48, 50–56, 58, 59, 62–64, 71, 74
**Distribution:** Vic, NSW, WA
**Synonym:** *Danthonia pallida*

*Chionochloa pallida*

# Chloris — Poaceae

## Chloris truncata — Windmill Grass, Umbrella Grass

**Size:** 15–45 cm high — F
**Habitat:** Grassy wetland, plains grassland, red gum and box woodland
**Form:** Open or compact tufted grass forming a dense low crown, occasionally with short, branched stolons
**Foliage:** Bluish-green, narrow, rough leaves to 15 cm long x 2–5 mm
**Flowers:** Panicle of 6–9 purplish to black spikes, each to 13 cm long, radiating horizontally from a common point at maturity; November to January. Spikelets arranged in 2 rows along each radiating arm.
**Requirements:** Well drained conditions, preferring heavy soils
**Comments:** An attractive grass for a sunny rockery, with very decorative seed heads. Similar to the weed 'Couch' (*Cynodon dactylon*), but larger with the exception of the length of the stolons. The flowering season may be extended through Autumn with additional watering.
**Propagation:** Seed
**Localities:** Richmond, Basalt Plains, 2, 5, 7, 9, 12, 15–18, 21, 23–26, 29, 34A, 67#
**Distribution:** All states except Tas

*Chloris truncata*

# Chorizandra — Cyperaceae

## Chorizandra cymbaria — Heron Bristle Rush

**Size:** 0.3–1 m high — F P
**Habitat:** Swamp scrub, valley sclerophyll forest, tea-tree heath
**Form:** Erect rush-like perennial herb with fleshy partitioned stems
**Foliage:** Fleshy basal leaves
**Flowers:** Solitary, dense, terminal flowerheads, globular dark brown with spikelets to 5 mm long, female floret terminal, surrounded by many male florets. Flowerhead sheathed by an erect bract which appears to extend the stem. September to November.
**Requirements:** Moist soils, tolerating drying out for short periods over Summer, but responds to Summer watering
**Comments:** An attractive rush-like plant for bog gardens. Flowerhead appears to burst out of the sides of the stems.
**Propagation:** Seed, division
**Localities:** 42, 54, 70, 73, 74
**Distribution:** All states except SA; N Cal

*Chorizandra cymbaria*

# Cladium — Cyperaceae

## Cladium procerum — Leafy or Tall Twig-rush

**Size:** 0.9–1.8 m high — F P
**Habitat:** Riparian scrub, wattle tea-tree scrub
**Form:** Large grass-like tufting perennial herb with creeping rhizomes, stems hollow, erect, with several nodes
**Foliage:** Flat leaves with minutely toothed margins and keels
**Flowers:** Terminal panicle dense and leafy, to 30 cm x 80 mm; November to December
**Requirements:** Moist to wet soils
**Comments:** Usually found near watercourses, tolerating saline conditions.
**Propagation:** Seed, division
**Localities:** Mouth of Yarra#, 36, 67, 74
**Distribution:** All states except Tas; N Cal

SCALE: 0 1cm 2cm — The scale representation on each plant equals 2cm.

*Cymbopogon refractus*

# Cymbopogon **Poaceae**

## Cymbopogon refractus

Barb-wire or Turpentine Grass

**F P**

**Size:** stems to 1.2 m high
**Habitat:** Escarpments in plains grassland
**Form:** Tufted perennial grass
**Foliage:** Basal leaves to 50 cm long, narrow and rough, rolled in bud, stem leaves shorter. Nodes on stems purple.
**Flowers:** Narrow panicle comprising distant clusters of glabrous racemes with reddish sheathing bract at the base; racemes bending downwards and becoming darker at maturity. Lemmas either awnless or with twice bent awn to 15 mm long; October to January
**Requirements:** Well drained soil, usually growing on rocky ground
**Comments:** The reflexed heads on this grass provide an interesting effect. The leaves have a gingery aroma when crushed. Known from one site within the Melbourne area.
**Propagation:** Seed, division of rhizome
**Localities:** Sunbury
**Distribution:** Vic, Qld, NSW, NT; N Cal, Poly

# Cyperus **Cyperaceae**

Annual or perennial sedges with creeping rhizomes, found in moist, boggy sites. Leaves of the local species are long and grass-like, at the base of the triangular flowering stems. The terminal inflorescence is an umbel or ball of several-flowered flattened spikelets, with bracts spreading from the base and usually exceeding the infloresence. Glumes overlap in 2 opposite rows. Plants are propagated by division or seed but are rarely available through nurseries.

*Cyperus brevifolius*

## Cyperus brevifolius

Globe Kyllinga

**F**

**Size:** 15–30 cm x spreading
**Habitat:** Known only from cultivated situations; usually on river flats
**Form:** Perennial herb with slender scattered or tufted stems
**Foliage:** Soft, limp leaves similar in length to the stems, to 3 mm wide; bracts 3–4, always longer than the flowerhead
**Flowers:** Tightly clustered, green or yellowish, single-flowered spikelets in a single rounded head. Flowers November to May.
**Requirements:** Permanently moist soils
**Comments:** Known only from lawn sites in Melbourne, it may be introduced.
**Localities:** Balwyn, Melbourne, ?69A
**Distribution:** Vic, Qld, NSW, SA; NZ, PNG, N Cal, N & S Am, Indon, Af, Philippines, S. Asia

## Cyperus gunnii

Flecked Flat Sedge

**P**

**Size:** 0.6–1 m x spreading
**Habitat:** Riparian scrub
**Form:** Densely tufted perennial herb with rigid, sharply triangular stems as long as the leaves
**Foliage:** Leaves folded flat, with parallel ridges and finely toothed margins, to 1 m x 5 mm; 2 to 4 bracts similar to leaves, longest pair 0.3–1 m long
**Flowers:** Simple or compound umbel with 5–8 primary branches. Numerous spikelets in dense, flattened clusters; bright brown spikelets are 6–15-flowered and linear, to 10 mm long; October to April
**Requirements:** Moist to boggy soils
**Localities:** 56
**Distribution:** Vic, Qld, NSW, Tas, SA

## Cyperus lhotskyanus

Flat Sedge

**P**

**Size:** 0.1–0.6 m high
**Habitat:** Moist depressions in box woodland on the basalt plains
**Form:** Tufted perennial herb with finely ridged stems as long as the leaves, and a short creeping rhizome
**Foliage:** Flat, rigid leaves to 60 cm long; several bracts like the leaves, the lower bract longer than the flowerhead
**Flowers:** Umbel of 3–7 spikes to 110 mm long; reddish-brown spikelets in dense, 8–12-flowered oblong-lanceolate heads; October to January.
**Requirements:** Wet soil
**Localities:** Toolern Vale, 14
**Distribution:** Vic, NSW, SA
**Synonym:** *C. rutilans*

*Illustrated p. 255*

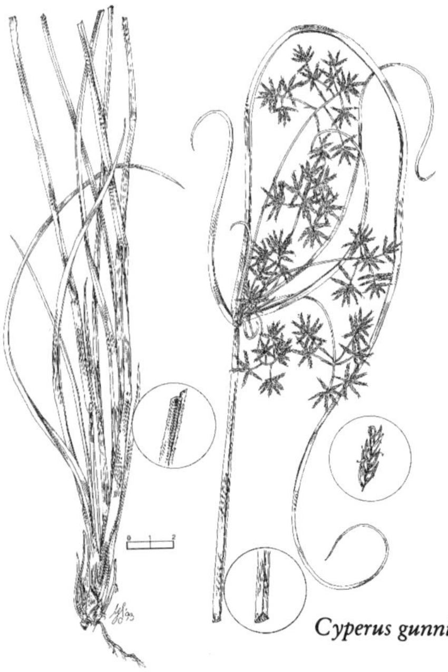

*Cyperus gunnii*

*Amphibromus fluitans*

*Amphibromus nervosus*

*Austrofestuca hookeriana*

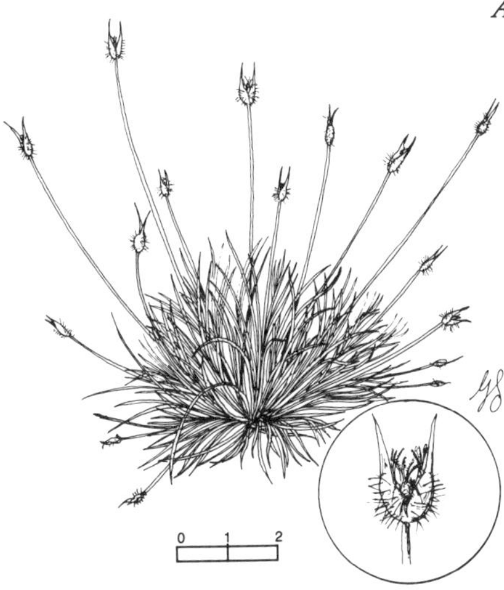

*Centrolepis fascicularis*

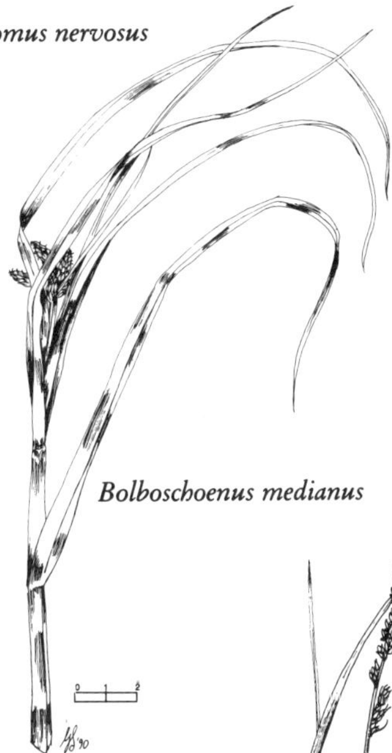

*Bolboschoenus medianus*

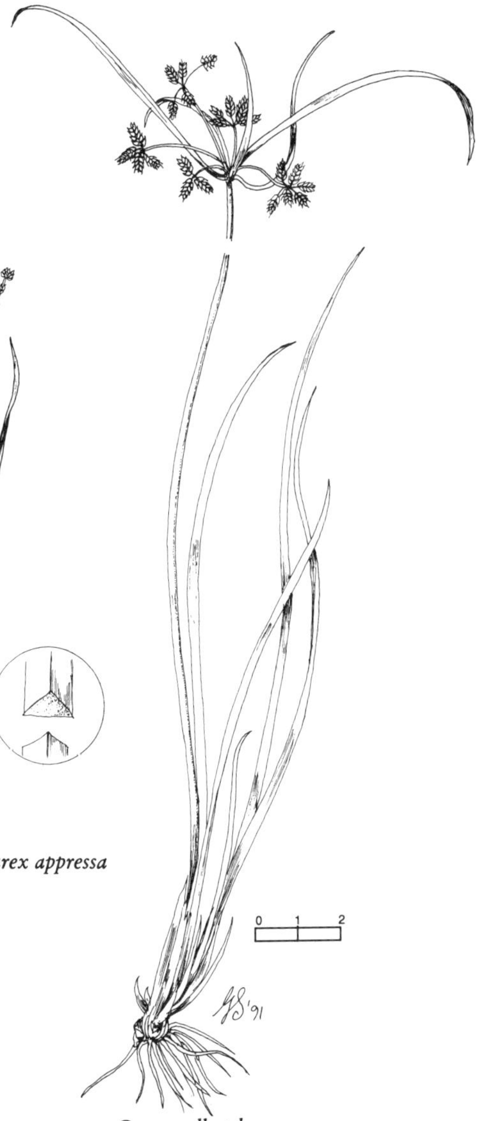

*Cyperus lhotskyanus*

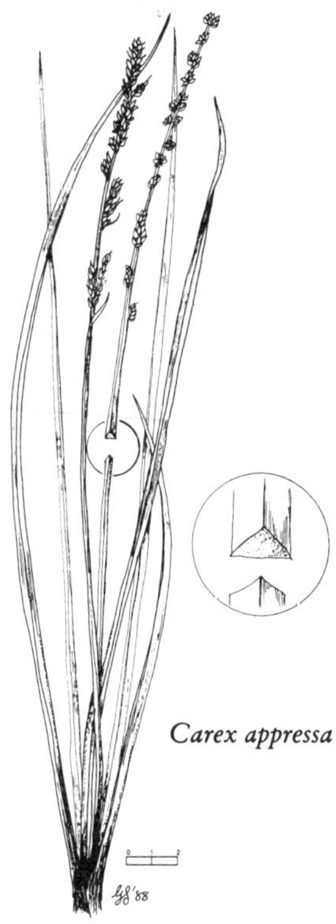

*Carex appressa*

*Baumea juncea*

*Carex iynx*

SCALE: 0 1cm 2cm The scale representation on each plant equals 2cm.

*Cyperus lucidus*

### Cyperus lucidus

Leafy Flat Sedge

**Size:** 0.6–1.5 m high **P**
**Habitat:** Around swamps and water courses of dry, valley and damp sclerophyll forests
**Form:** Robust tufted perennial herb with sharply triangular stems
**Foliage:** Shiny dark green, flat or channelled leaves with rough margins and many shallow ridges, often longer than the stems, to 15 mm wide; 3–6 large bracts 5–15 cm long
**Flowers:** Compound umbel of up to 12 rigid branches to 20 cm long, divided again into secondary, dense cylindrical brown spikes to 5 cm long, loosely arranged spikelets 5–10-flowered; October to February.
**Requirements:** Wet soils or boggy conditions
**Comments:** Will grow as an aquatic plant.
**Localities:** 30, 34A, 35, 42, 46, 47, 55, 59, 61
**Distribution:** Vic, Qld, NSW, Tas, SA

### Cyperus sanguinolentus

Dark Sedge

**Size:** 10–40 cm high **P**
**Habitat:** Riparian scrub
**Form:** Tufted annual herb
**Foliage:** Variable dark green leaves 10–40 cm x 3 mm, sometimes very short and thread-like; 3–4 bracts 8–10 cm long
**Flowers:** Umbel of few short spikes to 40 mm long, on slender stems as long as the leaves. Spikelets clustered, oblong and 8–20-flowered, to 14 mm long; glumes green with dark margins; November to April.
**Requirements:** Moist or wet soils with a sheltered aspect
**Comments:** May become a weedy in cultivation.
**Localities:** 29
**Distribution:** Vic, Qld, NSW, SA; PNG, Asia, Af

*Damasonium minus*

## Damasonium — Alismataceae

### Damasonium minus

Star Fruit

**Size:** to 0.5 m high **F P**
**Habitat:** Grassy wetlands, swamps and depressions in box woodland
**Form:** Emergent aquatic or semi-aquatic annual tufted herb
**Foliage:** Erect, long-stalked ovate-lanceolate to cordate leaves, blades to 100 mm long
**Flowers:** Stiff erect panicle with whorls of several 1-flowered branches held high above the foliage on leafless stems. Tiny pale pink 3-petalled flowers; October to July, followed by star-like fruit with 6–10 triangular sections.
**Requirements:** Shallow water in a sunny position
**Comments:** An interesting feature for a shallow pond. Plants may be perennial in permanent water, developing into dense, long-flowered colonies. May be confused with *Alisma* spp.
**Propagation:** Division
**Localities:** Melton, Yarra R., 6, 24, 26
**Distribution:** All states

Genus in Australia has been reviewed & most are now in Austrodanthonia & some are in Joycea & Rhytidosperma

## Danthonia — Poaceae

Danthonias (or Wallaby Grasses) are common perennial grasses of open plains and lightly forested slopes, forming discrete clumps or tussocks varying in height from 0.1–1 m. Seventeen species occur within the Melbourne metropolitan area. They may be distinguished from other grasses by the characteristic arrangements of hairs, either covering the lemma, in tufts or in rows. The lemma is distinctly lobed along the top and has a central awn rising from the sinus between 2 pointed or awned lobes. These features also help to identify individual species of *Danthonia*.

Most species are excellent contrast plants in a native landscape requiring full sun or semishaded positions with well drained soil, often thriving on neglect.

Most *Danthonia* spp. show different characteristics in response to changes in their environment. They become much smaller if trampled or grazed regularly and the infloresence may become quite purple after fire. Older plants can be readily rejuvenated by either lifting and dividing or (in some species) by burning the tussock. In the wild, after fire, the first rains induce the plants to spring into life very quickly. Danthonias are an important component of natural pastures and several species have considerable agricultural potential. Other species make excellent lawns if infrequently mown.

Propagation is by seed or by division.

## Danthonia auriculata

Lobed Wallaby Grass

**Size:** to 10 cm high; stems to 30 cm high **F P**
**Habitat:** Red gum woodland, plains grassland
**Form:** Slender tufted perennial grass
**Foliage:** Short, fine, densely hairy leaves to 10 cm long, margins inrolled
**Flowers:** Dense compact panicle to 40 mm long, spikelets usually 6-flowered. Immature spikelets pale green, often tinged purple, ageing to straw coloured. Outer edges of lobes on lemma end in triangular protuberances (auricles); lemma broad with 3 transverse rows of hair-tufts, hairs of upper ring to 6 mm long. Central awn to 15 mm long, strongly twisted 2 or more times near base. October to December
**Requirements:** Well drained clay or loam soils
**Comments:** A species which is well adapted to grazing and natural disturbance.
**Localities:** 2, 4, 7, 12, 16, 21, 25, 26, 35, 44, 45
**Distribution:** Vic, NSW, SA

*Danthonia auriculata*

## Danthonia caespitosa

Common Wallaby Grass

**Size:** 20–40 cm x 40 cm; stems to 1.2 m high **F P**
**Habitat:** Grassy wetland, plains grassland, red gum woodland, valley sclerophyll forest, tea-tree heath
**Form:** Variable perennial grass often forming a dense tussock, but sometimes reduced to small tufts
**Foliage:** Leaves glabrous or hairy, flat or loosely rolled, and often blue-green on upper surface; vary substantially under dry conditions
**Flowers:** Panicles usually oblong to 15 cm, with 10–30 spikelets of 4–9 flowers. Lemma pale, to 6 mm long, with 3 transverse rows of hair-tufts, central awn usually dark, to 25 mm long, twisted 3–5 times near base. October to January, March
**Requirements:** Moist well drained soils
**Comments:** Widespread and common throughout the region and often a dominant component of the ground flora. Mature flowerheads appear bleached and fluffy. May persist under mowing or grazing. Large variability between forms may warrant future subdivision of this species.
**Localities:** Montrose, 2, 4–9, 12, 14–18, 21, 24–26, 28, 29, 32, 34A, 35, 50A, 54, 56, 58, 62, 63, 67, 68, 69A, 71, 74, 76
**Distribution:** Vic, NSW, Tas, SA, WA; NZ

*Danthonia caespitosa*

## Danthonia carphoides var. carphoides

Short Wallaby Grass

**Size:** to 10 cm high; stems to 30 cm high **F**
**Habitat:** Plains grassland, red gum woodland
**Form:** Small tufted perennial grass
**Foliage:** Short fine leaves 3–20 cm long, inrolled
**Flowers:** Dense ovoid raceme to 40 mm long with few 3–5-flowered greenish spikelets soon turning straw-coloured. Lemma to 6 mm long with hairs scattered over the back, lateral lobes tiny and awnless to 2 mm long, central awn short and thick. Glumes are broad and strongly boat-shaped. October to December
**Requirements:** Well drained soils, tolerating short wet periods in Winter
**Comments:** A rare and adaptable small tuft, often associated with *D. auriculata*. Ideal for filling small spaces between rocks. May persist under mowing or grazing.
**Localities:** Preston, 2, 4, 7, 21, 25, 26, 28
**Distribution:** Vic, NSW, Tas, SA

*Danthonia carphoides* var. *carphoides*

## Danthonia duttoniana

Brown-back Wallaby Grass

**Size:** to 40 cm x 50 cm; stems to 1 m high **F P**
**Habitat:** Grassy wetland, low lying areas in plains grassland and red gum woodland
**Form:** Erect, densely tufted glabrous perennial grass
**Foliage:** Leaves typically 20 cm x 2–4 mm
**Flowers:** Numerous 5–7-flowered pale green spikelets, clustered towards ends of flowering branches in panicles to 120 mm long. Lower branches long, becoming reflexed. Lemma golden brown, lobes paler, with 2 rows of sparse hair-tufts. Central awn bent and twisted 2–3 times. October to January
**Requirements:** Moist, poorly drained soils, in basalt and peat
**Comments:** One of several *Danthonia* spp. which tolerate poorly drained conditions. The flowerheads are often distinctly reddish before maturity, making them most attractive.
**Localities:** 2, 4–7, 12, 16–18, 21, 24–26, 28, 35, 36
**Distribution:** Vic, NSW, SA

*Danthonia duttoniana*

SCALE: 0 1cm 2cm — The scale representation on each plant equals 2cm.

Danthonia eriantha

## Danthonia eriantha

Hill Wallaby Grass

**Size:** to 20 cm x 30 cm; stems to 50 cm high **F P**
**Habitat:** Red gum woodland, plains grassland, dry sclerophyll forest
**Form:** Erect, densely tufted perennial grass
**Foliage:** Leaves fine, stiff, inrolled, to 25 cm long, hairy at least when young
**Flowers:** Short, dense, lanceolate to ovate panicle, 20–70 mm long, 4–15 greenish spikelets, on downy branches. 2 rows of hair tufts on lemma. Central awn of lemma to 15 mm long, twisted 3–4 times, long lateral lobes. Anthers yellow to pale orange-yellow. October to December
**Requirements:** Well drained soils on exposed sites
**Comments:** A widespread and common grass, often co-extensive with other *Danthonia* spp., particularly *D. caespitosa*. Persists in rough lawns.
**Localities:** Montmorency, 2, 4, 6, 12, 17#, 18, 24–26, 35, 41, 42, 47, 48, 74
**Distribution:** Vic, NSW, SA

## Danthonia geniculata

Kneed Wallaby Grass

**Size:** to 15 cm x 20 cm; stems to 30 cm high **F P**
**Habitat:** Box and red gum woodland, dry sclerophyll forest, tea-tree heath
**Form:** Tufted perennial grass with slender stems, often bent near base
**Foliage:** Very fine hairy inrolled leaves to 15 cm long
**Flowers:** Dense panicle to 50 mm long, with pale, 4–6-flowered spikelets. Lemma hairs scattered, lateral lobes pointed, 2–4 times length of body, central awn bent at base, no longer than side lobes; October to December, March
**Requirements:** Tolerant, adapting to most soils
**Comments:** Useful in lawns but slow to establish.
**Localities:** Kilsyth#, Eltham, Prahran, 5, 6, 16–18, 21, 23–26, 32, 34A–36, 42, 44, 45, 49, 66–71, 74–76
**Distribution:** Vic, SA

Danthonia geniculata

## Danthonia induta

Yellow Anther Wallaby Grass

**Size:** to 30 cm high; stems 0.6–1.2 m high **F P**
**Habitat:** Box woodland
**Form:** Robust erect tufted perennial grass
**Foliage:** Leaves to 30 cm long, flat to 4 mm wide or inrolled
**Flowers:** Loose, large, spreading panicle to 18 cm long bearing many green or straw-coloured spikelets with yellow-orange anthers. Lemma hairs scattered, increasing in size towards the apex, central awn strongly twisted at base.
**Requirements:** Heavy well drained soils
**Comments:** An attractive, vigorously growing grass which is never abundant. Responds to extra moisture during Summer, producing flowers over an extended period.
**Localities:** Keilor, 25, 26, 35, 42
**Distribution:** Vic, Qld, NSW

*Illustrated p. 262*

## Danthonia laevis

Wallaby Grass

**Size:** to 40 cm x 40 cm; stems to 0.7 m high **F P**
**Habitat:** Plains grassland, red gum and box woodlands, dry and valley sclerophyll forests, tea-tree heath
**Form:** Erect densely tufted perennial
**Foliage:** Leaves to 30 cm long, inrolled in bud, becoming flat or loosely inrolled
**Flowers:** Panicle to 50 mm long with 4–7 greenish spikelets. Lemma has complete ring of long hair-tufts near the top and only small tufts (rather than a ring) below; lateral lobes broad, to 10 mm long ending in fine awns, central awn long to 30 mm, bent at 4 mm. Anthers yellow-orange to purple. October to January
**Requirements:** Moist soils
**Comments:** Extremely adaptable to different soils providing moisture is adequate. Often found bordering wet depressions with *D. duttoniana* on basalt or *D. semiannularis* elsewhere.
**Localities:** Silvan, Montmorency, 25, 26, 28, 34A, 35, 42, 44, 47, 48, 50A, 54–56, 63, 69A, 70, 73, 75
**Distribution:** Vic, NSW, Tas, SA

*Illustrated p. 262*

Danthonia lepidopoda

## Danthonia lepidopoda

**Size:** to 15 cm high; stems to 0.6 m high **P**
**Habitat:** Sclerophyll woodland
**Form:** Weakly tufted perennial grass, with long scaly rhizomes
**Foliage:** Flat or channelled leaves to 15 cm long, glabrous or sparsely hairy
**Flowers:** Narrow to narrowly ovate panicle to 80 mm long, with few, purplish, 3–4-flowered spikelets. Lemma covered in hairs which are weakly tufted, lobes ending in short bristly tips, central awn to 10 mm long, weakly twisted.
**Requirements:** Sandy soil
**Comments:** A newly named species known from 2 collections on the south-eastern slopes of the Dandenong Rges. More common in the Otways and Grampians. Differs from all other *Danthonia* spp. in that it develops a long creeping rhizome.
**Localities:** Belgrave Sth
**Distribution:** Endemic to Vic

## Danthonia linkii var. linkii
Wallaby Grass

**Size:** to 40 cm x 40 cm; stems to 0.6 m high — F P
**Habitat:** Red gum woodland, dry and valley sclerophyll forest
**Form:** Variable perennial grass, densely tufted and erect
**Foliage:** Leaves firm, 20–70 cm x 1–2.5 mm wide
**Flowers:** Loose, sometimes sparse, panicle to 120 mm long with pale green spikelets, crowded and 5–6-flowered. Lemma pale, fluffy, densely hairy at base, usually a row of fine long hairs around the top, back covered in hairs. Lateral lobes twice as long as lemma body, central awn twisted for 1.5 mm from base. Palea (or upper bract) lanceolate and pointed. Flowers mostly from November to March, but in a good season flowering may continue all year. Panicles ripen slowly from the top down, often showing the full range of stages of development.
**Requirements:** Heavy soils
**Comments:** High rates of foliage production may be useful for fodder or specimen plants, but very poor for lawns.
**Localities:** Heyington, Hurstbridge, 18, 21, 23, 25, 26, 34A–36, 39, 41, 42, 44, 51, 53, 56, 69A
**Distribution:** Vic, Qld, NSW

## Danthonia linkii var. fulva
Wallaby Grass

**Habitat:** Box woodland, dry and valley sclerophyll forests
**Comments:** Generally more robust, with broader leaves (to 3 mm), than *D. linkii* var. *linkii*. The central awn is brown and more strongly twisted for 3–4 mm at the base. Both varieties commonly occur together.
**Localities:** Keilor basalt plains, Montmorency, Dandenong Rges, 62, 63
**Distribution:** Vic, NSW, SA

*Danthonia linkii* var. *fulva*

## Danthonia pallida = Chionochloa pallida

## Danthonia penicillata
Slender Wallaby Grass

**Size:** to 20 cm x 30 cm; stems to 0.6 m high — P
**Habitat:** Dry, valley and wet sclerophyll forests, riparian woodland and scrub, grassy low open forest
**Form:** Fine, sparsely hairy, weeping perennial grass forming sparse open tussocks,
**Foliage:** Narrow flat or inrolled leaves to 15 cm long
**Flowers:** Slender, arching, furrowed stems to 60 cm high, bearing narrow weak racemose panicles to 15 cm long with scattered 4–6-flowered spikelets. Lateral lobes longer than lemma body; separate hair-tufts on sides and back of lemma; October to January.
**Requirements:** Moist well drained soil
**Comments:** A graceful distinctive little grass that is easily overlooked, and is ideal for establishing in shaded areas. Persists in unimproved pasture and rough lawn. Distinguished from *D. pilosa* by arching stems and short twisted part of awn.
**Localities:** 8, 16, 18, 25, 26, 29, 34A–36, 41, 42, 44, 47–50A, 54–57, 60, 63–67, 74, 76
**Distribution:** Vic, NSW, Tas; NZ

*Danthonia penicillata*

## Danthonia pilosa var. pilosa
Velvet Wallaby Grass

**Size:** to 20 cm x 30 cm; stems to 0.5 m high — P
**Habitat:** Sclerophyll woodland, valley sclerophyll forest, grassy low open forest, plains grassland
**Form:** Erect or flattened tufted perennial grass, depending on its environment
**Foliage:** Flat or inrolled, hairy linear leaves usually to 12 cm x 2 mm
**Flowers:** Short, ovate, stiff panicle to 70 mm long, contracting at maturity. 5–9-flowered pale spikelets crowded and overlapping. Top hair-tufts on lemma separated; dark central awn strongly twisted near base, coarse and long; November to February.
**Requirements:** Moist soils
**Localities:** 16, 23, 26, 28, 34A, 36, 40–42, 44, 47–50A, 54–56, 58, 60, 63, 64, 69A, 74–76
**Distribution:** All states except NT; NZ

*Illustrated p. 262*

## Danthonia procera
Tall Wallaby Grass

**Size:** to 0.5 m x 0.7 m; stems to 1.4 m high — F P
**Habitat:** Red gum woodland, grassy low open forest
**Form:** Large, robust perennial tussock grass
**Foliage:** Fine, inrolled, glabrous leaves to 30 cm x 3 mm, bluntly pointed; similar to *Chionochloa pallida*
**Flowers:** Loose, open panicle to 15 cm long. Lemma pale with hair-tufts in rows, short awns on lateral lobes.
**Requirements:** Well drained soils, although very tolerant once established
**Comments:** Regionally rare and uncommon throughout its range.
**Localities:** 23, 28, 35, 74, 76,
**Distribution:** Vic, NSW, Tas

*Danthonia procera*

SCALE: 0 1cm 2cm — The scale representation on each plant equals 2cm.

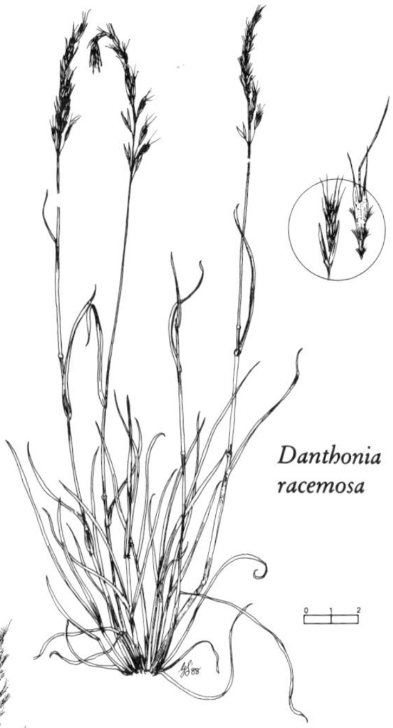

*Danthonia racemosa*

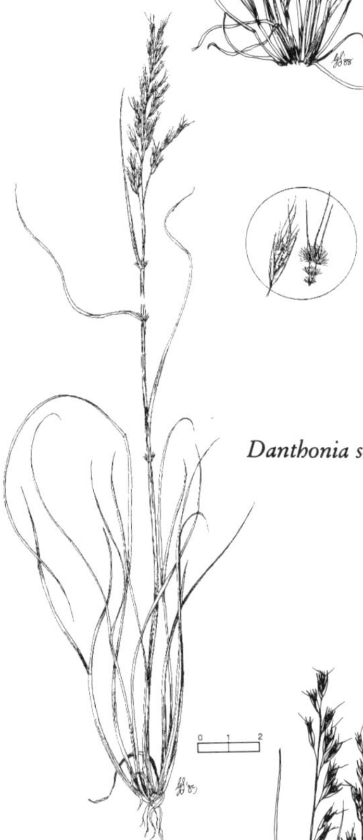

*Danthonia setacea* var. *setacea*

*Danthonia tenuior*

## Danthonia racemosa

Clustered Wallaby Grass

**Size:** to 20 cm high; stems 0.1–0.6 m high **F P**
**Habitat:** Widespread in plains grassland, riparian, box and red gum woodland, dry and valley sclerophyll forests
**Form:** Variable; slender tufts or dense tussock-forming perennial grass
**Foliage:** Fine light green leaves to 15 cm long, flat to strongly inrolled
**Flowers:** Slender compact racemose panicle to 15 cm long with 6–10-flowered pale greenish spikelets. Lemma narrow, parallel-sided, to 7 mm long, with densely hairy basal tuft, top whorl has only marginal and central hair-tufts, lemma tip elongated into a prominent point. Flowers most of the year, peaking Summer.
**Requirements:** Dry to moist well drained soils
**Comments:** Very common throughout its range, usually growing with other *Danthonia* spp. or *Microlaena stipoides*. A useful component for native lawns.
**Localities:** Montrose, 5, 8, 14–16, 18, 21, 23–26, 28, 29, 32, 34A–36, 39, 41, 42, 44, 45, 47, 50A, 51, 53, 55, 56, 69A, 71, 74, 76
**Distribution:** All states except NT; NZ

## Danthonia semiannularis

Tasmanian Wallaby Grass

**Size:** to 30 cm x 40 cm; stems to 0.6 m high **F P**
**Habitat:** Valley sclerophyll forest, red gum and sclerophyll woodland, grassy low open forest, tea-tree heath
**Form:** Erect tussock-forming perennial grass
**Foliage:** Fine grey-green leaves to 15 cm long, usually inrolled
**Flowers:** Profuse flowering stems with oblong panicles to 80 mm long, dense and fine-textured, with many 6-flowered greenish-purple spikelets. Lemma tiny, sparsely hairy with row of longer hairs near the top, lateral lobes shortly pointed and minutely awned, level with the lemma hairs, brown central awn to 9 mm long, twisted for 3 mm, then bent. October to January.
**Requirements:** Moist soils
**Localities:** Boronia, 4, 31, 34A, 35, 42, 45, 49, 50A, 51, 54–56, 60, 62, 63, 69A, 70, 74–76
**Distribution:** Vic, NSW, Tas, SA, WA; NZ *Illustrated p. 262*

## Danthonia setacea var. setacea

Small-flowered or Bristly Wallaby Grass

**Size:** to 30 cm x 40 cm; stems to 0.6 m high **F P**
**Habitat:** Widespread in plains grassland, red gum and box woodlands, dry and valley sclerophyll forests, tea-tree heath
**Form:** Erect, slender, densely tufted perennial
**Foliage:** Leaves fine, inrolled, to 20 cm long
**Flowers:** Dense, oblong to lanceolate, dappled silvery panicle to 75 mm long with pale to purplish 4–10-flowered spikelets. Lemma to 3 mm long, 2 whorls of hair-tufts, the upper ring longer and often incomplete, awns very slender, often purplish. October to January
**Requirements:** Adaptable, ranging from sandy to heavy soils
**Comments:** Often confused with *D. caespitosa* which has longer collar hairs on broader leaves, coarser glumes and longer lemma body, lateral lobes and awns.
**Localities:** Eltham, 4, 5, 7, 9, 12, 14, 16–18, 21–23, 25, 26, 28, 32, 34A, 36, 41, 42, 44, 45, 47–49, 50A–56, 61, 63, 64, 67–71, 73–76
**Distribution:** Vic, NSW, Tas, SA, WA

## Danthonia tenuior

Wallaby Grass

**Size:** to 30 cm x 40 cm; stems 0.7–0.8 m high **F P**
**Habitat:** Dry and valley sclerophyll forests, red gum woodland
**Form:** Erect tufted perennial grass
**Foliage:** Leaves flat in lower half, becoming loosely inrolled
**Flowers:** Loose, oblong panicle, usually to 10 cm long, with 4–6-flowered spikelets. Lemma, to 3.5 mm long, has 3 transverse rows of hair-tufts, awns of lateral lobes (if present) much shorter than flat section of lobes, central awn to 12 mm long, loosely twisted near base. Margins of side lobes usually prominently purple Anthers dark orange-yellow. October to January.
**Requirements:** Well drained soil
**Comments:** A very common grass, similar to *D. linkii*. It is distinguished by the panicle shape and size, lemma size and the purplish lobe margins.
**Localities:** Research, 23, 26, 32, 36, 41, 42, 44, 47, 48, 50A, 54, 55, 63, 74
**Distribution:** Vic, Qld, NSW, SA; NZ
**Synonym:** *D. purpurascens*

# Deyeuxia

**Poaceae**

Deyeuxias (or Bent Grasses) are erect, sparsely-tufted perennials up to 1.5 m high, with mostly basal leaves, flat or convolute and tapering to a fine point. The inflorescence is an open or dense terminal panicle with numerous and often densely overlapping 1-flowered spikelets. An awn rises from the back of the lemma similar to *Agrostis* but the lemma is narrower and papery rather than transparent. None of the species are important as pasture grasses. Propagation is by seed or division of the tuft. Seed should be stored 6–12 months before sowing.

## Deyeuxia contracta

Bent-grass

**Habitat:** Wet sclerophyll forest
**Form:** Leaves and flowering stems lax
**Flowers:** Panicle to 7–20 cm long, loose with spreading branches with green or purplish spikelets to 2.5 mm long. Lemma rough with minute awn attached to the upper third
**Requirements:** Moist rich soils
**Localities:** 57
**Distribution:** Vic, NSW, Tas *Illustrated p. 262*

## Deyeuxia densa

Bent-grass

**Size:** 0.5–1 m high **F P**
**Habitat:** Tea-tree heath, swamp scrub
**Form:** Stout or slender tufted perennial grass
**Foliage:** Leaves narrow, flat or folded
**Flowers:** Dense spike-like panicle to 120 mm long. Spikelet stalk ends in feathery bristle. Awn, to 5 mm long, rises from near the middle of the lemma. October to January
**Requirements:** Moist soil
**Comments:** Locally rare.
**Localities:** Clayton, Kilsyth, 74
**Distribution:** Vic, Tas, SA

## Deyeuxia minor

Bent-grass

**Size:** to 0.4 m high **F P**
**Habitat:** Coastal banksia woodland, dry sclerophyll forest
**Comments:** A slender tufted perennial similar to *D. densa* but smaller with very fine leaves. Panicle is shorter to 40 mm and bristle is absent. A coastal grass which may be extinct within Melbourne.
**Localities:** Warrandyte[#], 67[#]
**Distribution:** Vic, NSW, Tas, SA

## Deyeuxia quadriseta

Reed Bent-grass

**Size:** to 15 cm x 40 cm; stems 1 m or more high **F P**
**Habitat:** Widespread in damp, valley and dry sclerophyll forests, sclerophyll woodland, grassy low open forest, plains grassland
**Form:** Open, sparsely tufted perennial tussock
**Foliage:** Flat to slightly inrolled leaf to 30 cm long
**Flowers:** Dense spike-like panicle to 15 cm long with pale green spikelets, awn almost basal; September to May
**Requirements:** Well drained soils
**Comments:** Although a coarse grass, large clumps are attractive in flower.
**Localities:** Research, 7, 26, 28, 34A–36, 41, 42, 44–48, 50–58, 61–64, 67, 68, 69A, 70, 74–76
**Distribution:** Vic, NSW, Tas, SA, WA *Illustrated p. 262*

## Deyeuxia rodwayi

Bent-grass

**Size:** to 10 cm high; stems 0.75 m high **P N**
**Habitat:** Wet, damp and valley sclerophyll forests
**Form:** Weakly ascending slender tussock
**Flowers:** Narrow but loose panicle to 15 cm long. Glumes widely gaping at maturity. Awn attached to the middle of the lemma, curved and longer than lemma.
**Requirements:** Moist soils
**Localities:** Belgrave, 55, 57
**Distribution:** Vic, Qld, NSW, Tas *Illustrated p. 262*

*Deyeuxia densa*

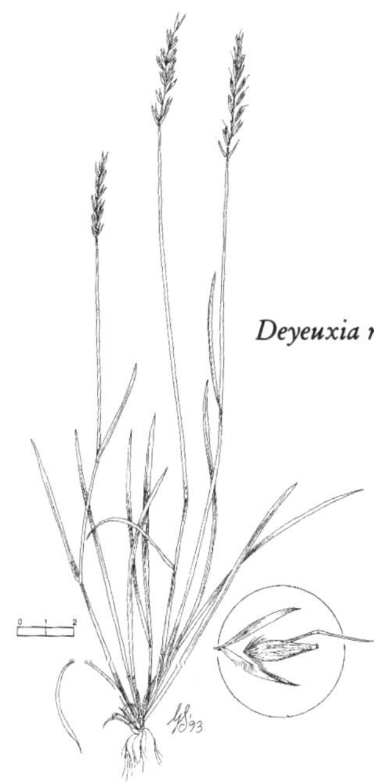

*Deyeuxia minor*

# Dichanthium — Poaceae

## Dichanthium sericeum

Queensland or Silky Blue-grass

**Size:** 30–50 cm high; stems to 0.8 m high **F**
**Habitat:** Plains grassland
**Form:** Upright grassy clump, nodes of flowering stems densely bearded with silky white hairs
**Foliage:** Flat, bluish leaves to 100 mm x 2–4 mm, often with purple tinges
**Flowers:** Groups of 2–6 sessile racemes to 50 mm long, densely covered in silky white hairs, radiating from the end of the flowering stem. Spikelets in very unequal pairs; awn of fertile lemma dark, twisted and bent, to 30 mm long. Flowering most of the year.
**Requirements:** Well drained heavy clay soils
**Comments:** Highly ornamental when growing vigorously because of its distinctive blue-grey appearance. Responds positively to extra water in Summer and also to hard cutting back.
**Propagation:** Seed or division
**Localities:** Sydenham, Sunshine, Footscray, Rockbank, Broadmeadows, 10, 15, 16, 25, 26, 29
**Distribution:** All mainland states; PNG, N Cal, Philippines

*Dichanthium sericeum*

SCALE: 0 1cm 2cm The scale representation on each plant equals 2cm.

*Danthonia semiannularis*

*Danthonia laevis*

*Danthonia induta*

*Danthonia pilosa* var. *pilosa*

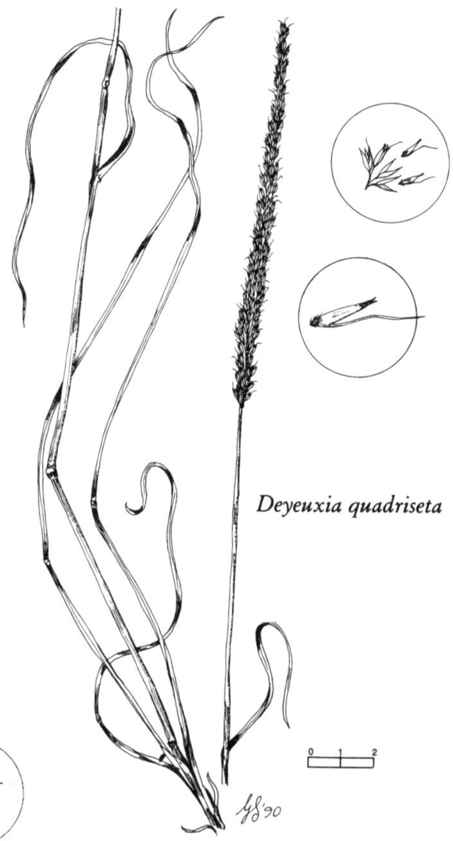

*Deyeuxia quadriseta*

*Deyeuxia contracta*

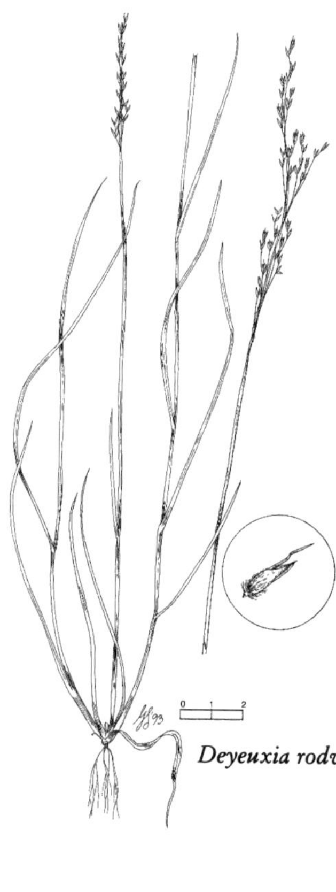

*Deyeuxia rodwayii*

# Dichelachne **Poaceae**

Sparsely tufted perennial grasses with fine, hairy flowerheads. Spikelets are single-flowered. Fine awns become curly at maturity. They are attached close to the top of the bilobed lemma and are more than twice the length of the lemma. Propagate by seed or division.

## Dichelachne crinita — Long-hair Plume-grass

**Size:** to 10 cm high; stems 0.5–1 m high **F P**
**Habitat:** Widespread in plains grassland, red gum, box and riparian woodland, dry and valley sclerophyll forest, grassy low open forest, tea-tree heath
**Form:** Sparse open tuft
**Foliage:** Flat, green to bluish-green leaves to 20 cm long
**Flowers:** Dense spike-like panicle to 20 cm long, of more than 100 overlapping spikelets, awn to 55 mm long, thread-like and curly rather than bent; October to December
**Requirements:** Well drained soils
**Comments:** An ornamental grass which seeds readily. Seeds may irritate the skin. Distinguish from *D. micrantha* by the usually denser flowerheads and lack of a bend in the awn.
**Localities:** 2, 4, 7, 8, 12, 13, 15–18, 20, 21, 24–26, 28, 31, 32, 33–36, 41, 42, 44, 49, 61, 67–69, 74, 76
**Distribution:** All states except NT; NZ
**Synonym:** *D. longiseta*

*Dichelachne crinita*

## Dichelachne micrantha — Short-hair Plume-grass

**Size:** to 10 cm high; stems 0.3–1.1 m high **F P**
**Habitat:** Dry sclerophyll forest, plains grassland
**Foliage:** Flat, dark green, mainly basal leaves to 20 cm long
**Flowers:** Dense panicle to 20 cm long, often looser at the base. Spikelets separated, from 30–100+. Awn to 20 mm long, slightly thickened and twisted in lower half, then often weakly bent. July to October
**Requirements:** Well drained soils
**Comments:** An ornamental grass
**Localities:** Research, 28, 36, 41, 42, 47, 48–49, 50A, 51, 53–58, 62, 74
**Distribution:** All states except NT; NZ
**Synonym:** Separated from *D. sciurea*

*Dichelachne micrantha*

## Dichelachne sieberiana — Plume-grass

**Size:** to 10 cm high; stems to 1.3 m high **P**
**Habitat:** Dry sclerophyll forest
**Form:** Tufted perennial grass, nodes of flowering stems hairy
**Foliage:** Narrow, flat, rough leaves
**Flowers:** Sparse, loose open panicles to 25 cm long; awns up to 4 times length of lemma, very tightly twisted. Summer
**Requirements:** Well drained soils
**Propagation:** Seed or division
**Localities:** 36, 42, 44, 46–48, 56, 64
**Distribution:** Vic, NSW, Tas
**Synonym:** Segregated from *D. rara* *Illustrated p. 273*

# Distichlis **Poaceae**

## Distichlis distichophylla — Australian Salt-grass

*Distichlis distichophylla*

**Size:** 5–20 cm x spreading **F**
**Habitat:** Saltmarshes, primary dune scrub
**Form:** Perennial grass with ascending branches and long-creeping rhizomes, forming a spreading lightly prickly clump
**Foliage:** Thin, rigid leaves 25–50 mm long, sometimes with short tips, alternate in 2 rows on opposite sides of the stem
**Flowers:** Dioecious; short terminal spike or raceme to 25 mm long, spikelets green to straw-coloured, flat, 2–4 on male plants; often more, crowded together on female plants and 6–14-flowered. Lemma keeled, awnless, stamens or styles often persisting. October to December
**Requirements:** Moist saline soils
**Comments:** Useful for soil stabilisation.
**Propagation:** Division of creeping stems
**Localities:** 2–5, 16, 21, 25, 26, 67, 69A, 71, 74
**Distribution:** Vic, NSW, Tas, SA

SCALE: 0 1cm 2cm — The scale representation on each plant equals 2cm.

*Dryopoa dives*

# Dryopoa — Poaceae

## Dryopoa dives — Giant Mountain Grass

**Size:** stems to 1.5–4.5 m high — P N
**Habitat:** Cool temperate rainforest, wet sclerophyll forest
**Form:** Biennial, erect grass with cane-like flowering stems to 25 mm thick, forming large, loose tussocks
**Foliage:** Flat, arching, bright green leaves to 1.5 m x 25 mm, often with prominent white midrib
**Flowers:** Straw-coloured loose, spreading panicle to 50 cm x 20 cm, spikelets to 8 mm long. Lemma awnless, with 5 prominent dorsal ribs; October to March
**Requirements:** Moist, well drained, fertile soils
**Comments:** A graceful fast-growing grass which flowers in the second year. It has considerable ornamental appeal used amongst ferns and by watercourses.
**Propagation:** Fresh seed
**Localities:** 49, 57
**Distribution:** Vic, NSW, Tas
**Synonym:** *Festuca dives*

# Echinopogon — Poaceae

## Echinopogon ovatus — Forest or Common Hedgehog Grass

**Size:** to 10 cm high; stems 15–40 cm high — P
**Habitat:** Riparian scrub, valley and damp sclerophyll forests, often in rocky areas
**Form:** Slender erect or ascending rhizomatous grass
**Foliage:** Very rough, flat bright green leaves, 2–16 cm x 8 mm, tapering to a point
**Flowers:** Bright green, bristly, ovoid spike-like panicle, spikelets to 5 mm long. Straight, stout awn to 15 mm long, rising from the top of the lemma, lemma bearded at the base. October to January
**Requirements:** Moist well drained soils
**Propagation:** Seed, division of rhizomes
**Localities:** Sunshine, Deer Park, Research, 26, 28, 29, 31, 32, 34A, 41, 42, 44, 50A, 55, 57, 58, 60, 62, 63
**Distribution:** All states except NT; NZ — *Illustrated p. 273*

# Eleocharis — Cyperaceae

Annual or perennial leafless sedges growing in water or in heavy soils which receive periodic flooding, spreading by creeping rhizomes or stolons. Stems are hollow and tufted with 1 or more sheathing bracts at the base (instead of leaves) and a single erect flower spike at the tip. Glumes overlap in a spiral arrangement. They provide interest planted around dams and ponds and a safe habitat for waterbirds and fish. Propagation is by seed using the bog method, or division carried out in Spring and early Summer.

*Eleocharis acuta*

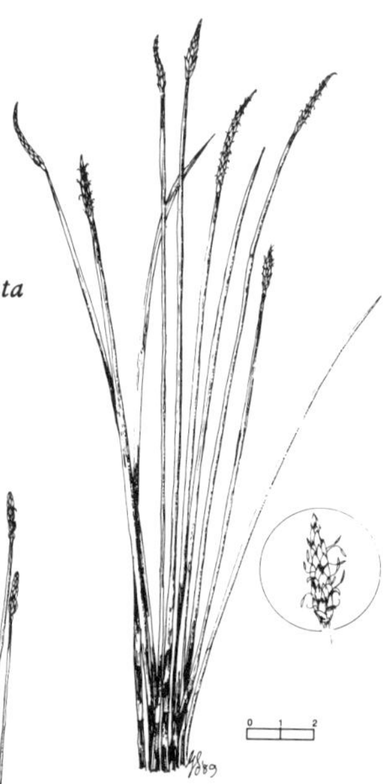

*Eleocharis gracilis*

## Eleocharis acuta — Common Spike-rush

**Size:** 0.3–0.9 m high — F
**Habitat:** Widespread in swampy areas of grassy wetlands, red gum and box woodland, riparian scrub, tea-tree heath, grassy low open forest
**Form:** Perennial aquatic herb with tufts of foliage at intervals along slender rhizomes
**Flowers:** Small linear dark brown spikelet to 30 mm x 3–7 mm; glumes ovate-lanceolate to pointed; September to April
**Comments:** Easy to establish in a dam or heavy damp soils, but may spread rapidly in shallow water.
**Localities:** The Basin, 1, 2, 4, 6, 7, 9, 14–18, 24–26, 28, 29, 34A, 35, 60, 67#, 69A–72, 74, 76
**Distribution:** All states; Norfolk Is, NZ

## Eleocharis atricha — Tuber Spike-rush

**Size:** 3–40 cm high — P
**Habitat:** Aquatic, grassy wetlands
**Form:** Tufted herb bearing small tubers along the stolons
**Flowers:** Lanceolate to linear chestnut brown spikelet to 20 mm x 3 mm; glumes oblong; January
**Comments:** The Aborigines may have eaten the tubers.
**Localities:** Laverton Nth, Keilor, 11, 60
**Distribution:** Vic, Qld, NSW, SA — *Illustrated p. 273*

## Eleocharis gracilis — Slender Spike-rush

**Size:** 20–30 cm high — P
**Habitat:** Riparian scrub and moist depressions
**Form:** Slender, erect or curved tufted stems on creeping rhizomes
**Flowers:** Ovoid spikelet to 9 mm long, glumes ovate-oblong, rounded at the tip; October to March, June
**Localities:** 16, 62, 63, 69A
**Distribution:** All states except NT; Norfolk Is, NZ

### Eleocharis minuta
Variable Spike-rush

**Size:** to 10 cm high **F**
**Habitat:** Grassy wetlands
**Flowers:** Spikelet 2–7 mm long, glumes less than 2 mm long
**Comments:** An endangered plant of Victoria which is now extinct within the Melbourne region.
**Localities:** Port Melbourne at the mouth of the Yarra R, 1892
**Distribution:** Vic, Qld, NSW; E Af and its major islands

*Eleocharis pallens*

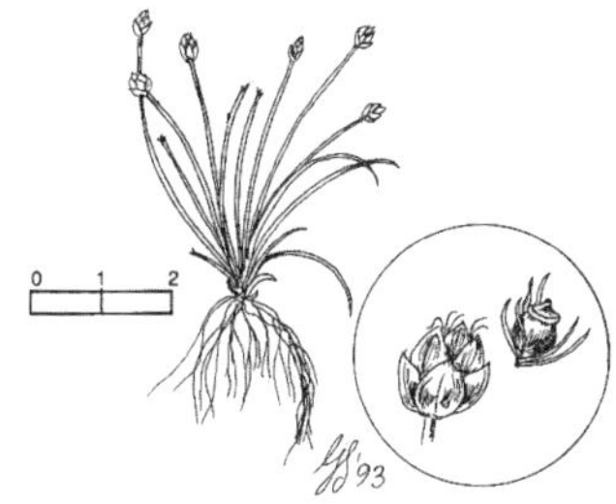
*Eleocharis minuta*

### Eleocharis pallens
Pale Spike-rush

**Size:** to 50 cm high **F P**
**Habitat:** Grassy wetlands
**Form:** Densely tufted herb with short rhizomes and prominently longitudinally ribbed, bluish culms
**Flowers:** Pale-coloured, linear cylindrical spikelet to 20 mm long; glumes shedding, ovate with triangular point; May, September
**Requirements:** Moist heavy soil, drying out in Summer
**Comments:** A vulnerable plant in Victoria, it was recently discovered in 2 sites in Melbourne, one of which was subsequently destroyed. Distinguish from *E. acuta* by the paler flowers, shorter rhizome and large pitted cells between the ribs on the stem.
**Localities:** Laverton Nth#, Rockbank
**Distribution:** All mainland states

### Eleocharis pusilla
Small Spike-rush

**Size:** 2 mm to 25 cm high **F P**
**Habitat:** Aquatic in saltmarshes, grassy wetlands, box woodland
**Form:** Tiny perennial herb with thread-like stems and rhizomes
**Flowers:** Tiny few-flowered ovate to linear spikelet to 7 mm long; December to April
**Comments:** Readily grown and easily controlled around margins of a pond or as a bog plant.
**Localities:** Laverton Nth, Deer Park, Sunshine, 4, 6, 11, 17#, 18, 24, 26
**Distribution:** All states; NZ

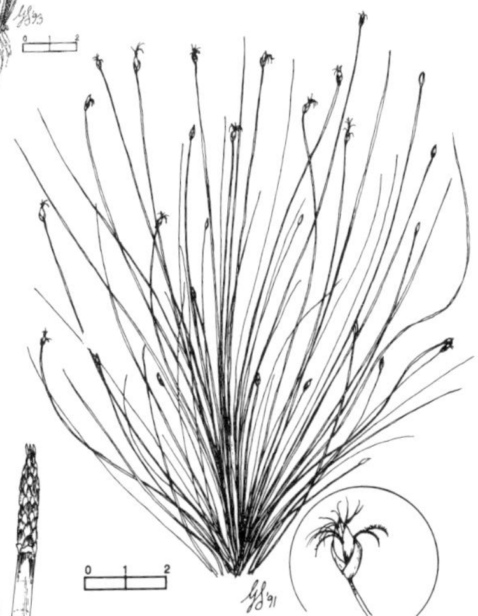
*Eleocharis pusilla*

### Eleocharis sphacelata
Tall Spike-rush

**Size:** 0.5–2 m high **F P**
**Habitat:** Shallow freshwater swamps of red gum woodland, swamp scrub
**Form:** Robust perennial aquatic herb with stout, ribbed stems to 12 mm wide along a thick woody rhizome
**Flowers:** Large cylindrical spike to 60 mm long, flowers changing from light green to light brown, glumes stiff and flattish to 8 mm long; November to February. Differs from the other species in that the spike is narrower than the stem.
**Comments:** Provides protection for fish in shallow dams or ponds but may become invasive. The Aborigines used the stems whole for making mats and split for weaving bags.
**Localities:** Moorabbin#, 29, 34A, 35, 60–63, 67#, 69A, 71, 74, 75
**Distribution:** All states; NZ, PNG

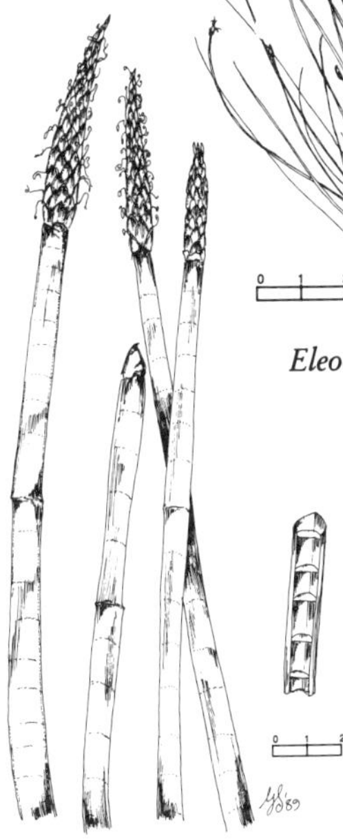
*Eleocharis sphacelata*

# Elymus
**Poaceae**

### Elymus scabrus
Common Wheat-grass

**Size:** 20 cm x 40 cm; stems to 0.8 m high **F P**
**Habitat:** Widespread throughout plains grassland, red gum woodland, dry and valley sclerophyll forests
**Form:** Open-tufted perennial grass with slender erect or drooping stems, trailing to several metres in some grassland forms
**Foliage:** Flat green or bluish leaves to 80 cm x 2–4 mm, but usually less, tapering gradually to a fine point. Upper surface and margins are rough.
**Flowers:** Slender spike with 1–10 large, distant 6–12-flowered spikelets. Lemmas taper into awns to 50 mm long which curve outwards as the spikelet matures; October to January.
**Requirements:** Adaptable to all well drained soils, including gravelly soils and alluvial clays
**Comments:** There are probably several species within *E. scabrus*. Bluish-leafed forms make attractive garden plants. Sought after by butterflies as a food source.
**Propagation:** Seed
**Localities:** St Albans, Deer Park, 2, 4, 7, 12, 13, 15–18, 21, 23–26, 28, 32–36, 38, 39, 41, 42, 44, 49, 50A, 53–55, 57, 58, 67#, 69A
**Distribution:** All states except NT; NZ
**Synonym:** *Elymus scaber*, *Agropyron scabrum*

*Elymus scabrus*

SCALE: 0 1cm 2cm The scale representation on each plant equals 2cm.

# Empodisma — Restionaceae

## Empodisma minus

Spreading Rope-rush

F P

**Size:** stems 0.5 x 2 m long
**Habitat:** Wet areas in sclerophyll woodland, swamp and wattle tea-tree scrub
**Form:** Dense, matted, perennial herb with bright green, wiry, curved branches, spreading from extensive rhizomes. May climb to 2 m when growing in dense undergrowth, developing into a tangled mass.
**Foliage:** Leafless; tiny pale green scales or sheathing bracts to 10 mm long along stems
**Flowers:** Dioecious; brown spikelets are sessile, axillary, solitary and distant, male spikelets 2–3-flowered to 4 mm long with 2 linear floral bracts and loose, narrow glumes; female spikelets single flowered with longer tight rigid glumes. September to March
**Requirements:** Ample moisture in poorly drained soils
**Propagation:** Seed, possibly division
**Localities:** Montrose, 50, 54, 56, 59, 60, 67#, 73–75
**Distribution:** Vic, Qld, NSW, Tas, SA; NZ
**Synonym:** *Calorophus lateriflorus*

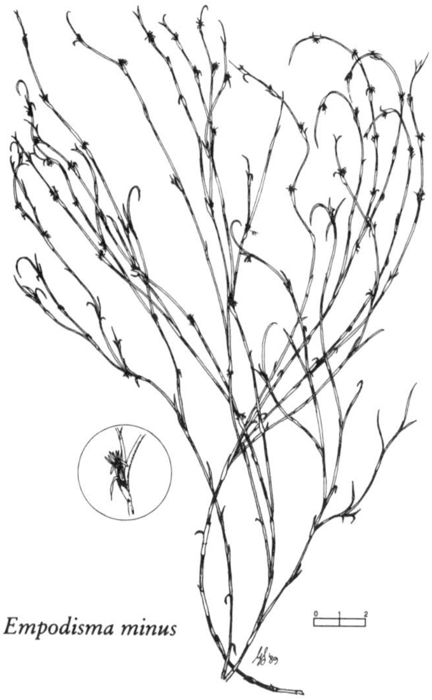

*Empodisma minus*

# Enneapogon — Poaceae

## Enneapogon nigricans

Niggerheads

F

**Size:** 20 cm x 30 cm; stems to 40 cm high
**Habitat:** Plains grassland, box woodland
**Form:** Tufted perennial grass, forming sparse tussocks
**Foliage:** Grey-green, downy, stiff narrow leaves to 15 cm x 2–3 mm, usually inrolled
**Flowers:** Dense terminal spike-like panicle 30–50 mm long, spikelets 3–4-flowered, becoming fluffy and dark purplish or black as seed matures and finally straw-coloured. Lemmas covered in silky hairs, with thin vertical ribs, each 9-awned, awns to 6 mm long. Mostly Summer, sporadically throughout the year.
**Requirements:** Dry, well drained soils, preferring exposed positions and no competition
**Comments:** Although plants are slow growing and short-lived, they are most attractive. Flowerheads are an ornamental feature.
**Propagation:** Seed, division
**Localities:** Sunbury, Essendon, Melton, Rockbank, Deer Park, 9, 16, 25, 26
**Distribution:** Vic, Qld, NSW, SA, WA

*Enneapogon nigricans*

# Entolasia — Poaceae

## Entolasia marginata

Bordered Panic

F P

**Size:** to 0.3 m high
**Habitat:** Grassy low open forest
**Form:** Wiry, erect to decumbent perennial grass with straggly stems which often branch
**Foliage:** Soft, flat lanceolate leaves 3–10 cm long, finely ribbed with numerous rough veins and pale, nerve-like margins
**Flowers:** Very narrow panicle to 75 mm long with few erect or slightly spreading branches, spikelets crowded. Upper glume prominently nerved, fertile lemma covered in silky hairs. October to December
**Requirements:** Tolerates infertile soils
**Comments:** Locally rare
**Propagation:** Seed, which may need to be stored first
**Localities:** 74
**Distribution:** Vic, Qld, NSW; PNG

## Entolasia stricta

F P

**Size:** to 0.6 m high
**Habitat:** Grassy low open forest
**Requirements:** Sandy soils
**Comments:** Differs from *E. marginata* in that stems are mostly erect and leaves are narrow and inrolled. Panicles are narrower with shorter branches.
**Propagation:** Seed
**Localities:** 74
**Distribution:** Vic
**Synonym:** Recently separated from *E. marginata*

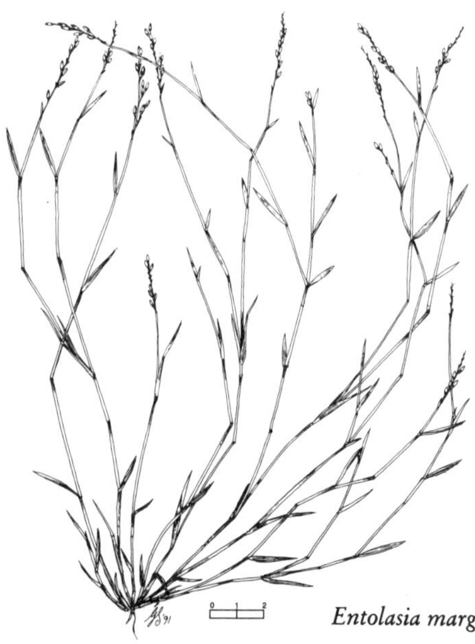

*Entolasia marginata*

# Eragrostis — Poaceae

## Eragrostis brownii — Common or Brown's Love-grass

**Size:** to 15 cm x 30 cm; stems to 40 cm high **F P**
**Habitat:** Widespread and common in riparian scrub, red gum woodland, dry and valley sclerophyll forests, tea-tree heath
**Form:** Coarse, tufted perennial grass, leaves spreading rather than erect
**Foliage:** Slender, flat, bright green leaves to 80 mm long, margins smooth
**Flowers:** Very open pyramid-shaped panicle 2–20 cm long, branches often widely spaced. Spikelets dense, to 15-flowered, olive green to greyish, thin and flat or curved with a longitudinal furrow between the 2 rows of flowers. September to April
**Requirements:** Adaptable to most soils
**Comments:** A decorative grass for rockeries, especially when well watered. Suitable for use as a low maintenance lawn in areas not subject to heavy use.
**Propagation:** Seed
**Localities:** 16, 21, 25, 26, 28, 34A–36, 39, 41, 42, 44–47, 50A, 51, 54–56, 58, 60, 65–71, 74, 75
**Distribution:** All states

*Eragrostis brownii*

## Eragrostis elongata — Close-headed Love-grass

**Size:** 10–20 cm high; stems 30–65 cm high **F P**
**Habitat:** Grassy wetlands
**Form:** Tufted perennial grass
**Foliage:** Leaves inrolled or flat, to 20 cm long, margins and upper surface rough
**Flowers:** Narrow, cylindrical spike-like panicle to 20 cm. Flat, white to purplish spikelets are sessile in dense clusters. December to March
**Requirements:** Moist heavy soils tolerating inundation
**Comments:** An isolated occurrence in Melbourne, more commonly found north of the Divide.
**Propagation:** Seed
**Localities:** Laverton Nth
**Distribution:** All mainland states; PNG

*Eragrostis elongata*

## Eragrostis infecunda — Cane Grass

**Size:** to 0.7 m high **F**
**Habitat:** Grassy wetland
**Form:** Rhizomatous or stoloniferous wiry perennial grass
**Foliage:** Stiff, inrolled or folded leaves to 100 mm long, margins thickened and rough
**Flowers:** Panicle narrow, elliptical-lanceolate to 10 cm long, spikelets very loosely 5–11-flowered, the small stem visible between the florets. December to March
**Requirements:** Tolerates saline clays, heavy soils and temporary inundation
**Propagation:** Seed, division
**Localities:** Altona
**Distribution:** Vic, SA *Illustrated p. 273*

## Festuca hookeriana = Austrofestuca hookeriana

## Festuca littoralis = Austrofestuca littoralis

# Gahnia — Cyperaceae

*Gahnia* species are perennial sedges with woody rhizomes which may form very large tussocks. They are generally found in situations where the soil is moist or wet for much of the year, and often subject to periodic inundation. Leaves are strap-shaped and/or inrolled, sheathing around the stem and spirally arranged up the stem. Leaf margins, and sometimes longitudinal ridges, have very sharp bristles on some species. The terminal dark brown to black panicle is composed of many branched spikes each with a leaf-like bract. The last 2 glumes of each spikelet contain flowers, the others are empty. Flowers are cream and fruit are hard nuts.

The Aborigines ground the nuts to provide food. Gahnias are important, (sometimes exclusive) food plants for caterpillars of several butterflies and provide safe habitat for small birds. Both foliage and flower spikes provide attractive contrast in cut and dried floral arrangements. Plants are useful in stabilising soil and creek erosion but may be difficult to propagate. Seed is the most successful, division and transplanting having limited success with some species.

SCALE: 0 1cm 2cm The scale representation on each plant equals 2cm.

Gahnia filum

Gahnia radula

Gahnia trifida

## Gahnia clarkei

Brickmaker's Sedge, Tall Saw-sedge

**Size:** 1.5–4.5 m x 1–3 m **P N**
**Habitat:** Damp and valley sclerophyll forests
**Form:** Tall perennial palm-like tussock, stems woody to 30 mm wide
**Foliage:** Very long, narrow, inrolled leaves with sharp margins
**Flowers:** Panicle initially golden brown, turning dark brown, to 60 cm x 60 mm. Numerous spikelets have more than 12 tightly overlapping, blunt glumes, the lower ones without awn-like points. The upper glumes contain 2 flowers. Nuts shiny bright red. September to February
**Requirements:** Moist to wet soils, tolerating permanent inundation
**Comments:** Slow growing but can eventually form impenetrable thickets. Becoming very rare in Melbourne with only 1 plant remaining in one location.
**Localities:** 51, 58, 67#
**Distribution:** Vic, Qld, NSW, SA

## Gahnia filum

Chaffy Saw-sedge

**Size:** 1–1.2 m high **F**
**Habitat:** Coastal saltmarshes and grassy wetlands
**Form:** Perennial leafy tussock
**Foliage:** Long, inrolled leaves tapering into slightly rough points; lower leaf-like bracts smooth, reddish-purple below, smaller floral bracts very broad and chaff-like
**Flowers:** Narrow pale brown panicle to 20–35 cm x 10–20 mm. Densely clustered, 1-flowered spikelets most of the year. Nuts pale brown.
**Requirements:** Moist sandy soil, tolerating saline situations
**Comments:** Attractive fruits
**Localities:** Seaholme, 3–5, 74
**Distribution:** Vic, NSW, Tas, SA, WA

## Gahnia radula

Thatch Saw-sedge

**Size:** 1–2 m x 0.5–2 m **F P**
**Habitat:** Dry and valley sclerophyll forests, sclerophyll woodland, grassy low open forest, wattle tea-tree scrub
**Form:** Clumping perennial herb spreading by underground rhizomes
**Foliage:** Long, flat, raspy, deep green leaves with very sharp inrolled margins
**Flowers:** Ovate, loosely-arranged chestnut brown panicle to 60 cm x 70 mm, becoming black and pendulous; most of the year. Nuts dark brown to black.
**Requirements:** Moist soils, tolerating dryness once established
**Comments:** Plants are tolerant of *Phytophthora cinnamomi.* The vigorous rhizomes are useful in controlling soil erosion. The foliage has been used for roof thatching in the past.
**Propagation:** Difficult to propagate as it sets little seed and transplants do not adapt well. Young plants seem to suffer transplanting the least provided minimum root damage occurs.
**Localities:** 26, 34A, 37, 40–44, 46–49, 50A, 51, 53–56, 59–67, 70, 71, 73–76
**Distribution:** Vic, Qld, NSW, Tas, SA

## Gahnia sieberiana

Red-fruited Saw-sedge

**Size:** 1.5–3 m x 2–3 m **F P N**
**Habitat:** Valley and damp sclerophyll forest, wattle tea-tree scrub
**Form:** Clumping perennial sedge
**Foliage:** Long, channelled, rough spreading leaves, becoming pendulous
**Flowers:** Dense, yellowish-brown compound panicle becoming almost black, to 65 cm x 70 mm. Spikelets differ from *G. clarkei* in that there are up to 10 loosely overlapping pointed glumes, all with awns. Bright shining red nuts dot the panicle. October to January.
**Requirements:** Moist soils for most of the year, but tolerates drier soils once established
**Propagaton:** Easily grown from seed which has been stored for 12 months.
**Localities:** 50–51, 53–59, 70, 71, 73–75
**Distribution:** Vic, Qld, NSW, Tas, SA; PNG, N Cal

*Illustrated p. 273*

## Gahnia trifida

Coast Saw-sedge

**Size:** 0.6–1.5 m x 0.6–1 m **F P**
**Habitat:** Coastal, in moist tea-tree heath
**Form:** Upright perennial clumping herb
**Foliage:** Long narrow leaves with very rough inrolled margins and pendulous, thin tips
**Flowers:** Erect, narrow, interrupted dark brown panicle to 40 cm long with densely clustered 1-flowered spikelets; October to December. Nut shiny and black.
**Requirements:** Moist well drained sandy soils, tolerating extended dry periods once established and alkaline soils
**Comments:** Differs from *G. filum* in that lower leaf bracts are sharply toothed and the smaller floral bracts are narrow. Flower stems are cut commercially.
**Propagation:** Seed or division but difficult to establish
**Localities:** 74
**Distribution:** Vic, Tas, SA, WA

# Glyceria **Poaceae**

## Glyceria australis
Australian Sweet-grass, Manna Grass

**Size:** 0.6–1 m x spreading **F P**
**Habitat:** Freshwater swamps and watercourses of riparian scrub, valley sclerophyll forest, grassy low open forest
**Form:** Perennial grass with individual stems arising from creeping rhizomes, lower part of the stems usually horizontal and rooting
**Foliage:** Flat leaves with rough edges and prominent veins
**Flowers:** Narrow panicle to 15 cm long with few pale green 6–20-flowered spikelets. Lemmas awnless with 7 distinct, parallel veins and tapered tips. October to February
**Requirements:** Moist or wet soils
**Comments:** May be used as a semi-aquatic grass in swampy ground. Easily confused with the introduced *G. declinata*, which has ragged-toothed tips of the lemmas.
**Propagation:** Seed, division
**Localities:** Yarra R, Plenty R, 35, 41, 42, 44, 51, 56, 74, 76
**Distribution:** Vic, Qld, NSW, Tas, SA

*Glyceria australis*

# Hemarthria **Poaceae**

## Hemarthria uncinata
Mat Grass

**Size:** to 40 cm x spreading **F P**
**Habitat:** Red gum woodland, swamp scrub, valley sclerophyll forest, tea-tree heath around swamps
**Form:** Perennial, rigid, matted grassy sward
**Foliage:** Flat, narrow-linear leaves to 20 cm long, keeled along the midrib
**Flowers:** Terminal, rigid, often curved spike to 130 mm long; 1-flowered spikelets paired, to 10 mm long, in 4 indistinct rows, appearing sessile, 1 of each pair in a hollow along the rhachis; December to February
**Requirements:** Moist often heavy clay soils
**Comments:** Vigorous plants may grow to 1 m high on sheltered sites, and spread may be up to 3 m. May be confused with the introduced *Hainardia* and *Parapholis* grasses, whose spikelets form 2 distinct rows on opposite sides of the stems.
**Propagation:** Seed, division
**Localities:** 25, 28, 34A–36, 51, 53, 54, 60, 69A–71, 74, 76
**Distribution:** All states except NT

*Hemarthria uncinata*

# Homopholis **Poaceae**

## Homopholis proluta
Rigid Panic

**Size:** 0.2–1 m high **F P**
**Habitat:** Red gum woodland, plains grassland, in low-lying areas
**Form:** Tufted, erect, leafy perennial grass
**Foliage:** Green or blue-green flat leaves (sometimes with margins inrolled), 2–15 cm x 2–7 mm
**Flowers:** Much branched, loose, very open panicle to 20 cm long, lower branches whorled. Short pointed spikelets to 4 mm long, greenish, tinged purple, on long thread-like stalks. September to April. The lower glume is as long as the spikelet, separating this genus from *Panicum* where the glume is never more than half as long.
**Requirements:** Heavy clay soils, plants forming extensive dense stands after wet winters. Drought resistant once established
**Comments:** Seeds are gathered by Aborigines and ground to make a kind of flour. Attractive used in dry arrangements.
**Propagation:** Seed
**Localities:** St. Albans, Laverton Nth, 4, 7, 11, 16–18, 21, 26
**Distribution:** Vic, Qld, NSW, SA, WA
**Synonym:** *Panicum prolutum* *Illustrated p. 273*

# Hypolaena **Restionaceae**

## Hypolaena fastigiata
Tassel Rope-rush

**Size:** 20–50 cm x 0.5–1.5 m **F P**
**Habitat:** Tea-tree heath
**Form:** Spreading grey-green rush-like rhizomatous perennial herb with slender, erect, much branched stems bearing up to 12 distinct longitudinal mealy-white ridges
**Foliage:** Leaves reduced to pointed dark sheathing bracts to 10 mm long
**Flowers:** Dioecious; many, stalked, pendulous reddish-brown male spikelets to 8 mm long, in a loose terminal panicle; few, solitary erect 1-flowered female spikelets to 10 mm long. Flowers and/or fruit most of the year.
**Requirements:** Moist sandy soils
**Comments:** Attactive foliage plant beside pools.
**Propagation:** Division of the creeping rhizome
**Localities:** 67–70, 73–75
**Distribution:** All states except NT

*Hypolaena fastigata*

SCALE: 0 1cm 2cm The scale representation on each plant equals 2cm.

*Imperata cylindrica*

# Imperata — Poaceae

## Imperata cylindrica

Blady Grass

**Size:** 0.3–0.7 m high; stems to 1 m high — **F P**
**Habitat:** Grassy wetland, damp, valley and dry sclerophyll forests, tea-tree heath
**Form:** Erect coarse, vigorously spreading, rhizomatous perennial grass
**Foliage:** Pale mottled green linear leaves to 70 cm long, often with reddish parts, with a prominent midrib. The margins having tiny serrations.
**Flowers:** Dense, fluffy, cylindrical spike-like panicle to 20 cm long. Paired spikelets have stalks of different lengths, the whole spikelet covered with long, silvery silky hairs; November to March
**Requirements:** Moist, well drained soils
**Comments:** The seed heads are extremely ornamental but are mainly seen after fire or slashing. Regenerates rapidly after fires and may be unwelcome in cultivated areas; however it provides an important habitat for many small animals, and should be maintained in remnant vegetation.
**Propagation:** Seed, division
**Localities:** Altona, 5, 50–51, 54–56, 59, 60, 74, 75
**Distribution:** All states; NZ, PNG, N Cal, Asia

# Isolepis — Cyperaceae

Annual or perennial sedges, *Isolepis* spp. may be rhizomatous, tufted, single-stemmed, creeping or floating. They are usually found in damp, swampy ground. Leaves are basal and are either small or reduced to sheaths on the nodeless stem. The brownish-yellow flowers are generally borne in dense terminal heads of 1 to many spikelets, a stiff erect terminal bract continuing the stem. Fruit are nuts. Some species are suited to garden culture, particularly in bog gardens. Propagation is by seed or division of the clumps.
**Synonym:** Previously included in the genus *Scirpus*

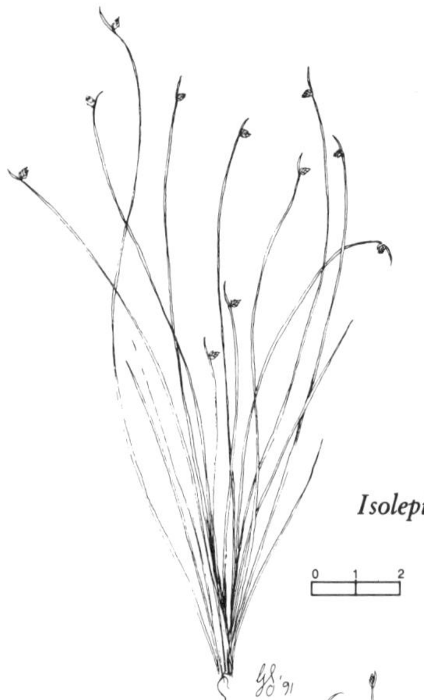

*Isolepis cernua*

## Isolepis cernua

Nodding, Low or Grassy Club-rush

**Size:** to 10 cm high — **F P**
**Habitat:** Grassy wetlands, saltmarshes, riparian scrub and woodland, frequently coastal
**Form:** Small erect tufted annual rush with leaves reduced to a small pointed sheath or non-existent; floral bract glume-like to 7 mm long
**Flowers:** Solitary broadly ovoid spikelet to 3 mm long; most of the year. Nuts are flattened and become dark brown or black on maturity.
**Localities:** 2, 4, 5, 8, 14, 16, 35, 36, 63, 67#, 69A, 70, 74
**Distribution:** All states except NT; most of the world
**Synonym:** *Scirpus cernuus*

*Isolepis fluitans*

## Isolepis fluitans

Floating Club-rush

**Size:** to 15 cm high — **P**
**Habitat:** Swamps or aquatic in tea-tree heath, valley sclerophyll forest
**Form:** Weak, many-noded, fine-stemmed perennial rush, creeping or floating in water, or tussock-forming on drier ground
**Foliage:** Pale green thread-like leaves, shorter than stems
**Flowers:** Solitary ovoid spikelet to 5 mm long on a long stalk, glumes pale green to straw-coloured; October to February. Nut is ovoid.
**Comments:** An attractive feature in a water garden.
**Localities:** Research, 34A, 35, 46, 47, 54, 70, 72, 74–76
**Distribution:** All states except NT; Eur, Af, Asia
**Synonym:** *Scirpus fluitans*

## Isolepis hookeriana

**Size:** to 15 cm high, often smaller — **P**
**Habitat:** Swamp scrub, depressions in grassy wetlands, valley sclerophyll forest
**Form:** Small tufted annual rush with fine, erect, bristly stems, leaves very fine to 3 mm long but usually absent
**Flowers:** Cluster of 1–2 slanted, oblong, several-flowered spikelets, to 3.5 mm long; September to February. Nut straw coloured to brown or black, longitudinally ridged and wrinkled.
**Localities:** Laverton Nth, Montrose, 5, 31, 44, 46, 50A, 54, 74, 75
**Distribution:** All states except NT
**Synonym:** *Scirpus hookeranus*, *S. calocarpus*

*Illustrated p. 273*

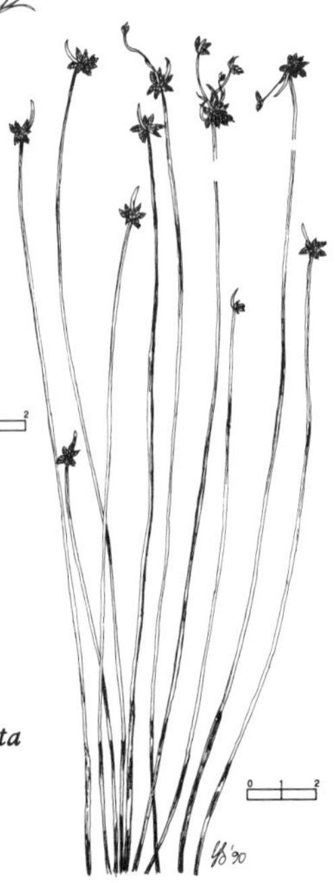

*Isolepis inundata*

## Isolepis inundata

Swamp Club-rush

**Size:** 5–30 cm x 10–40 cm — **P**
**Habitat:** Wattle tea-tree and riparian scrub, valley and damp sclerophyll forests
**Form:** Tufted perennial rush, stems stiffly erect or arching, with offshoots proliferating from flowerheads; leaves reduced to a point, erect bract longer than spikelets
**Flowers:** Cluster of 3–12 oblong, angled spikelets to 6 mm long, often becoming dark purplish brown. Glumes have a green keel; October to April
**Comments:** A widespread and useful tufting plant for moist to wet soils, tolerating periodic inundation, but can be a problem in nurseries.
**Localities:** 8, 14, 16, 34A–36, 42, 44, 50–58, 60, 63, 64, 67#, 69A, 72, 74–76
**Distribution:** All states except NT; NZ, S Am, SE Asia
**Synonym:** *Scirpus inundatus*

## Isolepis marginata

Tiny or Coarse Club-rush

**Size:** 25–100 mm high **F P**
**Habitat:** Wattle tea-tree and riparian scrub, tea-tree heath, saltmarshes, valley sclerophyll forest
**Form:** Tiny tufted leafy annual with bristly stems
**Foliage:** Well developed bristly leaves shorter than the stems; floral bract erect or spreading, longer than spikelets
**Flowers:** Single sessile cluster of 1–6 spikelets to 6 mm long. Rigid green glumes boat-shaped with a bright reddish or purplish-brown patch on sides; September to January
**Comments:** An adaptable species which tolerates drier conditions than most *Isolepis* spp. Often appears in gardens.
**Localities:** Montrose, The Basin, 2, 4, 12, 24, 26, 29, 32, 34A, 35, 44, 47, 48, 50A–52, 54–56, 69–71, 74, 75
**Distribution:5** Vic, NSW, Tas, SA, WA; NZ, S Af
**Synonym:** *Scirpus marginatus*

*Isolepis marginata*

## Isolepis nodosa

Knobby Club-rush

**Size:** 0.5–1.5 m x 0.6–2 m **F P**
**Habitat:** Coastal banksia woodland, riparian, swamp and primary dune scrub, saltmarshes, tea-tree heath, grassy low open forest
**Form:** Tall, coarse, wiry densely tufted perennial rush with creeping rhizomes, leaves reduced to sheaths
**Flowers:** Dense globular cluster to 20 mm wide of sessile brown spikelets to 5 mm long; most of the year. Short erect floral bract behind flowerhead appears to continue the stem making the cluster lateral.
**Requirements:** Moist soils, tolerating dry periods once established
**Comments:** Widespread in Melbourne, its upright habit provides interest in the landscape as a contrast plant. It is of some benefit in binding soils in moist areas.
**Localities:** 2–4, 8, 14, 26, 28, 31, 32, 61–64, 67–69A, 71, 72, 74, 76
**Distribution:** All states except NT; NZ, S Af, S Am
**Synonym:** *Scirpus nodosus*

*Isolepis nodosa*

## Isolepis platycarpa

Club-rush

**Size:** to 70 mm high **F P**
**Habitat:** Swamp and riparian scrub, damp sclerophyll forest
**Form:** Tiny tufted annual rush
**Comments:** Similar to *I. hookeriana* but leaves are short and well developed, to 30 mm long. Spikelets are often solitary, but sometimes paired, broader and less angled than *I. hookeriana*. Nut is shiny dark brown and wrinkled but not ridged. September to April.
**Localities:** The Basin, Montrose, 17#, 18, 28, 50A, 54, 58, 70, 72, 74
**Distribution:** Vic, NSW, Tas, SA, WA; NZ
**Synonym:** *Scirpus platycarpus*

## Isolepis producta

**P**

**Habitat:** Aquatic, in riparian scrub
**Comments:** Similar to *I. fluitans* differing in that *I. producta* has purplish leaf sheaths and glumes, and stems are always floating. Flowers October to March
**Localities:** Yarra R., 36
**Distribution:** Vic, NSW, Tas, SA, WA
**Synonym:** *Scirpus productus*

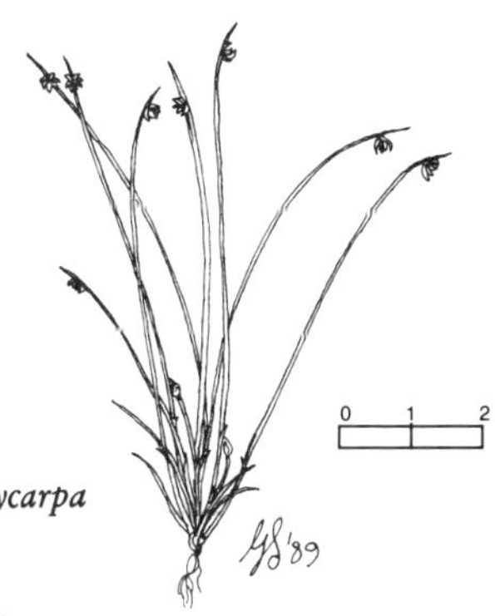

*Isolepis platycarpa*

## Isolepis stellata

Star Club-rush

**Size:** 60–100 mm high **P**
**Habitat:** Valley sclerophyll forest
**Form:** Slender tufted annual rush with rigid stems
**Foliage:** Leaves bristly to 25 mm long
**Flowers:** Star-like clusters of 3–8 densely packed greenish spikelets to 4 mm long. Glumes spreading with broad keel and sharp point; October to February. Nut triangular and black
**Localities:** 62, 75
**Distribution:** Vic, NSW, Tas, SA, WA
**Synonym:** *Scirpus stellatus*

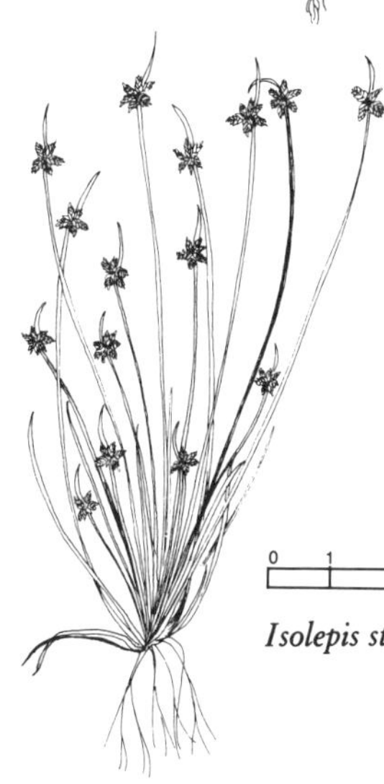

*Isolepis stellata*

## Isolepis subtilissima

**Size:** 5–12 cm high **P**
**Habitat:** Wet and valley sclerophyll forests
**Form:** Tufted bristly perennial rush with very slender stems and long-creeping rhizomes
**Foliage:** Very narrow channelled leaves sometimes longer than stems; floral bract erect and leaf-like to 30 mm long
**Flowers:** Single, or 2–3 ovoid, reddish spikelets to 3 mm long; March
**Localities:** 44, 57, 74, 76
**Distribution:** Vic, Qld, NSW, Tas, SA; NZ, PNG, Philippines
**Synonym:** *Scirpus subtilissimus*

*Illustrated p. 273*

SCALE: 0 1cm 2cm The scale representation on each plant equals 2cm.

*Isolepis victoriensis*

### Isolepis victoriensis

**Size:** 6–12 cm high **F**
**Habitat:** Damp depressions in plains grassland, box woodland
**Comments:** Tufted annual rush. Heads of 3–4 spikelets to 5 mm long with strongly keeled glumes; August to November. Nut globular and pale grey.
**Localities:** Werribee[#], Campbellfield[#], Rockbank, 7, 17[#], 18, 67[#]
**Distribution:** Endemic to Vic
**Synonym:** *Scirpus victoriensis*

## Juncus — Juncaceae

*Juncus* species are common plants of moist areas, and are often found in swamps and along watercourses. Several species are often found growing together. Many are adapted to growing in temporarily inundated areas, some being tolerant of saline conditions. They may be annual or perennial (perennials have a short-creeping rhizome), native or exotic, and mostly form robust tussocks of green stems. Flowers occur in terminal cymes or clusters varying in colour from greenish through browns to purple-black. The flowers usually appear to occur on one side of the stem, due to an erect bract which seems to continue the stem above the flowerhead. Fruit is a 3-celled capsule. Leaves are always glabrous and arranged in a basal rosette. They may be flat or terete, with or without spongy pith, or reduced to basal sheaths. *Juncus* spp. are useful in stabilising wet sloping soils, and for providing habitat for small birds and frogs as well as adding interest in a wet landscape area. Propagation is by fresh seed or division of rhizomes. Seed is produced prolifically and germinates readily. Some species may become invasive under certain conditions. Most species prefer a sunny to semi-shaded position.

*Juncus* species are difficult to tell apart. Features used here include the colour, stem length and diameter (measured midway between the rhizome and the base of the inflorescence); the appearance of the pith when cut longitudinally (solid, hollow or with air spaces); the length, colour and tightness of the basal sheaths; the arrangement of the inflorescence; the length of the capsule at maturity and the relative lengths of the capsule and the perianth (petals and sepals).

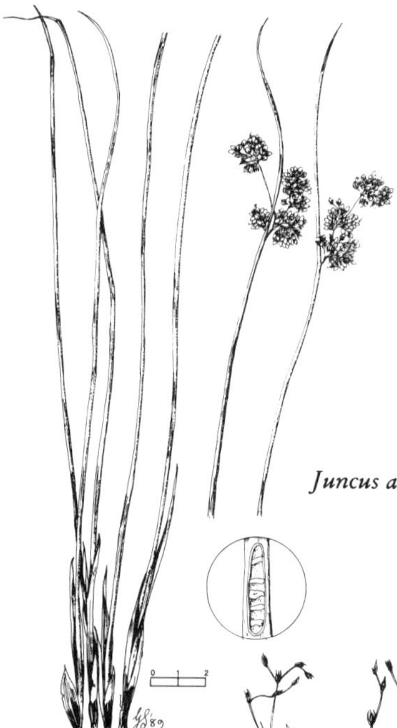

*Juncus amabilis*

### Juncus amabilis

**Size:** 0.2–1.2 m x 0.2–0.5 m **F P**
**Habitat:** Grassy wetlands, plains grassland, valley sclerophyll forest, wattle tea-tree scrub
**Form:** Rhizomatous tufted perennial rush with blue-green stems 1–2.5 mm wide, pith very open with large air spaces
**Foliage:** Leaves reduced to dark brown basal sheaths
**Flowers:** Compact panicle with light brown flowers clustered along the branches; November to February. Capsule 2–2.5 mm long, somewhat longer than the perianth.
**Comments:** Tolerates both periods of inundation and dryness once established.
**Localities:** Montrose, 16, 25, 34A, 42, 44, 47, 48, 53, 54, 56, 63, 64, 70, 71, 74, 76
**Distribution:** Vic, NSW, Tas, SA, WA

### Juncus australis — Austral Rush

**Size:** 0.6–1.2 m x 0.5–1 m **F P**
**Habitat:** Riparian scrub, valley sclerophyll forest
**Form:** Tufted perennial rush with short-creeping rhizome; stems stout, shiny, blue-green 1–3 mm wide, pith open with large air spaces
**Foliage:** Leaves reduced to shiny dark brown basal sheaths, longest sheaths 5–15 cm long, standing away from the stem and broadly dilated at the tip
**Flowers:** Compound panicles with almost globular clusters to 30 mm wide, of pale brown flowers; most of the year. Capsule usually shorter than the perianth.
**Requirements:** Prefers moist to wet soils but will tolerate short, dry periods.
**Localities:** 15, 18, 28, 34A, 36, 47
**Distribution:** Vic, NSW, Tas, SA; NZ

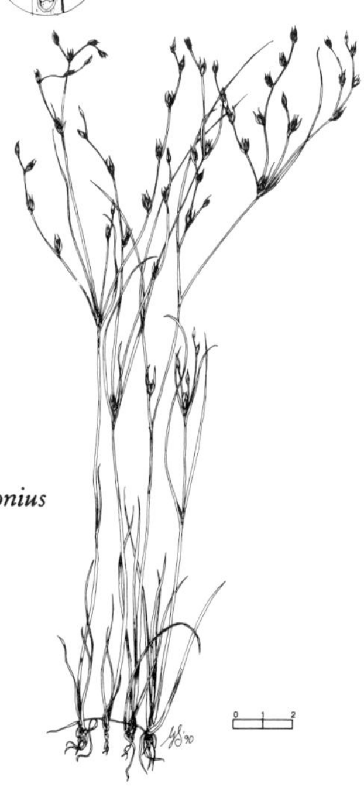

*Juncus bufonius*

### Juncus bufonius — Toad Rush

**Size:** 2–30 cm x 30 cm **F P**
**Habitat:** Widespread in grassy wetlands, plains grassland, red gum woodland, riparian and swamp scrub, wet, damp and valley sclerophyll forests, tea-tree heath
**Form:** Small slender annual herb
**Foliage:** 2–3 very narrow, solid, shiny flat channelled leaves on each stem
**Flowers:** Branched panicle to 120 mm long, of single pale green flowers scattered along the branches, bracts short and leafy; most of the year. Capsule shorter than perianth.
**Comments:** Considered by some botanists to be introduced to Australia. Often a weed, colonising anywhere sufficient soil moisture is available.
**Localities:** 2, 4, 5, 7, 12, 14–18, 24–26, 28, 31, 35, 41–44, 47, 49, 50A, 51, 53–58, 60, 61, 64, 67, 69A–72, 74, 75
**Distribution:** All mainland states; cosmopolitan

*Dichelachne sieberiana*

*Echinopogon ovatus*

*Eleocharis atricha*

*Eragrostis infecunda*

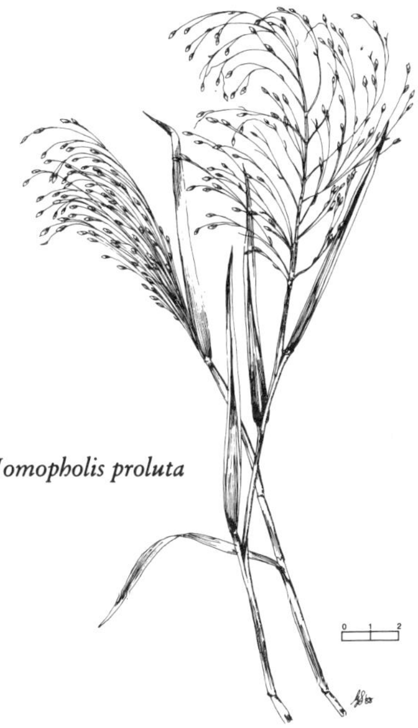

*Homopholis proluta*

*Gahnia sieberiana*

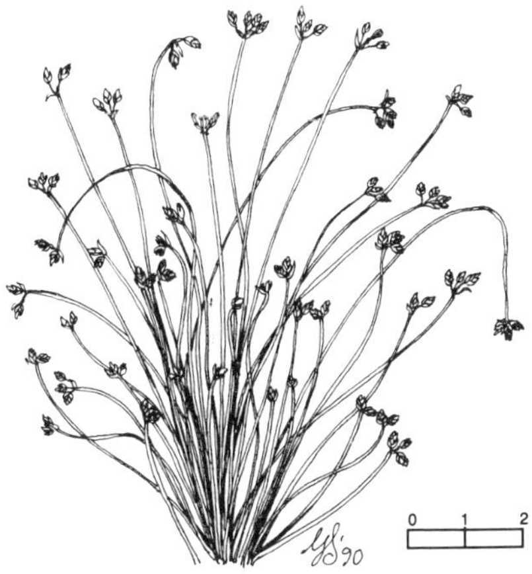

*Isolepis hookeriana*

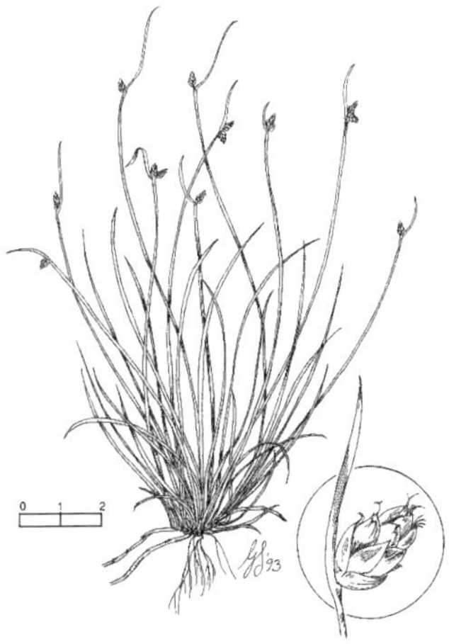

*Isolepis subtilissima*

*Juncus flavidus*

SCALE: 0 1cm 2cm — The scale representation on each plant equals 2cm.

*Juncus caespiticius*

*Juncus filicaulis*

## Juncus caespiticius

Grassy or Tufted Rush

**Size:** 10–40 cm x 20–50 cm — F P
**Habitat:** Marshy soaks in riparian scrub, tea-tree heath
**Form:** Rhizomatous tufted perennial rush
**Foliage:** Solid, flat narrow incurved leaves to 20 cm x 2–4 mm, tapering to fine points
**Flowers:** Cyme of up to 8 globular heads of brownish flowers; most of the year. Leafy bracts up to the length of the inflorescence.
**Localities:** 16, 29, 67, 69, 70, 74
**Distribution:** Vic, NSW, Tas, SA, WA

## Juncus filicaulis

Thread Rush

**Size:** 0.4–0.7 m x 0.3–0.5 m — F P
**Habitat:** Red gum woodland
**Form:** Tufted rhizomatous perennial herb with slender blue-green stems to 1.2 mm wide, pith cobwebby with very small air spaces
**Foliage:** Basal sheaths pale brown to 50 mm long, closely pressed to the stem
**Flowers:** Panicle of 1 to few small, almost round straw-coloured flower clusters; most of the year
**Requirements:** Moist soils, tolerating dry periods once established
**Localities:** Yarra R., 25, 26, 69A
**Distribution:** Vic, NSW, Tas, WA

## Juncus flavidus

**Size:** 0.4–1.2 m x 0.2–1 m — F P
**Habitat:** Saltmarshes, grassy wetland, box and red gum woodlands
**Form:** Tufted rhizomatous perennial rush with yellow-green stems to 3 mm wide, pith interrupted by small air spaces
**Foliage:** Basal sheaths dark reddish-brown
**Flowers:** Panicle with clusters of pale yellow flowers along branches, becoming bright yellow as they mature; November to May
**Comments:** More tolerant of dry soils than other *Juncus* spp.
**Localities:** Basalt plains, 2, 4–6, 17#, 18, 25, 34A, 47, 48, 69A, 70
**Distribution:** All states except NT

*Illustrated p. 273*

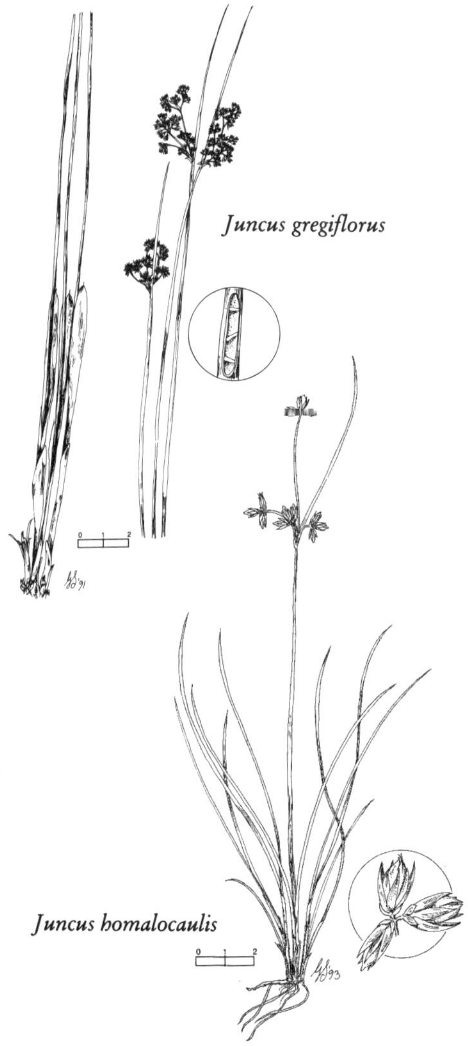

*Juncus gregiflorus*

*Juncus homalocaulis*

## Juncus gregiflorus

**Size:** 0.5–1.4 m x 0.6–1.5 m — P
**Habitat:** Valley and damp sclerophyll forest, riparian scrub
**Form:** Tufted rhizomatous perennial rush with bright green shining stems to 3 mm wide, pith interrupted
**Foliage:** Basal sheaths dark brown to dark reddish or purplish brown, top sheath usually loose
**Flowers:** Panicle of pale brown clustered flowers; November to March. Capsule flat-topped.
**Localities:** Montrose, 12, 24, 26, 34A, 36, 44, 50A, 53, 56, 63, 74
**Distribution:** Vic, NSW, Tas

## Juncus holoschoenus

Joint-leaved Rush

**Size:** 20–50 cm x 10–40 cm — F
**Habitat:** Grassy wetland, red gum woodland, valley sclerophyll forest, swamp scrub, tea-tree heath
**Form:** Open clumping rhizomatous perennial rush with leafy stems to 3 mm wide
**Foliage:** Many hollow, segmented, usually terete leaves as long as the stems
**Flowers:** Loose cymose panicle with 4–8 clusters each with 10–20 pale green flowers; November to March. Leafy bracts shorter than panicle.
**Comments:** A common, widespead plant often colonising garden areas. Similar to *J. prismatocarpus* which has multi-tubular partitions in the flattened leaves.
**Localities:** Research, 4, 6, 17#, 18, 25, 28, 34A, 35, 42, 44, 50A, 54–56, 58, 61–64, 69A–72, 74, 76
**Distribution:** All states except NT; NZ

*Illustrated p. 286*

## Juncus homalocaulis

Wiry Rush

**Size:** 2–15 cm x 5–30 cm — F
**Habitat:** Saltmarshes, grassy wetland, box woodland
**Form:** Dwarf, slender rhizomatous perennial rush, old leaf sheaths remaining around the dark green stems
**Foliage:** Very fine, solid, slightly flattened leaves to 20 cm long
**Flowers:** Few-flowered cyme of 1 central cluster, often with another cluster on the ends of 2 short recurved branches; 2–6 flowers per cluster; November to March. Capsule light reddish-brown, shorter than perianth. 2 leafy bracts, 1 longer than the inflorescence.
**Localities:** Dandenong Rges, Laverton Nth, Deer Park, 2, 11, 17#, 18, 24–26, 60
**Distribution:** All states except NT

## Juncus krausii

Sea Rush

**Size:** 0.6–2.3 m x 0.5–1.5 m F
**Habitat:** Marshes in coastal banksia woodland, primary dune scrub
**Form:** Rhizomatous perennial rush with terete stems, leaves and bracts, pith solid
**Foliage:** Few, dark green sharp-pointed leaves similar to and as long as the stems
**Flowers:** Loose or dense panicle with small clusters of dark brown flowers; November to March
**Requirements:** Brackish to saline areas
**Comments:** Plants are slow to re-establish after disturbance.
**Localities:** Footscray, 3, 4, 25, 67, 71, 72, 74
**Distribution:** All states except NT; NZ
**Synonym:** Previously included with *J. maritimus* which occurs in the Northern Hemisphere.

*Juncus krausii*

## Juncus pallidus

Pale Rush

**Size:** 0.5–2.3 m x 0.3–1 m F P
**Habitat:** Widespread in grassy low open forest, plains grassland, valley sclerophyll forest, tea-tree heath
**Form:** Tufted rhizomatous perennial rush, pith spongy and uninterrupted in pale green stems 3–8 mm wide
**Foliage:** Loose pale brown basal sheaths, darker brown below, to 15 cm long
**Flowers:** Erect, loose or dense panicles to 15 cm long with many pale yellow flowers; December to March. Capsule longer than perianth.
**Comments:** Vigorous grower adaptable to most situations, but requires periodic inundation for optimum growth.
**Localities:** Footscray, 16, 34A–36, 41, 42, 44, 46–48, 50–76
**Distribution:** Vic, NSW, Tas, SA, WA; NZ

*Juncus pallidus*

## Juncus pauciflorus

Loose-flower Rush

**Size:** 0.3–1 m x 0.2–0.6 m P
**Habitat:** Riparian and swamp scrub, valley sclerophyll forest, grassy low open forest
**Form:** Rhizomatous perennial rush forming a bright green, often arching tussock; pith spongy and continuous
**Foliage:** Tight, dark reddish brown basal sheaths
**Flowers:** Open, loose panicle to 80 mm long, with 20–60 reddish flowers (with 6 stamens) widely spaced along fine curved branches; December to March. Capsule longer than perianth.
**Comments:** An adaptable plant tolerating dryness once established.
**Localities:** 26, 27, 35, 36, 44, 47–49, 53–58, 60, 61, 63, 64, 67#, 74, 76
**Distribution:** Vic, NSW, Tas, SA, WA; NZ

*Juncus pauciflorus*

## Juncus planifolius

Broad-leaf rush

**Size:** 10–50 cm x 10–30 cm F P
**Habitat:** Valley and damp sclerophyll forests, swamp and wattle tea-tree scrub
**Form:** Annual or perennial tufting rush with slender leafless stems, lacking a rhizome
**Foliage:** Broad flat solid leaves 20–30 cm x 3–8 mm, in basal grass-like tuft
**Flowers:** Umbel-like cymes of brown flowers in many small round heads to 7 mm wide; September to March. Bract shorter than infloresence.
**Comments:** Tolerates permanently moist soils. Differs from *J. caespiticius* in that leaves are wider, and lacks rhizome. Local forms frequently die around January.
**Localities:** 35, 43, 44, 54–56, 58–61, 63, 64, 67#, 69A, 71, 74, 75
**Distribution:** All states except NT; NZ, S Am

*Juncus planifolius*

## Juncus prismatocarpus

Branching Rush

**Size:** 0.2–0.6 m x 10–30 cm P
**Habitat:** Wet areas in damp sclerophyll forest, wattle tea-tree scrub
**Form:** Prolifically-flowered rush similar to *J. holoschoenus*
**Foliage:** Leaves are flattened, hollow with multi-tubular partitions, to 25 cm long
**Flowers:** Long-branched, cymose panicles with clusters of 10–40 flowers distantly spaced, with a single small bract; November to March
**Comments:** Distinguished from *J. holoschoenus* by flattened leaves.
**Localities:** 58, 63, 67#
**Distribution:** All states except NT; NZ, Asia

*Illustrated p. 286*

SCALE: 0 1cm 2cm The scale representation on each plant equals 2cm.

*Juncus procerus*

## Juncus procerus

**Size:** 1–2 m x 0.6–1.5 m **P**
**Habitat:** Damp and valley sclerophyll forests
**Form:** Tufted rhizomatous perennial herb with soft, thick stems; large air spaces in pith (almost hollow), stems split lengthwise when compressed
**Foliage:** Large, broad, brown basal sheaths to 15 cm long, loosely surrounding the stem
**Flowers:** Compact panicle of single brown flowers; October to February. Capsule longer than perianth.
**Requirements:** Damp, well drained soils in sheltered situations
**Localities:** 34A, 35, 55, 56, 62, 63, 74, 76
**Distribution:** Vic, NSW, Tas, SA, WA; NZ

## Juncus radula — Hoary Rush

**Size:** 0.3–0.7 m x 0.2–0.6 m **F P**
**Habitat:** Plains grassland, red gum woodland, dry sclerophyll forest
**Form:** Tufted rhizomatous perennial rush with rough stems, branches of infloresences and flowers; greyish stems slender to 2.5 mm wide, pith solid
**Foliage:** Basal sheaths pale brown
**Flowers:** Loose panicle of single pale brown flowers spaced along the branches; October to March.
**Comments:** Often occurs in drier areas subject to periodic inundation.
**Localities:** Laverton Nth, Deer Park, 2, 5, 12, 17#, 18, 26, 35, 39
**Distribution:** Vic, NSW, SA, WA

*Juncus radula*

## Juncus revolutus — Creeping Rush

**Size:** 5–30 cm x 0.5–1.5 m **F**
**Habitat:** Saltmarshes
**Form:** Perennial rush, with small tufts at intervals along creeping rhizome, stems sometimes with 1 stem-leaf
**Foliage:** Solid, thin, flattened leaves to 25 cm long, margins incurved
**Flowers:** Short one-sided racemes of pale brown flowers, singly or in groups of 2–3 along branches; November to March
**Comments:** A rare plant in the Melbourne region, it has applications for areas with saline soil.
**Localities:** Mouth of Yarra R, 2, 4
**Distribution:** Vic, NSW, Tas

## Juncus sarophorus

**Size:** 0.6–2 m x 0.5–1 m **P**
**Habitat:** Wattle tea-tree and riparian scrub, valley sclerophyll forest
**Form:** Tufted rhizomatous perennial rush with hard blue-green stems, pith interrupted by air spaces
**Foliage:** Basal sheaths dark brown to reddish-brown, longest to 30 cm
**Flowers:** Dense, fan-shaped panicle of pale brown to whitish flowers on erect branches; November to March
**Comments:** Capsules much longer than perianth
**Localities:** Montrose, 8, 14, 16, 25, 34A, 42, 44, 47, 48, 54, 56, 62, 63, 70, 74
**Distribution:** Vic, NSW, Tas, SA; NZ

*Juncus sarophorus*

## Juncus sp. O.

**Size:** 1–1.6 m high **F P**
**Habitat:** Grassy wetlands
**Form:** Tufted rhizomatous perennial rush, pith usually interrupted in top half of stem, continuous below, stems glaucous, becoming bluish-green with maturity
**Foliage:** Loose, tapering dark brown basal sheaths to 40 cm long, reddish-brown to black at base
**Flowers:** Large panicle with flowers usually clustered on branchlets, erect bract to 50 cm long above panicle
**Requirements:** Heavy soil
**Comments:** A rare species in Melbourne, more commonly found north of the Divide.
**Localities:** Laverton, 25
**Distribution:** Vic, NSW *Illustrated p. 286*

## Juncus subsecundus — Finger Rush

**Size:** 0.5–1 m x 0.5–1 m **F P**
**Habitat:** Swampy depressions in plains grassland, riparian scrub, valley sclerophyll forest, grassy low open forest
**Form:** Tufted rhizomatous perennial rush with smooth, dull blue-green stems and air spaces in pith
**Foliage:** Basal sheaths pale to mid brown, longest to 120 mm long
**Flowers:** Loose or compact panicles of pale brown flowers, with a large cluster towards the base of the panicle and smaller terminal clusters; November to January, March to May.
**Comments:** May form large colonies in heavy, wet soils.
**Localities:** Research, 2, 4, 5, 16–18, 24–26, 28, 32, 34A, 36, 44, 47, 48, 58, 62, 63, 71, 74
**Distribution:** All states except NT

### Juncus usitatus

**Size:** 0.3–1.2 m x 0.3–1.5 m **F P**
**Habitat:** Plains grassland, red gum woodland, valley sclerophyll forest, often growing in shallow water
**Form:** Tufted rhizomatous perennial rush with slender yellow-green stems, pith interrupted, with air spaces and pithy segments almost equal
**Foliage:** Basal sheaths dark brown to dark reddish-brown, shining
**Flowers:** Loose fan-shaped panicles of pale brown flowers; November to March
**Comments:** A vigorous grower, particularly in poorly drained soils.
**Localities:** 12, 26, 33, 34A, 36, 58, 62, 63, 65, 67#
**Distribution:** Vic, Qld, NSW

### Juncus vaginatus

Clustered Rush

**Size:** 0.3–1 m x 0.5–1 m **F P**
**Habitat:** Wet depressions in red gum woodland
**Form:** Tufted rhizomatous perennial rush with robust stems to 8 mm wide, pith solid and spongy
**Foliage:** Basal sheaths loose, yellowish-green to light brown or greyish
**Flowers:** Large, much-branched and spreading panicle to 15 cm long, pale brown to deep orange-red flowers, with 6 stamens, in almost round clusters separated by bare primary branches; November to March
**Comments:** An attractive ornamental rush for wet gardens and poolside planting.
**Localities:** 25, 26
**Distribution:** Vic, Qld, NSW, Tas

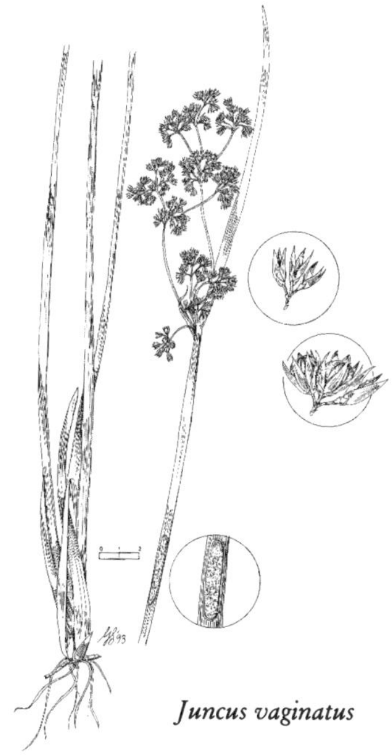

*Juncus vaginatus*

## Lemna

Lemnaceae

### Lemna disperma

Duckweed

**Size:** Individual plants 2–4 mm wide **F P N**
**Habitat:** Fresh, still water in grassy wetlands, riparian scrub, box woodland
**Form:** Floating mat plant, each plant consisting of a single leaf-like thallus and a single root 10–40 mm long, suspended in the water
**Foliage:** Bright green, flattened disk-like leaves
**Flowers:** Minute, almost insignificant white flowers on the margins of thallus; Spring
**Comments:** A rapid-growing and multiplying perennial aquatic plant providing good cover for fish and other aquatic animals. It is most prolific in Spring and Summer.
**Propagation:** Vegetative budding. Young thalli of *Lemna*, *Spirodela* and *Wolffia* spp. emerge from small pockets in the edges and eventually break away. Each species also occasionally produces minute seed.
**Localities:** 4, 6, 14, 16, 29, 34A–36, 42, 44, 49, 55, 60, 62, 63, 67#, 71, 75
**Distribution:** All states; NZ, cosmopolitan
**Synonym:** *L. minor*

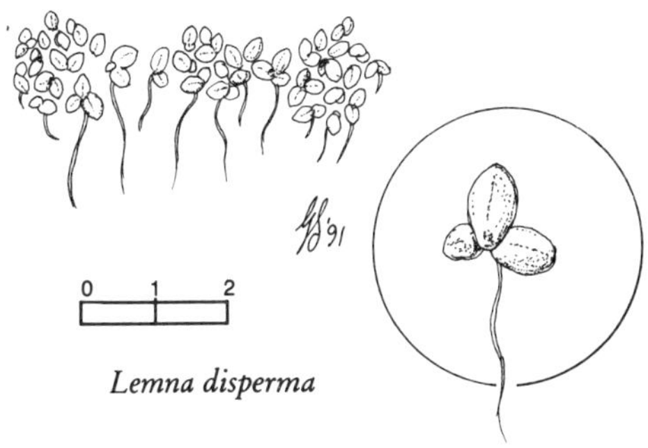


*Lemna disperma*

## Lepidosperma

Cyperaceae

*Lepidosperma* species, commonly called sword-sedges and rapier-sedges because of the shapes of their leaves, are a relatively abundant group of perennial sedges occurring in heaths and moist sites such as stream sides and the edges of swamps. Plants are tufted, spreading by creeping rhizomes. The erect stems and leaves are very similar in appearance and length, with all leaves basal and sometimes flatter than the stems. Flowers are in terminal panicles or spikes with the lowest bract leaf-like, sheathing the base of the panicle. The fruit is a nut, surrounded by 6 thickened whitish scales (not to be confused with stamen filaments).

The erect foliage and graceful panicles of flowers make many *Lepidosperma* spp. attractive plants for accent planting, particularly near water. The flower heads are most decorative in bold dried flower arrangements. Plants can be propagated by seed, but are generally propagated by division of the clumps. They are not fussy as to soil type and often grow well in very heavy soils as long as there is plenty of moisture available, and full or partial sun.

*Lepidosperma concavum*

### Lepidosperma concavum

Sand-hill or Hill Sword-sedge

**Size:** 0.6–1 m high **F P**
**Habitat:** Primary dune scrub, coastal banksia woodland, tea-tree heath
**Foliage:** Stems solid to 6 mm wide, flat on one side, convex the other side
**Flowers:** Short, widely pyramidal panicle 40–70 mm long with densely clustered blackish-brown spikelets; most of the year; bract broad. Nut dull brown, scales long and tapering.
**Requirements:** Moist well drained sandy soil
**Localities:** Clayton, 49, 67–71, 73–75
**Distribution:** Vic, Qld, NSW, Tas, SA

SCALE: 0 1cm 2cm The scale representation on each plant equals 2cm.

## Lepidosperma congestum

Clustered Sword-sedge

**Size:** 0.15–0.8 m high **F P**
**Habitat:** Tea-tree heath
**Foliage:** Leaves sticky at sheath margins; stems flat, to 5 mm wide, margins sharp and slightly rough
**Flowers:** Narrow elongated blackish panicle to 80 mm long, with primary branches very short, spikelets in dense golden rounded clusters; most of the year. Bracts sharply pointed, upper ones with long awns. Nut shiny olive-green.
**Requirements:** Moist sandy soil
**Localities:** Black Rock, 69A, 75
**Distribution:** Vic, SA

*Lepidosperma elatius* var. *elatius*

## Lepidosperma elatius var. elatius

Tall Sword-sedge

**Size:** 1–2.5 m x 1–2 m **P N**
**Habitat:** Wet, damp and valley sclerophyll forests, grassy low open forest in swamps and gullies
**Foliage:** Leaves to 10 mm wide, slightly shorter than stems which have sharp margins and a conspicuous mid-rib
**Flowers:** Loose panicle more than 100 mm long with numerous ovoid spikelets to 5 mm long; most of the year.
**Comments:** The Aborigines used the lower part of the leaf to obtain fibre for basket-making. A spectacular accent plant which should not be planted near walkways due to sharpness of leaves.
**Localities:** 34A, 46, 47, 50–52, 55–58, 60, 75
**Distribution:** Vic, NSW, Tas
**Synonym:** The var. *ensiforme* is now included within this variety.

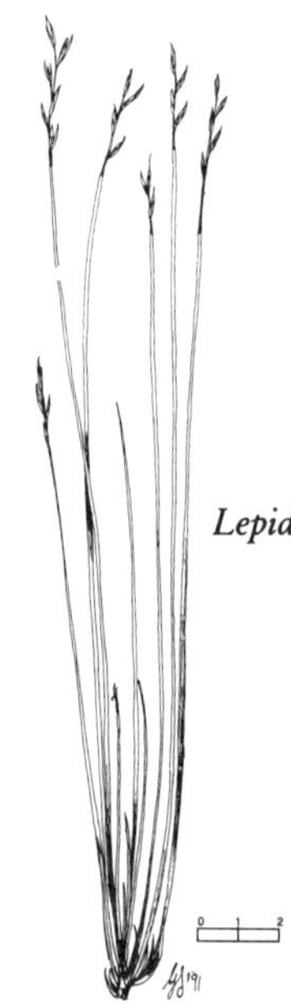

*Lepidosperma filiforme*

## Lepidosperma filiforme

Common Rapier-sedge

**Size:** to 0.9 m high **P**
**Habitat:** Swamp and wattle tea-tree scrub, sclerophyll woodland
**Foliage:** Leaves reduced to basal sheaths, stems terete and smooth to 1 mm wide
**Flowers:** Simple spike of brown or greyish flowers in up to 12 loosely arranged spikelets; Spring to Summer
**Comments:** Becoming rare in the Melbourne area.
**Localities:** Montrose, Clayton, 54, 67#, 74–76
**Distribution:** Vic, NSW, Tas; NZ

## Lepidosperma forsythii

Large-flowered Rapier-sedge

**Size:** 0.6–1 m high **F P**
**Habitat:** Tea-tree heath, in swamps
**Foliage:** Leafless, leaves reduced to bracts tapering to a point; stems terete to 1.5 mm wide, bracts large and broad, shorter than flowerhead
**Flowers:** Short flexuose panicle with 5 or more narrow, widely spreading spikelets
**Requirements:** Wet sandy soil
**Localities:** Beaumaris, 74
**Distribution:** Vic, NSW

*Lepidosperma gladiatum*

## Lepidosperma gladiatum

Sword Rush, Coast Sword-sedge

**Size:** 1–1.5 m high **F P**
**Habitat:** Primary dune scrub
**Form:** Similar to *L. concavum* but stems are convex on both faces, taller and broader, to 20 mm wide, with sharp edges and a prominent central rib (leaves flatter).
**Flowers:** The dense panicle is larger, to 15 cm long with pale brown spikelets. Nut is wrinkled and shiny brown, and scales beneath the nut are very broad.
**Comments:** Useful in dried floral arrangements. The tender leaf base was eaten by the Aborigines while the whole leaf was used for basket-making.
**Localities:** Seaford, 67#, 74
**Distribution:** All states except Qld, NT

## Lepidosperma gunnii

**Habitat:** Tea-tree heath
**Comments:** Similar to *L. laterale* from which it has recently been separated. It has narrower leaves and stems to 2–3 mm wide and shorter panicles to 75 mm long.
**Localities:** 66, 67, 73, 74–76
**Synonym:** *L. laterale* var. *angustum*

## Lepidosperma laterale var. laterale
Variable Sword-sedge

**Size:** 0.4–1 m high **P**
**Habitat:** Damp, valley and dry sclerophyll forests, sclerophyll woodland, swamp scrub, grassy low open forest
**Foliage:** Narrow, stiff shining dark green leaves and stems to 6 mm wide, hardly ribbed, flat to slightly concave on one side, margins sharp
**Flowers:** Narrow, loose panicle to 20 cm long with erect branches and narrow, brown spikelets to 8 mm long; September to March. Nut is wrinkled and shiny mid-brown.
**Requirements:** Very adaptable
**Comments:** As a species, *L. laterale* is the most common and widespread *Lepidosperma* in Victoria. It is very variable and several varieties have been described. Plants with narrow leaves occur at Diggers Rest, Warrandyte and Croydon. *L. lineare*, which does not occur within Melbourne, has been wrongly applied to some narrow leaf forms. A more robust variety is described below. *L. laterale* is currently undergoing revision.
**Localities:** Eltham, 26, 28, 32, 34A–36, 40–42, 44, 47, 48, 50–60, 62–66, 71, 74–76
**Distribution:** Vic, Qld, NSW, Tas, SA; NZ *Also illustrated p. 286*

*Lepidosperma laterale* var. *laterale*

## Lepidosperma laterale var. majus
Variable Sword-sedge

**Size:** Very variable, up to 2 m high **P**
**Habitat:** Valley and wet sclerophyll forests, grassy low open forest
**Comments:** Differs from *L. laterale* var. *laterale* in that it is more robust, with broad, very flat stems. The flowers are borne on a longer panicle to 30 cm. Large plants resemble *L. elatius*, which has smaller nuts with shorter scales, not drawn out into long fine points or bristles. The red leaf base was used as a contrast colour in basket-making by the Aborigines.
**Localities:** 44, 45, 47, 48, 54, 57, 74, 76
**Distribution:** Vic, NSW

*Lepidosperma laterale* var. *majus*

## Lepidosperma longitudinale
Pithy or Common Sword-sedge

**Size:** 0.6–1.7 m high **F P**
**Habitat:** Wattle tea-tree and swamp scrub, occurring in or near water
**Foliage:** Leaves and stems soft and pithy, sometimes hollowed, convex on both faces, to 8 mm wide, margins smooth
**Flowers:** Erect narrow panicle 10–30 cm long with densely crowded greyish-brown, 2–3-flowered spikelets to 7 mm long; most of the year. Nut shiny light brown to brown.
**Requirements:** Moist or wet soils
**Localities:** 60, 67, 69–71, 73–75
**Distribution:** All states except NT

*Lepidosperma longitudinale*

## Lepidosperma neesii
Stiff Rapier-sedge

**Size:** to 0.8 m high **P**
**Habitat:** Grassy low open forest, wattle tea-tree scrub
**Foliage:** Linear terete leaves, stems strongly angled
**Flowers:** Short, narrow blackish spike-like panicle to 25 mm long, with a rigid, very long bract beneath. Many spikelets to 8 mm long pressed flat against the branches.
**Requirements:** Swampy soils
**Localities:** 59, 74, 76
**Distribution:** Vic, NSW

*Lepidosperma neesii*

## Lepidosperma semiteres
Wire Rapier-sedge

**Size:** 0.3–1 m high **F P**
**Habitat:** Damp sclerophyll forest, tea-tree heath, grassy low open forest
**Foliage:** Leaves flat with thin margins; stems to 1 mm wide, strongly convex on both faces, sometimes with a deep groove
**Flowers:** Short, branched spike-like panicle to 60 mm long with 12 or more slender spikelets to 8 mm long, flowers brown or greyish; most of the year. Nut straw-coloured and wrinkled or smooth shiny olive-green.
**Localities:** 56, 67–70, 74, 76
**Distribution:** Vic, NSW, SA *Illustrated p. 286*

## Lepidosperma tortuosum
Tortuous Rapier-sedge

**Size:** to 40 cm high **P**
**Habitat:** Damp sclerophyll forest
**Foliage:** Leaves are convex on both faces, to 25 cm long, stems biconvex to terete, channelled with a ridge on the other side
**Comments:** Similar to *L. forsythii* but smaller and more slender. The 2–5 spikelets to 8 mm long are close together on the short spike; January. Bracts are inconspicuous.
**Localities:** 56
**Distribution:** Vic, NSW, Tas, SA *Illustrated p. 286*

SCALE: 0 1cm 2cm The scale representation on each plant equals 2cm.

# Lepilaena

## Zannichelliaceae

Aquatic annual or perennial rhizomatous grassy herbs, found in fresh or brackish water, some species tolerating salinities exceeding those of seawater. Rhizomes and stems are much-branched. Leaves are sheathed at the base with a well developed, cleft ligule (or growth between the sheath and leafblade). The unisexual flowers are inconspicuous and at first enclosed in a leaf sheath. The fruit and leaves are eaten by some seabirds.

*Lepilaena cylindrocarpa*

### Lepilaena bilocularis

Small-fruited or Feather-style Water-mat

**Size:** stems 5–15 cm long, occasionally to 1 m long **F P**
**Habitat:** Saltmarshes, lakes, in brackish water to 1 m deep
**Form:** Submerged slender perennial herb
**Foliage:** Fine, linear leaves to 20 mm long, tip truncate with 3 distinct points, borne at close intervals along thread-like stems
**Flowers:** Dioecious, flowers almost sessile; May to September
**Requirements:** Fresh to brackish water
**Localities:** Albert Park Lake, 4
**Distribution:** Vic, Qld, NSW, Tas, SA; NZ

### Lepilaena cylindrocarpa

Long-fruit Water-mat

**Size:** stems to 35 cm long **F**
**Habitat:** Saltmarshes, marine
**Comments:** Similar to *L. preissii*. Leaves end in a short point. Inconspicuous dioecious flowers are borne on stalks longer than the leaf sheaths. Petals are slightly notched. Stems are slender, branched and flexuose; stems on female plants shorter, to 25 cm high. Flowers August to November.
**Localities:** Western shores of Port Phillip, mouth of Yarra R, 2, 4
**Distribution:** Vic, Tas, SA, WA

### Lepilaena preissii

Slender Water-mat

**Size:** stems 5–50 cm long **F**
**Habitat:** Salt marshes, brackish lakes, marine
**Form:** Annual monoecious or dioecious herb
**Foliage:** Linear leaves to 80 mm long tapering to a sharp point
**Flowers:** Much-branched flowering stems with clusters of conspicuous white flowers (relative to other species) on short stalks no longer than the sheaths, petals deeply notched; September to February. Male and female flowers often together.
**Localities:** Western shores of Port Phillip, Albert Park, 4
**Distribution:** Vic, Tas, SA, WA

*Leptocarpus brownii*

# Leptocarpus

## Restionaceae

### Leptocarpus brownii

Coarse Twine-rush

**Size:** 0.5–1 m high **F P**
**Habitat:** Swamp scrub, tea-tree heath
**Form:** Rush-like rhizomatous perennial herb with erect unbranched stems and scale-like leaves or sheathing bracts to 7 mm long
**Flowers:** Dioecious, in panicles; male flowers in few short reddish-brown spikelets to 10 mm long, on white-woolly stalks; female flowers in numerous 3–6-flowered golden-brown spikelets to 6 mm long, sessile, clustered together; most of the year
**Requirements:** Moist soils, tolerating dryness once established
**Localities:** 61, 67#, 70, 71 74
**Distribution:** Vic, NSW, Tas, SA

### Leptocarpus tenax

Slender Twine-rush

**Size:** 0.5–1 m high **F P**
**Habitat:** Swamp scrub, wattle tea-tree scrub
**Foliage:** Sheathing bracts 5–10 mm long
**Flowers:** Differs from *L. brownii* in that male spikelets are smaller, to 4 mm long, and more numerous in a loose, often drooping panicle with glabrous stalks. Female spikelets fewer, to 10 mm long, in rigidly erect panicles; mostly October to June.
**Requirements:** Moist soil
**Comments:** Endangered in the Melbourne area. Very attractive as an accent plant, and also when used in dried arrangements.
**Localities:** Collected in Sandringham 1905; 61
**Distribution:** All states except NT

# Lepyrodia — Restionaceae

## Lepyrodia muelleri

Common or Erect Scale-rush

**Size:** 0.4–0.6 m high **P**
**Habitat:** Wattle tea-tree scrub
**Form:** Rush-like perennial herb with scaly rhizomes
**Foliage:** Leaves reduced to distant sheathing bracts to 25 mm long
**Flowers:** Dioecious on few stems, few of which are branched. Single, stalked pale brown flowers in a narrow panicle 30–120 mm long, glumes shorter than perianth segments; September to May. Floral bracts are very thin and fragile.
**Requirements:** Wet sandy soil
**Localities:** 59, 67, 68, 70, 73–75
**Distribution:** Vic, NSW, Tas, SA, WA

*Lepyrodia muelleri*

# Lomandra — Xanthorrhoeaceae (Dasypogonaceae)

Lomandras are hardy perennial rhizomatous herbs forming rush-like tufts. Their basal leaves are often held stiffly erect, longer than and sometimes hiding the dense clusters or panicles of small yellow or yellow-brown flowers. All species are dioecious. The flowers are composed of 3 outer and 3 inner perianth segments. Lomandras are important butterfly food plants. The Aborigines made drinks from the nectar and used fibre from the leaves of some species to weave bags and baskets. Propagation is by seed or division of clumps. Previously included in the family *Liliaceae.*

## Lomandra filiformis ssp. filiformis

Wattle Mat-rush

**Size:** 15–50 cm x 15–20 cm **F P**
**Habitat:** Widespread in valley and dry sclerophyll forests, sclerophyll and red gum woodland, grassy low open forest
**Form:** Sparsely tufted grass-like perennial
**Foliage:** Narrow channelled bluish-green to green leaves to 5 mm wide, flexible, with margins inrolled, tip minutely 1-3-pointed
**Flowers:** Panicle to 20 cm long (occasionally to 30 cm). The numerous stalked, yellow female flowers are small and globular and borne on fine short branches, looking like tiny wattle buds. Female panicles are slightly shorter than the male. Outer segments of male flowers are purplish and stems whitish; September to December.
**Requirements:** Moist, well drained clays or sands tolerating dry shady situations once established
**Comments:** An attractive, long-lived rockery plant grown for form rather than flowers. A form from Montrose is rhizomatous and spreading.
**Localities:** 12, 13, 16, 18, 21, 23–26, 28, 29, 31, 32, 34A–49, 50A–56, 58–68, 69A 71, 74–76
**Distribution:** Vic, Qld, NSW

*Lomandra filiformis* ssp. *filiformis*

## Lomandra filiformis ssp. coriacea

**Size:** 15–30 cm high **F P**
**Habitat:** Riparian scrub, valley and dry sclerophyll forest
**Foliage:** Leaves are flat, leathery and lightly raspy on the edges, tip entire.
**Flowers:** Male panicle has a few whorled branches and is distinctly rough at the axes; stem yellowish. Female panicle is smaller, axes not as rough.
**Comments:** The two subspecies are often found growing together in the wild.
**Localities:** Montrose, Sth Yarra, Brighton, 13, 21, 32, 34A, 36, 46, 47, 50–51, 53–55, 65, 69, 73–76
**Distribution:** Vic, Qld, NSW

*Lomandra longifolia*

## Lomandra longifolia

Spiny-headed Mat-rush

**Size:** 0.5–1 m x 0.5–1.2 m **F P N**
**Habitat:** Widespread in red gum woodland, dry and valley sclerophyll forests, grassy low open forest, tea-tree heath, coastal banksia woodland
**Form:** Large tussock plant
**Foliage:** Smooth bright green strap-like leaves to 12 mm wide, tip with 2 or 3 teeth
**Flowers:** Numerous clusters of sessile, scented yellowish flowers with purplish bases, whorled panicles to 15 cm long; panicles many-branched on male plants; narrow on female plants; September to December. Shiny greenish brown or brownish-orange capsules remain on plant for most of year. Spiky bracts to 30 mm long extend beyond each flower cluster.
**Requirements:** Well drained soils tolerating dry shade
**Comments:** Grows well under established trees.
**Localities:** 1, 12, 16, 26, 28, 29, 31, 32, 34A–36, 38, 40–56, 58–71, 73–76
**Distribution:** Vic, Qld, NSW, Tas, SA

SCALE: 0 1cm 2cm — The scale representation on each plant equals 2cm.

*Lomandra multiflora*

### **Lomandra micrantha** ssp. **micrantha** — Small-flowered Mat-rush

**Size:** 0.2–0.7 m x 0.5 m — **P**
**Habitat:** Plains grassland, box woodland
**Foliage:** Narrow, flexuose grass-like leaves to 2.5 mm wide, flat to terete, hiding flower stem
**Flowers:** Loose panicle with minutely-rough branches. Small, scattered flowers are stalked and greenish-yellow to dark red. Male flowers to 2 mm long, borne singly or in 2's, panicle branched; female flowers to 5 mm long, panicle may be unbranched; April to December. Shining pale yellow or greenish capsule.
**Requirements:** Suited to most soils
**Localities:** Sunshine, 17#, 18, 28, 38, 74
**Distribution:** Vic, NSW, SA, WA

### **Lomandra multiflora** — Many-flowered Mat-rush

**Size:** 0.3–0.5 m x 0.3–0.5 m — **P**
**Habitat:** Box woodland, valley sclerophyll forest, tea-tree heath
**Foliage:** Rigid, narrow greyish leaves to 3 mm wide, surfaces slightly rough
**Flowers:** Male flowers are borne in showy, densely whorled clusters on branched stems to 30 cm high. Each flower is 2–3 mm long on a stalk to 10 mm long. Petals are yellow, sepals purplish. Female flowers are slightly larger, to 4 mm, and sessile. Panicles are unbranched, stems to 25 cm long; capsules become black and wrinkled. June to January
**Requirements:** Well drained soils, tolerating extended dry spells once established
**Localities:** Montrose, St Albans, 34A, 42, 44, 54, 56, 62, 63, 67–70
**Distribution:** Vic, NT, Qld, NSW

*Lomandra nana*

### **Lomandra nana** — Pale Mat-rush

**Size:** 5–15 cm x 5–10 cm — **F P**
**Habitat:** Tea-tree heath, plains grassland
**Form:** Low densely tufted herb with flowers hidden at the base of the foliage
**Foliage:** Fine, rough, flat bluish-green leaves to 2 mm wide, erect or curved to one side
**Flowers:** Male panicle stout, unbranched, creamy-white flowers in sessile clusters. Female flowerhead almost stemless with flowers crowded into 1 to several sessile clusters. Flowers sessile; October to December.
**Requirements:** Well drained soils
**Localities:** Keilor, Melton, 63, 64, 68–69A, 74
**Distribution:** Vic, Tas, SA
**Synonym:** Segregated from *L. glauca*

## Luzula — Juncaceae

### **Luzula meridionalis** var. **meridionalis** — Field Woodrush

**Size:** to 10 cm x 15 cm; stem to 30 cm high — **P**
**Habitat:** Widespread in red gum woodland, dry, valley and wet sclerophyll forests, grassy low open forest
**Form:** Loosely tufted stoloniferous perennial herb
**Foliage:** Flat, linear grass-like basal leaves 2–5 mm wide, the edges fringed with long white hairs, denser towards the base, and blunt tips; 1–2 stem leaves
**Flowers:** Terminal umbel-like panicle to 10 mm wide, of 1–6 dense globular clusters of brown flowers subtended by 1–4 leaf-like bracts, held above the foliage; August to December. Flowers each have 6 tapering perianth parts and 6 stamens. Fruit brown to dark brown.
**Requirements:** Well drained soils, tolerant of both dry and wet periods, but not waterlogging
**Comments:** *Luzula* spp. often develop swollen onion-like bases.
**Propagation:** Seed, division
**Localities:** 35, 36, 39, 41, 42, 44, 46, 49, 50A, 51, 53, 55–57, 60, 62, 63, 64, 67, 74, 76
**Distribution:** Vic, NSW, Tas, SA, WA
**Synonym:** All varieties were originally combined under *L. campestris.*

*Luzula meridionalis* var. *densiflora*

### **Luzula meridionalis** var. **densiflora** — Woodrush

**Size:** stem 25–40 cm high — **F P**
**Habitat:** Valley sclerophyll forest, wattle tea-tree scrub
**Requirements:** Moist well drained soils
**Comments:** Differs from the typical species in that leaves are narrower, to 3 mm wide and margins densely fringed. Panicle consists of 1 or more rigidly branched sessile clusters. Fruit is brown to dark red-brown. July to December
**Localities:** Eltham, 42, 44, 45, 56, 74, 75
**Distribution:** Vic, NSW, Tas, SA, WA
**Synonym:** *L. densiflora*

### Luzula meridionalis var. flaccida

Woodrush

**Size:** 15–25 cm high — P N
**Habitat:** Tea-tree heath, valley sclerophyll forest
**Comments:** Differs from the above varieties in that leaves are very narrow, 1–2 mm wide, lax and almost glabrous. The flowers are smaller and yellowish-brown on limp stems; August to November.
**Localities:** 32, 34A, 36, 44, 47, 48, 67, 69A, 74
**Distribution:** Vic, Qld, NSW, Tas, SA
**Synonym:** *L. flaccida*

*Illustrated p. 286*

# Microlaena — Poaceae

### Microlaena stipoides

Weeping Grass

**Size:** Highly variable; to 30 cm x 0.6 m; stems to 1 m high — F P
**Habitat:** Widespread in plains grassland, dry and valley sclerophyll forests, sclerophyll woodland, grassy low open forest, swamp, riparian and wattle tea-tree scrub
**Form:** Sparse, shortly rhizomatous perennial grass with gracefully arching, erect or semi-prostrate stems which may develop roots at the basal nodes
**Foliage:** Narrow flat leaves to 80 mm long
**Flowers:** Narrow nodding racemes or panicles to 20 cm long on slender stems. Spikelets (to 30 mm long) composed of minute glumes, 2 sterile lemmas with awns to 20 mm long and 1 fertile, awnless lemma; September to November
**Requirements:** Moist, well drained soils
**Comments:** One of the best lawn grasses for shady sites, where it is often found as a remnant of the original vegetation of Melbourne. Regularly mown or grazed plants form a fine sward with stems as short as a few cms. In rural areas, grazing by macropods (kangaroos etc) sometimes creates weeping grass lawns.
**Propagation:** Seed, division
**Localities:** Laverton Nth, Sunshine, Deer Park, 8, 10, 16, 18, 21, 23–26, 28, 29, 31–36, 38–42, 44, 45, 47–60, 62–64, 67–71, 74–76
**Distribution:** All states; NZ

*Microlaena stipoides*

# Ottelia — Hydrocharitaceae

### Ottelia ovalifolia

Swamp Lily

F P

**Habitat:** Freshwater pools and slow streams in water to 0.6 m deep
**Form:** Aquatic perennial herb with a tuft of floating leaves rising on long stalks from the base. Plants root in mud.
**Foliage:** Elliptic to ovate leaves, blades 50–150 mm long, with 7 prominent veins; juvenile leaves linear
**Flowers:** Large, showy single white flowers with dark reddish-purple base, to 75 mm wide, held just above water level; plants also bear unopened submerged flowers which are self-pollinating; November to March. Seeds ripen and are released under water.
**Requirements:** Fresh water, tolerating seasonal drying out of pool
**Comments:** An ornamental pond plant with attractive flowers and foliage. An alternative to water-lilies. Locally rare.
**Propagation:** Division
**Localities:** Melton West, Warrandyte, 9, 11, 35, 63, 67#
**Distribution:** All mainland states; N Cal

*Ottelia ovalifolia*

# Panicum — Poaceae

### Panicum decompositum

Australian Millet or Umbrella Grass

**Size:** 0.3–0.5 m high, stems to 1.5 m high — F P
**Habitat:** Moist depressions in grassy wetlands
**Form:** Coarse, densely tufted perennial grass
**Foliage:** Erect, flat glabrous blue-green leaves to 50 cm x 12 mm gradually tapering to a fine point; edges whitish
**Flowers:** Large, very open, much branched panicle to 30 cm long, held above foliage on stout, hollow stems. Lower branches clustered. Solitary spikelets along the rhachis of branches. First glume up to 30% length of spikelet. Flowers mostly in Summer.
**Requirements:** Heavy soils that do not dry out
**Comments:** Although perennial, plants may be short lived, suffering adversely during dry periods. Very attractive in dried arrangements. The flour from ground seeds was made into cakes by the Aborigines. A rare plant both in Melbourne and the rest of Victoria.
**Propagation:** Seed
**Localities:** Braybrook 1905; Laverton, Sunshine
**Distribution:** All mainland states

*Panicum decompositum*

SCALE: 0 1cm 2cm — The scale representation on each plant equals 2cm.

*Panicum effusum* var. *effusum*

*Paspalidium constrictum*

*Philydrum lanuginosum*

### Panicum effusum var. effusum

Hairy Panic

F P

**Size:** 20–60 cm high
**Habitat:** Grassy wetlands, plains grassland, box and red gum woodland
**Form:** Tufted perennial grass with thickened, hairy base
**Foliage:** Sparsely hairy, flat, dull greyish-green leaves 5–25 cm x 2–6 mm, nodes bearded with silky hairs
**Flowers:** Differs from *P. decompositum* in that panicle is smaller, to 20 cm long, with usually 2 purplish spikelets near the ends of thread-like rhachis. First glume pointed, about half the length of spikelet. October to May
**Requirements:** Heavy clay soils
**Comments:** The old flowerheads of *Panicum* spp. soon become brittle and break away. Blown along by the wind they are often seen in large numbers caught along fences.
**Propagation:** Seed
**Localities:** Sunshine, 6, 9, 15, 17#, 18, 21, 25, 26
**Distribution:** All mainland states

### Panicum prolutum = Homopholis proluta

## Paspalidium

Poaceae

### Paspalidium constrictum

Knotty-butt or Box Grass

F P

**Size:** 30–60 cm high
**Habitat:** Rocky escarpments
**Form:** Tufted perennial herb
**Foliage:** Narrow, rough, flat or folded leaves to 15 cm long
**Flowers:** Narrow, slender spike-like panicle to 12 cm long, borne slightly above the leaves. Spikelets irregularly arranged on short panicle branches, lower glume downy; September to April
**Requirements:** Well drained rocky soils
**Comments:** A restricted species due to habitat destruction, it is now known from only 2 sites within the Melbourne area.
**Propagation:** Seed
**Localities:** Melton
**Distribution:** All mainland states

## Pentapogon

Poaceae

### Pentapogon quadrifidus

Five-awned Spear-grass

F P

**Size:** 8–18 cm high, stems 30–60 cm high
**Habitat:** Plains grassland, red gum woodland, dry and valley sclerophyll forests
**Form:** Tufted erect annual or short-lived perennial grass
**Foliage:** Leaves 8–18 cm long, inrolled, glabrous or pubescent
**Flowers:** Open to moderately dense light brownish panicle to 15 cm long with narrow 1-flowered spikelets. Lemma 2-lobed, 2 small awns per lobe, longer central awn to 25 mm long, rigid and twisted in first 10 mm. November to February.
**Requirements:** Adaptable to all but very wet soil conditions
**Propagation:** Seed
**Localities:** Laverton Nth, 5, 12, 25, 26, 28, 36, 42, 44, 47, 48, 50A, 54, 55, 74
**Distribution:** Vic, NSW, Tas, SA

*Illustrated p. 286*

## Philydrum

Philydraceae

### Philydrum lanuginosum

Woolly Water Lily

F P

**Size:** 0.5–1 m x 0.5 m
**Habitat:** Freshwater pools and swamps
**Form:** Erect aquatic perennial herb
**Foliage:** Smooth, pointed, reddish-green strap-shaped leaves 15–45 cm x 5–10 mm, reducing in length up the flowering stem to become ovate floral bracts to 25 mm long
**Flowers:** Long spike of up to 20 sessile, velvety yellow, hooded flowers, on a stout woolly flowering stem to 0.9 m high. Each flower has just 1 perfect stamen and 2 staminodes and lasts for one day; Summer.
**Requirements:** Ample fresh water
**Comments:** Attractive flowering pond plant which appears to be extinct within the Melbourne area. Aboriginal women used the leaves for girdles.
**Propagation:** Division
**Localities:** Cheltenham#, 67#
**Distribution:** Vic, Qld, NSW, WA, NT; SE Asia

# Phragmites — Poaceae

## Phragmites australis — Common Reed

**Size:** 1–3 m high — F P
**Habitat:** Riparian and swamp scrub, coastal banksia woodland
**Form:** Semi-aquatic, fast growing bamboo-like perennial grass with strongly creeping rhizomes, flowering stems leafy and unbranched
**Foliage:** Glabrous grey-green leaves 20–60 cm x 1–3 cm, tapering to a fine point
**Flowers:** White to purplish panicles, fluffy and plume-like to 40 cm long. Spikelets 2–6-flowered, to 15 mm long, lemmas glabrous, ending in long points, surrounded by many silky hairs. January to April
**Requirements:** Wet and poorly drained soils, tolerating brackish sites
**Comments:** An attractive water-side plant for dams and large ponds but must be kept contained or else it can take over. Provides food and nesting for waterbirds. Important to the Aborigines who used the stems for spears and ornaments and leaves to weave baskets.
**Propagation:** Division, seed
**Localities:** 1, 3, 4, 8, 14, 16, 18, 21, 25–30, 32, 33, 34A–36, 42, 44, 46, 47, 49, 51, 53, 55, 56, 59–67, 69A–76
**Distribution:** All states; all continents

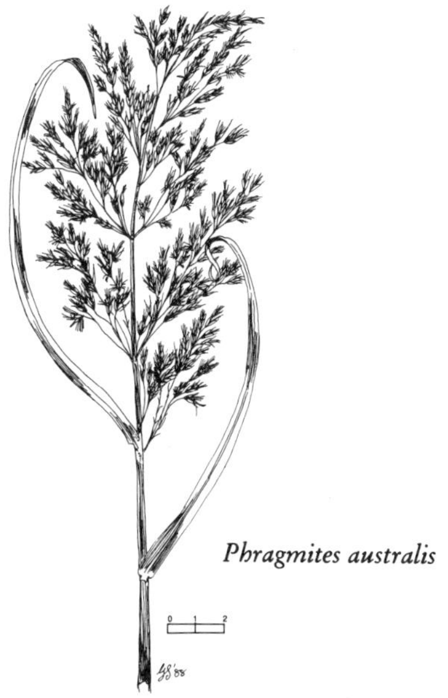

*Phragmites australis*

# Poa — Poaceae

Native *Poa* species found in Melbourne are perennial grasses with flowerheads arranged mostly in loose panicles. The lemmas are 5–9-veined while the glumes are 1–3-veined.

Some particularly attractive species are suited to ornamental horticulture. Their soft, graceful form makes them a very useful accent plant in a variety of landscape styles.

Propagation is from seed, or division (which is quicker and easier). The Aborigines used the leaves and stems of the larger species for string or basket-making.

Classification and identification of Poas into species is very difficult. Recognition must sometimes rely on a composite of rather variable features, and intermediates occur between what we presently call separate species.

## Poa clelandii

**Size:** to 25 cm high; stems to 0.75 m high — F P
**Habitat:** Plains grassland, tea-tree heath, valley sclerophyll forest
**Form:** Loose or dense tussocks with creeping rhizomes, new tufts forming at a distance from the parent
**Foliage:** Leaves flat, inrolled on drying, to 25 cm long, lower leaf sheaths purplish
**Flowers:** Narrow panicle to 20 cm long, with 3–5-flowered, purplish, greenish or straw-coloured spikelets on upper ends of branches. September to January
**Localities:** Beaumaris, 17#, 18, 51, 54, 74, 75
**Distribution:** Vic, SA

*Poa clelandii*

## Poa ensiformis — Purple-sheath Tussock Grass

**Size:** 0.3–0.75 m x 1 m; stems to 1.5 m high — P N
**Habitat:** Damp and valley sclerophyll forests
**Form:** Rhizomatous, forming loose to rather dense tussocks
**Foliage:** Long, narrow, flat dark green leaves to 75 cm x 6 mm, rough, especially on margins and back. Purple sheaths around lower leaves during most months. Old leaves inrolled and straw-coloured.
**Flowers:** Erect, spreading panicle to 30 cm long; October to March
**Comments:** Vigorous grass, ideally suited to moist, shaded sites. Useful for controlling erosion.
**Localities:** 34A, 36, 42, 44, 47, 48, 50A, 51, 53–57
**Distribution:** Vic, NSW
**Synonym:** Previously included with, and confused with *P. helmsii* which is found in the mountains of the Dividing Range. *Illustrated p. 286*

## Poa labillardieri — Cool season grass — Common Tussock Grass

**Size:** 0.3–0.8 m high; stems to 1.2 m high F P
**Habitat:** Widespread in red gum woodland, riparian scrub, plains grassland, dry and valley sclerophyll forests
**Form:** Large, coarse densely tufted tussock
**Foliage:** Rough, green, greyish-green or blue-green leaves to 80 cm x 1–3.5 mm, flat to tightly inrolled, tips sometimes pungent-pointed.
**Flowers:** Open panicle to 25 cm long, branches spreading. Green or purplish 3–4-flowered flattened spikelets along the length of upper branches but only along upper half of lower branches. October to February
**Requirements:** Adapts to moist or slightly dry soils
**Comments:** A very vigorous grass with many forms. Colour may change with season or age.
**Localities:** 1, 2, 4–6, 8, 11, 14, 16–18, 25–28, 32, 34A–36, 39, 41, 42, 44, 50A, 51, 53–55, 58, 60, 62, 63, 67, 68, 69A–71, 74, 76
**Distribution:** Vic, Qld, NSW, Tas, SA

*Poa labillardieri*

SCALE: 0 1cm 2cm — The scale representation on each plant equals 2cm.

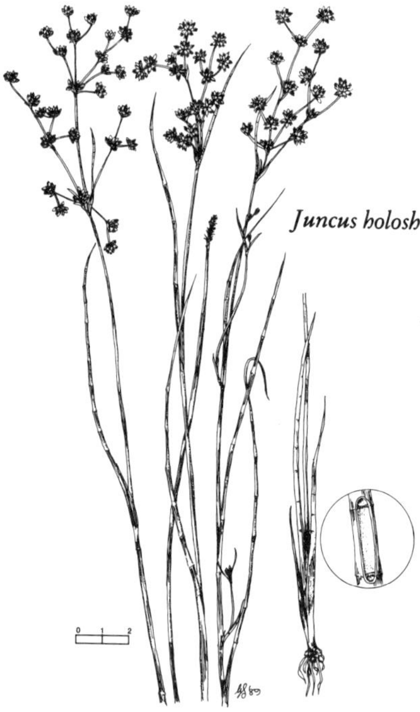

*Juncus holoshoenus*

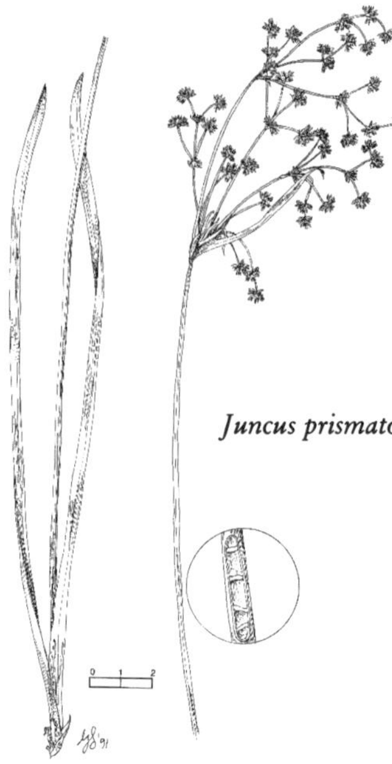

*Juncus prismatocarpus*

*Juncus* sp. O.

*Lepidosperma laterale* var. *laterale*

*Lepidosperma tortuosum*

*Lepidosperma semiteres*

*Luzula meridionalis* var. *flaccida*

*Poa ensiformis*

*Pentapogon quadrifidus*

## Poa morrisii

Velvet Tussock Grass

**Size:** to 30 cm high, stems 0.5–0.9 m high **F P**
**Habitat:** Dry and valley sclerophyll forests, sclerophyll woodland, grassy low open forest
**Form:** Soft dense greyish-green tussock-forming grass
**Foliage:** Loosely inrolled, soft, lightly to densely hairy leaves to 30 cm long, lower sheaths pale
**Flowers:** Open panicles to 25 cm long. Lemmas densely hairy on keel, soft downy hairs between veins. October to January
**Requirements:** Moist well drained soils
**Comments:** A most attractive grass, suited to a range of horticultural uses.
**Localities:** Eltham, 26, 28, 34A–36, 41, 42, 44, 47, 48, 50–51, 53–56, 59, 62–64, 70, 74, 76
**Distribution:** Vic, SA
**Synonym:** Previously included with *P. australis* sp. agg. a very variable group which has now been separated into several species.

*Poa morrisii*

## Poa poiformis var. poiformis

Coast or Blue Tussock Grass

**Size:** 0.2–0.9 m x 1 m **F P**
**Habitat:** Primary dune scrub, coastal banksia woodland
**Form:** Densely tufting grass
**Foliage:** Rigidly erect, thickened terete or angular bluish leaves about as long as flowering stems
**Flowers:** Dense panicles to 30 cm long, at first narrow, becoming loose. Branches in whorls of 3–5. Large, pale green to straw-coloured spikelets almost to the base of the branches. Lemmas rough on upper keel and hairy on lower keel veins and margins. September to January
**Requirements:** Well drained sandy soil, tolerating saline soil and salt spray
**Comments:** Sometimes hard to distinguish from some coastal forms of *P. labillardieri*, whose distribution may overlap. Features which need to be considered are: proximity to sea, stem height relative to leaves, compactness of seed heads.
**Localities:** Seaford, 2, 4, 5, 62, 67, 69A, 71, 74
**Distribution:** Vic, NSW, Tas, SA, WA

*Poa poiformis* var. *poiformis*

## Poa rodwayi

**Size:** 0.25–0.6 m high **F P**
**Habitat:** Grassy low open forest, valley sclerophyll forest, box woodland
**Comments:** Green to greyish-green tussock-forming perennial.
**Localities:** Tarneit, Sth of Melton, 62, 64, 74, 75
**Distribution:** Vic, NSW, Tas

## Poa sieberiana var. sieberiana

Tussock Grass

**Size:** 15–30 cm x 40 cm; stems to 0.9 m high **F P**
**Habitat:** Widespread in grassy low open forest, wattle tea-tree scrub, plains grassland, red gum woodland, dry sclerophyll forest
**Form:** Dense green to greyish-green tuft
**Foliage:** Fne, almost glabrous terete leaves
**Flowers:** Panicle to 25 cm long, with green, purplish or straw-coloured spikelets; branches mostly in pairs. Lower part of lemma hairy between the veins.
**Comments:** Extremely variable species with different forms from coastal to alpine. Very adaptable in cultivation, this attractive grass establishes well under existing eucalypts. Grows larger in shaded forests.
**Localities:** Eltham, 2, 10, 12 18, 23, 25, 26, 28, 31, 32, 34A–36, 40–42, 44, 47, 50A, 51, 53, 56, 58, 60, 65, 67#, 69–71, 74–76
**Distribution:** Vic, Qld, NSW, Tas
**Synonym:** Previously included with *P. australis* sp. agg. a very variable group which has now been separated.

## Poa sieberiana var. hirtella

**Habitat:** Valley sclerophyll forest
**Comments:** Distinguish from the typical form by the sparsely hairy leaves.
**Localities:** Montrose, 36
**Distribution:** Vic, Qld, NSW

## Poa tenera

Slender Tussock Grass

**Size:** 5–20 cm high **P N**
**Habitat:** Damp and valley sclerophyll forests, swamp scrub, grassy low open forest
**Form:** Bright green, soft, trailing stoloniferous grass, sometimes forming open tussocks
**Foliage:** Soft, glabrous, inrolled linear leaves 5–20 cm long
**Flowers:** Delicate open panicle to 120 mm long with few pale green 2–4-flowered spikelets towards the end of the fine branches. Lower branches paired. October to January
**Requirements:** Moist well-drained soils
**Comments:** Very effective when trailing down embankments.
**Localities:** 34A, 41, 42, 44, 47, 48, 50A, 51, 53–58, 61, 74–76
**Distribution:** Vic, NSW, Tas, SA *Illustrated p. 294*

*Poa sieberiana* var. *sieberiana*

SCALE: 0 1cm 2cm — The scale representation on each plant equals 2cm.

# Potamogeton

**Potamogetonaceae**

*Potamogeton* species are perennial, rhizomatous aquatic herbs commonly known as pondweeds. They thrive in both still backwaters and lagoons and flowing streams, as long as the water is fresh. They may be totally submerged except for the flowers, or partly emerged with floating leaves. Flowers are brownish and generally inconspicuous, borne in spikes held above the water. After pollination spikes become submerged, where the fruit develops. Plants are attractive in ornamental ponds; however their growth can become quite vigorous and must be kept in check. Propagation is by division of rhizomes or setting the nut-like fruit into the muddy base of ponds.

*Potamogeton crispus*

*Potamogeton ochreatus*

*Potamogeton pectinatus*

## Potamogeton crispus

Curly Pondweed

**Size:** stems to 4 m long **F P**
**Habitat:** Still or flowing water to 4.5 m deep in riparian and red gum woodland
**Form:** Submerged aquatic
**Foliage:** Sessile, narrow-lanceolate to linear-oblong, green to reddish-brown translucent leaves to 70 mm long, margins distinctly wavy, with tiny teeth
**Flowers:** Short, emergent few-flowered spike, held on stems to 70 mm above the water. Fruit to 5 mm x 3 mm with a long pointed beak, in clusters 10–20 mm long at the end of the spike; November to May
**Localities:** 8, 35, 42, 67#
**Distribution:** All mainland states; NZ, Eur, Af, Asia

## Potamogeton ochreatus

Blunt Pondweed

**Size:** stems to 4.5 m long **F P**
**Habitat:** Still or flowing water to 4.5 m deep
**Comments:** Differs from *P. crispus* in that leaves are entire, flat and linear, to 90 mm long, with stipule-like nerves on the leaf sheath. Flowering spike elongated, to 25 mm in fruit; August to March peaking late Spring
**Localities:** 4, 11, 29#, 34A, 42, 74, 76
**Distribution:** All states; NZ

## Potamogeton pectinatus

Fennel Pondweed

**Size:** stems to 3 m long **F P**
**Habitat:** Fresh to saline, still or flowing water to 3 m deep
**Foliage:** Stipular sheath is joined to the stem and leaves at the base, free above; leaves entire, narrow-linear and grass-like to 15 cm long
**Flowers:** Spikes with several isolated clusters of 2–3 flowers, to 40 mm long in fruit
**Localities:** Altona, Footscray, 2, 4, 11, 67#
**Distribution:** All states except NT; NZ, world-wide

## Potamogeton tricarinatus

Floating Pondweed

**Size:** stems to 2.7 m long **F P**
**Habitat:** Still and floating fresh water, mud beside receding water, grassy wetlands
**Form:** Aquatic or semi-aquatic, emergent leaves floating
**Foliage:** Long-stalked, thick and leathery, ovate to oblong emergent leaves 15–100 mm x 25 mm, with up to 16 parallel veins; submerged leaves thin, narrower, 5–23 cm long
**Flowers:** Flowering spike short and dense to 40 mm long at the end of stout, erect stalks. Fruit cluster to 45 mm long; September to April
**Localities:** Rockbank, Melton, 6, 7, 29, 35, 61, 67#, 71
**Distribution:** All states *Illustrated p. 294*

# Puccinellia

**Poaceae**

*Puccinellia stricta* var. *perlaxa*

## Puccinellia stricta var. stricta

Australian Saltmarsh-grass

**Size:** 15–60 cm high **F**
**Habitat:** Saltmarshes, grassy wetland
**Form:** Stiffly erect tufted annual grass
**Foliage:** Slender, bristle-like, erect leaves, sheaths loose, the uppermost leaf clasping the base of the panicle
**Flowers:** Very narrow erect panicle to 15 cm long with narrow 6–12-flowered spikelets to 10 mm long, pale green to greyish. September to January
**Requirements:** Poorly drained saline soils
**Propagation:** Seed
**Localities:** Laverton, Port Melbourne, 2–4, 26, 74
**Distribution:** Vic, NSW, Tas, SA, WA; NZ *Illustrated p. 294*

## Puccinellia stricta var. perlaxa

**Size:** to 50 cm high **F**
**Habitat:** Saltmarsh
**Comments:** Annual or perennial grass with pale green to bluish-green folded to inrolled leaves to 30 cm long. Panicles are larger than the typical form, to 30 cm long, and widely spreading. Spikelets are smaller, to 8 mm long, often purplish and 4–5-flowered.
**Localities:** Altona, Werribee, Footscray, 4, 5
**Distribution:** Vic, Tas

# Restio — Restionaceae

## Restio tetraphyllus — Tassel Cord-rush

F P

**Size:** to 1.5 m high
**Habitat:** Swamps in wattle tea-tree scrub, tea-tree heath
**Form:** Tall bright green feathery perennial tussock, stems smooth and terete with brown sheathing bracts at regular intervals
**Foliage:** Fine soft much-divided leaves or branchlets from a node just below the inflorescence
**Flowers:** Dioecious; loose tassel-like panicles of many brown or reddish ovoid to round spikelets; glumes with awn-like points; October to February
**Requirements:** Permanently boggy sandy soils
**Comments:** An attractive gracefully arching but non-invasive sedge-like plant for planting in or around ponds. The foliage is useful in floral arrangements. Very rare in the Melbourne area.
**Propagation:** Seed, division
**Localities:** 73–75
**Distribution:** Vic, Qld, NSW, Tas, SA

*Restio tetraphyllus*

# Ruppia — Ruppiaceae/Potamogetonaceae

## Ruppia maritima — Sea Tassel or Ditch Grass

F

**Size:** to 15 cm high
**Habitat:** Saltmarsh
**Form:** Rhizomatous, aquatic perennial with fine, erect branched stems to 50 mm long
**Foliage:** Clusters of thread-like leaves to 15 cm long, sheathed broadly at base
**Flowers:** 1 or 2 flowers on a straight stalk to 50 mm long, flowers pollinated below the surface; July to December
**Requirements:** Brackish water
**Comments:** Waterbirds eat the seeds, leaves and starchy leaf bases of all species.
**Propagation:** Division of rhizome, setting ripe fruit in mud
**Localities:** Port Phillip Bay, Kororoit Ck
**Distribution:** All states; NZ, cosmopolitan

## Ruppia megacarpa — Widgeon Grass

F

**Size:** stems to 1.5 m high but usually 20–30 cm high
**Habitat:** Estuarine aquatic, in water to 0.3 m deep, salt pans
**Comments:** More robust perennial herb than the other 2 Melbourne species. Bright to olive green thread-like leaves 5–25 cm long with 2-lobed tip, sheath to 40 mm long. Flowers rise to the surface on stalk 0.3–1 m long, stalk becoming coiled after pollination drawing the flowers below the surface, where the fruit ripens. August to December.
**Localities:** Altona, 5
**Distribution:** Vic, NSW, SA, WA; NZ

## Ruppia polycarpa — Widgeon Grass

F

**Size:** stems 20–40 cm high
**Habitat:** Saltmarshes and estuaries
**Comments:** Annual or delicate perennial similar to *R. maritima* but stems crowded. Dark green leaves to 15 cm long ending in a blunt tip. Flower stalks to 50 cm long each with 2 flowers, pollinated above the water surface but drawn underwater as fruit ripens. June to November
**Localities:** 2, 4
**Distribution:** Vic, NSW, Tas, SA, WA; NZ

*Ruppia maritima*

# Schoenoplectus — Cyperaceae

## Schoenoplectus dissachanthus — Club-rush

F P

**Size:** stems to 30 cm high
**Habitat:** Grassy low open forest
**Form:** Slender annual rush-like herb with a creeping rhizome and tufted triangular stems; leaves reduced to sheaths
**Flowers:** Terminal cluster of 1–3 sessile, cylindrical yellow or greenish-yellow, many-flowered spikelets to 15 mm long with scales at the base of nut. Bract at base of cluster to 100 mm long, appears to extend stem. A single female flower is found in a sheath at the base of the flowering stem. March to April
**Requirements:** Moist to wet sandy soil
**Comments:** Extremely rare in the Melbourne area.
**Localities:** 74
**Distribution:** All mainland states
**Synonym:** *Scirpus dissachanthus*

SCALE: 0 1cm 2cm — The scale representation on each plant equals 2cm.

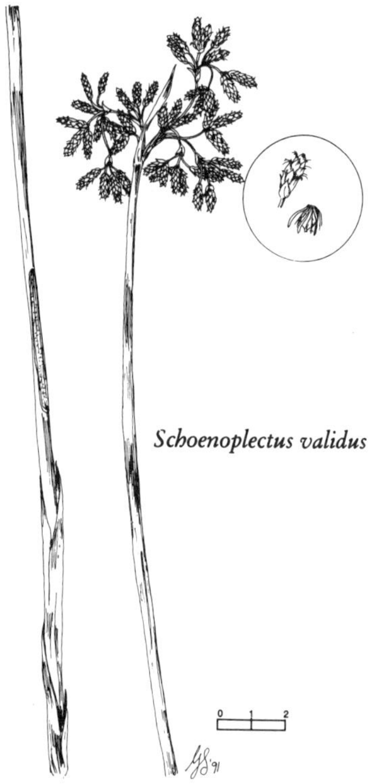

Schoenoplectus validus

### Schoenoplectus pungens
American Club-rush, Sharp Leaf-rush

**Size:** stems 30–60 cm high **F P**
**Habitat:** Grassy wetlands
**Form:** Robust tufted rhizomatous herb, stems stout and sharply triangular
**Foliage:** Bluish grass-like basal leaves with long sheaths shorter than the stems
**Flowers:** Cluster of 1–6 usually sessile dark brown spikelets to 10 mm long. Bract at base of cluster to 60 mm long appears as a continuation of the stem. October to March
**Requirements:** Wet or swampy soils
**Comments:** Has become rare as a result of development in Melbourne.
**Propagation:** Seed
**Localities:** Rockbank, Broadmeadows
**Distribution:** Vic, SA, WA; NZ, Eur, N Am, S Am
**Synonym:** *Scirpus americanus* *Illustrated p. 294*

### Schoenoplectus validus
River Club-rush

**Size:** stems to 2 m high **F P**
**Habitat:** Riparian and swamp scrub
**Form:** Robust tufted perennial rhizomatous herb, stems soft, terete and grooved
**Flowers:** Loose umbel-like panicle to 70 mm long, with numerus many-flowered ovoid spikelets to 10 mm long, glumes keeled and hairy along margins; bract shorter than flowerhead. November to April
**Requirements:** Wet soil
**Comments:** The stems have been used for weaving mats or baskets.
**Localities:** Sunshine, Deer Park, 11, 14, 16, 26, 28, 35, 36, 42, 72
**Distribution:** All states except NT

# Schoenus — Cyperaceae

*Schoenus*, or bog sedges, are tufted annual or perennial plants with a short creeping rhizome, occurring in moist areas which are often subject to regular inundation. Leaves are usually basal, sometimes reduced to a sheath. Flowers are in a cluster of spikelets, the rhachilla (or spikelet stem) strongly zig-zagged and curved over each nut. Plants propagate readily by division or seed.

*Schoenus apogon*

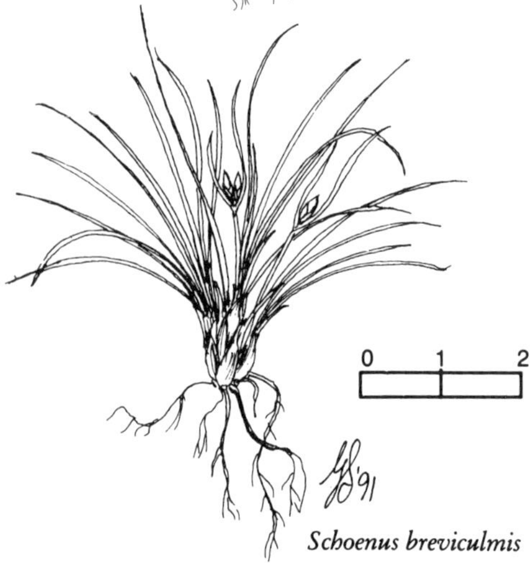

*Schoenus breviculmis*

### Schoenus apogon
Common Bog-rush

**Size:** 5–30 cm high **F P**
**Habitat:** Sclerophyll and red gum woodland, plains grassland, valley sclerophyll forest, wattle tea-tree heath
**Form:** Slender perennial tufted herb, stems terete, longitudinally grooved with 1–3 nodes below the flowerhead
**Foliage:** Fine inrolled leaves as long as the stems, or shorter in tall plants
**Flowers:** 2–5 clusters of dense, blackish, 2–3-flowered spikelets to 7 mm long; 1 cluster terminal, the others at a node in the axil of a leafy bract. Bracts longer than clusters. October to January
**Requirements:** Moist or wet soils
**Comments:** A widespread and variable species giving a grass-like appearance to a moist area. Often found growing in gardens.
**Localities:** 2, 4, 6, 7, 12, 13, 16–18, 25, 26, 28, 29, 32, 34A–36, 41, 42, 44, 47–59, 62, 64, 67#, 69–71 73–76
**Distribution:** All states except WA; PNG, NZ

### Schoenus breviculmis
Matted Bog-rush

**Size:** to 50 mm high **F P**
**Habitat:** Wattle tea-tree scrub
**Form:** Dwarf matting perennial herb with short erect stems hidden amongst the dense foliage, leaves flat
**Flowers:** Terminal cluster of 2 several-flowered spikelets to 10 mm long, surrounded by 2–3 floral bracts. Flowers most of the year.
**Requirements:** Moist peaty soil
**Localities:** 67, 68
**Distribution:** Vic, SA, WA

### Schoenus brevifolius
Zig-zag or Short-leaf Bog-rush

**Size:** 0.3–0.6 m high **F P**
**Habitat:** Tea-tree heath, grassy low open forest
**Form:** Large leafless herb, leaves reduced to basal, glabrous brown sheaths to 10 mm long
**Flowers:** Long panicle with loose, branched clusters of 3–5-flowered spikelets to 10 mm long, rhachillas very flexuose; bracts much smaller than spikelets.
**Requirements:** Moist sandy soil
**Localities:** 68–70, 73–75
**Distribution:** Vic, Qld, NSW, Tas; NZ, N Cal *Illustrated p. 294*

## Schoenus latelaminatus

Medusa Bog-rush

**P**

**Size:** 10–25 cm high
**Habitat:** Valley sclerophyll forest, red gum woodland
**Form:** Annual pale-green grass-like herb similar to *S. apogon* but stems are lax and do not have nodes
**Flowers:** Clusters of 1–3 tiny pale spikelets in the bract axils
**Requirements:** Temporarily wet sites
**Comments:** A very rare plant in the Melbourne area.
**Localities:** Epping, 42
**Distribution:** Vic, NSW, Tas, SA

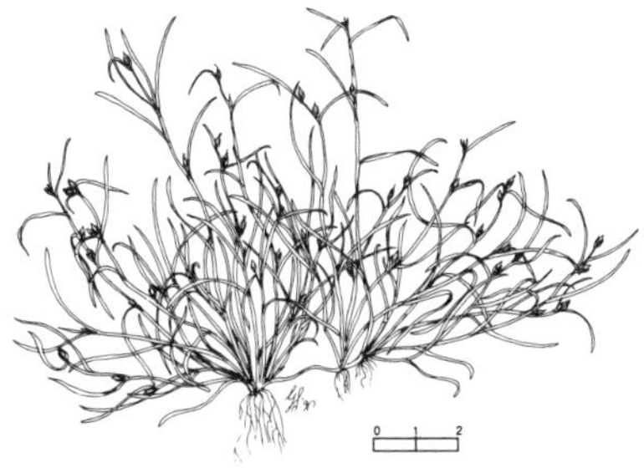

*Schoenus maschalinus*

## Schoenus maschalinus

Leafy Bog-rush

**P**

**Size:** stems to 25 cm long
**Habitat:** Valley sclerophyll forest, wattle tea-tree scrub
**Form:** Weak slender herb with prostrate to ascending fine leafy stems, sometimes branched
**Foliage:** Very narrow weak flat leaves to 30 mm long
**Flowers:** Clusters of 1–3 tiny, 1–2-flowered spikelets to 3 mm long in the upper axils; October to February; bracts longer than spikelets
**Requirements:** Moist heavy soils
**Comments:** Forms a fine bright green mat along the water's edge of ponds.
**Localities:** Diamond Ck, 44, 51, 53, 55–57, 67#, 69, 69A, 74, 75
**Distribution:** All states; NZ, PNG, Philippines

## Schoenus nitens

Shiny Bog-rush

**F P**

**Size:** 3–15 cm high
**Habitat:** Saltmarsh
**Form:** Slender-stemmed perennial, few short basal leaves with glabrous shining sheaths
**Flowers:** Terminal cluster of 2–4 tiny ovate, blackish, 1–3-flowered spikelets; September to March
**Requirements:** Moist soils
**Comments:** A coastal plant tolerating brackish conditions.
**Localities:** 4, 5, 67#, 69A
**Distribution:** All states except NT; NZ, PNG

*Illustrated p. 294*

## Schoenus tenuissimus

Slender Bog-rush

**F P**

**Size:** 10–25 cm high
**Habitat:** Wattle tea-tree and swamp scrub
**Form:** Slender wiry herb with leaves reduced to basal sheaths
**Flowers:** Single, 1-flowered, narrow blackish terminal spikelet to 10 mm long; most of the year.
**Localities:** Montrose, 50, 54, 67, 69, 70, 73–75
**Distribution:** Vic, NSW, Tas, SA

*Illustrated p. 294*

*Schoenus tesquorum*

## Schoenus tesquorum

Bog-rush

**P**

**Size:** 15–40 cm high
**Habitat:** Swamp and riparian scrub, red gum woodland
**Comments:** Similar to and confused with *S. apogon*, but with coarser stems (to 40 cm long) and leaves (to 18 cm long). Stems have a single node below the flowerhead. A rare plant.
**Localities:** Craigieburn, Clayton, 54, 61, 74, 75
**Distribution:** Vic, SA

# Sparganium — Sparganiaceae

## Sparganium subglobosum

Floating Burr-reed

**F P**

**Size:** 0.3–0.6 m x 0.3 m
**Habitat:** Aquatic in fresh, still or moving water to 0.5 m deep
**Form:** Rhizomatous grass-like plant
**Foliage:** Few, soft but stout, erect narrow leaves to 5 mm wide, floating or submerged, lowest leaves longest, to 60 cm, rising from the rhizome
**Flowers:** Unbranched stems to 30 cm high with a flexuose terminal spike of 2–9 sessile, globular flowerheads; male flowers at the top of the spike, female below. Burr-like clusters larger in fruit, to 15 mm wide, each nut beaked; Summer
**Requirements:** Permanent moisture
**Localities:** Yarra R. near Melbourne#, Cranbourne
**Distribution:** Vic, Qld, NSW; NZ
**Synonym:** S. antipodum

*Sparganium subglobosum*

SCALE: 0 1cm 2cm — The scale representation on each plant equals 2cm.

*Spinifex sericeus*

# Spinifex — Poaceae

## Spinifex sericeus

Hairy Spinifex

F

**Size:** to 50 cm high
**Habitat:** Primary dune scrub, beaches
**Form:** Strong perennial grass with long-creeping stolons
**Foliage:** Limp, silvery, flat densely hairy leaves 10–40 cm long, inrolled in bud, tufts of leaves produced along the stolons at thickened nodes
**Flowers:** Dioecious; terminal clusters of spikes; stalked, orange-brown male spikelets to 12 mm long in erect racemes to 50 mm long; female plant has large silvery-green to straw-coloured head-like clusters to 22 cm wide of sessile 1–2-flowered spikelets, each at the base of a long bristle; September to January
**Requirements:** Excellent drainage in sandy soil, with deep moisture
**Comments:** A very useful plant for stabilising sand dunes. Seed heads on female plants snap off whole, like a "tumble weed".
**Propagation:** Seed
**Localities:** Altona, Seaford, 5, 67, 69A, 74
**Distribution:** Vic, Qld, NSW, Tas, SA
**Synonym:** Segregated from *S. hirsutus* which only occurs in WA

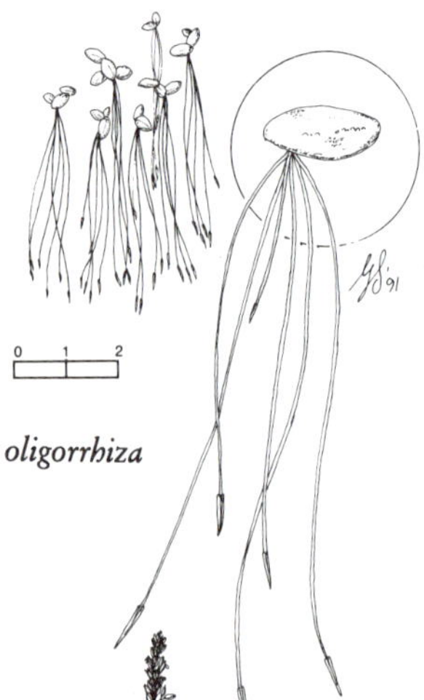

*Spirodela oligorrhiza*

# Spirodela — Lemnaceae

## Spirodela oligorrhiza

Thin Duckweed

F P

**Size:** thallus 3–6 mm long x 2–4 mm wide
**Habitat:** Fresh still water in riparian scrub, tea-tree heath
**Form:** Floating aquatic forming dense mats on the surface of still water. Individual plant consists of single thin ovate leaf-like thallus, often tinged purplish below, with 3–8 roots to 40 mm long.
**Propagation:** As for *Lemna disperma*
**Localities:** Yarra R., 36, 74, 76
**Distribution:** All mainland states; NZ, N Cal, Asia
**Synonym:** *Lemna oligorrhiza*

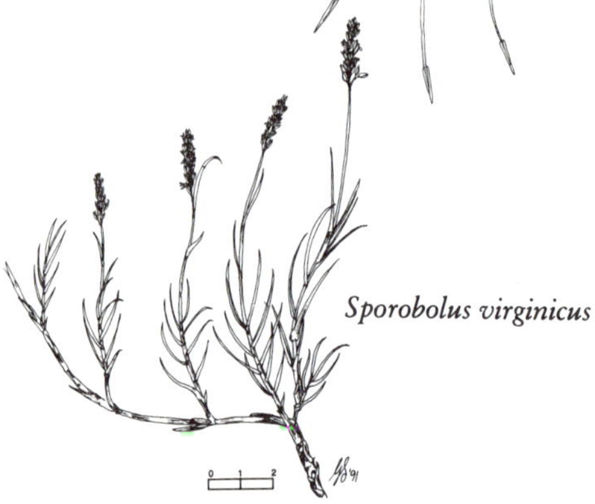

*Sporobolus virginicus*

# Sporobolus — Poaceae

## Sporobolus virginicus

Salt or Sand Couch

F

**Size:** 10–30 cm x spreading
**Habitat:** Estuaries, saltmarshes, primary dune scrub, coastal cliffs
**Form:** Rhizomatous perennial grass with leafy stems and thick, extensively creeping scaly rhizomes
**Foliage:** Pairs of rigid, wiry leaves with overlapping sheaths and short blades (to 50 mm long) and inrolled margins
**Flowers:** Dense, narrow spike-like panicle to 60 mm long with tiny, dark 1-flowered spikelets; all year.
**Requirements:** Moist sandy soils, tolerating saline conditions
**Comments:** A useful grass for salt-laden soils.
**Propagation:** Seed, division of rhizome
**Localities:** Port Melbourne, Altona, Seaford, around Port Phillip, 5, 74
**Distribution:** All states; NZ, PNG, Pacific, S Am, S Af

# Stipa — Poaceae

- Genus has been reviewed + all Australian species are now in the genus Austrostipa

*Stipa* species are tufted perennial grasses, with 200 species occurring in the Melbourne region. Several species found on the grasslands of the basalt plains are now quite rare. They are commonly known as spear grasses as the mature seeds can penetrate the skin, mouths and eyes of stock. Leaves are overlapping and hairy, often terete. The inflorescence is a panicle of single, stalked 1-flowered spikelets which have a hard cylindrical lemma sharply-pointed at the base (or callus). The terminal awn twists and bends on drying and may be very long. The lower section is termed the column and twists, the section above the bend does not twist and is called the bristle. Glumes remain on the plant after the seeds have fallen.

As horticultural subjects they can add a graceful accent to a bushland garden or rockery and are generally at their best in Spring and Summer when the soft new growth and feathery flowering panicles are present. Hard pruning of tussocks after flowering is recommended to maintain vigour and to remove dead leaves and flower stems. Plants will grow in full sun to semi-shade.

Some species such as *S. elegantissima* are often available from specialist nurseries. Plants are usually readily grown from seed which is easily collected, but be wary of introduced *Nassella* spp. which appear very similar and are major weeds. Often a high proportion of florets do not produce viable seed. Division of clumps is also successful.

## Stipa aristiglumis

Plump or Bristly Spear-grass

**Size:** to 40 cm high; stems 0.5–1 m high **F P**
**Habitat:** Plains grassland, red gum woodland
**Form:** Erect, densely tufted, stout or slender grass
**Foliage:** Narrow, glabrous, smooth leaves to 40 cm x 3–6 mm, inrolled from the base or only in the upper part
**Flowers:** Loose few-flowered panicle 10–40 cm long with long branches, lemmas swollen, awn twice bent, to 40 mm long, callus is hooked and very short in relation to lemma length, glumes prominently green-veined; October to February.
**Requirements:** Heavy soils
**Comments:** Becoming less common as habitat disappears.
**Localities:** Tottenham, St Albans, 5, 6, 7, 13, 26, 28
**Distribution:** Vic, Qld, NSW, SA

*Stipa aristiglumis*

## Stipa bigeniculata

Tall Spear-grass, Plains Grass

**Size:** to 20 cm high; stems 0.3–1.2 m high **F P**
**Habitat:** Plains grassland, grassy wetland, red gum and box woodland
**Form:** Slender, erect tussock
**Foliage:** Narrow ridged leaves 8–20 cm long, lower leaves rough
**Flowers:** Very similar to *S. blackii*. Panicle 25–45 cm long; lemma to 10 mm long, hair tuft at the top to 2 mm long, awn to 60 mm long strongly twice bent, callus curved, longer in relation to lemma length; September to February.
**Requirements:** Heavy clay soils on flood plains
**Localities:** 4, 6, 12, 14, 15, 17#, 18, 21, 22, 23, 26
**Distribution:** Vic, NSW

*Illustrated p. 294*

## Stipa blackii

Crested Spear-grass

**Size:** to 20 cm high; stems 0.45–1 m high **F P**
**Habitat:** Red gum woodland, plains grassland, box woodland
**Form:** Densely tufted erect grass
**Foliage:** Rough, stiff, linear leaves to 20 cm x 2–5 mm, lower leaves hairy
**Flowers:** Similar to *S. aristiglumis* and *S. bigeniculata*. Panicle narrower, 10–25 cm long with purplish or straw-coloured spikelets. Lemma covered in white spreading hairs, becoming yellow to orange, with a conspicuous tuft of hairs to 5 mm long on apex; awn to 40 mm long and twice bent. September to November.
**Requirements:** Moist heavy soils
**Comments:** Disappearing from Melbourne areas as habitat is destroyed.
**Localities:** Toolern Vale, Melton Sth, Sunshine, Deer Pk, Sth Yarra, Mt Cottrell, 7, 12, 13, 15, 20, 26, 32
**Distribution:** Vic, NSW, SA, WA

*Illustrated p. 294*

*Stipa curticoma*

## Stipa curticoma

Spear-grass

**Size:** to 30 cm high; stems to 1.2 m high **F P**
**Habitat:** Grassy wetlands, plains grassland, red gum woodland
**Form:** Open tufted grass
**Foliage:** Leaves to 30 cm x 3–5 mm
**Flowers:** Sparse open panicle 15–30 cm long. Glumes broad. Lemma hairs white, with tiny hair-tufts at top of lemma, awn 45–65 mm long with dense, fine spreading hairs on twice-bent column. September to December.
**Requirements:** Moist heavy soils, intolerant of dry conditions
**Comments:** A rare grass in Victoria.
**Localities:** Diggers Rest, Laverton, Sunshine, Derrimut, Deer Park, St Albans, 2, 10, 17#
**Distribution:** Vic, SA

## Stipa densiflora

Foxtail Spear-grass

**Size:** to 20 cm high; stems 0.5–1 m high **F P**
**Habitat:** Box and box ironbark woodland, dry sclerophyll forest
**Foliage:** Lax, hairy leaves to 130 mm long, flat or inrolled, tapering to a fine point
**Flowers:** Dense compact panicle, to 30 cm long. Glumes are downy, lemma to 15 mm long, awn to 45 mm long, column twice-bent, with feathery hairs. September to January.
**Requirements:** Well drained gravelly soils
**Localities:** Diamond Ck, 14, 26, 36, 42, 55
**Distribution:** Vic, Qld, NSW, SA

*Stipa densiflora*

SCALE: 0 1cm 2cm — The scale representation on each plant equals 2cm.

*Puccinellia stricta* var. *stricta*

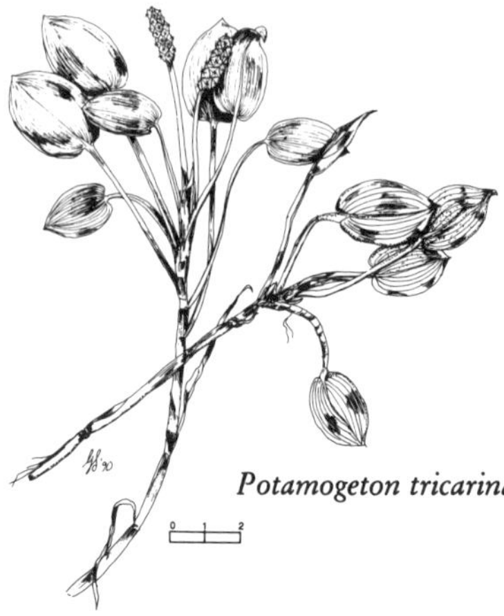

*Potamogeton tricarinatus*

*Poa tenera*

*Schoenus brevifolius*

*Schoenus nitens*

*Schoenoplectus pungens*

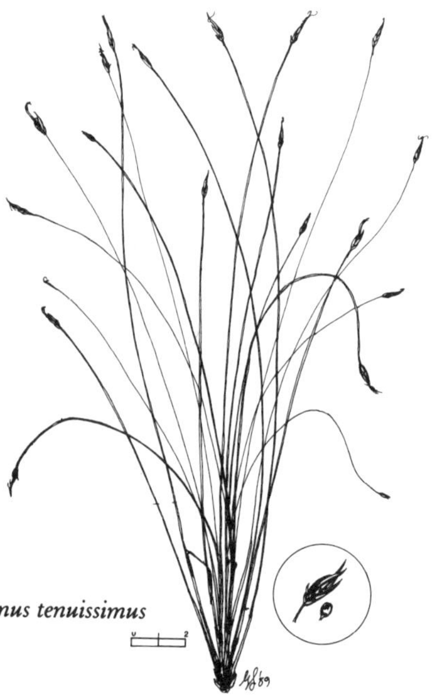

*Schoenus tenuissimus*

*Stipa blackii*

*Stipa bigeniculata*

## Stipa elegantissima
Feather Spear-grass, Spider Grass

**Size:** 0.5–0.8 m high; stems to 1 m high **F P**
**Habitat:** Box and red gum woodland
**Form:** Rhizomatous tussock, cane-like stems glabrous and often branched
**Foliage:** Rough, narrow, inrolled leaves 50–80 cm long,
**Flowers:** Large, very loose open panicle to 25 cm long, branches whorled, with long feathery, sometimes pinkish, hairs on branches and stalks. Glumes 3-ribbed, purplish with scattered hairs, lemma black at maturity. Awn to 50 mm long and strongly bent, column to 15 mm long, once-bent, with rough hairs. August to January
**Requirements:** Dry sandy or open soil
**Comments:** A particularly ornamental grass.
**Localities:** Altona, St Albans, Sth of Melton, Sunbury, 22, 23, 25, 26, 32, 74, 76
**Distribution:** Vic, NSW, SA, WA

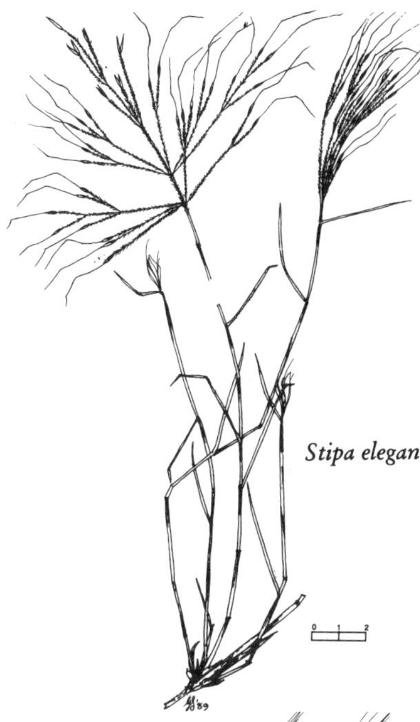

*Stipa elegantissima*

## Stipa exilis

**Size:** stems to 0.6 m high **F P**
**Habitat:** Woodland
**Form:** Small densely tufted perennial grass, stems erect, very fine with downy nodes
**Foliage:** Downy, flexuose, very narrow leaves to 20 cm long, tightly inrolled
**Flowers:** Narrow, few-flowered panicle to 20 cm long. Glumes purplish, narrow and tapering, unequal to 10 mm long; dark brown lemma lightly covered in whitish hairs with a short tuft of hairs at the apex, palea has band of silky hairs along centre back. Awn to 55 mm long, twice bent. August to November
**Requirements:** Well drained heavy soil
**Comments:** A rare grass in Victoria with a disjunct distribution in Melbourne.
**Localities:** Melton
**Distribution:** Vic, SA, WA *Illustrated p. 304*

## Stipa flavescens
Spear-grass

**Size:** 0.5–0.7 m high; stems to 1.2 m high **F P**
**Habitat:** Primary dune scrub, saltmarshes
**Form:** Open, shortly rhizomatous tufted grass, often branching after the season's first seeds begin to mature
**Foliage:** Narrow flexuose leaves to 70 cm x 7 mm
**Flowers:** Dense narrow many-flowered panicle to 40 cm long with erect branches. Glumes to 15 mm long, 3-ribbed, purplish with scattered hairs. Awn is slender to 70 mm long, column twice-bent, densely and softly hairy. September to February
**Requirements:** Moist well drained soils, tolerant of salt spray
**Comments:** Useful grass for front-line coastal planting and stabilisation of sand dunes.
**Localities:** Sandringham-Mt. Eliza, Altona, Mt Cottrell, 2–5, 49, 62, 63
**Distribution:** Vic, NSW, Tas, SA, WA
**Synonym:** *S. elatior*

*Stipa flavescens*

## Stipa gibbosa
Spear-grass

**Size:** to 30 cm high; stems to 1.5 m high **F P**
**Habitat:** Plains grassland, red gum woodland
**Foliage:** Few ridged leaves to 30 cm x 2.5 mm
**Flowers:** Narrow panicle on a rigid, erect stem. Differs from *S. bigeniculata* in that the lemma, to 6 mm long, is distinctly swollen, awn is shorter to 35 mm long; October to January.
**Requirements:** Moist soils which do not dry out
**Comments:** Becoming very rare in Melbourne as its habitat is destroyed for housing. Also rare throughout Victoria.
**Localities:** Laverton, Altona, Mt Cottrell, Rockbank, 4, 16–18
**Distribution:** Vic, NSW, SA *Illustrated p. 305*

## Stipa mollis
Soft Spear-grass

**Size:** to 30 cm high; stems to 1.5 m high **F P**
**Habitat:** Dry sclerophyll forest, red gum woodland, grassy low open forest
**Form:** Robust, erect tufted grass
**Foliage:** Limp, downy, tightly inrolled leaves to 30 cm x 1–3 mm
**Flowers:** Differs from *S. semibarbata* in that hairs on column are longer to 4 mm and spiral up the awn towards the end of the bristle, glume shorter, to 20 mm long; October to January
**Requirements:** Moist soils
**Comments:** A more robust species than *S. semibarbata*.
**Localities:** Alamein, Brighton, 26, 32, 34A, 36, 58, 59, 66–70, 74, 76
**Distribution:** Vic, NSW, Tas, SA, WA
**Synonym:** *S. semibarbata* var. *mollis;* segregated from *S. hemipogon*

*Stipa mollis*

SCALE: 0 1cm 2cm — The scale representation on each plant equals 2cm.

*Stipa muelleri*

## Stipa muelleri

Wiry Spear-grass

**F P**

**Size:** to 1 m high x spreading
**Habitat:** Sclerophyll woodland
**Form:** Bright green wiry rhizomatous grass forming dense tangles, usually branching at the nodes, leaves scale-like
**Flowers:** Sparse unbranched panicle consisting of 1–3 large spikelets to 30 cm long, lemmas hairy with 2 erect lobes, awn rigid, 50–100 mm long; October to December.
**Requirements:** Well drained clay soil
**Localities:** Olinda, Montrose, 56, 59
**Distribution:** Vic, SA

## Stipa nodosa

Spear-grass

**F P**

**Size:** to 30 cm high; stems 0.5–1 m high
**Habitat:** Red gum woodland, plains grassland
**Form:** Tufted grass without rhizomes, nodes on stems much wider than stem
**Foliage:** Coarse, flat or inrolled leaves to 30 cm long, many leaves also on stems
**Flowers:** Sparse, spreading panicle to 50 cm long, awns to 100 mm long with sickle-shaped bristles; August to September
**Requirements:** Moist well drained soils
**Localities:** Sth Yarra, 16, 20, 21, 31, 32, 34A, 35, 50A, 65
**Distribution:** Vic, Qld, NSW, SA
**Synonym:** Segregated from *S. variabilis*

*Illustrated p. 305*

*Stipa pubinodis*

## Stipa oligostachya

Spear-grass

**F P**

**Size:** to 20 cm high; stems to 1 m high
**Habitat:** Plains grassland, grassy wetlands, red gum woodland
**Foliage:** Erect, tightly inrolled leaves to 20 cm long
**Flowers:** Sparse, spreading panicle to 25 cm long, pale lemma warty, only hairy on veins, awn to 70 mm long, twice bent then slightly curved; November
**Requirements:** Moist heavy soils
**Localities:** Sunbury, Deer Park, St. Albans, 4, 6, 25
**Distribution:** Vic, SA

*Illustrated p. 305*

## Stipa pubinodis

Spear-grass

**P**

**Size:** to 35 cm high; stems 0.6–1.5 m high
**Habitat:** Valley and dry sclerophyll forests, grassy low open forest
**Foliage:** Rough, ridged tightly inrolled leaves to 35 cm x 1–2 mm
**Flowers:** Sparse panicle 15–25 cm long with more than 20 flowers. Lemma white hairy and granular, to 12 mm long, with a stout, rigid twice-bent awn to 100 mm long, column to 60 mm long; tip of glume appears cut off and soon tears; September to January.
**Requirements:** Moist soils
**Comments:** May be confused with *S. rudis* which has a smaller lemma and glume and much shorter, more slender awn column. It also prefers damper shadier positions although the two do overlap in habitat.
**Localities:** 16, 31, 32, 42, 44, 47, 48, 50A, 51, 54–56, 58, 61–64, 69A, 70, 74, 76
**Distribution:** Vic, Tas, SA
**Synonym:** Segregated from *S. pubescens* which does not occur in Victoria.

*Stipa rudis* ssp. *rudis*

## Stipa rudis ssp. rudis

Veined Spear-grass

**P N**

**Size:** to 40 cm high; stems to 1.2 m high
**Habitat:** Sclerophyll and red gum woodland, dry and valley sclerophyll forests, grassy low open forest
**Foliage:** Erect inrolled leaves to 40 cm long, woolly to slightly rough on one side; terete slightly ridged stems with 2–4 downy nodes, bent near the base
**Flowers:** Loose spreading many-flowered panicle to 40 cm x 15–40 mm. Lemma to 10 mm long, with fawn to golden hairs, awn twice bent, to 60 mm long with column to 25 mm long.
**Requirements:** Moist soils
**Localities:** Lilydale, Montmorency, 4, 12, 15, 26, 36, 42, 44, 47–49, 50A, 51, 53–56, 62–64, 69A, 71, 75
**Distribution:** Vic, Qld, NSW
**Synonym:** *S. nervosa* var. *neutralis*

## Stipa rudis ssp. australis

Spear-grass

**F P**

**Size:** stems to 1 m high
**Habitat:** Dry sclerophyll forest, grassy low open forest
**Comments:** Leaves are densely woolly and flower parts are larger than in ssp. *rudis*. Lemma is straw-coloured or purple with an awn 60–90 mm long, column to 45 mm long.
**Localities:** Wantirna, 49, 73, 74
**Distribution:** Vic, Tas

## *Stipa rudis* ssp. *nervosa*

Spear-grass

**F P**

**Size:** stems 0.5–1.2 m high
**Habitat:** Valley sclerophyll forest
**Comments:** Very similar to ssp. *rudis*. The panicle is larger and narrower (to 50 cm x 30 mm). Awn and column are smaller (to 45 mm and to 20 mm respectively) and the lemma has white hairs.
**Localities:** Eltham, 45, 47, 50A, 55, 60, 61, 64
**Distribution:** Vic, Qld, NSW
**Synonym:** *S. nervosa* var. *nervosa*

*Stipa rudis* ssp. *nervosa*

## *Stipa scabra* ssp. *scabra*

Rough Spear-grass

**F P**

**Size:** to 25 cm high; stems 0.3–0.5 m high
**Habitat:** Plains grassland, box and red gum woodland
**Form:** Slender, tufted perennial grass with erect stems
**Foliage:** Fine, ridged inrolled or folded leaves to 25 cm long with short stiff hairs
**Flowers:** Dense narrow few-flowered panicle to 30 cm x 10–20 mm, with erect branches and spikelets; awn to 70 mm long, sickle-shaped with a short column to 15 mm long; glumes usually purplish; August to October
**Requirements:** Well drained soils
**Localities:** 9, 12, 14, 18, 21–23, 26, 29, 49
**Distribution:** All states
**Synonym:** Segregated from *S. variabilis*

*Illustrated p. 305*

## *Stipa scabra* ssp. *falcata*

Slender Spear-grass

**F P**

**Habitat:** Plains grassland, box woodland
**Flowers:** Erect panicle to 25 cm long with spreading branches; August to December.
**Comments:** This subspecies has a more spreading panicle than the typical species. A form from Keilor is intermediate between the 2 subspecies.
**Localities:** 4, 10, 16, 17#, 24–26, 32
**Distribution:** Vic, NSW, SA
**Synonym:** *S. falcata*

*Stipa scabra* ssp. *falcata*

## *Stipa semibarbata*

Fibrous Spear-grass

**F P**

**Size:** to 30 cm high; stems 0.3–1 m high
**Habitat:** Plains grassland, red gum and box woodland, valley sclerophyll forest, grassy low open forest
**Form:** Erect, tufted grass
**Foliage:** Rough leaves 30 cm x 1–3 mm (occasionally to 7 mm), becoming torn and fibrous with age
**Flowers:** Large, dense stiffly erect panicle to 32 cm long with more than 30 spikelets. Awn to 90 mm long, twice bent, with short feathery hairs to 1 mm covering the stout column (but not present on the bristle) glumes to 25 mm long, tapering. August to November
**Requirements:** Well drained soil
**Comments:** This grass is suited to dry sandy soils. It is often confused with *S. mollis*.
**Localities:** Eltham, 2, 4, 12, 21, 24–26, 30, 32, 34A, 58, 65, 66, 71, 74–76
**Distribution:** Vic, NSW, Tas, SA

*Illustrated p. 305*

## *Stipa setacea*

Corkscrew Grass

**F P**

**Size:** to 30 cm high; stems 0.3–0.8 m high
**Habitat:** Red gum woodland, box woodland
**Form:** Slender tufted perennial grass
**Foliage:** Fine, rough or glabrous inrolled ridged leaves to 30 cm long
**Flowers:** Narrow but loose few-flowered panicle to 20 cm long, branches rough, the uppermost leaf sheath enveloping the base of the panicle. Lemma black with downy white hairs, awn to 40 mm long, twice bent; September to November.
**Requirements:** Heavy, well drained soil of basalt plains
**Localities:** Sunbury, Altona, Mt Cottrell, Rockbank, 16, 26, 49
**Distribution:** Vic, Qld, NSW, SA

*Stipa setacea*

## *Stipa stipoides*

Coast or Prickly Spear-grass

**F**

**Size:** to 1 m high
**Habitat:** Primary dune scrub, coastal cliffs, saltmarshes
**Form:** Hemispherical, densely-tufted rush-like tussock
**Foliage:** Long, prickly, glabrous, terete leaves to 70 cm long
**Flowers:** Tight narrow panicle to 25 cm long, mostly within the tussock. Glumes broad and erect to 20 mm long, lemmas covered in white to yellowish hairs, with 2 small hairy lobes, awn to 40 mm long, twice bent; October to March.
**Requirements:** Well drained sandy soil, tolerant of salt spray
**Localities:** 2, 3, 67, 69A, 74, 76
**Distribution:** Vic, NSW, Tas, SA
**Synonym:** *S. teretifolia*

*Illustrated p. 305*

SCALE: 0 1cm 2cm — The scale representation on each plant equals 2cm.

*Stipa stuposa*

**Stipa stuposa** — Tasmanian Spear-grass

**F P**

**Size:** stems to 1.2 m high
**Habitat:** Plains grassland
**Form:** Robust tufted perennial grass, stem nodes downy
**Foliage:** Leaves to 40 cm long, glabrous or hairy
**Flowers:** Narrow, sparse to many-flowered panicle to 35 cm long. Glumes tapering, downy, to 25 mm long; lemma covered in white to yellow hairs, distinct tuft of hairs at the apex. Long awn to 70 mm, twice bent, column covered in soft hairs to 1.5 mm. November
**Requirements:** Well drained heavy soil
**Localities:** 29
**Distribution:** Vic, NSW, Tas, SA

# Tetraria — Cyperaceae

*Tetraria capillaris*

## Tetraria capillaris — Hair-sedge, Bristle Twig-rush

**F P**

**Size:** 20–50 cm high
**Habitat:** Swamp scrub, damp and valley sclerophyll forest
**Form:** Slender rhizomatous perennial rush-like herb with many limp, thread-like, bright to yellow-green stems, terete with a groove along one side
**Foliage:** Leaves reduced to downy basal sheaths, sometimes with a bristly blade to 7 mm long
**Flowers:** Tiny loose panicle to 25 mm long of 1–8 narrow, reddish-brown 1–2-flowered spikelets to 6 mm long; March to June
**Requirements:** Moist soils
**Propagation:** Seed, division
**Localities:** Olinda, Montrose, 54, 56, 61
**Distribution:** All states except NT; NZ

# Tetrarrhena — Poaceae

*Tetrarrhena distichophylla*

## Tetrarrhena distichophylla — Hairy Rice-grass

**F P**

**Size:** stems to 0.6 m long
**Habitat:** Grassy low open forest
**Form:** Decumbent rhizomatous perennial grass with hairy stems, sometimes forming mats
**Foliage:** Flat or inrolled rough leaves to 60 mm long in pairs along the stem, paler green above
**Flowers:** Short erect terminal spike-like raceme to 20 mm long, spikelets downy, sterile lemmas rough and pointed; September to March
**Requirements:** Moist sandy soil
**Propagation:** Seed, division
**Localities:** Baxter, 60, 73, 74
**Distribution:** Vic, SA, Tas

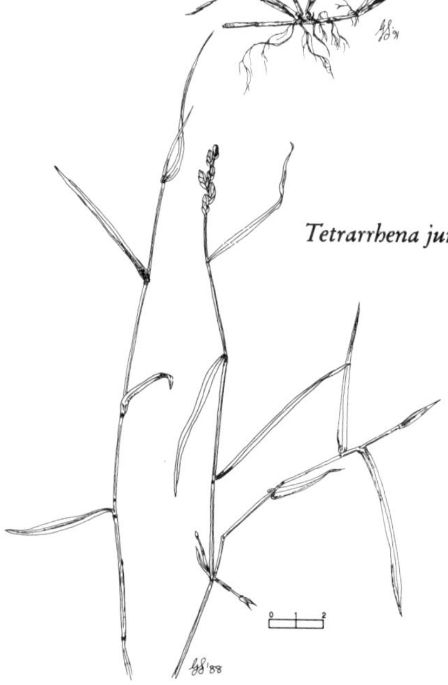

*Tetrarrhena juncea*

## Tetrarrhena juncea — Forest Wire Grass

**P N**

**Size:** stems scrambling 3–5 m long
**Habitat:** Wet, damp and valley sclerophyll forests, grassy low open forest
**Form:** Wiry, many-branched scrambling perennial grass with minutely hooked stems and extensive rhizomes
**Foliage:** Flat, rough leaves to 80 mm long, spread distantly and alternately up the stem
**Flowers:** Raceme with glabrous spikelets, glumes rounded and smooth
**Requirements:** Moist soils, tolerating drying out in Summer
**Comments:** This important component of forested areas is quick to re-establish after fire, preventing soil erosion until other plants regenerate. May cover large areas in moist environments, forming impenetrable barriers.
**Propagation:** Division
**Localities:** 43, 44, 49–52, 54–58, 62, 64, 75
**Distribution:** Vic, Qld, NSW, Tas

# Themeda **Poaceae**

## Themeda triandra

Kangaroo Grass

*Themeda triandra*

| | |
|---|---|
| **Size:** | to 40 cm x 0.75 m, stems 0.7–0.9 m high **F P** |
| **Habitat:** | Widespread in plains grassland, sclerophyll, red gum and box woodlands, grassy wetland, grassy low open forest, dry and valley sclerophyll forests |
| **Form:** | Soft, erect or sprawling perennial tussock |
| **Foliage:** | Narrow-linear, limp, green, purple or sometimes quite blue, leaves to 30 cm x 3–4 mm, keeled below, sometimes rough above |
| **Flowers:** | Leafy interrupted panicles 10–20 cm long on slightly arching stems; glossy-brown spikelets in groups of 6–8, central spikelet fertile and awned, each group surrounded by 4 sessile bract-like spikelets, each cluster sheathed in a leafy bract to 100 mm long; September to February. |
| **Requirements:** | Adaptable to most soils which do not remain wet |
| **Comments:** | A very common grass in many areas of remnant vegetation around Melbourne. The stems and leaves were used as string by the Aborigines to make nets. |
| **Propagation:** | Seed, division |
| **Localities:** | 2–7, 10, 12–18, 20, 21, 23–26, 28–32, 34A–36, 38–42, 44–48, 50–56, 58–71, 74–76 |
| **Distribution:** | All States; PNG, N Cal, SE Asia, Af |
| **Synonym:** | *T. australis* |

# Thismia **Burmanniaceae (Thismiaceae)**

## Thismia rodwayi

Fairy Lanterns

| | |
|---|---|
| **Size:** | stem 10–30 mm high **N** |
| **Habitat:** | Cool temperate rainforest, wet sclerophyll forest |
| **Form:** | Leafless saprophyte, often growing and flowering below the leaf litter, on a creeping mycorrhizal root system to 15 cm long; erect stem has 6 white scale-like bracts |
| **Flowers:** | Single, terminal amber and red to rosy pink lantern-shaped flower to 20 mm long with 12 vertical coloured stripes, outer 3 petals recurved, inner 3 arching inwards. |
| **Requirements:** | Damp humus |
| **Comments:** | A very rare plant known from few sites in Victoria and 1 site in the Melbourne region. |
| **Propagation:** | Impossible |
| **Localities:** | 57 |
| **Distribution:** | Vic, Tas; NZ |

# Tricostularia **Cyperaceae**

## Tricostularia pauciflora

Needle Bog-rush

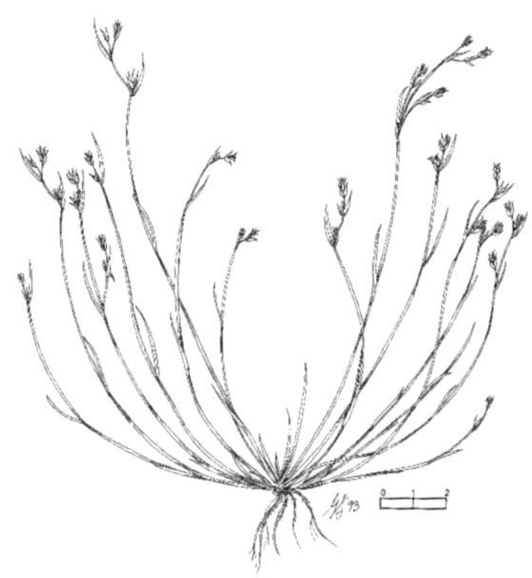

*Tricostularia pauciflora*

| | |
|---|---|
| **Size:** | 10–30 cm high **P** |
| **Habitat:** | Tea-tree heath |
| **Form:** | Tufted shortly rhizomatous perennial rush-like sedge, stems very fine, ridged and curved |
| **Foliage:** | Few basal leaves, reduced to reddish sheaths, sometimes with a green tapering point to 15 mm long |
| **Flowers:** | Short terminal spike to 25 mm long with 1–4 broad, brown 2-flowered spikelets to 6 mm long; December to February |
| **Requirements:** | Damp sandy or peaty soil |
| **Propagation:** | Seed, division |
| **Localities:** | Mordialloc–Mentone, 67, 69 |
| **Distribution:** | Vic, NSW, Tas, SA |

# Triglochin **Juncaginaceae**

## Triglochin centrocarpa

Dwarf Arrowgrass

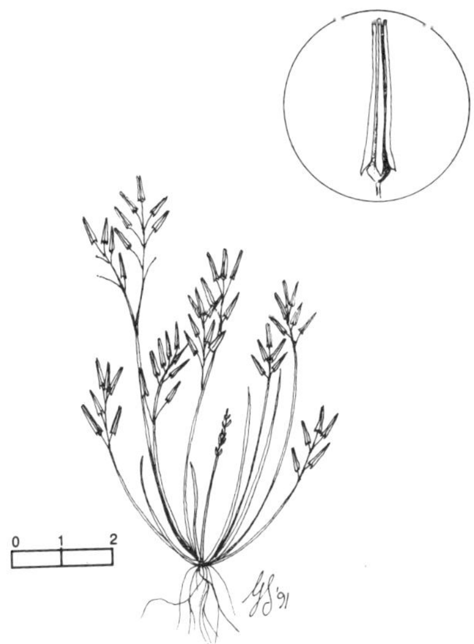

*Triglochin centrocarpa*

| | |
|---|---|
| **Size:** | stem 5–15 cm high **F P** |
| **Habitat:** | Tea-tree heath |
| **Form:** | Slender annual herb |
| **Foliage:** | Thread-like basal leaves to 50 mm long |
| **Flowers:** | Terminal racemes to 60 mm long of 6–23 minute greenish flowers. Fruit narrow-pyramidal, 2–5 mm long with inconspicuous basal spurs; May to October. |
| **Requirements:** | Damp sandy soil, intolerant of permanent water |
| **Comments:** | Presumed extinct within the Melbourne area. |
| **Propagation:** | Seed |
| **Localities:** | 67# |
| **Distribution:** | All States |
| **Synonym:** | *T. centrocarpum* |

SCALE: 0 1cm 2cm The scale representation on each plant equals 2cm.

## Triglochin minutissima

Tiny Arrowgrass

**F P**

**Size:** stems 8–12 cm high
**Habitat:** Saltmarshes
**Requirements:** Moist saline soils
**Comments:** Similar to *T. centrocarpa* but smaller. Leaves 15–40 mm long, flowers are distant in the 4–11-flowered raceme and almost sessile. Fruit are minute to 1.5 mm long, basal spurs deflexed points; August to October. A rare Victorian plant which has almost disappeared in Melbourne.
**Propagation:** Seed
**Localities:** Altona, Port Melbourne#, Brighton#
**Distribution:** Vic, NSW, SA, WA
**Synonym:** *T. minutissimum*

*Triglochin minutissima*

## Triglochin mucronata

Prickly Arrowgrass

**F P**

**Size:** stem 2–10 cm
**Habitat:** Saltmarshes
**Requirements:** Moist saline soil
**Comments:** Differs from *T. centrocarpa* in that the raceme is 1–7-flowered and up to 15 mm long and fruit, to 3 mm long, is broadly top-shaped with a 3-beaked tip; August to October. Very rare in the Melbourne area.
**Propagation:** Seed
**Localities:** Altona, Seaholme, Williamstown, Port Melbourne
**Distribution:** Vic, SA, WA
**Synonym:** T. mucronatum

*Triglochin procera*

## Triglochin procera

Water-ribbon

**F P**

**Size:** stems 20–50 cm high
**Habitat:** Common in shallows of fresh water lakes and streams of riparian and swamp scrub, riparian woodland, damp and valley sclerophyll forests, grassy low open forest
**Form:** Variable and robust aquatic or amphibious perennial herb with a thick rhizome ending in numerous tubers
**Foliage:** Ribbon-like, erect or floating leaves 5–40 mm wide and 0.1–2 m long
**Flowers:** Dense terminal raceme to 30 mm long and 20 mm thick with 60–200 small greenish flowers held erect above the water. Spikes are shorter if not growing in permanent water. Fruit is broad elliptic 10–20 mm long; August to April.
**Requirements:** Submerged in freshwater to 1.5 m deep or areas subject to regular inundation
**Comments:** An attractive plant for water gardens, the long ribbon-like leaves flowing freely with the water currents. Leaves are erect in still water. Small forms also occur in areas of periodic inundation. The tubers were an important food source for the Aborigines and were either roasted or eaten raw.
**Propagation:** Division
**Localities:** Altona, Kororoit Ck, 1, 4, 8, 14, 16, 19, 26, 28, 29, 32, 33, 34A, 35, 42, 44, 47, 54, 56, 61–63, 65–67#, 69A, 71, 72, 74–76
**Distribution:** All States
**Synonym:** *T. procerum*

*Triglochin striata*

## Triglochin striata

Streaked Arrowgrass

**F P**

**Size:** stems 3–25 cm high
**Habitat:** Saltmarshes, swamp and riparian scrub, coastal banksia woodland
**Form:** Slender semi-aquatic perennial herb with small rhizomes
**Foliage:** Fine terete or flattened erect leaves 2–30 cm x 1–3 mm in tufts along the rhizome
**Flowers:** Slender, loose terminal racemes to 20 cm long with 2–100 tiny greenish-yellow flowers. Fruit globular to 3 mm; August to May.
**Requirements:** Moist, poorly drained soils
**Propagation:** Division
**Localities:** Altona, Montrose, 2–4, 11, 17#, 18, 25, 26, 29#, 32, 34A, 35, 42, 44, 47, 54, 56, 58–64, 67#, 69A–72, 74, 76
**Distribution:** All States; NZ, S Af, N Am
**Synonym:** *T. striatum*

*Also illustrated p. 21*

# Tripogon — Poaceae

## Tripogon loliiformis

Rye Beetle-grass, Five-minute Grass

**Size:** 5–25 cm high — F P
**Habitat:** Box woodland, plains grassland on very shallow soil above underlying rock
**Form:** Tufted annual or biennial grass
**Foliage:** Basal, narrow-linear or inrolled, glabrous or silky-hairy leaves to 50 mm long, with white sheaths
**Flowers:** Terminal open spike to 100 mm long. Narrow, erect 6–14-flowered spikelets, sessile in 2 rows along one side of the rhachis; lemma notched with a short awn and a short bearded callus; July to September.
**Requirements:** Dry soils on the basalt plains
**Comments:** A rare Victorian grass, found in few sites in Melbourne. Its common name refers to the speed with which it 'resurrects' after rain.
**Propagation:** Seed
**Localities:** Sunbury, Deer Park, St Albans, Melton, 7, 16–18, 26
**Distribution:** All mainland states

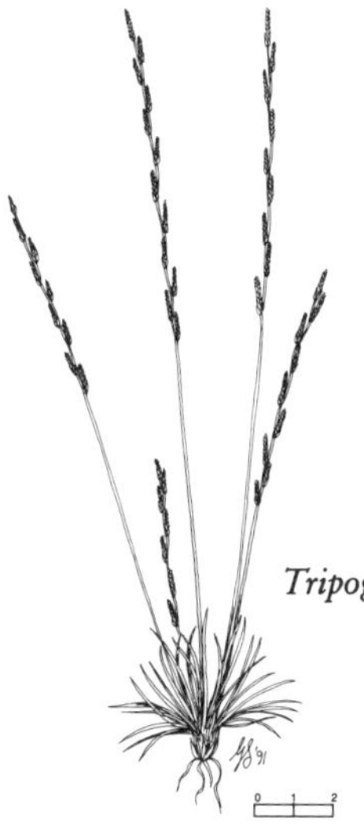

*Tripogon loliiformis*

# Trithuria — Hydatellaceae

## Trithuria submersa

Trithuria

**Size:** stems to 50 mm high — F P
**Habitat:** Grassy wetlands
**Form:** Tiny reddish annual herb
**Foliage:** Tuft of fine pointed leaves to 40 mm long
**Flowers:** Terminal head of reddish flowers lacking petals, each head surrounded by 4–6 spreading bracts on erect, unbranched, thread-like flowering stems; September to November
**Requirements:** Mud
**Comments:** Grows in Winter on the margins of seasonal pools on the Basalt Plains.
**Propagation:** Seed
**Localities:** Laverton
**Distribution:** Vic, NSW, Tas, SA, WA

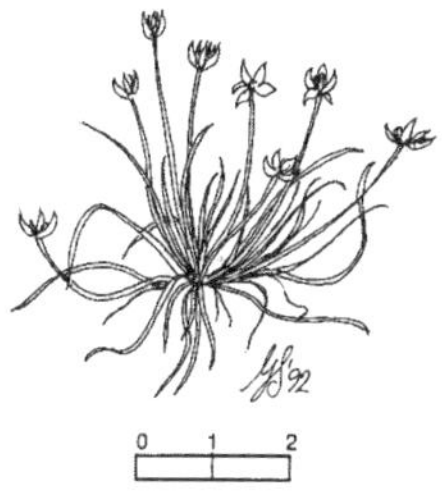

*Trithuria submersa*

# Typha — Typhaceae

## Typha domingensis

Bullrush or Cumbungi

**Size:** 1–3 m high x spreading — F P
**Habitat:** Fresh water lakes and pools in riparian scrub and woodland, box and red gum woodland, valley sclerophyll and grassy low open forest
**Form:** Vigorous, semi-aquatic perennial rush-like plant, spreading by rhizomes, stems erect and cane-like
**Foliage:** Stiff, flat narrow-linear leaves to 2 m x 4–10 mm, in pairs near the base of the stem
**Flowers:** Dense cylindrical spikes on tall stems above the foliage. Male flowers clustered 20–50 mm above the female spike which is usually 10–24 cm long and 5–15 mm wide. The stigmas are linear. Flowers all year.
**Requirements:** Grows in water along the shoreline and will not tolerate drying out
**Comments:** Often becomes invasive, particularly on farm dams and ornamental waterways. (See Chapter 5 on wetland regeneration for methods of containment.) Will not grow in water deeper than about a metre. The Aborigines roasted or steamed the rhizome, used stem and leaf fibres for string and wove baskets from the leaf.
**Propagation:** Seed, division
**Localities:** 1, 4, 16, 24, 26, 28, 33, 35, 36, 42, 44, 62–64, 67#, 71, 72, 74–76
**Distribution:** All States; N Af, SE Asia
**Synonym:** Both species were originally combined under *T. angustifolia*

*Typha domingensis*

## Typha orientalis

**Size:** 2–4.5 m high — F P
**Habitat:** Riparian and swamp scrub, saltmarshes
**Comments:** The main differences from *T. domingensis* are that the leaves are often broader, from 6–20 mm wide, the separation between the 2 spikes is less than 20 mm and the female spike is larger at 7–30 cm x 15–20 mm. The stigmas are narrow-ovate. Distinguishing *T. orientalis* from the introduced *T. latifolia* is less clear and requires microscopic examination for certainty. The flower spikes are often a darker brown in the latter. The introduced species dies down in Winter, while the native spp. remain green.
**Localities:** 4, 16, 26, 29, 35, 61
**Distribution:** All states except NT; NZ, SE Asia

*Typha orientalis*

SCALE: 0 1cm 2cm — The scale representation on each plant equals 2cm.

*Uncinia tenella*

# Uncinia
**Cyperaceae**

## Uncinia tenella
Delicate Hook-sedge

P N

**Size:** stems to 15 cm high
**Habitat:** Cool temperate rainforest, wet sclerophyll forest
**Form:** Dwarf rush-like sedge
**Foliage:** Very fine grass-like leaves
**Flowers:** Dioecious; single terminal spikelet to 15 mm long, nut enclosed in a bag which ends in a long, projecting hooked awn
**Requirements:** Moist humus-rich soils where terrestrial
**Comments:** Frequently found growing on rotting logs and treefern trunks as well as beside watercourses. It has a very restricted range in the Melbourne area.
**Propagation:** Seed
**Localities:** 57
**Distribution:** Vic, NSW, Tas

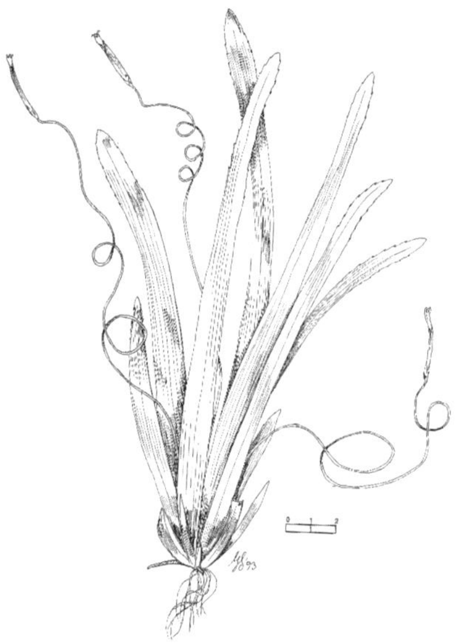

*Vallisneria spiralis*

# Vallisneria
**Hydrocharitaceae**

## Vallisneria spiralis
Eel or Ribbon Weed, Eel Grass

F P

**Size:** submerged x 30 cm
**Habitat:** Fresh water streams and lakes
**Form:** Submerged perennial stoloniferous herb
**Foliage:** Basal tufts of linear strap-like leaves 15–30 cm long with 5–9 longitudinal veins and tiny teeth along the margins towards the top of the leaf
**Flowers:** Dioecious; many minute male flowers in a head enclosed within a shortly-stalked bract to 20 mm long, flowerhead breaks off before pollen is released and floats on water surface; female flowers solitary to 25 mm wide enclosed in a bract on a very long and spirally-coiled stalk which lifts the flower to open on the surface of the water; November to May
**Requirements:** Fresh water
**Comments:** An excellent oxygenator for pools, providing habitat for water animals.
**Propagation:** Division
**Localities:** 35, 36, 60
**Distribution:** All states; NZ, cosmopolitan

# Wolffia
**Lemnaceae**

## Wolffia australiana
Duck Weed

F P

**Size:** 1 mm long x 2 mm deep
**Habitat:** Fresh water aquatic of shallow ponds and streams to 30 cm deep
**Form:** Swollen rootless perennial herb, thallus elliptic above, convex below, most of plant submerged
**Flowers:** Single male and female flower erupt from a hollow on upper surface of thallus, rarely flowers
**Requirements:** Fresh or stagnant water
**Comments:** Many individual plants form a dense mat on the surface of the water. The smallest flowering plant in Australia.
**Propagation:** As for *Lemna disperma*. Budding is from a pit in the upper surface.
**Localities:** 4, 16, 60, 72
**Distribution:** Vic, NSW, SA; NZ

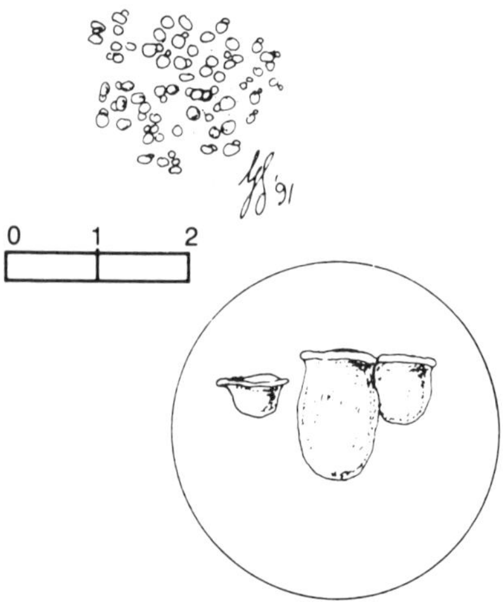


*Wolffia australiana*

# Xanthorrhoea
**Xanthorrhoeaceae**

## Xanthorrhoea australis
Austral Grass Tree

F P

**Size:** to 3 m high (occasionally more)
**Habitat:** Grassy low open forest
**Form:** A slow-growing stout perennial plant with a thick woody trunk or butt surmounted by a grassy tuft of leaves and, if not burnt, skirted with old leaves
**Foliage:** Four-sided rigid bluish-green leaves to 1 m x 1–3 mm
**Flowers:** Dense clusters of creamy white, strongly scented flowers in a terminal spike 1–1.8 m x 50–80 mm, much longer than the stem (to 50 cm long); July to December
**Requirements:** Well drained soils tolerating dry conditions once established
**Comments:** One of the outstanding features of the Australian bush, particularly *en masse*. Prominent after bushfires which encourage bright green new leaves and prolific flowering. The *Xanthorrhoea* species were very useful to the Aborigines. The nectar provided drinks and the root was eaten. Adhesive was obtained from the resin while the flowering stem provided spears and firesticks. Very restricted distribution within the Melbourne area, although it occurs frequently just outside it to the north-east and south-east.
**Propagation:** Seed
**Localities:** 76
**Distribution:** Vic, NSW, Tas, SA

*Illustrated p. vi and 304*

### Xanthorrhoea minor ssp. lutea

Small Grass Tree

**Size:** 0.6 m x 1 m — **F P**
**Habitat:** Widespread in dry and valley sclerophyll forests, sclerophyll woodland, grassy low open forest, tea-tree heath
**Form:** Slow growing, stout perennial plant with a subterranean, often branched, woody trunk. Superficially looks like a dense grassy clump.
**Foliage:** Arching, triangular leaves to 1 m x 2–4 mm
**Flowers:** Dense clusters of yellowish, strongly-scented flowers in a terminal spike 5–20 cm long x 10–40 mm; slender stems (30–90 cm long), much longer than spike. Plants may bear several spikes; November to February.
**Requirements:** Well drained soils, tolerating dryness once established
**Comments:** An attractive grassy clump in the bush, particularly when the plant displays several flowering stems. Easily grown. The most obvious distinguishing feature of this subspecies is the very broad yellow flower. *X. minor* ssp. *minor* does not occur within the Melbourne area.
**Propagation:** Fresh seed
**Localities:** Research, 40, 42–44, 46–56, 58–64, 66–71, 73–76
**Distribution:** Vic, SA

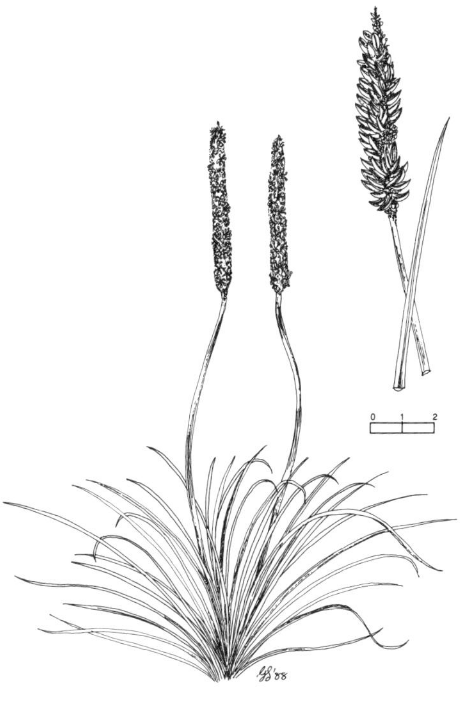

*Xanthorrhoea minor* ssp. *lutea*

# Xyris — Xyridaceae

### Xyris gracilis

Slender Yellow-eye

**Size:** 15–45 cm x 30 cm — **F P**
**Habitat:** Tea-tree heath
**Form:** Erect perennial tufted herb with slender leafless stems
**Foliage:** Rush-like, basal pale green leaves to 30 cm x 1–2 mm
**Flowers:** Single terminal spike to 5 mm wide, composed of 6–12 conspicuous brown bracts with large pale centres, each holding a 3-petalled bright yellow flower, each flower lasting only 1 day; November to January.
**Requirements:** Damp peaty soils
**Comments:** An inconspicuous plant until in flower. Suitable for bog gardens. Almost extinct within the Melbourne area.
**Propagation:** Seed, division
**Localities:** Clayton, Moorabbin, 67#, 73–75
**Distribution:** Vic, NSW, Tas

### Xyris operculata

Tall Yellow-eye

**Size:** to 0.6 m high, stems 0.5–1.5 m high — **P**
**Habitat:** Tea-tree heath
**Form:** Large clumping perennial
**Foliage:** Terete leaves to 60 cm long, sheaths shiny
**Flowers:** Large globular head of many, often fringed, broad dark brown bracts in 5 rows, lower bracts empty. The yellow flowers open one at a time; November to January
**Comments:** Both species used to grow together but this species is now presumed extinct within Melbourne.
**Localities:** Eastern Port Phillip Bay, Macclesfield
**Distribution:** All states except WA

*Xyris gracilis*

# Zostera — Zosteraceae

### Zostera muelleri

Dwarf Grass-wrack

**Habitat:** Marine aquatic of mud flats and shallow coastal areas
**Form:** Perennial sea grass with creeping rhizomes forming extensive meadows at times, emergent at low tide
**Foliage:** Short vegetative shoots with 2–6 narrow-linear leaves to 100 mm long with notched end, deciduous, leaving old sheath
**Flowers:** Flowers enclosed in a shortly-stalked sheathing structure, 2–3 bracteoles on the rhachis; October to March.
**Comments:** Other sea-grasses occurring in Port Phillip Bay are:
*Heterozostera tasmanica*—similar to above but with rounded leaves and no bracteoles on rhachis, fertile shoots branched; found in coastal waters to 8 m deep.
*Amphibolus antarctica (Zannichelliaceae)*—submerged with wiry upright stems to 60 mm long, rigid, shiny leaves broad-linear to 50 mm long on the top half of the stem, flowers sessile, either dioecious or on separate shoots.
**Propagation:** Division
**Localities:** All occur along the coast of Port Phillip Bay within the study area.
**Distribution:** Vic, Tas, SA; NZ. Latter 2 also in NSW, WA (not NZ)

SCALE: 0 1cm 2cm — The scale representation on each plant equals 2cm.

# Zoysia **Poaceae**

## Zoysia macrantha

Prickly Couch

**F**

**Size:** stem to 25 cm x spreading
**Habitat:** Saltmarshes, primary dune scrub
**Form:** Low rhizomatous mat-forming perennial grass
**Foliage:** Leaves small to 5 mm long, more than 2 mm wide with many smooth veins, ending in long pungent points
**Flowers:** Terminal narrow spike-like raceme of small purplish 1-flowered spikelets to 5 mm long, single glume pointed
**Requirements:** Moist to dry sandy soil, tolerating saline conditions
**Propagation:** Seed, division
**Localities:** Seaholme, Port Melbourne, 67
**Distribution:** Vic, Qld, NSW, Tas

*Xanthorrhoea australis*

*Stipa exilis*

*Stipa nodosa*

*Stipa oligostachya*

*Stipa gibbosa*

*Stipa semibarbata*

*Stipa stipoides*

*Stipa scabra* ssp. *scabra*

SCALE: 0 1cm 2cm The scale representation on each plant equals 2cm.

# Ferns and Fern Allies

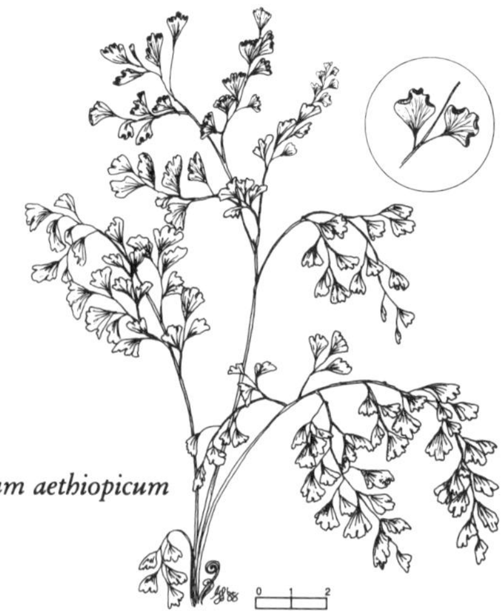

*Adiantum aethiopicum*

## Adiantum — Adiantaceae

### Adiantum aethiopicum

Common Maidenhair

**P**

**Size:** 10–45 cm high x spreading
**Habitat:** Red gum woodland, wet, damp, valley and dry sclerophyll forests, grassy low open forest, riparian scrub in crevices of rocky outcrops
**Form:** Carpeting groundcover suckering by vigorous underground rhizomes
**Foliage:** Delicate bipinnate to tripinnate fronds, pinnules lobed and wedge-shaped, light yellow-green but darkening with age
**Requirements:** Moist, well drained soils. Becomes dormant during warm dry spells, but reshoots quickly from underground rhizomes once the soil moisture increases.
**Comments:** A well known and reliable fern which can cover large areas in ideal conditions. Suitable also as a container or basket plant if given ample moisture and protection from drying winds.
**Propagation:** Spore, division
**Localities:** Sunshine, St Albans, 1, 25, 26, 28–30, 34A–36, 38, 40–45#, 47, 49–51, 53, 55–65, 67, 74, 76
**Distribution:** All states; Lord Howe Is, NZ, Af

*Allantodia australis*

## Allantodia — Athyriaceae

### Allantodia australis

Austral Lady Fern

**P N**

**Size:** 0.5–2 m tall
**Habitat:** Cool temperate rainforest, wet sclerophyll forest
**Form:** Erect tufted fern, with a small woody trunk in old plants
**Foliage:** Brittle, arching, bipinnate to tripinnate light green lacy fronds to 2 m long. Fronds are broadly triangular-shaped, pinnules to 25 mm long, prominently toothed.
**Requirements:** Cool, moist to wet well drained soils
**Comments:** An elegant fern for a moist outdoor area protected from wind. Old well grown plants resemble short-trunked tree ferns. Dormant during Winter. Suited to growing in a large tub indoors provided high humidity levels are maintained.
**Propagation:** Spores
**Localities:** 57
**Distribution:** Vic, Qld, NSW, Tas; NZ, N Cal
**Synonym:** *Diplazium australe*, *Athyrium australe*

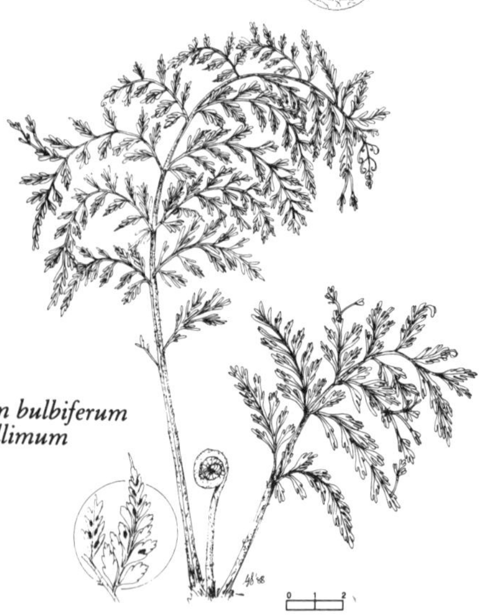

*Asplenium bulbiferum* ssp. *gracillimum*

## Asplenium — Aspleniaceae

### Asplenium bulbiferum ssp. gracillimum

Mother Spleenwort

**P**

**Size:** 0.5–1.2 m x 1 m
**Habitat:** Cool temperate rainforest, wet and damp sclerophyll forests, epiphytic on logs, rocks and tree fern trunks
**Form:** Tufted perennial spreading clump with erect or semi-weeping fronds
**Foliage:** Thin, dark green shining fronds bipinnate to tripinnate, often with small plantlets borne near the tips of pinnae
**Requirements:** Moist, very well drained soils that do not dry out, preferably high in humus
**Comments:** A relatively common and easy to grow fern, hardy in a protected situation provided soil moisture can be maintained without becoming wet. The slowly growing plantlets on old fronds should be allowed to develop on the plant until the frond withers, when plantlets may be removed and potted individually until large enough to plant out.
**Propagation:** Plantlets produced on the fronds, spores
**Localities:** 57, 58
**Distribution:** Vic, Qld, NSW, Tas, SA; NZ

### Asplenium flabellifolium

Necklace fern

**Size:** 5–30 cm x spreading **P**
**Habitat:** Moist rocky outcrops of riparian scrub and woodland, red gum woodland, wet, damp and valley sclerophyll forests also epiphytic on tree fern trunks, logs and moist rocks
**Form:** Small procumbent or prostrate fern forming large spreading colonies by creeping rhizomes
**Foliage:** Delicate, pale green pinnate fronds to 30 cm long with plantlets borne on the tip of the frond. Pinnae are wedge- or fan-shaped.
**Requirements:** Rich light moist open soil
**Comments:** This dainty little fern is useful for terrariums and hanging baskets, but must have ample moisture and dappled shade. In nature, it can be epiphytic, and is often found in sunny situations, protected in crevices among boulders and grasses. Plantlets take root as the tip of the frond touches the ground, and a large colony is soon formed in humus rich soil.
**Propagation:** Spore, plantlets, division of the rhizomes
**Localities:** St Albans, Rockbank, 8, 16, 25, 26, 28–30#, 34A–36, 42, 54–58
**Distribution:** All States; NZ

*Asplenium flabellifolium*

## Azolla

## Azollaceae

### Azolla filiculoides

Red Azolla or Pacific Azolla

**Size:** Prostrate-15 mm x 10–30 mm **F P**
**Habitat:** Billabongs and slow-moving watercourses in riparian scrub, grassy wetlands, box and red gum woodland, valley sclerophyll forest, grassy low open forest
**Form:** Free-floating fern forming carpets over still fresh water. Unbranched roots dangle below the water surface.
**Foliage:** Overlapping, irregularly divided sessile fronds with crinkled appearance, oblong-triangular in shape, to 50 mm long, green to deep red in colour
**Requirements:** Fresh still water
**Comments:** Can be very fast spreading but dies back in cold weather. Sometimes appears temporarily in lakes and slow flowing streams. Good cover for fish in ponds.
**Propagation:** Natural division as older plants break up
**Localities:** Maribyrnong Valley, Diamond Ck, 4, 6, 14, 24, 26, 27, 34A–36, ?42, 44, 48, 60, 67#, 71, 72
**Distribution:** All states; NZ, N Am, S Am

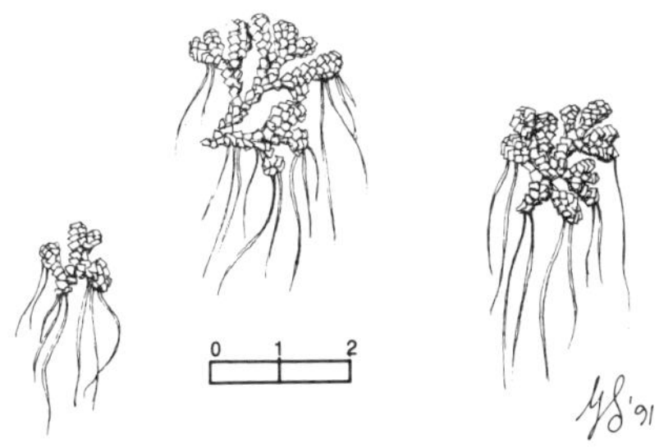

*Azolla filiculoides*

### Azolla pinnata

Ferny Azolla

**P**

**Habitat:** Billabongs and ponds in damp and valley sclerophyll forests
**Comments:** Very similar in form to *A. filiculoides*. Branching is more regular (pinnate) and fronds are triangular in appearance. The lower surface of the fronds is rough while *A. filiculoides* is smooth all over. The main distinguishing feature is the feathery roots.
**Localities:** 34A, 42, 48, 59
**Distribution:** All States except Tas; PNG, Asia, Af

*Azolla pinnata*

## Blechnum

## Blechnaceae

### Blechnum cartilagineum

Gristle Fern

**Size:** 0.5–1.5 m x 1 m **P**
**Habitat:** Damp and valley sclerophyll forests
**Form:** Upright tufting fern with short creeping stoloniferous rhizomes, forming spreading patches
**Foliage:** Very broad, slightly arching, leathery light green to yellow-green pinnate fronds to 1.5 m long, new growth often pinkish. Pinnae are longest at the base of fronds.
**Requirements:** Moist well drained soils, tolerating drier conditions once established
**Comments:** This attractive fern is readily available and quite easy to grow. It is the only Victorian *Blechnum* in which the fertile and barren fronds are similar in shape.
**Propagation:** Spore, division
**Localities:** Montrose, 42, 55–58, 60
**Distribution:** Vic, Qld, NSW, Tas; Philipines

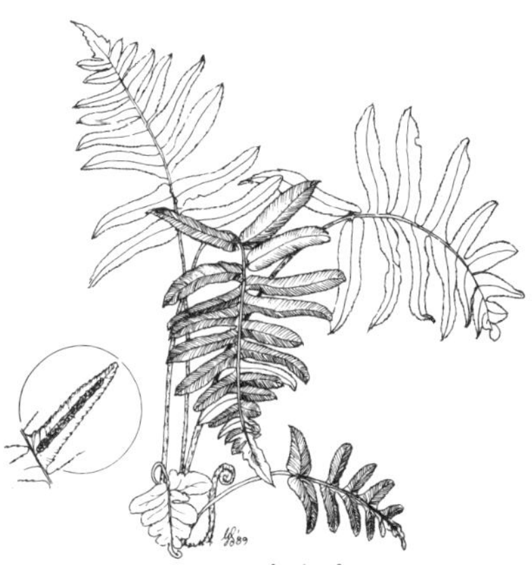

*Blechnum cartilagineum*

SCALE: 0 1cm 2cm The scale representation on each plant equals 2cm.

*Blechnum chambersii*

*Blechnum fluviatile*

## Blechnum chambersii

Lance Water Fern

**P N**

**Size:** 0.6 m x 0.6 m
**Habitat:** Wet and damp sclerophyll forests
**Form:** Semi-erect clump-forming fern
**Foliage:** Rough thin-textured dark green arching pinnate fronds to 60 cm long, but usually less. Pinnae small and broad at base of fronds, becoming longer towards the centre, then reducing again towards the tip. Fertile pinnae very narrow.
**Requirements:** Ample moisture in cool dark situations
**Comments:** Slow growing fern needing heavy shade and intolerant of dryness, or hot and cold winds. Locally rare.
**Propagation:** Spore, division
**Localities:** ?42, 57, 58, 60
**Distribution:** Vic, NSW, Tas, SA; NZ, Fiji, Norfolk Is

## Blechnum fluviatile

Ray or Star Water Fern

**P N**

**Size:** 20–40 cm x 0.5–1 m
**Habitat:** Wet and damp sclerophyll forests
**Form:** Prostrate to spreading rosetted fern, colonising small areas by underground stolons
**Foliage:** Thin light green pinnate fronds, spreading or prostrate, to 80 cm long. Pinnae sessile, the longest towards the centre of the frond, the lower ones rounded, the upper ones oblong. Fertile pinnae short, narrow and curved, on erect fronds.
**Requirements:** Well drained soil with ample moisture
**Comments:** Fast growing attractive small fern in ideal conditions but will not tolerate dryness.
**Propagation:** Spore, division
**Localities:** 47, 57, 58, 74
**Distribution:** Vic, NSW, Tas; NZ

*Blechnum minus*

## Blechnum minus

Soft Water Fern

**F P**

**Size:** 0.5–1.2 m x 0.5–1 m
**Habitat:** Widespread in wet, damp and valley sclerophyll forests, grassy low open forest
**Form:** Dense erect clump, forming spreading patches from underground stolons
**Foliage:** Erect or arching, soft shiny green deeply pinnate fronds to 1.2 m long. Pinnae stalked, linear and well spaced along the rhachis, (or midrib), the lowest ones smallest. The new fronds are pinkish. Pinnae on fertile fronds very narrow.
**Requirements:** Likes humus rich soil but quite adaptable. Must have ample moisture, tolerating wet soils
**Comments:** Will tolerate exposure to the sun if the roots are well watered. Old plants produce short trunks.
**Propagation:** Spore, division
**Localities:** 41, 42, 55–57, 74–76
**Distribution:** Vic, Qld, NSW, Tas, SA; NZ

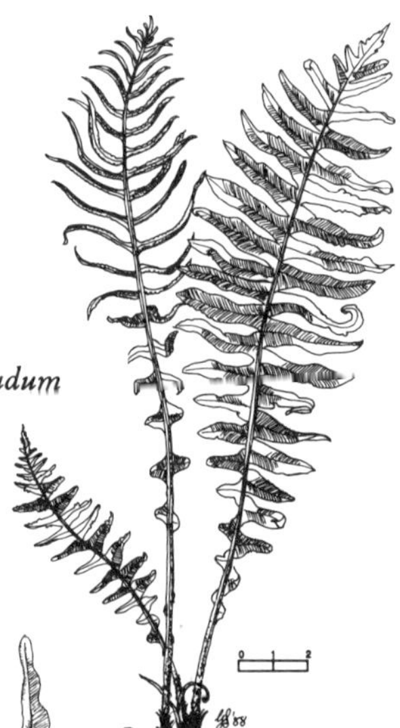

*Blechnum nudum*

## Blechnum nudum

Fishbone Water Fern

**P N**

**Size:** 0.5–1.2 m x 0.5–1 m
**Habitat:** Widespread in wet, damp and valley sclerophyll forests, grassy low open forest
**Form:** Individual plants form an erect rounded clump, with a small black fibrous trunk to 1 m in old plants. Plants spread by underground stolon to form large colonies.
**Foliage:** Erect or spreading dark green lanceolate pinnate fronds to 1.2 m long, young fronds yellow-green. Pinnae straight, held at right angles to the rhachis. Fertile fronds have much narrower pinnae.
**Requirements:** Very moist well-drained soil rich in humus. Adapts well to heavy soils.
**Comments:** Its attractive arching habit makes this fern an excellent long lived tub or garden plant but it must be protected from hot sun and drying winds.
**Propagation:** Spore, division
**Localities:** 42, 55, 56, 58, 75
**Distribution:** Vic, Qld, NSW, Tas, SA

*Blechnum patersonii*

## Blechnum patersonii

Strap Water Fern

**P N**

**Size:** 10–40 cm x 40 cm
**Habitat:** Cool temperate rainforest, wet sclerophyll forest
**Form:** Small erect fern forming open clumps
**Foliage:** Erect, leathery dark green fronds to 40 cm long, entire and strap-like or sometimes irregularly lobed. Young fronds reddish-pink. Fertile fronds narrower.
**Requirements:** Rich moist well drained soil
**Comments:** An attractive and hardy fern in a well protected moist position. The fertile fronds are simple when the plant has only simple fronds, but are divided and lobed when the plant is likewise.
**Propagation:** Spore
**Localities:** 57
**Distribution:** Vic, Qld, NSW, Tas

### Blechnum wattsii
Hard Water Fern

**Size:** to 1 m x spreading — **P N**
**Habitat:** Cool temperate rainforest, damp and wet sclerophyll forests
**Form:** Vigorous rhizomatous creeping fern, generally forming extensive dense colonies, often excluding other ground species
**Foliage:** Upright dark green coarsely pinnate fronds to 1 m long. Pinnae leathery and slightly serrated. Young fronds an attractive pink-bronze. Scales occur on all but the upper portion of the rhachis. Fertile fronds much narrower.
**Requirements:** Moist, well drained soil, rich in humus
**Comments:** Often the dominant ground cover in moist mountain forests, it is fairly hardy in cultivation, in a protected position .
**Propagation:** Spore, division
**Localities:** 49, 57, 58
**Distribution:** Vic, Qld, NSW, Tas, SA
**Synonym:** Previously referred to as *B. procerum* which does not occur in Australia.

*Blechnum wattsii*

# Botrychium — Ophioglossaceae

### Botrychium australe
Austral Moonwort

**Size:** 10–40 cm x 10–30 cm
**Habitat:** Moist depressions in plains grassland
**Form:** Small open fern with thick fleshy roots and a non-creeping rhizome
**Foliage:** Sterile fronds bipinnate to 4-pinnate, fleshy, bright green and lacy. The generally taller fertile fronds are borne as offshoots of the sterile fronds and bear grape-like clusters of spore cases at their tips.
**Requirements:** Cool, acidic moist well drained soil, rich in humus
**Comments:** A vulnerable plant throughout Victoria, presumed extinct in the Melbourne area. Plants grow in a mycorrhizal association with an invading fungus.
**Propagation:** Division
**Localities:** 25#
**Distribution:** All States except WA; NZ, S Am

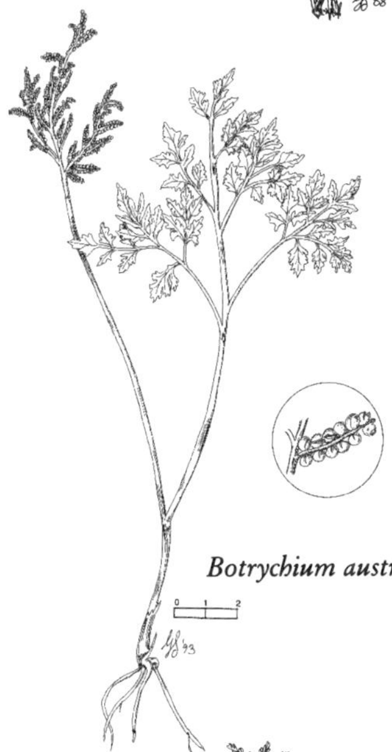
*Botrychium australe*

# Calochlaena — Dicksoniaceae

### Calochlaena dubia
False Bracken, Rainbow Fern

**Size:** 0.5–2 m x spreading — **P N**
**Habitat:** Riparian scrub, damp and valley sclerophyll forests
**Form:** Coarse open spreading fern, sometimes growing in dense clumps, with a stout, vigorous creeping rhizome
**Foliage:** Large, broadly triangular, yellow-green 3–4 pinnate fronds on strong erect stalks (or stipes) to 2 m long (rarely to 4 m long)
**Requirements:** Adaptable to most soils and exposed sites but thrives in moist clay soils
**Comments:** This fern is often confused with common bracken (*Pteridium esculentum*). It is distinguished by its softer fronds, and a groove running along the main rib. Plants are quite vigorous once established and need room to spread. Growth may be controlled by pruning of creeping rhizomes.
**Propagation:** Spore, division of rhizomes
**Localities:** Montrose, 16, 18, 26, 42, 43, 50A, 51, 53, 55–60, 62, 63, 74
**Distribution:** Vic, Qld, NSW, Tas
**Synonym:** *Culcita dubia*

*Calochlaena dubia*

# Cheilanthes — Adiantaceae

### Cheilanthes austrotenuifolia
Rockfern

**Size:** 0.2–0.5 m x spreading — **F P**
**Habitat:** Widespread in red gum and box woodlands, damp, valley and dry sclerophyll forests, tea-tree heath
**Form:** Dwarf fern forming open clumps, rhizomes short and branched
**Foliage:** Fronds erect bipinnate to tripinnate, shiny, bright green and almost glabrous, to 30 cm long but often shorter. Fertile and sterile fronds similar.
**Requirements:** Open position in well drained soil with root protection
**Comments:** A surprisingly hardy fern in natural conditions, the fronds tolerant of direct sunshine and drying winds. They do eventually die down to the creeping rhizome in Summer but are renewed in Autumn. Plants resent disturbance and are difficult to establish, but have proved successful in sunny rockeries where the rhizomes are protected from extremes in temperature. In dry areas fronds may only reach 30–50 mm long.
**Propagation:** Spore, division
**Localities:** Tottenham, Sydenham, Tullamarine, 4, 7, 12, 15–18, 25, 26, 29, 30, 32, 34A–36, 42, 44, 46, 49, 50A, 51, 55, 56, 58–61, 67, 76
**Distribution:** All States; NZ, PNG
**Synonym:** Previously called *C. tenuifolia* var. *tenuifolia*

*Cheilanthes austrotenuifolia*

SCALE: 0 1cm 2cm — The scale representation on each plant equals 2cm.

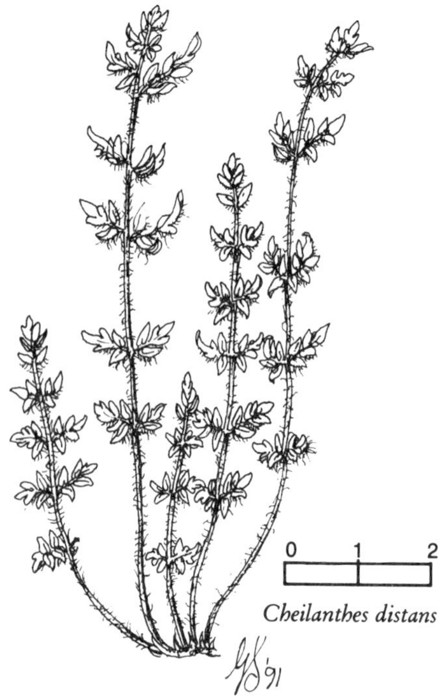

*Cheilanthes distans*

### Cheilanthes distans
Bristly Cloak Fern

**Size:** to 30 cm high **F**
**Habitat:** Uncommon in riparian scrub, escarpments, crevices in rocky outcrops
**Form:** Tiny fern with short, much-branched rhizomes, spreading to form small, compact, separate clumps
**Foliage:** Small, slender upright pinnate to bipinnate fronds, to 30 cm long, but usually much less, covered with hair-like scales
**Requirements:** Warm open position, preferably in a well drained rockery situation. Resents disturbance
**Comments:** One of the easiest species to grow, this plant resents shade and overwatering. Often grows in similar habitats to *C. austrotenuifolia* but is easily distinguished by the dense covering of hair-like scales. Restricted range in the Melbourne area.
**Propagation:** Spore, division
**Localities:** Keilor, Diggers Rest, Rockbank, 16, 18, 26
**Distribution:** All States except Tas; Norfolk Is, NZ, N Cal

### Cheilanthes sieberi ssp. sieberi
Mulga Fern

**Size:** to 0.4 m high **F**
**Habitat:** Box woodland, plains grassland, often amongst rocks
**Form:** Dwarf fern forming open spreading clumps
**Foliage:** Narrow erect, shiny bright green glabrous bipinnate to tripinnate fronds
**Requirements:** Warm well drained soils, resenting shade, but rhizomes need protection from temperature extremes.
**Comments:** Similar to *C. austrotenuifolia*, differing mainly in the narrower fronds and shorter rhizomes. Some water during the driest periods is beneficial.
**Propagation:** Spore, division
**Localities:** 14, 28, 34A
**Distribution:** All states except Tas; NZ, N Cal
**Synonym:** *C. tenuifolia* var. *sieberi*

*Ctenopteris heterophylla*

## Ctenopteris — Grammitidaceae

### Ctenopteris heterophylla
Gipsy Fern

**Size:** 10–30 cm high **N**
**Habitat:** In mossy crevices or on rocks or mossy tree trunks in wet sclerophyll forest, cool temperate rainforest
**Form:** Pendant, tufted or shortly rhizomatous, epiphytic fern
**Foliage:** Variable, lobed or pinnate, rarely entire, lanceolate fronds to 30 cm long, leathery and dull green
**Requirements:** Extremely humus rich moist soil or preferably mossy rock, logs or tree trunks in a constantly humid situation
**Comments:** Very slow growing. May be suited to terrarium cultivation but has never been successfully grown.
**Propagation:** Spore, division
**Localities:** 57
**Distribution:** Vic, Tas; NZ

### Culcita dubia = Calochlaena dubia

*Cyathea australis*

## Cyathea — Cyatheaceae

### Cyathea australis
Rough Tree Fern

**Size:** 5–15 m high **P N**
**Habitat:** Cool temperate rainforest, wet, damp and valley sclerophyll forests, banks of creeks and sheltered sites in drier areas
**Form:** Upright fibrous trunk with a head of long arching fronds. Base of trunk is often buttressed.
**Foliage:** Dark green tripinnate fronds, 2–5 m long x 1 m wide. Base of the stalks are covered with hard, pointed projections
**Requirements:** Moist soils tolerating dryness and some sun once established
**Comments:** Often confused with *Dicksonia antarctica*, the Soft Tree Fern—the chief vegetative difference being the rasp-like frond bases compared with the soft hairy frond bases of *D. antarctica*.
**Propagation:** Spore
**Localities:** 42, 51, 55–58, 60, 66#, 73–76
**Distribution:** Vic, Qld, NSW, Tas

### Cyathea cunninghamii

Slender Tree Fern

N

**Size:** 10–20 m high
**Habitat:** Wet sclerophyll forest
**Form:** Slender upright trunk with a small head of arching fronds
**Foliage:** Soft tripinnate fronds to 3 m long, dull, dark green above, paler below
**Requirements:** Ample moisture, particularly during Summer, with humus rich soil. Intolerant of wind and hot sun.
**Comments:** A most attractive treefern with a very slender, almost black trunk. Plants to 20 m in height are known with trunk diameters of only 15 cm. A rare plant in Victoria known from 1 site in the Melbourne region.
**Propagation:** Spore
**Localities:** Dandenong Ranges
**Distribution:** Vic, Qld, NSW, Tas; NZ

*Dicksonia antarctica*

## Dicksonia — Dicksoniaceae

### Dicksonia antarctica

Soft Tree Fern

P N

**Size:** 2–15 m high
**Habitat:** Cool temperate rainforest, wet, damp and valley sclerophyll forests, often in pure stands
**Form:** Robust, often buttressed trunk with a head of arching fronds. Trunk covered in soft brown fibrous roots.
**Foliage:** Tripinnate fronds to 5 m long, dark green and glossy above, dull green below.
**Requirements:** Soils rich in humus and ample moisture. Will tolerate some sun once established.
**Comments:** This is the fern commonly sold as bare trunks in plant nurseries. All sale plants should carry a licence tag. Trunks are excellent hosts for small epiphytic ferns.
**Propagation:** Should be propagated by spore only, to protect plants in the wild.
**Localities:** 55–58, 60, 66#, 74, 76
**Distribution:** Vic, Qld, NSW, Tas, SA

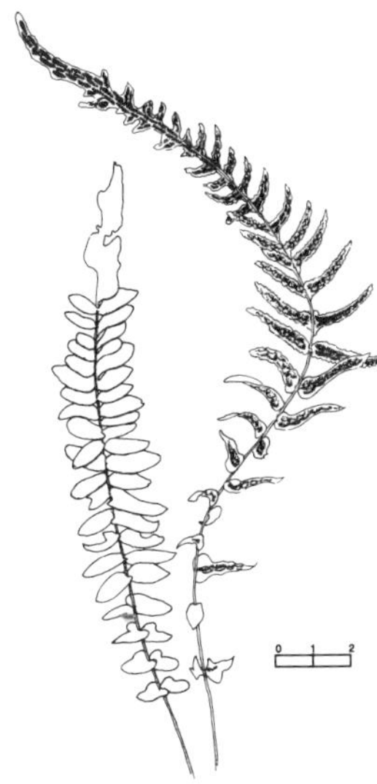

*Doodia caudata*

## Doodia — Blechnaceae

### Doodia caudata

Small Rasp Fern

P N

**Size:** 10–30 cm x 30–50 cm
**Habitat:** Embankments and rock crevices in riparian scrub
**Form:** Small erect or weeping fern, producing separate tufted clumps on short creeping rhizomes
**Foliage:** Narrow, dark green, pinnate fronds, generally having terminal pinnae united into a tail. Fertile fronds more erect and about twice as long as sterile fronds.
**Requirements:** Acidic rich peaty soils that do not dry out. Will tolerate quite a sunny position with ample moisture.
**Comments:** An adaptable fern suited to garden, pot or hanging basket culture. This fern is both variable and widespread but is locally rare.
**Propagation:** Spore, division
**Localities:** Broadmeadows
**Distribution:** Vic, Qld, NSW, Tas, SA; NZ, N Cal

*Doodia media* var. *australis*

### Doodia media ssp. australis

Common Rasp Fern

P N

**Size:** 0.3–0.6 m x 0.3–0.5 m
**Habitat:** Valley sclerophyll forest, riparian scrub
**Form:** Small erect fern forming spreading tufted clumps along short creeping rhizomes
**Foliage:** Pinnate, dark green lanceolate fronds to 60 cm long, new growth often purplish-pink. Stipe is covered in tiny projections giving it a rough texture.
**Requirements:** Humus rich, moist well drained soils
**Comments:** A variable species which will tolerate sunshine in protected situations, provided it is kept moist.
**Propagation:** Spore or division
**Localities:** Somerton, Dandenong Ranges 1863, 42, 74
**Distribution:** Vic, Qld, NSW, Tas; NZ

SCALE: 0 1cm 2cm — The scale representation on each plant equals 2cm

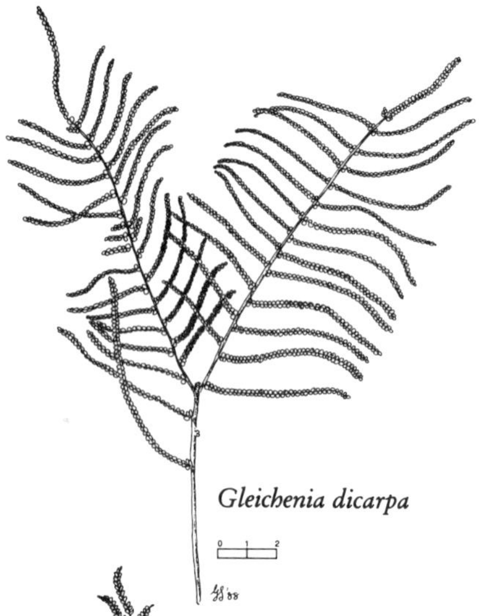

*Gleichenia dicarpa*

# Gleichenia — Gleicheniaceae

## Gleichenia dicarpa

Pouched Coral Fern, Tangle Fern

**Size:** to 3 m in scrubby thickets but usually lower — **F P**
**Habitat:** Bogs and soaks in damp sclerophyll forest, wattle tea-tree scrub
**Form:** Scrambling or climbing fern forming extensive tangled colonies, with vigorous much-branched wiry rhizomes
**Foliage:** Erect, many branched and very wiry, light green fronds. Pinnae are up to 50 mm long and pinnatifid. Margins of rounded pinnules are strongly rolled under forming a pouch.
**Requirements:** Acidic, damp or wet soils. Prefers its foliage in the sun.
**Comments:** Sometimes forms impenetrable thickets. Roots resent disturbance and will not tolerate drying out.
**Propagation:** Spores, occasionally division of the creeping rhizomes
**Localities:** Bentleigh$^{\#}$, 58, 66$^{\#}$, 67$^{\#}$, 74, 75
**Distribution:** Vic, Qld, NSW, Tas; NZ, N Cal, Mal
**Synonym:** *G. circinnata*

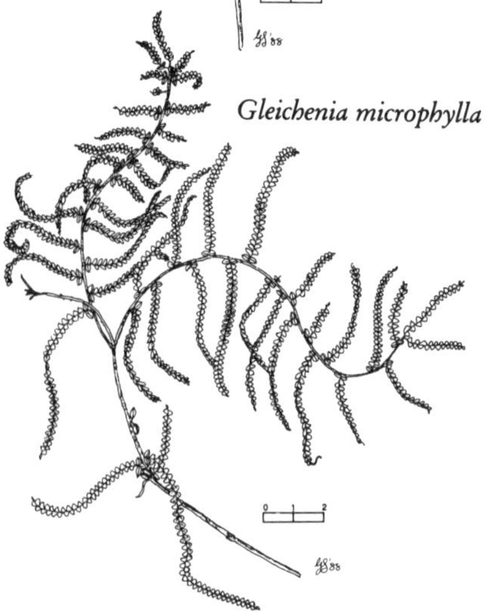

*Gleichenia microphylla*

## Gleichenia microphylla

Scrambling Coral Fern, Umbrella Fern

**Size:** to 4 m high in scrubby thickets — **P N**
**Habitat:** Wet, damp and valley sclerophyll forests, wattle tea-tree scrub
**Comments:** Often confused with *G. dicarpa. G. microphylla* is a stronger-growing species, prefering cooler, shaded sites. Fronds are dark green to yellowish with almost triangular, flat pinnules, whereas the pinnules of *G. dicarpa* are distinctly pouched.
**Propagation:** Spores, occasionally division
**Localities:** Montrose, 56, 57, 67$^{\#}$, 73–76
**Distribution:** All states except NT; NZ, N Cal, Mal

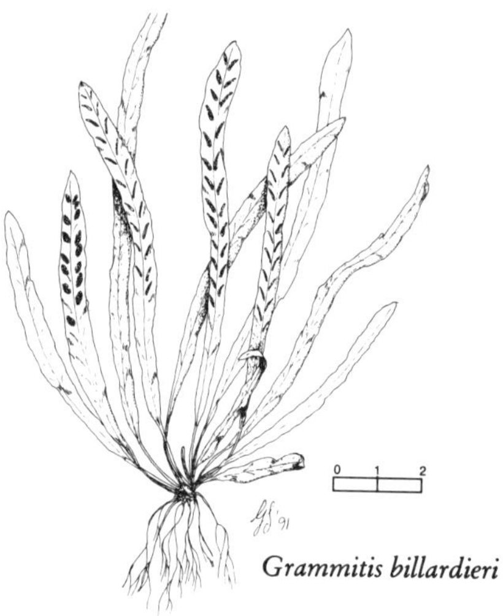

*Grammitis billardieri*

# Grammitis — Grammitidaceae

## Grammitis billardieri

Common Finger Fern

**Size:** to 15 cm high — **N**
**Habitat:** Cool temperate rainforest, wet and damp sclerophyll forests
**Form:** Tiny erect epiphytic or lithophytic fern with a short creeping rootstock
**Foliage:** Few leathery, dark green linear fronds (rarely lobed near the tip) to 15 cm long, often smaller
**Requirements:** Moist, heavily shaded position with high humidity
**Comments:** This inconspicuous fern is common on mossy rocks, logs and tree fern trunks in cool, moist forests. It has proved difficult to maintain in cultivation.
**Propagation:** Spore
**Localities:** 57, 58
**Distribution:** Vic, NSW, Tas; NZ

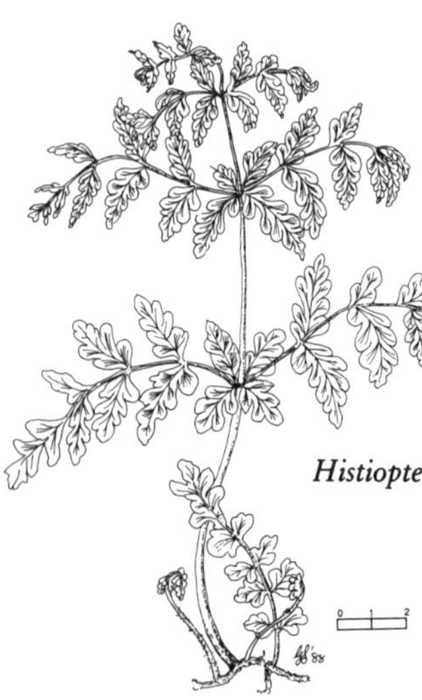

*Histiopteris incisa*

# Histiopteris — Dennstaedtiaceae

## Histiopteris incisa

Bat's Wing Fern

**Size:** 0.5–2 m high — **P N**
**Habitat:** Cool temperate rainforest, wet and damp sclerophyll forests
**Form:** Robust upright fern spreading by vigorous underground rhizomes
**Foliage:** Soft pale green or glaucous bipinnate to tripinnate fronds, triangular in outline, held erect on stout stalks, to 2 m high. Pinnae and pinnules in opposite pairs.
**Requirements:** Moist well drained soils, tolerating dryness once established
**Comments:** In the right conditions this plant can be extremely vigorous, forming spreading colonies. May also establish in more open areas such as escarpments where the young ferns are able to get sufficient moisture. It makes an interesting foliage contrast in a fernery, where its bluish fronds can be shown to advantage.
**Propagation:** Spore, division
**Localities:** 42, 56–58, 66$^{\#}$, 67$^{\#}$
**Distribution:** All States except WA; NZ, S Af, S Am, W Ind

# Hymenophyllum — Hymenophyllaceae

## Hymenophyllum australe — Austral Filmy Fern

**Size:** Fronds to 15 cm long — N
**Habitat:** Cool temperate rainforest
**Form:** Lithophytic or epiphytic filmy fern with wiry creeping rhizomes
**Foliage:** Erect, membranous dark green bipinnate fronds with secondary pinnae deeply divided into linear segments
**Requirements:** Constant high humidity and protection from sun and drying winds
**Comments:** Can form dense blankets on mossy rocks, logs and tree fern trunks. All species of *Hymenophyllum* have a restricted range in Melbourne.
**Propagation:** Spore, division
**Localities:** 57
**Distribution:** Vic, Qld, NSW, Tas

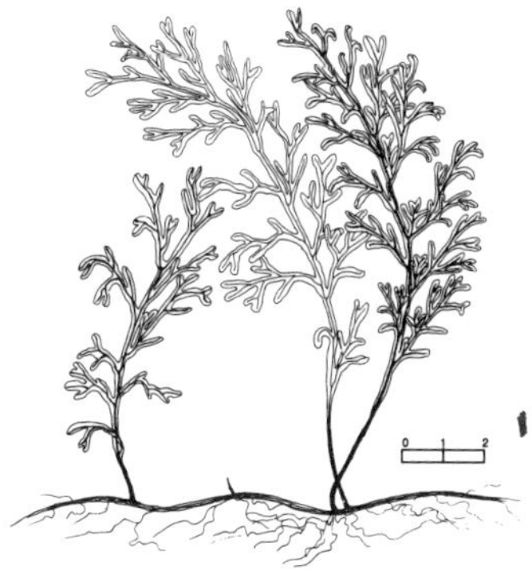

*Hymenophyllum australe*

## Hymenophyllum cupressiforme — Common Filmy Fern

**Size:** Fronds to 10 cm long — N
**Habitat:** Cool temperate rainforest, wet and damp sclerophyll forests
**Comments:** Lithophytic or epiphytic filmy fern with threadlike creeping rootstock. The main distinguishing feature is the form of the fronds. They are small, weakly erect or arching, dark green and membranous, with toothed margins.
**Localities:** 42, 57, 58
**Distribution:** Vic, Qld, NSW, Tas; NZ

## Hymenophyllum flabellatum — Shiny Filmy Fern

**Size:** Fronds to 25 cm long — N
**Habitat:** Cool temperate rainforest, wet and damp sclerophyll forests
**Comments:** Epiphytic fern with long creeping threadlike rootstock, often found growing on trunks of *Dicksonia antarctica*. Differs from *H. cupressiforme* in that the pale green pendent fronds are much longer, and from *H. australe* in that the fronds are usually pendent.
**Localities:** 57, 58
**Distribution:** Vic, Qld, NSW, Tas; NZ, Samoa, Tahiti

*Hymenophyllum flabellatum*

# Hypolepis — Dennstaedtiaceae

## Hypolepis glandulifera — Downy Ground Fern, Sticky Ground Fern

**Size:** to 2.5 m high — F P
**Habitat:** Wet sclerophyll forest
**Form:** A coarse vigorous spreading fern with a thick creeping much-branched rhizome, quickly developing into large colonies
**Foliage:** Large ovate to triangular, mostly 4-pinnate fronds, dull green and bracken-like on slender stalks, to 2 m long. Stalks, rhachises and young fronds covered densely with crooked sticky hairs
**Requirements:** Tolerant of most soils and positions but must have ample moisture in sunny positions
**Comments:** The fronds are less sticky if growing in sun. An easily grown fern which spreads quickly, but may be controlled by pruning the rhizome.
**Propagation:** Spore, division
**Localities:** 57
**Distribution:** Vic, Qld, NSW; S India, Sri Lanka to PNG and N Cal
**Synonym:** Previously referred to as *H. punctata.*

## Hypolepis muelleri — Harsh Ground Fern

**Size:** 0.5–1.5 m high — F P
**Habitat:** Wet and damp sclerophyll forests
**Form:** Coarse, vigorous spreading fern with a slender creeping rhizome, often forming extensive colonies
**Foliage:** Large ovate tripinnate fronds, dark green and bracken-like on slender stems, to 1.5 m long
**Requirements:** Very moist, swampy soils
**Comments:** May become quite invasive under ideal conditions.
**Propagation:** Spore, division
**Localities:** 57, 58
**Distribution:** Vic, Qld, NSW, Tas

SCALE: 0 1cm 2cm — The scale representation on each plant equals 2cm.

*Hypolepis rugosula*

### Hypolepis rugosula — Ruddy Ground Fern

**Size:** 0.5–1.5 m high — **F P**
**Habitat:** Wet and damp sclerophyll forests, swamps in wattle tea-tree scrub
**Form:** A coarse vigorously spreading fern with strong branching rhizomes, forming large clumps
**Foliage:** Erect broadly-triangular tripinnate fronds, dull green and bracken-like with reddish-brown warty stems, to 1.5 m long
**Requirements:** Moist or wet soils
**Comments:** May become invasive. As with other *Hypolepis* species, suitable only for large gardens where its vigorous habit is not a problem. Looks best in shaded moist situations, but adaptable to full sun if ample moisture is available.
**Propagation:** Spore, division
**Localities:** 57, 58, 67[#], 74, 75
**Distribution:** All States except NT

*Isoetes drummondii var. drummondii*

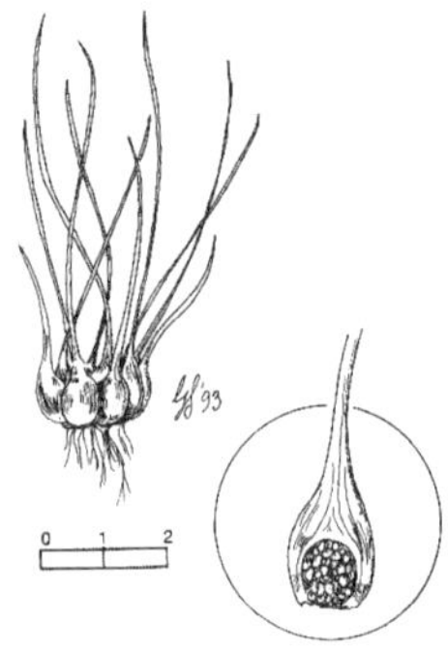

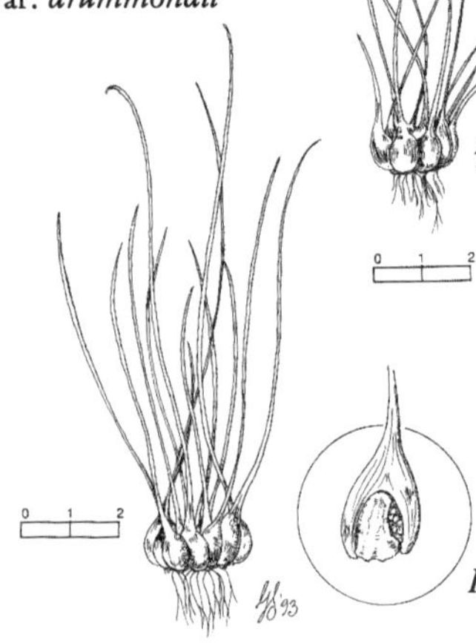

*Isoetes muelleri*

## Isoetes — Isoetaceae

### Isoetes drummondii var. drummondii — Plain Quillwort

**Size:** 3–20 cm high — **F P**
**Habitat:** Red gum woodland
**Form:** Small, erect or spreading, tufted annual fern ally, corm-like rootstock usually 3-lobed
**Foliage:** Narrowly cylindrical fronds 3–20 cm long, bases abruptly wider and overlapping. Broadly elliptical spore-containing cases at base of fronds, liver brown becoming shiny when mature; membranous veil absent.
**Requirements:** Poorly drained soil or shallow water
**Comments:** A rare in plant in Melbourne.
**Propagation:** Unknown
**Localities:** Dandenong, Langwarrin
**Distribution:** Vic, Tas, SA, WA

### Isoetes muelleri

**Size:** 5–30 cm high — **F P**
**Habitat:** Grassy wetlands
**Form:** Aquatic tufted fern ally with 2- or 3-lobed rootstock
**Foliage:** Erect or lax cylindrical fronds 50–150 mm long, becoming wider towards the base. Bases abruptly wider, flattened and loosely overlapping. Thin, rounded to elliptical, pale coloured spore-containing cases entirely or partly covered by pale membranous veil.
**Requirements:** Moist heavy soils subject to inundation
**Comments:** Known from one site in the Melbourne area.
**Localities:** Rockbank
**Distribution:** All states
**Synonym:** Previously confused with *I. humilior*

*Lastreopsis acuminata*

## Lastreopsis — Dryopteridaceae

### Lastreopsis acuminata — Shiny Shield Fern

**Size:** 0.5–0.8 m x 0.5–1 m — **P N**
**Habitat:** Wet, damp and valley sclerophyll forests
**Form:** Open, tufted or shortly rhizomatous fern
**Foliage:** Bipinnate (sometimes tripinnate) arching, glossy, dark green narrowly triangular fronds to 0.8 m long
**Requirements:** Moist, humus rich, well drained soils
**Comments:** An easily grown fern, its glossy fronds add interest to shaded gardens. Fronds become smaller in open or sunny situations, even with plenty of moisture. Locally rare.
**Propagation:** Spore, division
**Localities:** 57, 58, 62
**Distribution:** Vic, Qld, NSW, Tas, SA
**Synonym:** *L. shepherdii*

### Lastreopsis hispida — Bristly Shield Fern

**Size:** 0.1–0.8 m x 0.3–0.6 m — **P N**
**Habitat:** Cool temperate rainforest, wet sclerophyll forest
**Form:** Open tufted or shortly rhizomatous fern
**Foliage:** Thin glossy broadly triangular dark green 3- to 4-pinnate fronds, erect or arching, to 0.8 m long. Bristly hairs on the stems and rhachis.
**Requirements:** Moist to wet, humus rich, well drained soils
**Comments:** Sometimes grows on rotting logs and mossy trunks. It is quite rare both in Victoria and in the region.
**Propagation:** Spore, division
**Localities:** Olinda
**Distribution:** Vic, NSW, Tas; NZ

# Lindsaea — Lindsaeaceae

## Lindsaea linearis — Screw Fern

**F P**

**Size:** 5–30 cm x 20 cm
**Habitat:** Widespread in damp and valley sclerophyll forests, tea-tree heath, grassy low open forest
**Form:** Small erect fern with a short wiry creeping rhizome
**Foliage:** Pinnate, pale green slender fronds to 30 cm long forming an arching rosette. Pinnae paired and fan-shaped to 10 mm long. Fertile fronds more slender and erect than sterile fronds.
**Requirements:** Moist soils, often in open sites when the rhizome is protected
**Comments:** The pinnae on the narrow fertile fronds are obliquely bent down, giving rise to the common name of screw fern. It is difficult to establish and maintain in cultivation.
**Propagation:** Spore
**Localities:** 41, 43, 44, 47–51, 53–56, 58, 59, 61–64, 67, 68, 70, 71, 73–76
**Distribution:** All States; Norfolk Is, NZ, N Cal

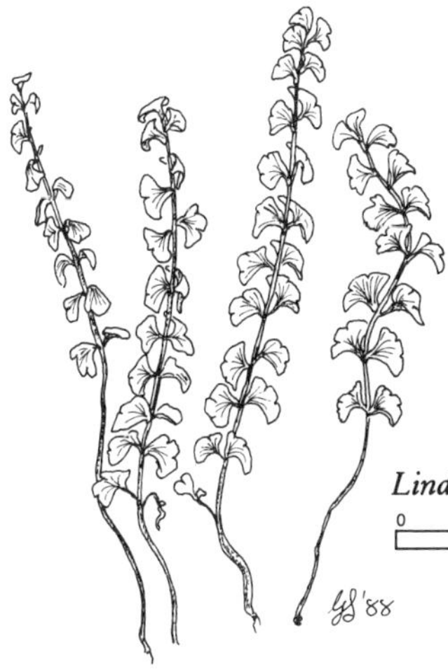

*Lindsaea linearis*

# Lycopodiella — Lycopodiaceae

## Lycopodiella laterale — Slender Clubmoss

**P N**

**Size:** Branches 10–40 cm long
**Habitat:** Peaty swamps amongst sedges in wattle tea-tree scrub
**Form:** Weakly erect or trailing single or few-stemmed non-woody plant, with thick branched underground rhizomes. Stems may be once or twice divided.
**Foliage:** Soft very narrow leaves to 6 mm long, sessile with incurved margins, crowded along the stems
**Requirements:** Moist to wet peaty, often sandy soils, protected by other scrubby plants
**Comments:** Small cone-like fruiting bodies (or stroboli) 10–15 mm long, are sessile and scattered along the stems. May be suited to terrarium culture. Possibly now extinct within the Melbourne area.
**Propagation:** Spore
**Localities:** Clayton#, 67#, 74
**Distribution:** All states except NT; NZ, N Cal
**Synonym:** *Lycopodium laterale*

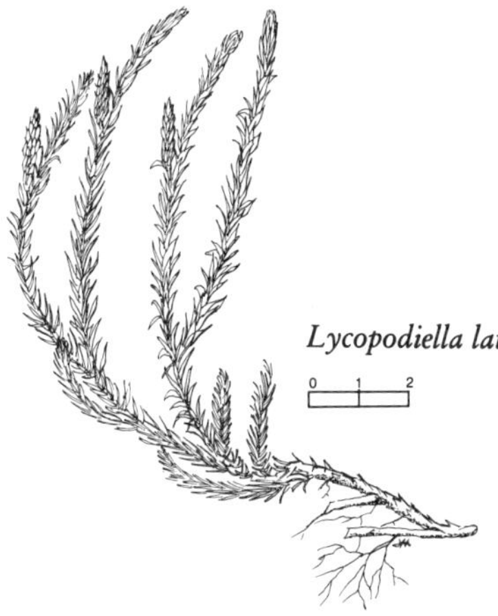

*Lycopodiella laterale*

# Marsilea — Marsileaceae

## Marsilea drummondii — Common Nardoo

**F P**

**Size:** Stalks 2–30 cm high
**Habitat:** Flood plains, bogs and swamps in box and red gum woodland, grassy wetlands, riparian scrub
**Form:** An aquatic with floating fronds and a creeping rhizome, glabrous on aquatic plants, covered in dense orange-brown hairs when terrestrial. Nodes on rhizome are close together.
**Foliage:** Silky, grey-green, "four-leaf clover" fronds on thin stalks to 30 cm long, held above the water surface. As the frond ages the stalk becomes more pliable and the rigid leaf floats. Glabrous wedge-shaped leaflets to 25 mm long (smaller and hairy on terrestrial plants).
**Fruit:** Light brown, rounded sporocarps to 10 mm wide singly on short stalks at the base of the frond. Fruiting generally occurs as water recedes, the sporocarps detaching from the plant as the soil dries out. They split following flooding to release the spores.
**Requirements:** Boggy soil, subject to inundation. Shallow water no deeper than 0.3 m.
**Comments:** Attractive in a pond or aquarium, plants may cover large areas in ideal situations, without being invasive. Tends to become dormant in cooler months. Spores form in hard structures the size of a pea, called sporocarps. They were an important food source for the Aborigines who either roasted them or ground them into flour.
**Propagation:** Spore or division
**Localities:** Sydenham, Brooklyn, Kew, 1, 6, 7, 9, 11, 25, 35, 36, ?42
**Distribution:** All mainland states

*Marsilea drummondii*

## Marsilea hirsuta — Short-fruit Nardoo

**F P**

**Size:** Stalks 2–17 cm high
**Habitat:** Grassy wetlands on muddy sites beside swamps
**Comments:** Similar to *M. drummondii*, and although often with long silky hairs on top of the fronds, usually less hairy than that species. Stalks of fronds shorter, to 17 cm long. Narrow wedge-shaped leaflets to 15 mm long. Sporocarps to 5 mm, borne singly or in small groups, held horizontally on very short stalks. Nodes on rhizomes usually distant.
**Localities:** Laverton, 7
**Distribution:** All mainland states

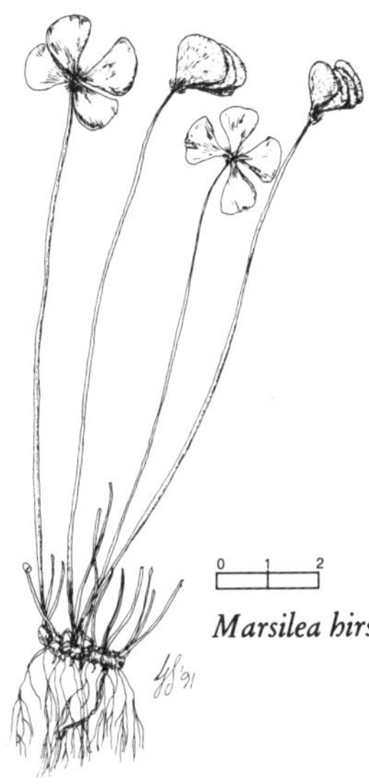

*Marsilea hirsuta*

SCALE: 0 1cm 2cm — The scale representation on each plant equals 2cm.

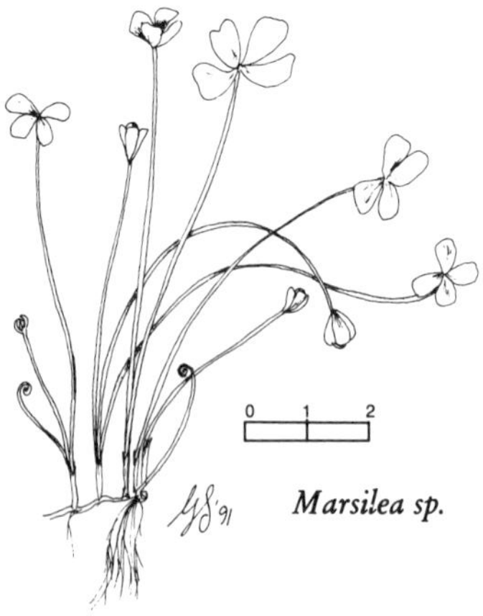

*Marsilea sp.*

### *Marsilea* sp. — Narrow-leaf Nardoo

**Size:** Stalks to 10 cm high **F P**
**Habitat:** Boggy areas subject to inundation in grassy wetlands, plains grassland
**Foliage:** Leaflets are narrowly obovate to 8 mm long, asymmetrically arranged, and sparsely hairy to glabrous
**Fruit:** Solitary ribbed sporocarps to 3 mm long on tiny stalks
**Comments:** Similar to *M. drummondii* and *M. hirsuta* but rhizomes are glabrous, stalks grow in clusters and plant is much more miniature in form. Plants form mats in boggy soils, or very shallow water. Much less common than previous species and locally rare.
**Localities:** Laverton, 17#, 18, 35
**Distribution:** All mainland states
**Synonym:** Previously confused with *M. angustifolia* which only occurs in tropical northern Australia.

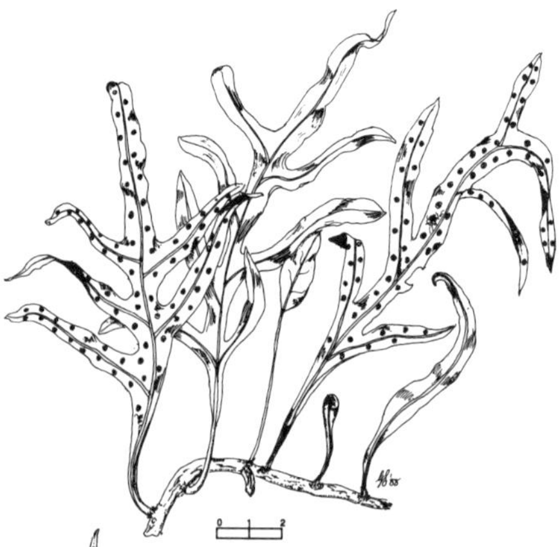

*Microsorum pustulatum*

# Microsorum — Polypodiaceae

## Microsorum pustulatum — Kangaroo Fern

**Size:** Scrambling or climbing fern **P N**
**Habitat:** Cool temperate rainforest, wet sclerophyll forest
**Form:** Vigorous epiphytic or lithophytic fern with a creeping rhizome, favouring *Dicksonia antarctica* as a host
**Foliage:** Variable, linear, irregularly lobed or pinnate, leathery fronds to 60 cm long, but usually much less. Margins are wavy and lobes vary from 5–25 cm long.
**Requirements:** Shaded protected site with high humidity
**Comments:** An attractive creeping fern growing on mossy rocks, treefern trunks and logs in cool, moist forests. This fern is quite adaptable to cultivation in well watered, protected fern gardens.
**Propagation:** Spore, division
**Localities:** 55, 57, 58
**Distribution:** Vic, Qld, NSW, Tas; Norfolk Is, NZ, N Cal
**Synonym:** *M. diversifolium*

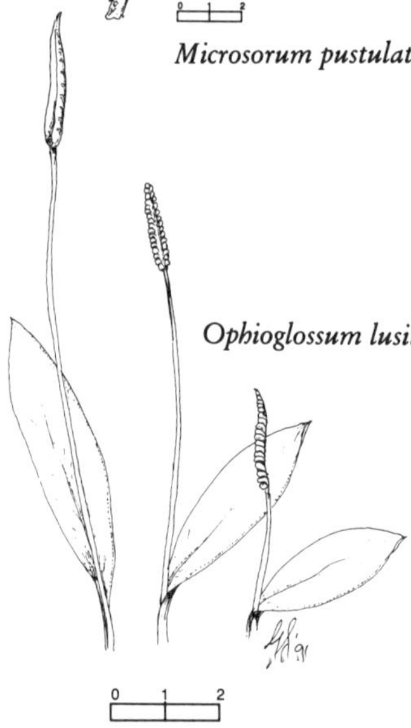

*Ophioglossum lusitanicum*

# Ophioglossum — Ophioglossaceae

## Ophioglossum lusitanicum — Adder's Tongue

**Size:** Fertile frond to 15 cm high **F P**
**Habitat:** Valley sclerophyll forest, tea-tree heath, wattle tea-tree scrub, plains grassland
**Form:** Very small sparse fern with tuber-like rhizome
**Foliage:** Each plant consists of a solitary, erect broadly elliptic to lanceolate sterile frond to 35 mm long and a fertile spike to 15 mm long rising from the base of the sterile frond on a thread-like stalk to 70 mm long
**Requirements:** Protected situation with moist, well drained soil, tolerating some flooding
**Comments:** Difficult to maintain in cultivation but may occur in dense clumps in its natural habitat. Very rare within the Melbourne area.
**Propagation:** Spore
**Localities:** Laverton, Hurstbridge, 34A, 67#
**Distribution:** All States; almost cosmopolitan
**Synonym:** *O. coriaceum*

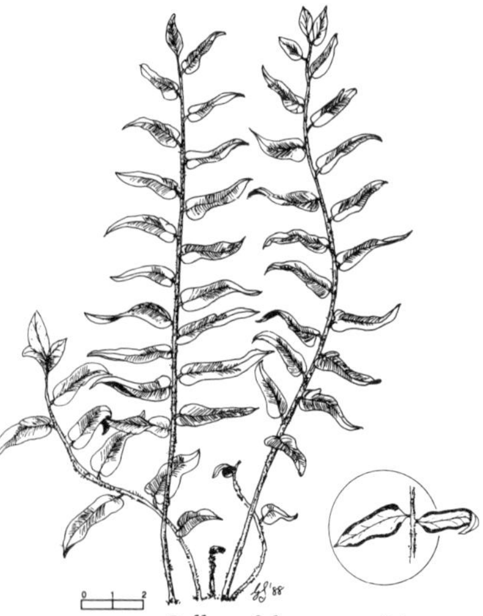

*Pellaea falcata* var. *falcata*

# Pellaea — Adiantaceae

## Pellaea falcata var. falcata — Sickle Fern

**Size:** 0.3–0.8 m x 0.3–0.6 m **P N**
**Habitat:** Rocky outcrops in riparian scrub and woodland, wet and damp sclerophyll forests
**Form:** Rigid tufted fern with short creeping rhizomes
**Foliage:** Erect, linear-lanceolate, pinnate, shiny dark green leathery fronds to 50 cm long, rhachis dark brown. Pinnae 20–50 mm long, ovate to lanceolate, often curved upwards.
**Requirements:** Moist, well drained organic soils with root protection (such as rock crevices). Tolerates wet periods and some sun.
**Comments:** A reliable fern for pot or garden culture.
**Propagation:** Spore, division
**Localities:** Deer Park, 8, 16, 26, 29#, 42, 57, 58, 60
**Distribution:** Vic, Qld, NSW, Tas; NZ, Mal, India

# Phylloglossum — Lycopodiaceae

## Phylloglossum drummondi

Pigmy Club Moss

P N

**Size:** 1–5 cm high
**Habitat:** Swamp and wattle tea-tree scrub
**Form:** Tiny colony-forming fern with a basal tufted rosette and a central erect fleshy white stalk bearing a cone-like sporing structure
**Foliage:** Few, fleshy terete basal fronds to 20 mm long
**Requirements:** Wet peaty soils
**Comments:** The only species in the genus. Leaves die back as the ground dries, and resprout from a tiny white tuber during Winter and Spring. Locally rare.
**Propagation:** Spore
**Localities:** 49, 67#, 74
**Distribution:** Vic, NSW, Tas, SA, WA; NZ

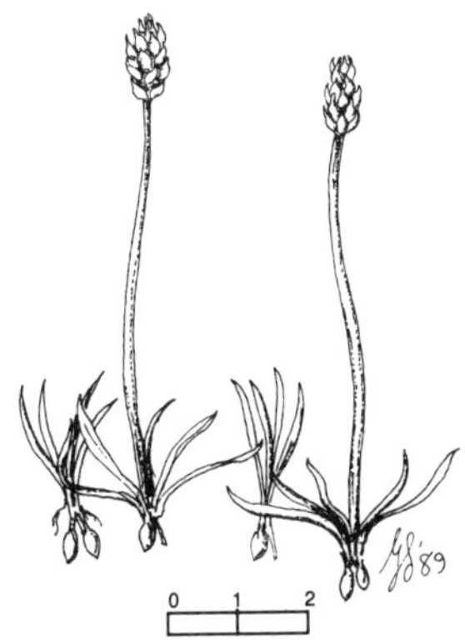

*Phylloglossum drummondi*

# Pilularia — Marsileaceae

## Pilularia novae-hollandiae

Austral Pillwort

F P

**Size:** 2–7 cm high
**Habitat:** Swamps and drying mud of temporary shallow water in box woodland
**Form:** Small perennial fern forming dense tufts along thread-like, glabrous green rhizomes
**Foliage:** Deep green grass-like sterile fronds to 70 mm long. Fertile fronds are buried in mud, and end in tiny globular sporocarps.
**Requirements:** Soft mud, tolerating temporary inundation
**Comments:** Growing amongst grasses and sedges it is easily overlooked. Recognised by the more or less flexuose fronds and the way they unfurl. A restricted occurrence.
**Propagation:** Spores, division
**Localities:** Keilor, Laverton Nth, Deer Park, Rockbank, 36
**Distribution:** Vic, NSW, Tas, SA, WA

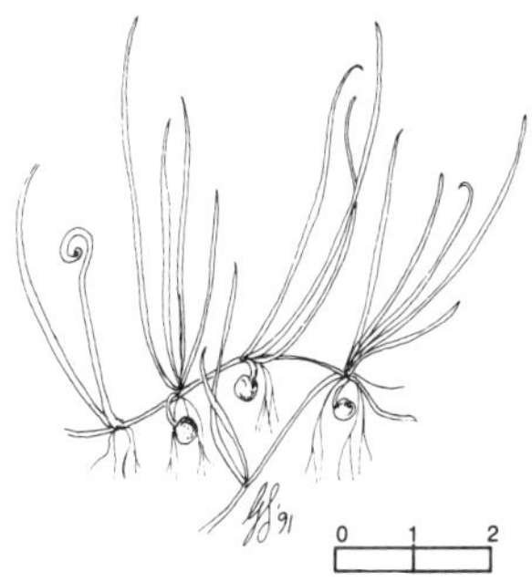

*Pilularia novae-hollandiae*

# Pleurosorus — Aspleniaceae

## Pleurosorus rutifolius

Blanket Fern

F

**Size:** to 17 cm high
**Habitat:** Rocky sites in riparian woodland and scrub, box woodland
**Form:** Small upright fern with a very short creeping rhizome covered in dark brown scales
**Foliage:** Erect, pinnate dark green fronds to 17 cm long, pinnae fan-shaped, notched or lobed, to 15 mm long
**Requirements:** Adaptable to most soils but resents overwatering and humidity. Like many Cheilanthes species, this fern prefers a sunny position with root protection such as afforded by rock crevices.
**Comments:** An attractive little fern which is quite drought resistant. It shrivels up at the height of Summer but resurrects with the onset of rain. Locally rare.
**Propagation:** Spore, division
**Localities:** Sunbury, Werribee, Altona, St Albans, 10, 11, 16, 26, 29, 36
**Distribution:** All States; NZ

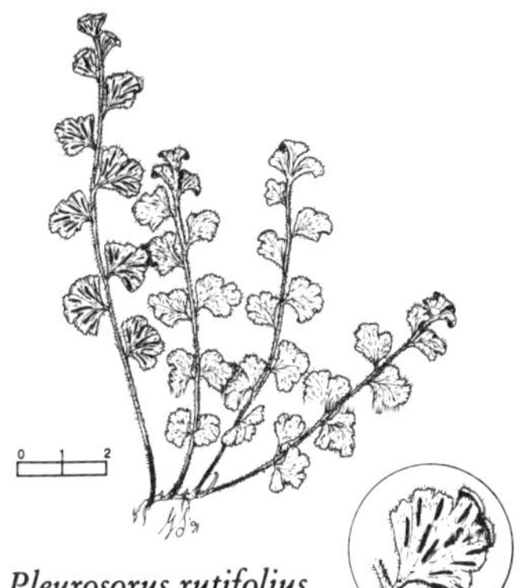

*Pleurosorus rutifolius*

# Polyphlebium — Hymenophyllaceae

## Polyphlebium venosum

Veined Bristle Fern

P N

**Size:** Fronds to 15 cm long
**Habitat:** Cool temperate rainforest, wet and damp sclerophyll forests
**Form:** A delicate rhizomatous epiphytic fern, the rhizome thread-like and much branched
**Foliage:** Delicate translucent, pendulous, light green, lobed to irregularly pinnate fronds, with prominent, branched veins
**Requirements:** High humidity
**Comments:** This is the most common filmy fern in the outlying moist hills around Melbourne, preferring, and often completely covering the trunks of *Dicksonia antarctica*.
**Propagation:** Spore, division
**Localities:** 57, 58
**Distribution:** Vic, NSW, Tas; NZ

*Polyphlebium venosum*

SCALE: 0 1cm 2cm — The scale representation on each plant equals 2cm.

*Polystichum proliferum*

# Polystichum — Dryopteridaceae

## Polystichum proliferum

Mother Shield Fern — P N

**Size:** 1–1.5 m x 1–1.5 m
**Habitat:** Cool temperate rainforest, wet and damp sclerophyll forests
**Form:** Tufted fern with long arching fronds, often occurring in large spreading colonies. Rhizomes are thick and become trunk-like in old plants.
**Foliage:** Young fronds bright light green, becoming dull dark green as they mature. At first erect, then arching, to 1 m long, bipinnate, often bearing bulbils near the tips. Stalks and rhachis densely covered with dark scales.
**Requirements:** Moist soils, tolerating dry periods once established
**Comments:** An excellent and reliable fern for garden culture that is readily propagated by layering the bulbils.
**Propagation:** Spore, bulbils
**Localities:** Deer Park, 42, 49, 55–58, 60, 74
**Distribution:** Vic, NSW, Tas

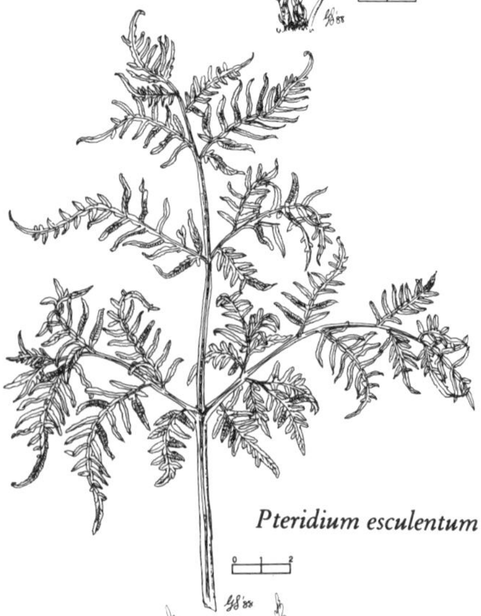

*Pteridium esculentum*

# Pteridium — Dennstaedtiaceae

## Pteridium esculentum

Austral Bracken — F P

**Size:** to 1.5 m high
**Habitat:** Widespread in wet, damp and valley sclerophyll forests, sclerophyll and red gum woodland, grassy low open forest, tea-tree heath, wattle tea-tree scrub, coastal banksia woodland
**Form:** Single erect frond, spreading vigorously from a strong creeping rhizome
**Foliage:** Dark green, broadly-triangular, leathery, 3- to 4-pinnate fronds, on rigid upright stalks, to 1.5 m long
**Requirements:** Moist or dry soils
**Comments:** An invasive fern that is poisonous to stock. Although this plant spreads readily by means of its branching underground rhizomes, it is difficult to propagate vegetatively. The sap from young fronds is a useful insect bite antidote. The rhizome was roasted in the hot ashes by the Aborigines, then beaten to a paste.
**Propagation:** Spore
**Localities:** 15, 16, 26, 29, 30, 32, 34A–36, 40–44, 46–76
**Distribution:** All States; NZ, South Pacific Is

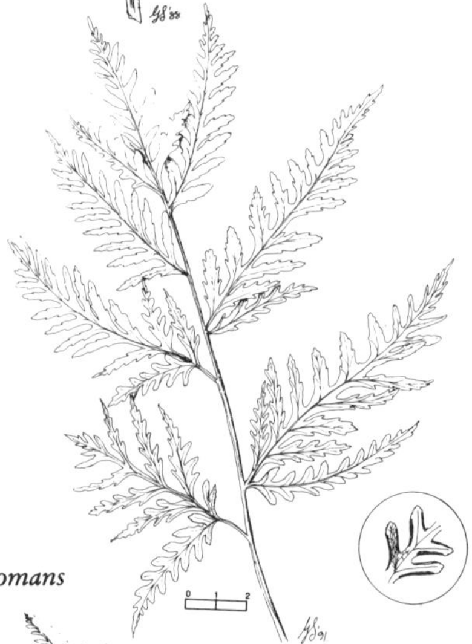

*Pteris comans*

# Pteris — Pteridaceae

## Pteris comans

Netted Brake — N

**Size:** to 1.5 m high
**Habitat:** Fern gullies of damp and wet sclerophyll forests
**Form:** Erect tufted fern
**Foliage:** Erect, dark green, 3- to 4-pinnate, broadly triangular lacy fronds to 1.5 m long. Pinnules are irregularly toothed and net veined. The grooved stalks are yellowish.
**Requirements:** Ample moisture and rich organic soils
**Comments:** This fern is rare in Victoria and endangered in the Melbourne region. It is difficult to maintain in cultivation as it resents disturbance.
**Propagation:** Spore
**Localities:** 57, 58
**Distribution:** Vic, Qld, NSW, Tas; NZ, South Pacific Is

## Pteris tremula

Tender Brake — P N

**Size:** 0.6–1.5 m high
**Habitat:** Damp sclerophyll forest; isolated moist sites elsewhere
**Form:** Tufted fern sometimes with multiple crowns on old plants
**Foliage:** Erect, fine lacy, 2- to 4-pinnate, pale green, triangular fronds to 1.5 m long, on grooved, shiny brown stalks. Pinnules are oblong-linear, usually with fine teeth on margins. Veins are not netted.
**Requirements:** Moist to wet humus rich soils
**Comments:** A very attractive fern for a large container or protected fernery. In ideal conditions it freely colonises in gardens. Locally rare.
**Propagation:** Spore, division
**Localities:** 57, 58, 74
**Distribution:** Vic, Qld, NSW, Tas, SA; Norfolk Is, NZ, Poly

*Pteris tremula*

# Pyrrosia — Polypodiaceae

## Pyrrosia rupestris

Rock Felt Fern

**P N**

**Size:** Prostrate-10 cm x 20 cm
**Habitat:** Wet sclerophyll forest
**Form:** An epiphytic or lithophytic mat-forming fern. The slender rhizome much branched and often exposed.
**Foliage:** Pale green, fleshy rounded to ovate button-like sterile fronds to 20 mm long, with lanceolate fertile fronds to 100 mm long. Covered in hairs when young, only on lower surface of frond when older.
**Requirements:** Ample moisture and humidity
**Comments:** This unusual fern is very rare in the Melbourne area, known only from two sites in the Dandenongs. It is more common in warm temperate and subtropical climates. Makes an interesting specimen in an open hanging basket on a mossy or fibrous host provided it has plenty of moisture and protection from winds. In times of drought fronds will shrivel, but swell again after rain.
**Propagation:** Spore, division
**Localities:** Wonga Park, 57
**Distribution:** Vic, Qld, NSW; PNG

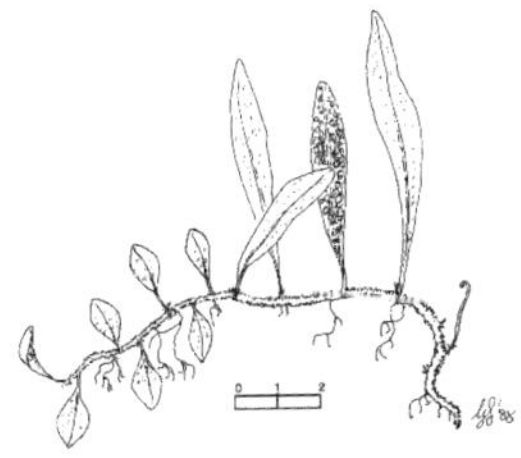

*Pyrrosia rupestris*

# Rumohra — Davalliaceae

## Rumohra adiantiformis

Leathery Shield Fern

**P N**

**Size:** Creeping or climbing epiphytic
**Habitat:** Cool temperate rainforest
**Form:** Epiphytic on treefern trunks, mossy logs and boulders, or rarely terrestrial fern, sparsely foliaged with a robust hairy creeping rhizome
**Foliage:** Large broadly-triangular light green to shiny dark green, bipinnate to tripinnate erect or arching fronds to 50 cm long, on long grooved stalks
**Requirements:** Moist organic soils or a fibrous host, in humid conditions
**Comments:** Common on trunks of *Dicksonia antarctica.*
**Propagation:** Spore, division
**Localities:** 57,58
**Distribution:** Vic, Qld, NSW, Tas; NZ, Central & S Am, S Af

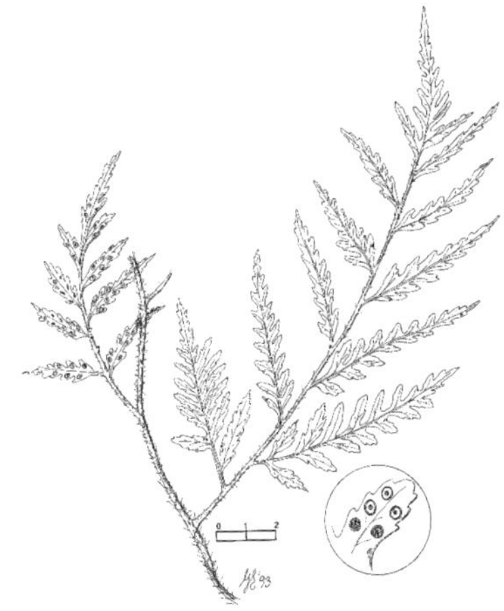

*Rumohra adiantiformis*

# Schizaea — Schizaeaceae

## Schizaea asperula

Rough Comb-fern

**P**

**Size:** 10–40 cm high
**Habitat:** Wattle tea-tree scrub, tea-tree heath
**Form:** Upright tufted fern, with wiry stems and short creeping rhizome
**Foliage:** Distinctly rough, erect fan-shaped fronds forked several times. Sterile fronds 10–15 cm long, fertile fronds 15–30 cm long, with end segments consisting of pinnately arranged sori-bearing branches, resembling a comb.
**Requirements:** Dry to moist, well drained, peaty acidic soils
**Comments:** Does not tolerate root disturbance. It requires a symbiotic fungal relationship for successful growth. Once more widespread in Melbourne but now very restricted in its range. The three *Schizaea* species were often found growing together.
**Propagation:** Spores, division
**Localities:** Grew in Cheltenham to Sandringham area last century; 74
**Distribution:** Vic, NSW, Tas, SA

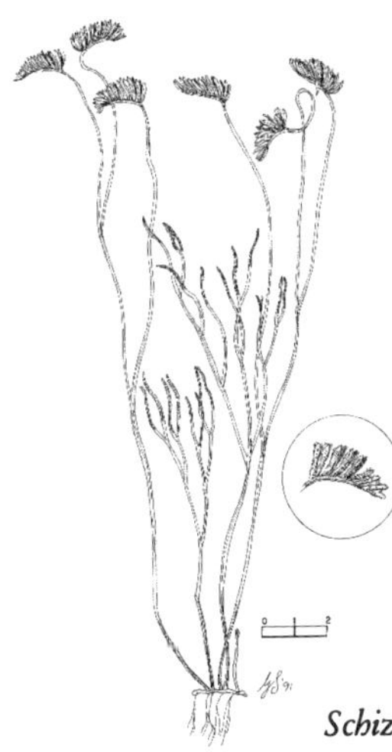

*Schizaea asperula*

## Schizaea bifida

Forked Comb-fern

**P**

**Size:** to 35 cm high
**Habitat:** Swamps in wattle tea-tree scrub, tea-tree heath
**Form:** Erect tufted fern with wiry stems and short rhizomes covered in glossy brown hairs
**Foliage:** Erect sterile fronds to 10 cm long, with simple flattened thread-like segments. Stems of fertile fronds wider, flattened and taller, to 35 cm long, similar to *S. asperula.*
**Requirements:** Moist to wet peaty soils
**Comments:** A slowly spreading fern now rare in the Melbourne area.
**Propagation:** Spores, division
**Localities:** 67#, 74
**Distribution:** Vic, Qld, NSW, Tas, SA; NZ

## Schizaea fistulosa

Narrow Comb-fern

**P**

**Size:** 5–30 cm high
**Habitat:** Tea-tree heath, wattle tea-tree scrub beside swamps, often on mossy mounds
**Comments:** Similar to and often found with *S. bifida.* Rhizomes dark brown. Sterile fronds are simple, grass-like and few in number. Stems of fertile fronds simple and terete with a single terminal comb-shaped spore-bearing head, narrower than *S. bifida.*
**Localities:** Cranbourne, 67#
**Distribution:** All states except NT; NZ, N Cal, Mal, Madagascar, S Am

SCALE: 0 1cm 2cm — The scale representation on each plant equals 2cm.

*Schizaea fistulosa*

*Schizaea bifida*

*Selaginella gracillima*

*Selaginella uliginosa*

# Selaginella — Selaginellaceae

## Selaginella gracillima — Tiny Selaginella or Clubmoss

**Size:** 2–6 cm high — P N
**Habitat:** Wet peaty depressions in wattle tea-tree scrub
**Form:** Almost moss-like wispy annual fern ally with erect stems, unbranched or branching once or twice
**Foliage:** Tiny pale green lanceolate leaves to 1 mm long scattered along the stem
**Requirements:** Ample moisture and humidity in water-holding peaty soils
**Comments:** Unknown in cultivation.
**Propagation:** Spore
**Localities:** Heatherton, 49, 67#, 74, 75
**Distribution:** All States except NT

## Selaginella uliginosa — Swamp Selaginella or Clubmoss

**Size:** 5–25 cm x 5–30 cm — F P
**Habitat:** Moist depressions in swamp and wattle tea-tree scrub
**Form:** A soft erect rhizomatous fern-like plant, with slender pinnately branched stems
**Foliage:** Tiny scale-like, pale yellowish-green leaves turning red with stress. Spore cases bunched together in cone-like structures on the tips of some stems.
**Requirements:** Poorly drained, moist organic soils
**Comments:** An interesting plant, each branched stem somewhat resembling a tiny pine tree.
**Propagation:** Spore, division
**Localities:** Montrose, 49, 50A, 54, 67, 73–76
**Distribution:** All States except SA

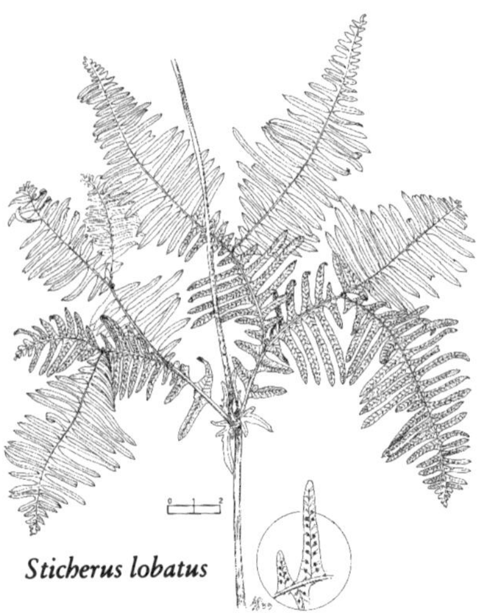

*Sticherus lobatus*

# Sticherus — Gleicheniaceae

## Sticherus lobatus — Spreading Fan Fern

**Size:** to 1 m high — P N
**Habitat:** Cool temperate rainforest
**Form:** An erect fern with long creeping rhizomes and tough wiry stalks
**Foliage:** Dull green fronds, to 1 m long, are repeatedly branched, bipinnate and fan-shaped on slender stems. Pinnules, with entire margins, are arranged almost at right angles to the rhachis.
**Requirements:** Moist soils in a protected position
**Comments:** Plants generally resent disturbance and drying out. Restricted distribution in Melbourne. When growing along creeks, stems may reach 2 m high, supported by surrounding growth, and plants may cover large areas.
**Propagation:** Spore
**Localities:** Dandenong Ranges
**Distribution:** Vic, NT, Qld, NSW, Tas

## Sticherus tener — Silky Fan Fern

**Size:** to 2 m high — P N
**Habitat:** Cool temperate rainforest, wet sclerophyll forest
**Comments:** Pinnules are entire, somewhat silky on the underside and arranged at an angle to the rhachis. The two local species of *Sticherus* are similar and easily confused. The angle of attachment of the pinnules to the rhachis is the major distinguishing feature. *S. tener* is much easier to cultivate, but will not tolerate drying out.
**Propagation:** Spore, division
**Localities:** 57
**Distribution:** Vic, NSW

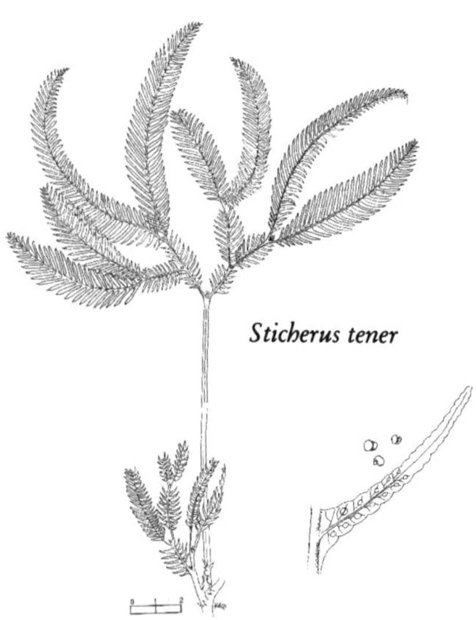

*Sticherus tener*

# Tmesipteris — Psilotaceae

## Tmesipteris billardieri — Long Fork Fern

**Size:** stems to 0.6 m long — N
**Habitat:** Cool temperate rainforest
**Form:** Fleshy rhizomatous epiphyte with pendulous leafy stems growing almost entirely on trunks of *Dicksonia antarctica*. The rhizomes are often deeply buried in the trunk.
**Foliage:** Weak flattened stems to 60 cm long are borne along the rhizome at regular intervals. The stems bear broad leaves to 25 mm long, with rounded tips and short point.
**Requirements:** Abundant water and high humidity
**Comments:** *Tmesipteris* spp. are not true ferns but a very primitive type of plant which lacks true roots. Well grown specimens can completely clothe the tops of large treeferns. This species is currently undergoing botanical revision.
**Propagation:** Spore
**Localities:** 55, 57
**Distribution:** Vic, NSW, Tas; Norfolk Is.

*Illustrated p. 321*

### Tmesipteris parva

Small Fork Fern

N

**Size:** stems to 10 cm long
**Habitat:** Cool temperate rainforest
**Form:** Fleshy rhizomatous epiphyte with pendulous leafy stems, usually growing on trunks of *Cyathea australis*
**Foliage:** Fleshy tapering leaves to 15 mm long
**Requirements:** Abundant moisture and high humidity
**Comments:** Superficially similar to *T. billardieri*, but smaller in all parts. It may be distinguished by the leaves, which are narrow and curved and taper to an acute point.
**Propagation:** Spore
**Localities:** Silvan, 57
**Distribution:** Vic, NSW, King Is; Philippines

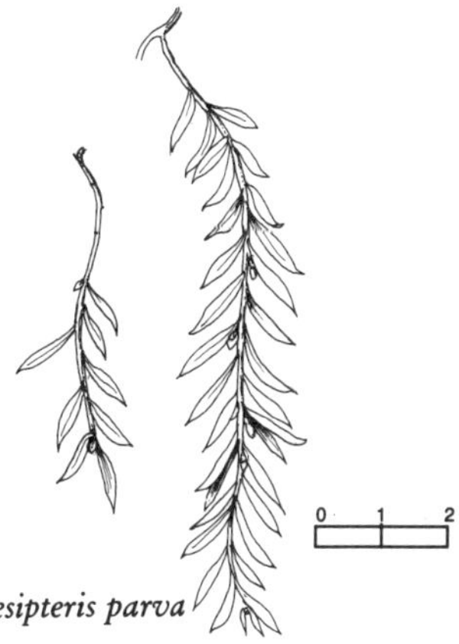

*Tmesipteris parva*

## Todea — Osmundaceae

### Todea barbara

King Fern

F P N

**Size:** 2–3 m x 4 m
**Habitat:** Wet and damp sclerophyll forests
**Form:** Spreading treefern-like plant with a dense crown of fronds, often with multiple crowns, from a stout, almost barrel-shaped trunk
**Foliage:** Upright to arching, dark green, glossy, leathery bipinnate fronds to 2.5 m long
**Requirements:** Moist, humus-rich soils
**Comments:** Very old plants can develop multi-crowned trunks rarely more than 1 m high but often up to 1.5 m wide. They are slow growing but extremely long lived. May be distinguished from other treeferns by the bipinnate rather than tripinnate fronds. Locally rare.
**Propagation:** Spore, transplant
**Localities:** 57, 58
**Distribution:** Vic, Qld, NSW, Tas, SA; NZ, S Af

*Todea barbara*

*Tmesipteris billardieri*

# Conifer

## Callitris — Cupressaceae

### Callitris glaucophylla

White Cypress-pine

F

**Size:** 7–20 m x 5–10 m
**Habitat:** Plains grassland, isolated rocky scarps in box woodland
**Form:** Slender erect tree with a straight trunk
**Foliage:** Glaucous or green, leaves to 3 mm long are raised segments with rounded edges, forming branchlets
**Cones:** Tiny male cones borne in terminal clusters, female cones usually single or in pairs on stalks to 15 mm long, smooth, rounded to 25 mm wide, with 6 woody scales (or segments) splitting almost to the base as cone opens; Spring to early Summer. Cones usually fall once mature.
**Requirements:** Well drained soils, tolerating extended dry periods
**Comments:** An ornamental slow growing tree well known for its termite-resistant wood. Useful as a shade tree or windbreak, and for binding sand. It is depleted within Victoria mainly as a result of timber cutting and rabbit grazing. It has a very restricted range in the Melbourne area.
**Propagation:** Seed
**Localities:** Keilor, Sydenham, Sth of Melton, 15, 16, 27
**Synonym:** Previously included in *C. columellaris*, also confused with *C. hugelii*
**Distribution:** All mainland states

*Callitris glaucophylla*

SCALE: 0 1cm 2cm The scale representation on each plant equals 2cm.

# *Appendix 1. Cross Reference for Easy Plant Selection*

The cross reference section should be used as an aid to find a plant or a group of plants quickly to fulfill a specific requirement. More information on each plant may then be obtained from the descriptions in the previous section.

## Explanation of Key to the Cross Reference Section

### Size and habit of plant

Sizes are an average and may vary according to the conditions in which a plant is grown.
The term 'tree' refers to a large woody plant with a distinct trunk.

TREE: Small—less than 8 metres high
Medium—8–15 metres high
Tall—more than 15 metres high

The term 'shrub' refers to a woody perennial plant which has several major branches growing from near the base of the plant.

SHRUB: Small—up to 1 metre high
Medium—1–2.5 metres high
Large—more than 2.5 metres high

A Herb does not develop a woody stem. It may be annual or perennial.
An Annual is a plant which flowers, seeds and dies in the one year.
A Tuft is a clumping plant where the leaves rise from the base of the plant. It is the common habit of grasses, lilies, irises and ferns.
A Groundcover is low and spreading, although it is not necessarily completely prostrate. It is always wider than its height and may be a shrub or herb, spreading from low branches or by stolons or rhizomes.
A Climber scrambles or twines up and/or over other plants.
An Epiphyte grows on other plants but is not a parasite.

### Flowering Times

Flowering times are given in seasons: Summer, Autumn, Winter, Spring, Y—most of the year. The months of flowering, where known, are indicated in the individual descriptions.

### Bird Attracting Plants

Some species of plants provide a rich source of nectar while others have sought-after seeds. Birds will seek insects from most plants and protection in prickly bushes. Waterbirds eat leaves, stems and/or roots of most water plants.

N—plants providing nectar
S—plants providing seeds
B—plants providing berries

### Butterfly Attracting Plants

About 60 species of butterflies exist in the Melbourne area. The number of moth species is unknown but it would be in the thousands.

Indigenous vegetation is the key to the conservation of butterflies and moths. Their survival is under threat from heavy use of insecticides, clearing of native vegetation for agricultural and urban use and the invasion of native vegetation by weeds.

Butterflies require plants at 2 separate stages of their life cycle. As an adult they seek nectar-rich flowers as an energy source for flying and mating. As a caterpillar they require food for growth. This is usually, but not always, leaves.

This list is by no means complete and further information is welcome.

F—food plant for the caterpillar
N—a nectar plant for the butterfly or moth

### Aspect

F—requires or tolerates full sun
P—prefers partial sun or dappled shade
N—prefers or tolerates full shade
A—will cope with all aspects

### Screen or Windbreak Plants

S—may be used to screen unwanted vistas
W—will tolerate reasonable wind force and protect less sturdy plants

### Soil and Drainage

D—dry, well drained soils—may be sandy, clayish or stony soils which do not accumulate water
M—moist, well drained soils—never dry out, holding water for short periods only
S—moist, swampy soils—poorly drained, includes areas subject to periodic inundation which may dry out in Summer
W—wet, boggy or waterlogged soils—remain wet all year. Includes semi-aquatic plants.
A—aquatic, in shallow or deep water

### Coastal

1—plants which can withstand the full brunt of salt-laden winds
2—plants which require some protection from the strong, salty winds, either behind the front row, behind the sand dunes or a little further from the coast

### Vegetation Communities

Covers the habitats found within the Melbourne region. While some plants are found in very specific locations, others may occur in a variety of habitats. (See Chapter 1 for further information.)

1 Aquatic
2 Cool temperate rainforest
3 Wet sclerophyll forest
4 Damp sclerophyll forest
5 Sclerophyll woodland
6 Swamp scrub
7 Grassy low open forest
8 Tea-tree heath
9 Wattle tea-tree scrub
10 Coastal banksia woodland
11 Primary dune scrub
12 Saltmarsh
13 Grassy wetland
14 Red gum woodland
15 Plains grassland
16 Riparian scrub
17 Chenopod rocky open scrub
18 Box ironbark woodland
19 Box woodland
20 Riparian woodland
21 Dry sclerophyll forest
22 Valley sclerophyll forest

### Aboriginal Uses

* The Aborigines had a range of uses for plants. These include: food, medicine, weapons, containers, utensils, transportation.

# Plant Selection List

## Dicotyledons

| | Tree | Shrub | Other | Flowers | Bird | Butterfly | Sun Aspect | Screen Windbreak | Drainage | Coast | Vegetation Communities | Aboriginal Uses |
|---|---|---|---|---|---|---|---|---|---|---|---|---|
| *Acacia acinacea* | | S-M | | W-Sp | S | | F P | S | D | | 14 17-19 21 | |
| *Acacia aculeatissima* | | | G | W-S | S | | P | | D | | 21 | |
| *Acacia brownei* | | S | | W-Sp | S | | P | | D | 2 | 8 21 | |
| *Acacia dealbata* | M-T | | | W-Sp | S | F | F P | W | M | | 3 4 7 14 16 22 | * |
| *Acacia genistifolia* | | M | | A-Sp | S | | F P | | D M | 2 | 21 22 | |
| *Acacia gunnii* | | S | | W-Sp | S | | F P | | D | | 18 21 | |
| *Acacia implexa* | S-M | | | S-A | S | | F P | S | D M | 2 | 14 19-22 | |
| *Acacia lanigera* | | S-M | | A-Sp | S | | F P | | D | | 18 21 | |
| *Acacia leprosa* | S-M | L | | W-Sp | S | | P N | S | M | | 4 22 | |
| *Acacia mearnsii* | S-T | | | Sp-S | S | F | A | W | D M | 1 | 7 14-16 20-22 | * |
| *Acacia melanoxylon* | M-T | | | W-Sp | S | F | F P | S W | M | 1 | 2 3 7 14-16 21 22 | * |
| *Acacia mucronata* | | M-L | | Sp | S | | F P | S | D M | | 4 22 | |
| *Acacia myrtifolia* | | M | | W-Sp | S | | A | | D M | | 4 5 22 | |
| *Acacia oxycedrus* | | M | | W-Sp | S | | F P | | M | 2 | 8 21 | |
| *Acacia paradoxa* | | M-L | | W-Sp | S | | F P | S | D | | 7 14 15 21 | |
| *Acacia pycnantha* | S-M | L | | W-Sp | S | | F P | S W | D | 2 | 4 14 15 18 21 22 | * |
| *Acacia retinodes* | S | L | | Y | S | | F P | S W | D M S | 2 | 10 16 20 | |
| *Acacia sophorae* | | S-L | | W-Sp | S | | F | S W | D | 1 | 7 10 11 | * |
| *Acacia stricta* | | M-L | | W-Sp | S | | A | W | D | 1 | 7 8 10 22 | |
| *Acacia suaveolens* | | M | | A-Sp | S | | F P | W | D | 2 | 8 9 | |
| *Acacia ulicifolia* | | M | | A-Sp | S | | P | S | M | 2 | 7 8 22 | |
| *Acacia verniciflua* | | L | | W-S | S | | F P | S | D M S | 2 | 4 7 22 | * |
| *Acacia verticillata* | S | M-L | | W-S | S | | P | | M S | 2 | 3 4 9 16 22 | * |
| *Acaena agnipila* | | | H | Sp | | | A | | D M | | 13-15 21 | |
| *Acaena echinata* | | | H | Sp | | | F P | | D M | | 13-15 22 | |
| *Acaena novae-zelandiae* | | | H G | Sp-S | | | A | | M | 1 | 3 6 14-16 20 22 | |
| *Acaena ovina* | | | H | Sp | | | F P | | D M | 1 | 4 7 14 15 20 21 | |
| *Acrotriche prostrata* | | | G | W | N | | P N | | M | 2 | 4 5 7 22 | * |
| *Acrotriche serrulata* | | | G | W-Sp | N | | P N | | D M | 1 | 5 7 21 22 | * |
| *Actites megalocarpa* | | | H | Sp-A | | | F | | D | 1 | 11 | |
| *Ajuga australis* | | | H A | S | | | A | | D | 2 | 14 21 22 | * |
| *Allocasuarina littoralis* | S | | A | S | | | F P | W | D | 2 | 7 14 21 22 | * |
| *Allocasuarina luehmannii* | M | | S | S | | | F | | D | | 19 | * |
| *Allocasuarina paludosa* | | S | | A-Sp | S | | F | | M | 2 | 5 7 8 22 | * |
| *Allocasuarina paradoxa* | | S | | W-Sp | S | | P | S | D M | 2 | 8 | * |
| *Allocasuarina verticillata* | M | | | A-Sp | S | | F | W | D | 1 | 11 14 17-19 | * |
| *Almaleea subumbellata* | | S | | Sp-S | | | P | | M | | 6 | |
| *Alternanthera denticulata* | | | H | Y | | | F P | | M S | 2 | 13-15 19 20 22 | |
| *Alyxia buxifolia* | | M | | Sp-S | B | | F | | D | 1 | 11 | |
| *Amperea xiphoclada* | | S | | Sp-S | | | F | | M | 2 | 8 21 | |
| *Amyema linophyllum* ssp. *orientale* | | | | Sp-S | B | F | P N | | | | 19 | * |
| *Amyema miquelii* | | | | Y | B | F | P N | | | | 14 18 19 | * |
| *Amyema pendulum* | | | | S | B | F | P N | | | 1 | 3 4 5 7 22 | * |
| *Amyema preissii* | | | | Sp-A | B | F | P N | | | | 11 | * |
| *Amyema quandang* | | | | Y | B | F | P N | | | | 22 | * |
| *Angianthus preissianus* | | | H A | Sp | | | F | | D M | 1 | 11 12 | |
| *Aotus ericoides* | | M | | Sp | | | F P | | D M | 2 | 8 | |
| *Apium annuum* | | | H A | Sp | | | F | | M | 1 | 7 11 15 | * |
| *Apium prostratum* ssp. *prostratum* var. *filiforme* | | | H | S-A | | | F | | M W | 1 | 10 16 | * |
| *Apium prostratum* ssp. *prostratum* var. *prostratum* | | | H | S-A | | | F P | | M W | 1 | 10 16 | * |

| | Tree | Shrub | Other | Flowers | Bird | Butterfly | Sun Aspect | Screen Windbreak | Drainage | Coast | Vegetation Communities | Aboriginal Uses |
|---|---|---|---|---|---|---|---|---|---|---|---|---|
| *Asperula conferta* | | | H G | Sp-S | | | P | | D M | 2 | 11 13-15 22 | |
| *Asperula euryphylla* | | | H | Sp | | | P N | | M | | 3 4 | |
| *Asperula scoparia* | | | H G | Sp-S | | | P N | | M | 2 | 4 13 15 21 | |
| *Asterolasia asteriscophora* | | M | | W-Sp | | | P | | M | | 4 22 | |
| *Astroloma humifusum* | | S | G | W-Sp | N | B | F P | | D | 2 | 7 8 14 18 21 | * |
| *Atherosperma moschatum* | T | | | A-W | | F | P N | | M | | 2 3 | |
| *Atriplex billardieri* | | | H G | | | | F | | D | 1 | 11 | * |
| *Atriplex cinerea* | | M | | Sp-A | | | F P | | D | 1 | 11 | * |
| *Atriplex paludosa* ssp. *paludosa* | | S | | Y | | | F | | W | 1 | 12 14 | * |
| *Atriplex semibaccata* | | S | | S | | | F P | | D | 1 | 11 16 19 | * |
| *Atriplex suberecta* | | | H A | Y | | | F | | W | 1 | 12 | * |
| *Australina pusilla* ssp. *muelleri* | | | H G | S | | | P N | | M | | 2-4 | |
| *Avicennia marina* | S | | | S | B | | F | | A | 1 | 1 12 | * |
| *Ballantinia antipoda* | | | H A | Sp | | | F P | | D M | | 13 | |
| *Banksia integrifolia* | T | | | S-Sp | N | N | F | W | D | 1 | 10 11 | * |
| *Banksia marginata* | S | M-L | | Sp-A | N | N | F P | S W | D M | 1 | 5 7 8 16 22 | * |
| *Banksia spinulosa* var. *cunninghamii* | | M-L | | A-W | N | | F P | S W | M | 2 | 5 22 | * |
| *Bauera rubioides* | | S-M | G | Y | | | P N | | M S | | 3 4 | |
| *Bedfordia arborescens* | S | L | | Sp-S | | | P N | | M | | 2 3 22 | |
| *Beyeria viscosa* | | L | | W-S | | | P N | S | M | | 4 | |
| *Billardiera longiflora* | | | C | W-S | N B | | P | | M | | 2 3 7 22 | * |
| *Billardiera scandens* var. *brachyantha* | | | C | W-S | N B | | P | | D M | | 21 | * |
| *Billardiera scandens* var. *scandens* | | | C | W-S | N B | | A | | D M | 2 | 4 5 7 21 22 | * |
| *Boerhavia dominii* | | | C G | Sp-S | | | F P | | D | 2 | 12 | * |
| *Boronia parviflora* | | S | | Sp | | | P | | M W | 2 | 8 | |
| *Bossiaea cinerea* | | M | | Sp | | | P | | D M | 2 | 8 10 | |
| *Bossiaea cordigera* | | | G C | S | S | | F P | | M | | 5 | |
| *Bossiaea prostrata* | | | G | Sp | S | | A | | D M | 2 | 5 7 14 15 21 22 | |
| *Brachyloma ciliatum* | | S | | Sp | | | F P | | M | 2 | 8 | * |
| *Brachyscome aculeata* | | | H | Sp-A | | N | F P | | M | | 16 | |
| *Brachyscome basaltica* var. *gracilis* | | | H | Sp | | N | F | | D M | | 13 15 | |
| *Brachyscome cardiocarpa* | | | H T | W-Sp | | N | F P | | S | 2 | 6 8 13 | |
| *Brachyscome ciliaris* | | | H | W-S | | N | F | | D | | 15 | |
| *Brachyscome decipiens* | | | H | Sp-S | | N | F P | | M S | 2 | 14 15 21 22 | |
| *Brachyscome dentata* | | | H | Sp | | N | F | | D M | | 13 15 16 19 | |
| *Brachyscome diversifolia* var. *diversifolia* | | | H T | Sp-S | | N | F P | | D | 1 | 10 11 21 | |
| *Brachyscome exilis* | | | H | Sp-S | | N | F P | | S | | 13 | |
| *Brachyscome graminea* | | | H | Y | | N | F P | | S | 2 | 8 13 16 | |
| *Brachyscome leptocarpa* | | | H A | Sp | | N | F | | D | | 15 | |
| *Brachyscome multifida* var. *multifida* | | | H G | Y | | N | F P | | D M | | 19 21 | |
| *Brachyscome parvula* var. *parvula* | | | H T | Sp-S | | N | F | | S | 1 | 11 | |
| *Brachyscome perpusilla* | | | H A | W-Sp | | N | F | | M | | 13 | |
| *Brachyscome scapigera* | | | H T | Sp-S | | N | F | | D M | | 21 | |
| *Brachyscome spathulata* | | | H | Sp-A | | N | F P | | D M | | 16 | |
| *Brachyscome trachycarpa* | | | H | Y | | N | F P | | D M | | 15 | |
| *Brachyscome uliginosa* | | | H | Sp-S | | N | F | | S | | 22 | |
| *Bracteantha bracteata* | | | H A | Sp-A | | N | F | | D M | 2 | 8 21 | |
| *Bracteantha viscosa* | | | H | Sp-A | | N | F P | | D | | 14 19 21 | |
| *Brunonia australis* | | | H | Sp-S | | N | F P | | D M | 2 | 8 14 15 21 22 | |
| *Bursaria spinosa* var. *macrophylla* | S | M-L | | S | | N F | F P | S | D M | | 15 16 | |
| *Bursaria spinosa* var. *spinosa* | S | M-L | | S | | N F | F P | S | D M | 1 | 7 11 14 16 20-22 | |
| *Calandrinia calyptrata* | | | A | W-Sp | | | F | | D | 1 | 10 11 14 17 19 | * |
| *Calandrinia eremaea* | | | A | W-Sp | | | F | | D | | | |
| *Callistemon sieberi* | S | L | | S-A | N S | N | F P | S | M S | | 16 | |
| *Callitriche brachycarpa* | | | H | Sp-S | | | P | | W | | 16 22 | |
| *Callitriche cyclocarpa* | | | H | Sp-S | | | P | | A | | 1 | |
| *Callitriche muelleri* | | | H | Sp-S | | | P N | | W A | | 1 | |
| *Callitriche palustris* | | | H | Sp-S | | | P | | A | | 1 16 22 | |
| *Callitriche umbonata* | | | H | Sp-S | | | P | | W A | | 1 13 | |
| *Calocephalus citreus* | | | H | Sp-S | | N | F P | | D M | | 14 15 | |
| *Calocephalus lacteus* | | S | | Sp-S | | N | F P | | D M | | 14 15 21 | |
| *Calotis anthemoides* | | | H | Sp | | | F P | | M S | | 15 | |
| *Calotis cymbacantha* | | | A | A W-Sp | | | F P | | D | | 15 | |
| *Calotis lappulacea* | | | H | Y | | | F P | | D | | 19 | |
| *Calotis scabiosifolia* var. *scabiosifolia* | | | H | Sp | | | F P | | M | | 14 15 | |
| *Calotis scabiosifolia* var. *integrifolia* | | | H | Sp | | | P | | M | | 21 | |
| *Calotis scapigera* | | | H | Y | | | F P | | M | | 13 14 | |
| *Calystegia marginata* | | | C | Sp-S | | | P N | | M S | | 4 22 | |
| *Calystegia sepium* | | | C | Sp-S | | | P | | M | | 13 16 22 | |

| | Tree | Shrub | Other | Flowers | Bird | Butterfly | Sun Aspect | Screen Windbreak | Drainage | Coast | Vegetation Communities | Aboriginal Uses |
|---|---|---|---|---|---|---|---|---|---|---|---|---|
| *Calytrix tetragona* | | M | | Sp | | N | F P | | D | | 16 | |
| *Cardamine gunnii* | | | H | Sp-S | | | P | | M W | | 2 8 | * |
| *Cardamine paucijuga* | | | H | W-Sp | | | P | | D M | | 6 16 | * |
| *Cardamine tenuifolia* | | | H | S | | | P | | W S | | 15 22 | * |
| *Carpobrotus modestus* | | | G | Sp-S | | | F | | D | | 17 | * |
| *Carpobrotus rossii* | | | G | Sp-S | | | F P | | D | 1 | 10 11 | * |
| *Cassinia aculeata* | | L | | S | | | P | | M | 1 | 3-5 7 21 22 | |
| *Cassinia arcuata* | | M | | S | | | F P | | D | 2 | 7 18 19 21 22 | |
| *Cassinia longifolia* | | L | | S | | | P N | | M | 2 | 7 14 18 19 21 22 | |
| *Cassinia trinerva* | | L | | S | | | P | | M | | 3 4 | |
| *Cassytha glabella* | | | C | Sp | | F | F P | | | 2 | 4 8 9 22 | * |
| *Cassytha melantha* | | | C | W-Sp | | F | F P | | D | 2 | 4 10 19 21 22 | * |
| *Cassytha pubescens* | | | C | S-A | | F | F P | | | 2 | 4 8 14 21 22 | * |
| *Centella cordifolia* | | | G | W-S | | | F P | | S | | 4 6 7 16 21 22 | |
| *Centipeda cunninghamii* | | | H | S-A | | | F P | | M S | 2 | 7 13 16 | * |
| *Centipeda minima* | | | H | S-A | | | P | | S | | 6 14 22 | * |
| *Ceratophyllum demersum* | | | H | S | | | | | A | | 1 | |
| *Chenopodium desertorum* ssp. *microphyllum* | | | H | W-Sp | | F | F | | S W | | 17 | |
| *Chenopodium glaucum* | | | A | A-W | | F | F P | | M | 1 | 11 12 15 | |
| *Chenopodium pumilio* | | | A | S-W | | F | F | | S W | | 13 16 | |
| *Chrysocephalum apiculatum* | | | H | Y | | N | F | | D M | 2 | 8 14 19 | |
| *Chrysocephalum semipapposum* | | | H | S-A | | N | F P | | D M | | 4 14 15 19 21 | |
| *Ciclospermum leptophyllum* | | | A | S | | | P N | | M | | 16 | |
| *Clematis aristata* | | | C | W-S | | | A | | D M | | 3 4 7 22 | |
| *Clematis microphylla* | | | C | W-Sp | | | F P | | D M | 1 | 7 10 11 15 18-21 | |
| *Comesperma calymega* | | S | | S | | | F P | | D M | 2 | 8 | |
| *Comesperma ericinum* | | M | | S | | | P | | D M | 2 | 8 22 | |
| *Comesperma polygaloides* | | S | | Sp-S | | | F | | D | | 14 15 | |
| *Comesperma volubile* | | | C | W-Sp | | | P | | D M | 2 | 7 8 21 22 | |
| *Convolvulus erubescens* | | | C G | Sp-S | | | F | | D | 2 | 4 8 14 15 21 | * |
| *Convolvulus remotus* | | | G | Sp-S | | | F | | D | | 14 15 | |
| *Coprosma hirtella* | | M | | Sp | B | | P | | M | | 3 4 22 | |
| *Coprosma quadrifida* | | M | | Sp | B | | P N | | M | | 2-6 | * |
| *Correa alba* | | M | | Y | N | | F P | S | D | 1 | 11 | |
| *Correa glabra* | | M | | Y | N | | A | S | D | | 16 | |
| *Correa reflexa* | | M | | A-W | N | | A | | D M | 2 | 4 5 7 8 21 22 | |
| *Correa reflexa* x *glabra* | | M | | A-W | N | | F P | | D | | 21 | |
| *Cotula australis* | | | H G | Sp-S | | | P N | | M | 1 | 5 10 14 16 21 | |
| *Cotula vulgaris* var. *australasica* | | | A | Sp | | | F P | | W | 2 | 8 12 | |
| *Craspedia variabilis* | | | T | Sp | | N | F P | | M S | 2 | 5 7 8 13 14 16 21 22 | |
| *Crassula colorata* var. *acuminata* | | | A | Sp | | | F | | W | | 7 15 17 | |
| *Crassula decumbens* var. *decumbens* | | | A | Sp | | | P | | W | 1 | 8 11 14 22 | |
| *Crassula helmsii* | | | H | S-A | | | F | | W A | | 1 7 16 20 22 | |
| *Crassula pedicellosa* | | | A | Sp | | | F P | | M | | 8 13 | |
| *Crassula peduncularis* | | | A | S | | | P N | | S | 1 | 10 19 | |
| *Crassula sieberiana* ssp. *sieberiana* | | | H | Sp-W | | | F P | | S W | 2 | 8 13-17 19 | |
| *Crassula sieberiana* ssp. *tetramera* | | | A | Sp | | | P | | W | 2 | 8 22 | |
| *Cryptandra amara* var. *amara* | | S | | W-Sp | | | F P | | D | | 19 | |
| *Cuscuta tasmanica* | | | C | S-A | | | F | | | 1 | 12 | |
| *Cymbonotus lawsonianus* | | | H | Sp | | | F | | M | | 14 15 22 | |
| *Cymbonotus preissianus* | | | H | Sp | | | P N | | M | 2 | 4 7 14 21 | |
| *Cynoglossum australe* | | | H | Sp-S | | | F P | | D | | 15 | |
| *Cynoglossum latifolium* | | S | | Y | | | P | | M S | | 3 4 | |
| *Cynoglossum suaveolens* | | | H | Y | | | F P | | D M | 2 | 4 10 15 21 | |
| *Dampiera stricta* | | | H G | A | | | F P | | M | | 8 21 | |
| *Daucus glochidiatus* | | | A | Sp | | | F P | | D | 1 | 10 15 21 22 | |
| *Daviesia latifolia* | | M | | Sp | | | F P | | D M | 2 | 4 7 21 22 | |
| *Daviesia leptophylla* | | M | | Sp-S | | | F P | | D | 2 | 5 10 21 22 | |
| *Daviesia ulicifolia* | | M | | W-Sp | | | P | | D M | 2 | 8 16 21 | |
| *Derwentia derwentiana* | | S | | S | | | P N | | M | | 3 16 22 | |
| *Desmodium gunnii* | | S | G | Y | | | P | | M | | 4 13 22 | |
| *Desmodium varians* | | S | | W-Sp | | | F P | | D | | 15 21 | |
| *Dichondra repens* | | | H G | Sp | | | P N | | M | 2 | 7 10 15 16 20-22 | |
| *Dillwynia cinerascens* | | M | | W-Sp | | | P | | D | 2 | 8 15 21 22 | |
| *Dillwynia glaberrima* | | M | | Sp | | | F P | | D M | 2 | 7 8 22 | |
| *Dillwynia hispida* | | S | | Sp | | | P | | M | 2 | 8 | |
| *Dillwynia juniperina* | | M | | Sp | | | P | | D | | 14 | |
| *Dillwynia phylicoides* | | M | | Sp | | | P | | D | | 21 | |
| *Dillwynia sericea* | | M | | Sp | | | F P | | D | 2 | 5 8 21 | |

| | Tree | Shrub | Other | Flowers | Bird | Butterfly | Sun Aspect | Screen Windbreak | Drainage | Coast | Vegetation Communities | Aboriginal Uses |
|---|---|---|---|---|---|---|---|---|---|---|---|---|
| *Disphyma crassifolium* ssp. *clavellatum* | | | G | Sp-S | | | F | | D | 1 | 10 12 | * |
| *Dodonaea viscosa* ssp. *spatulata* | | M | | Sp | | | F P | S | D M | | 7 16 19 21 | |
| *Drosera binata* | | | H | Sp-A | | | F P | | W | 2 | 8 22 | |
| *Drosera glanduligera* | | | H | W-Sp | | | F P | | S | 2 | 8 14 | |
| *Drosera macrantha* | | | H | W-Sp | | | F P | | S | 2 | 4-8 22 | |
| *Drosera peltata* ssp. *auriculata* | | | H | Sp | | | F P | | S | 2 | 5 7-9 21 22 | |
| *Drosera peltata* ssp. *peltata* | | | H | Sp | | | F P | | S | 2 | 8 14 15 21 22 | |
| *Drosera pygmaea* | | | H | W-S | | | F | | S | 2 | 6 7 | |
| *Drosera spathulata* | | | H | Y | | | P | | S | 2 | 6 8 | |
| *Drosera whittakeri* | | | H | W-Sp | | | F P | | S | 2 | 7 8 14 15 21 22 | |
| *Eclipta platyglossa* | | | H | Sp-S | | | F P | | S | | 13 | |
| *Einadia hastata* | | S | | S | | | F P | | D | 1 | 8 11 14 16-20 | |
| *Einadia nutans* ssp. *nutans* | | | G | S-A | | | F P | | D | 1 | 11 14 16 17 19-21 | |
| *Einadia trigonos* | | | G | S | | | P | | M | | 14 | |
| *Elatine gratioloides* | | | A G | Y | | | F | | W A | | 1 13-15 19 | |
| *Enchylaena tomentosa* | | S | G | Sp-S | | | F P | | D M | 1 | 11 14-17 19 | * |
| *Epacris gunnii* | | S | | A-Sp | N | | P | | M | | 6 22 | |
| *Epacris impressa* | | S | | A-Sp | N | | P | | D M | 1 | 4 5 7-9 21 22 | |
| *Epacris obtusifolia* | | M | | W-Sp | N | N | P | | M S | 2 | 8 | |
| *Epilobium billardierianum* ssp. *billardierianum* | | | H | S | | | F P | | M S | 2 | 6 9 14 16 21 22 | |
| *Epilobium billardierianum* ssp. *cinereum* | | | H | S | | | F P | | M S W | 2 | 4 6 22 | |
| *Epilobium hirtigerum* | | | H | S | | | F P | | M S | | 5 7 14-16 22 | |
| *Eremophila deserti* | | L | | A-Sp | B | | F | | D | | 19 | |
| *Erodium crinitum* | | | H | W-Sp | | | F P | | D | 2 | 8 10 15 | * |
| *Eryngium ovinum* | | | H | Sp-S | | | F | | M | | 14 15 | |
| *Eryngium vesiculosum* | | | H | Sp-S | | | F P | | S | | 6 13 22 | |
| *Eucalyptus albens* | M-T | | | A-W | N S | F | F P | W | D | | 19 | * |
| *Eucalyptus baueriana* | M-T | | | Sp-S | N S | F | F | W S | M | | 16 19 | * |
| *Eucalyptus baxteri* | M-T | | | S-A | N S | F | F P | W | D M | | 4 22 | * |
| *Eucalyptus blakelyi* | T | | | Sp-S | N S | F | F P | W S | M | | 18 | * |
| *Eucalyptus camaldulensis* | M-T | | | S | N S | F | F | | D M S | 2 | 14-16 20 | * |
| *Eucalyptus camphora* | M | | | A | N S | F | F P | W | S | | 22 | * |
| *Eucalyptus cephalocarpa* | M-T | | | A-Sp | N S | F | F P | W | M S | 2 | 4 5 7 9 22 | * |
| *Eucalyptus cypellocarpa* | M-T | | | S-W | N S | F | F P | | M | | 3 4 | * |
| *Eucalyptus dives* | M-T | | | Sp | N S | F | F P | W S | D M | | 5 22 | * |
| *Eucalyptus globoidea* | M-T | | | Y | N S | F | F P | | M S | | 22 | * |
| *Eucalyptus goniocalyx* | M-T | | | A-W | N S | F | F P | W | D | | 4 21 22 | * |
| *Eucalyptus ignorabilis* | M-T | | | S-A | N S | F | F | W | M | | 22 | * |
| *Eucalyptus leucoxylon* ssp. *connata* | M-T | | | W-Sp | N S | F | F P | W | D M | | 14 17-19 21 | * |
| *Eucalyptus macrorhyncha* | M-T | | | S-A | N S | F | F P | | D | | 18 21 22 | * |
| *Eucalyptus melliodora* | M-T | | | Sp-A | N S | F N | F P | W | D M | 2 | 14 19 21 22 | * |
| *Eucalyptus microcarpa* | M-T | | | S-W | N S | F | F | W | D | | 17 19 | * |
| *Eucalyptus nortonii* | M | | | Sp | N S | F | F P | W | M | | 14 | * |
| *Eucalyptus obliqua* | S-T | | | S-A | N S | F | F P | W | M | 2 | 3-5 7 21 22 | * |
| *Eucalyptus ovata* | M-T | | | A-W | N S | F | F P | W S | S | 2 | 4 6 7 14-16 20 22 | * |
| *Eucalyptus pauciflora* ssp. *pauciflora* | S-M | | | Sp-S | N S | F | F P | W | D M | 2 | 14 21 | * |
| *Eucalyptus polyanthemos* ssp. *vestita* | S-T | | | Sp-S | N S | F | F P | W S | D | | 21 22 | * |
| *Eucalyptus pryoriana* | S-M | | | A-W | N S | F | F | W S | D | 1 | 7 8 | * |
| *Eucalyptus radiata* | M-T | | | Sp-S | N S | F | F P | W | D M | | 4 5 7 22 | * |
| *Eucalyptus regnans* | T | | | S-W | N S | F | F | | M | | 2 3 | |
| *Eucalyptus rubida* ssp. *rubida* | M-T | | | Sp-S | N S | F | F P | | D | | 21 22 | |
| *Eucalyptus tricarpa* | M-T | | | W-S | N S | F | F P | W | D | | 18 21 | * |
| *Eucalyptus viminalis* | M-T | | | S-A | N S | F | F P | W | M | 2 | 3 6 7 20 22 | * |
| *Eucalyptus yarraensis* | M | | | S | N S | F | F P | W S | M S | | 22 | * |
| *Euchiton gymnocephalus* | | | H | Y | | | F P | | M | 2 | 9 16 22 | |
| *Euchiton involucratus* | | | H | Y | | | F P | | M | 1 | 10 16 19 21 22 | |
| *Euchiton sphaericus* | | | A H | W-S | | | F P | | M | | 15 19 22 | |
| *Euphorbia drummondii* | | | A G | Sp | | | F | | D | | 12 15 19 | |
| *Euphrasia collina* ssp. *collina* | | | H | W-S | | | F P | | M | 2 | 5 8 | |
| *Euphrasia collina* ssp. *muelleri* | | | H | W-Sp | | | F P | | M | | 8 | |
| *Euphrasia collina* ssp. *paludosa* | | | H | Sp-S | | | P | | S | | 6 22 | |
| *Euphrasia collina* ssp. *trichocalycina* | | | H | W-Sp | | | F | | M | | 21 | |
| *Euphrasia scabra* | | | H | Sp-A | | | F P | | M | | 11 12 | |
| *Eutaxia diffusa* | | S | | Sp | | | F | | D | 1 | 11 19 | |
| *Eutaxia microphylla* var. *microphylla* | | S | G | Sp | | | A | | D M S | | 14 15 | |
| *Exocarpos cupressiformis* | L | | | Y | | F | F P | S | D M | 1 | 4 7 8 14 18 19 21 22 | * |
| *Exocarpos stricta* | M | | | W-Sp | | F | F P | | D | 1 | 4 7 14 21 22 | |
| *Frankenia pauciflora* | | S | G | Y | | | F P | | D | 1 | 11 12 | |
| *Galium australe* | | | H | S | | | P | | M | | 7 16 22 | |
| *Galium binifolium* | | | A | Sp | | | P | | M | | 4 | |
| *Galium ciliare* | | | H | Sp-S | | | P | | D M | | 5 | |

| | Tree | Shrub | Other | Flowers | Bird | Butterfly | Sun Aspect | Screen Windbreak | Drainage | Coast | Vegetation Communities | Aboriginal Uses |
|---|---|---|---|---|---|---|---|---|---|---|---|---|
| *Galium compactum* | | | H | Sp-S | | | F | | D | | 21 | |
| *Galium gaudichaudii* | | | H | Sp-S | | | P | | M | | 3 4 7 16 22 | |
| *Galium migrans* | | | H | Sp | | | P | | M | | 16 20 22 | |
| *Galium propinquum* | | | H | Sp-S | | | P N | | M | 2 | 7 10 19 20 22 | |
| *Geranium potentilloides* | | | H | Sp-A | | | P N | | M | | 4 5 22 | * |
| *Geranium retrorsum* | | | H | W-S | | | P | | M | | 14 15 19 20 | |
| *Geranium solanderi* | | | H | Sp-S | | | P | | M S | | 7 14 16 21 22 | * |
| *Glossostigma elatinoides* | | | H G | S-A | | | F | | W A | | 19 | |
| *Glycine clandestina* | | | H C | Sp-S | | F | P N | | M | 2 | 8 14 21 22 | |
| *Glycine latrobeana* | | | G | Sp-S | | F | F P | | D | | 15 19 21 | |
| *Glycine microphylla* | | | C | Sp-A | | F | P N | | M | | 4 | |
| *Glycine tabacina* | | | C | S-A | | F | F P | | D | | 14-16 21 | |
| *Glycyrrhiza acanthocarpa* | | | H | Sp-S | | | F P | | M | | | |
| *Gnaphalium indutum* | | | H | Sp-S | | | F | | D | 2 | 7 10 13 14 | |
| *Gompholobium huegelii* | | S | | Sp-S | | | P | | D M | 2 | 5 8 21 | |
| *Gonocarpus elatus* | | | H | Sp-S | | | F P | | D M | | 19 | |
| *Gonocarpus humilis* | | S | | Sp-S | | | P | | S | | 4 7 22 | |
| *Gonocarpus micranthus* | | | G | S | | | P | | S W | 2 | 6 8 9 | |
| *Gonocarpus tetragynus* | | | H | S | | | F P | | D M | 2 | 5 7-9 14 15 21 22 | |
| *Goodenia blackiana* | | | H | Sp-S | | | F P | | D M | | 21 22 | |
| *Goodenia elongata* | | | H | Sp-S | | | P | | M | 2 | 6 8 22 | |
| *Goodenia geniculata* | | | H | W-S | | | F P | | M S | 2 | 7 8 | |
| *Goodenia gracilis* | | | H | Sp-S | | | F | | S | | 13 | |
| *Goodenia heteromera* | | | H | Sp-A | | | F P | | S | | 13 | |
| *Goodenia humilis* | | | H | Sp-S | | | F | | M S | 2 | 6 9 | |
| *Goodenia lanata* | | | H | Sp-S | | | F P | | D M | | 4 5 22 | |
| *Goodenia ovata* | | M | | W-S | | | A | | M S | 1 | 3 4 7 14 19 22 | |
| *Goodenia pinnatifida* | | | H | Sp-A | | | F | | M | 2 | 13-15 | |
| *Goodia lotifolia* | | M-L | | Sp | S | F | F P | | D M | | 4 21 | |
| *Gratiola peruviana* | | | H G | Sp-A | | | P N | | S | 2 | 7 16 22 | |
| *Gratiola pubescens* | | | H G | Sp-S | | | P | | S | | 6 | |
| *Grevillea alpina* | | M | | W-S | N | | P | | D M | | 21 | |
| *Grevillea infecunda* | | S | G | Sp S | N | | F P | | D | 2 | 8 | |
| *Grevillea rosmarinifolia* | | M | | W-Sp | N | | F P | | D M | | 15 21 | |
| *Gynatrix pulchella* | | M-L | | W-Sp | | | P | | M | | 4 15 16 22 | * |
| *Hakea nodosa* | | M-L | | A-W | N | | A | S | M S | 2 | 5 6 22 | |
| *Hakea sericea* | | L | | W-Sp | N | N | F P | | D | | 5 19 22 | |
| *Hakea teretifolia* ssp. *hirsuta* | | L | | Sp-S | N | | F | S | M | | 5 | |
| *Hakea ulicina* | | M-L | | W-Sp | N | | F P | | D M | 2 | 5 8 21 22 | |
| *Haloragis aspera* | | | H | Sp-A | | | F P | | M | | 15 19 | |
| *Haloragis heterophylla* | | | H | Sp-S | | | F P | | S | | 15 16 | |
| *Halosarcia halocnemoides* ssp. *halocnemoides* | | S | | S-A | | | F | | W | 1 | 12 | |
| *Halosarcia pergranulata* ssp. *pergranulata* | | S | | S-A | | | F | | W | 1 | 12 | |
| *Hardenbergia violacea* | | | C | W-Sp | | | F P | | D | 2 | 14 21 22 | |
| *Hedycarya angustifolia* | S | L | | Sp | | | N | | M | | 2-4 | * |
| *Helichrysum leucopsideum* | | | H | Sp-A | | N | F P | | D M | | 5 10 | |
| *Helichrysum rutidolepis* | | | H | Sp | | N | F P | | M | | 13-15 21 | |
| *Helichrysum scorpioides* | | | H G | Sp-S | | N | F P | | D M | 2 | 5 8 21 | |
| *Hemichroa pentandra* | | | H | Sp | | | F | | W | 1 | 12 | |
| *Hibbertia acicularis* | | S | | Sp-S | | F | F P | | D M | 2 | 5 8 | |
| *Hibbertia empetrifolia* | | S | G | Sp | | F | F P | | M | | 22 | |
| *Hibbertia obtusifolia* | | S | | W-S | | F | P | | D | | 18 21 | |
| *Hibbertia prostrata* | | S | | Sp | | F | F P | | M | 2 | 8 | |
| *Hibbertia riparia* | | S | | Sp-S | | F | F P | | M | | 5 7 8 22 | |
| *Hibbertia sericea* var. *sericea* | | S | | Sp | | F | F P | | M | 2 | 8 | |
| *Hibbertia stricta* var. *stricta* | | S | | W-Sp | | F | A | | M | 2 | 5 8 22 | |
| *Hibbertia virgata* var. *virgata* | | S | | W-Sp | | F | F P | | M | | 8 | |
| *Hovea linearis* | | S | | Sp | | | P N | | D M | 2 | 8 21 22 | |
| *Hydrocotyle callicarpa* | | | A | Sp | | | A | | M | 1 | 7 11 22 | |
| *Hydrocotyle capillaris* | | | A | Sp | | | F P | | M | | 12 15 21 | |
| *Hydrocotyle foveolata* | | | A | Sp | | | P | | M S | | 8 22 | |
| *Hydrocotyle geraniifolia* | | | H G | S | | | P N | | M S | | 3 22 | |
| *Hydrocotyle hirta* | | | H G | S | | | P | | S | | 3 4 6 9 22 | |
| *Hydrocotyle laxiflora* | | | H G | Sp | | | P N | | S | 2 | 20-22 | |
| *Hydrocotyle medicaginoides* | | | A | Sp | | | P | | S | | 9 | |
| *Hydrocotyle muscosa* | | | H G | Sp-A | | | P | | S W | | 7 | |
| *Hydrocotyle sibthorpioides* | | | H G | Y | | | P | | S | | 13 22 | |
| *Hydrocotyle tripartita* | | | H | S-A | | | P | | S W | 2 | 7 19 | |
| *Hydrocotyle verticillata* | | | H | S-A | | | P | | S W | 1 | 9 16 22 | |
| *Hymenanthera dentata* | | M-L | | Sp | | | F P | | M | | 14-16 19 | |
| *Hymenanthera* sp. aff. *dentata* | | S | | Sp | | | F | | D | | 15 | |

| | Tree | Shrub | Other | Flowers | Bird | Butterfly | Sun Aspect | Screen Windbreak | Drainage | Coast | Vegetation Communities | Aboriginal Uses |
|---|---|---|---|---|---|---|---|---|---|---|---|---|
| *Hypericum gramineum* | | | H | Sp-S | | | F P | | M S | 2 | 5 7 9 14 15 21 22 | |
| *Hypericum japonicum* | | | H | Sp-S | | | P | | S | 2 | 4 6 9 | |
| *Indigofera australis* | | M | | Sp | | F | A | | D M | | 4 7 16 22 | * |
| *Isoetopsis graminifolia* | | | A T | W-Sp | | | F | | D | | 15 | |
| *Isopogon ceratophyllus* | | S | | Sp | | | F P | | M | 2 | 8 | |
| *Isotoma fluviatilis* ssp. *australis* | | | G | Sp | | | F P | | S W | | 9 15 16 | |
| *Ixiolaena leptolepis* | | | H | Sp | | N | F | | D | | 15 | |
| *Ixiolaena* sp. | | S | | Sp-S | | | F | | M | | 15 | |
| *Kennedia prostrata* | | | G | A-S | | F | F P | | D | 2 | 7 8 14-16 21 22 | * |
| *Kunzea ericoides* | | L | | S | | N | F P | S | D M | | 21 22 | |
| *Lagenifera gracilis* | | | H | Sp-S | | | F P | | M | | 7 9 21 22 | |
| *Lagenifera huegelii* | | | H | Sp-A | | | F P | | M | | 19 21 | |
| *Lagenifera stipitata* | | | H | Sp-A | | | F P | | M | 2 | 3 4 7-9 22 | |
| *Lasiopetalum baueri* | | M | | Sp-S | | | F | | D | 1 | 11 | |
| *Lavatera plebeia* var. *plebeia* | | S-M | | W-Sp | | | F | | D | 1 | 12 16 | |
| *Lavatera plebeia* var. *tomentosa* | | S-M | | Y | | | F | | D | 1 | 11 | |
| *Lawrencia spicata* | | S | | Sp-S | | | F | | S | 1 | 8 12 | |
| *Lepidium aschersonii* | | | H | Sp | | | F | | S | 2 | 12 | * |
| *Lepidium pseudohyssopifolium* | | | H | W-Sp | | | F P | | D M | 1 | 15 16 19 | * |
| *Lepidium pseudotasmanicum* | | | H | W-Sp | | | P | | M | | 16 | * |
| *Leptinella reptans* | | | H G | Y | | | F P | | W | 1 | 10 16 | |
| *Leptorhynchos elongatus* | | | H A | Sp | | | F | | D | | 21 | |
| *Leptorhynchos squamatus* | | | H | Sp-S | | | F P | | M | | 14 15 22 | |
| *Leptorhynchos tenuifolius* | | | H | Sp-S | | | F P | | D M | 2 | 7 8 14 15 21 22 | |
| *Leptospermum continentale* | | M-L | | Sp-A | | N | F P | | M S | 2 | 5-9 22 | * |
| *Leptospermum grandifolium* | S | M-L | | Sp-S | | N | F P | S | S W | | 4 | * |
| *Leptospermum laevigatum* | S | L | | Sp | | N | F P | S W | D M | 1 | 7 8 10 11 | * |
| *Leptospermum lanigerum* | S | M-L | | Sp-S | | N | F P | S | S | 2 | 8 16 20 22 | * |
| *Leptospermum myrsinoides* | | S-M | | Sp | | N | F P | | M S | 2 | 5 8 22 | |
| *Leptospermum obovatum* | | M-L | | S | | N | P | S | S | | 16 | |
| *Leptostigma reptans* | | | H G | S | | | P N | | S W | | 22 | |
| *Leucochrysum albicans* ssp. *albicans* var. *albicans* | | | H | Sp-A | | N | F P | | D | | 21 | |
| *Leucophyta brownii* | | S | G | Sp-S | | | F P | | D | 1 | 11 | |
| *Leucopogon australis* | | S | | Sp | | | P | | M | 1 | 8 | * |
| *Leucopogon ericoides* | | S | | W-Sp | | | P | | D | | 8 21 | * |
| *Leucopogon parviflorus* | | S-M | | W-Sp | | | F P | S W | D | 1 | 10 11 | * |
| *Leucopogon virgatus* var. *virgatus* | | S | | W-S | | | F P | | D | 2 | 7 8 21 | * |
| *Levenhookia dubia* | | | H | Sp | | | P | | M | 2 | 8 13 22 | |
| *Levenhookia sonderi* | | | H | Sp | | | F P | | M S | 2 | 8 13 22 | |
| *Lilaeopsis polyantha* | | | H | Sp-A | | | F P | | W A | 1 | 1 11 12 | |
| *Limonium australe* | | | H | | | | F | | S W | 1 | 12 | |
| *Limosella australis* | | | G H | Sp | | | F | | S W | | 6 | |
| *Linum marginale* | | S | | Sp-A | | | F P | | M | 2 | 8 13 14 21 | * |
| *Lissanthe strigosa* | | S | | Sp | | | P | | D | | 21 | * |
| *Lobelia alata* | | | H | Y | | | F P | | M S | 2 | 5 8 16 22 | |
| *Lobelia gibbosa* | | | A | S | | | P | | M | 2 | 8 | |
| *Lobelia pratioides* | | | H G | S | | | P | | W | | 13 15 22 | |
| *Lobelia rhombifolia* | | | A G | Sp | | | P | | S W | | 4 | |
| *Lobelia simplicicaulis* | | | H | S | | | F P | | M | 2 | 8 21 22 | |
| *Lomatia fraseri* | S | L | | S | | | P N | | M | | 3 4 | |
| *Lomatia ilicifolia* | | S-M | | S | | | F P | | D | | 4 5 7 21 | |
| *Lomatia myricoides* | S | M | | S | | | P N | | M | | 16 | |
| *Lotus australis* | | | H | S-A | | F | F | | D M | | 15 | |
| *Lycopus australis* | | | H | S-A | | | P N | | S W | | 10 16 22 | |
| *Lysiana exocarpi* | | | | S-Sp | B | F | F P | | | | 19 | * |
| *Lythrum hyssopifolia* | | | A | S-A | | | F P | | M S | 2 | 6 7 13 14 16 22 | |
| *Lythrum salicaria* | | S-M | H | S | | | F | | S | | 16 22 | |
| *Maireana brevifolia* | | S | | S-W | | | F | | D | | 16 19 | |
| *Maireana decalvans* | | S | | S-A | | | F P | | S | | 15 19 | |
| *Maireana enchylaenoides* | | | H | Sp-S | | | F | | D | | 19 | |
| *Maireana pentagona* | | | H | Sp | | | F P | | M | | 15 | |
| *Mazus pumilio* | | | H G | Sp-S | | | A | | S W | | 8 21 | |
| *Melaleuca ericifolia* | S | M-L | | Sp | N | N | F P | S W | M S | 2 | 6 9 10 14 22 | * |
| *Melaleuca lanceolata* | S | L | | Sp-S | N | N | F | S W | D M S | 1 | 11 16 20 | * |
| *Melaleuca parvistaminea* | | M-L | | Sp | N | N | F P | S W | S | | 6 16 | * |
| *Melaleuca squarrosa* | | M-L | | Sp-S | N | N | F P | | S W | 2 | 4 6 9 22 | * |

| | Tree | Shrub | Other | Flowers | Bird | Butterfly | Sun Aspect | Screen Windbreak | Drainage | Coast | Vegetation Communities | Aboriginal Uses |
|---|---|---|---|---|---|---|---|---|---|---|---|---|
| *Mentha australis* | | | H G | Sp-S | | | A | | S W | | 16 22 | * |
| *Mentha diemenica* var. *diemenica* | | | H | S | | | A | | S | | 13 14 | * |
| *Mentha laxiflora* | | | H G | Sp-S | | | P N | | S | | 3 4 22 | * |
| *Mentha satureoides* | | | H | S | | | P N | | S | | 13 14 | * |
| *Microseris lanceolata* | | | H T | W-Sp | | N | P | | D M | 2 | 14 15 21 | * |
| *Millotia myosotidifolia* | | | A | W | | | F P | | D | | 14 | |
| *Millotia tenuifolia* | | | A | Sp | | | F | | D | 1 | 11 | |
| *Mimulus repens* | | | H G | Sp-A | | | F P | | S W | 2 | 12 13 16 | |
| *Minuria leptophylla* | | | H | Y | | | F | | D | | 15 | |
| *Mitrasacme distylis* | | | A | Sp | | | | | | 1 | | |
| *Mitrasacme paradoxa* | | | A | Sp | | | P | | D | 1 | 11 21 | |
| *Monotoca scoparia* | | M | | A | | | F P | | M | 2 | 8 21 | * |
| *Montia fontana* | | | H T | Sp | | | P | | S W | | 6 | |
| *Muehlenbeckia adpressa* | | | C | Sp | B | | F | | D | 1 | 10 11 | * |
| *Muehlenbeckia florulenta* | | M | | Y | | | F | | S | 2 | 12 13 16 | |
| *Muellerina eucalyptoides* | | | C | Sp-S | B | F | F P | | | | 3 4 7 14 19 22 | |
| *Myoporum insulare* | | M-L | | Sp | B | | F | S W | D | 1 | 10 11 | * |
| *Myoporum parvifolium* | | | G | S | B | | A | | D | 2 | 12 13 | * |
| *Myoporum viscosum* | | S-M | | Sp | B | | F | | D | 1 | 11 16 17 | * |
| *Myosotis australis* | | | A | Sp | | | F | | D | | 7 21 | |
| *Myosotis exarrhena* | | | H | Sp | | | | | D | | | |
| *Myriocephalus rhizocephalus* | | | A | Sp | | | P | | M | | 13 14 | |
| *Myriophyllum amphibium* | | | H | S | | | P | | S W | | 9 14 21 | |
| *Myriophyllum caput-medusae* | | | H | S | | | F P | | A | | 1 16 | |
| *Myriophyllum crispatum* | | | H | Sp-A | | | F P | | W A | | 7 13 22 | |
| *Myriophyllum integrifolium* | | | A | Sp | | | P | | S W | | 13 | |
| *Myriophyllum muelleri* | | | A | Sp | | | F P | | A | | 1 13 | |
| *Myriophyllum salsugineum* | | | H | Sp-S | | | F P | | A | | 1 | |
| *Myriophyllum simulans* | | | H | Sp | | | F P | | A | | 1 9 13 | |
| *Myriophyllum variifolium* | | | H | Sp-S | | | F P | | A | | 1 8 13 22 | |
| *Myriophyllum verrucosum* | | | H | Sp-A | | | F P | | A | | 1 4 13 | |
| *Neopaxia australasica* | | | H | G S | | | F P | | S W A | 1 | 1 6 8 13 | |
| *Nicotiana maritima* | | | A | Y | | | | | D | | 14 | |
| *Nicotiana suaveolens* | | | A | A-W | | | F | | D | | 16 19 | |
| *Notelaea ligustrina* | S | M-L | | Sp | B | | P N | S | M | | 2 3 | |
| *Olearia argophylla* | S | M-L | | Sp | | | P N | | M | | 2-4 22 | |
| *Olearia asterotricha* | | S-M | | S | | | P | | M S | | 7 22 | |
| *Olearia axillaris* | | S-M | | S-A | | | F | | D | 1 | 11 | |
| *Olearia ciliata* | | S | | Sp | | | F P | | D | 2 | 8 | |
| *Olearia decurrens* | | S-M | | S | | | F P | | D | | 15 19 | |
| *Olearia erubescens* | | S | | Sp | | | F P | | D | | 4 21 | |
| *Olearia glandulosa* | | S-M | | S-A | | | F P | | S | | 6 | |
| *Olearia glutinosa* | | M | | S | | | F | | D | 1 | 11 | |
| *Olearia lirata* | | M-L | | Sp | | | P N | | M | | 4 7 22 | |
| *Olearia myrsinoides* | | S-M | | Sp-S | | | F P | | D M | | 21 22 | |
| *Olearia phlogopappa* var. *phlogopappa* | | S-M | | Sp-S | | | F P | | D | | 22 | |
| *Olearia ramulosa* var. *ramulosa* | | S-M | | Sp-A | | | F P | | D | 2 | 4 7 8 11 14 | |
| *Olearia rugosa* | | S-M | | Sp | | | P | | M | | 4 | |
| *Opercularia ovata* | | | H | Sp | | | F P | | M | 2 | 7 14 21 22 | |
| *Opercularia varia* | | | H | Sp | | | P N | | M | 2 | 7-9 21 22 | |
| *Oreomyrrhis eriopoda* | | | H | Sp | | | F P | | D | | 19 21 | |
| *Oxalis exilis* | | | H | Y | | | F P | | M S | | 4 7 16 19 22 | |
| *Oxalis perennans* | | | H | Y | | | F P | | M S | | 13-15 22 | |
| *Oxalis radicosa* | | | H | A-Sp | | | F P | | D | | 19 | |
| *Ozothamnus ferrugineus* | S | M-L | | Sp-S | | | F P | | M | 2 | 3 4 6 7 9 14 22 | |
| *Ozothamnus obcordatus* | | S-M | | Sp-S | | | F P | | D | | 18 19 21 | |
| *Ozothamnus rosmarinifolius* | | M-L | | S | | | F P | | M | | 6 | |
| *Ozothamnus turbinatus* | | S-M | | A | | | F | | D | 1 | 11 | |
| *Pandorea pandorana* | | | C | Sp | | | F P | | M | | 3 4 7 21 22 | |
| *Parietaria debilis* | | | A | Sp | | | A | | M | 1 | 10 16 22 | |
| *Parsonsia brownii* | | | C | Sp | | | P N | | M | | 2 3 | |
| *Pelargonium australe* | | | H | Sp-S | | | F P | | D | 2 | 4 10 11 16 22 | |
| *Pelargonium inodorum* | | | A | Sp-S | | | F P | | D | | 7 21 22 | |
| *Pelargonium rodneyanum* | | | H | S | | | F P | | D | | 15 | |
| *Persicaria decipiens* | | | A | Y | S | | A | | W A | 2 | 4 8 14 16 20 22 | |
| *Persicaria hydropiper* | | | A | S | S | | F P | | W A | | 16 19 22 | * |
| *Persicaria lapathifolia* | | | A | Y | S | | F P | | S W | | 22 | |
| *Persicaria praetermissa* | | | H | S-A | S | | P | | S W | | 22 | |
| *Persicaria prostrata* | | | H | Sp-W | S | | F P | | S W | | 13 15 22 | |
| *Persicaria subsessilis* | | | H/A | | S | | F P | | S W | | 16 | |

| | Tree | Shrub | Other | Flowers | Bird | Butterfly | Sun Aspect | Screen Windbreak | Drainage | Coast | Vegetation Communities | Aboriginal Uses |
|---|---|---|---|---|---|---|---|---|---|---|---|---|
| *Persoonia juniperina* | | S-M | | S | N | | F P | | D M | 2 | 7 8 21 22 | * |
| *Pimelea axiflora* ssp. *axiflora* | | M | | Sp | | N | P N | | M | | 3 4 22 | * |
| *Pimelea curviflora* var. *sericea* | | S | | Sp-S | | N | F P | | D M | | 4 14 15 21 22 | |
| *Pimelea flava* ssp. *flava* | | S-M | | Sp | | N | P N | | M | | 4 22 | |
| *Pimelea glauca* | | S | | W-S | | N | F P | | D | 2 | 11 15 | |
| *Pimelea humilis* | | S | | Sp-S | | N | F P | | M | 2 | 7 8 14 21 22 | |
| *Pimelea ligustrina* ssp. *ligustrina* | | S-M | | Sp | | N | P N | | M | | 3 4 | |
| *Pimelea linifolia* ssp. *linifolia* | | S | | W-S | | N | F P | | D | | 16 22 | |
| *Pimelea micrantha* | | S | | Sp | | N | F P | | D | | 16 | |
| *Pimelea octophylla* ssp. *octophylla* | | S | | Sp | | N | F P | | D | 2 | 5 8 | |
| *Pimelea pauciflora* | | M | | Sp | | N | F P | | M | | 22 | * |
| *Pimelea phylicoides* | | S | | Sp-S | | N | F | | D | 2 | 5 8 | |
| *Pimelea spinescens* ssp. *spinescens* | | S | | A-Sp | | N | F | | D | | 15 | |
| *Pittosporum bicolor* | S-M | L | | Sp | S | | P N | | M | | 3 4 | |
| *Plantago debilis* | | | H | Y | | F | P | | M | | 4 7 22 | |
| *Plantago gaudichaudii* | | | H | Sp | | F | F | | M | | 14 15 | |
| *Plantago hispida* | | | H | Sp | | F | F | | D | | 19 | |
| *Plantago varia* | | | H | Y | | F | F P | | D | 2 | 4 7 14 21 22 | |
| *Platylobium formosum* | | S-M | | Sp | | | F P | | M | | 4 8 22 | |
| *Platylobium obtusangulum* | | S | | Sp | | | F P | | D | 2 | 5 7 8 14 21 22 | |
| *Platysace heterophylla* | | | H | Sp-S | | | P | | M | 2 | 7 8 | |
| *Platysace lanceolata* | | S-M | | S | | | P | | D | | 21 | |
| *Podolepis jaceoides* | | | T | Sp | | N | F P | | D | | 8 13-15 22 | * |
| *Podotheca angustifolia* | | | A | Sp | | | F | | D | 2 | 10 | |
| *Polygonum plebeium* | | | A | S | S | | F P | | W | | 13? | |
| *Polyscias sambucifolius* | S | M-L | | Sp-S | | | P N | | M | | 3 4 22 | |
| *Pomaderris aspera* | M | L | | Sp-S | | F | P N | | M | | 3 4 16 22 | * |
| *Pomaderris elachophylla* | | M-L | | Sp | | | P | | M | | 22 | |
| *Pomaderris elliptica* | | M-L | | Sp | | | P | | M | | 22 | |
| *Pomaderris ferruginea* | | M-L | | Sp | | | F P | | M | | 22 | |
| *Pomaderris lanigera* | | M | | Sp | | | F P | | M | | 21 22 | |
| *Pomaderris paniculosa* ssp. *paralia* | | S-M | | Sp | | | F P | | D | 1 | 10 | |
| *Pomaderris prunifolia* | | M-L | | Sp | | | A | | M | | 22 | |
| *Pomaderris racemosa* | | M-L | | Sp | | | A | | D M | | 16 | |
| *Pomaderris vacciniifolia* | | M-L | | Sp | | | F P | | M | | 22 | |
| *Poranthera microphylla* | | | A | Sp | | | F P | | D M | 2 | 7 14 19 21 22 | |
| *Portulaca oleracea* | | | A G | Sp-S | | | F | | D M | 1 | 4 8 11 15 | * |
| *Pratia concolor* | | | H G | S | | | P N | | S W | | 13 15 | |
| *Pratia irrigua* | | | H G | S | | | F P | | S W | | 12 16 | |
| *Pratia pedunculata* | | | H G | S | | | F P | | S W | | 8 13-15 22 | |
| *Prostanthera lasianthos* var. *lasianthos* | S-M | L | | S | | | A | S | M | | 3 4 22 | |
| *Prostanthera melissifolia* | | M | | Sp-S | | | P N | S | M | | 3 | |
| *Prunella vulgaris* | | | H G | Sp-S | | | A | | M S | 2 | 3 4 6 8 14 16 22 | |
| *Pseudognaphalium luteo-album* | | | A | Sp-A | | | A | | D M | 2 | 4 6 7 12 15 22 | * |
| *Psoralea adscendens* | | | H G | Sp-S | | | P | | M | | 15 | |
| *Psoralea parva* | | | H | Sp | | | F | | M | | 15 | |
| *Psoralea tenax* | | | H | Sp-S | | | F P | | M S | | 15 | |
| *Ptilotus macrocephalus* | | | H | Sp | | | F | | D M | | 14 15 19 | |
| *Ptilotus spathulatus* | | | T | Sp A | | | F | | D | 1 | 11 15 19 | * |
| *Pultenaea daphnoides* | | M | | Sp | | | A | | M | | 7 21 | |
| *Pultenaea dentata* | | S | | Sp | | | P | | M | | 7 22 | |
| *Pultenaea gunnii* | | S-M | | Sp | | | F P | | D M | | 4 5 21 22 | |
| *Pultenaea hispidula* | | M | | Sp | | | F P | | D | | 4 | |
| *Pultenaea humilis* | | S | G | Sp | | | P | | M | | 22 | |
| *Pultenaea juniperina* var. *mucronata* | | M | | Sp | | | F P | | M | | 4 22 | |
| *Pultenaea muelleri* var. *muelleri* | | M | | Sp-S | | | P | | M | | 22 | |
| *Pultenaea paleacea* var. *sericea* | | S | | Sp | | | F P | | M | 2 | 8 | |
| *Pultenaea pedunculata* | | S | G | Sp | | | F P | | D | | 21 | |
| *Pultenaea scabra* | | M | | Sp | | | P | | M | | 22 | |
| *Pultenaea stricta* | | S | | Sp | | | F P | | M | 2 | 8 22 | |
| *Pultenaea tenuifolia* | | S | | Sp | | | F P | | D | 1 | 10 11 | |
| *Pultenaea weindorferi* | | S | | Sp | | | P N | | M | | 22 | |
| *Pycnosorus chrysanthes* | | | T | Sp A | | N | F P | | S | | 13 | |
| *Pycnosorus globosus* | | | T | Sp-S | | N | F | | S | | 13 | |
| *Ranunculus amphitrichus* | | | H | Y | | | F P | | W | 2 | 9 14 22 | |
| *Ranunculus glabrifolius* | | | H | Sp | | | P | | S W | | 6 22 | |
| *Ranunculus inundatus* | | | H | Sp-S | | | F P | | S W | | 14 22 | |
| *Ranunculus lappaceus* | | | H | Sp | | | F P | | M S | 2 | 3 4 9 13 14 21 22 | |
| *Ranunculus pachycarpus* | | | H | W-Sp | | | F P | | M S | | 22 | |
| *Ranunculus papulentus* | | | H | S | | | F P | | W | | 16 | |
| *Ranunculus plebeius* | | | H | Sp-S | | | P N | | S W | | 3 4 16 22 | |
| *Ranunculus pumilio* var. *pumilio* | | | A | Sp | | | F P | | S W | 2 | 9 10 15 | |
| *Ranunculus robertsonii* | | | H | Sp | | | F P | | S | | 15 | |

| | Tree | Shrub | Other | Flowers | Bird | Butterfly | Sun Aspect | Screen Windbreak | Drainage | Coast | Vegetation Communities | Aboriginal Uses |
|---|---|---|---|---|---|---|---|---|---|---|---|---|
| *Ranunculus scapiger* | | | H | Sp-S | | | P N | | M | | 21 | |
| *Ranunculus sessiliflorus* var. *sessiliflorus* | | | A | W-Sp | | | P | | S | | 9 16 | |
| *Rapanea howittiana* | S | M-L | | W-S | | | P N | S | M | | 16 22 | |
| *Rhagodia candolleana* ssp. *candolleana* | | M | | S-A | | F | F | | D | 1 | 10-12 | * |
| *Rhagodia parabolica* | | S-M | | S | | F | F P | | D | | 17 19 | |
| *Rhodanthe anthemoides* | | | H | Sp-A | | N | F P | | M | | 15 | |
| *Rhodanthe corymbiflora* | | | H | W-Sp | | N | F | | D | 2 | 10 15 | |
| *Ricinocarpos pinifolius* | | M | | Sp | | | F P | | M | 2 | 8 | |
| *Rorippa dictyosperma* | | | H | Y | | | P N | | S | | 3 4 | * |
| *Rorippa laciniata* | | | H | Y | | | P | | S | | 16 | * |
| *Rubus parvifolius* | | S | | Sp-S | B | | P N | | D M | | 3 4 16 20 22 | * |
| *Rumex bidens* | | | H | Sp-S | | | F P | | A | | 1 | |
| *Rumex brownii* | | | H | Sp | | | F P | | S | | 9 14 16 19 20 22 | |
| *Rumex dumosus* | | | H | Sp-S | | | F P | | S | | 12-15 | |
| *Rumex tenax* | | | H | A Sp | | | F P | | S | | 13 | |
| *Rutidosis leptorrhynchoides* | | | H | Sp-A | | | F P | | M | | 13 15 | |
| *Rutidosis multiflora* | | | A | Sp | | | F P | | D | 2 | 10 11 21 | |
| *Sambucus gaudichaudiana* | | S | H | Sp | B | | P | | M | | 3 4 16 20 22 | * |
| *Samolus repens* | | | H T | Y | | | F P | | S | 1 | 11-14 | |
| *Sarcocornia blackiana* | | S | | Sp-S | | | F | | S W | 1 | 11 12 | |
| *Sarcocornia quinqueflora* ssp. *quinqueflora* | | S | G | S | | | F | | S W | 1 | 11-13 | |
| *Sclerolaena diacantha* | | S | G | Y | | | F P | | D | | 15 | |
| *Sclerolaena muricata* var. *muricata* | | S | | Y | | | F | | S | 2 | 11 12 16 19 | |
| *Sclerolaena muricata* var. *villosa* | | S | | Y | | | F P | | S | | 16 | |
| *Sclerostegia arbuscula* | | M | | Y | | | F | | S W | 1 | 12 | |
| *Scutellaria humilis* | | | H G | Sp-S | | | A | | M | 2 | 8 19 | |
| *Sebaea albidiflora* | | | A | Sp | | | P | | S W | 2 | 9 12 22 | |
| *Sebaea ovata* | | | A | Sp | | | P N | | S W | 2 | 4 8 13-15 22 | |
| *Selliera radicans* | | | H G | Sp | | | A | | M S | 2 | 10 12 13 16 | |
| *Senecio biserratus* | | | A | Sp-S | | F | F P | | S | 1 | 7 11 16 | |
| *Senecio cunninghamii* var. *cunninghamii* | | S | | Sp-A | | F | F P | | S W | | 15 | |
| *Senecio glomeratus* | | | A | Sp-S | | F | F P | | S | 2 | 3 4 7 8 14 21 22 | |
| *Senecio hispidulus* var. *hispidulus* | | | H | Sp-S | | F | F P | | M S | 2 | 8 10 15 21 22 | |
| *Senecio hispidulus* var. *dissectus* | | | H | Sp-S | | F | F P | | D | | 21 | |
| *Senecio lautus* | | S | H | Sp-S | | F | F P | | D M | 2 | 4 10 16 22 | |
| *Senecio linearifolius* | | | H | Sp-S | | F | P N | | S | | 3 4 22 | |
| *Senecio macrocarpus* | | | H | Sp | | F | F | | D | | 15 | |
| *Senecio minimus* | | | A | S-A | | F | P N | | M | | 4 6 8 14 20 22 | |
| *Senecio odoratus* | | | H | Sp-S | | F | F P | | D | | 7 13 22 | |
| *Senecio quadridentatus* | | | H | Sp-S | | F | F | | M S | 2 | 8 14-16 20-22 | |
| *Senecio runcinifolius* | | | A | Sp | | F | F | | S | | 12 | |
| *Senecio spathulatus* | | | H | Sp-S | | F | F | | M | 1 | 11 | |
| *Senecio squarrosus* | | | H | S | | F | F P | | D | | 7 15 | |
| *Senecio tenuiflorus* | | | A | Sp-S | | F | F P | | S | | 8 21 22 | |
| *Senecio vagus* | | | H | Sp | | F | P N | | M | | 3 | |
| *Senecio velleioides* | | | H | Sp-S | | F | P N | | M | | 3 4 | |
| *Senna artemisioides* ssp. *filifolia* | | M | | W-Sp | | | F P | | D | | 19 | |
| *Sigesbeckia orientalis* | | | A | Sp-S | | | P | | M S | | 3 4 22 | |
| *Solanum aviculare* | | M | | S | B | | A | | D M | | 3 4 7 14 22 | * |
| *Solanum laciniatum* | | M | | Sp-S | B | | A | | D M | | 14 16 19 22 | |
| *Solanum prinophyllum* | | S | A | Sp-S | | | P | | M | | 3 4 | |
| *Solanum vescum* | | M | | Sp-S | | | F P | | D M | | 4 | |
| *Solenogyne dominii* | | | H | Y | | | F P | | D | | 7 12 14 15 | |
| *Solenogyne gunnii* | | | H | Y | | | P | | M | | 7 14 15 21 | |
| *Sphaerolobium vimineum* | | S | | Sp-S | | | F P | | M | | 8 22 | |
| *Sprengelia incarnata* | | S-M | | Sp | N | | F P | | S W | 2 | 6 8 | |
| *Spyridium parvifolium* | | M | | W-Sp | | | P N | | D M | | 4 22 | |
| *Stackhousia monogyna* | | | H | Sp-S | | N | P N | | M | | 3 4 7 13-15 21 22 | |
| *Stackhousia spathulata* | | | H | W-S | | N | F P | | D | 2 | 11 | |
| *Stackhousia viminea* | | | H | Sp-S | | N | F P | | D | 2 | 8 | |
| *Stellaria flaccida* | | | H G | Sp | | | P N | | M S | | 3 4 22 | |
| *Stellaria multiflora* | | | A | Sp | | | F P | | M | 1 | 11 | |
| *Stellaria palustris* var. *palustris* | | | H | Sp-S | | | P | | S W | | 8 13 | |
| *Stellaria pungens* | | | H G | Sp-S | | | F P | | M | 2 | 4 11 14 16 19 22 | |
| *Stenopetalum lineare* | | | A | Sp | | | F | | D | 1 | 11 15 | * |
| *Stuartina muelleri* | | | A G | Sp | | | F | | D M | 1 | 8 11 17 | |

| | Tree | Shrub | Other | Flowers | Bird | Butterfly | Sun Aspect | Screen Windbreak | Drainage | Coast | Vegetation Communities | Aboriginal Uses |
|---|---|---|---|---|---|---|---|---|---|---|---|---|
| *Stylidium beaugleholei* | | | A | Sp-S | | | P | | S | 2 | 8 | |
| *Stylidium despectum* | | | A | Sp | | | F P | | W | | 16 | |
| *Stylidium graminifolium* var. *angustifolium* | | | H T | Sp | | | P | | M | | 3 4 21 22 | |
| *Stylidium graminifolium* var. *graminifolium* | | | H T | Sp | | | F P | | M | 2 | 7 14 15 21 22 | |
| *Stylidium inundatum* | | | A | Sp | | | F P | | S | | 6 8 | |
| *Stylidium perpusillum* | | | A | Sp | | | F P | | S | | 8 | |
| *Suaeda australis* | | S | | Y | | | F | | S | 1 | 10 12 | |
| *Swainsona lessertiifolia* | | | H | W-Sp | | | F P | | D | 1 | 11 | |
| *Templetonia stenophylla* | | S | | W-Sp | | | F P | | D | | 19 | |
| *Tetragonia implexicoma* | | | C | Sp-S | | | F P | | D | 1 | 10-12 | * |
| *Tetragonia tetragonioides* | | | H G | Sp-S | | | F | | M | 1 | 11 | * |
| *Tetratheca bauerifolia* | | S | | Sp-S | | | P | | M | | 22 | |
| *Tetratheca ciliata* | | S | | W-S | | | P N | | D | | 4 7 21 22 | |
| *Teucrium corymbosum* | | | H | S | | | P N | | M | | 22 | |
| *Teucrium racemosum* | | | H | Sp-S | | | F | | M S | | 13 14 | |
| *Thesium australe* | | | H | S | | | F | | M S | | 15 | |
| *Threlkeldia diffusa* | | | H G | W-Sp | | | F | | S | 1 | 11 12 | |
| *Trachymene anisocarpa* | | | A | Sp-S | | | F P | | D M | 2 | 8 | |
| *Triptilodiscus pygmaeus* | | | A | Sp | | | F P | | D | | 14 19 21 | |
| *Urtica incisa* | | | H | W-S | | F | A | | M S | | 4 16 19 20 | * |
| *Utricularia australis* | | | H | Sp | | | P N | | A | | 1 8 14 | |
| *Utricularia dichotoma* | | | H | Sp-S | | | P N | | W | | 6 8 16 22 | |
| *Utricularia laterifolia* | | | A | Sp | | | P N | | W | | 8 | |
| *Utricularia* sp. aff. *dichotoma* | | | H | Sp | | | P N | | S W | | 13 | |
| *Utricularia tenella* | | | H | Sp | | | P | | W | | 8 | |
| *Velleia paradoxa* | | | H | Sp A | | | AA | | M S | | 13-16 | |
| *Veronica calycina* | | | H G | Sp | | | P N | | M S | | 4 7 16 21 22 | |
| *Veronica gracilis* | | | H | Sp-S | | | P N | | M | | 4 8 13 14 16 21 22 | |
| *Veronica notabilis* | | | H G | Sp-S | | | P N | | S | | 3 4 | |
| *Veronica plebeia* | | | H G | Sp-S | | | A | | M | | 4 7 20 22 | |
| *Villarsia exaltata* | | | H T | Sp | | | F P | | W A | | 1 8 | |
| *Villarsia reniformis* | | | H T | Sp | | | A | | W A | 2 | 1 6 8 | |
| *Viminaria juncea* | | M-L | | Sp-S | | | F P | | M S W | 2 | 6-8 14 16 22 | |
| *Viola betonicifolia* | | | H T | Sp-S | | | P N | | M | | 4 7 21 22 | |
| *Viola cleistogamoides* | | | H | Sp-S | | | P N | | S | | 6 8 | |
| *Viola hederacea* | | | H G | W-A | | | A | | S W | | 3 4 7 9 21 22 | |
| *Viola sieberiana* | | | H | Sp-S | | | P N | | S | | 3 4 8 | |
| *Vittadinia cervicularis* | | S | | Y | | | F | | D | | 14 19 | |
| *Vittadinia cuneata* var. *cuneata* | | S | | Y | | | F P | | D M | | 8 13-15 | |
| *Vittadinia gracilis* | | S | | Y | | | F P | | D M | | 13-15 | |
| *Vittadinia muelleri* | | S | | Y | | | F P | | D | | 14 15 21 | |
| *Wahlenbergia communis* | | | H | Sp-A | | | A | | M | | 14-16 19 | * |
| *Wahlenbergia gracilenta* | | | A | Sp | | | F P | | M | | 4 8 14 15 22 | * |
| *Wahlenbergia gracilis* | | | A/H | Sp | | | F P | | M | | 3 4 7 16 22 | * |
| *Wahlenbergia gymnoclada* | | | H | Sp-S | | | F P | | D | | 7 22 | * |
| *Wahlenbergia luteola* | | | H T | Y | | | F | | M | | 14 15 | |
| *Wahlenbergia multicaulis* | | | H T | Sp-A | | | F P | | D | | 7 14 21 | * |
| *Wahlenbergia stricta* | | | H | Sp-S | | | F P | | D | | 3 4 7 14 15 21 22 | * |
| *Wilsonia backhousei* | | S | G | Sp | | | F | | S W | 1 | 11 | |
| *Wilsonia humilis* | | S | S | Sp | | | F | | S W | 1 | 12 | |
| *Wilsonia rotundifolia* | | S | G | Sp | | | F | | S | 1 | 11-13 | |
| *Xanthosia dissecta* | | | H | Sp | | | P | | M | | 4 5 9 22 | |
| *Xanthosia pilosa* | | | H | Sp-S | | | P | | M | 2 | 7 | |
| *Xanthosia pusilla* | | | H | Sp | | | F P | | D | 2 | 7 8 | |
| *Ziera arborescens* | S | M-L | | Sp | | | P N | | M | | 3 | |
| *Zygophyllum billardieri* | | | H | Y | | | F | | D | | 11 16 | |
| *Zygophyllum glaucum* | | | A | W-Sp | | | F | | M | | 16 | |

## Lilies and Irises

| | Tree | Shrub | Other | Flowers | Bird | Butterfly | Sun Aspect | Screen Windbreak | Drainage | Coast | Vegetation Communities | Aboriginal Uses |
|---|---|---|---|---|---|---|---|---|---|---|---|---|
| *Arthropodium fimbriatum* | | | H T | Sp | | | F | | D | | 15 19 | * |
| *Arthropodium milleflorum* | | | H T | Sp-S | | | F P | | S | | 4 7 16 22 | * |
| *Arthropodium minus* | | | H T | Sp | | | F P | | D M | | 14 15 | * |
| *Arthropodium strictum* | | | H T | Sp | | | F P | | D | | 4 5 7 14 15 21 22 | * |
| *Bulbine bulbosa* | | | H T | Sp-S | | | F P | | M | | 14 15 21 22 | * |
| *Burchardia umbellata* | | | H T | Sp-S | | | F P | | M | | 5 7 8 10 14 15 19 | * |
| *Caesia calliantha* | | | H T | Sp-S | | | F P | | M S | 2 | 14 15 22 | * |
| *Caesia parviflora* var. *minor* | | | H T | Sp-S | | | F P | | M S | 2 | 14 | * |
| *Caesia parviflora* var. *parviflora* | | | H T | Sp-S | | | F P | | M | | 4 7 8 22 | * |

| | Tree | Shrub | Other | Flowers | Bird | Butterfly | Sun Aspect | Screen Windbreak | Drainage | Coast | Vegetation Communities | Aboriginal Uses |
|---|---|---|---|---|---|---|---|---|---|---|---|---|
| *Chamaescilla corymbosa* | | | H | Sp | | | F P | | S | 2 | 4 7 14 22 | |
| *Dianella caerula* | | | T | Sp-S | | | F P | | D M | | 22 | * |
| *Dianella longifolia* var. *longifolia* | | | T | Sp-S | | | F P | | M | | 7 15 19 21 22 | * |
| *Dianella revoluta* var. *brevicaulis* | | | T | Sp-S | | | F P | | D | 1 | 11 | * |
| *Dianella revoluta* var. *revoluta* | | | T | Sp-S | | | F P | | D M | 2 | 4 8 15 19 21 22 | * |
| *Dianella tasmanica* | | | T | Sp-S | | | P N | | S | | 3 4 22 | * |
| *Diplarrena moraea* | | | T | Sp | | | F P | | M | | 4 5 | |
| *Hypoxis glabella* var. *glabella* | | | H T | W-S | | | P | | M | | 7 14 15 21 22 | * |
| *Hypoxis hygrometrica* var. *hygrometrica* | | | H T | Sp-S | | | P | | M | | 6 | |
| *Hypoxis hygrometrica* var. *villosisepala* | | | H T | Sp-S | | | F P | | M | 2 | 15 22 | * |
| *Hypoxis vaginata* var. *brevistigmata* | | | H T | Sp-S | | | F P | | M | 2 | 8 | * |
| *Hypoxis vaginata* var. *vaginata* | | | H T | W-S | | | F P | | M | | 7 14 22 | * |
| *Laxmannia orientalis* | | | H T | Sp | | | F P | | M | | 8 | |
| *Libertia pulchella* | | | H T | Sp-S | | | P N | | M S | | 4 | |
| *Patersonia fragilis* | | | H T | Sp-S | | | F P | | S | 2 | 6 8 | |
| *Patersonia occidentalis* | | | H T | Sp-S | | | F P | | S | | 6 9 22 | |
| *Stypandra glauca* | | | H | W-Sp | | | F P | | D M | | 6 16 | |
| *Thelionema caespitosum* | | | H T | Sp-S | | | F P | | S | 2 | 6 8 9 22 | |
| *Thysanotus patersonii* | | | C | Sp | | | F P | | D M | 2 | 7 8 14-16 21 22 | |
| *Thysanotus tuberosus* ssp. *tuberosus* | | | H T | S | | | F P | | D | | 14 16 21 22 | * |
| *Tricoryne elatior* | | | H | Sp-S | | | F P | | D M | 2 | 7 8 14 15 19 21 22 | |
| *Wurmbea dioica* | | | H | W-Sp | | | F P | | M | 2 | 4 5 7 8 14 21 22 | * |

## Orchids

| | Tree | Shrub | Other | Flowers | Bird | Butterfly | Sun Aspect | Screen Windbreak | Drainage | Coast | Vegetation Communities | Aboriginal Uses |
|---|---|---|---|---|---|---|---|---|---|---|---|---|
| *Acianthus caudatus* | | | H | W-Sp | | | P | | M | 2 | 7 8 21 22 | * |
| *Acianthus pusillus* | | | H | A-W | | | P | | M | | 4 8 21 22 | * |
| *Arthrochilus huntianus* | | | H | Sp-S | | | P | | D | | 21 | |
| *Burnettia cuneata* | | | H | Sp-S | | | P | | S W | | 8 | * |
| *Caladenia aurantiaca* | | | H | Sp | | | P | | D | | 8 | * |
| *Caladenia australis* | | | H | Sp | | | P | | D M | | 21 | * |
| *Caladenia caerula* | | | H | Sp | | | P | | D | | 8 18 21 | * |
| *Caladenia cardiochila* | | | H | Sp | | | P | | D | 2 | 8 11 | * |
| *Caladenia carnea* var. *carnea* | | | H | Sp-S | | | F P | | D | | 5 8 14 21 | * |
| *Caladenia catenata* | | | H | A-Sp | | | P | | D | | 7 14 21 22 | * |
| *Caladenia clavigera* | | | H | Sp | | | F P | | M | | 21 22 | * |
| *Caladenia concinna* | | | H | Sp | | | P | | D | | 21 | * |
| *Caladenia congesta* | | | H | Sp-S | | | P | | D | | 9 22 | * |
| *Caladenia deformis* | | | H | W-Sp | | | F P | | D | 2 | 7 8 21 | * |
| *Caladenia dilatata* | | | H | Sp-S | | | F P | | D | | 4 8 14 21 22 | * |
| *Caladenia filamentosa* var. *filamentosa* | | | H | W-Sp | | | F P | | M | | 4 22 | * |
| *Caladenia flavovirens* | | | H | Sp-S | | | P | | M | | 22 | * |
| *Caladenia gracilis* | | | H | Sp | | | P | | D | | 7 21 22 | * |
| *Caladenia iridescens* | | | H | Sp | | | P | | D | | 7 22 | * |
| *Caladenia latifolia* | | | H | Sp | | | F P | | D | 2 | 8 10 | * |
| *Caladenia menziesii* | | | H | Sp | | | P | | S | 2 | 4 7 9 22 | * |
| *Caladenia oenochila* | | | H | Sp | | | P N | | M | | 4 22 | * |
| *Caladenia praecox* | | | H | Sp | | | P | | D | | 21 22 | * |
| *Caladenia pusilla* | | | H | Sp | | | P | | D | | 7 22 | * |
| *Caladenia robinsonii* | | | H | Sp | | | F P | | M | | 7 | * |
| *Caladenia rosella* | | | H | W-Sp | | | P | | D | | 21 | * |
| *Caladenia tentaculata* | | | H | Sp | | | P | | D | | 7 22 | * |
| *Caladenia thysanochila* | | | H | Sp | | | F P | | D | 2 | 7 | * |
| *Caladenia venusta* | | | H | Sp | | | F | | M | 2 | 8 | * |
| *Caleana major* | | | H | Sp-S | | | F P | | D | 2 | 7 21 | |
| *Calochilus campestris* | | | H | Sp-S | | | P | | M | | 4 7 22 | * |
| *Calochilus paludosus* | | | H | Sp-S | | | P | | M | | 21 22 | * |
| *Calochilus robertsonii* | | | H | Sp | | | P | | M | | 4 5 7 21 22 | * |
| *Chiloglottis cornuta* | | | H | Sp-S | | | P N | | S | | 3 | * |
| *Chiloglottis reflexa* | | | H | S-A | | | P | | M | | 7 8 21 | * |
| *Chiloglottis trapeziformis* | | | H | Sp | | | P | | D M | 2 | 7 21 | * |
| *Chiloglottis valida* | | | H | Sp-S | | | P | | M | | 4 7 22 | * |
| *Corybas aconitiflorus* | | | H | A-W | | | P N | | M | | 4 | * |
| *Corybas diemenicus* | | | H | W-Sp | | | N | | S | | 3 7 22 | * |
| *Corybas incurvus* | | | H | W | | | P | | S | | 7 8 22 | * |
| *Corybas unguiculatus* | | | H | A-Sp | | | P N | | S W | | 7 8 | * |
| *Cryptostylis leptochila* | | | H | S-A | | | P | | D M | | 4 7 22 | * |
| *Cryptostylis subulata* | | | H | Sp-A | | | P | | S | | 4 7 9 22 | * |
| *Cyrtostylis reniformis* | | | H | W-Sp | | | P | | D M | | 8 14 22 | * |
| *Dipodium roseum* | | | H | S-A | | | P | | M | | 7 14 21 22 | * |
| *Diuris brevissima* | | | H | Sp | | | F P | | D M | | 21 | |
| *Diuris corymbosa* | | | H | W-Sp | | | P | | D M | 2 | 7 14 21 22 | * |

| | Tree | Shrub | Other | Flowers | Bird | Butterfly | Sun Aspect | Screen Windbreak | Drainage | Coast | Vegetation Communities | Aboriginal Uses |
|---|---|---|---|---|---|---|---|---|---|---|---|---|
| *Diuris fragrantissima* | | | H | Sp | | | F | | M | | 15 | * |
| *Diuris lanceolata* | | | H | Sp | | | F P | | S | | 5 8 14 15 21 22 | * |
| *Diuris* x *fastidiosa* | | | H | Sp | | | F P | | S | | 15 | * |
| *Diuris palustris* | | | H | W-Sp | | | F P | | S W | | 8 15 | * |
| *Diuris pardina* | | | H | Sp | | | P | | D M | | 7 8 21 22 | * |
| *Diuris punctata* | | | H | Sp | | | F P | | S | | 5 7 14 19 | * |
| *Diuris sulphurea* | | | H | Sp | | | P | | M | | 8 14 19 21 22 | * |
| *Eriochilus cucullatus* | | | H | S-A | | | F P | | D M | | 5 7 14 21 22 | |
| *Gastrodia procera* | | | H | S | | | F P | | S | | 4 | |
| *Gastrodia sesamoides* | | | H | Sp-S | | | P | | S | | 4 7 22 | |
| *Genoplesium archeri* | | | H | S-A | | | P | | D M | 2 | 7 11 22 | * |
| *Genoplesium despectans* | | | H | S-A | | | P | | D | | 7 21 22 | * |
| *Genoplesium morrisii* | | | H | S-A | | | P | | S | | 7 22 | * |
| *Genoplesium rufum* | | | H | S-A | | | FP | | M | | 8 14 | * |
| *Glossodia major* | | | H | Sp | | | F P | | D M | | 7 8 14 19 21 22 | * |
| *Leporella fimbriata* | | | H | A-Sp | | | F P | | M | | 7 8 | * |
| *Lyperanthus nigricans* | | | H | Sp | | | P | | D M | 2 | 7 8 10 | * |
| *Lyperanthus suaveolens* | | | H | Sp | | | P | | D M | | 8 21 22 | * |
| *Microtis atrata* | | | H | Sp | | | P N | | S W | | 7 8 22 | * |
| *Microtis parviflora* | | | H | Sp-S | | | F P | | S | | 7 8 21 22 | * |
| *Microtis rara* | | | H | Sp-S | | | A | | S | | 22 | * |
| *Microtis unifolia* | | | H | Sp-S | | | F P | | M S | | 8 9 13 14 21 22 | * |
| *Orthoceras strictum* | | | H | Sp-S | | | F P | | S W | | 7 9 22 | * |
| *Prasophyllum australe* | | | H | Sp-S | | | P | | S W | | 4 22 | * |
| *Prasophyllum brevilabre* | | | H | Sp-S | | | P | | S | | 4 22 | * |
| *Prasophyllum colemanae* | | | H | Sp | | | P | | M S | | 22 | * |
| *Prasophyllum elatum* | | | H | Sp | | | F P | | M | | 4 8 9 | * |
| *Prasophyllum flavum* | | | H | Sp-S | | | P N | | S | | 3 | * |
| *Prasophyllum frenchii* | | | H | Sp-S | | | F P | | M | | 7 8 22 | * |
| *Prasophyllum fuscum* | | | H | Sp | | | F P | | S | | 4 9 15 | * |
| *Prasophyllum lindleyanum* | | | H | Sp-S | | | P | | D M | | 4 7 22 | * |
| *Prasophyllum odoratum* | | | H | Sp-S | | | P | | M | | 4 22 | * |
| *Prasophyllum patens* | | | H | Sp-S | | | F P | | M | 2 | 8 11 | * |
| *Prasophyllum pyriforme* | | | H | Sp-S | | | P | | M | | 22 | * |
| *Prasophyllum spicatum* | | | H | Sp-S | | | P | | M | | 8 | * |
| *Pterostylis alata* | | | H | A-W | | | P | | M | | 8 21 22 | * |
| *Pterostylis alpina* | | | H | Sp | | | P | | M | | 4 22 | * |
| *Pterostylis cncinna* | | | H | W-Sp | | | F P | | M | | 7 8 14 21 22 | * |
| *Pterostylis* x *toveyana* | | | H | W-Sp | | | P | | M | | 7 8 21 | * |
| *Pterostylis cucullata* | | | H | Sp | | | P | | M | | 8 | * |
| *Pterostylis curta* | | | H | W-Sp | | | P N | | M | | 4 8 16 22 | * |
| *Pterostylis cycnocephala* | | | H | Sp-S | | | F P | | D M | | 19 21 | * |
| *Pterostylis decurva* | | | H | Sp-S | | | P N | | S | | 4 22 | * |
| *Pterostylis foliata* | | | H | Sp-S | | | P N | | S | | 4 22 | * |
| *Pterostylis furcata* | | | H | Sp-S | | | P N | | S W | | 4 22 | * |
| *Pterostylis grandiflora* | | | H | W | | | P | | S | | 7 8 22 | * |
| *Pterostylis longifolia* | | | H | W-Sp | | | P | | M | | 7 8 21 22 | * |
| *Pterostylis* sp. aff. *longifolia* | | | H | W-Sp | | | P | | M | | 18 21 | * |
| *Pterostylis mutica* | | | H | W-S | | | F P | | D M | | 8 19 21 | * |
| *Pterostylis nana* | | | H | W-Sp | | | A | | D M | | 7 11 21 | * |
| *Pterostylis nutans* | | | H | A-Sp | | | P N | | S | | 7 8 14 21 22 | * |
| *Pterostylis* x *ingens* | | | H | Sp | | | P | | S | | 22 | * |
| *Pterostylis obtusa* | | | H | S-W | | | P | | S | | 4 8 22 | * |
| *Pterostylis parviflora* | | | H | S-W | | | P | | S | | 7 8 21 | * |
| *Pterostylis pedoglossa* | | | H | A-W | | | P N | | S | | 7 8 | * |
| *Pterostylis pedunculata* | | | H | W-Sp | | | P N | | S | | 7 8 21 22 | * |
| *Pterostylis plumosa* | | | H | Sp | | | P | | M | | 7 21 22 | * |
| *Pterostylis revoluta* | | | H | S-W | | | P | | D M | | 18 21 | * |
| *Pterostylis robusta* | | | H | W-Sp | | | P | | D M | | 4 21 | * |
| *Pterostylis rufa* | | | H | Sp-S | | | P | | D | | 21 | * |
| *Pterostylis sanguinea* | | | H | A-W | | | F P | | D | | 7 10 | * |
| *Pterostylis truncata* | | | H | S-W | | | P | | M | | 5 15 22 | * |
| *Sarcochilus australis* | | | E | Sp | | | P N | | | | 2 3 | |
| *Spiranthes sinensis* | | | H | Sp-S | | | F P | | S W | | 9 22 | * |
| *Thelymitra antennifera* | | | H | Sp | | | F P | | M | | 4 5 7 8 22 | * |
| *Thelymitra* x *macmillanii* | | | H | Sp | | | F P | | M | | 7 | * |
| *Thelymitra aristata* | | | H | Sp | | | F P | | S | | 14 15 21 | * |
| *Thelymitra carnea* | | | H | Sp | | | F P | | S | | 7 22 | * |
| *Thelymitra epipactoides* | | | H | Sp | | | F P | | S W | | 8 | * |
| *Thelymitra flexuosa* | | | H | Sp | | | F P | | S | | 7 9 22 | * |
| *Thelymitra ixioides* var. *ixioides* | | | H | Sp | | | F P | | W | | 7 21 22 | * |
| *Thelymitra ixioides* var. *subdifformis* | | | H | Sp | | | F P | | S | | 22 | * |
| *Thelymitra luteocilium* | | | H | Sp | | | F P | | M | | 9 22 | * |
| *Thelymitra media* var. *media* | | | H | Sp-S | | | F P | | S | | 3 4 22 | * |
| *Thelymitra megcalyptra* | | | H | Sp | | | F P | | D | | 10 15 | * |
| *Thelymitra nuda* | | | H | Sp | | | F P | | D M | | 8 21 | * |
| *Thelymitra pauciflora* | | | H | Sp-S | | | F P | | M | | 7 8 14 15 21 22 | * |
| *Thelymitra rubra* | | | H | Sp | | | P | | S | | 7 8 21 22 | |

## Grasses, Rushes, Sedges and Aquatics

| Species | Tree | Shrub | Other | Flowers | Bird | Butterfly | Sun Aspect | Screen Windbreak | Drainage | Coast | Vegetation Communities | Aboriginal Uses |
|---|---|---|---|---|---|---|---|---|---|---|---|---|
| *Agrostis adamsonii* | | | A T | Sp | | | | | | | | |
| *Agrostis aemula* var. *aemula* | | | A T | Sp-S | | | F P | | M S | | 13 15 22 | |
| *Agrostis aemula* var. *setifolia* | | | A T | Sp-S | | | F P | | S | | 13 | |
| *Agrostis avenacea* | | | A T | Sp-S | | | F P | | S | | 7 11 13 15 19 22 | |
| *Agrostis billardieri* var. *billardieri* | | | A T | Sp-S | | | F P | | S | 1 | 11 16 | |
| *Agrostis billardieri* var. *robusta* | | | A T | Sp-S | | | F P | | M | | 21/22 | |
| *Agrostis* aff. *hiemalis* | | | T | Sp-S | | | P N | | S | | 4 | |
| *Alisma plantago-aquatica* | | | H T | S | | | F P | | W A | | 1 13 16 | |
| *Amphibromus archeri* | | | T | Sp-S | | | F P | | S W | | 6 16 | |
| *Amphibromus fluitans* | | | T | Sp | | | P | | S W | | 7 16 | |
| *Amphibromus macrorhinus* | | | T | Sp | | | F | | S W | | 13 | |
| *Amphibromus neesii* | | | T | Sp-A | | | F P | | S W | | 9 13 15 | |
| *Amphibromus nervosus* | | | T | Sp-A | | | F P | | S W | | 13 | |
| *Amphibromus pithogastrus* | | | T | Sp | | | F P | | S W | | 13 | |
| *Aphelia gracilis* | | | A T | Sp | | | F P | | S | 2 | 6 8 | |
| *Aphelia pumilio* | | | A | Sp | | | F P | | S | 2 | 8 | |
| *Aristida ramosa* | | | H T | Sp | | | F | | D | | 15 19 | |
| *Austrofestuca hookeriana* | | | T | Sp-S | | | F P | | S | | 6 9 22 | |
| *Austrofestuca littoralis* | | | T | Sp | | | F | | D | 1 | 11 | |
| *Baumea acuta* | | | T | Sp-A | | F | F | | S | 2 | 8 | |
| *Baumea articulata* | | | T | S-A | | F | F P | | S | | 16 | |
| *Baumea gunnii* | | | T | Y | | F | F P | | S | | 6 9 | |
| *Baumea juncea* | | | T | Sp-A | | F | F P | | S | 2 | 8 10 | |
| *Baumea rubiginosa* | | | T | Y | | F | F | | S W | | 6 | |
| *Baumea tetragona* | | | T | Y | | F | F | | S | 2 | 6 8 9 | |
| *Bolboschoenus caldwellii* | | | T | Sp-A | | | F P | | W A | 1 | 12-14 | * |
| *Bolboschoenus medianus* | | | T | Sp-A | | | F P | | S W A | 1 | 7 16 | * |
| *Bothriochloa macra* | | | T | S-A Sp | | | F P | | D M | | 11 14 15 19 | |
| *Carex appressa* | | | T | Sp-S | | F | F | | S W | | 7 8 14 16 21 22 | * |
| *Carex bichenoviana* | | | T | Sp-S | | | F P | | S | | 14 15 | |
| *Carex breviculmis* | | | T | Sp-S | | | F P | | M-S | | 4 7 12-15 21 22 | |
| *Carex fascicularis* | | | T | Sp-A | | | P | | S W | | 22 | |
| *Carex gaudichaudiana* | | | T | Sp-S | | F | F P | | S W | | 6 22 | |
| *Carex incomitata* | | | T | Sp-S | | | F | | D | | 15 | |
| *Carex inversa* | | | T | Sp-A | | | F P | | M | | 7 8 13-15 19 22 | |
| *Carex iynx* | | | T | Sp | | | P | | S | | 22 | |
| *Carex pumila* | | | T | Sp-S | | | F | | S | 1 | 11 13 | |
| *Carex tasmanica* | | | T | Sp | | | F P | | S W | | 13 | |
| *Carex tereticaulis* | | | T | Sp-A | | | F P | | S W | | 15 16 22 | * |
| *Caustis restiacea* | | | H T | Sp-S | | | F P | | S | | 6 | |
| *Centrolepis aristata* | | | A T | Sp | | | P | | S | | 6 9 14 22 | |
| *Centrolepis fascicularis* | | | H T | S | | | P | | S | | 6 9 | |
| *Centrolepis polygyna* | | | A T | W-Sp | | | F P | | S | 1 | 8 12 | |
| *Centrolepis strigosa* | | | A T | Sp | | | F P | | S | | 8 9 14 22 | |
| *Chionochloa pallida* | | | T | Sp-S | | | F P | | D M | | 5 21 22 | |
| *Chloris truncata* | | | T | S | | | F | | D | | 13-15 19 | |
| *Chorizandra cymbaria* | | | T | Sp | | | F P | | S | | 6 8 22 | |
| *Cladium procerum* | | | T | Sp-S | | | F P | | S W | 1 | 9 16 | |
| *Cymbopogon refractus* | | | T | Sp-S | | | F P | | D | | 15 | |
| *Cyperus brevifolius* | | | T | Sp-A | | F | F | | S | | 16 | |
| *Cyperus gunnii* | | | T | Sp-A | | F | P | | S | | 16 | |
| *Cyperus lhotskyanus* | | | T | Sp-S | | F | P | | W | | 19 | |
| *Cyperus lucidus* | | | T | Sp-S | | F | P | | W A | | 4 21 22 | |
| *Cyperus sanguinolentus* | | | T A | Sp-A | | F | P | | S W | | 16 | |
| *Damasonium minus* | | | A T | Sp-W | | | F P | | W A | | 1 13 19 | |
| *Danthonia auriculata* | | | T | Sp-S | | | F P | | D | | 14 15 | |
| *Danthonia caespitosa* | | | T | Sp-S A | | | F P | | M | | 8 13-15 22 | |
| *Danthonia carphoides* var. *carphoides* | | | T | Sp-S | | | F | | M S | | 14 15 | |
| *Danthonia duttoniana* | | | T | Sp-S | | | F P | | S W | | 13-15 | |
| *Danthonia eriantha* | | | T | Sp-S | | | F P | | D | | 14 15 21 | |
| *Danthonia geniculata* | | | T | Sp-S | | | F P | | D M S | | 8 14 19 21 | |
| *Danthonia induta* | | | T | Sp-S | | | F P | | M | | 19 | |
| *Danthonia laevis* | | | T | Sp-S | | | F P | | S | | 8 14 15 19 21 22 | |
| *Danthonia lepidopoda* | | | T | Sp | | | P | | M | | 5 | |
| *Danthonia linkii* var. *fulva* | | | T | Sp-A | | | F P | | M S | | 19 21 22 | |
| *Danthonia linkii* var. *linkii* | | | T | Sp-A | | | F P | | M S | | 14 21 22 | |
| *Danthonia penicillata* | | | T | Sp-S | | | P | | M | | 7 16 20-22 | |
| *Danthonia pilosa* var. *pilosa* | | | T | Sp-S | | | P | | S | | 5 7 15 22 | |
| *Danthonia procera* | | | T | Sp-S | | | F P | | D M | | 7 14 | |
| *Danthonia racemosa* | | | T | Y | | | F P | | D M | | 14 15 19-22 | |
| *Danthonia semiannularis* | | | T | Sp-S | | | F P | | S | | 5 7 8 14 22 | |
| *Danthonia setacea* var. *setacea* | | | T | Sp-S | | F | F P | | D M | | 8 14 15 19 21 22 | |
| *Danthonia tenuior* | | | T | Sp-S | | | F P | | D | | 14 21 22 | |

| | Tree | Shrub | Other | Flowers | Bird | Butterfly | Sun Aspect | Screen Windbreak | Drainage | Coast | Vegetation Communities | Aboriginal Uses |
|---|---|---|---|---|---|---|---|---|---|---|---|---|
| *Deyeuxia contracta* | | | T | Sp | | | P N | | S | | 3 | |
| *Deyeuxia densa* | | | T | Sp-S | | | F P | | S | | 6 8 | |
| *Deyeuxia minor* | | | T | Sp-S | | | F P | | D | 1 | 10 21 | |
| *Deyeuxia quadriseta* | | | T | Sp-A | | | F P | | D M | | 4 5 7 15 21 22 | |
| *Deyeuxia rodwayi* | | | T | | | | P N | | S | | 3 4 22 | |
| *Dichanthium sericeum* | | | T | Y | | | F | | M | | 15 | |
| *Dichelachne crinita* | | | T | Sp-S | | | F P | | D | | 7 8 14 15 19-22 | |
| *Dichelachne micrantha* | | | T | W-Sp | | | F P | | D | | 15 21 | |
| *Dichelachne sieberiana* | | | T | S | | | P | | D | | 21 | |
| *Distichlis distichophylla* | | | G | Sp-S | | | F | | S | 1 | 11 12 | |
| *Dryopoa dives* | | | T | Sp-S | | | P N | | M | | 2 3 | |
| *Echinopogon ovatus* | | | T | Sp-S | | | P | | M | | 4 16 22 | |
| *Eleocharis acuta* | | | T | Sp-A | | | F | | W A | | 1 7 8 13 15 16 19 | |
| *Eleocharis atricha* | | | T | S | | | P | | A | | 1 13 | * |
| *Eleocharis gracilis* | | | T | Sp-S W | | | P | | W | | 16 | |
| *Eleocharis minuta* | | | T | S | | | F | | W | 1 | 13 | |
| *Eleocharis pallens* | | | T | A S | | | F P | | S W | | 13 | |
| *Eleocharis pusilla* | | | T | S-A | | | F P | | A | | 1 12-14 | |
| *Eleocharis sphacelata* | | | T | S | | | F P | | W A | | 1 6 14 | * |
| *Elymus scabrus* | | | T | Sp-S | | F | F P | | D M | | 14 15 21 22 | |
| *Empodisma minus* | | | G | Sp-A | | | F P | | S W | | 5 6 9 | |
| *Enneapogon nigricans* | | | T | S | | | F | | D | | 15 19 | |
| *Entolasia marginata* | | | T | Sp | | | F P | | M | | 7 | |
| *Entolasia stricta* | | | T | S | | | F P | | D M | | 7 | |
| *Eragrostis brownii* | | | T | Sp-A | | | F P | | D M | | 8 14 16 21 22 | |
| *Eragrostis elongata* | | | T | S-A | | | F P | | S W | | 13 | |
| *Eragrostis infecunda* | | | T | S | | | F | | S W | 1 | 13 | |
| *Gahnia clarkei* | | | T | Sp-S | | F | P N | | S W | | 4 22 | * |
| *Gahnia filum* | | | T | Y | | F | F | | S W | 1 | 12 13 | * |
| *Gahnia radula* | | | T | Y | | F | F P | | D M S | | 5 7 9 21 22 | * |
| *Gahnia sieberiana* | | | T | Sp-S | | F | A | | S | | 4 9 22 | * |
| *Gahnia trifida* | | | T | Sp-S | | F | F P | | M | 2 | 10 | * |
| *Glyceria australis* | | | G | Sp-S | | | F P | | S W | | 7 16 22 | |
| *Hemarthria uncinata* | | | T | S | | | F P | | S | | 6 8 16 22 | |
| *Homopholis proluta* | | | T | Sp-A | | | F P | | S | | 14 15 | * |
| *Hypolaena fastigata* | | | G | Y | | | F P | | S | 2 | 8 | |
| *Imperata cylindrica* | | | T G | S | | F | F P | | M | | 4 8 13 21 22 | |
| *Isolepis cernua* | | | A T | Y | | | F P | | S W | 1 | 12 13 16 20 | |
| *Isolepis fluitans* | | | T | Sp-S | | | P | | W A | | 1 6 8 22 | |
| *Isolepis hookeriana* | | | A T | Sp-S | | | P | | S | | 6 13 22 | |
| *Isolepis inundata* | | | T | Sp-A | | | P | | S W | | 4 9 16 22 | |
| *Isolepis marginata* | | | A T | Sp-S | | | F P | | M | 1 | 8 12 16 22 | |
| *Isolepis nodosa* | | | T | Y | | | F P | | S | 1 | 6-8 10 11 16 | |
| *Isolepis platycarpa* | | | A T | Sp-A | | | F P | | S | | 4 6 16 | |
| *Isolepis producta* | | | H | Sp-A | | | P | | A | | 1 16 | |
| *Isolepis stellata* | | | A T | Sp-S | | | P | | S | | 22 | |
| *Isolepis subtilissima* | | | T | A | | | P | | S | | 3 22 | |
| *Isolepis victoriensis* | | | A T | Sp | | | F | | S | | 15 19 | |
| *Juncus amabilis* | | | T | S | S | | F P | | M S | | 9 13 15 22 | |
| *Juncus australis* | | | T | Y | S | | F P | | S W | | 16 22 | |
| *Juncus bufonius* | | | T | Y | S | | F P | | S W | | 3 4 6 8 13-16 22 | |
| *Juncus caespiticius* | | | T | Y | S | | F P | | W | | 8 16 | |
| *Juncus filicaulis* | | | T | Y | S | | F P | | S | | 14 | |
| *Juncus flavidus* | | | T | S-A | S | | F P | | M | | 12-14 19 | |
| *Juncus gregiflorus* | | | T | Sp-S | S | | P | | S | | 4 16 22 | |
| *Juncus holoshoenus* | | | T | S | S | | F | | S | | 6 8 13 14 22 | |
| *Juncus homalocaulis* | | | T | S | S | | F | | S | | 12 13 19 | |
| *Juncus krausii* | | | T | S | S | | F | | S W | 1 | 10 11 | |
| *Juncus pallidus* | | | T | S | S | | F P | | S W | | 7 8 15 22 | |
| *Juncus pauciflorus* | | | T | S | S | | P | | M S | | 6 7 16 22 | |
| *Juncus planifolius* | | | A/T | Sp-S | S | | F P | | W | | 4 6 9 22 | |
| *Juncus prismatocarpus* | | | T | S | S | | P | | W | | 4 9 | |
| *Juncus procerus* | | | T | Sp-S | S | | P | | M | | 4 22 | |
| *Juncus radula* | | | T | Sp-S | S | | F P | | S | | 14 15 21 | |
| *Juncus revolutus* | | | T | Sp-S | S | | F | | S W | 1 | 12 | |
| *Juncus sarophorus* | | | T | Sp-S | S | | P | | S | | 9 16 22 | |
| *Juncus* sp. O | | | T | S | S | | F P | | S | | 13 | |
| *Juncus subsecundus* | | | T | S | S | | F P | | S W | | 7 15 16 22 | |
| *Juncus usitatus* | | | T | S | S | | F P | | S W | | 14 15 22 | |
| *Juncus vaginatus* | | | T | S | S | | F P | | S | | 14 | |
| *Lemna disperma* | | | | Sp | | | A | | A | | 1 13 16 19 | |
| *Lepidosperma concavum* | | | T | Y | S | F | F P | | M | 2 | 8 10 11 | |
| *Lepidosperma congestum* | | | T | Y | S | | F P | | S | | 8 | |
| *Lepidosperma elatius* var. *elatius* | | | T | Y | S | | P N | | S | | 3 4 7 22 | * |

| | Tree | Shrub | Other | Flowers | Bird | Butterfly | Sun Aspect | Screen Windbreak | Drainage | Coast | Vegetation Communities | Aboriginal Uses |
|---|---|---|---|---|---|---|---|---|---|---|---|---|
| *Lepidosperma filiforme* | | | T | Sp-S | S | | P | | S | | 5 6 9 | |
| *Lepidosperma forsythii* | | | T | Sp-S | S | | F P | | W | 2 | 8 | |
| *Lepidosperma gladiatum* | | | T | Sp-S | S | | F P | | D | 1 | 11 | * |
| *Lepidosperma gunnii* | | | T | Sp-S | S | | F P | | M | | 8 | |
| *Lepidosperma laterale* var. *laterale* | | | T | Sp-S | S | | P | | M S | | 4-7 21 22 | |
| *Lepidosperma laterale* var. *majus* | | | T | Sp-S | S | | P | | M S | | 3 7 22 | * |
| *Lepidosperma longitudinale* | | | T | Y | S | | F P | | S W | | 6 9 | |
| *Lepidosperma neesii* | | | T | Sp | S | | P | | S | | 7 9 | |
| *Lepidosperma semiteres* | | | T | Y | S | | F P | | S | | 4 7 8 | |
| *Lepidosperma tortuosum* | | | T | Sp-S | S | | P | | W | | 4 | |
| *Lepilaena bilocularis* | | | H | W-Sp | B | | F P | | A | 1 | 1 12 | |
| *Lepilaena cylindrocarpa* | | | H | Sp | B | | F | | A | 1 | 1 12 | |
| *Lepilaena preissii* | | | A | Sp-S | B | | F | | A | 1 | 1 12 | |
| *Leptocarpus brownii* | | | T | Y | | | F P | | S | | 6 8 | |
| *Leptocarpus tenax* | | | T | Sp-W | | | F P | | S | | 6 9 | |
| *Lepyrodia muelleri* | | | T | Sp-A | | | P | | W | | 9 | |
| *Lomandra filiformis* ssp. *coriacea* | | | T | Sp | S | F | F P | | M | | 16 21 22 | * |
| *Lomandra filiformis* ssp. *filiformis* | | | T | Sp | S | F | F P | | M | | 5 7 14 21 22 | * |
| *Lomandra longifolia* | | | T | Y | S | F | A | | D M | 2 | 7 8 10 14 21 22 | * |
| *Lomandra micrantha* ssp. *micrantha* | | | T | A-S | S | F | P | | M | | 15 19 | * |
| *Lomandra multiflora* | | | T | W-S | S | F | P | | D M | | 8 19 22 | * |
| *Lomandra nana* | | | T | Sp | S | F | F P | | D M | | 8 15 | * |
| *Luzula meridionalis* var. *densiflora* | | | T | W-S | | | F P | | M | | 9 22 | |
| *Luzula meridionalis* var. *flaccida* | | | T | Sp | | | P N | | M | 2 | 8 22 | |
| *Luzula meridionalis* var. *meridionalis* | | | T | Sp | | | P | | D M | | 3 7 14 21 22 | |
| *Microlaena stipoides* | | | T | Sp | | F | F P | | M | 2 | 5-7 9 15 16 21 22 | |
| *Ottelia ovalifolia* | | | H T | S | | | F P | | A | | 1 | |
| *Panicum decompositum* | | | T | S | | | F P | | M S | | 13 | * |
| *Panicum effusum* | | | T | Sp-A | | | F P | | M | | 13-15 19 | * |
| *Paspalidium constrictum* | | | T | Sp-A | | | F P | | D | | | |
| *Pentapogon quadrifidus* | | | A T | S | | | F P | | M S | | 14 15 21 22 | |
| *Philydrum lanuginosum* | | | H | S | | | F P | | A | | 1 | |
| *Phragmites australis* | | | T | S-A | | | F P | | W A | | 6 10 16 | * |
| *Poa clelandii* | | | T | Sp-S | S | F | F P | | M | | 8 15 22 | * |
| *Poa ensiformis* | | | T | Sp-A | S | F | P N | | M | | 4 22 | * |
| *Poa labillardieri* | | | T | Sp-S | S | F | F P | | M | 2 | 14 16 21 22 | * |
| *Poa morrisii* | | | T | Sp-S | S | F | F P | | M | | 5 7 21 22 | |
| *Poa poiformis* var. *poiformis* | | | T | Sp-S | S | F | F P | | D | 1 | 10 11 | * |
| *Poa rodwayi* | | | T | Sp | S | | F P | | M | | 7 19 22 | |
| *Poa sieberiana* var. *hirtella* | | | T | Sp-S | S | F | P | | M | | 22 | |
| *Poa sieberiana* var. *sieberiana* | | | T | Sp-S | S | F | F P | | M | | 7 9 14 15 21 | * |
| *Poa tenera* | | | T | Sp-S | S | F | P N | | M | | 4 6 7 22 | |
| *Potamogeton crispus* | | | H | Sp-A | | | F P | | A | | 1 14 16 | |
| *Potamogeton ochreatus* | | | H | Sp-A | | | F P | | A | | 1 13 | |
| *Potamogeton pectinatus* | | | H | Sp-A | | | F P | | A | | 1 13 | |
| *Potamogeton tricarinatus* | | | H | Sp-A | | | F P | | A | | 1 13 | |
| *Puccinellia stricta* var. *perlaxa* | | | A T | Sp-S | | | F | | W | 1 | 12 13 | |
| *Puccinellia stricta* var. *stricta* | | | A T | Sp | | | F | | W | 1 | 12 | |
| *Restio tetraphyllus* | | | T | Sp-S | | | F P | | W | | 8 9 | |
| *Ruppia maritima* | | | H | W-S | S | | F | | A | 1 | 1 12 | |
| *Ruppia megacarpa* | | | H | Sp-S | S | | F | | A | 1 | 1 12 | |
| *Ruppia polycarpa* | | | A/H | W-Sp | S | | F | | A | 1 | 1 12 | |
| *Schoenoplectus dissachanthus* | | | A T | A | | | F P | | S W | | 7 | |
| *Schoenoplectus pungens* | | | Y | Sp-A | | | F P | | S W | | 13 | |
| *Schoenoplectus validus* | | | T | S-A | | | F P | | W | | 6 16 | * |
| *Schoenus apogon* | | | T | Sp-S | | | F P | | S W | 2 | 5 7 9 14 15 22 | |
| *Schoenus breviculmis* | | | T | Y | | | F P | | S | | 9 | |
| *Schoenus brevifolius* | | | T | Sp-S | | | F P | | S | | 7 8 | |
| *Schoenus latelaminatus* | | | A | Sp | | | P | | S | | 14 22 | |
| *Schoenus maschalinus* | | | T | Sp-S | | | P | | S | | 9 22 | |
| *Schoenus nitens* | | | T | Sp-A | | | F P | | S | | 12 | |
| *Schoenus tenuissimus* | | | T | Y | | | F P | | S W | | 6 9 | |
| *Schoenus tesquorum* | | | T | Sp-S | | | P | | S W | | 6 16 | |
| *Sparganium subglobosum* | | | H | S | | | F P | | A | | 1 | |
| *Spinifex sericeus* | | | T G | Sp-S | | | F | | D | 1 | 11 | |
| *Spirodela oligorrhiza* | | | | | | | F P | | A | | 1 8 16 22 | |
| *Sporobolus virginicus* | | | T G | Y | | | F | | S | 1 | 11 12 | |
| *Stipa aristiglumis* | | | T | Sp-S | S | | F P | | S | | 14 15 | |
| *Stipa bigeniculata* | | | T | Sp-S | S | | F P | | S | | 13-15 19 | |
| *Stipa blackii* | | | T | Sp | S | | F P | | S | | 14 15 19 | |
| *Stipa curticoma* | | | T | Sp | S | | F P | | S | | 13-15 | |
| *Stipa densiflora* | | | T | Sp-S | S | | F P | | D | | 18 19 21 | |

| | Tree | Shrub | Other | Flowers | Bird | Butterfly | Sun Aspect | Screen Windbreak | Drainage | Coast | Vegetation Communities | Aboriginal Uses |
|---|---|---|---|---|---|---|---|---|---|---|---|---|
| *Stipa elegantissima* | | | T | Sp-S | S | | F P | | D | | 14 19 | |
| *Stipa exilis* | | | T | Sp | S | | F P | | M | | | |
| *Stipa flavescens* | | | T | Sp-S | S | | F P | | M | 1 | 11 12 | |
| *Stipa gibbosa* | | | T | Sp-S | S | | F P | | S | | 14 15 | |
| *Stipa mollis* | | | T | Sp-S | S | | F P | | S | | 7 14 21 | |
| *Stipa muelleri* | | | T | Sp | S | | F P | | D | | 5 | |
| *Stipa nodosa* | | | T | Sp | S | | F P | | M | | 14 15 | |
| *Stipa oligostachya* | | | T | Sp | S | | F P | | S | | 13-15 | |
| *Stipa pubinodis* | | | T | Sp-S | S | | P | | S | | 7 21 22 | |
| *Stipa rudis* ssp. *australis* | | | T | Sp-S | S | | F P | | S | | 7 21 | |
| *Stipa rudis* ssp. *nervosa* | | | T | Sp-S | S | | F P | | S | | 22 | |
| *Stipa rudis* ssp. *rudis* | | | T | Sp-S | S | | P N | | S | | 5 7 14 21 22 | |
| *Stipa scabra* ssp. *falcata* | | | T | Sp | S | | F P | | D M | | 15 19 | |
| *Stipa scabra* ssp. *scabra* | | | T | Sp | S | | F P | | D M | | 14 15 19 | |
| *Stipa semibarbata* | | | T | Sp | S | F | F P | | D M | | 7 14 15 19 22 | |
| *Stipa setacea* | | | T | Sp | S | | F P | | M | | 14 19 | |
| *Stipa stipoides* | | | T | Sp-A | S | | F P | | D | 1 | 11 12 | |
| *Stipa stuposa* | | | T | Sp | S | | F P | | M | | 15 | |
| *Tetraria capillaris* | | | T | A-W | | | F P | | S | | 4 6 22 | |
| *Tetrarrhena distichophylla* | | | G | Sp-A | | | F P | | S | | 7 | |
| *Tetrarrhena juncea* | | | G | Sp-A | | | P N | | M S | | 3 4 7 21 22 | |
| *Themeda triandra* | | | T | Sp-S | S | F | F P | | D M | | 5 7 13-15 19 21 22 | * |
| *Thismia rodwayi* | | | H | | | | N | | S | | 2 3 | |
| *Tricostularia pauciflora* | | | T | S | | | P | | S | | 8 | |
| *Triglochin centrocarpa* | | | A | Sp-A | | | F P | | S | | 8 | |
| *Triglochin minutissima* | | | A | Sp | | | F P | | S | 1 | 12 | |
| *Triglochin mucronata* | | | A | Sp | | | F P | | S | 1 | 12 | |
| *Triglochin procera* | | | H | Sp-A | | | F P | | A | | 1 4 6 7 16 20 22 | * |
| *Triglochin striata* | | | H | Sp-A | | | F P | | W | 1 | 6 10 12 16 | |
| *Tripogon loliiformis* | | | A T | W | | | F P | | D | | 15 19 | |
| *Trithuria submersa* | | | A | Sp | | | F P | | S W | | 13 | |
| *Typha domingensis* | | | T | Y | | | F P | | A | | 1 6 7 14 19 20 22 | * |
| *Typha orientalis* | | | T | Y | | | F P | | A | | 1 6 12 16 | * |
| *Uncinia tenella* | | | T E | | | | P N | | S | | 2 3 | |
| *Vallisneria spiralis* | | | H | S-A | | | F P | | A | | 1 13 | |
| *Wolffia australiana* | | | H | | | | F P | | A | | 1 | |
| *Xanthorrhoea australis* | S | | T | W-S | N | N | F P | | D | | 7 | * |
| *Xanthorrhoea minor* ssp. *lutea* | | | T | Sp-S | N | N | F P | | D M | | 5 7 8 21 22 | * |
| *Xyris gracilis* | | | T | S | | | F P | | S | | 8 | |
| *Xyris operculata* | | | T | S | | | P | | S | | 8 | |
| *Zostera muelleri* | | | H | Sp-A | | | F | | A | | 1 | |
| *Zoysia macrantha* | | | G | | | | F | | D M | 1 | 11 12 | |

## Ferns and Fern Allies

| | Tree | Shrub | Other | Flowers | Bird | Butterfly | Sun Aspect | Screen Windbreak | Drainage | Coast | Vegetation Communities | Aboriginal Uses |
|---|---|---|---|---|---|---|---|---|---|---|---|---|
| *Adiantum aethiopicum* | | | G | | | | P | | M | | 3 4 7 14 16 21 22 | |
| *Allantodia australis* | | | T | | | | P N | | M | | 2 3 | |
| *Asplenium bulbiferum* | | | T | | | | P | | M | | 2-4 | |
| *Asplenium flabellifolium* ssp. *gracillimum* | | | G E | | | | P | | M S | | 3 4 14 16 20 22 | |
| *Azolla filiculoides* | | | | | | | F P | | A | | 1 7 13 14 16 19 22 | |
| *Azolla pinnata* | | | | | | | F P | | A | | 1 4 22 | |
| *Blechnum cartilagineum* | | | T G | | | | P | | M | | 4 22 | |
| *Blechnum chambersii* | | | T | | | | P N | | S W | | 3 4 | |
| *Blechnum fluviatile* | | | T G | | | | P N | | M | | 3 4 | |
| *Blechnum minus* | | | T G | | | | F P | | S W | | 3 4 7 22 | |
| *Blechnum nudum* | | | T G | | | | P N | | M S | | 3 4 7 22 | |
| *Blechnum patersonii* | | | T G | | | | P N | | M | | 2 3 | |
| *Blechnum wattsii* | | | G | | | | P N | | M | | 2-4 | |
| *Botrychium australe* | | | | | | | | | M | | 15 | |
| *Calochlaena dubia* | | | T G | | | | P N | | M S | | 4 16 22 | |
| *Cheilanthes austrotenuifolia* | | | T G | | | | F P | | D | | 4 14 19 21 22 | |
| *Cheilanthes distans* | | | T | | | | F | | D | | 16 | |
| *Cheilanthes sieberi* ssp. *sieberi* | | | T G | | | | F | | D | | 15 19 | |
| *Ctenopteris heterophylla* | | | E | | | | N | | S | | 2 3 | |
| *Cyathea australis* | S-M | | | | | | P N | | M S | | 2-4 22 | |
| *Cyathea cunninghamii* | M-T | | | | | | N | | S W | | 3 | |
| *Dicksonia antarctica* | S-M | | | | | | P N | | S | | 2-4 22 | |
| *Doodia caudata* | | | T | | | | P N | | W | | 16 | |
| *Doodia media* ssp. *australis* | | | T G | | | | P N | | M | | 16 22 | |

| | Tree | Shrub | Other | Flowers | Bird | Butterfly | Sun Aspect | Screen Windbreak | Drainage | Coast | Vegetation Communities | Aboriginal Uses |
|---|---|---|---|---|---|---|---|---|---|---|---|---|
| *Gleichenia dicarpa* | | | C | | | | F P | | S W | | 4 9 | |
| *Gleichenia microphylla* | | | C | | | | P N | | S W | | 3 4 9 22 | |
| *Grammitis billardieri* | | | E | | | | N | | | | 2-4 | |
| *Histiopteris incisa* | | | G | | | | P N | | M | | 2-4 | |
| *Hymenophyllum australe* | | | E | | | | N | | | | 2 | |
| *Hymenophyllum cupressiforme* | | | E | | | | N | | | | 2-4 | |
| *Hymenophyllum flabellatum* | | | E | | | | N | | | | 2-4 | |
| *Hypolepis glandulifera* | | | G | | | | F P | | M S | | 3 | |
| *Hypolepis muelleri* | | | G | | | | F P | | S W | | 3 4 | |
| *Hypolepis rugosula* | | | G | | | | F P | | S W | | 3 4 9 | |
| *Isoetes drummondii* var. *drummondii* | | | T | | | | F P | | S W | | 14 | |
| *Isoetes muelleri* | | | T | | | | F P | | S W | | 13 | |
| *Lastreopsis acuminata* | | | T | | | | P N | | M | | 3 4 22 | |
| *Lastreopsis hispida* | | | T | | | | P N | | M S | | 2 3 | |
| *Lindsaea linearis* | | | G | | | | F P | | M S | | 4 7 8 22 | |
| *Lycopodiella laterale* | | | | | | | P N | | S W | | 9 | |
| *Marsilea drummondii* | | | G | | | | F P | | W A | | 13 14 16 19 | * |
| *Marsilea hirsuta* | | | G | | | | F P | | W | | 13 | |
| *Marsilea* sp. | | | G | | | | F P | | W A | | 13 15 | |
| *Microsorum pustulatum* | | | E C | | | | P N | | | | 2 3 | |
| *Ophioglossum lusitanicum* | | | T | | | | F P | | M S | | 8 9 15 22 | |
| *Pellaea falcata* var. *falcata* | | | T | | | | P N | | M S | | 3 4 16 20 | |
| *Phylloglossum drummondi* | | | T | | | | P N | | S | | 6 9 | |
| *Pilularia novae hollandiae* | | | T | | | | F P | | S W | | 19 | |
| *Pleurosorus rutifolius* | | | T | | | | F | | D M | | 16 19 20 | |
| *Polyphlebium venosum* | | | E | | | | P N | | | | 2-4 | |
| *Polystichum proliferum* | | | T G | | | | P N | | M S | | 2-4 | |
| *Pteridium esculentum* | | | | | | | F P | | D M | 2 | 3-5 7-10 14 22 | * |
| *Pteris comans* | | | T | | | | N | | S | | 3 4 | |
| *Pteris tremula* | | | T | | | | P N | | S W | | 4 | |
| *Pyrrosia rupestris* | | | E | | | | P N | | | | 3 | |
| *Rumohra adiantiformis* | | | E | | | | P N | | | | 2 | |
| *Schizaea asperula* | | | T | | | | P | | D M | 2 | 8 9 | |
| *Schizaea bifida* | | | T | | | | P | | S W | 2 | 8 9 | |
| *Schizaea fistulosa* | | | T | | | | P | | S W | 2 | 8 9 | |
| *Selaginella gracillima* | | | H | | | | P N | | S W | | 6 9 | |
| *Selaginella uliginosa* | | | H | | | | F P | | S W | | 6 9 | |
| *Sticherus lobatus* | | | | | | | P N | | S | | 2 | |
| *Sticherus tener* | | | | | | | P N | | S | | 2 3 | |
| *Tmesipteris billardieri* | | | E | | | | N | | | | 2 | |
| *Tmesipteris parva* | | | E | | | | N | | | | 2 | |
| *Todea barbara* | S | | | | | | A | | S | | 3 4 | |

## Conifers

| | Tree | Shrub | Other | Flowers | Bird | Butterfly | Sun Aspect | Screen Windbreak | Drainage | Coast | Vegetation Communities | Aboriginal Uses |
|---|---|---|---|---|---|---|---|---|---|---|---|---|
| *Callitris glaucophylla* | M-T | | | | S | | F | W | D | | 15 | |

# *Appendix 2. Reserves and Localities*

These locations are representative of the plant communities that were once found across the greater Melbourne area and are referred to in the individual plant descriptions. Each is indicated on the map of the Melbourne Study on end papers.

Visits to these areas will provide an idea of the way plants relate to each other in a community. Unfortunately all are affected, to some degree, by invading weeds.

It must be stressed that removal of plant material should only occur with the permission of the land owner and that plants should never be transplanted from public land.

1. Weeroona Garden
   Set up by Brunswick Electricity Supply and Victorian Indigenous Nurseries Co-operative
   CERES, Lee St, East Brunswick
   Plants used by the local Wurundjeri tribe
2. Merrett Rifle Range, Williamstown
   Kororoit Creek Rd. Williamstown
   Communities: Primary Dune Scrub, Saltmarsh, Grassy Wetlands, Plains Grassland, Coastal and Marine vegetation
   Owner: Dept. Conservation & Environment
   Whilst the bay side of this area has been retained as a reserve (now known as Jawbone Conservation Reserve), the grasslands area was bulldozed for a housing estate in 1991.
3. Kororoit Creek mouth
   Altona Coastal Park
   Communities: Saltmarsh, Marine
   Owner: managed by Cities of Altona & Williamstown
4. Cherry Lake—Truganina Swamp
   Millers Rd. Altona
   Community: Saltmarsh
   Owner: Melbourne Water
5. Point Cook Metropolitan Park
   Point Cook—Homestead Rd, Werribee
   Communities: Primary Dune Scrub, Grassy Wetland, Saltmarsh, Plains Grassland
   Owner: Melbourne Water
6. Laverton RAAF Base Swamp
   Princes Highway, Laverton
   Community: Grassy Wetlands
   Owner: RAAF (prohibited access) & Public Transport Corporation (small area)
7. Altona Grasslands (including Laverton North Grassland Reserve)
   South and East of Princes Highway to Melbourne Water Drainage Basin, Altona
   Communities: Plains Grassland, Grassy Wetland, Saltmarsh
   Owner: Both private and public land.
   Latrobe University have reintroduced some rarer grassland species back into the Laverton North Grassland Reserve.
   A site of National significance, containing many rare and depleted species.
8. Cobbledicks Ford
   Cobbledicks Ford Rd, Werribee
   Community: Riparian Woodland
   Owner: Public land
9. Exford woodlands
   Melton South
   Community: Box Woodland (State significance)
   Owner: Private
10. North-western Rail Reserve
    Melton Reservoir through Melton
    Community: Plains Grassland
    Owner: Public Transport Corporation
11. Deans Marsh
    East of Rockbank, Sydenham West
    Community: Grassy wetland
    Owner: Mostly private
12. Evans St. Sunbury, Rail Reserve
    Community: Plains Grassland (State significance)
    Owner: Public Transport Corporation, Shire of Bulla
    Contains a wide range of grasses.
13. Melbourne–Bendigo Rail Reserve
    Sites between Sunbury and Diggers Rest railway stations
    Community: Plains Grassland (National significance)
    Owner: Public Transport Corporation
14. Jacksons Creek, Sunbury
    Communities: Box Woodland, Riparian Scrub, Chenopod Rocky Open Scrub
    Owner: Mostly private
15. Diggers Rest Rail Reserve Grasslands
    Between Calder Raceway and Diggers Rest
    Community: Plains Grassland (National significance)
    Owner: Public Transport Corporation
16. Organ Pipes National Park
    Calder Highway, Sydenham
    Communities: Riparian Scrub (mostly original flora); Box Woodland, Plains Grassland
    Owner: Managed by Dept. Conservation & Environment
    A revegetation project where many locally threatened species have been collected from nearby areas and reintroduced into the Park.
17. Taylors Lakes Grassland
    Between Keilor-Melton Rd and railway line, Taylors Lakes
    Community: Plains Grassland (National significance)
    Owner: Has now been cleared for housing, leaving tiny pockets of remnant vegetation which will ultimately disappear.
18. Taylors Creek
    St. Albans Rd to Calder Park Dve, Taylors Lakes
    Communities: Box Woodland, Riparian Scrub, Plains Grassland
    Owner: Mostly private; Surveys prepared by Melbourne Water

19. Steele Creek Grassland
West of confluence with Maribyrnong R., Avondale Heights
Communities: Riparian Scrub, Plains Grassland, escarpment vegetation
Owner: Melbourne Water
20. Braybrook Rail Reserve Grassland
West of Tottenham railway station, between Sunshine Rd and Duke St.
Community: Plains Grassland
Owner: Public Transport Corporation
21. Royal Park West
Manningham St, Parkville
Community: Red Gum Woodland
Owner: City of Melbourne
22. Union St. Brunswick, along Moonee Ponds Ck
Community: Riparian Scrub
Owner: Public
23. Napier Park, Essendon
Community: Red Gum Woodland
Owner: City of Essendon
24. Radar Hill, Tullamarine
Lancefield and Perimeter Rds, Melbourne International Airport
Community: Box Woodland
Owner: Commonwealth Government
25. Gellibrand Hill State Park
Oaklands and Somerton Rd, Oaklands Junction
Communities: Red Gum Woodland, Box Woodland
Owner: Managed by Dept. Conservation & Environment
26. Shire of Bulla
The vascular flora of the Shire of Bulla
Communities: Red Gum Woodland, Box Woodland, Plains Grassland, Riparian Scrub, Chenopod Rocky Open Scrub
27. Deep Creek environs
Konagaderra Rd, East of Sunbury
Community: Riparian Scrub
Owner: Private & public
28. Merri Creek, Cooper St, Somerton
Communities: Plains Grassland, Riparian Scrub
Owner: Public & Private; a small section along the Merri Ck has recently been acquired by the State Government to become part of the Merri Ck Metropolitan Park. At the time of writing the rest of this area is under threat of industrial development.
29. Merri Creek Valley
North Coburg to Campbellfield
Communities: Riparian Scrub, Plains Grassland
Owner: Public and private
Two lists were used: Rupp's species list of 1896 and Merri Ck Co-ordinating Committee study 1984
30. Yarra River to Studley Pk 1856
Community: Riparian Scrub and others
Early list by Hannaford of plants along the Yarra River.
31. Heyington Railway Reserve 1945
Communities: Escarpment, Red Gum Woodland
Owner: Public Transport Corporation
Early list of Yarra R. cliffs and the railway reserve at Heyington
32. Studley Park & Yarra Bend, Kew
Communities: Red Gum Woodland, Riparian Scrub, Dry Sclerophyll Forest
Owner: Some private, most managed by Melbourne Water
33. Heidelberg area
Communities: include Red Gum Woodland, Riparian Scrub
34. Templestowe area
Communities: Red Gum Woodland, Dry Sclerophyll Forest
34A. Yarra Valley Metropolitan Park
Communities: Riparian Scrub, Valley Sclerophyll Forest
Owner: Melbourne Water
35. Gresswell Forest
Bendoran Cres, Bundoora
Community: Red Gum Woodland
Owner: managed by Latrobe University Wildlife Reserve
36. Plenty Gorge
Plenty River from Wilton Vale Rd to Plenty River Ave
Communities: Valley and Dry Sclerophyll Forests, Box Ironbark Woodland
Owner: Mostly public, managed by Melbourne Water and Dept. Conservation & Environment
37. Yandell Reserve
St. Helena Rd, Greensborough
Community: Dry Sclerophyll Forest
Owner: Managed by Shire of Diamond Valley
38. Meruka Park
Karingal Dve, Eltham
Community: Dry Sclerophyll Forest
Owner: Managed by Shire of Eltham
39. Eltham Lower Park
Homestead Rd, Eltham
Community: Dry and Valley Sclerophyll Forests
Owner: Managed by Shire of Eltham
40. Bemboka Reserve
Bemboka Rd, Warranwood
Community: Dry Sclerophyll Forest grading to Valley Sclerophyll Forest
Owner: Managed by City of Croydon
41. Professors Hill
Research-Warrandyte Rd, Warrandyte Nth
Communities: Dry and Valley Sclerophyll Forests
Owner: Managed by Shire of Eltham
42. Warrandyte State Park
Several areas including 4th & 5th Hill, Jumping Ck Reserve
Communities: Dry and Valley Sclerophyll Forests, Riparian Scrub
Owner: Managed by Dept. Conservation & Environment
43. Koolunga Reserve
Forest Rd, Ferntree Gully
Community: Valley Sclerophyll Forest
Owner: Managed by City of Knox
44. Mullum Mullum/Koonung Valley
Doncaster to Ringwood
Communities: Dry and Valley Sclerophyll Forests, Riparian Scrub
Owner: Public and private land
45. Wattle Park 1917, 1958
Riversdale Rd, Wattle Park
Community: Left over remnant of Red Gum Woodland
Owner: Managed by Melbourne Water
46. Blackburn Lake
Central Rd, Blackburn
Community: Valley Sclerophyll Forest, some Riparian and Aquatic vegetation
Owner: Managed by City of Nunawading
47. City of Nunawading
Communities: Dry and Valley Sclerophyll Forests
48. Antonio Park
Maroondah Highway and Deep Creek Rd, Mitcham
Community: Dry Sclerophyll Forest
Owner: Managed by City of Nunawading
49. Mitcham area
General lists by J W Audas, 1937 and F Rogers 1970
Communities: Dry and Valley Sclerophyll Forests
50. 'Uambi'
Heathmont
Community: Valley Sclerophyll Forest
Owner: Private
50A. Ringwood
Lists of several reserves within the City of Ringwood personal knowledge by members of SGAP
Communities: Dry and Valley Sclerophyll Forests
51. Croydon
Lists of several reserves within the City of Croydon
Communities: Dry and Valley Sclerophyll Forests, Riparian scrub

52. Birts Hill
Alice St, Croydon
Community: Dry Sclerophyll Forest
Owner: Managed by Dept. Conservation & Environment
53. Hotchkins Ridge Flora Reserve
Exeter Rd, Croydon North
Community: Valley Sclerophyll Forest
Owner: City of Croydon & Dept. Conservation & Environment
54. Tereddan Dve Reserves
Kilsyth South
Communities: Valley Sclerophyll Forest, Swamp Scrub
Owner: Shire of Lillydale, Dept. Conservation & Environment & Melbourne Water
55. Mt. Dandenong National Park—west face
Mt. Dandenong Tourist Rd. to Glasgow Rd, Montrose
Communities: Dry and Valley Sclerophyll Forests, pockets of Damp Sclerophyll Forest
Owner: Managed by Dept. Conservation & Environment
56. Mt Evelyn area
Communities: Valley and Damp Sclerophyll Forests
57. Sherbrooke Forest
Dandenong Ranges
Communities: Cool Temperate Rainforest, Wet Sclerophyll Forest
Owner: Managed by Dept. Conservation & Environment
58. Ferntree Gully National Park
Burwood Highway & Mt Dandenong Tourist Rd, Ferntree Gully
Communities: Wet and Damp Sclerophyll Forests
Owner: Managed by Dept. Conservation & Environment
59. Baluk Willam Flora Reserve
Courtnays Rd, Belgrave South
Community: Sclerophyll Woodland
Owner: Managed by Dept. Conservation & Environment
60. Lysterfield Lake Park
Wellington Rd, Lysterfield
Communities: Grassy Low Open Forest, Wetland and Aquatic vegetation
Owner: Managed by Dept. Conservation & Environment
61. Churchill National Park
Churchill Park Dve, Rowville
Communities: Damp and Valley Sclerophyll Forests, Swamp Scrub
Owner: Dept. Conservation & Environment
62. Dandenong Valley Metropolitan Park
City of Waverley
Communities: Valley Sclerophyll Forest, Swamp Scrub, Aquatic
Owner: Melbourne Water
63. Waverley District
Communities: Valley Sclerophyll Forest, Swamp scrub
64. Valley Reserve, Mt. Waverley
Wills Ave, Mt. Waverley
Community: Valley Sclerophyll Forest
Owner: Managed by City of Waverley
65. Ashburton Wood 1931
High St. Rd, Gardiners Creek, Warrigal Rd
Community: Red Gum Woodland
Now totally developed.
66. Oakleigh area
General list of original species. Areas mostly developed
Community: Red Gum Woodland
67. City of Sandringham
Many plants noted as extinct within the area.
Communities: Tea-tree Heath, Wattle Tea-tree Scrub, Grassy Low Open Forest, Red Gum Woodland, Coastal Banksia Woodland, Primary Dune Scrub, Coastal Cliffs
68. Beaumaris High School
Balcombe & Reserve Rds, Beaumaris
Community: Tea-tree Heath
Owner: Ministry of Education & Training
69. Royal Melbourne Golf Course
Reserve & Cheltenham Rds, Black Rock
Communities: Primary Dune Scrub, Tea-tree Heath
Owner: Royal Melbourne Golf Club
69A. City of Mordialloc
Communities: Grassy Low Open Forest, Tea-tree Heath, Red Gum Woodland, Swamp Scrub, Primary Dune Scrub
70. The Grange Heathland Reserve
Osborne Ave, Clayton South
Community: Tea-tree Heath
Owner: City of Oakleigh
71. Braeside Metropolitan Park
Lower Dandenong Rd, Braeside
Communities: Tea-tree Heath, Grassy Low Open Forest, Red Gum Woodland, Swamp Scrub, Wattle Tea-tree Scrub
Owner: Melbourne Water
72. Edithvale-Seaford Wetlands
West of Wells Rd, Edithvale to Seaford
Community: Swamp Scrub, sub-saline wetlands
Owner: Dandenong Valley & Western Port Authority
73. Cranbourne Annexe
Ballarto Rd, Cranbourne; Annexe of Royal Botanic Gardens
Communities: Tea-tree Heath, Wattle Tea-tree scrub, Grassy Low Open Forest
Owner: Managed by Dept. Conservation & Environment
74. City of Frankston
General list including The Pines
Communities: Tea-tree Heath, Wattle Tea-tree Scrub, Grassy Low Open Forest, Primary Dune Scrub, Coastal Banksia Woodland
75. Langwarrin Flora and Fauna Reserve
Robinsons Rd, Langwarrin
Communities: Tea-tree Heath, Wattle Tea-tree Scrub, Grassy Low Open Forest
Owner: Managed by Dept. Conservation & Environment
76. Mooroodue Quarry Flora & Fauna Reserve & Mooroodue Plains
Between Mooroodue & Canadian Bay Rds, Mooroodue
Community: Grassy Low Open Forest
Owner: Managed by Dept. Conservation & Environment

# *Appendix 3. A. Associations & Contacts for Australian and Indigenous Plants*

1. Society for Growing Australian Plants Maroondah Inc. (SGAP)
   PO Box 33 Ringwood 3134

Caters for people from Kew to Montrose interested in growing and conserving Australian plants. Activities include monthly meetings, garden visits, field trips, camps and displays.

2. Society for Growing Australian Plants Victoria Inc.
   C/o 17 Craig Crt, Heathmont 3135

The co-ordinating committee for the 24 district branches of SGAP, both city and country. All membership enquiries should be directed to them.

3. Indigenous Flora and Fauna Association Inc. (IFFA)
   C/o 69 Spenseley St., Clifton Hill 3068

An organisation of professionals and amateurs from diverse backgrounds who are committed to the conservation of indigenous flora and fauna through preservation, restoration and management of habitat. Their objective is to both research and provide information.

4. Australasian Native Orchid Society, Victoria Inc. (ANOS)
   Secretary: PO Box 285, Cheltenham 3192

5. Conservation Council of Victoria, 247 Flinders Lane Melbourne 3000 Ph 654 4833

6. 'Friends' activities published in newsletters of IFFA (Indigenotes) and Victorian National Parks Association. (VNPA contact number (03) 650 8296)

Local government departments can provide information on groups managing local bushland reserves or local environment groups.

7. Australian Trust for Conservation Volunteers

Co-ordinate plantings throughout Australia during the week and on weekends, predominantly using indigenous plants.

National Headquarters in Ballarat (053) 32 7490
Melbourne (03) 532 8446

8. A project is underway to set up a database which will provide very detailed information on the indigenous plants of Melbourne and their management. A pilot program is currently being prepared. For further information contact Melbourne Water on (03) 615 4619.

9. Indigenous Nurseries Network (IFFA). Members represent more than 60 indigenous nurseries throughout Victoria. Contact: Sue Mills (03) 383 2937.

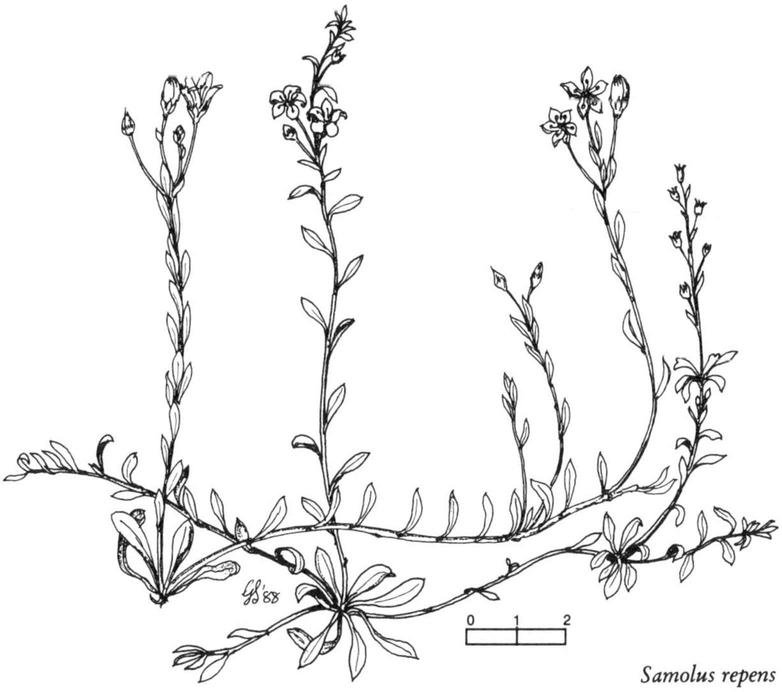

*Samolus repens*

# Appendix 3. B. Nurseries Selling Indigenous plants of Melbourne

Locating plants from local provenances has been a difficulty experienced by many people involved in both large and small scale regeneration projects. Some large projects such as Nunawading Indigenous Plants Project (878 5998) and Friends of Warrandyte State Park grew their own to solve this problem. Fortunately, as the realisation of the benefits of growing indigenous plants increases, nurseries and individuals are beginning to cater for the demand.

The nurseries below are not an exhaustive list. Some nurseries also sell a general range of Australian plants. It is up to the buyer to check labelling on stock to ensure that the plants are indigenous to their area of need.

Many local councils are now setting up indigenous nurseries for their own areas. Some of these also have plants available to the public.

Austraflora Nursery, Belfast Rd, Montrose 3765
Ph (03) 728 1222 Proprietors: Bill Molyneux & Sue Forrester
Limited stock of indigenous plants.

Best Plant Brokers, 141 Stud Rd, Dandenong 3175
Ph (03) 791 8387 Proprietor: Darren Wallace
Tubestock to advanced sizes of species derived from provenance material within the Greater Knox region, also Lysterfield, Belgrave to Montrose.

Bimbadeen Nursery, 104 Webb St, Warrandyte 3113
Ph (03) 844 3906 Proprietor: Peter Hansen
Specialises in species from Warrandyte.

Bindelong Nursery (Yarra Valley Tubes), 55 Railway Ave, Wandin Nth 3139 Ph (059) 64 4974 Proprietor: Doug Pocock
Wholesale supplier of plants of the Yarra Valley.

Bush Garden Native Plants, 3 Burswood Close, Frankston 3199 (Melway 102 F10) Ph (059) 71 2585 Proprietor: Bev Courtney
Specialises in understorey species of the Mornington Peninsula, including Frankston, Langwarrin and Cranbourne. Species include grasses, orchids and wildflowers, also groundcovers, shrubs and small trees.

Bushland Flora, PO Box 312 Mt Evelyn 3796 Ph (03) 736 4364 Proprietors: Andrew Paget & Ian Shimmen
Contact re retail outlets throughout Melbourne.
Contract, retail and wholesale in tubestock. Species Melbourne-wide.

Bush Life Nursery, cnr Canterbury & Wantirna Rds, Ringwood 3134 Ph (03) 870 0016 Proprietor: Greg Jacobs
Specialises in plants from the Outer Eastern suburbs from Box Hill to Dandenong Ranges, also Warrandyte.

Chris Fletcher, 58 King St, Yarra Glen 3775 Ph (03) 730 1517 (by appointment only)
Species from Yarra Valley.

Dragonfly Aquatics, RMB AB 366, Colac 3250 Proprietor: Nick Romanowski
Specialises in water and moisture-loving plants, with the largest range in Australia. Send 2 stamps to the above address for a catalogue.

E.J. Fenton, Larapinta, Box 570 Hamilton 3300 Ph (055) 73 4555 Proprietor: Elizabeth Fenton
Seed and tubestock from the Hamilton district.

Geelong Indigenous Nursery, Ph (052) 298 087 Proprietor: Mark Trengove

Kareelah Bush Nursery, Lot 7 Luxton Dve, Balnarring 3926
Ph (059) 89 5801 Proprietor: Kathie Strickland
Tubestock from grasses to trees, from Mornington Peninsula. Retail and contract growing.

Kuranga Native Nursery, 393 Maroondah Highway Ringwood 3134 Ph (03) 879 4076 Proprietors: Evan Clucas & Leanne Weston
Specialises in tubestock, mostly from the Eastern suburbs, especially Ringwood. The source of all indigenous stock is marked on the label.

Local Native Flora, 4 Gull St, Rye 3941 Ph (059) 85 1122
Proprietor: Mark Adams
Southern Mornington Peninsula species and contract growing.

Otways Indigenous Nursery, 14 Tenth Ave, Anglesea 3230
Ph (052) 63 1630 (by appointment only) Proprietor: Geoff Clark
Specialises in coastal and heathland plant propagation from seeds and cuttings, from Point Addis to Apollo Bay to Colac and Winchelsea.

Put It Back Nursery, 1/51 Spenser St, St Kilda 3182 Ph (03) 534 3154 AH. Proprietor: Rob Scott
Contract seed collection and growing of plants throughout Melbourne.

Sandringham Community Nursery, cnr Reserve & Talinga Rds, Sandringham Ph (03) 598 8111 (Sandringham City Council) Open Saturdays.
Specialises in Sandringham provenances.

Victorian Indigenous Nurseries Co-op Ltd (VINC) CERES, Lee St., Brunswick East 3057 Ph (03) 387 4403
Collects seeds in and grows plants for the Northern, Western and Inner suburbs of Melbourne. Wholesale and retail nursery. Also contract growing and revegetation assistance.

Wildlife Reserve Nursery, Latrobe University, Bundoora 3083. Take Ring Rd at Latrobe University to the North East end of the campus, opposite Car Park 7
Specialises in plants from Heidelberg, Preston, Whittlesea, Diamond Valley and Eltham municipalities; also wetland plants of the Yarra Valley.

Wyeena Nursery, Smiths Gully Rd, Smiths Gully 3760.
Ph (03) 710 1340 Proprietors: Julie & Kahn Franke
Specialises in plants from Eltham Shire, Plenty Gorge, and Yarra Valley.

# *Appendix 4. Glossary*

**abaxial** Away from the axis or stem.

**achene** A small dry one-seeded fruit which does not split on maturity (sometimes referred to as a cypsela).

**annual** A plant that completes its life cycle in one season.

**anther** The pollen bearing part of a stamen.

**apex** The free end or tip of a leaf.

**appressed** Pressed flat against something.

**ascending** Directed upwards.

**auricle** An ear-shaped lobe or appendage.

**awn** A bristle like appendage, e.g. on the seeds of many grasses.

**axil** The upper angle between two dissimilar parts e.g. that formed by a leaf in relation to the stem.

**axillary** Of buds or flowers arising in an axil.

**basal rosette** A cluster of leaves radiating from a common point at the base of the stem.

**berry** A fleshy fruit often containing many seeds.

**bilobed** Two lobed.

**bipinnate** Twice pinnately divided.

**bract** A leaflike structure which subtends a flower stem or inflorescence.

**bracteole** A small bract.

**bulbil** A specialized bud produced at the junction of main veins on the fronds of some ferns.

**burr** A prickly fruit.

**callus (pl. calli)** A thickening, eg the glandular, warty structure often found on the labellum in orchids; hardened, often hairy base of grass florets.

**calyx** Outer envelope of the flower, consisting of the sepals.

**carpel** Female reproductive organs.

**column** A fleshy growth in the flowers of orchids formed by the union of the stigmas and stamens.

**compound** Consisting of 2 or more parts.

**concave** Curves inward.

**convex** Curves or bulges outwards.

**convolute** Rolled around and overlapping, as in petals in a bud or lemmas in a spikelet.

**coppice** Wood of small trees grown for periodic cutting; growth arising from dormant buds in trunk, usually after lopping.

**cordate** Heart-shaped.

**corolla** The petals; whorl of floral segments inside the calyx.

**crenulate** Bordered by very small round teeth.

**cuneate** Wedge-shaped.

**cyme** An inflorescence where each branch is opposite and ends in a flower, the flowers opening sequentially downwards, and further development is by lateral branches.

**decumbent** Reclining on the ground with the tips of branches ascending.

**decussate** Of opposite leaves, each successive pair at right angles to the pair below.

**dicotyledon** The group of flowering plants which bear 2 seed leaves at the seedling stage.

**dioecious** Bearing male and female flowers on separate plants.

**diffuse** Widely spreading and much branched; of open growth.

**discoid** Flowerheads which only have central tubular disc florets (esp. Asteraceae).

**drupe** A fruit with a 'stone' or seed surrounded by a fleshy layer.

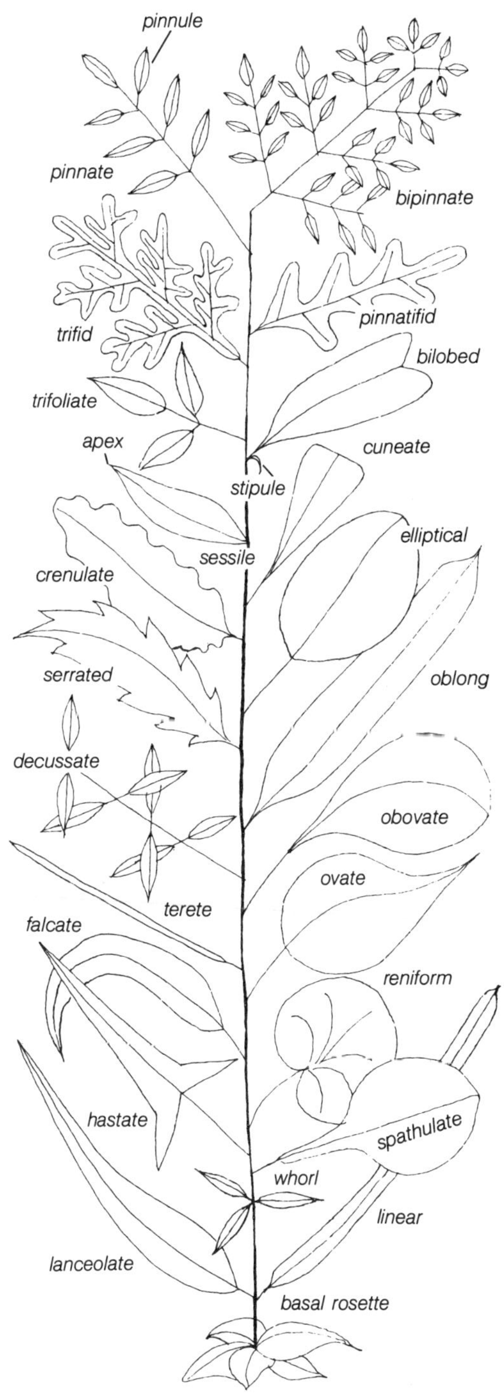

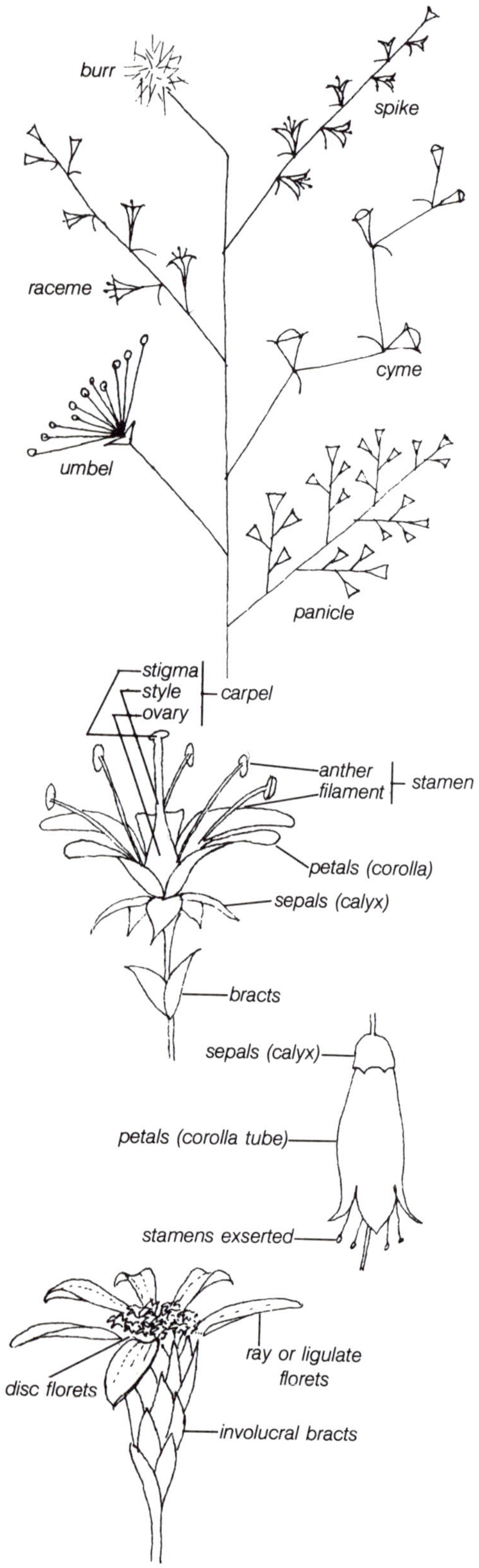


**elliptical** Of a flat surface in the shape of an ellipse.
**endemic** Found only in a specified location.
**entire** Whole, not toothed or divided in anyway.
**ephemeral** A plant completing its life cycle within a very short period e.g. 3–6 months.
**epiphytic** A plant growing on another but not parasitic.
**exserted** Protruding beyond adjacent parts.
**falcate** Sickle-shaped.
**filament** The stalk of a stamen supporting the anther..
**filiform** Long and very slender; thread-like.
**flexuose** Stem bent alternately in opposite directions.
**floret** The smallest unit of a compound flower.
**frond** Leaf of a fern.
**frontline coastal** The closest to the water's edge, able to withstand salt-laden winds.
**glabrous** Smooth, without hairs.
**glaucous** Covered with a bloom, giving a bluish lustre.
**globose** Like a globe; round.
**glume** One of 2 bracts at the base of the spikelets of grasses and sedges.
**hastate** Like an arrow head with spreading basal lobes.
**herb** A plant that produces a fleshy rather than woody stem.
**herbaceous** A perennial plant which dies down each year after flowering.
**inflorescence** The flowering structure of a plant.
**internode** Portion of stem between two adjacent nodes.
**juvenile leaf** A seedling leaf that differs markedly from the adult leaf.
**labellum** A lip; in orchids the petal in front of the column.
**lanceolate** Lance-shaped; narrow and tapering at each end especially the apex.
**lateral** Arising from the main axis; at the side of.
**lemma** The lower of 2 bracts enclosing each grass flower.
**lignotuber** A woody, usually underground, swelling at the base of a trunk, containing dormant buds.
**ligulate** Of daisy flowers, a strap-like corolla lobe (ray floret).
**linear** Long and narrow, with parallel sides.
**linear-lanceolate** Long, narrow lance-shaped.
**lithophyte** Of a plant growing on rock.
**lobe** A rounded projecting part of a leaf, petal or sepal.
**mallee** A shrub or tree with several trunks or stems rising from a lignotuber (esp. *Eucalyptus* spp.).
**mealy** Covered with a flour-like powder.
**monocotyledon** The group of flowering plants which bear a single seed leaf at the seedling stage.
**monoecious** Bearing male and female flowers on the same plant.
**mycorrhiza** A beneficial relationship between the roots or seeds of a plant and soil borne fungi resulting in a nutrient exchange system. Some plants (e.g. orchids) cannot grow without such a relationship.
**node** The place on a stem marked by the attachment of a leaf or bracts.
**oblanceolate** Lanceolate with the broadest part above the middle.
**oblong** Greater in breadth than length.
**obovate** Ovate with the broadest part above the middle.
**obtuse** Blunt or rounded at the apex.
**ovate** Egg-shaped and attached by the broad end.
**palea** The upper of 2 bracts enclosing each grass flower.
**pappus** A tuft of bristles, hairs or scales representing the calyx on the seeds of Asteraceae.
**panicle** A much branched inflorescence.
**parasitic** Of a plant living on another plant.
**pendent** Hanging, swinging.
**perennial** Living for more than two years.
**perianth** Floral parts or segments (petals and sepals) of monocotyledons.
**phyllode** A stem or stalk modified to act as a leaf.
**pinna** A primary segment of a divided leaf.
**pinnate** Once divided with the divisions extending to the midrib.
**pinnatifid** Once divided with the divisions not extending to the midrib.
**pinnule** The segment of a compound leaf divided more than once.
**procumbent** Flat along the ground without rooting.
**provenance** Source.
**pungent** Very sharply pointed; also, smelling strongly.
**quadrangular** Four-sided and angular.
**raceme** A long unbranched inflorescence.

**radiate** A compound flowerhead (esp. in Asteraceae) comprising central tubular disc florets surrounded by ray florets.
**ray floret** One of the ligulate florets around the edge of inflorescence, as in daisies.
**recurved** Curved backwards.
**reflexed** Bent backwards and downwards.
**reniform** Kidney-shaped.
**rhachilla** The main axis of a grass spikelet.
**rhachis** The main axis of an inflorescence, compound leaf or frond.
**rhizome** An underground stem.
**riparian** Of or on a watercourse or drainage line.
**saprophyte** A plant which derives its food from dead or decaying matter.
**scarified seed** Seed that has had the surface scratched or lacerated (e.g. by sandpapering).
**sclerophyll** Referring to a forest dominated by eucalypts with an understorey of shrubs with small hard leaves.
**scrub** A plant community dominated by shrubs.
**second line** The second grouping of plants (usually used in reference to coastal plants), those behind the 'front line'.
**sedge** Waterside plant resembling grass.
**sepal** Free segment of the calyx or outer whorl of the perianth.
**serrated** Toothed with sharp, forward pointing teeth.
**sessile** Without a stalk.
**simple** Undivided, of one piece.
**spathulate** Spoon-shaped.
**spike** An inflorescence of sessile flowers on an unbranched axis.
**spikelet** Small spikes of the grass infloresence, bearing 2 glumes and one or more florets.
**spore** A simple asexual reproductive unit of ferns.
**sporocarp** A thick wall covering the spore cases.
**stalk** A support for a flower or leaf.
**stamen** The male part of a flower producing pollen, consisting of an anther and a filament.
**stellate** Star-shaped or of starlike form.
**stigma** Part of the style receptive to pollen.
**stipule** Small bract-like appendages borne in pairs at the base of the stalk.
**stolon** A basal stem growing above or just below the ground surface and rooting at intervals.
**stoloniferous** Bearing stolons; spreading by stolons.
**style** The elongated part of a carpel between the ovary and the stigma.
**sward** Expanse of short grass.
**symbiotic** Of a mutually beneficial partnership.
**terete** Cylindrical and slender.
**terrestrial** Growing in the ground.
**terminal** At the end or extremity.
**trifid** Divided almost to the middle into three parts.
**tripinnate** Divided three times.
**trifoliate** A leaf with three leaflets.
**truncate** With the apex appearing as if cut off.
**tuberoid** The swollen end of an underground root.
**tuberous** Resembling a tuber.
**umbel** An inflorescence in which all the stems arise at the same point and the flowers lie at the same level, more or less umbrella-shaped.
**unisexual flowers** Having only one sex.
**valves** A segment of the case of a woody fruit.
**whorl** An arrangement of three or more parts at the same level around an axis.

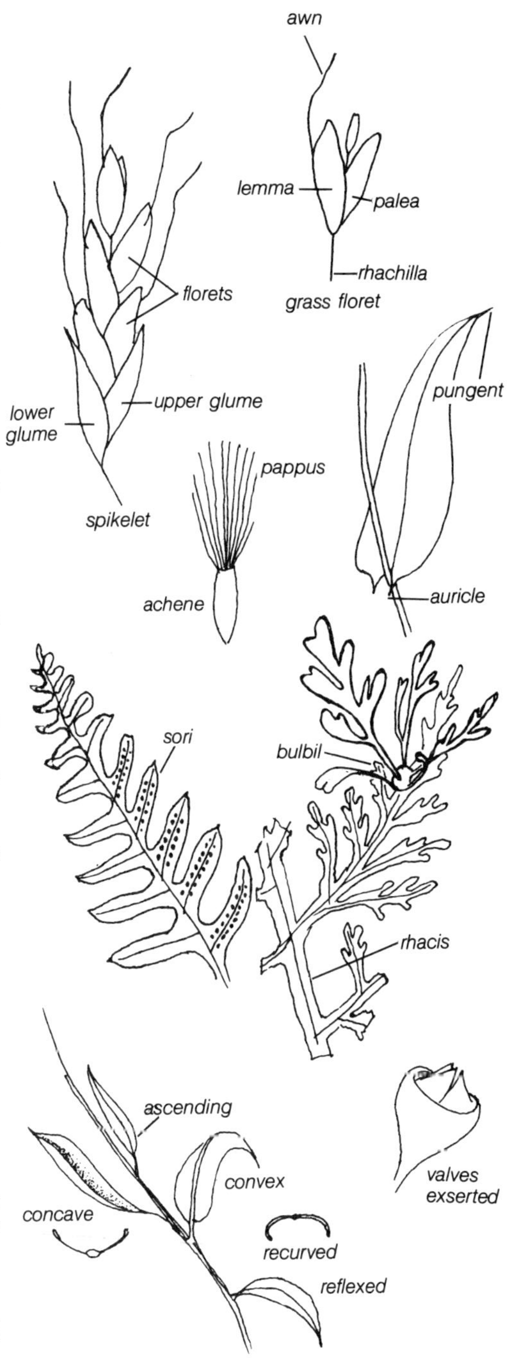


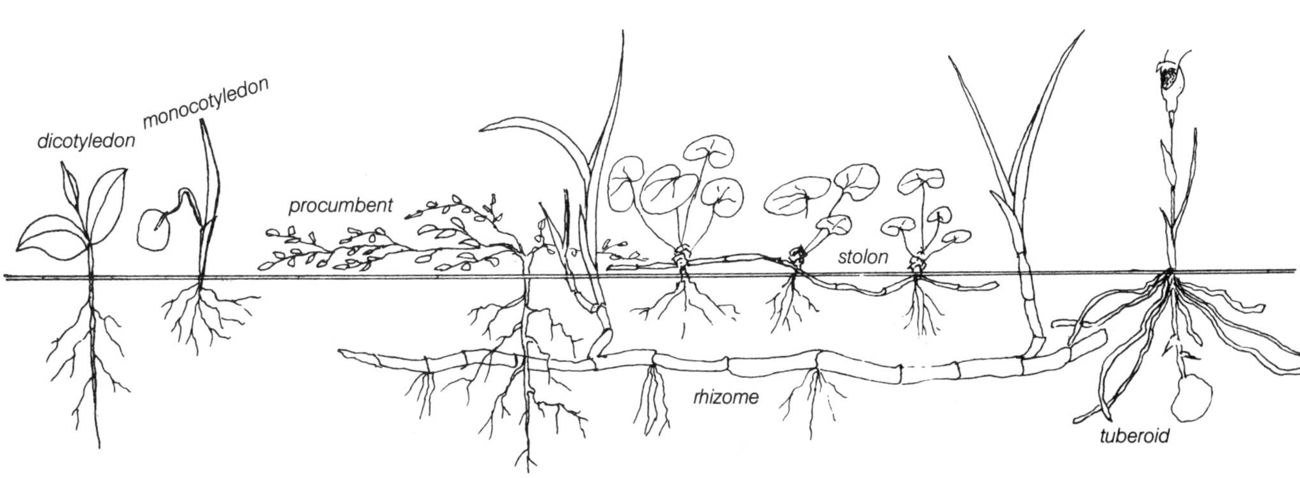

# *Appendix 5. Bibliography*

Other references are listed under Further Reading after each chapter.


Albrecht, D. (unpubl) *A Review of Leafless Species of Juncus in Victoria.*

Aston, H.I. (1973) *Aquatic Plants of Australia*, Melbourne University Press.

Audas, J.W. (1937) *The Australian Bushland*, Robertson & Mullens Ltd, Melbourne.

Australian Daisy Study Group (1987) *Australian Daisies for Gardens and Floral Art*, Lothian Publishing Co., Melbourne.

Australian Plant Study Group (1987) *Grow What Small Plant*, Thomas Nelson (Australia) Ltd, Melbourne.

Australian Plant Study Group (1982) *Grow What Wet*, Thomas Nelson (Australia) Ltd, Melbourne.

Australian Plant Study Group (1980) *Grow What Where*, Thomas Nelson (Australia) Ltd, Melbourne.

Australian Systematic Botany Society (1981) *Flora of Central Australia*, (Ed. J. Jessop), A.H. & A.W. Reed, Sydney.

Australian Systematic Botany Society, (1982) Newsletter Nos. 32, 33 (Ed. G.P. Guymer), Queensland Herbarium, Indooroopilly, Qld

*Austrobaileya* Vol. 2 (1984), Periodical Journal of Queensland Herbarium, Dept. of Primary Industries, Brisbane, Government Printer, Qld.

Beadle, N.C.W. (1976) *Student's Flora of Northeastern New South Wales*, Pt. 3, University of New England, Armidale.

Beadle, N.C.W., Evans O.D., Carolin R.C. (1972) *Flora of the Sydney Region*, A.H. & A.W Reed, Sydney.

Beardsell, C. & D. (1983) *A Year of Orchids*, Richard Griffin, South Melbourne.

Beauglehole, A.C. (1983) *The Distribution and Conservation of Vascular Plants in the Melbourne area, Victoria*, Western Victorian Field Naturalists Clubs Association, Portland Vic.

Bentham, G. & Mueller, F. (1863–78) *Flora Australiensis* Pts 1–7, Lovell Reeve & Co., London.

Black, J.M. (1943–57) *Flora of South Australia*; 4th edition (1986) Pts 1–4 (Ed. J.P. Jessop & H.R. Toelken) Government Printer, Adelaide.

Blombery, A.M. (1967) *A Guide to Australian Native Plants*, Angus & Robertson, Sydney.

Bridgewater, P.B., Rosser, C. & de Corona, A. (1981) *The Saltmarsh Plants of Southern Australia*, Botany Dept., Monash University, Melbourne.

Brooker, M.I.H. & Kleinig, D.A. (1983) *Field Guide to Eucalypts* Vol. 1, Inkata Press Pty Ltd, Melbourne

*Brunonia*, (1979–87) Vols 2:1, 4:2, 5:1, 8:3, 9:2, Periodical Journal of the Herbarium Australiense, CSIRO, Melbourne.

Buckhorn, R. et al. (1989) *Urban Forestry Handbook*, Dept. Conservation, Forests & Lands, Melbourne.

Burbidge, N.T. (1984) *Australian Grasses*, revised by Jacobs, S.W.L., National Herbarium, Royal Botanic Gardens, Sydney, Angus & Robertson, Sydney.

Burbidge, N.T. & Gray, M. (1970) *Flora of the Australian Capital Territory*, Australian National University Press, Canberra.

Cady, L. & Rotherham, E.R. (1970) *Australian Native Orchids*, A.H. & A.W. Reed, Sydney.

Calder, W. (1975) *Langwarrin Military Reserve: Ecological Evaluation for the Shire of Cranbourne*, Report, Centre for Environmental Studies, University of Melbourne, Melbourne.

Carr, G.W. (1982) *Vegetation of Exeter Road Ridge, Croydon North, and its significance*, Report to Croydon Conservation Society, Croydon, Victoria.

Carr, G.W. (1983) *Report on the Vegetation and Management of the Proposed Eltham Lower Park Flora Reserve*, Society for Growing Australian Plants Yarra Yarra & Eltham Shire Council, Melbourne.

Carr, G.W. (1991) *New Taxa in Caladenia, Chiloglottis and Gastrodia (Orchidaceae) from South Eastern Australia*, Indigenous Flora & Fauna Association, Melbourne.

Carr, G.W. (1980) *Proposed development of the Vegetation of LaTrobe University Wildlife Reserve*, Wildlife Reserves Management Committee, LaTrobe University, Melbourne.

Carr, G.W., Reid, J. & Albrecht, D. (1987) *The Vegetation, Fauna and Management of Antonio Park*, City of Nunawading, Victoria.

Chandler, C. (1975) *Two Hundred Wattles for Gardens*, David G. Stead Memorial, Wildlife Research Foundation of Australia, Sydney.

Cheal, D.C. (1988) *Botanical Assessment of Grasslands, Merri Creek, Somerton, Cooper Street*, Dept. Conservation & Environment, Melbourne.

Cheal, D.C., et al. (unpubl.) *Vegetation Survey and Sites of Botanical Significance in the Melbourne Area*, Dept. Conservation & Environment, Melbourne.

City of Box Hill (undated) *Box Hill: An Introduction to the Natural Vegetation*, City of Box Hill, Victoria.

Clucas, R.D. (1980) *Peripheral Vegetation of Lysterfield Lake*, Ministry for Conservation, Victoria.

Cochrane, R.G., Fuhrer, B.A., Rotherham, E.R. & Willis, J.H. (1968) *Flowers and Plants of Victoria*, A.H. & A.W. Reed, Sydney.

Conabere, E. & Garnett, J.R. (1987) *Wildflowers of South-Eastern Australia*, Greenhouse Publications, Richmond.

*Contributions from the New South Wales National Herbarium*, (1962) No. 102, Dept. of Agriculture, NSW.

Costermans, L. (1981) *Native Trees and Shrubs of South-Eastern Australia*, Rigby, Adelaide.

Cribb, A.B. & Cribb, J.W. (1974) *Wild Food in Australia*, Collins, Sydney.

Croydon Conservation Society, (1988) *Trees and Wildflowers of Croydon*, Croydon Conservation Society, Croydon, Victoria.

Cunningham, G.M., Mulham, W.B., Milthorpe, P.C. & Leigh, J.H. (1981) *Plants of Western New South Wales*, Soil Conservation Service of New South Wales.

Curtis, W.M. (1956–67) *The Student's Flora of Tasmania* Pts 1–3, Government Printer, Tas.

Dandenong Valley & Western Port Authority (1991) *Edithvale-Seaford Wetlands*, Dandenong Valley & Western Port Authority, Melbourne.

Dashorst, G.R.M. & Jessop, J.P. (1990) *Plants of the Adelaide Plains and Hills*, State Herbarium of South Australia, Kangaroo Press, Kenthurst, Sydney.

Dockrill, A.W. (1969) *Australian Indigenous Orchids*, Vol 1 Society for Growing Australian Plants, Sydney.

Dunk, G. (1983) *Ferns for the Home and Garden*, Angus & Robertson, Sydney.

Ecological Horticulture Pty. Ltd. 'Appendix F' in Riedel, Byrne & Wilson Sayer Core (1987) *Merrett Rifle Range Environmental Effects Statement* Vol 2, Urban Land Authority, Melbourne.

Edwards, J., Reid, J. et al. (1988) *Mt. Evelyn's Original Garden: Plants of the Northern Dandenongs*, Mt. Evelyn Environment Protection & Progress Association, Melbourne.

Elliot, G. (1979) *Australian Plants for Small Gardens & Containers*, Hyland House, Melbourne.

Elliot, R. (1975) *An Introduction to the Grampians Flora*, Algona Guides, Melbourne.

Elliot, R. & Blake, T. (1984) *Common Native Plants of Sherbrooke Forest & Dandenong Ranges*, Pioneer Design Studio, Lilydale, Vic.

Elliot, R. & Jones, D.L. (1980-90) *Encyclopaedia of Australian Plants* Vol 1-5, Lothian, Melbourne.

Erickson, R. (1975) *Triggerplants*, Paterson Brokensha Pty. Ltd. Perth.

Fairley, A. & Moore, P. (1989) *Native Plants of the Sydney District*, Kangaroo Press, Kenthurst, Sydney.

Fisher, D. (1988) *Vegetation Survey and Sites of Botanical and Geological Significance: Taylors Creek*, Melbourne & Metropolitan Board of Works, Melbourne.

Fitzsimons, P.F. (1983) *Vegetation Management Strategy for Werribee Shire Coastline* (Unpubl. thesis), Monash University.

*Flora of Australia* (1981–91) (Ed. A.S. George), Vols 3, 4, 18, 45, 46, Australian Government Publishing Service, Canberra.

Friends of the Valley Reserve (undated) *The Valley Reserve, History, Plants, Birds*, Friends of the Valley Reserve & Waverley City Council, Melbourne.

Galbraith, J. (1977) *Collin's Field Guide to the Wildflowers of South-East Australia*, Collins, Sydney.

*Geobotany*, (1982) Acta Phytotax. Vol 33.

Gott, B. (unpubl.) 'Plant Species used by Koories, found in the Melbourne LCC Area', *Vicuse Database.*

Gullan, P.K. et al. (1989) *Rare or Threatened Plants in Victoria*, Dept. Conservation & Environment, Melbourne.

Gullan, P. & Walsh, N. (1986) *Ferns and Fern Allies of the Upper Yarra Valley & Dandenong Ranges*, National Herbarium of Victoria, Dept. Conservation & Environment, Melbourne.

Harrington, C.D. (1977) *How to Identify Grasses and Grass-like Plants*, Ohio University, Ohio.

Harris, T.Y. (1977) *Gardening with Australian Plants—Shrubs*, Thomas Nelson (Australia) Ltd., Melbourne.

*Indigenotes* (1990–91) Newsletter of Indigenous Flora & Fauna Association, Melbourne.

Jones, D.L. (1988) *Native Orchids of Australia*, Reed Books Pty. Ltd, Sydney.

Jones, D.L. & Clemesha, S.C. (1976) *Australian Ferns and Fern Allies*, A.H. & A.W. Reed, Sydney.

Jones, D.L. & Gray, B. (1988) *Climbing Plants in Australia*, Reed Books, Sydney.

*Journal of the Adelaide Botanic Gardens* (1979–81) Adelaide Botanic Gardens, S.A.

Kemp, B. (1987 unpubl.) *Organ Pipes National Park Flora.*

Lamp, C & Collet, F. (1976) *Weeds in Australia*, Inkata Press, Melbourne.

Lamp, C.A., Forbes, S.J. & Cape, V.W. (1990) *Grasses of Temperate Australia*, Inkata Press, Melbourne.

Land Conservation Council (1991) *Melbourne Area District 2 Review — Descriptive Report*, Land Conservation Council, Melbourne.

Launceston Field Naturalists Club (1981) *Guide to Flowers and Plants of Tasmania*, (Ed. M. Cameron), A.H. & A.W. Reed, Sydney.

Leigh, J., Boden, R. & Briggs, J. (1984) *Extinct and Endangered Plants of Australia*, Macmillan Co. Aust. Pty. Ltd., Melbourne.

McCubbin, C. (1971) *Australian Butterflies*, Thomas Nelson (Australia) Ltd., Melbourne.

McDougall, K. (1987) *Sites of Botanical Significance in the Western Region of Melbourne*, Dept. Conservation, Forests & Lands,Melbourne.

Marchant, N.G. et al. (1987) *Flora of the Perth Region*, Pts 1&2, Western Australian Herbarium, Dept. of Agriculture, Perth.

Morton, J. (1984 unpubl.) *The Design of Nature Parks on the Keilor Plains*, Final Project, Landscape Architecture, RMIT, Melbourne.

Miscellaneous species lists as part of published reports for various management committees, state and local government departments (see Appendix 2)

Miscellaneous unpublished species lists (refer to Appendix 2 for many of these).

*Muelleria* (1986–) Periodical Journal of the National Herbarium of Victoria, Government Printer, Vic.

Nicholls, W.H. (1964) *Orchids of Australia*, Complete Edition, (Ed. D.L. Jones & T.B. Muir), Thomas Nelson (Australia) Ltd, Melbourne.

*Nuytsia* (1984) Bulletin of the Western Australian Herbarium, Dept. Conservation & Land Management, Government Printer, W.A.

Pearson, L.G.C. (undated) *The Indigenous Plants of the Melbourne Region*, L.G.C. Pearson, Melbourne.

Port Phillip Authority (1982 unpubl.) *The Coastal Vegetation of Port Phillip Bay*, Port Phillip Authority, Melbourne.

Prescott, A. (1988) *It's Blue with Five Petals: Wildflowers of the Adelaide Region*, A Prescott, Prospect, S.A.

Raynor, C. (unpubl. photographic record) *Flora of the Keilor Plains*, Society for Growing Australian Plants, Keilor Plains.

Robinson, R.W. et al. (1986) *Vegetation Survey of the Melbourne Metropolitan Area* (preliminary Report), Dept. Conservation, Forests & Lands, Melbourne.

Rogers, F.J.C. (1978) *A Field Guide to Victorian Wattles*, Revised Edition, F.J.C. Rogers.

Ross, J.H. (1993) *A Census of the Vascular Plants of Victoria*, 4th Edition, National Herbarium of Victoria, Dept. Conservation & Environment, Melbourne.

*Sandringham Environment Series* (1979–89) Nos 3, 4, 7. Sandringham City Council, Sandringham, Victoria.

Sharpe, P.R. (1986) *Keys to Cyperaceae, Restionaceae and Juncaceae of Queensland*, Botany Branch, Dept. Primary Industries, Brisbane.

Simon, B.K. (1990) *A Key to Australian Grasses* Botany Branch, Dept. Primary Industries, Brisbane.

Society for Growing Australian Plants (1983–90) *Australian Plants*, Quarterly, Society for Growing Australian Plants, Sydney.

Society for Growing Australian Plants, Maroondah Inc. (Rev. 1991) *Attracting Birds to Native Gardens*, Society for Growing Australian Plants, Maroondah Inc. Ringwood.

Society for Growing Australian Plants, Maroondah Inc. (1978) *500 Australian Native Plants*, Society for Growing Australian Plants, Maroondah Inc., Ringwood.

Stanley, T.D. & Ross, E.M. (1983–86) *Flora of South-Eastern Queensland* Vol 2, Queensland Dept. of Primary Industries, Brisbane.

Sutton, C.S. (1916) 'A Sketch of the Keilor Plains Flora', *Victorian Naturalist*, Victorian Field Naturalists Club, Victoria.

*Telopea* (1981–89) Contributions from the National Herbarium of New South Wales, Government Printer, NSW.

*Victorian Naturalist* (1928–90) Newsletter of Victorian Field Naturalists Club, Victoria.

Waddell, W. (1976) *Wildflower Diary*, Native Plants Preservation Society.

Wheeler, D.J.B., Jacobs, S.W.L. & Norton, B.E. (1982) *Grasses of New South Wales* University of New England, Armidale, NSW.

Willis, J.H. (1962–72) *A Handbook to Plants in Victoria* Vols 1 & 2, Melbourne University Press, Melbourne.

Willis, J.H. (1964) *Vegetation of the Basalt Plains in Western Victoria*, Proceedings of the Royal Society of Victoria, Vol 77, Victoria.

Woolcock, C. and D. (1984) *Australian Terrestrial Orchids*, Thomas Nelson (Australia) Ltd, Melbourne.

Wrigley, J.W. & Fagg, M. (1979) *Australian Native Plants*, William Collins, Sydney.

# *Index of Photographs*

# *Index of Botanical and Common Names*

Current botanical names are written in bold type. Synonyms, common names and weed species are in medium type. Common names based directly on generic names have not been listed. Page numbers in bold type refer to the plant description while numbers preceded by Col refer to Colour Plates. * represents species introduced to Melbourne.

Most are now Austrodanthonia

Melicytus dentatus - Syn Hymenanthera dentata

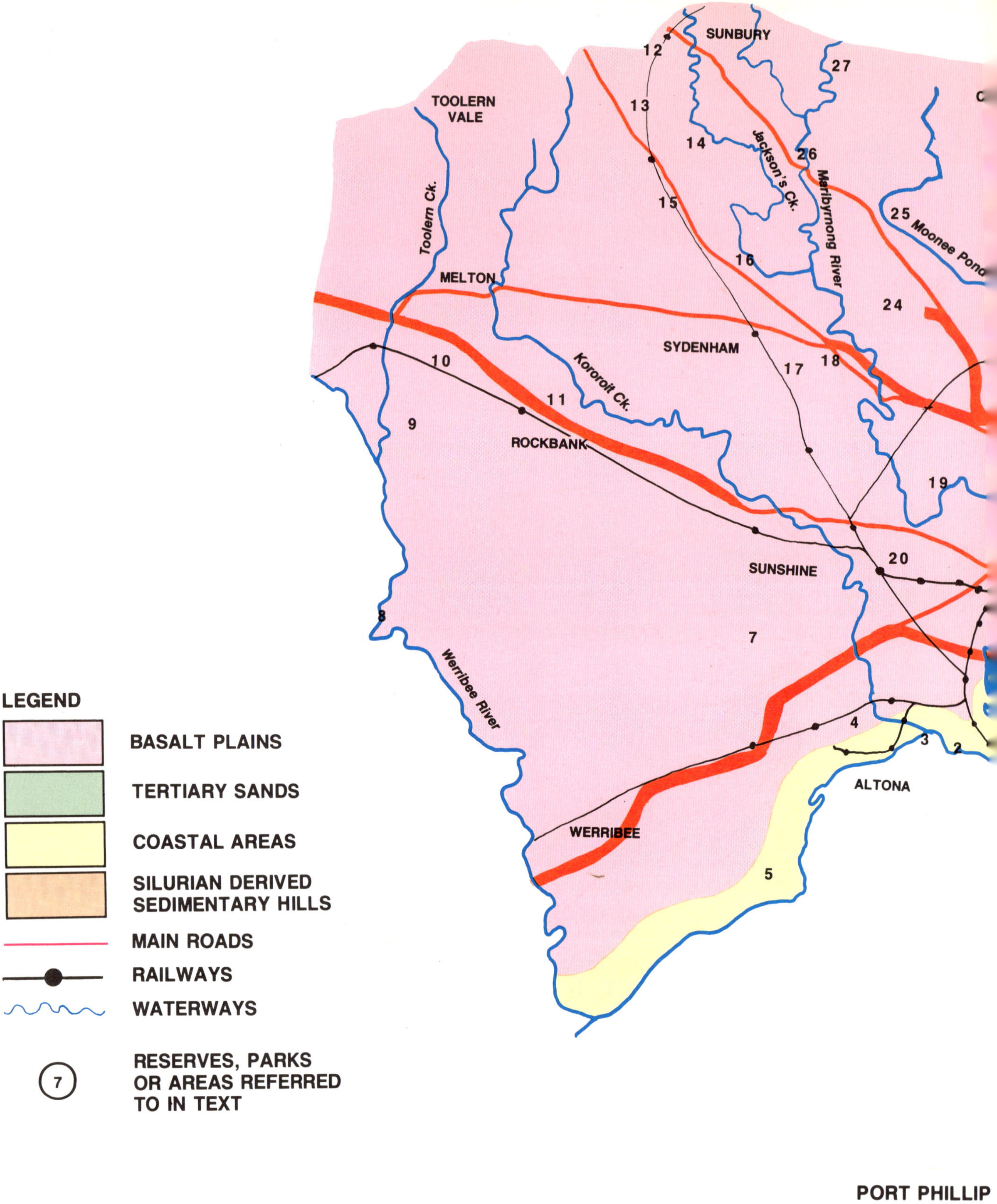


# REFERENCE MAP FOR FLORA of MELBOURNE